Guo-Fan Hong (Ed.)

The Nitrogen Fixation and its Research in China

With 198 Figures

Springer-Verlag Berlin Heidelberg GmbH

Guo-Fan Hong

Shanghai Institute of Biochemistry
Academia Sinica, 320 Yue-Yang Road
Shanghai 200031, PR China

ISBN 978-3-662-10387-6

Library of Congress Cataloging-in-Publication Data

The Nitrogen fixation and its research in China / Guo-Fan Hong.
p. cm.
ISBN 978-3-662-10387-6   ISBN 978-3-662-10385-2 (eBook)
DOI 10.1007/978-3-662-10385-2

1. Nitrogen—Fixation. 2. Nitrogen—Fixation—Research—China.
I. Hong, Guo-Fan. QR89.7.N558 1992 589.9'0133—dc20 91-35220 CIP

Typesetting: Thomson Press (India) Ltd., New Delhi;

2151/3020-543210 – Printed on acid-free paper

Preface

Studies on nitrogen fixation have now been recognised as a topic of high priority in Chinese scientific programmes, since nitrogen fixation research is considered essential for helping to solve the food problem in this country.

In China, more than half of the total nitrogen fertilizer applied to cultivated fields comes from biological nitrogen fixation sources. The area of cultivated lands involving symbiotic associations between nitrogen-fixing bacteria and plants has been estimated at 300 000 000 mu (1 mu = 0.1647 acres), and these lands produced more than 1 600 000 tons of biologically fixed nitrogen fertilizer, equal to the total amount of chemical fertilizer used in 1970 in China. In addition, the amount of nitrogen fixed by free-living nitrogen-fixing bacteria came to 7 000 000 tons per year.

Chinese scientists have been successful in applying various forms of biological nitrogen fixation as green manure and green feed to agriculture, which has markedly increased production. They have also been working on the nitrogen-fixing actinomycete *Frankia* and non-leguminous woody plants, and their research has also been fairly fruitful. About 44 species in 8 genera among 6 families of non-leguminous woody plants have been reported to have actinomycete-induced nodulation and nitrogen fixation, about 20 of which were identified for the first time.

Studies on the chemistry of nitrogen fixation with special emphasis on the mechanism and chemical modelling of nitrogenase catalysis have been conducted vigorously in this country for many years. Readers will find in this book a detailed overview of this subject by Prof. Jia-Xi Lu. The book deals with a wide range of topics, including biochemistry, molecular genetics and biology of nitrogen fixation, that have been studied in this country. Readers should obtain a clear picture of what Chinese scientists have been pursuing in various aspects of nitrogen fixation research.

Another prominent feature of this book is the contributions from leading scientists in this field who are active in the international community. The articles by Drs. P. Boistard, J.A. Downie, A.W.B. Johnston, W. Klingmüller, A. Puhler, J.P.W. Young and their colleagues describe in detail their respective research activities which have brought forth some recent advances in various aspects of biological nitrogen fixation research.

I would very much like to thank Ms. Wen-Wen Shi for her great patience and painstaking efforts in typing a mountain of manuscripts; without her help it would have been impossible to complete the editing in time.

Shanghai, November 1991   Guo-Fan Hong

List of Contributors

Article numbers are in parentheses following the names of contributors. Affiliations listed are current.

Xue-Liang Bai (19)
Laboratory of Molecular Genetics, Guangxi Agricultural College, Nanning 530005, China

Pierre Boistard (15)
Ministère de l'Agriculture, Centre de Recherches d'Auzeville, Laboratoire de Biologie Moleculaire des Relations Plantes-Microorganismes, CNRS-INRA, Chemin de Borde Rouge-BP 27, 31326 Castanet-Tolosan, Cedex-France

Long-Xiang Cai (30)
Fujian Institute of Subtropical Botany, Xiamen. 361009, China

Keqiang Cai (30)
Fujian Institute of Subtropical Botany, Xiamen. 361009, China

Hui-Min Cao (10)
Shanghai Institute of Biochemistry, Chinese Academy of Sciences, Shanghai 200031, China

Y.Z. Cao (23)
Laboratory of Biological Nitrogen Fixation
Huazhong Agricultural University, Wuhan, Hubei. 430072, China

Guan-Xiong Chen (28)
Institute of Applied Ecology, Academia Sinica, Shenyang 110015, China

H.K. Chen (23)
Laboratory of Biological Nitrogen Fixation
Huazhong Agricultural University, Wuhan, Hubei. 430072, China

Li-Mei Chen (19)
Laboratory of Molecular Genetics, Guangxi Agricultural College, Nanning 530005, China

Pencheng Chen (7)
Fujian Institute of Research on the Structure of Matter, Fuzhou Research Laboratory of Structural Chemistry, Chinese Academy of Sciences, Fuzhou, Fujian 350002, China

Yu-Hai Chui (19, 29)
Institute of Applied Ecology, Academia Sinica, Shenyang 110015, China

X.P. Dai (12)
Institute of Applied Ecology, Academia Sinica, Shenyang 110015, China

Jian Ding (29)
Institute of Applied Ecology, Academia Sinica, Shenyang 110015, China

J.A. Downie (9, 16)
John Innes Institute, John Innes Centre for Plant Science Research, Norwich NR4 7UH, U.K.

A. Economou (16)
John Innes Institute, John Innes Centre for Plant Science Research, Norwich NR4 7UH, U.K.

C.S. Fan (22)
Nanjing Agricultural University, Nanjing 210014, China

Gui-Lan Guan (31)
Xinjiang Institute of Biology, Pedology and Desert Research, Chinese Academy of Sciences, China

Xiang-Ju Gu (21)
Institute of Botany, Chinese Academy of Sciences, Beijing 100044, China

G.F. Hong (9, 10)
Shanghai Institute of Biochemistry, Chinese Academy of Sciences, Shanghai 200031, China

Jia-Bin Huang (28)
Institute of Applied Ecology, Academia Sinica, Shenyang 110015, China

Liang-Ren Huang (7)
Fujian Institute of Research on the Structure of Matter,

Fuzhou Research Laboratory of Structural Chemistry, Chinese Academy of Sciences, Fuzhou, Fujian 350002, China

Wei-Nan Huang (30)
Fujian Institute of Subtropical Botany, Xiamen 361009, China

Ya-Li Huang (29)
Institute of Applied Ecology, Academia Sinica, Shenyang 110015, China

Jian-De Jiang (28)
Institute of Applied Ecology, Academia Sinica, Shenyang 110015, China

Yu-Xiang Jing (21)
Institute of Botany, Chinese Academy of Sciences, Beijing 100044, China

Andrew W.B. Johnston (14)
School of Biological Sciences, University of East Anglia, Norwich NR4 7TJ, U.K.

Bei-Sheng Kang (6)
Fujian Institute of Research on the Structure of Matter, Chinese Academy of Sciences, Fuzhou, Fujian 350002, China

Walter Klingmüller (17)
Genetics Department, University of Bayreuth, 8580 Bayreuth, Federal Republic of Germany

An-Jian Lan (7)
Fujian Institute of Research on the Structure of Matter, Fuzhou Research Laboratory of Structural Chemistry, Chinese Academy of Sciences, Fuzhou, Fujian 350002, China

Gu Lan (30)
Fujian Institute of Subtropical Botany, Xiamen 361009, China

F.D. Li (23)
Laboratory of Biological Nitrogen Fixation Huazhong Agricultural University, Wuhan, Hubei 430072, China

Guang-Shan Li (21)
Institute of Botany, Chinese Academy of Sciences, Beijing 100044, China

Shang-Hao Li (S.H. Ley) (26)
Institute of Hydrobiology, Chinese Academy of Sciences, Wuhan, 430072, P. R. China

Wei-Guang Li (29)
Institute of Applied Ecology, Academia Sinica, Shenyang
110017, China

Zhong-Wei Li (25, 29)
Institute of Applied Ecology, Academia Sinica, Shenyang
110015, China

Yuan-Dan Li (29)
Institute of Applied Ecology, Academia Sinica, Shenyang
110015, China

Ji-Shang Lin (25)
Institute of Applied Ecology, Academia Sinica, Shenyang
110015, China

Yong-Qi Lin (11)
Dept. of Molecular Biology, Jilin University, Changchun, China

Chun-Wan Liu (8)
Fujian Institute of Research on the Structure of Matter,
Chinese Academy of Sciences, Fuzhou, Fujian 350002, China

Chung-Chu Liu (27)
Fujian Academy of Agricultural Sciences, Fuzhou, Fujian
350002, China

Hui-Chang Liu (28)
Institute of Applied Ecology, Academia Sinica, Shenyang
110015, China

Qiu-Tian Liu (6)
Fujian Institute of Research on the Structure of Matter,
Chinese Academy of Sciences, Fuzhou, Fujian 350002, China

Jia-Xi Lu (1, 6, 7, 8)
Fujian Institute of Research on the Structure of Matter,
Fuzhou Research Laboratory of Structural Chemistry Chinese
Academy of Sciences, Fuzhou, Fujian, 350002, China

You-Yi Lu (10)
Shanghai Institute of Biochemistry, Chinese Academy. of
Sciences, Shanghai 200031, China

Qing-Sheng Ma (9, 19)
Laboratory of Molecular Genetics, Guangxi Agricultural
College, Nanning 530005, China

F.D. Ni (9, 10)
Shanghai Institute of Biochemistry, Chinese Academy of
Sciences, Shanghai 200031, China

Shu-Yun Niu (3)
Department of Chemistry, Jilin University, Changchun, China

I.M. Pretorius-Güth (13)
Lehrstuhl für Genetik, Fakultät für Biologie, Universität Bielefeld, Postfach 8640, 4800 Bielefeld 1, Federal Republic of Germany

A. Pühler (13)
Lehrstuhl für Genetik, Fakultät für Biologie, Universität Bielefeld, Postfach 8640, 4800 Bielefeld 1, Federal Republic of Germany

Bo Qi (25)
Institute of Applied Ecology, Academia Sinica, Shenyang 110015, China

Min Qin (19)
Laboratory of Molecular Genetics, Guangxi Agricultural College, Nanning 530005, China

Xue-Qin Shan (21)
Institute of Botany, Chinese Academy of Sciences, Beijing 100044, China

L.A. Sharypova (13)
Lehrstuhl für Genetik, Fakultät für Biologie, Universität Bieleffeld, Postfach 8640, 4800 Bielefeld 1, Federal Republic of Germany

Shan-Min Shen (28)
Fujian Academy of Agricultural Sciences, Fuzhou, Fujian, China

B.V. Simarov (13)
All-Union Research Institute for Agricultural Microbiology, Podbelsky Shanssee 3, Leningrad-Pushkin 6, 189620, USSR

Hong-Yu Song (20)
Shanghai Institute of Plant Physiology, Academia Sinica, 200032, China

Wei Song (5, 24)
Institute for Application of Atomic Energy, CAAS. Beijing, 100094, China

Feng-Yan Su (29)
Institute of Applied Ecology of Academia Sinica, Shenyang, China

Hui-Jun Sun (29)
Institute of Applied Ecology of Academia Sinica, Shenyang, China

Ao-Qing Tang (2, 3)
Institute of Theoretical Chemistry, Jilin University, Changchun, China

K.R. Tsai (4, 5)
Nitrogen Fixation Research Group and Laboratory of Catalysis, Xiamen University, Xiamen, Fujian 361005, China

H.L. Wan (4)
Nitrogen Fixation Research Group and Laboratory of Catalysis, Xiamen University, Xiamen, Fujian 361005, China

Fei Wang (11)
Dept. of Molecular Biology, Jilin University, Changchun, China

Hui-Xian Wang (24)
Institute for Application of Atomic Energy, CAAS. Beijing, 100094, China

Shu-Jin Wang (25)
Institute of Applied Ecology, Academia Sinica, Shenyang 110015, China

Yan-Ling Wang (19)
Laboratory of Molecular Genetics, Guangxi Agricultural College, Nanning 530005, China

Yang Wu (29)
Institute of Applied Ecology of Academia Sinica, Sheyang, China

Yi-De Wu (30)
Fujian Institute of Subtropical Botany, Xiamen, 361009, China

Guang-Ming Xu (25)
Institute of Applied Ecology, Academia Sinica, Shenyang 110015, China

Ji-Qing Xu (2, 3)
Department of Chemistry, Jilin University, Changchun, China

Qing-De Xu (29)
Institute of Applied Ecology, Academia Sinica, Shenyang 110015, China

L.S. Xu (4)
Nitrogen Fixation Research Group and Laboratory of Catalysis, Xiamen University, Xiamen, Fujian 361005, China

Zhi-Wei Xu (30)
Fujian Institute of Subtropical Botany, Xiamen 261009, China

De-Lin Xue (25)
Institute of Applied Ecology, Academia Sinica, Shenyang 110015, China

Hui-Fan Yang (28)
Institute of Applied Ecology, Academia Sinica, Shenyang 110015, China

J.S. Yang (12)
Institute of Applied Ecology, Academia Sinica, Shenyang 110015, China

Chong-Biao You (5, 24)
Institute for Application of Atomic Energy, CAAS, Beijing 100094, China

J.P.W. Young (18)
John Innes Institute, Colney Lane, Norwich NR4 7UH, U.K.

A-dong Yu (28)
Institute of Applied Ecology, Academia Sinica, Shenyang 110015, China

D. Zeng (5)
Nitrogen Fixation Research Group, Xiamen University, Xiamen Fujian 361005, China

W.-Y. Zeng
North Western China's Agricultural University, Wugong County, Shan X1 Province, China

C.G. Zhang (12)
Institute of Applied Ecology, Academia Sinica, Shenyang 110015, China

Dao-Hai Zhang (29)
Institute of Applied Ecology, Academia Sinica, Shenyang 110015, China

H.T. Zhang (4)
Nitrogen Fixation Research Group and Laboratory of Catalysis, Xiamen University, Xiamen, Fujian 361005, China

Xian-Wu Zhang (25)
Institute of Applied Ecology, Academia Sinica, Shenyang 110015, China

Zhi-Gui Zhang (3)
Department of Chemistry, Jilin University, Changchun, China

Zhong-Ze Zhang (29)
Institute of Applied Ecology, Academia Sinica, Shenyang
110015, China

Zhen-Ying Zhao (28)
Institute of Applied Ecology, Academia Sinica, Shenyang
110015, China

Xiao-Ten Zhao (7)
Fujian Institute of Research on the Structure of Matter,
Fuzhou Research Laboratory of Structural Chemistry, Chinese
Academy of Sciences, Fuzhou, Fujian 350002, China

Wei-Wen Zheng (27)
Fujian Academy of Agricultural Sciences, Fuzhou, Fujian
350002, China

Hua Zou (29)
Institute of Applied Ecology, Academia Sinica, Shenyang
110015, China

Xiao-Lu Zou (30)
Fujian Institute of Subtropical Botany, Xiamen 361009, China

Bo-Tao Zhuang (7)
Fujian Institute of Research on the Structure of Matter,
Fuzhou Research Laboratory of Structural Chemistry, Chinese
Academy of Sciences, Fuzhou, Fujian 350002, China

Table of Contents

Section I. Chemistry of Nitrogen Fixation

Section II. Biochemistry and Molecular Genetics of Nitrogen Fixation

Section III. Biology of Nitrogen Fixation

Section I:
Chemistry of Nitrogen Fixation

Research on the Chemical Modelling of Biological Nitrogen Fixation in the New China— An Overview of Research Carried out at the Fujian Institute During the 1970s and the Early 1980s

Lu Jiaxi (Chia-Si Lu)

In the New China, the summer of 1971 witnessed the gradual resumption of more-or-less peaceful working conditions after the first destructive phase of the catastrophic Cultural Revolution. Professor Guo Xingxian, then the deputy director of the Provisional Bureau of Biological Sciences, Chinese Academy of Sciences (CAS), began to recognize the academic significance of chemical modelling of biological nitrogen fixation, which was then attracting the attention of a great number of chemists and biochemists. He therefore proposed to organize a mini-conference consisting of a number of professors and research scientists for a preliminary evaluation of the feasibility of drawing up a research program in this connection. This received at once very warm response from chemists working in the field of the chemistry of transition-metal clusters and looking forward to their possible applications to catalysis problems. As a matter of fact, it is already very much well-known that strongly acidified cuprous compounds play an important role in the catalytic dimerization of acetylene to vinyl acetylene. Again, several research groups in the New China were then already engaged, with certain degrees of success, in research projects organized for the catalytic cyclic trimerization of acetylene to benzene with supported chromic oxide catalysts, and even the catalytic "tetramerization" of acetylene to styrene (via the oligomerization of vinyl acetylene with two molecules of acetylene) with similar chromium catalysts. (Perhaps it is worth pointing out that China was then rather deficient in hydrocarbon resources!) There had been sufficient reasons to speculate that these cyclic trimerizations took place through the catalytic effect of trinuclear chromium clusters. Meanwhile, the molybdenum iron protein (the so-called MoFe-pr), known to be responsible for the very efficient dinitrogen fixation of the enzyme nitrogenase, has been found to contain the transition metals molybdenum and iron. The enthusiastic response of the Fujian Institute of Research on the Structure of Matter, CAS (Fuzhou, Fujian), actively involved in research on the structural chemistry of transition-metal clusters, as well as the Department of Chemistry, Xiamen (Amoy) University (Xiamen, Fujian), well-known for its catalytic research under the capable leadership of Professor K.R. Tsai (Cai Qirui), is therefore quite understandable.

The Nitrogen Fixation
and its Research in China
Editor: Guo-fan Hong
© Springer-Verlag Berlin Heidelberg 1992

Also, the Department of Chemistry, Jilin (Kirin) University (Changchun, Jilin), again well-known for its research work in quantum chemistry under the eminent leadership of Professor A.C. Tang (Tang Aoqing) has also been very much interested in this joint research project. However, without the hearty collaboration of chemists and biochemists in other Chinese research institutions, such an interdisciplinary research program would be doomed to failure.

The first national symposium on the chemical modelling of biological nitrogen fixation was a mini-conference organized by the Provisional Division of Biological Sciences, CAS, and held in Changchun, Jilin in February of 1972. A research program was drawn up mainly along the three following directions: (1) isolation, purification, and even crystal growth of nitrogenase; investigation of its structure-function relationships (provided that circumstances permit) and measurements of certain physicochemical and biochemical properties so as to provide necessary information for chemical modelling from the angles of structural chemistry and coordination catalysis; (2) syntheses of dinitrogen complexes of transition metals and bonding properties of the dinitrogen-transition-metal bonds so as to provide necessary information for formulation of structural criteria for chemical activation of the unusually inert dinitrogen molecule, thus enabling us to propose structural models for the active center of nitrogenase and measure their catalytic reduction capabilities under quasi-bioassaying conditions; (3) development of electron-donor–acceptor (EDA) type catalysts for dinitrogen fixation as well as improvement of the Haber–Bosch synthetic ammonia iron catalysts for use under milder conditions, say at temperatures 350–250 °C and under pressures ca. 150 atm.

The second national symposium was held in Xiamen (Amoy), Fujian, in March of 1973. A good number of helpful and useful review papers were read, leading to a series of interesting and enlightening discussions as well as formulation of attacking strategies for this research program. Thus in view of the unusual inertness of the N≡N triple bond in the dinitrogen molecule, the principal difficulties in the problem of chemical activation of such a small molecule were analyzed in detail, leading thus to a series of qualitative quantum-chemical discussions of N_2 complexes involving d electrons under typical octahedral coordination conditions. We have therefore been led to the following conclusions [1], valid in particular for trinuclear vs mononuclear and binuclear dinitrogen complexes and confirmed by quantum-chemical calculations [2, 3] (systematically carried out in Jilin University):

1) There are some essential similarities. First, when the total number of d electrons is quite large and the energy levels of the d orbitals of the "donor" transition-metal atom lie quite high up, the N atoms tend to become more negative; such complexes will favor electrophilic reactions. On the other hand, if the total number of d electrons is rather small and the energy levels of the d orbitals of the "acceptor" transition-metal atom lie quite low, the N atoms tend to become somewhat positive; such complexes may then be preferentially reduced via a nucleophilic reaction. Secondly, the exo-N atom is in general more negative than the endo-N atom.

2) There are, however, some striking differences worthy of note. For both mononuclear and binuclear dinitrogen complexes, the end-on mode of coordination is in general more stable. For trinuclear dinitrogen complexes, however, coordination complexes involving both end-on and side-on modes of coordination to the transition-metal atoms turn out to be more stable in general than the all-end-on modes of coordination. This conclusion is true as long as the total number of d electrons is not less than 12. Intuitively, this conclusion seems to hold true even for other polynuclear (i.e., higher than trinuclear) transition-metal dinitrogen complexes having a sufficient number of d electrons. Moreover, mononuclear and binuclear dinitrogen complexes are stable provided that either the energy levels of the d orbitals lie rather high up when the total number of d electrons is sufficiently large or the energy levels of the d orbitals lie rather low when the total number of d electrons is sufficiently small. On the contrary, trinuclear dinitrogen complexes are unstable under such circumstances. We have therefore reasons to believe that neither mononuclear nor binuclear N_2 complexes will lead to satisfactory chemical models for the active center of nitrogenase. In other words, the active center for catalytic nitrogen fixation in nitrogenase must be at least a trinuclear transition-metal cluster involving both end-on and side-on coordinations of N_2 onto the transition-metal atoms.

This qualitative discussion leads at once to the following structural criteria for activation of the dinitrogen molecule via coordination. First, end-on coordination is absolutely necessary to stretch the "abnormal" N≡N triple bond to restore its "normal" chemical behavior. One such end-on coordination bond may suffice, provided that it is strong enough. If it proves to be not sufficient, enough space must then be provided for a second end-on coordination to the substrate. Secondly, side-on coordination is also absolutely necessary. In general, one such side-on bond will not be sufficient for N_2 with its mutually orthogonal π orbitals. It is therefore desirable to have several such bonds in synergic action; indeed these may only be counted upon as a double side-on coordination. Thirdly, it is absolutely necessary to protect any such side-on coordinations against possible isomerization into coordinations of the end-on type. The syntheses and the crystal structure determinations of the stable lithium–nickel dinitrogen complexes: (A) $[(PhLi)_6Ni_2N_2(Et_2O)_2]_2$ (Fig. 1a) and (B) $[Ph\{Na(OEt_2)\}_2(Ph_2Ni)_2N_2NaLi_6(OEt)_4 \cdot (OEt_2)]_2$ (Fig. 1b) by Jonas (1973), Kruger and Tsay (1973), and Jonas et al. (1976) [4] provide unequivocal support for the proposed structural criteria discussed above. The Li atoms working in synchronization provide metal atoms for double end-on coordinations to N_2, and the pair of Ni atoms provides two centers for a double side-on coordination. Taken together, these suffice to protect the side-on coordinations against isomerization into end-on modes. The N–N bond of this complex is thus stretched to 1.35 Å; the stretching frequency is thus also brought down to below $1,550\,cm^{-1}$. As a matter of fact, the N_2 ligand can be quantitatively displaced by C_2H_4 without any difficulty. Such a complex has been shown to liberate ammonia readily in the presence of water (Fischler and von Gustorf (1975) [4]).

Thus, a satisfactory structural model for the active center of nitrogenase must be a polynuclear transition-metal cluster containing at least four metal atoms. It is also believed that these proposed structural criteria may also help in the mechanistic elucidation of the activation of dinitrogen by the industrial Haber–Bosch iron catalysts.

It was in August of 1973, only within a few months of the closing of the second national symposium, that two structural models, the Xiamen model X-1

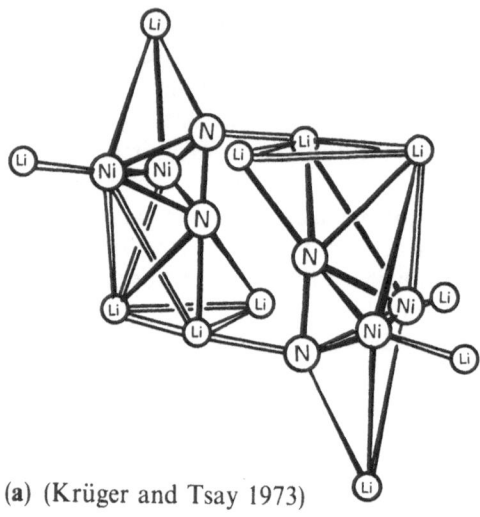

(a) (Krüger and Tsay 1973)

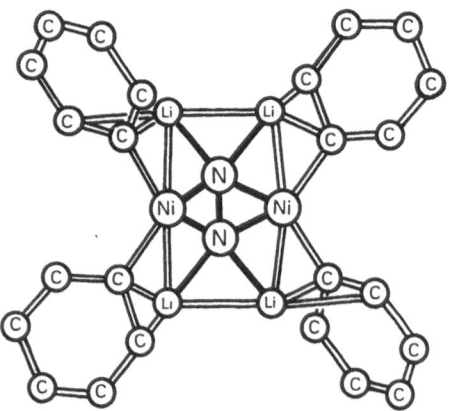

(b) (Jonas 1976)

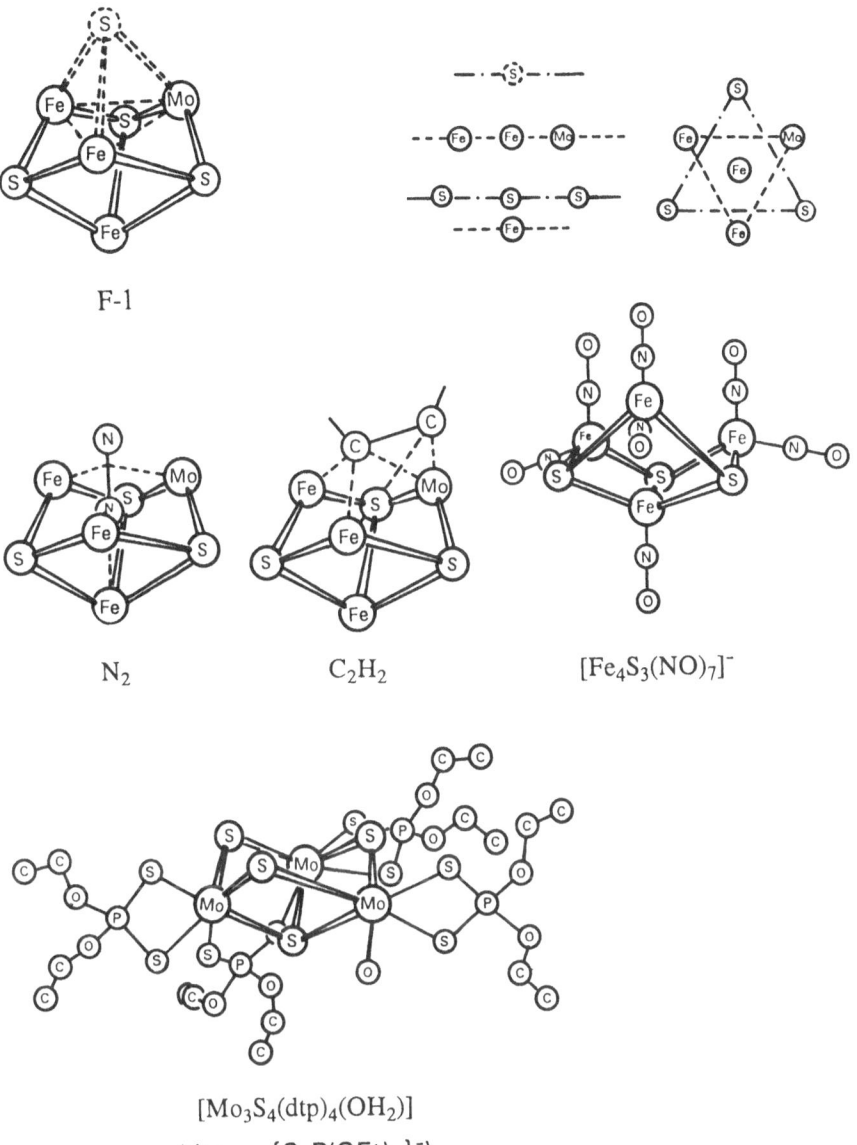

F-1

N₂   C₂H₂   [Fe₄S₃(NO)₇]⁻

[Mo₃S₄(dtp)₄(OH₂)]

a   (dtp ≡ {S₂P(OEt)₂}⁻)

Fig. 2 a, b. Early structural models proposed for the active center of nitrogenase and modes of coordination activation for the substrates N₂ and C₂H₂: **a** Fuzhou model F-1; **b** Xiamen model X-1

Fig. 1. Coordination geometry of N₂ to the Li and Ni atoms. **a** in A: [(PhLi)₆Ni₂N₂(Et₂O)₂]₂; **b** in B: [Ph{Na(OEt₂)}₂ (Ph₂Ni). NaNaLi₆(OEt)₄ (OEt₂)]₂

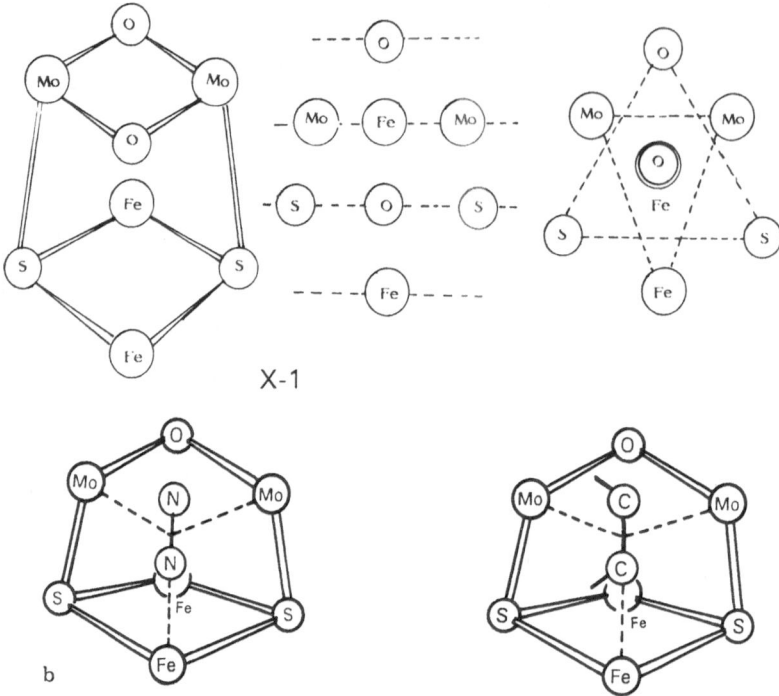

X-1

b

Fig. 2. (*Continued*)

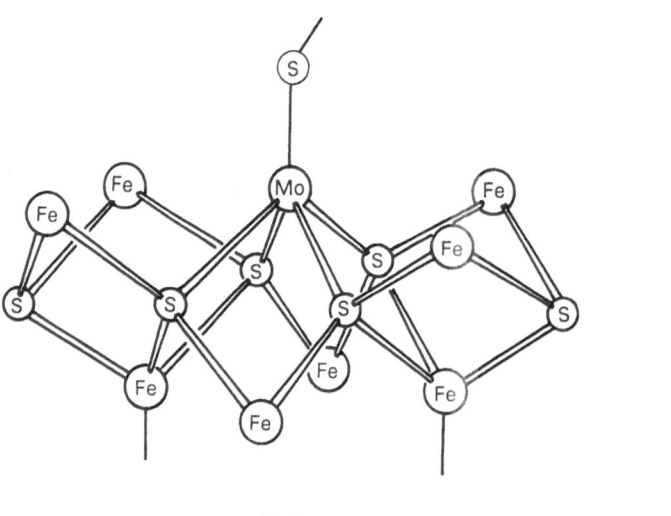

a   F-2

Fig. 3 a, b. Further developed structural models proposed for the active center of nitrogenase: **a** Fuzhou model F-2; **b** Xiamen models X-1, 1′, 1″, 2, 3, 4, 5, with N_2 as substrate occupying the position of an O or N atom in a labile (O or N)-containing ligand

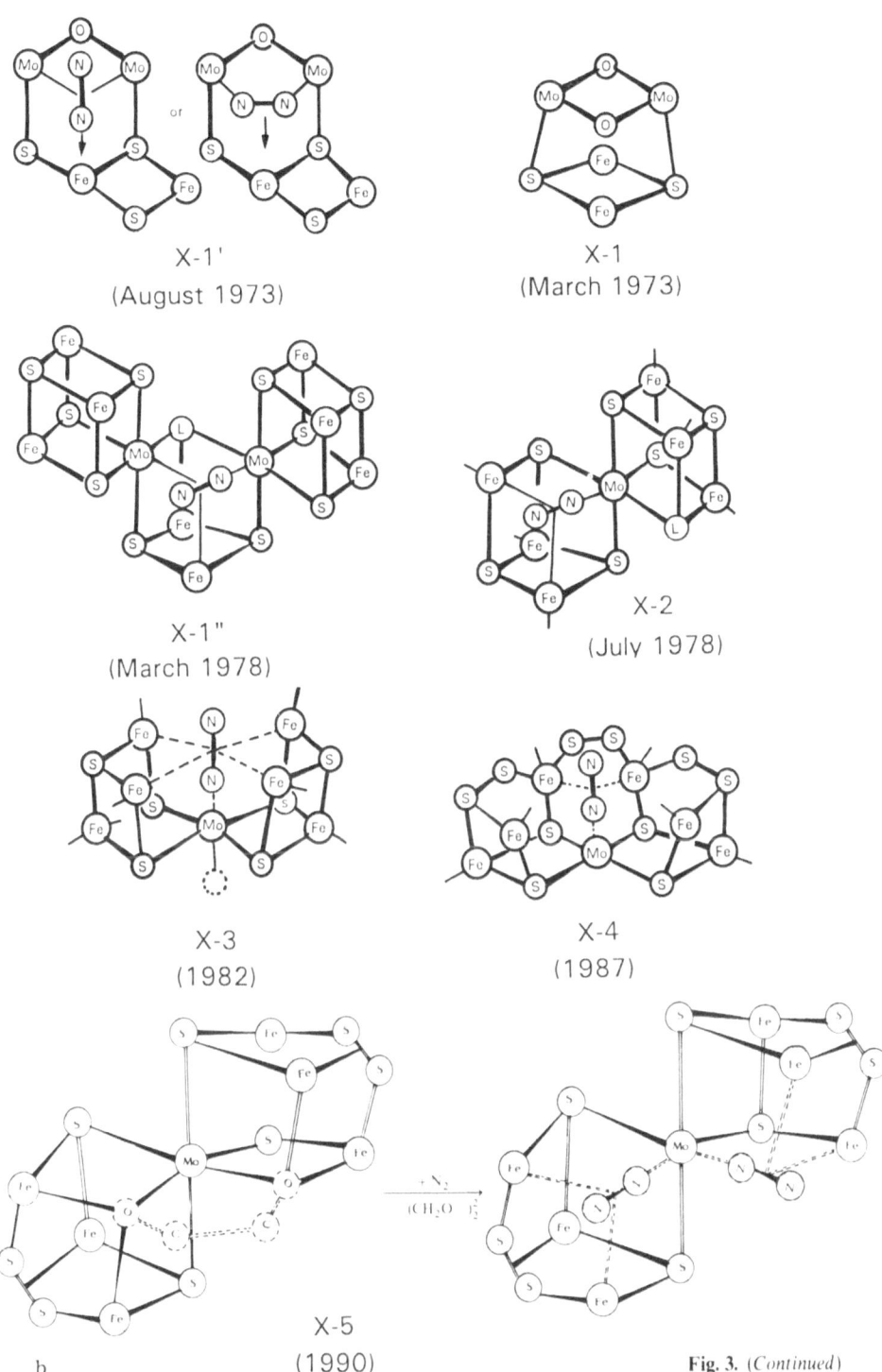

X-1'
(August 1973)

X-1
(March 1973)

X-1"
(March 1978)

X-2
(July 1978)

X-3
(1982)

X-4
(1987)

X-5
(1990)

b

Fig. 3. (Continued)

[5]: $Fe_2S_2Mo_2O_2$ or $Fe(S_2O)(Mo_2Fe)O$, a "twin-sited" *closo* cubane-like cluster (Fig. 2b) and the Fuzhou model F-1 [1]: $Fe_{bot}(S_3)_{waist}(MoFe_2)_{neck}$, a "string-bag" *nido* cubane-like cluster (Fig. 2a), were independently and almost simultaneously proposed (and further developed later into a few more "mature" structural models: Xiamen models X-1′, 1″, 2–5 and Fuzhou model F-2 by the Xiamen University and the Fujian Institute nitrogen fixation research groups, respectively (Fig. 3a and b). Their differences are: (1) Mo_2Fe_2-$Fe_{bot}(Mo_2Fe)_{neck}$ *vs* $MoFe_3$-$Fe_{bot}(M_3)_{neck}$ ($M = \frac{1}{3}(Mo + 2Fe)$; (2) 2 labile O *vs* 1 S vacancy; (3) modes of coordination: N_2—end-on to Fe, double side-on to 2 Mo (or double end-on to 2 Mo, side-on to Fe) *vs* end-on to Fe_{bot}, "triple" side-on to 3 M; C_2H_2—same as for N_2 *vs* end-on to M, double side-on to 2 M′. The similarity would be enhanced to some extent if the "bottom" atom Fe and the "neck" atom Mo in Fuzhou model F-1 should be reversed. However, the more satisfactory agreement between the calculated EXAFS $MoK\alpha$ intensity curve for the F-1 and the experimental curve in the case of FeMo-co would at once be spoiled. It is interesting to note that the Xiamen model X-2 and the Fuzhou model F-2 do not seem to differ that much provided that the two Fe atoms and the S atom not belonging to either of the two "string-bags" in the latter are not taken into consideration. As a matter of fact, all those structural models characterized by a Siamese-twinned dicubane-like cluster joined through a

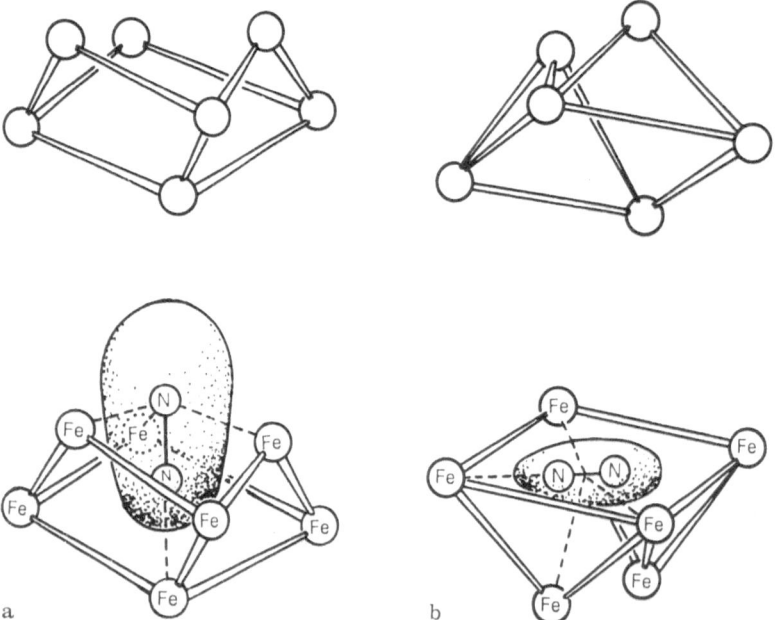

Fig. 4 a, b. Structural models proposed for the active center of Haber-Bosch iron catalyst for ammonia synthesis and models of coordination activation for N_2; **a** 7-Iron-atom Nanjing University "open-mouth pot" model; **b** 6-Iron-atom Xiamen University model

corner-sharing Mo atom can be easily shown to be essentially related to each other in the relative orientation of the two cubane moieties.

Incidentally, there is reason to believe that from the mechanistic point of view, there may exist some similar features in the structural models of the active centers as well as the modes of coordination of the dinitrogen molecule in the two cases of nitrogen fixation: the nitrogenase and the Haber–Bosch catalyst. As a matter of fact, Dai and his nitrogen fixation research group in the Nanjing (Nanking) University proposed a 7-iron-atom $Fe_{bot}(Fe_3)_{waist}(Fe_3)_{neck}$ "open-mouth pot" structural model [6] (Fig. 4a), very much similar to the "string-bag" Fuzhou model F-1, with N_2 coordinated to it in a similar "diving" mode. On the other hand, Huang and his coworkers in the Department of Chemistry, Xiamen University, proposed a 6-iron-atom $Fe_{bot}(Fe)_{waist}(Fe_3)_{neck}(Fe)_{cheek}$ structural model [7] (Fig. 4b), somewhat similar to the "twin-sited" Xiamen model X-1, with N_2 coordinated to it in a similar "cannon mounting" mode. This similarity is by no means accidental.

Perhaps it is not out of place to mention in this connection that a lot more structural models have been proposed by the Xiamen University research group for the active center of nitrogenase. They have certainly kept up fast with new developments and data available. On the other hand, the Fujian Institute research group has focussed more attention to attempted syntheses of modelling compounds and measurements of the cooperative phenomena in these multi-nuclear cluster frameworks, without losing sight of their N_2 and C_2H_2 reduction activities, regardless of how small or even negligible they are, provided that they can be established beyond experimental uncertainty.

The ATP-driven electron transfer in nitrogenase catalysis is an important electron- and energy-transfer in biological nitrogen fixation and other biological processes. Research work in this connection has been discussed in detail in the review article of C.-B. You, K.R. Tsai et al. [8] in this monograph, and thus will not be discussed any further in our article.

Most of the cluster-type structural models proposed so far for the active center of nitrogenase have the structure of an imperfect cubane-type cluster (e.g., M_3X_3M' or a double cubane thereof) or a perfect cubane-like structure with at least a loose ligand L occupying one of the corners (e.g., $M_3X_3M'L$ or a double cubane thereof). Thus the Fuzhou model F-1: $FeS_3(MoFe_2)$ has a structure of the former type, very much similar to the structure of the black Roussinate anion $[Fe_4S_3(NO)_7]^-$ (Fig. 5a); while each of the cubane-like cluster in the Xiamen model X-2: $[SFe_3S_2L]Mo[L'S_2Fe_3S]$ has a structure of the latter type.

Since the black Roussinate anion looks somewhat like a Chinese string-bag (Fig. 6) in its overall appearance, the author has chosen to call the Fuzhou model F-1 a "string-bag" structure, and the other a "capped" variation. Apparently, such a proposed "string-bag" configuration, regardless of whether it is "open" or "capped", is not at all accidental. After all, the 4Fe–4S ferredoxins are very common electron-transfer enzymes, and thus either the open neck or the easily removable cap of the "string-bag" configuration provides an easy

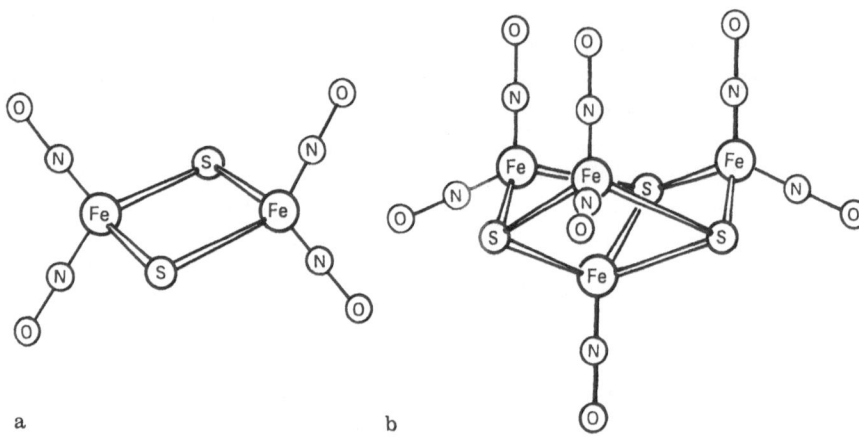

Fig. 5 a, b. Molecular structures of the Roussinate anions. **a** Black Roussinate monanion $[\{(ON)Fe\}S_3\{Fe(NO)_2\}_3]$; **b** Red Roussinate dianion $[\{(ON)_2Fe\}S_2Fe(NO)_2\}]$

access to most small attacking substrate molecules. It is therefore understandable that the nitrogen fixation research group in our Fujian Institute of Research on the Structure of Matter chose to start our research work with the black Roussinate anion as the first modelling compound.

Two questions arise at once in this connection. The first question is how we can proceed to synthesize a modelling compound with a string-bag frame-

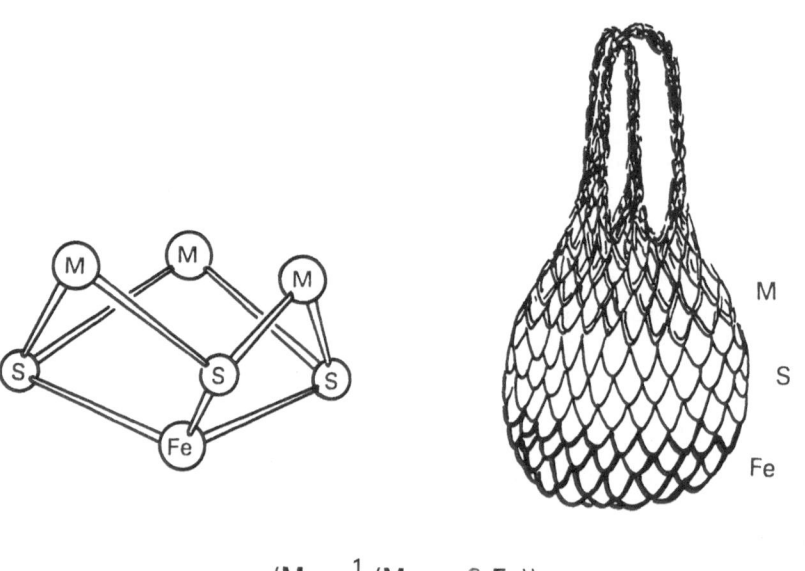

$$(M = \tfrac{1}{3} (Mo + 2 Fe))$$

Fig. 6. Fuzhou model F-1 *vs* Chinese "string bag"

work structure as required by the Fuzhou model F-1. It seems rather awkward that there exists quite a good number of tetranuclear Fe–S and Mo–S clusters with a core structure S_3M_3S having a configuration which is just the geometrical "inverse" of M_3S_3M while only very few clusters with a string-bag structure similar to the black Roussinate anion are known so far as we are aware of. It had been rather fortunate that as the first modelling compound in question, we had an opportunity to synthesize batches of the alkali black Roussinate $M[Fe_4S_3(NO)_7]$ (i.e., $M[\{(ON)Fe\}S_3\{Fe(NO)_2\}_3]$ (Fig. 5a) and the alkali red Roussinate $M_2[Fe_2S_2(NO)_4]$ (i.e., $M[\{(ON)_2Fe\}S_2\{Fe(NO)_2\}])$ (Fig. 5b) as well according to the standard procedures described in Brauer's: Handbuch der preparativen anorganischen Chemie, Bd.II. The synthesis of the black Roussinate is a "spontaneous self-assembly" reaction from $FeSO_4, M_2S, MNO_2$ and MOH:

$$26Fe^{2+} + 21NO^- + 17S^{2-} + 21H_2O \rightarrow 3[Fe_4S_3(NO)_7]^- + 14Fe(OH)_3\downarrow + 8S\downarrow,$$

and is usually rather tricky. The synthesis of the red Roussinate by converting the black Roussinate back into the red salt in a strong alkali solution

$$2[Fe_4S_3(NO)_7]^- + 4OH^- + H_2O \rightarrow 3[Fe_2S_2(NO)_4]^{2-} + N_2O\uparrow + 2Fe(OH)_3\downarrow$$

is even worse. However, in our joint ESR research project on the Roussinates with Professor W.C. Lin of the Department of Chemistry, University of British Columbia (Vancouver, B.C., Canada), then on sabbatical leave from U.B.C. (1976–77), it was found that both the black salt and the red salt gave rise practically to same ESR fission fragments in our work [1]. As a matter of fact, a casual examination of their structural models show clearly that the "black" monanion consisted of three edge-sharing rhomboidal fragments very much similar to the core of the "red" dianion. Indeed, if an appropriate volume of aqueous solution of the red Roussinate was introduced into a separatory funnel with an equal volume of ether on top of it (thus the ethereal layer remained colorless, as the red salt was insoluble in ether), and then a few drops of a dilute weak acid (e.g., acetic acid) were added, the etheral layer turned dark black at once on account of the practically instantaneous formation of the corresponding black salt, which was very soluble in ether:

$$4[Fe_2S_2(NO)_4]^{2-} + 6H^+ \rightarrow 2[Fe_4S_3(NO)_7]^- + N_2O\uparrow + 2H_2S\uparrow + H_2O.$$

A few drops of a sulfide-ion precipitant of medium strength (e.g., $FeCl_2, CoCl_2$ or $NiCl_2$) would do the same trick:

$$4[Fe_2S_2(NO)_4]^{2-} + 2M^{2+} + 2H^+ \rightarrow 2[Fe_4S_3(NO)_7]^- + N_2O\uparrow + 2MS\downarrow + H_2O$$

$$(M = Fe(II), Co(II), Ni(II)).$$

(It is worth noting here that a strong precipitant such as $HgCl_2$ would lead at

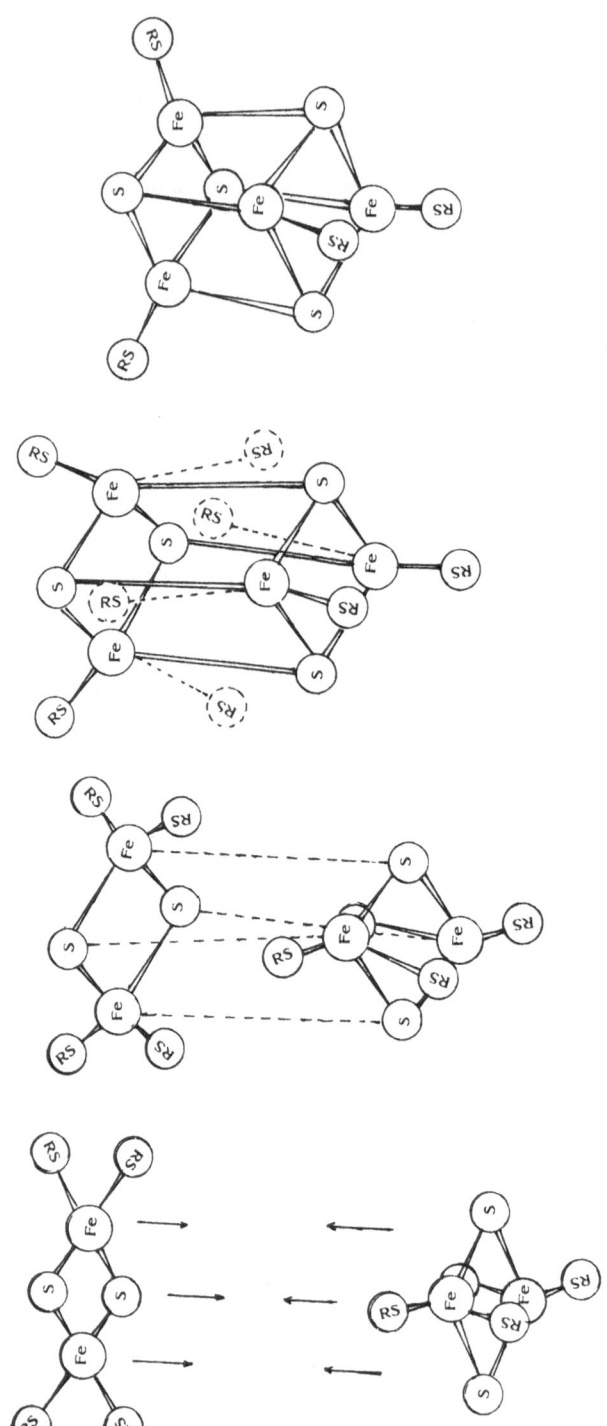

Fig. 7. Two cases of dimeric condensation of the following types: $2 [Fe_2S_2] \rightarrow [Fe_4S_4]$ or $[Fe_4S_3] + H_2S\uparrow$ or $S\downarrow$: **a** Formation of 4Fe-4S ferredoxin from 2Fe-2S ferredoxin: **b** Formation of black Roussinate from red Roussinate

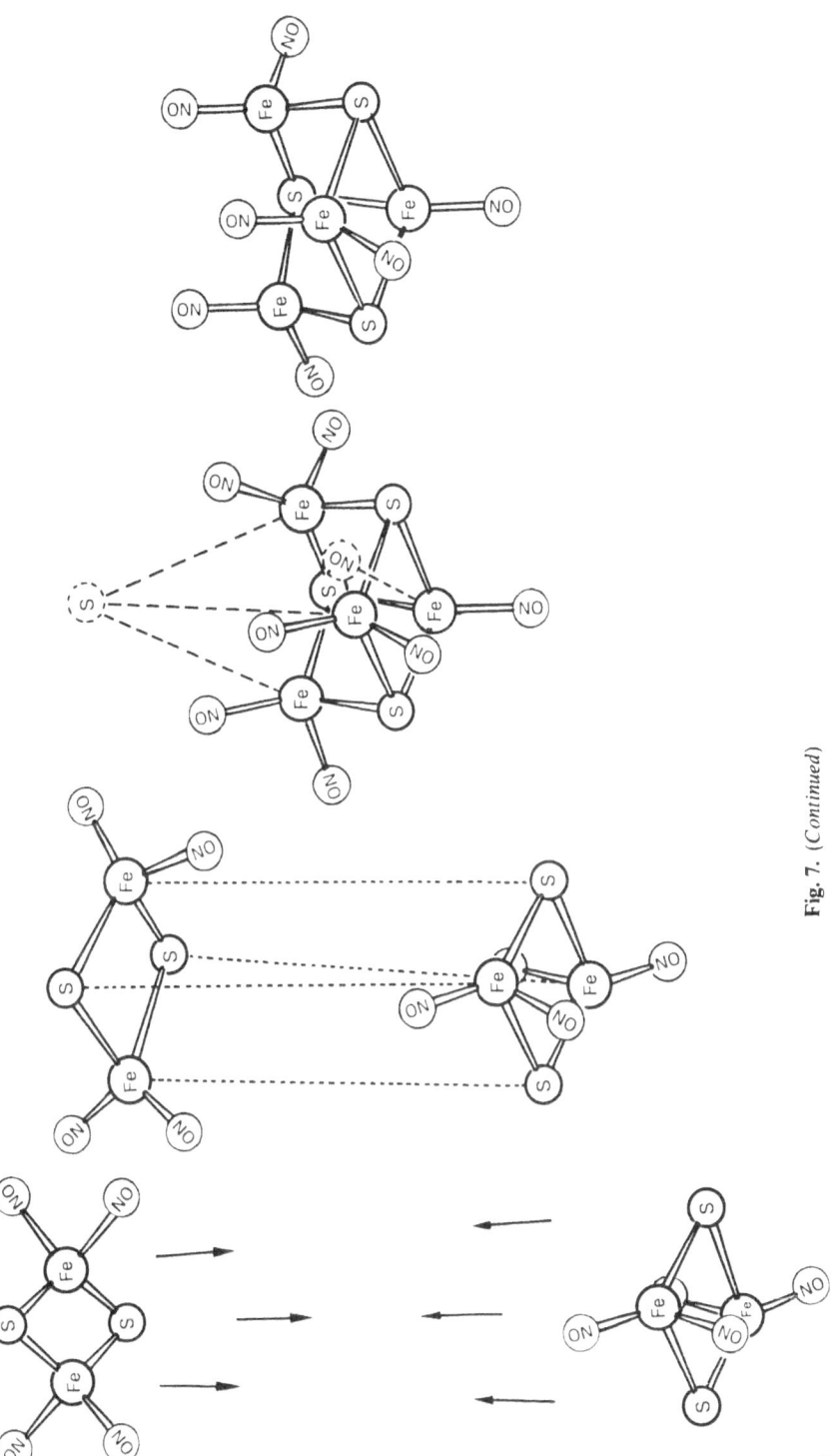

Fig. 7. (*Continued*)

b

once to a complete rupture of the red Roussinate framework; on the other hand, a very mild precipitant such as $ZnCl_2$ would not lead to any appreciable change at all in the red Roussinate.) It has been well-known that the 4Fe–4S ferredoxin $[Fe_4S_4(SR)_4]^{2-}$ can be easily formed from the 2Fe–2S ferredoxin $[Fe_2S_2(SR_4)]^{2-}$ by dimeric condensation (Fig. 7a). Thus these simple and yet interesting experiments led Lin to suggest that the formation of the black Roussinate from the red Roussinate should be another case of dimeric condensation (Fig. 7b). The only difference was that in this case one of the bridging S atoms turned out to be expelled before one of the eight NO ligands was split off. In other words, the Fe–S bond is weaker than the Fe–NO bond in the case of the red Roussinate, while for the 2Fe–2S ferredoxin, it is stronger than the Fe–SR bond. This has been confirmed in a quantum-chemical analysis carried out by Liu et al. [10]. In the meanwhile, Lin developed in Canada in 1979 a very neat "sureshot" method [11a] of synthesizing the black Roussinate by dissolving $FeSO_4$, MNO_2, M_2S, and MOH in stoichiometric proportions in ion-exchanged water; introducing the solution into a clean auto-clave; leading in pure CO to

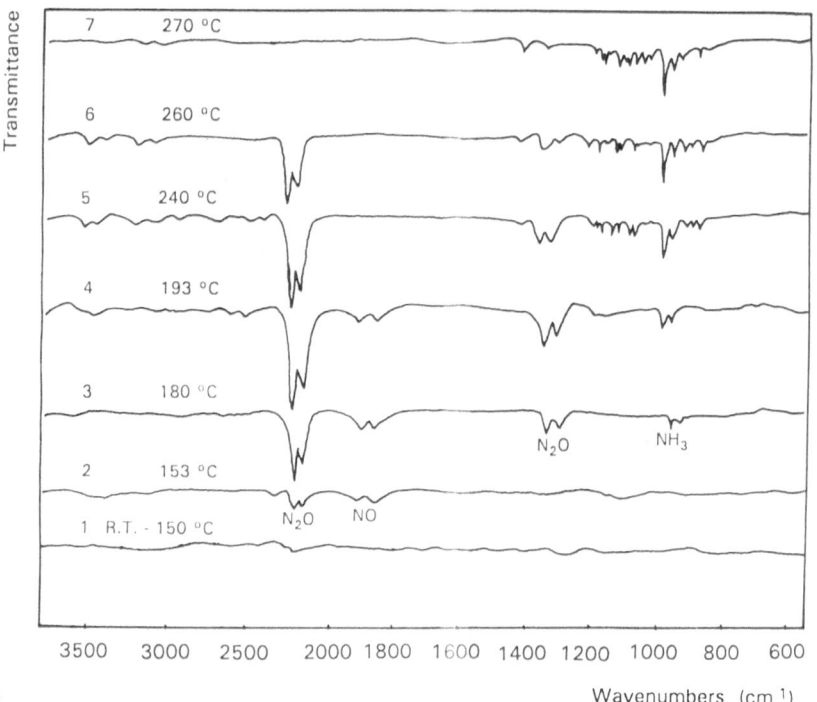

Fig. 8 a–d. IR spectra of gaseous phase alone in the course of reduction of the gaseous pyrolytic products NO, N_2O, and N_2 from pyrolyzed Roussinates by gaseous H_2: **a** $K[Fe_4S_3(NO)_7]$; **b** $[(CH_3)_4N]_2[Fe_2S_2(NO)_4]$; **8c, d** spectra of gaseous phase + solid phase

a pressure of ca. 30 atm.; heating the mixture up to 95–100 °C for an hour or so; and then allowing the whole system to cool down to room temperature. Upon letting off the pressure and opening the autoclave a whole batch of shiny needle-shaped black crystals is obtained in very high yield and can be easily recrystallized from ether to very high purity:

$$4Fe^{2+} + 7NO_2^- + 3S^{2-} + 6OH^- + 5CO$$

$$\xrightarrow[\sim 95\,°C]{\sim 30\,atm} [Fe_4S_3(NO)_7]^- + 5CO_3^{2-} + 3H_2O.$$

On the basis of this interesting finding, which has been confirmed by our research group in Fuzhou, we have been able to modify the stoichiometric proportions and develop this method to make the red Roussinate with similar

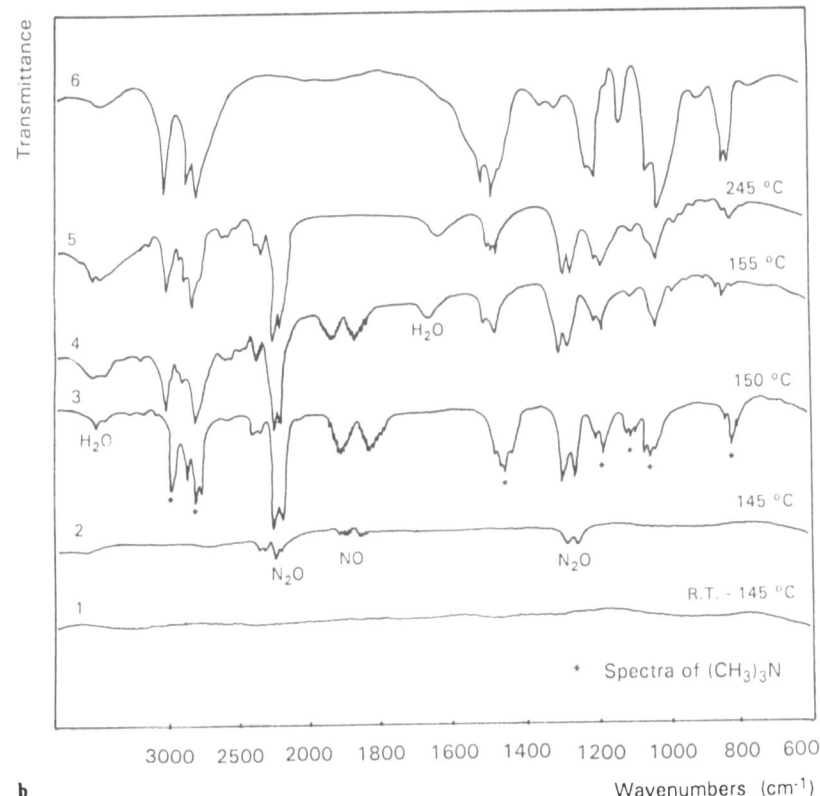

b

Fig. 8. (*Continued*)

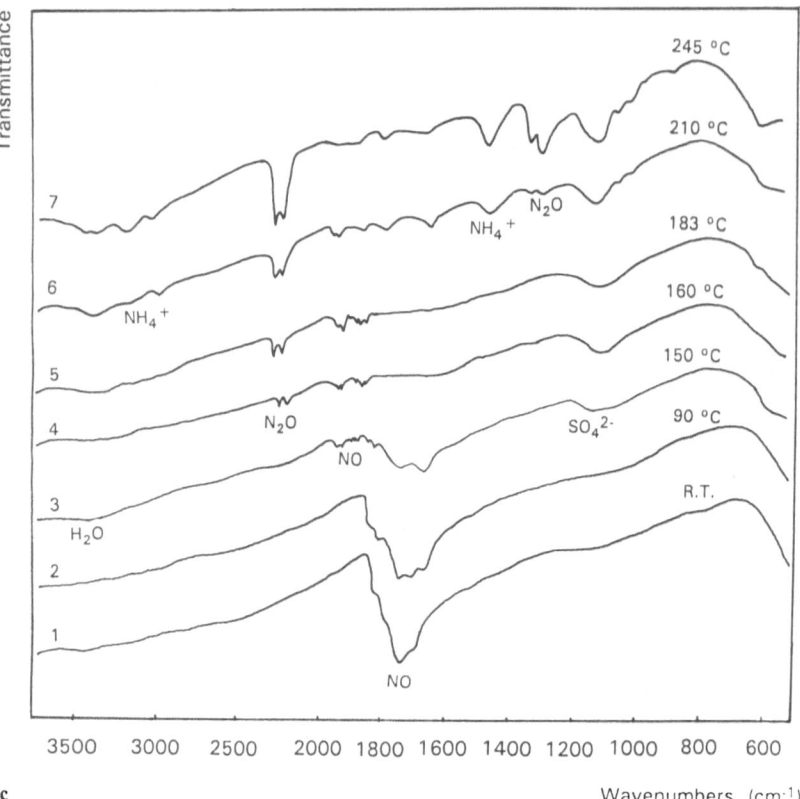

Fig. 8. (*Continued*)

high yield and high purity [11b]:

$$2Fe^{2+} + 4NO_2^- + 2S^{2-} + 4OH^- + 3CO$$

$$\xrightarrow[\sim 95\ C]{\sim 30\,atm} [Fe_2S_2(NO)_4]^{2-} + 3CO_3^{2-} + 2H_2O.$$

We have therefore been led to the development of a "modular assembly" or "unit construction" approach [12] as a method of rational synthesis of cubane-like transition-metal clusters. As an example, it is worth noting that we have succeeded in synthesizing $[Mo_xFe_{4-x}S_4(R_2dtc)_n]$, $(x = 0, 1, 2; n = 4, 5, 6)$, not by a "spontaneous self-assembly" reaction, but by a bimolecular modular assembly reaction of the type $[Fe_2S_2] + [Fe_2S_2] \rightarrow [Fe_4S_4]$ [13]:

$$2[Mg(DMF)_6][Fe_2S_2Cl_4] + 4R_2dtc^- \longrightarrow [Fe_4S_4(R_2dtc)_4]$$

$$(R_2 = Me_2, Et_2, -(CH_2)_4 -, -(CH_2)_5 -); \quad (1)$$

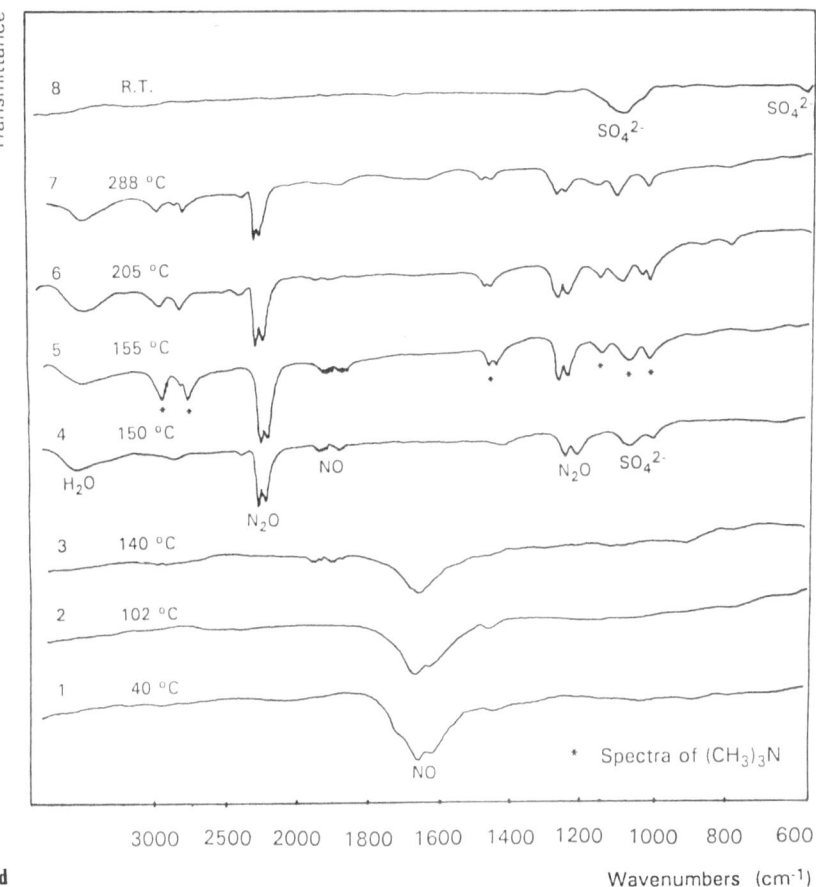

Fig. 8. (*Continued*)

$$[Mg(DMF)_6][Fe_2S_2Cl_4] + [Mg(DMF)_6][Cl_2FeS_2MoS_2] + 5R_2dtc$$
$$\longrightarrow [MoFe_3S_4(R_2dtc)_5] \quad (R_2 = Et_2, -(CH_2)_5-, Bz_2); \quad (2)$$

$$2[Mg(DMF)_6][Cl_2FeS_2MoS_2] + 6R_2dtc^- \longrightarrow [Mo_2Fe_2S_4(R_2dtc)_5]$$
$$+ [Mo_2Fe_2S_4(R_2dtc)_6] \quad (R_2 = -(CH_2)_4-). \quad (3)$$

The second question we have in mind is how much the cooperative effect of the string-bag structure of Fuzhou model F-1 contributes to the reduction of simple substrates such as N_2 and C_2H_2. (Ms.) L.-N. Zhang made a simple variable-temperature attachment [14a] for use on the Perkin–Elmer Model 577 IR grating spectrophotometer so as to follow changes in the IR absorption spectra of reaction intermediates and products in gaseous phase alone or in

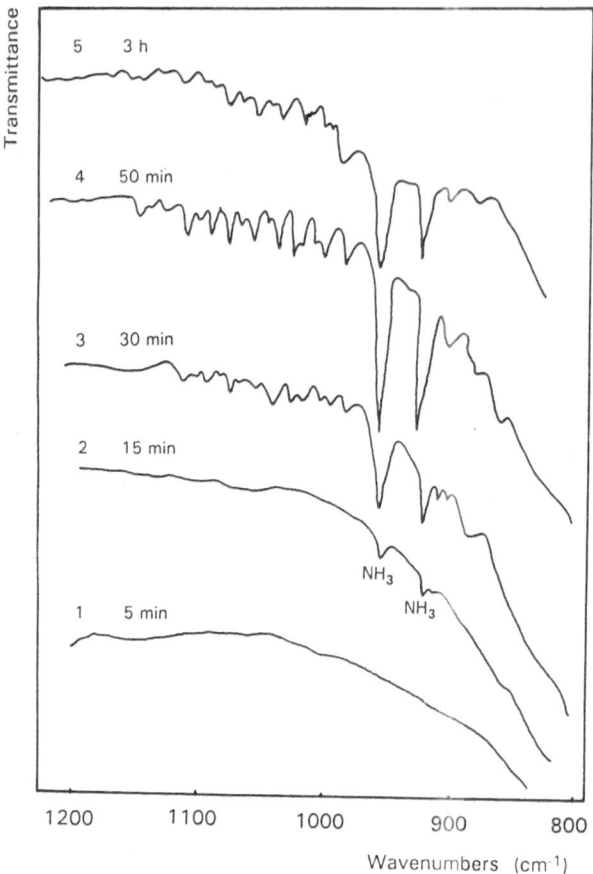

Fig. 9 a, b. IR spectra of gaseous phase alone in the course of reduction of the decomposition products of Roussinates by KBH_4 in $Na_2B_4O_7$-NaOH buffer (time measured from the instant of dropping in the buffer: **a** $K[Fe_4S_3(NO)_7]$; **b** $[(CH_3)_4N]_2[Fe_2S_2(NO)_4]$

gaseous phase + solid phase at a series of pre-assigned temperature steps up to ca. 700 °C and thus track down the reaction path which a pyrolyzed solid phase and its decomposition products undergo. In collaboration with her co-workers, she made a rather detailed comparative study of the black Roussinate $K[Fe_4S_3(NO)_7]$ vs the red Roussinate $[(CH_3)_4N]_2 [Fe_2S_2(NO)_4]$ [14b–e] under the following conditions: (1) reduction of their gaseous pyrolytic products NO, N_2O, and N_2 by gaseous H_2 (Figs. 8a–d); (2) reduction of the decomposition products of Roussinates by KBH_4 in a $Na_2B_4O_7$–NaOH buffer (Fig. 9); (3) reduction (in the presence of a Roussinate) of C_2H_2 to C_2H_4 by KBH_4 in a $Na_2B_4O_7$–NaOH buffer (Fig. 10). As a blank test for reaction condition (1), they even ran the reduction of gaseous N_2 by gaseous H_2 in the presence of a mixture

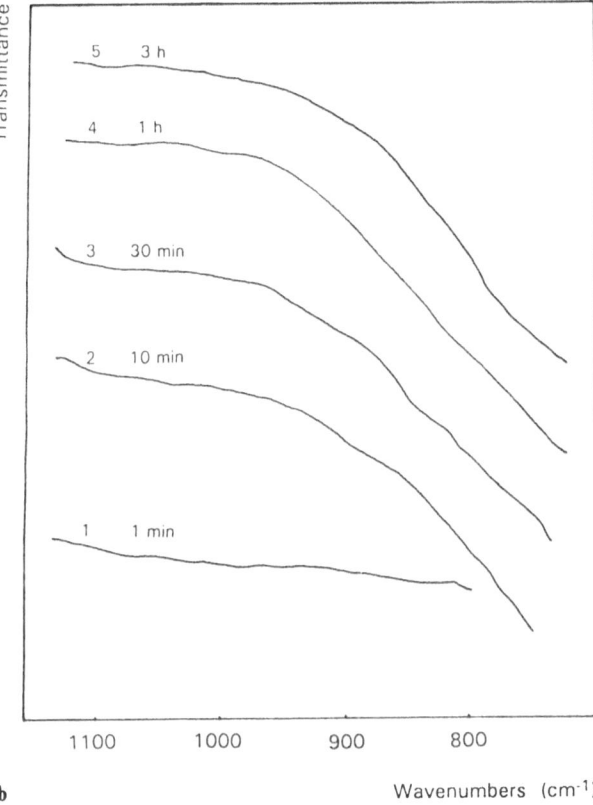

b   Wavenumbers (cm⁻¹)

Fig. 9. (*Continued*)

of presumed solid pyrolytic products $FeSO_4$, $Fe_2(SO_4)_3$, K_2SO_4, Fe_2O_3, and FeS of the Roussinates plus KNO_2 (Fig. 11). Under reaction conditions (1) and (2), NH_3 and/or NH_4^+ were found to be formed in the case of black Roussinate. This seems to imply that the "string-bag" structure is indeed a satisfactory prerequisite for the reduction of N_2, and thus confirms to a good extent our qualitative structural criteria for the reduction of such a molecule of unusual inertness. This is further supported by the IR spectra for reaction condition (1) for the black Roussinate, extended to the far IR region of ca. $200\,cm^{-1}$ (Fig. 12), which shows clearly that NH_3 does not cease to be formed until the "string-bag" framework structure of $[Fe_4S_3]$ collapses altogether, leaving Fe_2O_3 or Fe_3O_4 without exhibiting noticeable cooperative effect of the "string-bag" configuration any longer. On the other hand, under reaction condition (3), C_2H_4 was formed from both Roussinates. This is understandable, as C_2H_2 can be coordinated double side on either to two of the three "neck" Fe atoms (indeed plus one single end-on to the third "neck" Fe atom) in the case of the black Roussinate or to the rhomboidal $[Fe_2S_2]$ (three in the case of black Roussinate, but one only in the case of red Roussinate).

The plausibility of the "string-bag" Fuzhou model F-1 as a basic structural unit in the active center of nitrogenase has been further studied by attempted synthesis of "string-bag" modelling compounds along two different synthetic routes, leading to two series of new crystals, the "G" series and the "F" series, some of which have been shown to exhibit N_2 and C_2H_2 reduction capability under quasi-bioassaying conditions beyond experimental uncertainty.

Anaerobic reaction of EtMgBr (or ButMgCl), $(NH_4)_2MoS_4$, and $FeCl_3$ (in molar ratio 11:1:4) in ether-THF leads to the formation of a black precipitate,

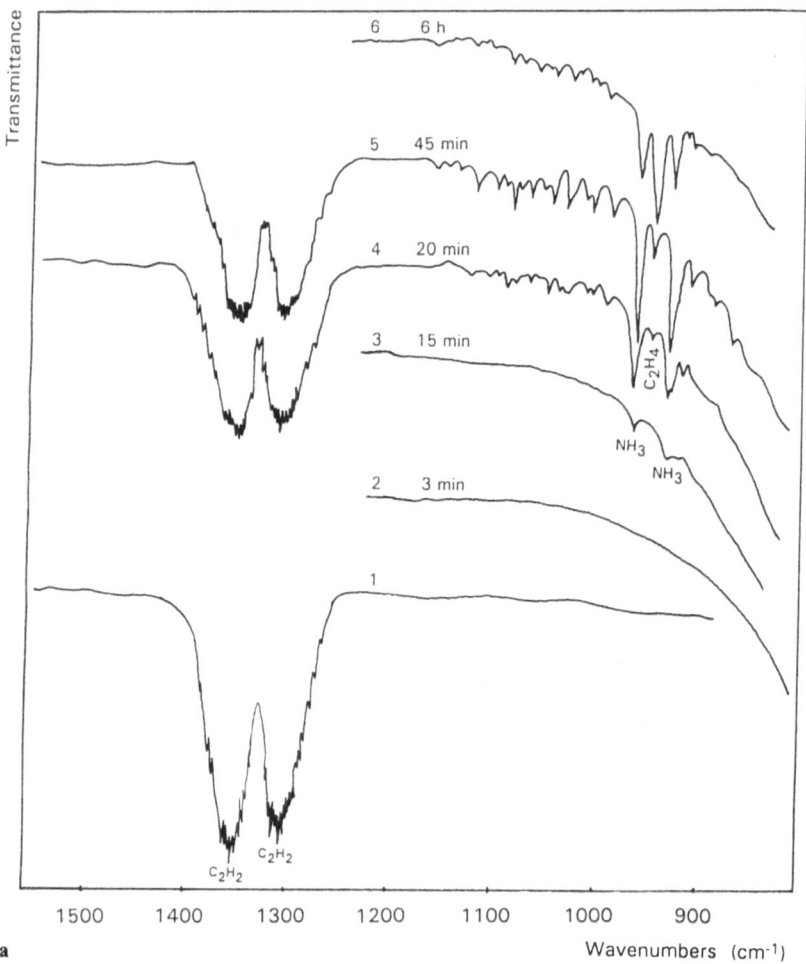

Fig. 10 a, b. IR spectra of gaseous phase alone in the course of reduction of C_2H_2 to C_2H_4 in the presence of Roussinates by KBH_4 in $Na_2B_4O_7$-NaOH buffer (time measured from the instant of dropping in the buffer: **a** $K[Fe_4S_3(NO)_7]$; **b** $[(CH_3)_4N]_2[Fe_2S_2(NO)_4]$

from which three different species of crystals GA, GB, GC of the new "G" series [15b] (with Mo contents ca. 10, 2.5, and 0.4% by weight, respectively) have been separated by fractional crystallization using DMF or DMF-THF. Elemental analyses of a typical GB crystal (designated as G-2) give $[Mg(DMF)_6]$ $(MoFe_{10}S_7Cl_8Br_{10})$. The anionic nature of the Fe-containing cluster is confirmed by paper electrophoresis. Thus, this cluster had been once mistaken to be the first modelling compound with the structure of Fuzhou model F-2. After combination with crude extracts from the *Azotobacter vinelandii* UW45 mutant (whose activity had not been checked by any standard assaying), this specimen reduced C_2H_2 to C_2H_4 in the presence of $Na_2S_2O_4$ and an ATP-generating system, with a specific activity of 1.1×10^{-2} nmol C_2H_4/nmol Mo/min. Under similar assaying conditions, reduction of N_2 to NH_3 was demonstrated using isotopically labelled N_2. In contrast, neither GA nor GC have exhibited any such C_2H_2 or N_2 reduction activity under similar quasi-bioassaying conditions. However, in the presence of KBH_4 as a reducing agent, all samples of GA, GB, and GC crystals reduced C_2H_2 to C_2H_4, with GB

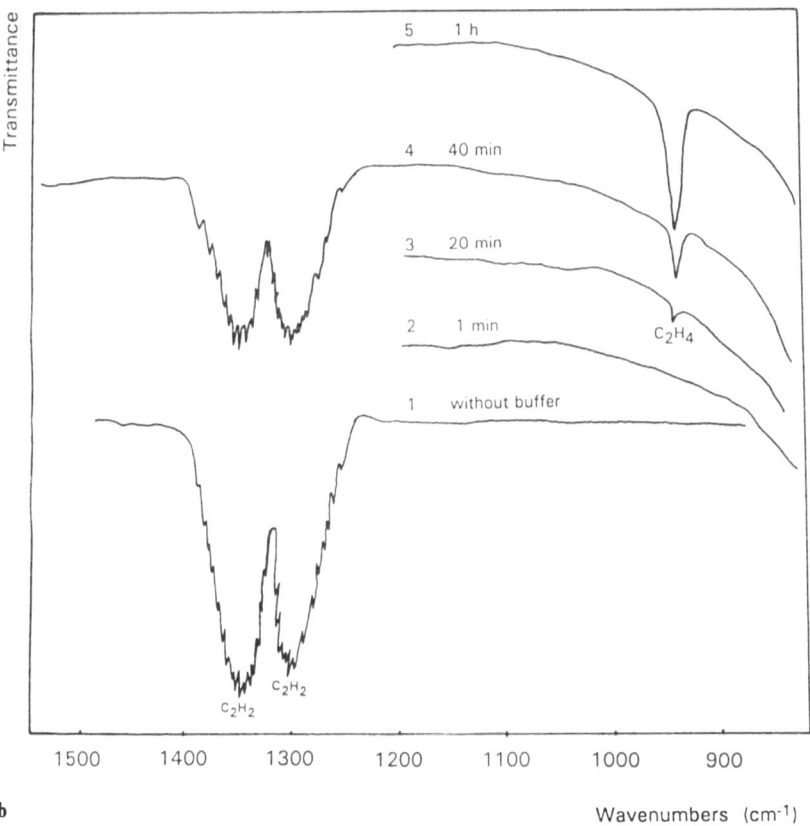

b

Wavenumbers (cm⁻¹)

Fig. 10. (*Continued*)

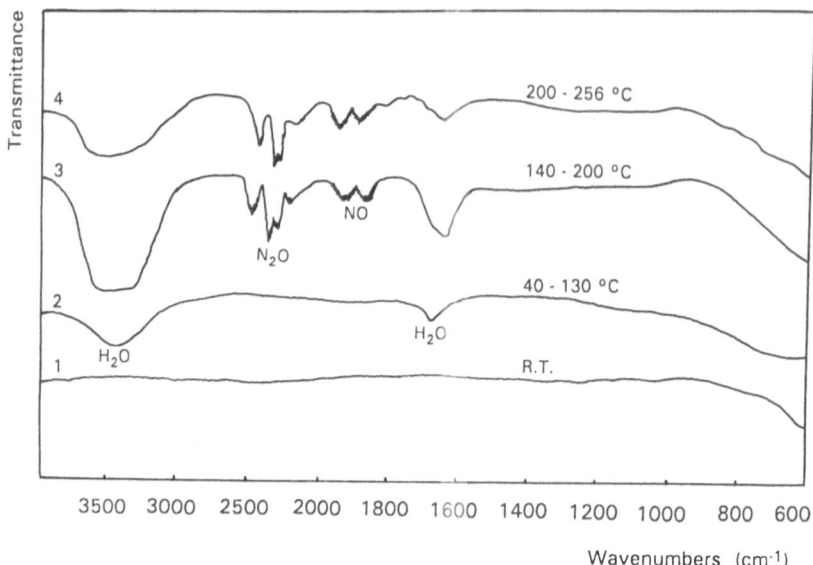

Fig. 11. IR spectra of gaseous phase alone in the course of reduction of gaseous N_2 by gaseous H_2 (admixed with N_2 in ratio ca. 1:3 by volume) in the presence of the presumed solid pyrolytic products $FeSO_4$, $Fe_2(SO_4)_3$, K_2SO_4, Fe_2O_3, FeS of the Roussinates $K[Fe_4S_3(NO)_7]$ and $K_2[Fe_2S_2(NO)_4]$ plus KNO_2

exhibiting the highest activity of 7.19 nmol C_2H_4/nmol Mo/min and a selectivity for C_2H_4 of over 90%.

X-ray structure analyses reveal beyond our expectation that all the "G series" crystals obtained after recrystallization from DMF or DMF-THF have practically the same crystallographic parameters: $a = 11.27\,\text{Å}$, $b = 9.47\,\text{Å}$, $c = 9.25\,\text{Å}$, $\alpha = 87.2°$, $\beta = 71.2°$, $\gamma = 74.9°$; space group $C_1^i - P1$; $Z = 1$ (referred to Mg). For a typical GA crystal with Mo:Fe:S = 1:1.3:4.1, the structural formula for the cluster dianion has been found to be K_A:

$$\left[\begin{smallmatrix} S \\ \\ S \end{smallmatrix} \!\!>\!\! Mo \!\!<\!\! \begin{smallmatrix} S \\ \\ S \end{smallmatrix} \!\!>\!\! Fe \!\!<\!\! \begin{smallmatrix} Cl \\ \\ Cl \end{smallmatrix} \right]^{2-},$$

with a molecular configuration identical with that reported by Tieckelmann et al. [16]. Very careful refinement of this structure determination to $R = 0.050$ has shown that the cluster anion possibly contains 10% of K_D:

$$\left[\begin{smallmatrix} Cl \\ \\ Cl \end{smallmatrix} \!\!>\!\! Fe \!\!<\!\! \begin{smallmatrix} Cl \\ \\ Cl \end{smallmatrix} \!\!>\!\! Fe \!\!<\!\! \begin{smallmatrix} Cl \\ \\ Cl \end{smallmatrix} \right]^{2-}$$ statistically admixed with K_A. In contrast to it, for

two typical GB crystals with empirical formulas approximating $[Mg(DMF)_6]$ $[MoFe_{10}S_7Cl_8Br_{10}]$ and $[Mg(DMF)_6]$ $[MoFe_8S_7Br_{10}]$, there are at least three different types of cluster dianions, most likely with molecular configurations and sizes quite similar to K_A; namely, K_B: $\left[\begin{smallmatrix} X \\ \\ X \end{smallmatrix} \!\!>\!\! Mo \!\!<\!\! \begin{smallmatrix} S \\ \\ S \end{smallmatrix} \!\!>\!\! Fe \!\!<\!\! \begin{smallmatrix} X \\ \\ X \end{smallmatrix} \right]^{2-}$, K_c:

$\left[\begin{smallmatrix} X \\ \\ X \end{smallmatrix} \!\!>\!\! Fe \!\!<\!\! \begin{smallmatrix} S \\ \\ S \end{smallmatrix} \!\!>\!\! Fe \!\!<\!\! \begin{smallmatrix} X \\ \\ X \end{smallmatrix} \right]^{2-}$, K_D: $\left[\begin{smallmatrix} X \\ \\ X \end{smallmatrix} \!\!>\!\! Fe \!\!<\!\! \begin{smallmatrix} X \\ \\ X \end{smallmatrix} \!\!>\!\! Fe \!\!<\!\! \begin{smallmatrix} X \\ \\ X \end{smallmatrix} \right]^{2-}$ (X = Cl, Br). These

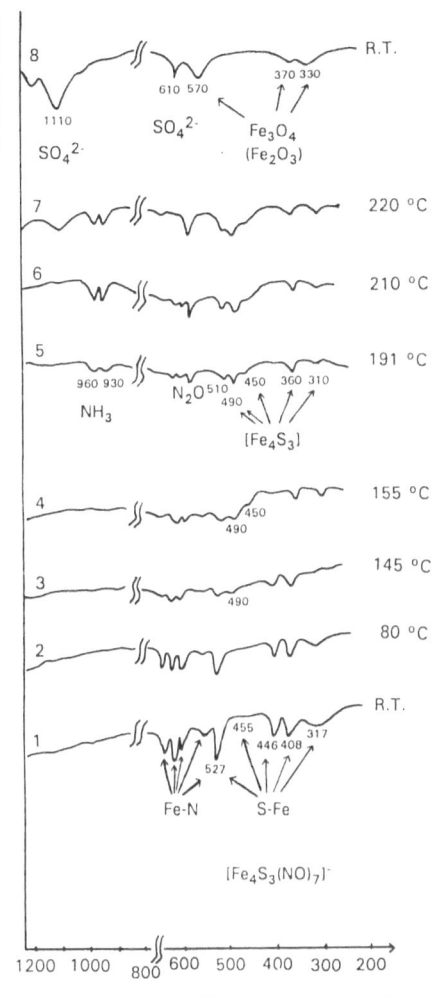

Fig. 12. Same as Fig. 8b (c), except that it is extended down to $200\,\mathrm{cm}^{-1}$ in the far IR region

determinations have been refined to R = 0.068. We are thus led to he conclusion that at least four different cluster dianions with similar structural formulas $\begin{bmatrix} T \diagdown \diagup B \diagdown \diagup T \\ T \diagup M \diagdown B \diagup M \diagdown T \end{bmatrix}^{2-}$ (M = Fe, Mo; T = S, Cl, Br) and practically identical configurations and sizes are statistically distributed within a clathrate structure, each of these dianions having the right shape and size to fit into the space inside the triclinic unit cell with the eight corners occupied by the $[\mathrm{Mg(DMF)}_6]^{2+}$ dications (Fig. 13) [17]. Apparently it is possible to take simple combinations of these cluster anions to make up Mo:Fe:S ratios approximating Shah and Brill's ratio for the FeMo-co. This seems to lend further support for the "unit

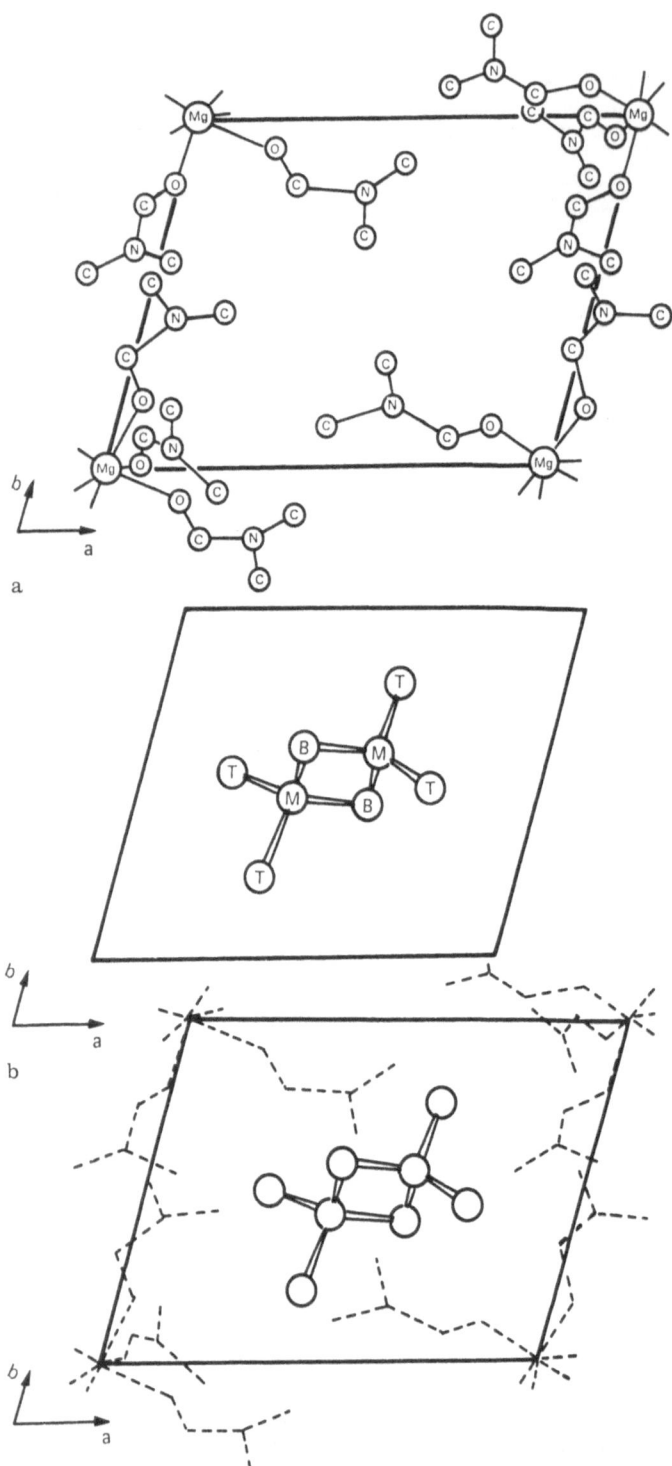

construction" approach for the rational syntheses of cubane-like Mo–Fe–S clusters mentioned above.

The "F series" crystals are dark-colored compounds, with a composition approximating $(Et_4N)_2$ $[Mo_xFe_{4-x}S_4(SBu^t)_4]$ ($x = 0.07 - 0.5$) from elemental analyses and IR spectra, obtained in methanol under anaerobic conditions from the reaction of $FeCl_2$ and $(NH_4)_2MoS_4$ with $NaSBu^t$ plus NaSH (in molar ratios 6:1:24:4) as a mixed reductant [16]. These compounds have demonstrated their ability to reduce C_2H_2 to C_2H_4 (with good selectivity) and N_2 to NH_3 in a quasi-bioassaying catalyzed reaction after combination with *Azotobacter vinelandii* UW45 mutant extracts plus an ATP-generating system. They have also exhibited good activity for reduction of C_2H_2 to C_2H_4 in the presence of KBH_4 alone. Preliminary X-ray structure analysis gives $a = 11.83$ Å, $c = 19.18$ Å; space group $I_4^2 - I_4^-$ or $D_{2d}^{11} - I_4^-2m$; $Z = 4$ (referred to Et_4N), with clear indication that the dianions are essentially *closo* cubane-like $[Fe_4S_4^*(SBu^t)_4]^{2-}$ clusters. The electron occupancy ratios $Z_{occ}^{Fe}/Z_{nom}^{Fe} = 1.06$ and $Z_{occ}^{s*}/Z_{nom}^s = 1.02$ at the Fe and S* sites have been rather carefully measured for a typical "F series" crystal containing ca. 3% Mo and 18% Fe (corresponding to atomic ratio 1:10). The occupancy ratio for Fe can be easily accounted for, insofar as $Z_{Mo}/Z_{Fe} = 1.61$. Thus there must be a statistical distribution of Mo atoms in the Fe sites. This interpretation has been confirmed by a preliminary Mössbauer spectra measurement. It is worth noting that the acid-labile sulfur S* is chemically very close to S^{2-} and thus Z_{s*}/Z_s 1.12. Thus, there may be some vacancies at the S* sites, suggesting that there may exist a small percentage of *nido* cubane-like clusters, some of which might even have the "string-bag" $Fe(S^*)_3$ $(MoFe_2)$ configuration of Fuzhou model F-1. This seems to add another piece of weak "evidence" for the "string-bag" structure as a possible basic structural unit for the active center of nitrogenase.

Before closing these introductory remarks, it is worth pointing out that our preliminary research work on the chemical modelling of biological nitrogen fixation in the New China during the past two decades has helped in paving new grounds in the field of bio-inorganic chemistry on one hand and in the development of the chemistry of transition-metal clusters (Fe–S, Mo–S, and Mo–Fe–S clusters in particular) on the other. We look forward to the day when further elucidation of the nature of chemical bonding in the quasi-aromaticity (a concept recently established on sound experimental and theoretical basis and studied in detail by us [18]) in the puckered $[Mo_3S_3]$ six-membered rings of certain $[Mo_3S_4]^{4+}$ clusters may reveal some new "modular assembly" approach for rational syntheses of chemical modelling cluster compounds designed according to our proposed structural models.

Fig. 13a–c. The clathrate structure of $\left[\begin{smallmatrix} T \\ T \end{smallmatrix}\!\!>\!\!M\!\!<^{\!\!B}_{\!\!B}\!\!>\!\!M\!\!<^{\!\!T}_{\!\!T}\right]^{2-}$ in the crystal unit cell of $[Mg(DMF)_6]$ $T_2MB_2MT_2]$: **a** $[Mg(DMF)_6]$ alone; **b** $[T_2MB_2MT_2]$ alone; **c** $[T_2MB_2MT_2]^{2-}$ fitted into the space inside the triclinic unit cell with the eight $[Mg(DMF)_6]^{2+}$ dications in position

We would like to acknowledge in this connection our indebtedness to Professor W.C. Lin of the University of British Columbia, Professor A.-Q. Tang of Jilin University, and Professor Q.-R. Cai of Xiamen University for enlightening discussions and helpful comments.

References

1. (a) Nitrogen Fixation Res Group, Fujian Inst of Res Struct of Matter, Chin Acad Sci (1975) A preliminary model for the active center of the catalytic nitrogen-fixing nitrogenase, with discussion on the structural criteria for the chemical activation of the dinitrogen molecule via coordination, Kexue Tongbao (Chinese Science Bull) 12:540 (in Chinese); (b) Lu JX (1980) Composite "string bag" cluster model for the active center of nitrogenase, in: Newton WE, Orme-Johnson WH (eds) Nitrogen fixation, vol I. University Park Press, Baltimore, p 343
2. Yatsimirskii KB (1971) Pure Appl Chem 27: 251
3. (a) Nitrogen Fixation Res Group, Dept of Chemistry, Kirin (Jilin) Univ (1974); (b) Hsu CC (Xu JQ) (1981) Scientia Sinica 17: 193; (c) Hsu CC (1980) A quantum-chemical theory of transition-metal-dinitrogen complexes, in: Newton WE, Orme-Johnson WH (eds) Nitrogen fixation, vol I, Univ Park Press, Bultimore, p 317; (d) Tang AQ, Xu JQ (1991) A chemical bond theory of transition-metal-dinitrogen complexes. This Monograph
4. (a) Jonas K (1973) Angew Chem Int Edn 12: 997; (b) Krüger C, Tsay YM (1973) Angew Chem Int Edn 12: 998; (c) Jonas K, Brauer DJ, Krüger C, Roberts PJ, Tsay YH (1976) J Amer Chem Soc 98: 74; (d) Fischler I, von Gustorf EK (1975) Naturwiss 62: 63
5. (a) Nitrogen-Fixation Res Group, Dept of Chemistry, Xiamen Univ (1974) Acta Univ Amoien (Nat Sci) 1: 111 (in Chinese); (b) Nitrogen-Fixation Res Group, Lab of Catalysis, Dept of Chem, Xiamen Univ (1976) A model of nitrogenase active-center and mechanism of nitrogenase catalysis. Scient Sinica 19: 460; (c) Tsai KR, Lin ST, Wan HL (1979) Acta Univ Amoien (Nat Sci) 18 (2): 30 (in Chinese, with English abstract); (d) Tsai KR (1980) Development of a model of nitrogenase active center and mechanism of nitrogenase catalysis, in: Newton WC, Orme-Johnson WH (eds) Nitrogen fixation, vol 1: 373; (e) Tsai KR, Wan HL, Zhang HT, Xu LS (1991) Studies on the mechanism of nitrogenase catalysis-substrates-cluster-coordination-chemistry approach. This Monograph
6. Nitrogen Fixation Group, Chem Dept, Nanking (Nanjing) Univ (1977) Studies on iron catalysts of ammonia synthesis promoted by alkali metal. II. Acta Chim Sinica 35: 141 (in Chinese, with English abstract)
7. Huang KH (1978) A model of active center on ammonia synthesis iron catalysts and coordination activation of N_2. Acta Univ Amoien (Nat Sci) 112 (in Chinese, with English abstract)
8. You CB, Song W, Zeng D, Tsai KR (1991) ATP binding to nitrogenase and ATP-driven electron transfer in nitrogen fixation. This Monograph:
9. Unpublished work
10. (a) Cao HZ, Liu CW, Lu JX (1986) The electronic structures of $[Fe_2S_2(SH)_4]^{2-}$ and $[Fe_4S_4(SH)_4]^{2-}$. Acta Chim Sinica 44: 1197 (in Chinese, with English abstract); (b) Liu CW, Cao HZ, Lu JX, Zheng SJ, Liu RZ (1987) The electronic structures of red Roussinate and black Roussinate. Acta Chim Sinica 1; (c) Liu CW, Cao HZ, Lu JX (1989) J Mol Struct (Theochem) 183: 1
11. (a) Personal Communication from Lin WC; (b) Lin XT, Zheng A, Lin SH, Huang JL, Lu JX (1982) Chin J Struct Chem 1: 79 (in Chinese, with English abstract)
12. Lu JX, Zhuang BT (1989) Chin J Struct Chem 8: 233 (in Chinese, with English abstract)
13. Liu QT, Huang LR, Lu JX (1990) Modular assembly synthesis of monocubane-like clusters $[Mo_xFe_{4-x}S_4]$ from heterobinuclear rhomboidal complexes and the crystal structure of $[Mo_2Fe_2S_4(C_4H_8dtc)_6]$ $3.5C_2H_2Cl_4$. Science in China XXX-XXX
14. (a) Zhang LN, Huang DR, Peng ZR (1980) A simple variable-temperature IR spectrophotometer attachment for use in tracking reaction intermediates and products in pyrolysis of a solid-phase reactant and its applications. Bull Fujian Inst Res Struct Matter 1: 71 (in Chinese); (b) Zhang LN, Huang DR, Peng ZR, He LJ, Zheng Y, Cai YK (1981) A comparative IR spectral study

of the coordination activity of the black Roussinate anion vs the red Roussinate anion. Bull Fujian Inst 1: 23 (in Chinese); (c) Zhang LN, Liu ZP, He LJ, Zheng Y, Huang DR (1983) Chin J Struct Chem 2: 11; (d) Zhang LN, Liu ZP, He LJ, Zheng Y, Hu YX (1984) Chin J Struct Chem 3: 161; (e) Zhang LN, He LJ, Liu ZP, Zheng Y, Huang DR (1985) Chin J Struct Chem 4: 331

15. (a) Lu JX (1981); (b) Huang WK, Tan Z, Cui YX, Zhao XT, Lin CZ, Jiang FL, Huang LR, Lu JX (1981); (c) Lu JX, Zhuang BT, Kang BS, Zhuang HH, Zhang MY (1981) in: Gibson AAH, Newton WE (eds) Current Perspectives in Nitrogen Fixation. Australian Acad Sci, Canberra, pp 50, 345, 346

16. Tieckelmann RH, Silvis HC, Kent TA, Huynh BH, Waszezak JV, Teo BK, Averill BA (1980) J Amer Chem Soc 102: 5550

17. Huang LR, Lu JX (1983) Structural Chemistry of G Series Modelling Compound for the Active Center of Nitrogenase. (a) Crystal and Molecular Structure of [Mg·6DMF] [MoFeS$_4$Cl$_2$]. Scient Sinica 3 B: 193; (b) Crystal and molecular structure of [Mg·6DMSO] [MoFeS$_4$Cl$_2$]. Scient Sinica 3 B: 199

18. (a) Huang JQ, Lu SF, Shang MY, Lin XT, Huang MD, Lin YH, Wu DM, Zhang JL, Lu JX (1987) Chin J Struct Chem 6: 219 (in Chinese, with English abstract); (b) Huang JQ, Huang JL, Shang, MY, Lu SF, Lin XT, Lin YH, Huang MD, Zhang HH, Lu JX (1988) Pure Appl Chem 60: 1185; (c) Lu JX (1989) Chin J Struct Chem 8: 327 (in Chinese, with English abstract); (d) Chen ZD, Li J, Cheng WD, Huang JQ, Lin CW, Lu JX (1990) A preliminary quantum-chemical analysis of the nature of quasi-aromaticity of the puckered [Mo$_3$S$_3$] rings in certain [Mo$_3$S$_4$]$^{4+}$ clusters. Chin Sci Bull (Kexue Tongbao) 35: 1698

CHAPTER 2
A Chemical Bond Theory
of Transition-Metal-Dinitrogen Complexes

Ao-Qing Tang and Ji-Qing Xu

The Nitrogen Fixation
and its Research in China
Editor. Guo-fan Hong
(C) Springer-Verlag Berlin Heidelberg 1992

$$\begin{array}{c} \diagdown \diagup \\ M'\text{-}3 \end{array}$$

$$\text{of the} \quad \begin{array}{c} \diagup \!\!\!^{N-2} \\ M\!-\!1 \diagdown \!\!\!\!\diagup M\!-\!1' \\ \diagdown \!\!\!_{N-3} \\ | \\ M'\!-\!4 \end{array} \quad \text{Type} \quad \dots \dots \dots \dots \dots \dots \quad 53$$

Abbreviations

Bu^t	*tert*-Butyl (C_4H_9)
bz	Benzyl (C_7H_7)
dmpe	Ethylene bisdimethyl phosphine ($Me_2PCH_2CH_2PMe_2$)
dppe	Ethylene bisdiphenyl phosphine ($Ph_2PCH_2CH_2PPh_2$)
en	Ethylene diamine ($C_2H_8N_2$)
Et	Ethyl (C_2H_5)
Me	Methyl (CH_3)
Me$_8$[16]aneS$_4$	3,3,7,7,11,11,15,15-Octamethyl-1,5,9,13-Tetrathiacyclohexadecane
Ph	Phenyl (C_6H_5)
Pr^i	Isopropyl (C_3H_7)
Pr^n	*n*-Propyl (C_3H_7)
Py	Pyridine (C_5H_5N)
THF	Tetrahydrofuran (C_4H_8O)
TMP	5,10,15,20-Tetramesitylporphyrine

1 Introduction

Hundreds of dinitrogen complexes of transition metals have been successfully synthesized since the first one was prepared (Allen and Senoff 1965). The electronic structures of dinitrogen complexes have been studied using various quantum-chemical methods, for example, semiexperienced (EHMO, CNDO, and INDO etc.), $X\alpha$, and ab initio (Chatt et al. 1978; Pelican and Boca 1984). In order to investigate the change tendency for stability and charge density on the nitrogen atoms of dinitrogen complexes with a number of d electron, using HMO and graphical methods (Tang and Kiang 1976, 1977), we have carried out molecular orbital calculation of mono-, bi-, and trinuclear dinitrogen complexes with a variety of reasonable coordination structures (Nitrogen Fixation Research Group 1974; Hsu CC 1980; Xu JQ et al. 1979, 1980a, 1980b, 1981).

2 Mononuclear Dinitrogen Complexes

2.1 Molecules with Single End-on Coordination Structure of the M-3-N-2-N-1 Type

We used a right-handed coordinate system with the z axis oriented in the direction of the molecular axis of the N_2. For this coordination type, two equivalent sets of bonding orbitals can be formed from the two different sets of atomic orbitals $(d_{xz}, p_x^{(2)}, p_x^{(1)})$ and $(d_{yz}, p_y^{(2)}, p_y^{(1)})$. These molecular orbitals (MOs) are twofold degenerate. Now consider the MO formed from the atomic orbital (AO) set $(d_{xz}, p_x^{(2)}, p_x^{(1)})$, and described by the linear combination of atomic orbitals (LCAO):

$$\psi = C_1\phi_1 + C_2\phi_2 + C_3\phi_3$$

where ϕ_1, ϕ_2, and ϕ_3 represent the AOs $p_x^{(1)}, p_x^{(2)}$, and d_{xz}, respectively. Using MO graphical theory, we obtained the following results. First, the MO graph can be written as:

$$
\begin{array}{ccc}
-\eta & -1 & \\
\bigcirc \text{———} \bigcirc \text{———} \bigcirc \\
x-\delta \quad\quad x \quad\quad x \\
3 \quad\quad 2 \quad\quad 1
\end{array}
$$

with $x = \dfrac{\varepsilon - \alpha}{\beta}, \delta = \dfrac{\alpha' - \alpha}{\beta'}$ and $\eta = \dfrac{\beta'}{\beta}$ where $\alpha(\alpha'), \beta(\beta')$, and ε denote, respectively, the AO energy of $p_x^{(1)}$ or $p_x^{(2)}(d_{xz})$, the interaction energy between ϕ_1 and ϕ_2 (ϕ_2 and ϕ_3) and the MO energy.

Second, the eigen equation corresponding to this MO graph is:

$$(x - \delta)(x^2 - 1) - \eta^2 x = 0$$

Third, the LCAO coefficients in this MO can be obtained explicitly through the equations:

$$C_1(x) = \frac{1}{N(x)} x(x - \delta) - \eta^2$$

$$C_2(x) = \frac{1}{N(x)}(x - \delta)$$

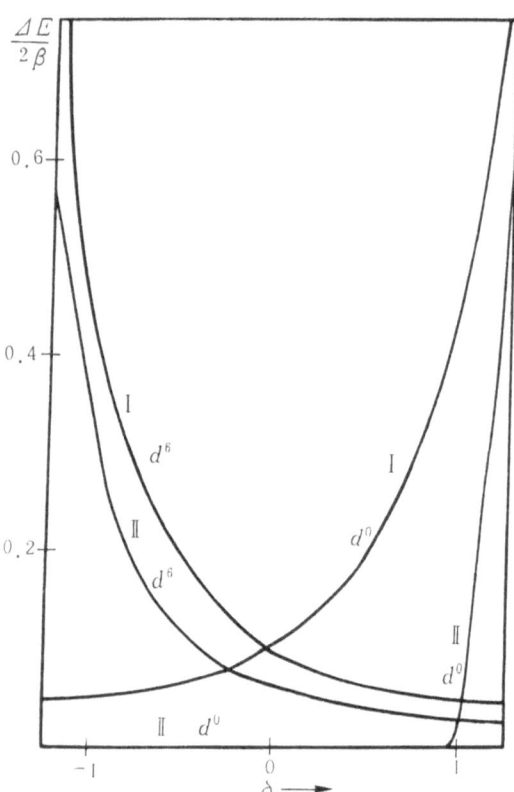

Fig. 1. Plots of the stabilization energy ($\Delta E/2\beta$) versus δ, which is a relative measure of the energy of the metal d orbital involved in π bonding with N_2, for mononuclear complexes. **I** represents coordination type, M–N-2–N-1, and **II** the coordination type M $\begin{array}{c} \diagup \text{N-1} \\ \diagdown \text{N-1}' \end{array}$ ($\eta^2 = 0.1$; $\eta'^2 = 0.025$; $\eta''^2 = 0.00625$). d^n represents the occupation (n) of the metal d orbitals

$$C_3(x) = \frac{1}{N(x)}\eta$$

$$N(x) = \{[x(x-\delta)-\eta^2]^2 + (x-\delta)^2 + \eta^2\}^{1/2}$$

where $N(x)$ is the normalization factor. Substituting the roots x into $\varepsilon = \alpha + \beta x$, we have been able to evaluate the stabilization energy ΔE (or $\Delta 1/2\beta$), which is very useful for discussion of the stability of such a dinitrogen complex. From the explicit form of $C_j(x)$, the charge density on the jth atom will be given by:

$$q_j = \sum_i n_i C_j(x_i) C_j(x_i)$$

The two quantum-chemical parameters, $\Delta E/2\beta$ and q_1, are plotted against δ, as shown in curves I of Figs. 1 and 2, respectively. For the d^6 configuration, these plots show that the stability of the end-on complex increases monotonously as δ decreases. The cases $\delta = 1$ and $\delta = -1$ correspond, respectively, to the π and π^* levels of the N_2 molecule. Hence, when tends toward -1 and even becomes more negative, i.e., the energy level α' of d_{xz} approaches the π^* level of N_2 from below and even gets beyond it, the complex tends to become more stable. With α' now lying rather high, the electron will tend to migrate from the d orbital to the AO of the nitrogen atoms, resulting in an increase in q_1. This situation favors attack by electrophilic groups such as H^+, and thus facilitates the reduction of N_2. For the d^0 configuration, in contrast, the complex becomes more stable only when α' approaches the level of N_2 from above and then goes further down. With α' now lying rather low, the N-1 atom will become more positive and thus the N_2 molecule will be reduced through attack by nucleophilic agents, such as H^-.

2.2 Molecules with Single Side-on Coordination Structure
of the M-2 $\overset{\diagup\text{N-1}}{\underset{\diagdown\text{N-1}'}{\mid}}$ Type

Here, the MO graph can be written as:

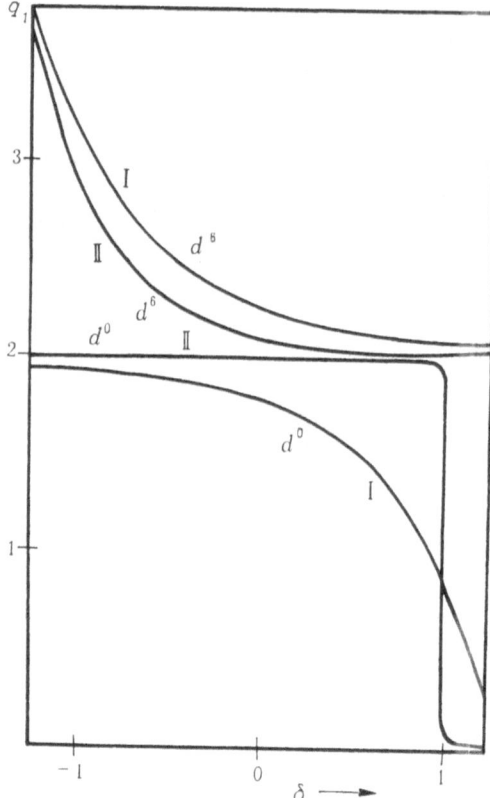

Fig. 2. Plots of charge density, q_1, on the *exo*-nitrogen (N-1) atom versus δ for mononuclear complexes. $\mathbf{I} = \mathrm{M-N-2-N-1}$ and $\mathbf{II} = \mathrm{M} \begin{array}{c} \mathrm{N\text{-}1} \\ \diagup \\ \diagdown \\ \mathrm{N\text{-}1'} \end{array}$ $(\eta^2 = 0.1;\ \eta'^2 = 0.025;\ \eta''^2 = 0.00625)$. d^n is the number (n) of d electrons available

For such a complex, two nonequivalent sets of π bonding orbitals can be formed from the two different AO sets $(d_{xz}P_x^{(1)}P_x^{(1')})$ and $(d_{yz}P_y^{(1)}P_y^{(1')})$. It is obvious from the accompanying diagram that molecular parameters, η' and η'', for this case cannot be equal to each other, nor can they be equal to the parameter η of the previous case (see Fig. 3). In order to discuss the two different LCAO for this side-on coordination structures on a basis comparable to the end-on case, we assume that the distance between the metal and N-2 in the first case is equal to that between the metal and the midpoint of the N–N bond in the present case, and the three parameters may be approximately interrelated by $\eta' = 1/2\,\eta$ and $\eta'' = 1/4\,\eta$.

Because there is a symmetry plane through the transition-metal ion M and perpendicular to the N–N bond, the MO graph can be reduced to a symmetric part plus an antisymmetric part.

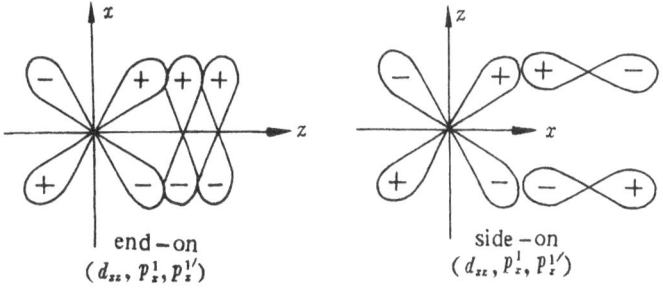

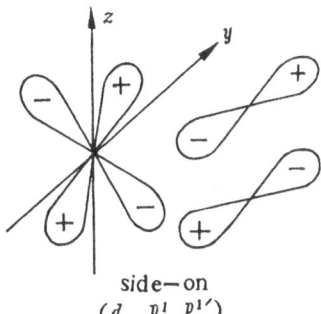

Fig. 3. Atomic orbitals involved in d–p π bonds in end-on and side-on coordinative dinitrogen complexes

For the symmetric graph:

$$\underset{x-1}{\bigcirc}$$

the eigen equation is:

$$x - 1 = 0$$

and the symmetric MO is given as:

$$\psi^{(s)} = \frac{1}{\sqrt{2}}(\phi_1 + \phi_{1'})$$

For the antisymmetric graph:

$$\overset{-\sqrt{2}\eta'}{\underset{\underset{1}{x+1} \qquad \underset{2}{x-\delta}}{\bigcirc\!\!-\!\!-\!\!-\!\!-\!\!\bigcirc}}$$

the eigen equation is:

$$(x - \delta)(x + 1) - 2\eta'^2 = 0,$$

and thus

$$\psi^{(A)} = C_1^{(A)} \frac{1}{\sqrt{2}} (\phi_1 - \phi_{1'}) + C_2^{(A)} \phi_2$$

where

$$C_1^{(A)} = \frac{1}{N(x)} (x - \delta)$$

$$C_2^{(A)} = \frac{1}{N(x)^2} \eta'$$

in which

$$N(x) = [(X - \delta)^2 + 2\eta'^2]^{1/2}.$$

It is obvious that these solutions hold good for the MO of $(d_{yz}, P_y^{(1)}, P_y^{(1)'})$, provided that the parameter η' is replaced by η''.

The stabilization energy $\Delta E/2\beta$ and the charge density q_1 (on the N-1 atom) can then be plotted against δ for the side-on coordination structure (curves II of Figs. 1 and 2). Figure 1 shows that, for the d^6 configuration, curve II is quite similar to curve I but is always lower in position, indicating that the side-on coordination structure is certainly less stable. For the d^0 configuration, however, the stabilization energy vanishes for $\delta < 1$ and thus no stable dinitrogen complexes can be formed at all. It is only when $\delta > 1$ (i.e., the energy level of the d orbital is equal to or lower than the level of N_2) that we can obtain stable dinitrogen complexes. This result helps to explain why no stable mononuclear side-on-coordinated dinitrogen complex has been synthesized to date. [RhCl(N$_2$)(PPr$_3^i$)$_2$] was reported as a dinitrogen complex possessing side-on coordination structure for nitrogen molecule (Busetto et al. 1977). Later on, however, the results of crystal structure determination of the complex and the ^{31}P- and ^{15}N-NMR spectra of a ^{15}N$_2$-labelled sample in solution obtained by Ibers et al. (1979) show that the complex contains a single end-on coordination structure for dinitrogen.

A comparison of curves I and II in Fig. 2 for the d^6 configuration shows that, since q_1 is always larger for end-on coordination, the end-on complex will be more easily reduced. For the d^0 configuration, only where $\delta > 1$ (i.e., the energy level of the d orbital is lower than the level of N_2) will the N_2 molecule be activated by a transition metal ion. Becoming more positive under this circumstance, the nitrogen atom is now in a more advantageous position of being successfully attacked by a nucleophilic group, such as H$^-$, and being reduced accordingly. In contrast, when $\delta < 1$, it is the end-on-coordinated N_2 molecule that is more easily activated.

2.3 Molecules with *trans*-Double-End-on Coordination Structure of the N-1–N-2–M-3–N-2′–N-1′ Type

The MO graph can be written as:

$$\underset{\underset{1}{x}}{\circ}\overset{-1}{\text{———}}\underset{\underset{2}{x}}{\circ}\overset{-\eta}{\text{———}}\underset{\underset{3}{x-\delta}}{\circ}\overset{-\eta}{\text{———}}\underset{\underset{2'}{x}}{\circ}\overset{-1}{\text{———}}\underset{\underset{1'}{x}}{\circ}$$

Thus, two equivalent sets of bonding MO orbitals can be formed from the two different AO sets $(p_x^{(1)}, p_x^{(2)}, d_{xz}, p_x^{(2')}, p_x^{(1')})$, and $(p_y^{(1)}, p_y^{(2)}, d_{xz}, p_y^{(2')}, p_y^{(1')})$, which are obviously two-fold degenerate. The charge densities q_1 and q_2 now plotted against δ (Fig. 4). Because q_1 is always larger than q_2 for all values of δ

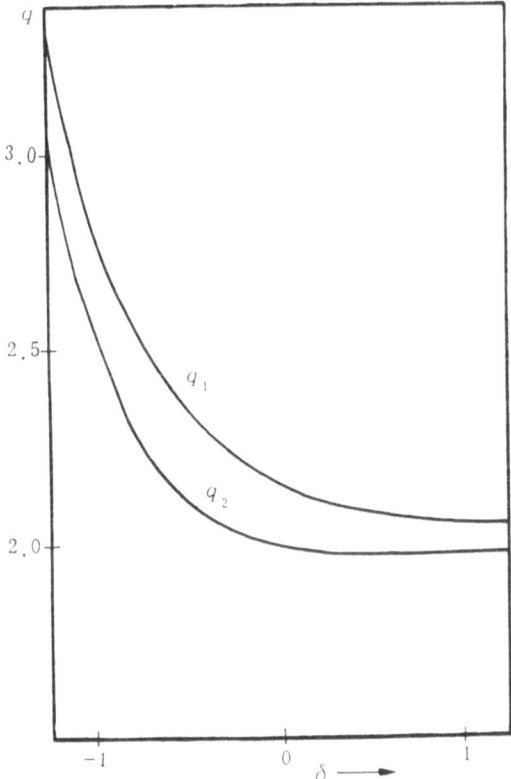

Fig. 4. Charge densities on both the *exo*-(N-1) and *endo*-(N-2) nitrogen atoms in mononuclear *trans*-double-end-on coordination compound (N-1–N-2–M-3–N-2′–N-1′) of a d^6 configuration as a function of δ. q_1 is the charge density on N-1 and q_2 is for N-2. $\eta^2 = 0.1$

(i.e., N-1 is always more negative than N-2), the N-2 atom is the first to be attacked by electrophilic groups, including H^+. The higher activity of the *exo*-N atom is entirely analogous to the single end-on-coordinated N_2 complexes, in agreement with the X-ray photoelectron spectroscopic studies (Brant and Feltham 1977).

2.4 Molecules with *cis*-Double-End-on Coordination Structure of the M-3–N-2–N-1

Three nonequivalent sets of bonding MOs can be formed from the AO sets $(p_x^{(1)}, p_x^{(2)}, d_{xz}, p_z^{(2')}, p_z^{(1'2)})$, $(p_y^{(1)}, p^{(2)}, d_{yz})$, and $(d_{xy}, p_y^{(2)}, p_y^{(1)})$, respectively. The bonding corresponding to the first AO set is analogous to that of the third case, whereas the other two sets are analogs of the first case.

For a comparison of the three end-on coordination type, q_1 is plotted against for the d^6 configuration as curves I (single N_2), II (*trans*-$(N_2)_2$) and III(*cis*-$(N_2)_2$) of Fig. 5. It is found that $q_1(I) > q_1(III) > q_1(II)$ for all values of δ. This result explains the fact that, in the reduction of bisdinitrogen complexes of molybdenum and tungsten, one of the N_2 molecules is easily replaced by a π-donating ligand, such as SO_4H^- or OH^- (Chatt et al. 1975, 1976), producing an intermediate simple dinitrogen complex. Obviously, such an intermediate complex with a more negative *exo*-N atom can be easily reduced.

2.5 Molecules with Quasi-Single End-on Coordination Structure of N-2–N-1 Type M-3

The atomic orbitals involved in $d - p\,\pi$ bond in such dinitrogen complexes are given in Fig. 6.

The procedure handling chemical bonds of the dinitrogen complexes with this coordination structure is similar to that for dinitrogen complexes with single end-on coordination structure. The difference lies in molecular parameter (η) changing for the former with a deviation angle (θ), but that for the latter is stationary. This can be clearly seen from Figs. 6 and 3, respectively.

From the result of qualitative calculation, we find that stability, charge density on the nitrogen atoms, and bond order (P_{NN}) for dinitrogen complexes with quasi-single end-on coordination structure are very close to that for dinitrogen complexes with single end-on coordination structure when θ is small

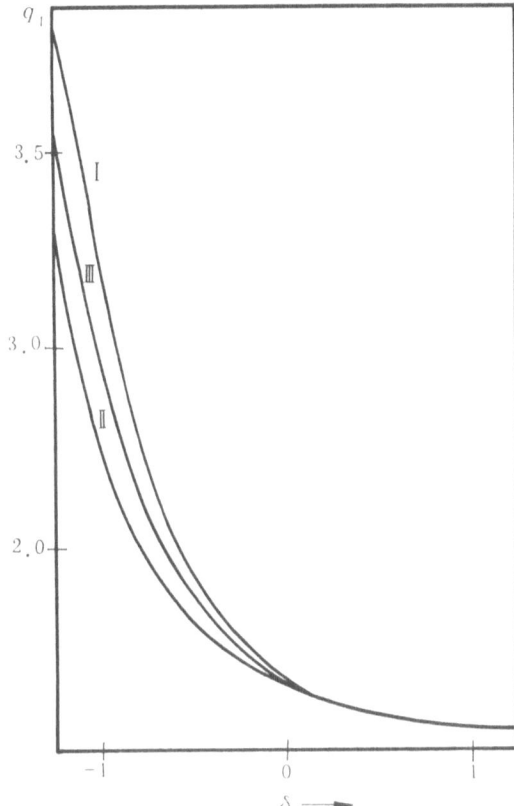

Fig. 5. A comparison of charge density (q_1) on the *exo*-nitrogen (N-1) atom for the single-end-on [M–N–N-1; **I**]. *trans*-double-end-on [N-1–N–M–N–N-1′; **II**], and *cis*-double-end-on [M–N–N-1′; **III**] coordination types. d^6 configuration and $\eta^2 = 0.1$

N
|
N-1

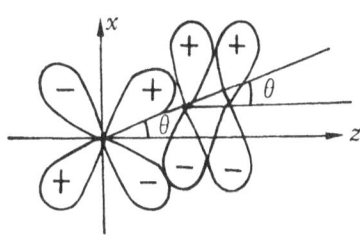

quasi—end—on

$(d_{zz} , p_z^1 , p_z^{11})$

Fig. 6. Atomic orbitals involved in d–p π bonds in quasi-end-on coordinative dinitrogen complexes

Table 1. Coordination structure types of some dinitrogen complexes

No	Compound	Coordination type	Angle(deg) M–N–N	Ref.
1	*trans*-[Cr(N$_2$)$_2$(dmpe)$_2$]	N–N / Cr / N–N	177.3(2)	A
2	[Mo(N$_2$)(PMe$_3$)$_5$]	N–N / Mo	–	B
3	[MoCl(N$_2$)(PMe$_3$)$_4$]	Mo–N–N	180.0	C1
4	Mo(dppe)$_2$(CO)(N$_2$)·$\frac{1}{2}$C$_6$H$_6$	N–N / Mo	177.0(12)	D
5	*trans*-[Mo(dppe)$_2$(N$_2$)$_2$]	N–N / Mo / N–N	176.6(5)	E1, E2
6	mer-[Mo(N$_2$)$_3$(PPr$^n_2$Ph)$_3$]	N–N–Mo–N–N / N / N	178.8(9) 178.3(8) 176.8(7)	F
7	*trans*-[Mo(N$_2$)$_2$(PMePh$_2$)$_2$ (PPh$_2$CH$_2$CH$_2$SMe)	Mo / \ N–N N–N	168(3) 174(3)	G
8	*trans*-Mo(N$_2$)$_2$Me$_8$[16]aneS$_4$	Mo / \ N–N N–N	176.7(5) 176.2(5)	H
9	*cis*-[Mo(N$_2$)$_2$(PMe$_3$)$_4$]	N–N–Mo / N / N	179(1) 177(1)	C$_2$
10	[WCl(N$_2$)(PMe$_3$)$_4$]	W–N–N	180.0	C1
11	*trans*-[ReCl(N$_2$)(PMe$_2$Ph)$_4$]	N–N / Re	177(1)	I1
12	mer-[Re(P(OMe)$_3$)$_3$Cl(CNMe)(N$_2$)]	Re–N–N	180.0	J

Table 1. (*Continued*)

No	Compound	Coordination type	Angle(deg) M–N–N	Ref.
13	*cis*-[Re(PMe$_3$)$_4$(NHPh)(N$_2$)]	N–N / Re	176.5(15)	K
14	*mer*-[Re(S$_2$PPh$_2$)(N$_2$)(CNMe)(PMe$_2$Ph)$_3$]	N–N / Re	174(1)	L1,L2
15	[Ru(NH$_3$)$_5$(N$_2$)]Cl$_2$	Ru–N–N	180	M1,M2
16	*trans*-[Ru(N$_3$)(en)$_2$(N$_2$)]PF$_6$	N–N / Ru	179.3(9)	I2
17	[Os(NH$_3$)$_5$(N$_2$)]Cl$_2$	N–N / Os	178.3(13)	N
18	[Ru(TMP)(THF)(N$_2$)]	N–N / Ru	177.6(16)	O
19	{OsCl$_2$(N$_2$)[N(OH)CHCO$_2$Et(PPh$_3$)$_2$}	N–N / Os	—	P
20	*mer*-[OsCl(SC$_6$F$_5$)(N$_2$)(PMe$_2$Ph)$_3$]	N–N / Os	176.1(5)	Q
21	[CoH(N$_2$)(PPh$_3$)$_3$]	N–N / Co	178(1)	13, 14
22	[CoH(PPh$_3$)$_3$(N$_2$)]Et$_2$O	N–N / Co	175(4)	R
23	K[Co(PMe$_3$)$_3$(N$_2$)]	N–N / Co	176.2 177.5 178.3	S
24	[RhH(PPhBu$^t_2$)$_2$(N$_2$)]	Rh–N–N	180	T
25	[RhCl(N$_2$)(PPr$^i_3$)$_2$]	N–N Rh	177.3	U
26	{Ti(η^5-C$_5$Me$_5$)$_2$]$_2$(μ-N$_2$)}	N–N / \ Ti Ti	176.8(4)[a] 178.1(4) 176.9(4)[b] 177.8(4)	V1
27	{[C$_5$H$_5$)$_2$Ti(PMe$_3$)]$_2$(μ-N$_2$)}	Ti \ N–N \ Ti	172.6(5) 169.0(5)	W

Table 1. (*Continued*)

No	Compound	Coordination type	Angle(deg) M–N–N	Ref.
28	$\{[C_5H_5)_2Ti(p\text{-}CH_3C_6H_4)]_2(\mu\text{-}N_2)\}$	Ti $\backslash$ N–N $\quad\backslash$ $\quad\quad$Ti N N	176.5	X
29	$\{(\mu\text{-}N_2)[Zr(N_2)(\eta^5\text{-}C_5Me_5)_2]_2$	Zr–N–N–Zr $\quad\quad\mid$ $\quad\quad$N $\quad\quad\mid$ $\quad\quad$N	177.9(5)[c] 176.7(3)[d] 177.4(3) 177.8(5)[c]	V²
30	$(\mu\text{-}N_2)\{[(o\text{-}Me_2NCH_2)C_6H_4]_2$ $V(Py)\}_2THF)_2$	N–N $/\quad\backslash$ V$\quad\quad$V	1,716(3) 1.713(3)	Y
31	$\{(\mu\text{-}N_2)[Ta(CHCMe_3)(CH_2CMe_3)$ $(PMe_3)_2]_2\}$	N–N $/\quad\backslash$ Ta$\quad\quad$Ta	171.4(6) 172.4(6)	Z, Ab1
32	$(\mu\text{-}N_2)[TaCl_3(P(bz)_3(THF)]_2$ $0.7CH_2Cl_2$	Ta $\backslash$ N–N $\quad\backslash$ $\quad\quad$Ta	178.9(4)	Ab2
33	$\{[Ta(S\text{-}2,6\text{-}C_6H_3\text{-}i\text{-}Pr_2)_3(THF)]_2$ $(\mu\text{-}N_2)\}$	N–N $/\quad\backslash$ Ta$\quad\quad$Ta	166.2(47) 165.9(48)	Bc
34	$\{[Ta(O\text{-}2,6\text{-}C_6H_3\text{-}i\text{-}Pr_2)_3(THF)]_2$ $(\mu\text{-}N_2)\}$	N–N $/\quad\backslash$ Ta$\quad\quad$Ta	176.6(6)	Bc
35	$\{[Cr(\eta^6\text{-}C_6Et_6)(CO)_2]_2(\mu\text{-}N_2)\}$	Cr–N–N–Cr	180.0(10)	Cd
36	$\{[Mo(\eta^6\text{-}C_6H_3Me_3)(dmpe)_2]_2$ $(\mu\text{-}N_2)\}$	Mo $\backslash$ N–N $\quad\backslash$ $\quad\quad$Mo	175.6(4)	De

Table 1. (*Continued*)

No	Compound	Coordination type	Angle(deg) M–N–N	Ref.
37	[(MeO)MoCl$_4$(μ-N$_2$)ReCl(PMe$_2$Ph)$_4$]	N–N / Re \ Mo	179.6(14)^e 178.7(9)^f	Ef1, Ef2
38	{[(η^5-C$_5$Me$_5$)W(CH$_3$)$_3$]$_2$(μ-N$_2$)}	N–N / W \ W	167.0 170.2	Ab3
39	{[W(C$_6$H$_5$C-CC$_6$H$_5$)(CH$_3$OCH$_2$CH$_2$OCH$_3$) Cl$_2$]$_2$(μ-N$_2$)}0.5CH$_3$OCH$_2$CH$_2$OCH$_3$	N–N / W \ W	176.4(10) 175.6(10)	Ab4
40	{[Mn(η^5-C$_5$H$_4$Me)(CO)$_2$]$_2$(μ-N$_2$)}	Mn \ N–N \ Mn	176.5(4)	Fg, Gh
41	{[Fe(CO)$_2$(P(OMe)$_3$)$_2$]$_2$(μ-N$_2$)}	Fe \ N–N \ Fe	175.8(7)	Hi
42	{[Ru(NH$_3$)$_5$]$_2$(μ-N$_2$)}(BF$_4$)$_4$	Ru \ N–N \ Ru	178.3(5)	Ij
43	[Ru$_2$H$_6$N$_2$(PPh$_3$)$_4$]	N–N / Ru–Ru	——	Jk
44	{[Ni(P(C$_6$H$_{11}$)$_3$)$_2$]$_2$(μ-N$_2$)}	N–N / Ni \ Ni	178.2 178.3	Kl
45	{[(PhLi)$_3$Ni]$_2$(N$_2$)·2Et$_2$O}$_2$	N / Ni | Ni^9 |/ N		Lm

Table 1. (*Continued*)

No	Compound	Coordination type	Angle(deg) M–N–N	Ref.
46	{Ph[NaOEt$_2$]$_2$(Ph$_2$Ni)$_2$(N$_2$) NaLi$_6$(OEt)$_4$OEt$_2$}$_2$	N ╱╲ Ni ∣ Ni[g] ╲╱ N		Mn
47	*trans*-{MoCl$_4$[(N$_2$)ReCl (PMe$_2$Ph)$_4$]$_2$}	Re–N–N–Mo–N–N–Re	178.6(21)[f] 177.1(22)[e]	No1, No2
48	{[Co(PMe$_3$)$_3$(N$_2$)]$_2$Mg(THF)$_4$}	Co–N–N ╲ Mg ╲ N–N–Co	180[h] 158[i]	Op
49	{[(Et$_2$O)C(CH$_3$)$_3$MgN$_2$Co (PMe$_3$)$_3$]$_2$}	Mg ╱ ╲ Co–N–N N–N–Co ╲ ╱ Mg	174.8(3)[h] 134.4(2) 129.5(2)	Pq

[a,b] Different structural units.
[c] Terminal angle M–N–N.
[d] Bridge angle M–N–N.
[e] The angle Re–N–N.
[f] The angle Mo–N–N.
[g] There are two structural units of the type in each molecule.
[h] The angle Co–N–N.
[i] The angle Mg–N–N.

Ref.: (A) Girolami GS et al. (1983). (B) Poveda ML et al. (1983). (C1,C2) Carmona E et al. (1983a, 1983b). (D) Sato M et al. (1978). (E1,E2) Uchida T et al. (1971, 1975). (F) Anderson SN et al. (1986). (G) Morris RH et al. (1984). (H) Yoshida T et al. (1984). (I1,I2,I3,I4) Davis BR et al. (1969a, 1969b, 1970, 1971). (J) Carvalho FN et al. (1982). (K) Chiu KW et al. (1982). (L1,L2) Pombeiro AJL et al. (1983, 1987). (M1,M2) Bottomey F et al. (1966, 1968). (N) Fergusson JE et al (1972). (O) Camenzind MJ et al. (1988). (P) Gallop MA et al. (1984). (Q) Cruz-Garrity D et al. (1988). (R) Enemark JH et al. (1968). (S) Klein HF et al. (1978). (T) Hoffman PR et al. (1976). (U) Thorn DL et al. (1979). (V1, V2) Sannar RD et al. (1976a, 1976b). (W) Berry DH et al. (1988). (X) Zeinstra JD et al. (1979). (Y) Edema JJH et al. (1989). (Z) Turner HU et al. (1980). (Ab1, Ab2, Ab3, Ab4) Churchill MR et al. (1981, 1982, 1984, 1986). (Bc) Schroch RR et al. (1988). (Cd) Denholm S et al. (1987). (De) Forder RA et al. (1974). (Ef1, Ef2) Mercer M et al. (1973, 1974). (Fg) Zeigler ML et al. (1976). (Gh) Weidenhammer K et al. (1979). (Hi Berke H et al. (1981). (Ij) Treitel IM et al. (1969). (Jk) Chaudret B et al. (1983). (K1) Jolly PW et al. (1971). (Lm) Kruger C et al. (1973). (Mn) Jonas K et al. (1976). (No1, No2) Cradwick PD et al. (1975, 1976). (Op, Hammer R et al. (1976). (Pq) Klein H-F et al. (1987)

(e.g. $\theta < 10^0$). In fact, only a few mononuclear dinitrogen complexes contain exact end-on coordination structure but most of them possess the quasi-end-on coordination structure (see Table 1). The steric effect of ligands in the dinitrogen complexes probably leads to this result.

3 Homo-Binuclear Dinitrogen Complexes

3.1 Molecules with All-End-on Coordination Structure of the M-1–N-2–N-2′–M-1′

Such a coordination structure gives rise to two equivalent sets of π bonding MOs corresponding to the two AO sets: $(d_{xz}^{(1)}, p_x^{(2)}, p_x^{(2')}, d_{xz}^{(1')})$ and $(d_{yz}^{(1)}, p_y^{(2)}, p_y^{(2')}, d_{yz}^{(1')})$.

3.2 Molecules with All-Side-on Coordination Structure of the M-1$\overset{\diagup\text{N-2}\diagdown}{\underset{\diagdown\text{N-2′}\diagup}{}}$M-1′ Type

This coordination type gives rise to two nonequivalent sets of bonding MOs arising from $(d_{xz}^{(1)}, p_x^{(2)}, p_x^{(2')}, d_{xz}^{(1')})$ and $(d_{yz}^{(1)}, p_y^{(2)}, p_y^{(2')}, d_{yz}^{(1')})$ respectively.

For molecules of both structure types, the stabilization energy, $\Delta E/2\beta$, and the charge density, q_2, can be plotted against δ (Figs. 7 and 8, respectively). The $\Delta E/2\beta$ curves are analogous to those for mononuclear complexes (Fig. 1), just as expected. Thus these curves indicate that all-end-on coordination type is more stable, which accounts for the paucity of binuclear dinitrogen complexes with all-side-on coordination (Krüger and Tsay 1973; Jonas et al. 1976). Moreover, it is worth noting from Fig. 6 that when $\delta > -1$ for d^{12} and $\delta < 1$ for d^0 it is easier to activate an end-on-coordination N_2 molecule, whereas when $\delta < -1$ for d^{12} (the superscript refers to the total number of d electrons) or $\delta > 1$ for d^0, a side-on-coordinated N_2 molecule will be activated more easily instead.

Similar considerations can be extended immediately to heterobinuclear dinitrogen complexes with essentially similar results.

3.3 Molecules with All-Quasi-End-on Coordination Structure of $\underset{\text{M-1}\qquad\text{M-1′}}{\diagup\text{N-2–N-2′}\diagdown}$ Type

Like mononuclear dinitrogen complexes, the stability, charge density on the nitrogen atoms, and bond order (P_{NN}) of the dinitrogen complexes with this

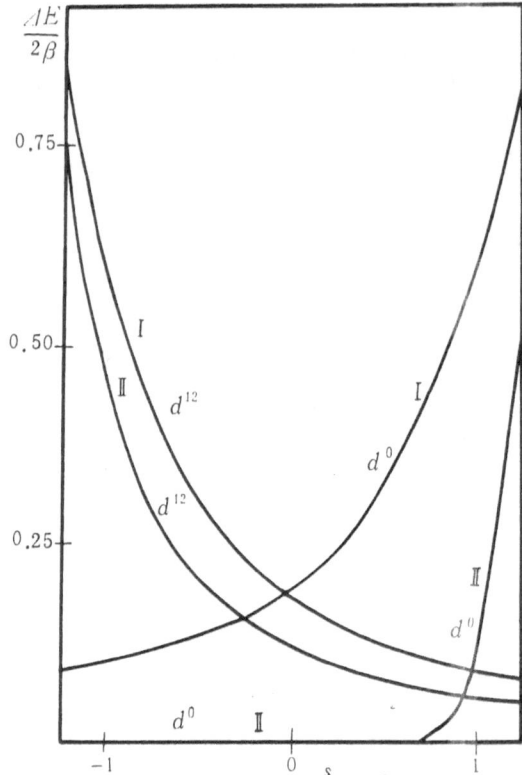

Fig. 7. Stabilization energy ($\Delta E/2\beta$) as a function of δ for homo-binuclear dinitrogen complexes oof the types M–N-2–N-2'–M' (I) and M⟨N-2, N-2'⟩M' (II). $\eta^2 = 0.1; \eta'^2 = 0.025; \eta''^2 = 0.00625$) with d^0 and d^{12} electron configurations

coordination structure are very close to that for the dinitrogen complexes with a precise all-end-on coordination structure in case where θ is very small.

4 Trinuclear Dinitrogen Complexes

4.1 All-*trans*-End-on Coordination Structure of the M-1–N-2–N-3–M'-4–N-3'–N-2'–M-1' Type

For this coordination type, the quantum-chemical parameters $\Delta E/2\beta$ and q_2 are again plotted against δ_2, as shown in Figs. 9 and 10, respectively (the superscript n in the figures refers to the total number of d electrons available in the

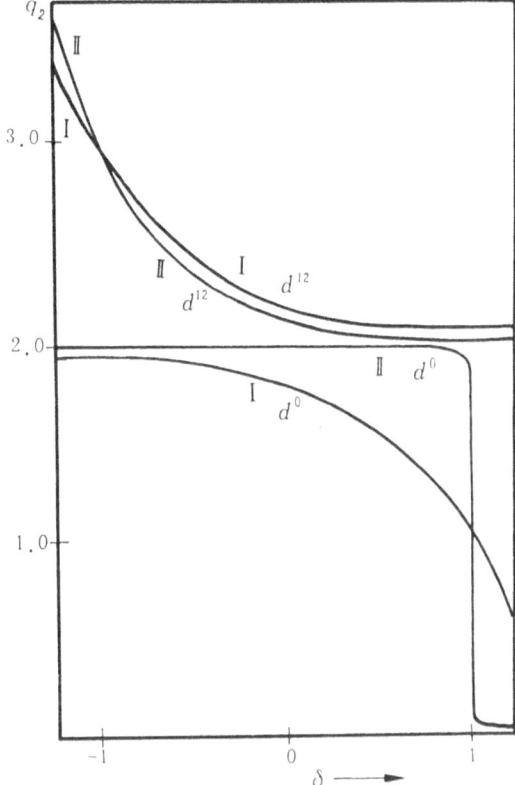

Fig. 8. Charge density (q_2) on the nitrogen atom as a function of δ for M–N-2–N-2–2′–M′ (**I**) and

$$
\begin{array}{c}
\text{N-2} \\
\text{M} \quad\quad \text{M′} \\
\text{N-2′}
\end{array}
$$

(**II**. $\eta^2 = 0.1$; $\eta'^2 = 0.025$; $\eta''^2 = 0.00625$) for both d^0 and d^{12} electron configurations

three transition-metal atoms). Figure 9 shows that, when $\delta_1 = 0.5$ and $-1.25 < \delta_2 < 0.5$, the dinitrogen complexes become more stable as n decreases, whereas in the case $\delta_1 = 0.5$ and $0.5 < \delta_2 < 1.25$, any decrease in n will result in decreased stability of these complexes. Thus, for the d^{16} configuration with $\delta_2 < 0.5$, no dinitrogen complex can be formed because $\Delta E/2\beta$ would then become negative. When $\delta_2 > 0.5$, however, the stability of these complexes will increase as δ_2 increases. This result is contrary to·that expected from the results obtained for mononuclear and binuclear dinitrogen complexes.

In contrast, Fig. 10 shows that the q_2 curves for these trinuclear complexes do bear similarities to those found for mononuclear and binuclear complexes. For example, when n is rather large (say, 16) and the energy level of the d orbital

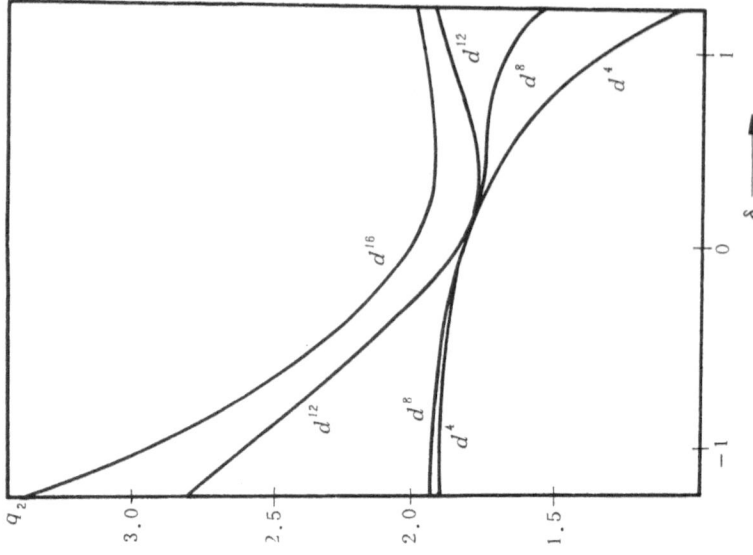

Fig. 10. Charge density (q_2) of the nitrogen (N-2) atoms in the complexes M–N-2′–N-3′–M′–N-3-N-2–M as a function of δ_2 and the total electron configurations d^4, d^8, d^{12}, and d^{16} (with $\delta_1 = 0.5$ and $\eta_1^2 = \eta_2^2 = 0.1$)

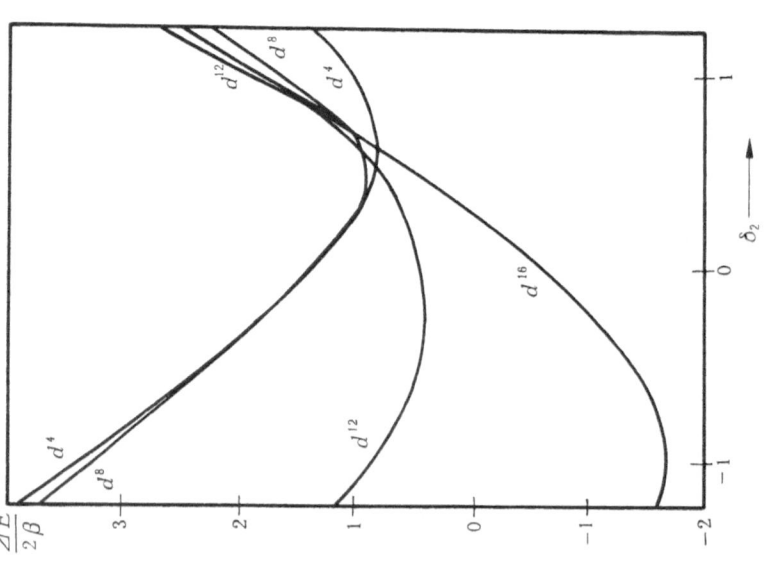

Fig. 9. Stabilization energy ($\Delta E/2\beta$) for the trinuclear complexes M–N-2′–N-3′–M′–N-3-N-2–M as a function of δ_2 (the relative energy of the d orbital on the M′ metal atom) for the total electron populations of d^4, d^8, d^{12}, and d^{16} (with $\delta_1 = 0.5$ and $\eta_1^2 = \eta_2^2 = 0.1$)

of the central atom M' approaches the π^* level of N_2 from below and even gets beyond it, d electrons will tend to migrate from M' toward the nitrogen atom, giving rise to an increase in q_2 and thus facilitating an electrophilic reaction. When n is small enough (say, 4) and the energy level of the d orbital of M' approaches the π level of N_2 from above and gets even further down, the N-2 atom will become somewhat positive, thus favoring a nucleophilic reaction.

4.2 Double-End-on Plus Single-Side-on Coordination Structure of the M-1–N-2–N-2'–M-1' Type

The plots of $\Delta E/2\beta$ and q_2 against δ for such a coordination type are shown in Figs. 11 and 12, respectively. Again, we find that the varition of the stabiliza-

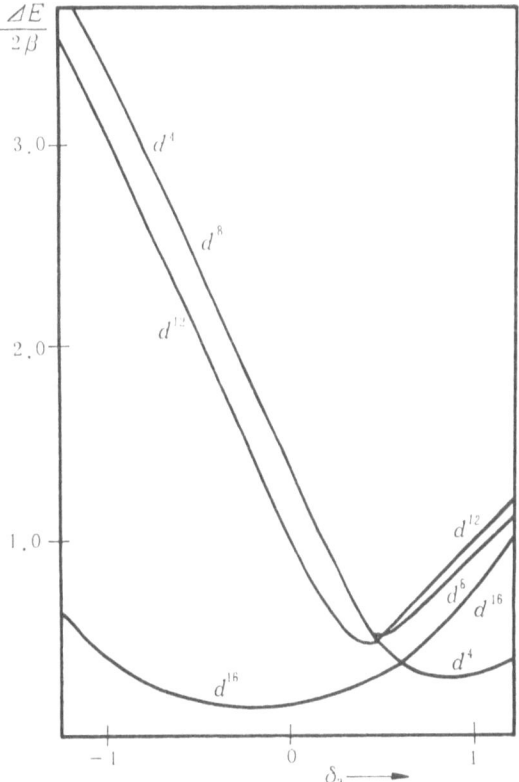

Fig. 11. Plots of $\Delta E/2\beta$ against δ_2 for the total electron populations d^4, d^8, d^{12}, and d^{16} for the complex M–N-2–N-2'–M (with $\delta_1 = 0.5$; $\eta_1^2 = \eta_2^2 = 0.1$; $\eta_{2'}^2 = 0.025$; $\eta_{2''}^2 = 0.00625$)

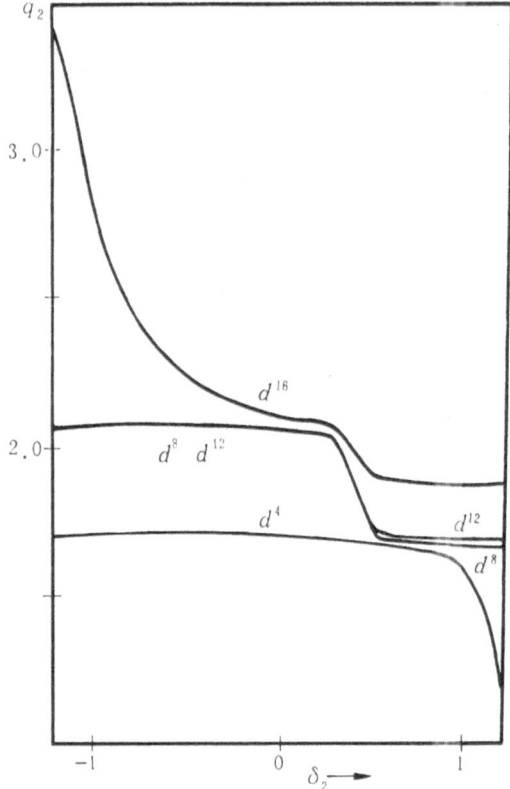

Fig. 12. Charge density (q_2) on the nitrogen atoms for the complexes M–N-2–N-2'–M for d^4, d^8,
$\underset{\diagdown M'\diagup}{}$
d^{12}, and d^{16} electron populations (with $\delta_1 = 0.5$; $\eta_1^2 = \eta_2^2 = 0.1$; $\eta_{2'}^2 = 0.025$; $\eta_{2''}^2 = 0.00625$)

tion energy is, in general, quite analogous to the case of all-*trans*-end-on coordination. For example, when $\delta < 0.5$, the stability of the dinitrogen complexes
increases as n decreases, so that complexes with d^{16} configuration would be
most unstable, yet they are more stable than the d^{16} complexes with all-*trans*-
end-on coordination. On the other hand, Fig. 12 shows that when n is rather
large (say, 16) and the energy level of the d orbital of M' lies rather high, the
nitrogen atoms tend to carry more negative charge. However, if n becomes
small enough (say, 4) and the energy level of the d orbital of M' lies rather low,
the nitrogen atoms would tend to become somewhat positive instead.

4.3 Single-End-on Plus Double-Side-on Coordination Structure

of the 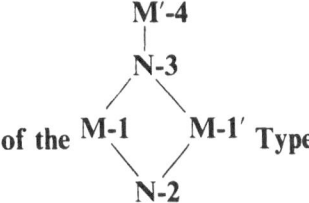 Type

The $\Delta E/2\beta$ versus δ_2 and q_2 versus δ_2 plots (Figs. 13 and 14) again show that the variation of both the stabilization energy and the charge density for this coordination structure follow patterns similar to the double-end-on plus single-side-on coordination type.

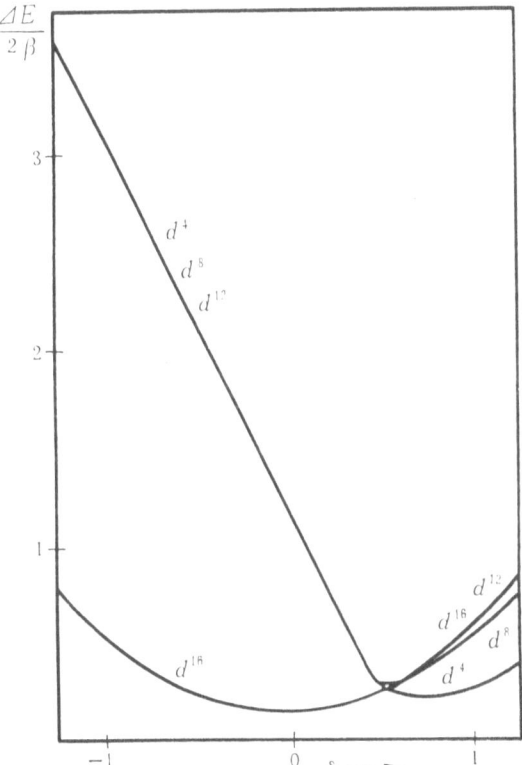

Fig. 13. Plots of stabilization energy ($\Delta E/2\beta$) against δ_2 for the complexes $M\overset{N-2}{\underset{N-3}{\diamond}}M$ as a function of total electron populations of d^4, d^8, d^{12}, and d^{16} (with $\delta_1 = 0.5$; $\eta_1^2 = \eta_2^2 = 0.1$; $\eta_{1'}^2 = 0.025$; $\eta_{1''}^2 = 0.00625$)

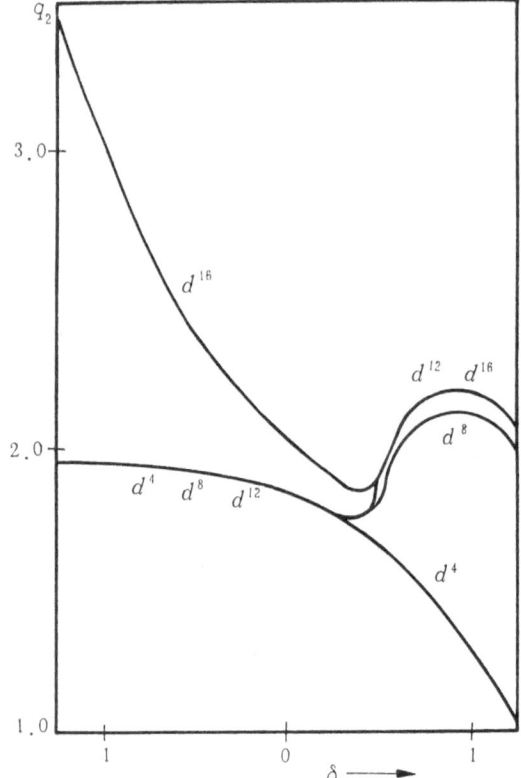

Fig. 14. Plots of charge density (q_2) on the nitrogen atom (N-2) against δ_2 as a function of total electron populations d^4, d^8, d^{12}, and d^{16} for the complexes $M\begin{smallmatrix} \text{N-2} \\ | \\ \text{N-3} \\ | \\ \text{M}' \end{smallmatrix}M$ (with $\delta_1 = 0.5$; $\eta_1^2 = \eta_2^2 = 0.1$; $\eta_{1'}^2 = 0.025$; $\eta_{1''}^2 = 0.00625$)

On comparison of the charge density curves for these three trinuclear coordination structures for the d^{16} configuration (Fig. 15), q_2 is in general somewhat larger for the third case, in conformity with the result of bond order (P_{NN}) calculations shown in Fig. 16. Again, the N-2 atom, which is *exo* with respect to M' carries more negative charge, in general, than the other *endo*-N atom (cf. Fig. 17).

5 Summary

We are now in a position to give a comparative discussion on dinitrogen complexes, particularly trinuclear versus mononuclear and binuclear, on the basis of our quantum-chemical calculations. There are some essential similarities.

First, when the total number of d electrons is quite large and the energy level of the d orbital of the "donor" transition metal atom lies quite high, the nitrogen atoms tend to become more negative, such complexes will favor an electrophilic reaction. On the other hand, if the total number of d electrons is small and energy level of the d orbital of the "acceptor" transition-metal atom lies rather low, the nitrogen atoms tend to become somewhat positive, such complexes may then be preferentially reduced via a nucleophilic reaction. Second, the exo-N atom is in general more negative than the $endo$-N atom.

However, there are also some striking differences worthy of note. First, for both mononuclear and binuclear dinitrogen complexes the stability order is in general of the end-on > quasi-end-on > side-on coordination type. For tri-nuclear dinitrogen complexes, however, Fig. 18 shows that, when $-1.25 < \delta_2 < 0.5$,

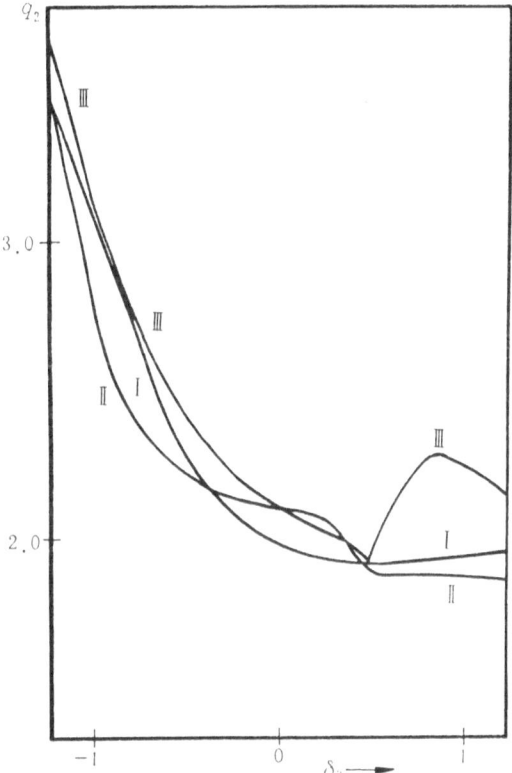

Fig. 15. Comparison of the charge density (q_2) curves for the d^{16} total electron configurations of the three trinuclear structures. M–N–N–M′–N–N–M (I), M–N–N–M (II), and M′ (III) with $\eta_1^2 = \eta_2^2 = 0.1$)

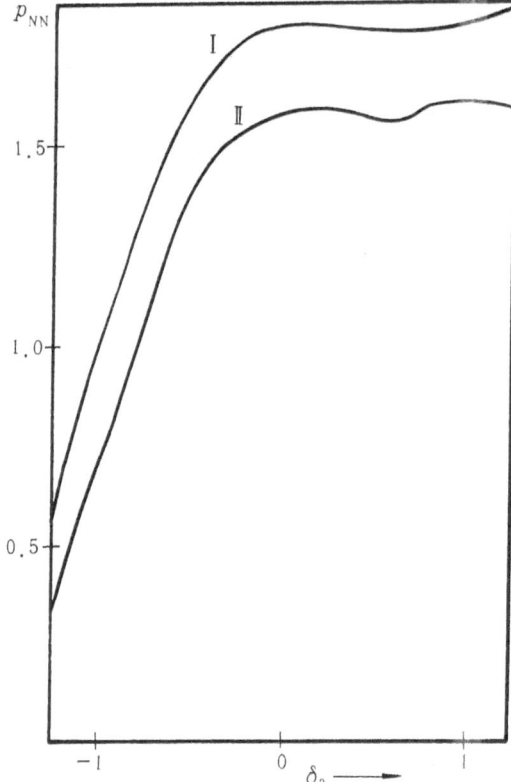

Fig. 16. Comparison of plots of bond order (P_{NN}) versus δ_2 for the d^{16} configurations in the trinuclear complexes M-N-N-M'-N-N-M (I) and M⟨N|N⟩M (II)

coordination structure involving both end-on and side-on coordination to transition-metal atoms, such as the second and third cases presented, turns out to be more stable than the all-end-on coordination type. This conclusion is true as long as the total number of d electrons is not less than 12. We therefore predict that the active center for catalytic nitrogen fixation in nitrogenase must be at least a trinuclear transition-metal cluster involving both end-on and side-on coordinations of N_2. The prediction stemming from the result of our qualitative calculation is just in accord with the elementary standpoint of the models proposed by Lu Jia-Xi (Nitrogen Fixation Research Group 1975; Lu 1980) and Cai Qi-Rui (Nitrogen Fixation Research Group 1976; Tsai 1980) in describing the catalytic behavior of nitrogenase.

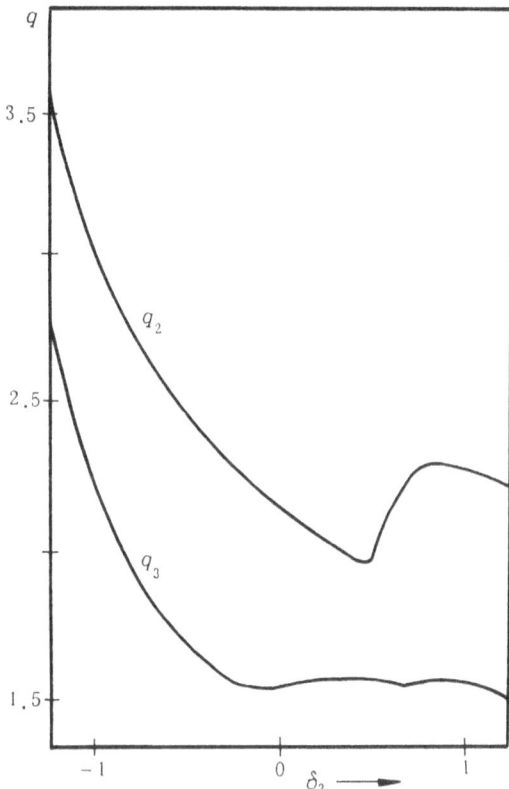

Fig. 17. A plot charge density (q) on the *exo* (q_2) and *endo* (q_3) nitrogen atoms in complexes

$$M\overset{N\text{-}2}{\underset{N\text{-}3}{<\!\!|\!\!\!\!\geqq}}M$$ as function of δ_2 (with $\delta_1 = 0.5$; $\eta_2^2 = 0.1$; $\eta_{1'}^2 = 0.025$; $\eta_{1''}^2 = 0.00625$)

$$\underset{M'}{|}$$

Second, mononuclear and binuclear dinitrogen complexes are stable provided that either the energy level of the d orbital lies rather high when the total number of electrons is large or the energy level of the d orbital lies rather low when the total number of d electrons is sufficiently small. On the contrary, trinuclear dinitrogen complexes are unstable under such circumstances. So far, very few trinuclear dinitrogen complexes have been successfully synthesized. Examples are $[MoCl_4\{(N_2)ReCl(PMe_2Ph)_4\}_2]$ (Cradwick et al. 1975, 1976) with a d^{14} configuration and $[Mg(THF)\{(N_2)Co(PMe_3)_3\}_2]$ (Hammer 1976) with a d^{18} configuration, both of which possess an all-*trans*-end-on coordination structure. For these complexes, the total number of d electrons is quite large. Thus, it is inferred that the energy level of the d orbital involved must lie in the range of low energies.

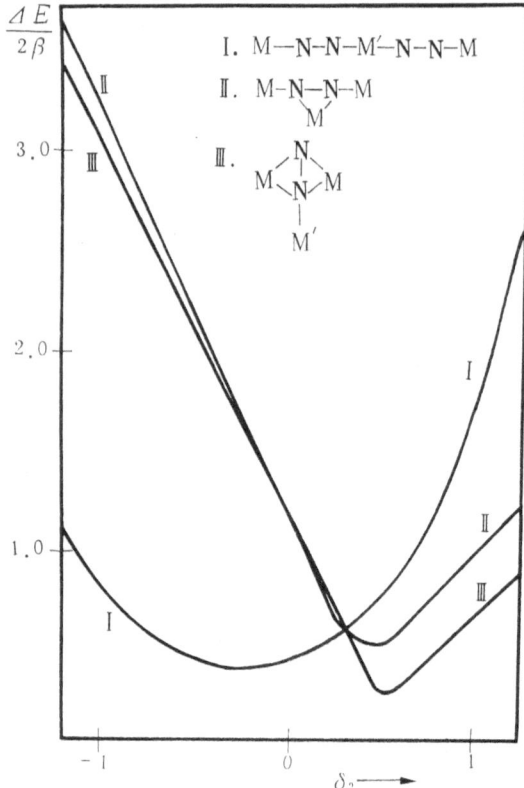

Fig. 18. Plots of stabilization energy ($\Delta E/2\beta$) against δ_2 for the d^{12} electron configurations in the trinuclear complexes M–N–N–M'–N–N–M (**I**), M–N–N–M, and M$\diagdown$N$\diagup$M (**III**) (with $\eta_1^2 = \eta_2^2 = 0.1$)

6 References

Allen AD, Senoff CV (1965) Nitrogenpentammineruthenium (II) complexes. Chem Commun 621
Anderson SN, Hughes DL, Richards RL (1986) Preparation of mer-[Mo(N$_2$)$_2$(PPr$_2^n$Ph)$_2$] from trans-[Mo(N$_2$)$_2$(PPr$_2^n$Ph)$_4$]: its X-ray crystal structure, spectroscopic properties, and protonation to give ammonia. J Chem Soc Dalton Trans 1591
Berke H, Kanhard W, Huttner G, von Seyerl I, Zsolan L (1981) Eisenkomplex als Modellverbindungen zur homogen Hydrierung von Kohlenmonoxid. Chem Ber 114: 2754
Berry DH, Procoplo LJ, Carrol PJ (1988) Molecular structure of {Cp$_2$Ti(PMe$_3$)}$_2$(μ-N$_2$). a titanocene dinitrogen complex. Organometallics 7: 570
Bottomey F, Nyburg SC (1966) Molecular nitrogen as a ligand: the crystal structure of nitrogenpentammineruthenium (II) dichloride. J Chem Soc Chem Commun 897
Bottomey F, Nyburg SC (1968) Molecular nitrogen as a ligand: the crystal structure of nitrogenpentammineruthenium (II) dichloride and related salts. Acta Crystallogr Sect B. 24: 1289

Brant P, Feltham RD (1977) X-ray photoelectron spectra of molybdenum dinitrogen complexes and their derivatives. J Less-Common Metals 54: 81

Busetto C, D'Alfonso A, Maspero F, Perego G, Lazzeta A (1977) Side on bonded dinitrogen and dioxygen complexes of rhodium(I). Synthesis and crystal structures of *trans*-chloro(dinitrogen)-, chloro(dioxygen)- and chloro(ethylene)-bis-(triisopropylphosphine)rhodium(I). J Chem Soc. Dalton Trans 1828

Camenzind MJ, James BR, Dolphin D, Sparapany JW, Ibers JA (1988) Molecular dinitrogen complexes of rutheniun (II) porphyrins. Inorg Chem 27: 3054

Carmona E, Marin JM, Poveda ML, Atwood JL, Rogers RD (1983a) Preparation and properties of dinitrogen trimethlyphosphine complexes of molybdenum and tungsten-II. Synthesis and crystal structures of $[MCl(N_2)(PMe_3)_4](M = Mo, W)$ and trans-$[MoCl_2(PMe_3)_4]$. Polyhedron. 2: 185

Carmona E, Marin JM, Poveda ML, Atwood JL, Rogers RD (1983b) Preparation and properties of dinitrogen trimethylphosphine complexes of molybdenum and tungsten. 4. Synthesis, chemical properties and X-ray structure of *cis*-$[Mo(N_2)_2(PMe_3)_4]$. The crystal and molecular structure of *trans*-$[Mo(C_2H_4)_2(PMe_3)_4]$ and trans, mer-$[Mo(C_2H_4)_2(CO)(PMe_3)_3]$. J Am Chem Soc 105: 3014

Carvalho FNN, Pombeiro AJL, Orama O, Schubert U, Pickett CJ, Richards RL (1982) Preparation of new dinitrogen complexes of Rhenium(I) with organophosphite and isocyanide ligands. X-ray structure of mer-$[ReCl(N_2)(CNMe)(P(OMe)_3)_3]$. J Organomet Chem 240: C18

Chatt J, Pearman AJ, Richards RL (1975) Reduction of monocoordinated molecular nitrogen to ammonia in a protic environment. Nature 253: 39

Chatt J, Pearman AJ, Richards RL (1976) Relevance of oxygen-ligands to reduction of ligating dinitrogen. Nature 259: 204

Chatt J, Dilworth JR, Richards RL (1978) Recent advances in the chemistry of nitrogen fixation. Chem Rev 78: 589

Chaudret B, Devillers J, Poilblance R (1983) Preparation, Characterization and X-ray crystal structure of $Ru_2H_6N_2(PPh_3)_4$, a compound containing four bridging hydrides and a ruthenium–ruthenium double bond. J Chem Soc Chem Commun 641

Chiu KW, Wang WK, Wilkinson G, Galase AMR, Hursihouse MB (1982) Reactions of phenyl-imidotrichlorobis(triphenylphosphine)rhenium(V). Reaction with trimethylphosphine and reduction of trimethylphospine complex to phenylamido complexes of rhenium (I, III): The X-ray crystal structures of phenylamido(dinitrogen)tetrakis(trimethylphos-phine)rhenium(I) and phenylamido (buta-1, 3-dien)tetrakis(trimethylphosphine)rhenium (I). Polyhedron 1: 37

Churchill MR, Wasserman HJ (1981) Crystal and molecular structure of $[Ta(=CHCMe_3)(CH_2CMe_3)(PMe_3)_2]_2(\mu-N_2)$, a molecule with a dinitrogen ligand behaving as a dinimido group in a Ta = NN = Ta bridge. Inorg Chem 20: 2899

Churchill MR, Wasserman HJ (1982) The $Ta(\mu-N_2)Ta$ system. 2. Crystal structure of $[TaCl_3(P(bz)_3)(THF)]_2(\mu-N_2)$ $0.7CH_2CH_2$. A binuclear diimido complex of octahedral tantalum(V). Inorg Chem 21: 218

Churchill MR, Li YJ, Theopold KH, Schock RR (1984) High oxidation state dinitrogen complexes. Synthesis and crystal structure of $[W(C_6H_5C = CC_6H_5)(CH_3OCH_2CH_2OCH_3)Cl_2](\mu-N_2)$ $0.5CH_3OCH_2CH_2OCH_3$. Inorg Chem 23: 4472

Churchill MR, Li YJ (1986) The $W(\mu-N_2)W$ system, II. Crystal structure of $[(5-C_5Me_5)W(CH_3)_3]_2(\mu-N_2)$. An organotungsten(VI) complex containing a W = N – N = W core. J Organomet Chem 301: 49

Crawick PD, Chatt J, Crabtree RH, Richards RL (1975) Preparation and X-ray structure of a trinuclear dinitrogen-bridged complex, trans-$[MoCl_4\{(N_2)ReCl(PMe_2Ph)_4\}_2]$. J Chem Soc Chem Commun 351

Cradwick PD (1976) Crystal structure of a trinuclear dinitrogen-bridged complex: Di- -dinitrogen-tetrachlorobis [chlorotetrakis(dimethylphenylphosphine)rhenio]molybdenumdichloromethane(1/1). J Chem Soc Dalton Trans 1934

Cruz-Garrity D, Gelove S, Torrens H, Leal J, Richards R (1988) Dinitrogen complexes of osmium(II) with thiolate co-ligand: X-ray structure of mer-$[OsCl(SC_6F_5)(N_2)(PMe_2PH)_3]$. J Chem Soc Datlton Trans 2393

Davis BR, Payne NC, Ibers JA (1969a) The geometry of coordinated molecular nitrogen. The structure of $Co(H)(N_2)(P(C_6H_5)_3)_3$. J Am Chem Soc 91: 1240

Davis BR, Payne NC, Ibers JA (1969b) The bonding of molecular nitrogen. I The crystal and molecular structure of hidridodinitrogentris(triphenylphosphine)cobalt(I). Inorg Chem 8: 2719

Davis BR, Ibers JA (1970) The bonding of molecular nitrogen. II The crystal and molecular structure of azidodinitrogenbis(ethylenediamine)ruthenium(II) hexafluorophosphate. Inorg Chem 9: 2768

Davis BR, Ibers JA (1971) The bonding molecular nitrogen. III The crystal and molecular structure of chlorodinitrogentetrakis(dimethylphosphine)rhenium(I). Inorg Chem 10: 578

Denholm S, Hunte G, Weakley TJR (1987) Dinitrogen complexes derived from tricarbonyl (η^6-hexaethylbenzene)chromium(0). Crystal and molecular structure of -dinitrogen bis[dicarbonyl(6-hexaethylbenzene)chromium(0)]toluene(1/1). J Chem Soc Dalton Trans 2789

Edema JJH, Meetsma A, Gambarotta S (1988) Divalent vanadium and dinitrogen fixation: The preparation and X-ray structure of $(\mu-N_2)\{[(o-Me_2NCH_2)C_6H_4]_2V(Py)\}_2$. J Am Chem Soc 111: 6878

Enemark JH, Davis BR, McGinnety JA, Ibers JA (1968) The structure of a molecular nitrogen compound of cobalt and evidence for $CoH(N_2)(PPh_3)_3$. J Chem Soc Chem Commun 96–97

Fergusson JE, Love JL, Robinson WT (1972) The crystal and molecular structure of dinitrogenpantamineosmium(II) chloride, $[Os(NH_3)_5N_2]Cl_2$, and rulated rethenium complexes. Inorg Chem 11: 1662

Forder RA, Prout K (1974) μ-Dinitrogen-bis{(π-mesitylene)[1,2-bis(dimethylphosphino)ethane]-molybdenum}. Acta crystallogr Sect B 30: 2778

Gallop RA, Richard CEF, Roper WR (1984) An unusual nitrosyldiazoalkane coupling reaction. Synthesis, structure and reactivity of the aldoxime-dinitrogen complex, $OsCl_2(N_2)[N(OH)-CHCO_2Et](PPh_3)_2$. J Organomet Chem 269: C21

Girolami GS, Salt JE, Wilkinson G (1983) Alkyl, hydride and dinitrogen 1,2-bis(dimethylphsphoin)-ethane complexes of cromium. Crystal structure of $Cr(CH_3)_2(dmpe)_2$, $CrH_4(dmpe)_2$ and $Cr(N_2)_2(dmpe)_2$. J Am Chem Soc 105: 5954

Hammer R, Klein HF, Schubert U, Frank A, Huttner G (1976) A novel hetero-bimetallic dinitrogen complex. Angew Chem Int Ed Engl 15: 612

Hoffman PR, Yoshida T, Okano T, Otsuka S, Ibers JA (1976) Crystal and molecular structure of hydrido(dinitrogen)bisphenyl(di-tert-butyl)phosphine]rhodium(I). Inorg Chem 15: 2462

Hsu CC (Xu, JQ) (1980) A quantum-chemical theory of transition metal-dinitrogen complexes. in "Nitrogen Fixation", Vol I, Newton WE, Orme-Johnson WH, Eds University Park Press, Baltimor, 317

Jolly PW, Jonas K, Kruger C, Tsay YH (1971) The preparation, reactions and structure of bis[bis(tricyclohexylphosphine)nickel]dinitrogen, $\{[C_6H_{11})_3P]_2Ni\}_2N_2$. J Organomet Chem, 33: 109

Jonas K, Brauer DJ, Kruger G, Roberts PJ, Tsay YH. (1976) "Side-on" dinitrogen-transition metal complexes. Molecular structure of $\{C_6H_5[NaO(C_2H_5)_2]_2[(C_6H_5)_2Ni]_2N_2NaLi_6(OC_2H_5)_4O-(C_2H_5)_2\}_2$, J Am Chem Soc 98: 74

Klein HF, Hammer R, Wenninger I, Friedrich P, Huttner G (1978) Eine neuartige assoziation zwischen kalium und komplex gebundenem distickstoff-struktur von $KN_2Co(PMe_3)_3$. Z Naturforsch Teil B 33: 1267

Klein HF, Konig H, Kopert S, Ellrich K, Riede J (1987) Cobalt diazenides of main-group I-III(1-3) metals: X-ray structure of a grignard compound containing(dinitrogen)(trimethylphosphane)-cobalt anions. Organometallics 6: 1341

Kruger C, Tsay YH (1973) Molecular structure of a dinitrogen-nickel-lithium complex. Angew Chem Int Ed Engl 12: 998

Lu JX (1980) Composite "string bag" Cluster model for the active center of nitrogenase. In "Nitrogen Fixation", Vol I Newton WE, Orme-Jhonson WH eds University Park Press, Baltimore, 343

Mercer M, Crabtree RH, Richards RL (1973) A μ-dinitrogen complex with a long N–N bond. X-ray crystal structure of $[(PMe_2Ph)_4 ClReN_2MoCl_4(OMe)]$. J Chem Soc Chem Commun 808

Mercer M (1974) Crystal structure of a dinuclear dinitrogen complex: Tetrachloro-{chloro-tetrakis[dimethyl(phenyl)phosphine]rhenium(1)}-μ-dinitrogenmethoxymolybdnum(v)ethanolhydrochloric acid. J Chem Soc Dalton Trans 1637

Morris RH, Ressner LM, Sawyer JF, Shiratian M (1984) A sulfur-ligated molybdenum complex that reduces dinitrogen to ammonia. The crystal and molecular structure of trans-$Mo(N_2)_2(PMePh_2)_2(PPh_2CH_2CH_2SMe)$. J Am Chem Soc 106: 3683

Nitrogen Fixation Research Group, Department of Chemistry, Kirin University (1974) Theory of chemical bonding in dinitrogen complexes. Scientia Sinia 17: 193

Nitrogen Fixation Research Group, Department of Chemistry, Xiamen University (1976) A model of the nitrogenase active center and the mechanism of nitrogenase catalysis. Scientia Sinica 19: 460

Nitrogen Fixation Research Group, Fujian Institute of Research on the Structure of Matter, Academia Sinica (1975) A preliminary model for the actve site of catalytic nitrogen fixation in nitrogenase. —Also on structural criteria for the activation of dinitrogen molecule via coordination. Kexue Tongbao 20: 540

Pelikan P, Boca R (1984) Geometric and electronic factors of dinitrogen activation on transition metal complexes. Coord Chem Rev 55: 55

Pombeiro AJL, Hitchcock PB, Richard RL (1983) Reaction of the dinitrogen complex [Re(η^2-S$_2$PPh$_2$)(N$_2$)(PMe$_2$Ph)$_3$] with methylisocyanide. Preparation and X-ray structure of the mixed dinitrogen-isocyanide complex mer-[Re(η^1-S$_2$PPh$_2$)(N$_2$)(CNMe)(PMe$_2$Ph)$_3$, Inorg Chem Acta 76: L225

Pombeiro AJL, Hitchcock PB, Richard RL (1987) Preparation of isocyanide and mixed dinitrogen-isocyanide complexes of rhenium(I) from reaction of trans-[ReCl(N$_2$)(PMe$_2$Ph)$_4$], mer-[Re(S$_2$PPh$_2$)(N$_2$)(PMe$_2$Ph)$_3$, or mer-[Re(S$_2$CNEt$_2$)(N$_2$)(PMe$_2$Ph)$_3$] with methyl isocyanide. Crystal structure of mer-[Re(S$_2$PPh$_2$)(N$_2$)(CNMe)(PMe$_2$Ph)$_3$]$^+$. J Chem Soc Dalton Trans 319

Poveda ML, Rogers RD, Atwood JL (1983) Preparation and properties of dinitrogen trimethyl-phosphine complexes of molybdenum and tungsten. 4. Synthesis, chemical properties and X-ray structure of cis[Mo(N$_2$)$_2$(PMe$_3$)$_4$]. The crystal and molecular structures of trans-[Mo(C$_2$H$_4$)$_2$] and trans, mer-[Mo(C$_2$H$_4$)$_2$(CO)(PMe$_3$)$_3$]. J Am Chem Soc 105: 3014

Sanner RD, Duggan DM, Mckenzie TC, Marsh RE, Bercaw JE (1976a) Structure and magnetism of μ-dinitrogenbis(bis(pentamethylcyclopentadienyl)titanium(II)).{(η^5–C$_5$(CH$_3$)$_5$)$_2$Ti}$_2$N$_2$. J Am Chem Soc 98: 8358

Sannar RD, Manriquez IM, Marsh RE, Bercaw JE (1976b) Structure of μ-dinitrogenbes-(bis(pentamethylcyclo-pentadienyl)nitrogenzirconium(II)), {(η^5-C$_5$(CH$_3$)$_5$)$_2$ZrN$_2$}$_2$N$_2$. J Am Chem Soc 98: 8351

Sato M, Tatsumi T, Kodama T, Hidai M, Uchida T, Uchida Y (1978) Preparation and properties of dinitrogen-molybdenum complexes. 6. Syntheses and molecular structures of a five-coordinate Mo(O) complex Mo(CO)(PH$_2$PCH$_2$CH$_2$PPh$_2$)$_2$ and a related six-coordinate complex Mo(CO)(N$_2$)(Ph$_2$PCH$_2$CH$_2$PPh$_2$)$_2$ 1/2C$_6$H$_6$. J Am Chem Soc 100: 4447

Schrock RR, Wesolek M, Liu AH, Wallace KC, Dewan JC (1988) Thiolato dinitrogen (or hydrazido(4-1)) complexes, [Ta(SAr)$_3$(THF)]$_2$(μ-N$_2$) (Ar = 2,6-C$_6$H$_3$-i-Pr$_2$, 2,4,6,-C$_6$H$_2$-i-Pr$_3$), and phenoxide analogues. Structural comparison of [Ta(S-2,6-C$_6$H$_3$-i-Pr$_2$)$_3$(THF)](μ-N$_2$) and [Ta(O-2,6-C$_6$H$_3$-i-Pr$_2$)$_3$(THF)]$_2$(μ-N$_2$). Inorg Chem 27: 2050

Tang AC, Kiang YS (1976) Graph theory of molecular orbitals. Scientia Sinica 19: 207

Tang AC, Kiang YS (1977) Graph theory of molecular orbitals.II. Symmetrical analysis and calculations of Mo coefficients. Scientia Sinica 20: 585

Thorn DL, Tulip TH, Ibers JA (1979) The structure of trans-chloro(dinitrogen)bis(tri-isopropylphosphine)rhodium(I); An X-ray study of the structure in the solid state and a nuclear magnetic resonance study of the structure in solution. J Chem Soc Dalton Trans 2022

Treitel IM, Flood MT, Marsh RE, Gray HB (1969) Molecular and electronic structure of μ-nitrogen-decaammine-diruthenium(II). J Am Chem Soc 91: 6512

Tsai KR (1980) in: Newton WE, Orme-Johnson WH (eds) Nitrogen fixation, vol I, University Park Press, Baltimore, p 373

Turner HU, Fellmann JD, Rocklage SM, Sckrock RR, Churchill MR, Wasserman HJ (1980) Tantalum complexes containing dimido bridging dinitrogen ligands. J Am Chem Soc 102: 7809

Uchida T, Uchida Y, Hidai M, Kodama T (1971) The crystal and molecular structure of trans-bis(dinitrogen)bis[1,2-bis(diphenylphosphino)ethane]molybdenum(0). Bull Chem Soc Jpn 44: 2883

Uchida T, Uchida Y, Hidai M, Kodama T (1975) trans-bis(dinitrogen)-bis-[1,2-bis(diphenylphos-phino)ethane]molybdenum(0). Acta Crystallogr Sect B. 31: 1197

Weidehammer K, Hermann WA, Ziegler ML (1979) Rontgenstrukturanalyse von μ-distickstoff-bis[dicarbonyl-5-me-thylcyclopentadienylmangan],[η^5-(C$_5$H$_4$CH$_2$)Mn(CO)$_2$]$_2$(μ-N$_2$). Z Anorg Allg Chem 457: 183

Xu JQ, Xu LJ, Liu XS, Zhang ZG (1979) On the theory of chemical bond in transition-metal dinitrogen complexes (I). Acta Scientiarum Nturalium Universitatis Jilinensis. (3): 51

Xu JQ (1980a) On the theory of chemical bond in transition-metal dinitrogen complexes (II). Scientiarum Nturalium Universitatis Jilinensis. (1): 83

Xu, JQ (1980b) On the theory of chemical bond in trinuclear transition-metal dinitrogen complexes. J Cat (CUIHUA XUEBAO) 1: 36

Xu JQ, Xu LJ, Lin XS, Zhang ZG (1981) A quantum-chemical theory of transition-metal dinitrogen complexes. Scientia Sinica. 24: 35

Yoshida T, Adachi T, Kaminaka M, Ueda T, Higuchi T (1988) A novel molybdenum(0) dinitrogen complex containing crown thioether as the sole auxiliary ligand: trans-Mo(N$_2$)$_2$Me$_8$[16]aneS$_4$ (Me$_8$[16]aneS$_4$ = 3,3,7,7,11,11,15,15-octamethyl-1,5,9,13-tetrathiocyclohexadecane). J Am Chem Soc 110: 4872

Zeinstra JD, Teuben JH, Jellinek F (1979) Structure of dinitrogenbis[p-tolyldicyclopentadienyl-titanium(III)], [(C$_5$H$_5$)$_2$Ti(P-CH$_3$C$_6$H$_4$)]$_2$N$_2$. J Organomet Chem 170: 39

Ziegler ML, Weidenhammer K, Zeiner H, Skell RS, Herman WA (1976) Stickstoff-Übertragung von einem diazoalkan auf ein metallzentrum. Angew Chem 88: 761

CHAPTER 3
Study on the Chemistry of Molybdenum–Iron–Sulfur and Iron–Sulfur Clusters

Ji-Qing Xu, Zhi-Gui Zhang, Shu-Yun Niu, and Ao-Qing Tang

The Nitrogen Fixation
and its Research in China
Editor· Guo-fan Hong
(C) Springer-Verlag Berlin Heidelberg 1992

1 Introduction

The occurrence and development of some clusters are closely related to chemical bionics, the chemistry of Fe–S and Mo–Fe–S clusters has been developed under the promotion of the biochemical investigations of ferredoxins and nitrogenase and investigations of chemical imitation. It is known that there are two or more than two metal clusters in proteins and enzymes, in which metal atoms bridged with sulfur atoms are connected by terminal groups with residual groups of amino acid in polypeptide chains, the 2-Fe and 4-Fe prosthetic groups of ferredoxin are examples of them. The biological clusters have been investigated thoroughly [Sweeney et al. 1980], in which an iron and four sulfur atoms form a tetrahedron configuration. Holm et al. prepared the imitators of biological clusters in very simple system in 1978, and investigated deeply their structures, redox properties and reaction chemistry (Holm, Laskowski 1978, 1979; Mascharak et al. 1981). Fe–S cluster chemistry has been developed further on the basis of above-mentioned work.

Chemists began to study on biological chemistry of nitrogenase in 60's in order to imitate the function of nitrogenase and realize fixed nitrogen at ambient temperature and pressure, and obtained breakthrough in 1977. Shah et al. separated out the FeMo prosthetic groups (FeMo–co) with molecular weight about 1500 from the MoFe protein with molecular weight about 220 000 in 1977. Cramer et al. investigated its structure by EXAFS (Extended X-ray Absorption Fine Structure) method, the result indicates that it is a new type of a biological cluster containing Mo–Fe–S, and considered to be a important component part of active centers of nitrogenase. Therefore, chemists began also to study on synthesis, structure and properties of imitators of active centers of nitrogenase, Mo–Fe–S cluster chemistry has been developed further on the basis of above-mentioned work.

Chemical investigation on Mo–Fe–S and Fe–S clusters in our lab are introduced as follows.

2 Syntheses

So far over 20 kinds of Mo(W)–Fe–S, over 10 kinds of Fe–S and Mo–S clusters have been prepared in our lab. They are listed in Tables 1 and 2, respectively.

The syntheses of M–Fe–S(M = Mo, W) and Fe–S clusters were mostly performed in purified nitrogen atmosphere under deoxygenated and anhydrous condition at ambient or moderate temperature.

2.1 The Syntheses of Mo–Fe–S Clusters

Mo–Fe–S clusters are usually prepared using MoS_4^{2-} as Mo source and iron compounds (e.g. $FeCl_2$, $FeCl_3$, etc.) as Fe source in the presence of ligand

Table 1. Properties of Mo(W)-Fe–S cluster compounds prepared in the authors' laboratory

No. of compound	P	St	IR	UV-V	V	Mb	Ms	Cat
1.	A1		A1	A1				A1
2.	B1		B1			B1		B1
3.	B1		B1	B1		B1		B1
4.	B1		B1	B1		B1		B1
5.	B1		B1	B1		B1		B1
6.	B1, B2, C1	B2, C1	B1, B2	B1, B2	D	B1, B2		B1, B2
7.	B1, E	B1, E	B1, E	B1, E	D	B1, E		B1, E
8.	A2, F1	A2, F1	A2	A2	A2	A2		A2
9.	G3	G3			G3	G3		G3
10.	G1, C2	G1, C2	G1	G1				G1
11.	G2	G2						
12.	B3		B3	B3				B3
13.	B3		B3	B3				B3
14.	B4	B4		B4		B4		B5
15.	B5, H	B6, H	H	H	H	H	H	H
16.	B7, B8	B7, B8		B7, B8		B8	B8	B8
17.	B9		B9	B9		B9		B9
18.	G4, F2	G4, F2	G4	G4				
19.	B10, I	B10, I	B10, I	B10, I	B10, I	I	I	I
20.	B11, I	B11, I	B11, I	B11, I	B11, I	I	I	I
21.	B12, I	B12, I	B12, I	B12, I	B12, I	I	I	I
22.	G5	G5	G5	G5	G5	G5		

Notes:

I. Compounds

1 $[Bu_4N][MoFe_2S_4(PhSH)_2(OCH_3)_4]$
2 $[Bu_4N]_2[MoFe_6S_8(PhCH_2SH)(CH_3O)_6]$
3 $[NH_4]_2[MoFe_2S_4Cl_4]\cdot 5DMF$
4 $[NH_4]_2[MoFe_2S_4Cl_4(PhCH_4S)]\cdot 5DMF$
5 $[NH_4]_2[MoFe_2S_4Cl_4(PhS)]\cdot 5DMF$
6 $[Et_4N]_3[Mo_2FeS_8O_2]$
7 $[Bu_4N]_3[Mo_2FeS_8O]$
8 $[Et_4N]_3[Mo_2FeS_8O]\cdot CH_3Cn$
9 $[Bu_4N]_3[Mo_2FeS_8O]HOCH_2CH_2OH$
10 $[Et_4N]_3\{[(SCH_2CH_2S)MoS_3]_2Fe\}$
11 $[Et_4N]_3\{[(SCH_2CH_2S)MoS_3]_2Fe\}\cdot 1/4CH_3CN$
12 $[Et_4N]_2[MoFeS_4(SCN)_2OCH_3)_2]\cdot 3CH_3OH$
13 $[Et_4N]_3[Mo_2FeS_8O(OCH_3)_2]$
14 $[Et_4N]_4[Mo_2Fe_2S_{10}]\cdot 2CH_3OH$
15 $[Et_4]_2[(CH_3)_2CCNH_2(CH_3)][Mo_2Fe_2S_{10}]$
16 $Mo_2Fe_2S_4(S_2CNEt_2)_5\cdot CH_3CN$
17 $[Et_4N]_4[Mo_2Fe_7S_{12}(PhS)_6\cdot 6CH_3OH$
18 $[Et_4N]_4[Mo_2Fe_4S_9(SCH_2CH_2S)_2]$
19 $[Et_4N]_3[Mo_2Fe_6S_{14}(OCH_3)_3]$
20 $[Et_4N]_3[Mo_2Fe_6S_{11}(SPh)_3(OCH_3)_3]$
21 $[Et_4N]_3[Mo_2Fe_6S_8(SPh)_6(OCH_3)_3]\cdot 1/2CH_3OH$
22 $[Et_4N]_3\{[(SCH_2CH_2S)WS_3]_2Fe\}$

II. Abbreviations used in this and following tables:
P: preparation
St: structure
IR: infrared spectrum
UV-V: ultraviolet-visible spectrum
V: voltammetry
Mb: Mössbauer spectroscopy
Ms: magnetic susceptibility
Cat: catalytic activity for reduction of C_2H_2

Table 1. (*Notes Continued*)

III. References:

A1: Liu XS et al. 1982	I: Cai H 1990
B1: Xu JQ et al. 1983	A2: Liu XS et al. 1985
B3: Xu JQ et al. 1984b	B2: Xu JQ et al. 1984a
B5: Xu JQ et al. 1988	B4: Xu JQ et al. 1989a
B7: Xu JQ et al. 1989b	B6: Xu JQ et al. 1990a
B9: Xu JQ et al. 1986	B8: Xu JQ et al. 1989c
B11: Xu JQ et al. 1990c	B10: Xu JQ et al. 1990b
C1: Wang FS et al. 1986	B12: Xu JQ et al. 1990d
D: Huang TB et al. 1985	C2: Wang FS et al. 1984
F1: Fan YG et al. 1983	E: Xu LJ et al. 1984
G1: Zhang ZG et al. 1984a	F2: Fan YG et al. 1988a
G3: Zhang ZG et al. 1989a	G2: Zhang ZG et al. 1988
G5: Zhang ZG et al. 1989b	G4: Zhang ZG et al. 1987
	H: Nan YM 1990

Table 2. Properties of Fe–S and Mo–S cluster compounds prepared in the authors' laboratory

No. of compound	P	St	IR	UV-V	V	Mb	Ms	Cat
1	B13, I	B13, I	B13, I	B13, I	B13, I	I	I	I
2	B14	B14	B14	B14				
3	B15	B15						
4	G6	G6	G6					
5	G6	G6	G6					
6	B16, J	B16, J	B16	B16	B16	B16		
7	B17, H	B17, H	B17, H	B17, H	H	H	H	
8	K	K						
9	B18, I	B18, I	B18, I	B18, I	B18, I	I	I	
10	G7, F3	G7, F3	G7, F3	G7, F3				
11	L	L	L	L				
12	G8	G8	G8	G8				

Notes:

I. Compounds:

1 $[Et_4N]_2[Fe_2S_2(SPh)_4$ (monoclinic)
2 $[Et_4N]_2[Fe_2S_2(SPh)_4$ (orthogonal)
3 $[Et_4N]_2[Fe_2(SCH_2CH_2S)_4]$
4 $[Et_4N]_2[Fe_3(CO)_9S]$
5 $[Bu_4N]_2[Fe_3(CO)_9S]$
6 $[Et_4N]_2Fe_4S_4(S_2CNEt_2)_4]$
7 $[Et_4N][Fe_4S_4(S_2CNEt_2)_4]$
8 $[Bu_4N]_2\{Fe_4S_4[SC(CH_3)_3]_4\}$
9 $[Bu_4N]_2[Fe_4S_4(SPh)_4]$
10 $[Et_4N]_4[Fe_6S_9(SCH_2CH_2OH)Cl]$
11 $[Et_4N]_2[[Mo_2S_6O_2]$
12 $[Bu_4N]_2[Mo_2S_6O_2]$

II. References:

B13: Xu JQ et al. 1990e
B15: Xu JQ et al. 1987
B17: Xu JQ et al. 1990g
F3: Fan YG et al. 1988b
G7: Zhang ZG et al. 1985a
J: Yang GD et al. 1989
L: Yan YZ et al. 1987
B14: Xu JQ et al. 1990f
B16: Xu JQ et al. 1989d
B18: Xu JQ et al. 1990h
G6: Zhang ZG et al. 1989c
G8: Zhang ZG et al. 1985b
K: Hu NH et al. 1990

compounds under alkaline condition. Mo–Fe–S clusters in our lab were mostly obtained by means of this method.

For example, the clusters $[MoFe_2S_4Cl_4(PhS)]^{2-}$ 5, $Mo_2Fe_2S_4(S_2CNEt_2)_5$ 16, $[Mo_2FeS_8O_2]^{3-}$ 6, and $[Mo_2Fe_4S_9(SCH_2CH_2S)_2]^{4-}$ 18 were obtained by following reactions, respectively.

5   $FeCl_2 + MoS_4^{2-} + PhSH + CH_3ONa \xrightarrow{DMF,EtOEt}$

$$[MoFe_2S_4Cl_4(PhS)]^{2-} + \text{other products} \quad (1)$$

6   $FeCl_3 + MoS_4^{2-} + CH_3ONa \xrightarrow{CH_3OH} [Mo_2FeS_8O_2]^{3-} +$

$$\text{other products} \quad (2)$$

16   $FeCl_3 + MoS_4^{2-} + NaS_2CNEt_2 + PhSH + CH_3ONa \xrightarrow{DMF,CH_3OH,CH_3CN}$

$$Mo_2Fe_2S_4(S_2CNEt_2)_5 + \text{other products} \quad (3)$$

18   $FeCl_3 + MoS_4^{2-} + HSCH_2CH_2SH + CH_3ONa \xrightarrow{CH_3OH}$

$$[Mo_2Fe_4S_9(SCH_2CH_2S)_2]^{4-} + \text{other products} \quad (4)$$

Practically, the clusters 3–13, 16 and 17 were prepared by similar method, but the clusters 1 and 2 were prepared using $MoCl_4(CH_3CN)_2$ as molybdenum source.

1   $MoCl_4(CH_3CN)_2 + FeCl_3 + NaHS + CH_3ONa + PhSH \xrightarrow{CH_3OH}$

$$[MoFe_2S_4(PhSH)_2(CH_3O)_4]^- + \text{other products} \quad (5)$$

Recently, we prepared a series of new clusters using simple iron or $Fe + FeCl_3$(or $FeCl_2$) as iron source, the clusters 14, 15 and 19–21 were prepared by this method.

14   $Fe + MoS_4^{2-} + PhSH + CH_3ONa \xrightarrow{DMF,CH_3OH}$

$$[Mo_2Fe_2S_{10}]^{4-} + \text{other products} \quad (6)$$

19   $FeCl_3 + Fe + MoS_4^{2-} + S + p\text{-Me-PhSH} + CH_3ONa \xrightarrow{DMF,CH_3OH}$

$$[Mo_2Fe_6S_{14}(OCH_3)_3]^{3-} + \text{other products} \quad (7)$$

20   $FeCl_3 + Fe + MoS_4^{2-} + S + PhSH + CH_3ONa \xrightarrow{DMF,CH_3OH}$

$$[Mo_2Fe_6S_{11}(SPh)_3(OCH_3)_3]^{3-} + \text{other products} \quad (8)$$

Sometimes the produced cluster does not contain the ligand related to the compound added to the reaction system, but in its absence the reaction does not occur, this shows that the compound participates in the reaction. For example, in reaction (3) cluster *16* was formed under the reduction condition, the total oxidation number of the core Mo_2Fe_2 in cluster *16* is 13, but that of Mo_2Fe_2 in the reactants is 18, this indicated that the total oxidation number reduces 5 units in the reaction process. Both PhS^- and Et_2NCS^- provide electrons by dimerisation as Reaction (9) and (10), they are all reductant.

$$2PhH^- \longrightarrow PhSSPh + 2e \tag{9}$$

$$2Et_2NCS_2^- \longrightarrow Et_2NC(S)SS(S)CNEt_2 + 2e \tag{10}$$

Since the coordination ability of $Et_2NCS_2^-$ is much higher than that of PhS^-, $Et_2NCS_2^-$ not only works as a reductant, but also is a ligand source, but PhS^- only is a reductant (Q.J. Xu et al. 1989b, c).

We discovered that alkalinity of reaction systems is an important factor affecting cluster structure, and influences ligand sort to a certain extent, but Fe/Mo mole ratio does not effect on cluster structure obviously. The results are consistent with the results of formation condition of Mo–Fe–S cluster which were researched by infrared spectra and UV-visible spectra (S.Y. Niu et al. 1987).

2.2 Syntheses of Fe–S Clusters

According to the source of bridging sulfur atoms synthetic methods may be divided into three kinds (J.Q. Xu 1989e). The first method is using HS^- as the source of bridging sulfur atom. T. Hersskovitz et al. synthesized the first analogue of the active center of Iron–Sulfur protein, $[Fe_4S_4(SCH_2Ph)_4]^{2-}$, by this method in 1972.

$$4FeCl_3 + 6RS^- + 4HS^- + 4CH_3O^- \xrightarrow{CH_3OH}$$

$$[Fe_4S_4(SR)_4]^{2-} + RSSR + 12Cl^- + 4CH_3OH \tag{11}$$

where R denotes Aryl or Alkyl.

The second method is using simple sulfur as the source of bridging sulfur atoms. The method requires sufficient thioalcoholate as a reductant, G. Cristou et al. synthesized a series of Fe–S clusters by this method.

$$4FeCl_3 + 14RS^- + 4S \xrightarrow{CH_3OH} [Fe_4S_4(SR)_4]^{2-} + 5RSSR + 12Cl^- \tag{12}$$

$$4FeCl_2 + 10RS^- + 4S \xrightarrow{CH_3OH} [Fe_4S_4(SR)_4]^{2-} + 3RSSR + 8Cl^- \tag{13}$$

We discovered that MS_4^{2-} (M = Mo, W or V) can redox in the suitable condition (J.Q. Xu et al. 1989b–e):

$$MS_4^{2-} (M = Mo, W \text{ or } V) \longrightarrow MS_3^{2-} + S \tag{14}$$

Therefore, the clusters of Fe–S can be prepared by MS_4^{2-} as the source of bridging sulfur atom, this is the third method of synthesizing Fe–S clusters.

Actually, the Fe–S clusters listed in Table 2 were all prepared by this method, for example:

$$2 \quad Fe + VS_4^{2-} + PhSH + CH_3ONa \xrightarrow{CH_3CN,CH_3OH}$$

$$[Fe_2S_2(PhS)_4]^{2-} + \text{other products} \quad (15)$$

$$4 \quad Fe(CO)_5 + MoS_4^{2-} + CH_3ONa \xrightarrow{CH_3OH}$$

$$[Fe_3S(CO)_9]^{2-} + \text{other products} \quad (16)$$

$$6 \quad FeCl_2 + WS_4^{2-} + NaS_2CNEt_2 \xrightarrow{CH_3OH,DMF}$$

$$[Fe_4S_4(S_2CNEt_2)_4]^{2-} + \text{other products} \quad (17)$$

$$7 \quad FeCl_3 + WS_4^{2-} + NaS_2CNEt_2 + PhSH + CH_3ONa \xrightarrow{DMF,CH_3OH,CH_3CN}$$

$$[Fe_4S_4(S_2CNEt_2)_4]^- + \text{other products} \quad (18)$$

$$10 \quad FeCl_2 + MoS_4^{2-} + HSCH_2CH_2OH + CH_3ONa \xrightarrow{CH_3OH}$$

$$[Fe_6S_9(SCH_2CH_2OH)Cl]^{4-} + \text{other products} \quad (19)$$

The mechanism of the reaction synthesizing Fe–S cluster **6** by the third method may be infered as follows: (J.Q. Xu et al. 1989d)

$$MoS_4^{2-}(M = Mo, W \text{ or } V) \longrightarrow MoS_3^{2-} + S$$

$$FeCl_2 + 4RS^- \longrightarrow [Fe(SR)_4]^{2-} + 2Cl^-$$

$$2Fe(SR)_4^{2-} + 2S \longrightarrow [Fe_2S_2(SR)_4]^{2-} + RSSR + 2RS^-$$

$$2Fe_2S_2(SR)_4]^{2-} \longrightarrow [Fe_4S_4(SR)]^{2-} + RSSR + 2RS^-$$

where $R = Et_2NC=S$.

3 Structures

3.1 M–Fe–S Clusters

Mo(W)–Fe–S clusters synthesized and structure-determined in our lab are classified into four types: linear (clusters 6, 11, 14, 15 and 22), single cubane (cluster 16), double cubane (clusters 19–21) and reticulate (cluster 18) geometric configuration, Tables 3 and 4 list strctural data of the four types of clusters.

For the linear type of Mo–Fe–S cluster we obtained the following conclusion: (1) All the linear Mo–Fe–S clusters contain FeS_2Mo structural unit with approximate plane configuration, the change range of $Fe–S_b–Mo$ bond angles is not large, the Mo–Fe distance depends on the bond angle to a great extent.

Table 3. Selected structural data for some linear cluster compounds[a]

No. of compound	6[c]	7[d]	10[i]	14[k]	22[n]
M–Fe	2.731	2.708[e] 2.768[f]	2.724	2.801	2.720
Fe–Fe				2.746	
M–Sb	2.299	2.287[e] 2.241[f]	2.314	2.255	2.270
Fe–Sb	2.241	2.227[g] 2.267[h]	2.222	2.279	2.210
M–St	2.388	2.377[e] 2.179[f]	2.139[j]	2.178	1.980[j]
Sb–M–Sb	101.5	100.2[e] 105.4[f]	99.4	104.5	100.2
Sb–Fe–Sb	105.2	105.1[g] 103.0[h]	105.1	102.9[l] 104.6[m]	104.9
M–Sb–Fe	73.9	73.8[g] 75.6[h]	73.8	76.5	75.1
M–Fe–M	160.5	162.9	155.4		165.8
Ref.	Cl	E	C2	B4	G5

Notes:
* The compounds of the numbers:
6 $[Mo_2FeS_8O_2]^{3-}$
7 $[Mo_2FeS_8O]^{3-}$
10 $[\{[(SCH_2CH_2S)MoS_3]_2Fe\}^{3-}$
14 $[Mo_2Fe_2S_{10}]^{4-}$
22 $\{[(SCH_2CH_2S)WS_3]_2Fe\}^{3-}$

[a] In this and succeeding table, b, bridging; t, terminal; distance, Å; angles, deg.; given bond distances and angles are all the mean value
[b] See Table 1
[c] Mo–O 1.718
[d] Mo–O 2.204
[e] Mo Containing oxygen ligand
[f] Mo without oxygen ligand
[g] Sb Linking Mo containing oxygen ligand
[h] Sb Linking Mo without oxygen ligand
[i] Mo–L, 2.423
[j] M=S
[k] Mo–Fe–Fe, 179.1
[l] Sb Linking Fe and Mo
[m] Sb Linking both Fe
[n] W–L, 24.20

(2) The distance of Mo–S* (S* denotes inorganic sulfur) relates to coordination number of Mo. When the coordination number of Mo is five, mean bond length of Mo–S* ranges between 2.26 and 2.34 Å and is obviously longer than that in MoS_4^{2-} (2.18 Å); when the coordination number of Mo is four, the bound length of Mo–S* (2.21 Å) is shorter. (3) The bond length of Mo–Fe being between 2.70–2.80 Å belongs to bonding range.

For the cubane clusters, clusters _19, 20_ and _21_ contain similar skeletons of double cubane and have close bond lengths to Mo–S_b (2.34–2.35 Å) and Fe–S_b (2.27 Å), but the differences among the bond lengths of Mo–u_2–O are obvious

Table 4. Selected structural data for some cubane-like and reticular cluster compounds

No. of compound	16	18	19	20	21
Mo–Mo	2.734		3.118	3.146	3.217
Fe–Fe	2.784	2.702	2.715	2.705	2.688
Mo–Fe	2.753	2.701	2.744	2.726	2.739
Mo–Sb	2.332	2.253[a]	2.344	2.337	2.352
		2.360[b]			
Mo–St	2.531	2.412			
Mo–O			2.113	2.147	2.201
Fe–Sb	2.256	2.219[c]	2.271	2.269	2.269
		2.262[d]			
		2.320[e]			
Fe–St	2.356		2.225	2.235	2.238
Ref.	B7	F2	I	I	I

Notes:
[a] $Mo(u_2\text{-}Sb)$
[b] $Mo(u_3\text{-}Sb)$
[c] $Fe(u_2\text{-}Sb)$
[d] $Fe(u_3\text{-}Sb)$
[e] $Fe(u_4\text{-}Sb)$
* The compounds of the numbers:
16 $Mo_2Fe_2(S_2CNEt_2)_5$
18 $[Mo_2Fe_4S_9(SCH_2CH_2S)_2]^{4-}$
19 $[Mo_2Fe_6S_{14}(OCH_3)_3]^{3-}$
20 $[Mo_2Fe_6S_{11}(SPh)_3OCH_3)_3]^{3-}$
21 $[Mo_2Fe_6S_8(SPh)_6(OCH_3)]^{3-}$

and they appear regular change: $19(2.11\,\text{Å}) < 20(2.15\,\text{Å}) < 21(2.20\,\text{Å})$, the bond lengths of Mo–Mo also undergo regular change: $19(3.12\,\text{Å}) < 20(3.15\,\text{Å}) < 21(3.22\,\text{Å})$. In the three kinds of clusters the metallic oxidation numbers are $19(\text{MoFe}_6)^{28+}$, $20(\text{Mo}_2\text{Fe}_6)^{25+}$ and $21(\text{Mo}_2\text{Fe}_6)^{22+}$, respectively. It is shown from the bond lengths of Fe–S and data of Mössbauer spectra that the valence states of the Fe atoms are close, but the valence states of the Mo atoms are distinct, they are $19\,\text{Mo}^{+5.9}$, $20\,\text{Mo}^{4.5}$ and $21\,\text{Mo}^{2.7}$, respectively. The difference results in the regular changes of bond lengths of Mo–S and Mo–Fe. Figures 1 and 2 show the structures of some Mo–Fe–S clusters with linear, cubane, and reticulate type, respectively.

3.2 Fe–S Clusters

Table 5 lists the structural data of some Fe–S clusters. It can be seen from the data in Table 5 that (1) the bond lengths of Fe–S_b and Fe–S(R) in the cluster anion $[Fe_4S_4(S_2CNEt_2)]^{2-}$ **6** are much longer than that in the cluster anion $[Fe_4S_4(S_2CNEt_2)_4]^-$ **7**, this results from the formal valence of Fe atoms ($+2.5$) in cluster **6** lower than that of Fe atoms ($+2.75$) in cluster 7; (2) The clusters $[Fe_4S_4(SC(CH_3)_3)_4)]^{2-}$ **8**, $[Fe_4S_4(SPh)_4]^{2-}$ **9**, and **6** all have cubane-like

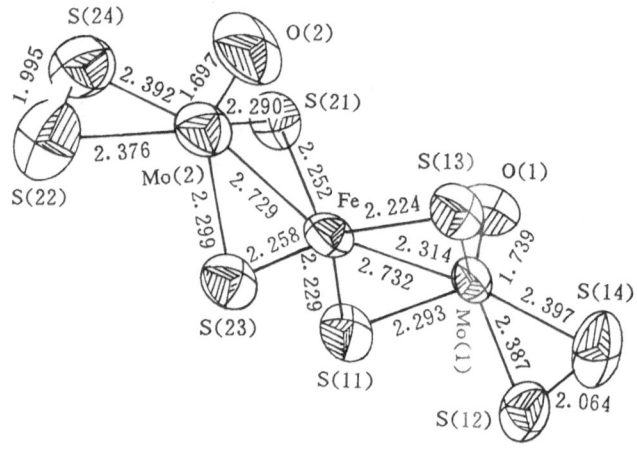

6

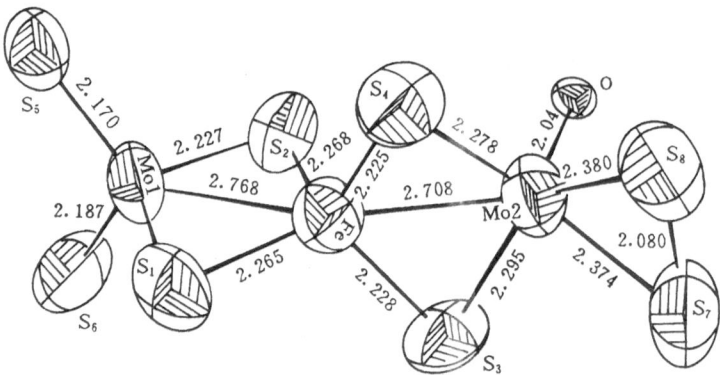

7

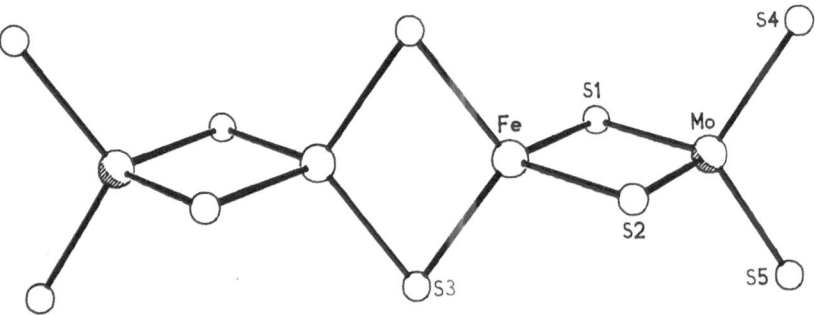

14

Fig. 1. Structure of some linear cluster compounds (6, 7, 14, 22)

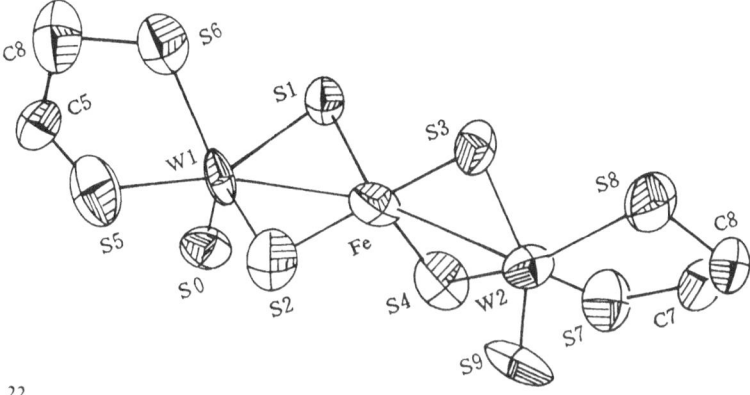

22

Fig. 1. (*Continued*)

structure and the same formal valence for their Fe atoms ($+2.5$), but the bond lengths of Fe–S_b and Fe–S(R) in cluster **6** are longer than those in clusters **8** and **9**, especially their differences on Fe–S(R) are more obvious, this may arise from different coordination numbers of Fe atoms. The coordination number of Fe atoms in cluster **6** is five, whereas four in clusters **8** and **9**.

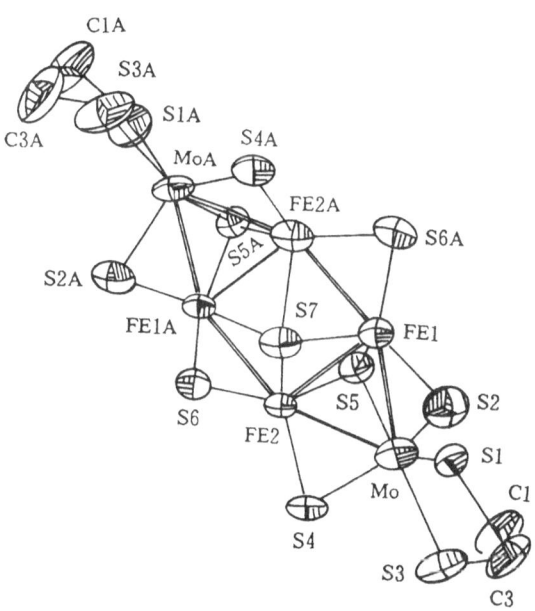

18

Fig. 2. Structure of some cubane and reticulated cluster compounds (*18, 19, 20, 21*)

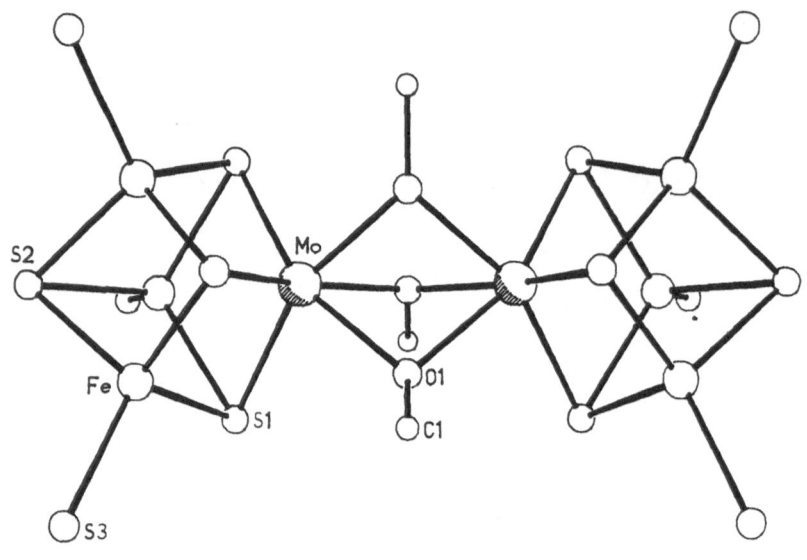

19

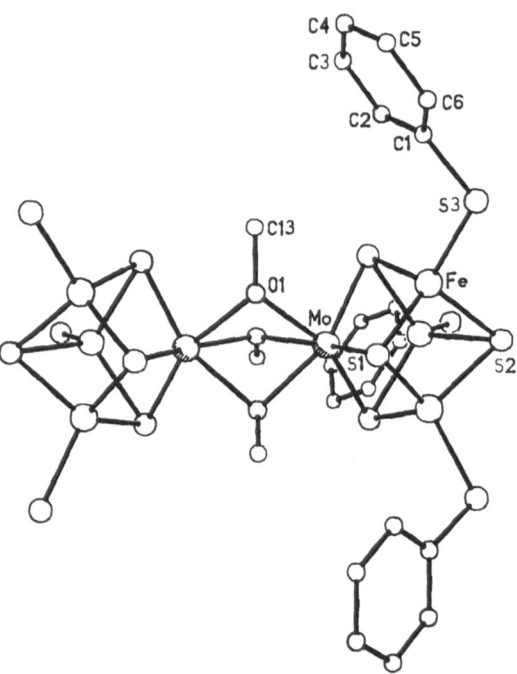

20

Fig. 2. (*Continued*)

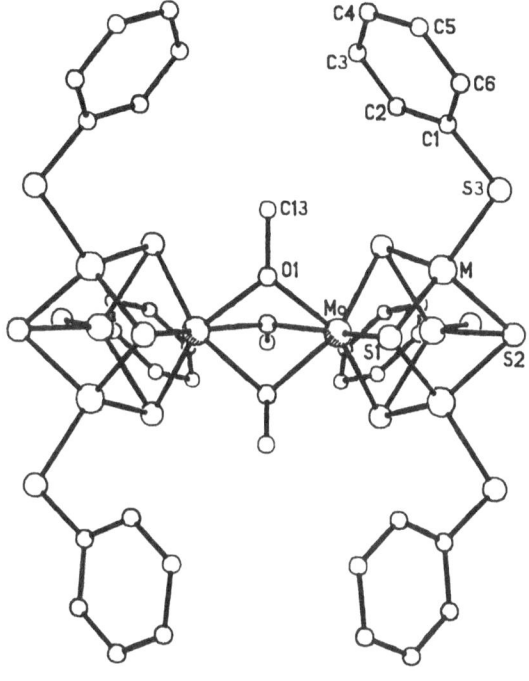

21

Fig. 2. (*Continued*)

Table 5. Selected structural data of some Fe–S clusters

No. of compound	Fe–Fe	Fe–Sb	Fe–S(R)	Ref.[a]
6	2.900	2.293	2.499	B16
7	2.800	2.257	2.390	H
8	2.780	2.287	2.254	K
9	2.744	2.282	2.267	I
10	2.735	2.342[b]	2.285	F3
		2.279[c]		
		2.234[d]		

Notes:
[a] See Table 2
[b] Fe–(u_4-S)
[c] Fe–(u_3-S)
[d] Fe–(u_2-S)

* The compounds of the numbers:
  6 $[Fe_4S_4(S_2CNEt_2)_4]^{2-}$
  7 $[Fe_4S_4(S_2CNEt_2)]^-$
  8 $\{Fe_4S_4[SC(CH_3)_3]_4\}^{2-}$
  9 $[Fe_4S_4(SPh)_4]^{2-}$
  10 $[Fe_6S_9(SCH_2CH_2OH)Cl]^{4-}$

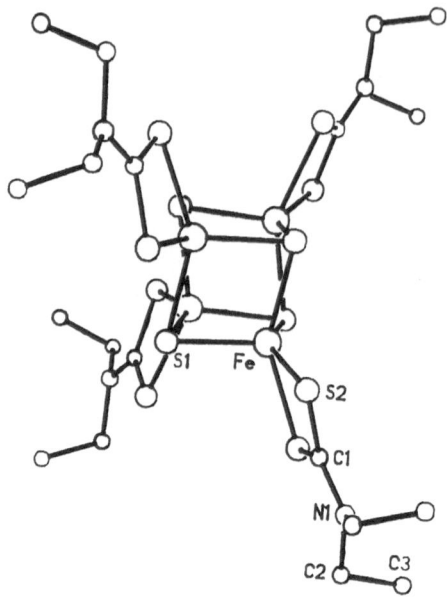

7

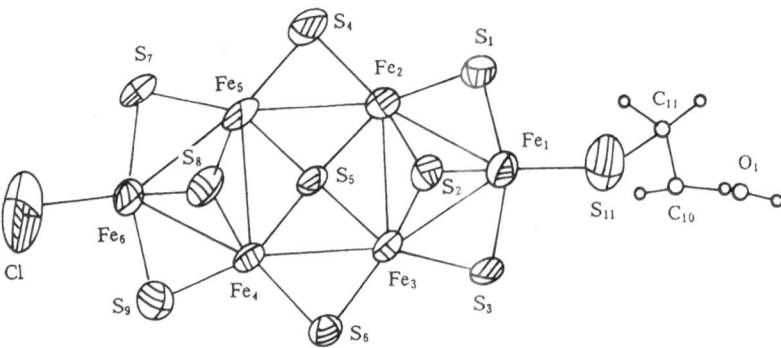

10

Fig. 3. Structure of clusters (**7, 10**) [Fe$_4$S$_4$(S$_2$CNEt$_2$)$_4$]$^-$ and [Fe$_6$S$_9$(SCH$_2$CH$_2$S)$_2$]$^{4-}$

The structures of [Fe$_4$S$_4$(S$_2$CNEt$_2$)$_4$]$^-$ **7** and [FE$_6$S$_9$(SCH$_2$CH$_2$OH)Cl]$^{4-}$ **10** are shown in Fig. 3. It is interesting that the anions of cluster **6** and **7** have the same component, but their valences are different, the result makes the latter have much higher symmetry compared with the former, the crystal of cluster **6** belongs to monoclinic system and Cc space group, the crystal of cluster **7** square system and I$\bar{4}$2m space group.

4 Properties

4.1 UV-Visible Spectra

The data of UV-Visible spectra of some Mo–Fe–S clusters are listed in Table 6. It can be seen from the data in Table 6 that the electronic spectra of linear M–Fe–S clusters generally have more than four narrow and strong absorption bands, and two of them mostly range in 400–510 nm. It is evident that the two absorption bands arise from the split of the electron transition energy of S→Mo in Mo–Fe–S clusters (Coucouvanis 1981).

The characteristic absorption peaks of UV-Vis absorption spectra of Mo–Fe–S clusters with cubane configuration are scarce and in general only there are two: one is about 450 nm, another about 300 nm, furthermore they are not obvious.

The UV-Vis spectra of Fe–S clusters with cubane-like configuration are similar to that of Mo–Fe–S with cubane-like configuration, their absorption peaks also are scarce and are all the results arised from electron transfer from sulfur atoms in the ligands to cubane skeleton.

4.2 Mössbauer Spectra and Magnetic Susceptibilities

Mössbauer spectra and magnetic susceptibilities of some Mo(W)–Fe–S clusters determined at room temperature are listed in Table 7. Mössbauer parameters

Table 6. UV-Vis spectra of some Mo(W)–Fe–S cluster compounds

No. of compound	Solvent	λ_{max}(nm)	Ref.[a]
6	DMF	660, 560, 570, 438, 365, 315, 275	M
7	DMF	625, 570, 505, 469, 343, 307, 275	M
8	DMF	663, 506, 436, 360, 316, 267	A2
	CH_3CN	445, 305, 229	A2
10	DMF	685, 575, 470, 360, 310	M
12	DMF	575 (shoulder), 475, 303	B2
13	DMF	570 (shoulder), 465, 335, 295	B3
14	DMF	470, 340, 304	B4
15	DMF	580, 500, 480, 350, 320, 265	H
16	DMF	348 (shoulder)	B8
17	DMF	470 (shoulder)	B9
18	DMF	585, 525, 360, 285	M
19	DMF	450, 310	I
20	DMF	430, 300	I
21	DMF	450, 310	I
22	DMF	620, 488, 355, 310	G5

Notes:
[a] See Table 1
M Niu SY et al. 1987b
* The compounds of the numbers: see Table 1

Table 7. Mössbauer parameters and magnetic susceptibilities of some Mo(W)–Fe–S cluster compounds

No. of compound	δ (mm/s)	ΔEq (mm/s)	$\mu(\mu_B)$	Ref.[a]
1	0.00	1.24		B1
	0.07	0.08		
2	0.31	0.97		B1
3	0.22	1.08		B1
4	0.08	0.48		B1
5	0.07	0.54		B1
6	0.31	0.00		B1
7	0.23	0.45		B1
8	0.73	0.14		A2
9	0.39	0.76		G3
10	0.05	0.85	5.01	G5
14	0.29	1.03	2.25	L
15	0.23	0.95	2.41	H
16	0.58	1.42	4.27	B8
	0.47	0.58		
17	0.37	0.72		B9
	0.32	0.81		
18	0.06	1.02	6.09	G5
	0.14	3.14		
19	0.27	0.77	5.37	I
20	0.27	0.10	5.23	I
	0.32	0.62		
21	0.25	0.78	5.57	I
22	0.26	0.00	4.13	G5

Notes:
[a] See Table 1
* The compounds of the numbers: see Table 1

in Table 7 refer to the result of α-Fe atom at room temperature. According to the data in Table 7 we can obtain the conclusions as follows: (1) The Mössbauer spectra of linear Mo–Fe–S clusters and double cubane clusters 19 and 21 mostly consist of one group of quadripole double lines. It shows that there is only one kind of valence state of Fe atom in the clusters. This is consistent with geometric configuration of the clusters. (2) The Mössbauer spectra of cluster 18 and 20 consist of two groups of quadripole double lines with the absorption intensity of 1:1, it shows that the clusters contain different valence of Fe atoms and both are identical. The conclusion is easily obtained from their molecular structures. (3) The Mössbauer spectra of cluster 16 consist of two groups of quadripole double lines. The result shows that cluster 16 has two kinds of Fe atoms with different valence, each Fe atom in cluster 16 is coordinated with five S atoms and has the same ligands. It seems that they should have the same valence, but, in fact, we found that the difference between Fe–S bond lengths of two Fe atoms is obvious. This indicates that they have different valence (Xu et al. 1989c).

4.3 Redox Properties

We determined cyclic voltammgrams of two kinds of Mo–Fe–S clusters, four kinds of Fe–S clusters and a kind of Fe–S complex. Their results are listed in Table 8. Redox properties and effecting factors of the compounds were obtained by comparison and analysis.

The experiments were performed in the degassed and anhydrous solution of dimethyl-formamide containing tetrabutylammonium perchlorate as the supporting electrolyte. A platinum plate was used as working electrode, its potential scan rate was 100 mv per second, the oxidation or reduction potentials are all referred to the potential of saturated calomel electrode.

We realized that the redox properties of the compounds relate to charge and geometric configuration of a reactant particle, charge density of the metal atom in the cluster core and the electron effect of the ligand and so on. The more negative the charge of a reactant particle, the more difficulty it gains an electron and more easily it loses an electron; the more positive the formal valence of metal atom in the cluster core, the more easily it gains an electron and the more difficulty it loses an electron; if the ligand is an electron-withdrawing

Table 8. Cyclic voltammetric data of six clusters and a coordinate compound

No. of compounds	$P_O(V)$	$P_R(V)$	Ref.[a]
2	0.530	0.420	N.B14
	−0.150	−0.730	
14	0.470	0.170	N.B4
	−0.100	−0.580	
6	0.460	0.390	N.B16
	0.030	−0.090	
	−0.375	−0.475	
		−0.725	
8	0.700	0.220	N.K
	−0.250	−0.630	
9	−0.050	−0.100	N.I
	−0.320	−0.800	
16	0.500	0.410	N.B8
	−0.350	−0.475	
Fe(DTC)₃	0.480	0.420	
	0.060	−0.070	
	−0.040	−0.400	
		−0.550	
		−1.120	

Notes:
[a] See Tables 1 and 2
N Zhang HB et al. 1990
P_O Potentials of oxidation peaks
P_R Potentials of reduction peaks
* The compounds of the numbers: see Tables 1 and 2

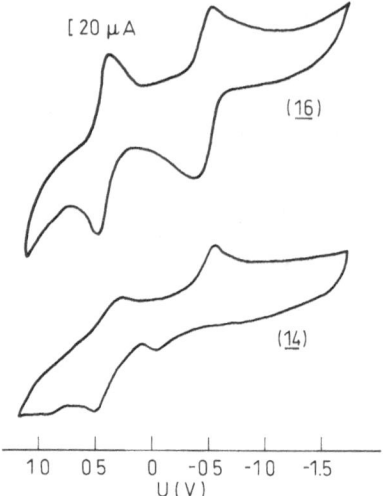

Fig. 4. Cyclic voltammogram of $[Fe_2Mo_2S_4(DTC)_5]$ **(16)** and $[Fe_2Mo_2S_{10}]^{4-}$ **(14)**

group, the reactant particle gains an electron easily, otherwise, if the ligand is an electron-donating group, it loses an electron easily; the geometric configuration in the structure, e.g. chain or cubane, also bring about different redox properties.

Figure 4 shows cyclic voltammograms of $[Et_4N]_4[Mo_2Fe_2S_{10}]$ and $Fe_2Mo_2S_4(DTC)_5$. The composition of their cluster cores is similar to each other, i.e. both contain Fe_2Mo_2 with inorganic sulfur atoms as bridges. But the chain structure of the former is different from the cubane structure of the latter with DTC group ligand. A great disparity between the two reactant particles is that the former is a cluster anion with a valence of -4, whereas the latter is a neutral molecule. Another difference is that the formal valence of iron atom in the former is $+2.7$ and $+2.5$ in the latter.

It can be seen from curves and data that the former has more negative value for oxidation and reduction peak potentials at the more positive position compared with the latter. Especially, in the reduction process the value of the former is much more negative than that of the latter. This indicates that the former is easier to be oxidized and more difficult to be reduced compared with the latter. It is known from the above analysis that there are two effecting factors: the charge of reactant particles and the formal valence of iron atoms. In addition, the effects of DTC group and geometric configuration cannot be negligible either.

4.4 Catalytic Activity

The activity of catalytic reduction of C_2H_2 for some Mo–Fe–S and Fe–S cluster compounds synthesized in our laboratory was determined (Xu et al. 1988, 1989f).

Table 9. Activity of Mo–Fe–S cluster and related compounds for catalytic reduction of C_2H_2

No. of compounds	Amount of M in system $(\mu mol)^b$	TN^a $(\mu$ mmol products/μ mol Mo min)^b		TTN	Selectivity to C_2H_4 (%)
		C_2H_4	C_2H_6		
Blank^c	0	0	0	—	—
14	0.187(Mo)	3.88	0.39	4.27	90.9
16	0.274(Mo)	0.07	0.01	0.08	87.5
7	0.385(Mo)	2.56	0.17	2.73	93.8
6	0.172(Mo)	2.18	0.10	2.28	95.6
12	0.283(Mo)	4.00	0	4.00	100
13	0.312(Mo)	3.00	0	3.00	100
17	0.321(Mo)	6.70	0.74	7.44	90.1
6	0.358(Fe)	0	0	—	—
8	0.340(Fe)	0.04	0.01	0.05	80.0
A	0.634(Mo)	0.70	0.07	0.77	90.9^d
B	0.634(Mo)	1.09	0.09	1.18	92.4^e
C	0.634(Mo)	0.41	0.05	0.46	89.1
D	0.706(W)	0	0	—	—
E	1.637(Fe)	0	0	—	—
F	1.247(Fe)	0	0	—	—
G^f	0.490(Mo)	2.54	0.14	2.68	94.8
H^g	0.490(Mo)	1.88	0.14	2.02	93.0
I^h	0.490(Mo)	2.56	0.16	2.27	94.1
J^i	0.490(Mo)	1.64	0.08	1.27	95.6
K^j	0.353(W)	0	0	—	—
L^k	0.353(W)	0	0	—	—

Notes:
^{a} TN = turnover number, average value in 30 min; TTN = total turnover number
^{b} M = Mo, Fe or W
^{c} Blank experiment was performed under the same experimental condition
^{d} Specific activities relative to Fe are 1.31 μ mol C_2H_4/μ mol Fe min and 0.12 μ mol C_2H_6/μ mol Fe min
^{e} Specific activities relative to Fe are 1.64 μ mol C_2H_4/μ mol Fe min and 0.14 μ mol C_2H_6/μ mol Fe min
^{f} Fe:Mo = 1.5:1 (by mole)
^{g} Fe:Mo:DTC = 1.5:1:1, DTC = NaS_2CNEt_2
^{h} Fe:Mo = 1:1
^{i} Fe:Mo:DTC = 1:1:1
^{j} Fe:W = 2:1
^{k} Fe:W = 1.5:1

^{l} Compounds:
A $(Bu_4N)_2Fe_4S_4(SC(CH_3)_3)_4 + (NH_4)_2MoS_4$
B $(Bu_4N)Fe_2S_2(SCH_2Ph)_2 + (NH_4)_2MoS_4$
C $(NH_4)_2MoS_4$
D $(NH_4)_2WS_4$
E $FeCl_2$
F $FeCl_3$
G $(NH_4)MoS_4 + FeCl_2$
H $(NH_4)_2MoS_4 + FeCl_2 + DTC$
I $(NH_4)_2MoS_4 + FeCl_3$
J $(NH_4)_2MoS_4 + FeCl_3 + DTC$
K $(NH_4)_2WS_4 + FeCl_2$
L $(NH_4)_2WS_4 + FeCl_3$
* The compounds of the numbers: see Tables 1 and 2

The catalytic activity measurements were carried out under anaerobic condition at 25 °C with KBH_4 as the reductant, and a solution of sodium borate (PH 9.6) as the buffer. The catalytic activity of some combination systems of relevant compounds was also determined for the sake of comparison. The catalytic activity data are given in Table 9.

Then again, we have determined catalytic activity of combination systems formed from $(NH_4)_2MoS_4$ and $FeCl_3$ with different mole ratio. Results obtained are listed in Table 10. The relation of catalytic activity (total turnover number) to the Fe/Mo ratio is shown in Fig. 5.

The following preliminary conclusions are obtained:

(1) Seven Mo–Fe–S cluster compounds, the catalytic activities of which were determined, all displayed obvious catalytic activity with the exception of compound 16 (see Table 9). And the catalytic activity increased with increasing Fe/Mo mole ratio in the cluster compounds (see Table 11).

(2) The catalytic activity of the combination systems of $(NH_4)_2MoS_4$ with $FeCl_3$ reached a maximum when Fe/Mo = 8. This ratio is in agreement with the Fe/Mo ratio (Fe/Mo = 6–8:1) in FeMo–co of nitrogenase. When this optimum value was surpassed, the catalytic activity decreased as the Fe/Mo ratio further increased.

(3) The catalytic activity for Mo–Fe–S cluster compounds is closely related to the coordination environment of the metal atoms, particularly that of the Mo atom. The metal ions in cluster compound **16** are coordinated by chelate ligands with strong coordination ability, and the ligands produce marked steric effects. This results in that the substrate molecules can not get close to the cluster core and thus can not be activated. This is why the catalytic

Table 10. Activity of combination systems with different compositions for catalytic reduction of C_2H_2

No.	Fe:Mo in system	Amount of Mo in system (μ mol)	Amount of Fe in system (μ mol)	TN^a (μ mol products/ μ mol Mo min)		TTN	Selectivity to C_2H_4 (%)
				C_2H_4	C_2H_5		
1	0.5:1	0.462	0.231	0.28	0.05	0.33	84.0
2	1:1	0.462	0.462	0.28	0.05	0.33	86.5
3	2:1	0.462	0.924	0.57	0.05	0.62	92.4
4	3:1	0.462	1.386	0.77	0.06	0.83	92.4
5	4:1	0.462	1.848	1.02	0.11	1.13	90.0
6	6:1	0.462	2.772	1.50	0.17	1.67	90.0
7	8:1	0.462	3.696	1.98	0.25	2.23	88.7
8	10:1	0.462	4.620	1.34	0.09	1.43	93.9
9	3:1:4 (DTC)^b	0.462	1.386	0.01	0	0.01	100

Notes:
^{a} TN = turnover number, average value in 30 min
TTN = total turnover number
^{b} In addition to $(NH_4)_2MoS_4$ and $FeCl_3$, sample 9 contains DTC, Fe:Mo:DTC = 3:1:4 by mole

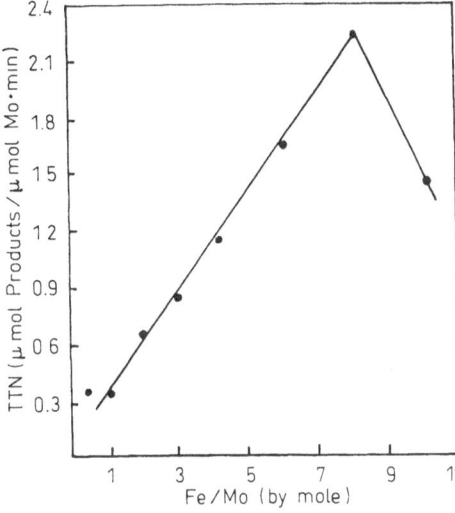

Fig. 5. The relation of activity of catalytic reduction of C_2H_2 to Fe/Mo by mole in combination system

Table 11. Dependence of activity of Mo–Fe–S cluster compounds for catalytic reduction of C_2H_2 on Fe/Mo in such compounds

Compounds	7	6	13	14	12	17
Fe:Mo	1:2	1:2	1.2	1:1	1:1	3.5:1
TTN	2.37	2.28	3.00	4.27	4.00	7.44

Notes:
Compounds: See Table 1

activity was greatly different between cluster compounds **14** and **16**, although they all contain two Mo and two Fe atoms. The catalytic activity of cluster compound **16** was very small.

(4) $(NH_4)_2MoS_4$ was found to have some catalytic activity, $FeCl_2$ and $FeCl_3$ had none, while Fe–S cluster compounds displayed very low activity (see Table 9). The catalytic activity was greatly enhanced, however, when $(NH_4)_2MoS_4$ was combined with iron compounds (including Fe–S cluster compounds), and the catalytic activity of the combination systems was clearly higher than the sum of the activity of the individual compounds. This appears to show that chemical reaction took place when starting compounds were combined, and Mo–Fe–S cluster compounds possessing comparatively high catalytic activity were probably formed as a consequence. It is believed that acetylene molecules are mainly activated by a Mo–Fe multinuclear action synergistically, but not by a mononuclear action. This is consistent with the active site of nitrogenase being a Mo–Fe–S cluster.

(5) $(NH_4)_2WS_4$ and its combination systems with iron compounds did not display activity of catalytic reduction of C_2H_2. This is in accord with WFe protein having no catalytic activity.

(6) The catalytic activity of combination systems decreased with increasing amounts of chelate ligands.

Acknowledgement. This research was supported by a grant from the National Science Foundation of China.

5 References

Cai H (1990) Master thesis, Jilin University, June

Coucoucanis D (1981) Fe–Mo–S complexes derived from $MS_4{}^{2-}$ anions (M = Mo, W) and their possible relevance as analogues of structure features in the Mo site of nitrogenase. Accts Chem Res 14: 201

Cristou G, Garner CD (1979) A convenient synthesis of tetrakis [thiolato-u_3-sulphido-iron] $(2-)$ clusters. J Chem Soc Dalton Trans (6): 1093

Cramer SP, Hodgson KO, Gillum WO, Mortenson LE (1978a) The molybdenum site of nitrogenase. Preliminary structural evidence from X-ray absorption spectroscopy. J Am Chem Soc 100: 3398

Cramer SP, Gillum WO, Hodgson KO, Mortenson LE, Stiefel EI, Chisnell JR, Brill WJ, Shah VK (1978b) The molybdenum site of nitrogenase. 2. A comparative study of Mo–Fe protein and the iron-molybdenum confactor by X-ray absorption spectroscopy. J Am Chem Soc 100: 3814

Fan YG, Lu PZ, Wang FS, Liu SX, Xu JQ (1983) Crystal structure of the cluster compound $[Mo_2FeS_8O_2][Et_4N]_3CH_3CN$. Acta Chimica Sinica 41: 776

Fan YG, Zhang ZG, Guo CX (1988) The crystal and electronic structure of reticular the Mo–Fe–S cluster compound $[Et_4N]_4[Mo_2Fe_4S_9(SCHCHS)_2]$. Chem J Chinese Univ (English Edition), 4: 61

Herskovitz T, Averill BA (1972) Structure and properties of a synthetic analogue of bacterial iron-sulfur proteins Proc Nat Acad Sci USA 69: 2437

Holm RH (1977) Synthetic approaches to the active sites of iron–sulfur proteins. Acc Chem Res 10: 427

Huang TB, Xu LJ, Guo CX (1985) The oxidation–reduction activities of the Mo–Fe–S cluster compound, $[MoS_4)_2FeL]^{3-}$ (L: O_2CH_3CN, O_2, O). FENZI KEXUE YU HUAXUE YANJIU (1): 93

Hu NH, Liu YS, Xu JQ, Yan YG (1990) Crystal structure of the cluster compound, $[Bu_4N]_2[Fe_4S_4 \cdot (S\text{-}t\text{-}Bu)_4]$. J Struct Chem (in press)

Laskovski EJ, Frankel RB, Gillum WO, Papaefthymiou GC, Renaud J, Ibers JA, Holm RH (1978) Synthetic analogues of the 4-Fe active sites of reduced ferredoxins. Electronic properties of the tetranuclear trianios $[Fe_4S_4(SR)_4]^{2-}$ and the structure trianious $[(C_2H_5)_3(CH_3)N]_3[Fe_4S_4 \cdot (SC_6H_5)]$. J Am Chem Soc 100: 5322

Laskovski EJ, Reynolds JG, Frankel RB, Foner S, Papaefthymiou GC, Holm RH (1979) Demonstration of the generality of the $[Fe_4S_4(SR)_4]^{2-}$ (compressed D_{2d})/$[Fe_4S_4(SR)_4]^{3-}$ (elongated D_{2d}) structural change in electrotransfer reactions of ferredoxin-4Fe site analogues. A model for unconstrained structural changes in ferredoxin proteins J Am Chem Soc 101: 6562

Liu XS, Xu JQ, Li J, Niu SY, Li SQ, Sun CT, Chen YY, Lin YQ, Yang SC, Zhao Y (1982) Synthesis and property study of a model compound of active center of nitrogenase. Chem J Chinese Univ 3: 555

Liu XS, Xu JQ, Niu SY, Li SQ (1985) Synthesis, properties and structure of cluster complex, $[Et_4N]_3[Mo_2Fe_2O_2]CH_2CN$. Chem J Chinese Univ 6: 258

Mascharak PK, Papaefthymiou GC, Frankel RB, Holm RH (1981) Evidence for the localized Fe(III)/Fe(II) oxidation state configuration as an intrinstic property of $[Fe_2S_2(SR)_4]^{3-}$ cluster. J Am Chem Soc 103: 6110

Nan YM (1990) Master thesis, Jilin University, June

Niu SY, Jiang Y, Xu JQ, Li SQ (1987a) Spectroscopic investigation of the condition for producing Mo–Fe–S clusters. GUANGPU XUE YU GUANGPU FENXI 7: 33

Niu SY, Zhang ZG, Pan M, Jiang Y, Wang TG, Lin XH (1987b) Investigation of molecular spectra for series of Mo–Fe–S clusters. KEXUE TONGBAO 32: 273

Pienkos PT, Shah VK, Brill WJ (1977) Molybdenum cofactors from molybdoenzymes and in vitro reconstitution of nitrogenase and nitrate reductase. Proc Natl Acad Sci US 74: 5468

Sweeney WV, Rabinowitz JC (1980) Proteins containing 4Fe–4S clusters: an overview. Annu Rev Biochem 49: 139

Wang FS, Zhang ZG, Fan YG, Shen C (1984) Structure of crystal of cluster compound, $[N(C_2H_5)_4]_3 \cdot \{[(SCH_2CH_2S)MoS_3]_2Fe\}$. Acta scientiarum naturalium Universitis Jilinensis (2): 94

Wang FS, Fan YG, Lu PZ, Xu JQ, Liu XS (1986) A cluster compound with di-end-oxygen ligands. Scientia Sinica (Series B) XXIX: 531

Xu JQ, Xu LJ, Liu XS, Zhang ZG, Niu SY, Li SQ, Fan YG, Wang FS, Lu PZ, Tang AQ (1983) Chemical modeling of active center of nitrogenase. J Mol Sci (Wuhan, China) 1: 13

Xu JQ, Liu XS, Niu SY, Fan YG, Wang FS, Lu PZ (1984a) Synthesis, structure and properties of the complex, $[(C_2H_5)_4N]_3[Mo_2FeS_8O_2]$. KEXUE TONGBAO 29: 344

Xu JQ, Liu XS, Zhou X, Liu LW, Li SQ, Wang TG (1984b) A study on Mo–Fe–S cluster compounds-synthesis and properties of $[Et_4N]_2[MoFeS_4(SCN)_2(OCH_3)_2]3CH_3OH$ and $[Et_4N]_2 \cdot [Mo_2FeS_8O(OCH_3)_2]$. Acta Scientiarum Naturalium Universitatis Jilinensis (4): 97

Xu JQ, Liu XS, Wang LN, Li SQ (1986) A study on synthesis and properties of cluster compound, $[Et_4N]_4[Mo_2Fe_7S_{12}(PhS)_6 \cdot 6CH_3OH$. Acta Scientiarum Naturatium Universitatis Jilinensis. (2): 121

Xu JQ, Fan YG (1987) Synthesis, crystal and molecular structure of the cluster compound, $[(C_2H_5)_4N]_2 \cdot [Fe_2(SCH_2CH_2S)_4]$. KEXUE TONGBAO 32: 1176

Xu JQ, Yan YZ, Wei Q (1988) A study on the activity of Mo–Fe–S cluster compounds for catalytic reduction of acetylene. Cuihua Xuebao 9: 218

Xu JQ, Yan YZ, Wei Q, Guo CX, Hu NH, Jin ZS, Liu YS (1989a) Synthesis on new molybdenum–iron–sulfur reaction system. I. Synthesis, structure, properties and quantum-chemical calculation of cluster compound, $[Et_2N]_4[Mo_2Fe_2S_{10}] \cdot 2CH_3OH$. In "proceedings of the Fifth China–Japan–US Symposium on Organometallic Chemistry", Chengdu, June 7–11, p 103

Xu JQ, Qian JS, Wei Q, Nu NH, Jin ZS, Wei GC (1989b) Synthesis and structure of a novel molybdenum–iron–sulfur cluster with Mo_2Fe_2 core and all-disulfur chelate ligand, $[Mo_2Fe_2(u_3-S) \cdot (S_2CNEt_2)_5 \cdot CH_3CN$. Inorg Chem Acta 164: 55

Xu JQ, Qian JS, Wei Q, Hu NH, Jin ZS, Wei GC (1989c) A study on the cluster compounds with all-dithio-chelate ligands. II. Synthesis, structure and properties of $[Mo_2Fe_2(u_3-S)_4(S_2CNEt_2)_5] \cdot CH_3CN$. Acta Chimica Sinica 47: 853

Xu JQ, Qian JS, Wei Q, Guo CX, Yang GD (1989d) Synthesis, structure, properties and quantum-chemical studies of cubane-like cluster compound with all-chelating ligands, $[(C_2H_5)_4N]_2\{Fe_4S_4 \cdot [S_2CN(C_2H_5)_2]_4\}$. Science in China (series B) 32: 927

Xu JQ (1989e) Synthetic chemistry of cluster compounds of transition metals. Chinese J Appl Chem 6: 1

Xu JQ, Yan YZ, Wei Q, Qian JS, Li SQ (1989f) A study on the activity of Mo–Fe–S cluster compounds for catalytic reduction of C_2H_2 and its relation to the composition and structure of such compounds. J Mol Cat 2: 229

Xu JQ, Nan YM, Cai H, Li SQ, Hu NH, Liu YS (1990a) Study on new reaction systems for Mo–Fe–S cluster compound synthesis, VIII structure and properties of $[(C_2H_5)_4N]_2[(CH_3)NCNH_2CH_3]_2 \cdot [Mo_2FeS_{10}]$. In "proceedings of the sixth symposium on organometallic chemistry of China", Changchun, Sept. 5–8, p 214

Xu JQ, Cai H, Nan YM, Li SQ, Hu NH, Liu YS (1990b) Study on new reaction systems for Mo–Fe–S cluster synthesis. III. Synthesis, structure and properties of $[Et_4N]_3[Mo_2Fe_6S_{14}(OCH_3)]$. In "Proceedings of the Third Symposium on Inorganic Solid State Chemistry and Synthetic Chemistry of China". Harbin, July 15–18, p 166

Xu JQ, Zhang XG, Cai H, Nan YM, Wang TG, Hu NH, Jin ZS (1990c) Study on new reaction systems for Mo–Fe–S cluster syntheses. IV. Synthesis, structure and properties of $[Et_4N]_3 \cdot [Mo_2Fe_6S_{11}(SPh)_3(OCH_3)_3]$. Ibid. p 168

Xu JQ, Cai H, Nan YM, Li SQ, Hu NH, Liu YS (1990d) Study on new reaction systems for Mo–Fe–S cluster syntheses. V. Synthesis, structure and properties of $[Et_4N]_3[Mo_2Fe_6S_8(SPh)_6((OCH_3)_3] \cdot 1/2CH_3OH$. Ibid. p 167

Xu JQ, Cai H, Nan YM, Li SQ, Hu NH, Liu YS (1990e) Study on new reaction systems for Mo–Fe–S cluster syntheses. VII. Synthesis, structure and properties of $[Et_4N]_2[Fe_2S_2(SPh)_4]$. Ibid. p 171

Xu JQ, Zeng J, Wang TL (1990f) Synthesis and structure of the cluster compound $[Et_4N]_2 \cdot [Fe_2S_2(SPh)_4]$ (to be published)

Xu JQ, Nan YM, Cai H, Wang TG, Hu NH, Liu YS, Jin ZS (1990g) Study on new reaction systems for Mo–Fe–S cluster syntheses. IX. synthesis, structure and properties of $[Et_4N][Fe_4S_4\text{-}(S_2CNEt_2)]_4$. Ibid. p 172

Xu JQ, Cai H, Nan YM, Li SQ, Hu NH, Liu YS (1990h) Study on new reaction systems for Mo–Fe–S cluster synthesis, structure and properties of $[Et_4N]_2[Fe_4S_4(SPH)_4]$. Ibid. p 170

Xu LJ, Wang FS, Lu PZ, Fan YG (1984) Investigation on synthesis, structure and properties of the cluster complex, $[(MoS_4)Fe(MoS_4O)][C_4H_9)_4N]_3$. Scientia Sinica (series B). XXVII: 877

Yang GD, Sun HL, Xu JQ, Qian JS, Wei Q (1989) Crystal and molecular structure of $[C_2H_5)_4N]_2 \cdot \{Fe_4S_4[S_2CN(C_2H_5)_2]_4\}$. Acta Chimica Sinica 47: 1

Yan YZ (1987) Master Thesis, Jilin University, June

Zhang ZG, Wang FS, Fan YG (1984a) Study on the synthesis structure and catalytic properties of a Mo–Fe–S cluster compound, $\{(C_2H_5)_4N\}_3\{[(SCH_2CH_2S)MoS_3]_2Fe\}$. KEXUE TONGBAO 29: 1486

Zhang ZG, Li X, Fan YG (1985a) Study on the synthesis, structure and catalytical activity of the Fe center model compound of nitrogenase. KEXUE TONGBAO 30: 1351

Zhang ZG, Zhang ZY (1985b) Study on synthesis, structure and properties of the cluster complex, $[(C_4H_9)_4N]_2[Mo_2S_6O_2]$. KEXUE TONGBAO 30: 764

Zhang ZG, Fan YG, Li Y, Niu SY, Li SQ (1987) Study on the synthesis, structure and properties of a novel type Mo–Fe–S cluster compound. KEXUE TONGBAO 32: 1405

Zhang ZG, Fan YG (1988) Crystal and molecular structure of cluster compound, $[Et_4N]_3[Mo_2FeS_6(SCH_2CH_2S)_2] \cdot 1/4CH_3CN$. Acta Scientiarum Naturalium Universitatis Jilinensis (2): 97

Zhang ZG, Sun WT, Niu SY, Li SQ (1989a) Studies on synthesis, structure and properties of the cluster compound, $[Bu_4N]_3[Mo_2FeS_8O](CH_2OH)_2$. Acta Chimica Sinica 47: 1061

Zhang ZG, Sun WT, Niu SY, Li SQ, Ye L, Fan YG, Guo CX (1989b) Study on syntheses, structures and properties of a series of Mo–Fe–S cluster compounds containing chelate group. J Inorg Chem 5: 35

Zhang ZG, Niu SY, Li SQ, Ye L, Fan YG (1989c) Studies on synthesis, structures and properties of Fe–S cluster compounds containing carbonyl group. In "Proceedings of the fifth China–Japan–US Symposium on Organometallic Chemistry", Changchun, China, Sept. 5–8, p 31

CHAPTER 4
Studies on the Mechanism of Nitrogenase Catalysis—Substrates-Cluster-Coordination-Chemistry Approach[1]

K.R. Tsai*, H.-L. Wan, H.-T. Zhang, and L.-S. Xu

[1] This work is supported by the National Natural Science Foundation of China and has also been supported by grants from the National Science & Technology Commission administered through Academia Sinica and through National Education Commission.
* To whom correspondence should be addressed.

The Nitrogen Fixation
and its Research in China
Editor: Guo-fan Hong
© Springer-Verlag Berlin Heidelberg 1992

1 Introduction

Nitrogenase is a complex metalloenzyme reversibly dissociable into two metallo-protein components: component 1, the MoFe-protein, a tetramer $(\alpha_2\beta_2)$ with 2 Mo, about 30 Fe and about the same number of acid-labile S, and a molecular weight of about 220 kDa, carries the binding sites for the reducible substrates; and component 2, the Fe-protein, a dimer (Γ_2), with a 4Fe–4S cluster and a molecular weight of about 60 kDa, serves as a specific electron carrier to transfer electrons to the MoFe-protein with the aid of the "electron activator" ATP, which goes into the enzyme complex as MgATP bound to the Fe-protein and is hydrolyzed into ADP and inorganic phosphate, Pi, concomitant with the electron transfer; each dimeric molecule of Fe-protein being able to bind reversibly one or two molecules of MgATP or MgADP. However, MgATP appears to be a nonreducible substrate of nitrogenase, its hydrolysis may, or may not be coupled to electron transfer in nitrogenase reactions, as to be dis-cussed later.

One of the striking features of nitrogenase catalysis is the extraordinarily wide variety of reducible substrates. Besides catalyzing the reduction of proton, H^+ (the only endogenous substrate of nitrogenase) to H_2, and N_2 (an exogenous substrate) to NH_3, nitrogenase also catalyzes the reduction of about 10 other types of exogenous substrates, the reduction products of which have been well characterized (Table 1). With the exception of H^+ and N_2H_4 (which is a very poor substrate at neutral pH), all other substrates of nitrogenase are compounds with terminal triple-bonds, or potential triple-bonds. The enzyme-catalyzed reduction of all of the exogenous substrates is strongly inhibited by CO, which, however, does not inhibit the ATP-dependent hydrogen-evolution reaction and the ATP hydrolysis reaction. The reduction of N_2 to ammonia is competitively inhibited by H_2, and this has been identified mechanistically with the reductive hydrogenation of D_2 to 2HD catalyzed specifically by N_2 and nitrogenase (Bulen 1976; Stiefel et al. 1980).

The mechanism of nitrogenase catalysis has been the subject of extensive investigation since the mid 1960s by many laboratories throughout the world, and many important progresses have been made in the last two decades (for more recent reviews, see Mortenson & Thorneley 1979; Burgess 1984; Orme-Johnson 1985; Smith et al. 1988). In China, organized researches on the chemical aspects of biological nitrogen fixation were started in the early 1970s, beginning with model studies led by quantum-chemical approach on mononuclear and multinuclear coordination activation of the sturdy $N\equiv N$ triple-bond (Tang et al. 1972, 1973) and on all the possible mononuclear and multinuclear bonding schemes of N_2 (Lu 1973, 1975). Another approach (the chemical-probes-and-coordination-catalysis approach) to the mechanistic problem of nitrogenase catalysis was started in the early 1973s and developed in later years (Nitrogen Fixation Research Group of Xiamen University 1974, 1976, 1982; Tsai et al. 1979; Tsai 1980; Tsai & Wan 1981; Cai (Tsai) et al. 1987, 1990). By regarding all the known exogenous substrates and inhibitors of nitrogenase as chemical probes

Table 1. Nitrogenase substrates, reaction products, K_m & EAC values

Substrate	Products	K_m(mM)	EAC	Reference
N_2	NH_3	0.06–0.12	0.7	Burns & Hardy (1975)
N_2, H_2 in D_2O	ND_3, 2HD			Hoch et al. (1960)
N_2O	H_2O, N_2, (NH_3)	~1.0	0.7	Hardy & Knight (1966)
N_3^-, HN_3	NH_3, N_2, (N_2H_4)	~1.1	0.7	Hardy & Knight (1966)
N_2H_4 (at pH8)	NH_3	10–20	0.2	Bulen (1976)
C_2H_2	C_2H_4	0.14–0.4	0.9	Burris (1971)
C_2H_2 (in D_2O)	cis-CHD=CHD (>99%)			Hardy et al. (1966)
$CH_3C{\equiv}CH$ (in D_2O)	cis- & $trans$- CH_3CD=CHD	30	0.4	Hardy & Jackson (1967)
CH_2=C=CH_2	CH_3CH=CH_2	8	0.7	Burns et al. (1975)
CN^-, HCN	CH_4, NH_3, CH_3NH_2, C_2H_4, C_2H_6	0.4–1.0	0.35	Hardy & Knight (1967)
CH_3NC	CH_3NH_2, CH_4, C_2H_4, C_2H_6, C_3H_6, C_3H_8	0.2–1.0	0.6	
CH_3CH_2NC	$CH_3CH_2NH_2$, CH_4, C_2H_4, C_2H_6	10–25	0.25	Kelly et al. (1967)
CH_3CH_2NC (in D_2O)	$CH_3CH_2NH_2$, CD_4, C_2D_4, C_2D_6			
CH_3CN	C_2H_6, NH_3	450	0.01	Hardy & Jackson (1967)
CH_3CH_2CN	C_3H_8, NH_3	500	0.01	Hardy & Jackson (1967)
CH_2=CHCN	NH_3, C_3H_6, C_3H_8	10–50	0.2	Fuchsman & Hardy (1967)
Cyclopropene	cis-(CH_2)(CHDCHD)	0.1	~0.6	McKenna et al. (1976, 1980)
(CH_2)(CH=CH)	cis- & $trans$- CH_3CD=CHD, CH_2DCH=CHD			
Diazirine	NH_3, CH_4, CH_3NH_2	0.1	~0.7	McKenna et al. (1982)
(CH_2)(N=N)	(ca. 3:1:1)			Orme-Johnson (1981)
H_2NCN	NH_3, CH_4, CH_3NH_2	0.8		Miller & Eady (1988)

and based upon the known coordination chemistry of some of the substrates and related compounds as ligands in mononuclear and multinuclear transition-metal complexes, valuable information can be obtained regarding the probable modes of coordination of these substrates on nitrogenase active-center; from this and from the known biophysical and biochemical properties of nitrogenase reported in the literature, as well as from the principles of coordination catalysis (Tsai 1965; Tsai & Wan 1981), important information about the structure and function of nitrogenase active-center can also be obtained. This has led us to the proposal and development of a monocubane-like cluster-structural model (Nitrogen Fixation Research Group of Xiamen University 1973, 1974, 1976) and a corner-sharing-twin-cubanes-like cluster-structural model (with two labile or labilizable ligands coordinated to adjacent sites on the Mo) of nitrogenase active-center (Tsai 1980). (Fig. 1a, d). Similar cluster-structural models, the "string-bag" and "composite-string-bag" cluster-structural models, of nitrogenase active-center were proposed independently at about the same time by Lu (1973, 1975, 1980) from slightly different approach (Fig. 1b, c).

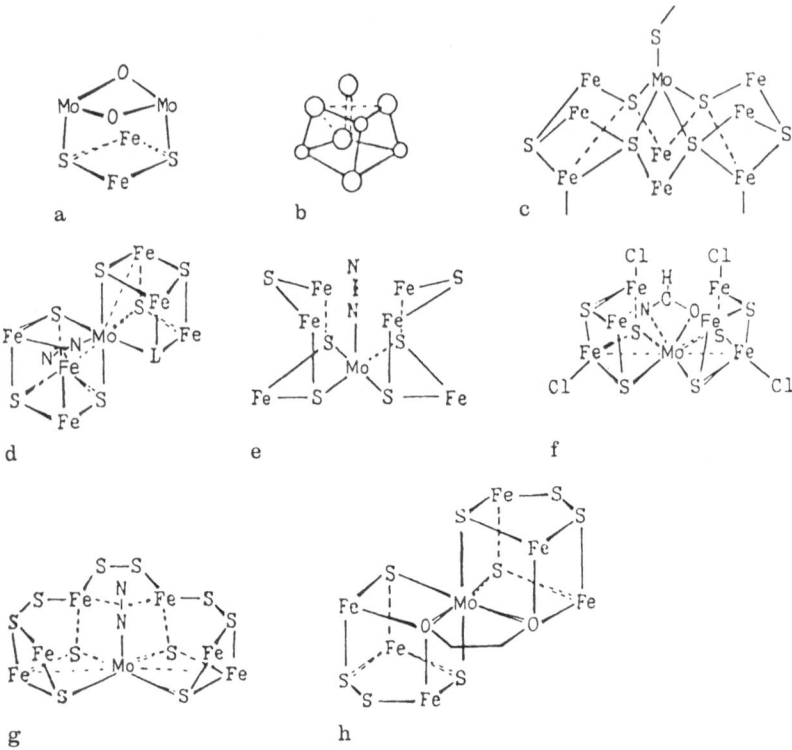

Fig. 1 a–h. Development of cluster-structural models for nitrogenase active-center. **a** Monocubane model, *X-1* (1973); **b** "String bag" model, *F-1* (1973); **c** Composite "string bag" model, *F-2* (1978); **d** Corner-sharing-twin-cubanes model, *X-2* (1978); **e** B.K. Teo model (1981); **f** Edge-sharing-twin-cubanes model, *X-3* (1982); **g** Edge-sharing-twin-cubanes model with μ_3-S_2 *endo*-ligands, *X-4* (1987); **h** Corner-sharing-twin-cubanes model with μ_3-S_2 *endo*-ligands, *X-5* (1990)

Subsequently, two edge-sharing-twin-cubanes-like cluster-structural models (Fig. 1f, g, the second one of which has two S_2 *endo*-ligands and one S_2 *exo*-ligand) have also been proposed (Nitrogen Fixation Research Group of Xiamen University 1982; Cai et al. 1987), and a slightly modified corner-sharing-twin-cubanes-like model (Fig. 1h, with 2 μ_3-S_2, instead of 2 μ_3-S, coordinated at the 2 Fe_3 corners) has been developed very recently (Tsai et al. 1990). These models have been used in the elucidation of the mechanisms of nitrogenase-catalyzed reduction of all the known substrates, and some of them have served as guides in the design of synthetic methods for the synthesis of the model compounds. Experimental basis and relevant literature for the development of these models and the proposed mechanisms of nitrogenase-catalyzed reactions, as well as the strategies of the synthetic methods and preliminary results from attempted synthesis of the corner-sharing-twin-cubanes-like cluster-structural model compounds will be reviewed and commented in this article.

In the same period of time beginning from 1973, studies on the mechanism of ATP-driven electron (and proton) transfer in nitrogenase catalysis have also been undertaken in our laboratory, and a mechanism of 2-step ATP-driven electron-transfer has been proposed and developed (Nitrogen Fixation Research Group of Xiamen University 1974, 1976, 1982; Tsai 1980; Cai et al. 1987; Wu et al. 1988). The essential points of the proposed mechanism are: (1) coordination of 2 MgATP to the reduced Fe-protein in the nitrogenase complex increases the ligand field acting on the Fe-protein 4Fe–4S center to drive the protein–protein electron-transfer with concomitant hydrolysis of 2 ATP into 2 ADP and 2 Pi, but no hydrogen evolution in the presteady state: (2) after each electron-transfer, the oxidized Fe-protein with the bound 2 MgADP and entrapped 2 Pi is obliged to dissociate from the enzyme complex for the replenishment of electron and ATP and for the release of the 2 MgADP and 2 Pi; (3) the fully-reduced MoFe-protein alone is still unable to reduce substrates (e.g., H^+) readily (Orme-Johnson & Munck 1980), so, in the steady state of the enzyme turnover, a step of ATP-driven protein–protein-substrate electron- (and proton-) transfer with concomitant ATP hydrolysis is required by complexation of the fully-reduced MoFe-protein with the reduced Fe-protein bound with 2 MgATP; and (4) electron backflow may take place whenever Fe-protein in the oxidized state with no, or only one bound adenine nucleotide molecule has a chance to complex with a fully reduced MoFe-protein tetramer. Some supports of the proposed mechanism have been obtained from model experiments involving synthetic 4Fe–4S cubane-like cluster complexes (Chen et al. 1985; Wu et al. 1988). The proposed mechanism has undergone some modifications very recently. A precursory mode of adenine uncleotides binding to Fe-protein before its complexation with MoFe-protein is postulated: the free Fe-protein binds 2 molecules of MgATP or MgADP on 2 equivalent sites *not in the coordination sphere* of the 4Fe–4S center, but close enough (about 5.5 to 6.0 Å from the center of the 4Fe–4S cluster) for appreciable electrostatic interaction with the effective core-charge of the 4Fe–4S center to account for most of the binding energy, and for a shift of at least $-100\,mV$ (caused by the adenine nucleotide binding) of the midpoint redox potential (E_m) of Fe-protein, which in the oxidized state with its higher effective core-charge of the 4Fe–4S center is expected to bind the adenine nucleotides, MgATP or MgADP, much more tightly than does Fe-protein in the reduced state, resulting in a decrease (by about 2 orders of magnitude) in the concentration ratio of Fe-protein in the oxidized state and that in the reduced state in the Nernst's equation in the presence of sufficient adenine nucleotides. A negative shift of the "intrinsic" redox potential (i.e., a lowering of the ionization potential) of the 4Fe–4S center by more than a hundred mV may also be estimated from a simple dipole-charge-interaction model. This precursory binding of MgATP will not lead to premature ATP hydrolysis since the 2 MgATP are not actually coordinated on the 4Fe–4S center of isolated Fe-protein. By formation of the enzyme complex with the MoFe-protein, further conformational change takes place which may allow each of the 2 MgATP bound to the Fe-protein to move closer to the 4Fe–4S center by about 1.3 to 1.8 Å to become a bridging chelate-ligand susceptible to hydrolysis by a mechanism analogous to that proposed by Haight for metal-ion catalyzed hydrolysis of phosphate anhydrides (for a review and relevant literature, see Haight 1987), and to increase the ligand field acting on the Fe-protein 4Fe–4S center to drive the protein-protein, or protein-protein-substrate electron- (and proton-) transfer, as proposed previously by us (for a review and more recent work: Cai et al. 1987; Wu et al. 1988). This model also implies that, with this mode of bridging coordination of the MgATP chelate-ligands on the 4Fe–4S center of the Fe-protein *in the enzyme complex*, ATP hydrolysis will readily take place even in the absence of electron transfer, but subsequent displacement of the MgADP bound to the oxidized Fe-protein by MgATP may be very slow. Since a more detailed discussion of this mechanism of ATP-driven electron (and proton) transfer and ATP hydrolysis is given in an accompanying article in this volume (You et al. 1990), it will not be discussed further in this paper.

2 Substrates and Inhibitors of Nitrogenase as Chemical Probes

The usefulness of the wide variety of reducible substrates of nitrogenase in the study of the structure and function of nitrogenase active-center was first recognized by Hardy and his co-workers; their pioneering works (for a review, cf. Hardy 1979) provide a wealth of valuable information about the products and K_m values of nitrogenase-catalyzed reduction of a variety of exogenous substrates, each of which has to compete with the ubiquitous endogenous substrate H^+ for electrons for its reduction, so there is also an electron availability coefficient (EAC) for each of the exogenous reducible substrates as listed in Table 1.

Since the rate-determining step of the enzyme turnover is known (Hardy 1979) to be independent of the substrates, the activation energy being about the same (14.6 kcal/mol of dithionite consumed in invitro experiments at room temperature above 20 °C) for all the reducible substrates, it may be assumed that the coordination and desorption of the substrates are fast in comparison with the rate of the enzyme turnover; thus $K_m = (k_{-1} + k)/k_1 \approx k_{-1}/k_1 = K_d = 1/K_b$, i.e., the K_m value of a reducible substrate is approximately equal to the inverse of the binding constant (K_b) of the substrate; the smaller the K_m value of a substrate, the stronger the binding of the substrate to the active center of nitrogenase.

From the known reduction products and the K_m values of the exogenous reducible substrates listed in Table 1, as well as from the known coordination chemistry of some of the substrates and related compounds (Table 2), important inferences can be made regarding the probable modes of coordination of these substrates (Nitrogen Fixation Research Group of Xiamen University 1974, 1976,

Table 2. Modes of coordination of chelate ligands with triple-bonds in binuclear and polynuclear transition-metal complexes

Ligand (L)	Metal complex mode of coord.	Bond length (Å)	Reference
PhC≡CPh	$Co_2(CO)_6 \cdot \mu_2(\eta^2)$-L	C–C, 1.37	Sly (1959)
PhC≡CPh	$Fe_3(CO)_9 \cdot \mu_3(\eta^2, w)$-L	C–C, 1.41	Blout et al. (1966)
(C–C, 1.21, 2.05 Å)		Fe–C, 1.95 Fe–Fe, 2.48, 2.58	
PhC≡CPh	$Fe_3(CO)_8 \cdot \mu_3(\eta^2, w, w')L_2$	C–C, 1.38	Dodge & Schomaker (1965)
RC≡CR	$Co_4(CO)_{10} \cdot \mu_4(\eta^2, w, w')$L	C–C, 1.44	Dahl & Smith (1962)
N≡N 1.099 Å	$ReCl_2L_4$-(η^2, w, w')-N_2-$MoCl_4(OCH_3)$	N–N, 1.21	Overcer et al. (1978)
N≡N	$(Ru^{II}(edta)_2)_2$-$\mu_2(\eta^2, w, w')$-N_2	N–N	Diamantis & Dubrawski (1981)
N≡N	Nb^{III}_2-$\mu_2(\eta^2, w, w')N_2$	N–N, 1.3	Rocklage & Schrock (1982)
N≡N	Ni_2-$\mu_2(\eta^2)$-N_2	N–N, 1.36 Ni–N, 1.97 Ni–Ni, 2.75	Kruger et al. (1973, 1975)
N≡N	$L'_n Ti^{II}_4$-$\mu_3(\eta^2, w)$-N_2	N–N, 1.30	Pez et al. (1982)
n-BuCN	$L'_n Fe_3$-$\mu_3(\eta^2, w)$-L	C–N, 1.26 C–N, 1.16 Fe–N, 1.78 Fe–Fe, 2.56	Andrews et al. (1979)
t-BuNC	$L'_n Ni_4$-$\mu_3(\eta^2, w, w')$-L	C–N	Thomas & Muetterties (1976)

1982; Tsai et al. 1979; Tsai 1980; Tsai & Wan 1981; Cai et al. 1987). Firstly, as substrates of nitrogenase, the mono-alkyl acetylenes, alkyl (and alkenyl) cyanides and isocyanides are all small straight-chained molecules subject to about the same chain-length restriction (not more than 2 to 3C atoms in the n-alkyl or alkenyl groups), and the corresponding α-methyl-substituted derivatives are not substrates of nitrogenase. This indicates that these substrates are most probably coordinated to the active center in about the same orientation; and since the n-alkyl isocyanides, like CO (a very potent inhibitor of the exogenous substrates, but not an inhibitor of the ATP-dependent H_2 evolution reaction), will most probably be coordinated C-end-on to Mo (which is most probably in the EPR-silent Mo^{IV} valence state; cf. Orme–Johnson et al. 1980, 1981), it may be inferred that the n-alkyl acetylenes and n-alkyl cyanides will most probably be coordinated C-end-on and N-end-on, respectively, also to the Mo^{IV}. Secondly, N_2, like the isoelectronic molecule CO with a strong tendency towards end-on coordination, may also be coordinated end-on to the Mo^{IV}; but single-end-on coordinated N_2 and single-end-on coordinated alkyl acetylenes and alkyl (or alkenyl) cyanides would all be very weakly bound, poor substrates; while the K_m values listed in Table 1 show that N_2 is the most strongly bound substrate, and methyl-acetylene and vinylcyanide have about the same binding affinity as ethyl-isocyanide; so it may be inferred from the known coordination chemistry of the acetylenes and N_2 (Table 2), that N_2 and the alkyl acetylenes, besides single-end-on coordination to Mo^{IV}, are mostly probably also coordinated double-side-on to 2 $Fe^{II,III}$, and that the alkyl cyanides and isocyanides may also be coordinated likewise (i.e., single-end-on to Mo and double-side-on to 2 Fe) since these substrates also have triple-bonds. As can be seen from Table 2, the predictions of these μ_3 modes of coordination for N_2 and for the alkyl cyanides and isocyanides have obtained supports later from successful syntheses of multi-nuclear transition-metal complexes containing these types of μ_3 ligands reported in the literature by other investigators. Thirdly, the isoelectronic substrate-molecules N_2O and N_3^- or HN_3 are, respectively, competitive inhibitors of N_2 and CH_3NC (and CN^- or HCN); as nitrogenase substrates, N_2O and N_3^- (or HN_3) are predominantly ($> 90\%$) reduced, respectively, to H_2O and NH_3, with liberation of N_2, indicating that these two substrates are most probably coordinated, respectively, with O-end-on and N-end-on to Mo^{IV} and weakly double-side-on to 2 $Fe^{II,III}$, and that O ("oxene") and NH (nitrene) are labilizable ligands. Fourthly, the modes of coordination of acetylene, cyclo-propane, which was first used by McKenna (1976, 1980) in his chemical-probes approach, and diazirine (a competitive inhibitor of acetylene) may be quite similar; but these modes may be somewhat different from that of the n-alkyl acetylenes, n-alkyl cyanides and isocyanides, in view of the relevant experimental data listed in Table 1. Thus the Mo-nitrogenase-catalyzed reductive deuteration of acetylene in D_2O yields practically 100% cis-dideutero-ethene, while the reductive deuteration of methylacetylene in D_2O gives a mixture of cis- and $trans$-dideutero-propene (implying sequential reductive hydrogenation); further-more, acetylene binds to the active center about 2 orders of magnitude more strongly than methylacetylene, as shown by their K_m values. This strongly indicates different modes of coordination. In view of the known experimental fact (Dahl et al. 1962, 1966) that, for the acetylenes in multinuclear transition-metal complexes, there are two possible modes of $\mu_3(\eta^2)$ coordination which require about the same internuclear M–M distances (ca. 2.6–2.7 Å) between the three metal nuclei, it may be inferred that $HC\equiv CH$ is most probably coordinated single-side-on to Mo^{IV} plus double C-end-on to 2 $Fe^{II,III}$, while methylacetylene takes the alternative μ_3 mode of coordination as discussed above; namely, C-end-on to Mo^{IV} plus double-side-on to two $Fe^{II,III}$. The reductive deuteration of cyclopropene in D_2O gives dideutero-cyclopropane ($> 90\%$ cis) and dideutero-propenes in 1:2 yields, the high cis content of the dideutero-cyclopropane indicating similar modes of coordination for cyclopropene and acetylene, and conceivably also for diazirine, another triangular-cyclic substrate molecule with structure very similar to that of cyclopropene, and a potent competitive inhibitor of acetylene reductive-hydrogenation. As monodentate ligands, cyclopropene and diazirine would be very weakly coordinated (in fact, no stable transition-metal complex of mono-dentate cyclopropene, or diazirine, has been known so far); but their very small K_m values (Table 1) show that their binding affinities to nitrogenase active-center are about as high as that of the most strongly coordinated substrate-molecule, N_2, and somewhat higher than that of acetylene. This again strongly implies multinuclear coordination of cyclopropene and diazirine at nitrogenase active-center. An EHMO study of the $\mu_3(\eta^2)$ trinuclear (Mo + 2Fe, cationic) bonding schemes for cyclopropene, acetylene, as well as dinitrogen, has been made by Wan and Tsai [1981]. For cyclopropene, this involves double-(C)-end-on to two metal nuclei (equivalent to the opening of the olefinic π-bond to form 2σ-M–C bonds, with the release of some of the bond strain known to be about 26 kcal/mol from the heat of hydrogenation of cyclopropene to cyclopropane) plus single-side-on to a third metal nucleus (conceivably involving the Walsh-type orbital of the quasi-double-bond of the triangular-cyclic molecule). The results of the EHMO calculation seem to justify the proposed trinuclear bonding schemes for these substrate-molecules. For single-side-on coordination to Mo^{IV} plus double-(C)-end-on coordination to 2 Fe ions, the results may be even better.

The fact that CO (a very potent inhibitor of the exogenous substrates) also has a strong triple-bond, but is itself not a substrate, requires some comment. Conceivably, a strongly C-end-on coordinated CO on Mo^{IV} has practically no tendency to form side-on coordinative bond with another *cationic* metal nucleus, according to the known coordination chemistry of CO. So, with CO bound to Mo^{IV}, the adjacent $Fe^{II,III}$ ion or ions may actually recede away a little from the Mo^{IV} and the coordinated CO ligand, and *cis*-insertion of a hydrido ligand, H, coordinated on $Fe^{II,III}$ to CO coordinated on Mo^{IV} will not be able to take place; however, this will not inhibit the reductive elimination of $2H$ on the Fe ion to give H_2 (or by some other pathway similar to other iron-sulfur-protein-type hydrogenases), thus the hydrogen-evolution reaction is not inhibited by CO. Note that MoFe-protein catalyzed reduction of the redox dye methylene blue by H_2 is also not inhibited by 0.1 atm pCO (Wang & Watt 1984), which is known to inhibit all other hydrogenases; probably, CO can not be coordinated on the Fe ion sites of the M-cluster due to steric hindrance.

The behaviors of coordination, reduction, and inhibition of the endogenous substrates in the case of vanadium nitrogenase appear to be closely similar to that in the case of molybdenum nitrogenase, but there are some minor differences between Mo-nitrogenase and V-nitrogenase in the distribution of the reductive-hydrogenation products for certain substrates and in the extent of inhibition by CO (e.g., reductive hydrogenation of C_2H_2 catalyzed by V-nitrogenase composed of $Ac1^V$ plus either $Ac2^V$ or $Ac2^{Mo}$ gives C_2H_4 and some C_2H_6, whereas $Ac1^{Mo}$–$Ac2^V$ cross yields only C_2H_4 with concomitant evolution of smaller proportion of hydrogen, according to Eady et al. (cf. Thorneley et al. 1989), implying the involvement of Mo, or V, in the binding of the endogenous substrates and of the inhibitor CO on the Mo- or V-nitrogenase active-center.

To recapitulate, the above discussions clearly illustrate that most, if not all, of the endogenous reducible substrates with terminal triple-bonds or potential triple-bonds are most probably coordinated to nitrogenase active-center in one of the two alternative modes of $\mu_3(\eta^2)$ coordination, which require approximately the same internuclear distances of the three metal nuclei.

3 Development of the Twin-Cubanes Cluster Models with Two Labile Ligands on the Mo

Noticing that an active site occupied by a labile or labilizable ligand, like a μ_2- or μ_3-"oxene" or nitrene endo-ligand, at one corner of a Mo–Fe–S(O) cubane cluster provides just about the right structural parameters (bond lengths and bond angles) for the above-mentioned $\mu_3(\eta^2)$ modes of coordination activation of N_2, acetylene, cyclopropene, n-alkylacetylenes, n-alkyl cyanides and other substrates with terminal triple-bonds or potential triple-bonds, a mono-cubane-like cluster-structural model $(X-1)$, $Mo_2O_2(\mu_3$-S$)_2Fe_2$ (Fig. 1a) was proposed in 1973 (Nitrogen Fixation Research Group of Xiamen University 1974, 1976). This model was modified in 1978 (Tsai et al. 1979; Tsai 1980) to a corner-sharing-twin-cubanes-like model $(X-2)$, $MoFe_6S_6(O, N)_2$, still retaining the feature of having 2 adjacent labile ligands on Mo (Fig. 1d), in view of the Mo-EXAFS work by Cramer et al. (1978) and the successful isolation of FeMo-cofactor by Shah & Brill (1977); these two works show that there is most probably only one Mo, associated with about 6–7 Fe and about 6 S, in an active center of nitrogenase. Note that, besides the similar cluster-structural models, the "string-bag model" $(F-1)$ and the "composite-string-bag model" $(F-2)$ (Fig. 1b, c) proposed independently by Lu (1973, 1975, 1980) as mentioned above, binuclear models, 2M, 2Mo, and Mo–Fe (linked by S), have been proposed previously by Brintzinger (1966), Shilov (1970), and Hardy et al. (1971), respectively (for a brief review of these binuclear models, see Hardy et al. 1979).

Subsequently, as modifications of model $X-2$ and the cluster model proposed by Teo (1981) (Fig. 1e), two edge-sharing-twin-cubanes-like models $(X-3, X-4)$ (Fig. 1f, g) have been proposed (Nitrogen Fixation Research Group of Xiamen University 1982; Cai et al. 1987), the second one $(X-4)$ with $2\mu_3$-S$_2$ endo-ligands, one μ_2-S$_2$ exo-ligand, and a Mo:Fe:S ratio of 1:6:10. As pointed out perviously (Cai et al. 1987), these edge-sharing-twin-cubanes models are structurally closely related to the corner-sharing-twin-cubanes model $(X-2)$. Thus in each of these three models, in the first coordination sphere of Mo^{IV}, there are 4S*, 2Fe (each chelate-bonded by 2 of the 4S*, as in a $MoFe_3S_3^*$ cubane-cluster), and two O-(or N-) containing adjacent labile ligands, together with probably 1 or 2 more Fe (also metal–metal bonded to Mo); and it has been shown (Cai et al. 1987) that these structural features are in excellent accord with the MoKa-EXAFS of MoFe-protein or FeMo-co (Newton et al. 1984), and in fairly good accord with the Fe-EXAFS of FeMo-co (Antonio et al. 1982) (Table 3).

The presence of 4S ligands bonded to Mo is also strongly supported by the chemical evidence that controlled degradation of FeMo-co (Newton et al. 1984), or successive treatments of MoFe-protein with acid and alkali (Zumft 1978) gives MoS_4^{2-} and MoS_3O^{2-} in the decomposition products, the latter anionic species being mostly probably produced by the ready partial-hydrolysis of the former anionic species in aqueous solution. The postulated presence of $2\mu_3$-S$_2$ endo-ligands in model $X-4$ (instead of $2\mu_3$-S as in Model $X-2$), each

Table 3. Structural Features of M-cluster, FeMo-co and Models X-4, X-2

Cluster	Bond type and numbers (n) average bond-length (Å)		EPR signals & spin-state	Reference
M-cluster	from Mo-EXAFS:	2.69	$(1_s):S = 3/2$;	Newton et al.
	Mo–Fe (3.5);	2.67	$g = 4.3, 3.7, 2.0$	(1984)
	Mo–S* (4.5);	2.37		Smith et al.
	Mo–O(N) (1.9);	2.17		(1988)
FeMo-co	from Mo-EXAFS:			
			$S = 3/2$;	Autonio et al.
	Mo–Fe (2.8);	2.69	$g = 4.6, 3.3, 2.0$	(1982)
	Mo–S* (3.0);	2.37		
	Mo–O(N) (3.0);	2.09		
	from Fe-EXAFS:			Smith et al.
	Fe–Fe (2.3 ± 0.9)	2.64		(1988)
	Fe–S* (Cl) (3.4 ± 1.6);	2.25		
	Fe–Mo (0.4 ± 0.1);	2.70		
	Fe–N(O) (1.2 ± 1.0);	1.81		
Model X-4	Mo–O(N) (ca. 2.7);	2.12		Tsai et al. (1987)
Model for	Mo–Fe (3–4);	2.70	$S = 3/2$ probable	
M-cluster	Mo–S* (4.0);	2.39		
(Fig. 1h)	Mo–O(N) (2);	2.10		
Model for	Mo–Fe (2);	2.70	$S = 3/2$ probable	
FeMo-co	Mo–S* (4);	2.39		
(Fig. 1i)	Mo–O(N) (3);	2.10		
	Fe–Fe (8/6 to 9/6);	2.66		
	Fe–S* (Cl) (20/6 = 3.3);	2.27		
	Fe–Mo (2/6 = 0.33);	2.70		
	Fe–O(N) (4/6 to 6/6);	1.90		
		2.10		
Model X-2	Mo–Fe (2);	2.75		Tsai et al. (1979)
	Mo–S* (4); 2.38,	2.40		
	Mo–O(N) (2)	2.10		
	Fe–Fe (2); 2.70, 2.70,	2.72		
	Fe–S* (Cl); 2.23, 2.33,	2.20		
	2.33, 2.20,	2.33		
	Fe–Mo (2/6);	2.70		
	Fe–O(N);	2.02		

bonded to 3 Fe at one of the two opposite corners of the twin-cubanes (Fig. 1g) is based mainly upon the following considerations: any μ_3-S bonded to 3 cationic Fe, as in the case of the Fd-type 4Fe–4S cubane clusters, which are known to collapse even in very weakly acid solution (at pH near or below 5), would be too acid-labile to withstand the treatment with dilute citric acid (at ca. pH 3) in the isolation of FeMo-co by Shah & Brill's (1977) method; while a μ_3-S_2 endo-ligand bonded to 3 cationic Fe should be much less acid-labile since HS_2^- as an acid is estimated to be stronger than HS^- by about 2 to 3 orders of magnitude. Furthermore, according to Kubas and Vergamini (1976, 1981), iron-sulfur monocubanes of the type $[L_4Fe_4S_3(S_2)]^{0,1+,2+}$, each with a μ_3-S_2 endo-ligand, also exhibit facile redox changes similar to that shown by the Fd-type 4Fe–4S cubane clusters. This may be one of the requisite properties of nitrogenase active-center.

However, these *edge-sharing*-twin-cubanes models were later found to be not in accord with the results of polarized X-ray absorption spectroscopy of MoFe-protein single-crystal (Flank et al. 1986), which indicate, from the change by as much as a factor of 2.5 in the Mo–Fe amplitude of the Mo-EXAFS with crystal orientation, that models of nitrogenase active-center with linear metal–metal-bonded Fe–Mo–Fe can be ruled out, while by similar method of inspection a linear S–Mo–S is compatible with the spectroscopic results. It then appears to us that model *X-2*, the *corner-sharing*-twin-cubanes-like model with 4S and 2 adjacent *O*- or *N*-containing ligands on the first coordination sphere of the pivotal Mo^{IV}, a linear S–Mo–S, and a metal–metal-bonded Fe–Mo–Fe angle of about 130° is in satisfactory agreement with the Mo-EXAFS, Fe-EXAFS, and the single-crystal polarized X-ray absorption spectroscopic data published in the literature (Cramer et al. 1978; Newton et al. 1984; Antonio et al. 1982; Flank et al. 1986); Furthermore, this model can provide at least 5 different micro-environments for the $3Fe^{II}$ & $3Fe^{III}$, which is indicated by the ^{57}Fe ENDOR (Hoffman et al. 1982) of MoFe-protein. So, model *X-2* appears to be a good basis for further refinement and development of a satisfactory model of nitrogenase active-center. Actually, only a slight modification is needed. For the same reason as discussed above, it is postulated as in the case of model *X-4*, that the two μ_3-S endo-ligands at the two opposite corners of model *X-2* should be replaced by 2 μ_3-S_2 endo-ligands. This leads us (Tsai et al. 1990) directly to the proposal of model *X-5* (Fig. 1h), with Mo:Fe:S = 1:6:8, falling into the range of 1:(6–8):(8–10) as reported by Nelson et al. (1983). It is known from the work of Kubas and Vergamini (1976, 1981) that iron-sulfur monocubanes with this type of μ_3-S_2 ligand are highly fluxional (Fig. 2d, e); this may account for the difficulty of characterizing FeMo-co by means of X-ray crystal structural methods. On the other hand, this structural fluxionality makes the internuclear distances involving the substrate-binding cationic Fe atoms slightly more flexible (Fe–Fe internuclear distances may vary between ca. 2.6 to 3.1 Å; cf. Kubas & Vergamini 1976), and so may be more adaptable to the binding of certain highly unstable transition-state intermediates in the reductive hydrogenation of N_2 and some other substrates.

As an indirect support of the plausibility of the proposed corner-sharing-twin-cubanes type of cluster-structural model, it may be mentioned that *five* examples of well-characterized compounds have already been reported in the literature, with similar corner-sharing-twin-cubanes-like-cluster structures (Fig. 2a, b, c), including three recently reported by Shibahara et al. (1987, 1988, 1989); namely, $\{[(H_2O)_9Mo_3\text{-}(\mu_3\text{-}S)(\mu\text{-}S)_3]_2M\}^{8+}$ (where the pivotal atom M = Mo, Hg, or Sn); and a twin-cubanes cluster, $[BuSn_3(\mu_3\text{-}O)(\mu_3\text{-}S)_3]_2Sn$, recently reported by Swamy et al. (1988), with 6 μ_3-S endo-ligands around the central Sn and 2 μ_3-O endo-ligands at the two opposite corners of the twin-cubanes (Fig. 2b), which would be the type of structure we were looking for if the 2 μ_3-O *endo*-ligands at the two opposite corners of the twin-cubanes could be made to exchange places with two adjacent μ_3-S *endo*-ligands cis-coordinated to the central Sn.

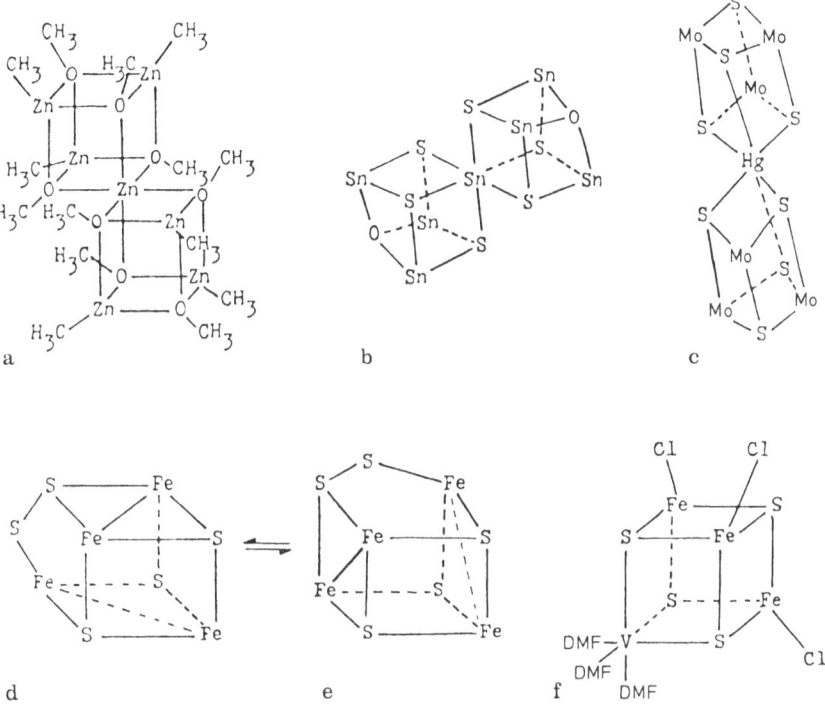

Fig. 2 a–f. Some known examples of corner-sharing-twin-cubanes compounds and relevant monocubane cluster compounds. **a** Heptanuclear zinc methoxide with μ_3-OCH$_3$ *endo*-ligands (Eisenhuth and Van Wazer 1968); **b** Heptanuclear Sn$_7$(μ_3-S)$_6$(μ_3-O)$_2$ twin-cubanes frame (Swamy et al. 1988); **c** Mo$_6$Hg(μ_3-S)$_8$ twin-cubanes frame (Ooi and Sykes 1989); **d** and **e** [Fe$_4$(μ_3-S)$_3$(μ_3-S$_2$)cp$_4$]$^{2+}$ cubane-frames, showing two of the three possible fluxional structures (Kubas and Vergamini 1981). **f** VFe$_3$S$_4$ cubane cluster formed by self-assembly in DMF (Kovacs and Holm 1986)

Note that the idea of having two adjacent labile, or labilizable O-containing-ligands (e.g., "oxene" ligands) was conceived originally to account for the peculiar mixed-type of inhibition exhibited by CO in the reductive hydrogenation of the endogenous substrates and by some of the substrates (e.g., N$_2$ and C$_2$H$_2$) in their mutual inhibitions; but it also serves to account for the *cis*-insertion, or *cis*-condensation side-reactions in the reductive hydrogenation of methylisocyanide (as well as CN$^-$, or ethylisocyanide), and for the fairly strong coordination (K$_m$ ca. 8 mM) of allene at the nitrogenase active-center, as illustrated with model *X-2* (Tsai et al. 1979; Tsai 1980), or with model *X-5* (Fig. 3g-2 to 3g-4; 3i-1).

Spectroscopic evidence in support of the presence of μ_3-S$_2$ in a MoFe-protein M-cluster or FeMo-co has recently been obtained by Zhang et al. [1988], who observed a low-temperature (24K) Raman peak at 448 cm^{-1} with carefully purified FeMo-co sample from Av MoFe-protein (Fig. 4); the assignment of this Raman peak to the S–S stretching frequency of μ_3-S$_2$ ligand coordinated

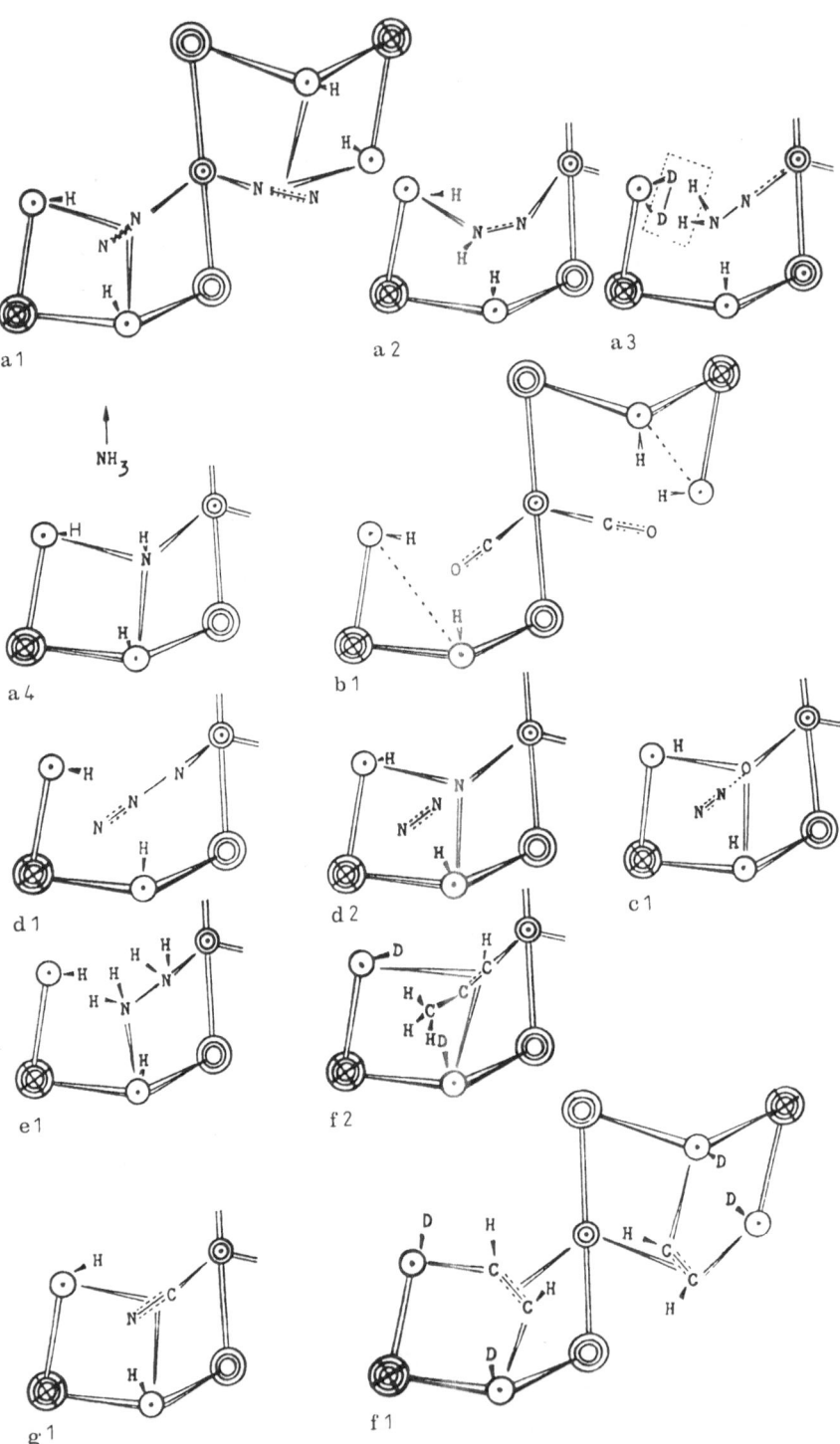

Fig. 3. (a1–f1)

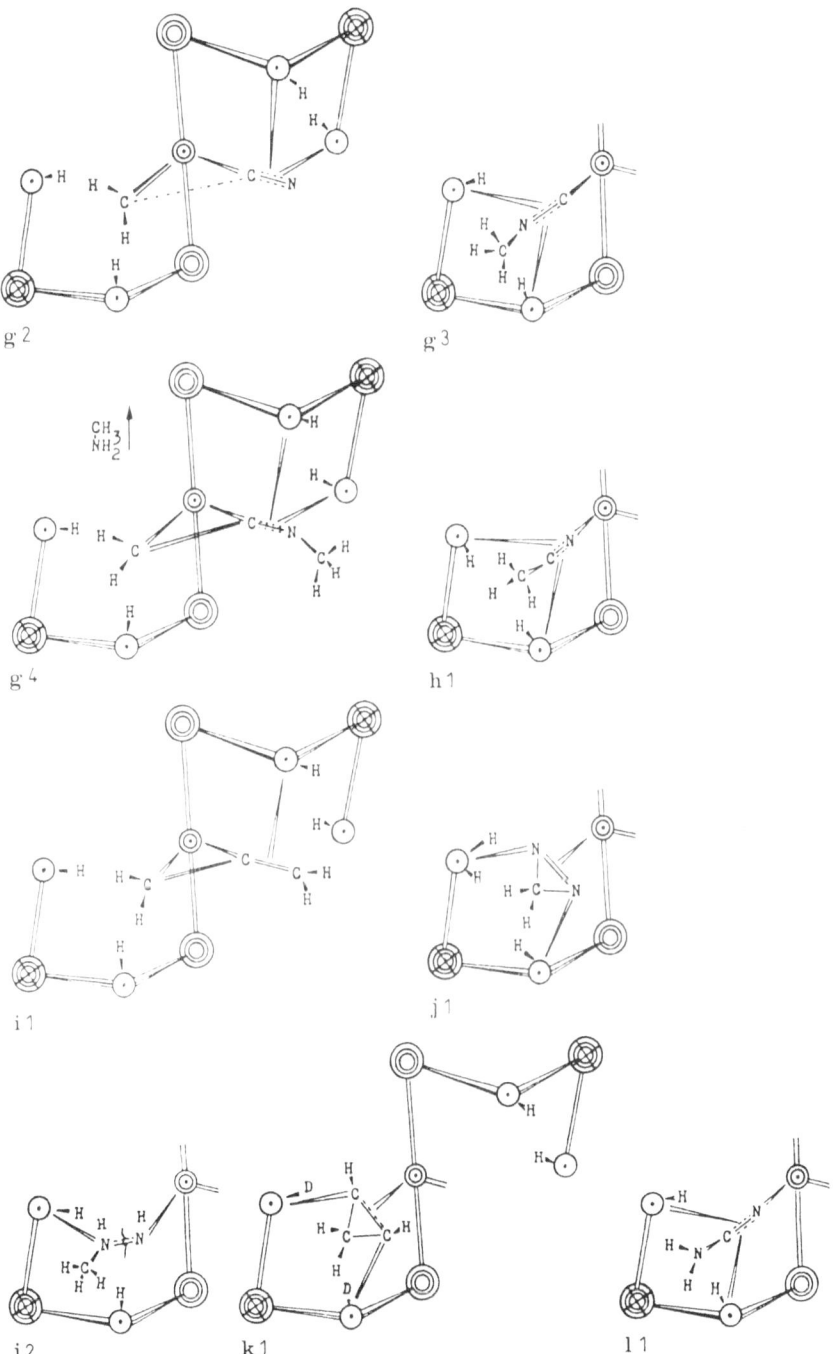

Fig. 3. Probable modes of coordination of substrates and of some of the N$_2$ase-catalyzed reaction intermediates on N$_2$ase active-center as illustrated with model X-2 or model X-5. (For clarity, those Fe and S atoms not directly involved in substrate-binding are omitted in the figure)

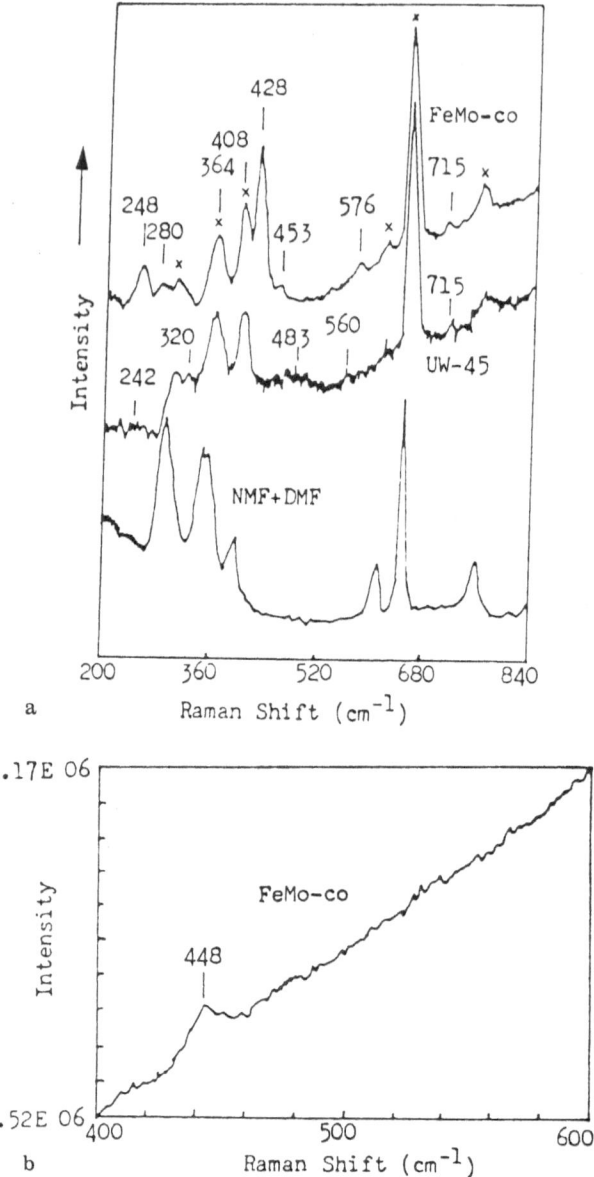

Fig. 4 a, b. Laser Raman spectra of FeMo-cofactor from *Av1*. **a** Raman spectrum of FeMo-co and of NMF extract of *Av1* from uw-45, obtained at room temperature with 488-nm Ar⁺ laser excitation, 14 scans, scanning rate 2 cm⁻¹/s. **b** Low temperature Raman spectrum of FeMo-co sample taken with 488-nm Ar⁺ laser excitation, 100 scans, scanning rate 2 cm⁻¹/s; sample at 24 K

Table 4. Some observed Raman peaks of FeMo-co vs. S–S stretching frequencies of some iron-sulfur compounds

Compound	Formulas	S_2-ligand μ_{s-s} cm^{-1}		Reference
1 (FeMo-co)	$Ln(S_2)_2(Fe^{II}_3 \cdot Fe^{III}_3)Mo^{IV}S_4$	μ_3-S_2(?)	448	Zhang et al. (1988)
2	$[cp_4(Fe^{II}_2Fe^{III}_2)(S)_3(S_2)]^{2-}$	μ_3-S_2	466	Kubas & Vergamini (1981)
3	$[cp_4(Fe^{II}Fe^{III}_3)(S)_3(S_2)]^-$	μ_3-S_2	520	Same
4	$cp_4(Fe^{III})_4(S)_2(S_2)_2$	μ_3-S_2	497	Same
			509	Same
5	$[cp_4Fe^{II}_2Fe^{III}_2(S)_2(S_2)_2]_2Ag^+$	μ_3-S_2	476	Same
			487	
6	$cp_2Fe^{II}_2(S_2)(SR)_2$	μ_2-S_2	~512	Same
7	Na_2S_2	S_2^{2-}	457	Zhang et al. (1988)

to 3 cationic Fe is based upon the known S–S stretching frequencies of disulfido endo-ligands in a variety of iron-sulfur compounds containing metal-bound disulfide moieties (Table 4), including two $Fe_4S_3(S_2)L_4$ cubane clusters each containing a μ_3-S_2 ligand. Some regularities about the variation of the S–S stretching frequencies with the numbers and valence states of the ligated transition-metal ions can be seen from Table 4. Thus for the same valence state of the transition-metal ions, the S–S stretching frequency of a tridentate disulfidoligand, μ_3-S_2, is lower than that of a bidentate disulfido-ligand, μ_2-S_2 (497 & 509 cm^{-1} for compound 4, vs. 507 & 516 cm^{-1} for compound 6), as expected; and for some average valence state of the Fe in the cubane clusters, the S–S stretching frequency of the one with 2 μ_3-S_2 endo-ligands (compound 5) is somewhat higher (by almost 10 to 20 cm^{-1}) than that with 1 μ_3-S_2 endo-ligand (compound 2); while for iron-sulfur cubane clusters of the same type and with the same number of metal nuclei ligated by each disulfido-ligand, the lower the average valence state of metal ions, the lower the S–S stretching frequency of the disulfide ligand (e.g., 466 cm^{-1} for compound 2, $[(Cp)_4(Fe^{II})_2(Fe^{III})_2(S)_3 (S_2)]^{2-}$, vs. 520 cm^{-1} for compound 3, $[(Cp)_4(Fe^{II})(Fe^{III})_3(S)_3(S_2)]^-$, both of the disulfido-ligands being μ_3-S_2 ligands; similar variation can also be seen from the difference in the S–S stretching frequencies of the μ_2-S_2 endo-ligands in cluster compounds 4 and 5, the S–S stretching frequency of the latter is about 20 cm^{-1} higher than that of the former); this seems to imply that the lower the redox potential of a disulfido-ligand-containing iron-sulfur clusters, the lower the S–S stretching frequency of the μ_3-S_2 ligand in it. Thus the S–S stretching frequencies of the disulfido-ligands appear to be sensitive indications for the modes of coordination of these ligands in iron-sulfur cluster compounds, as well as for the average valence state of the transition-metal ions in the clusters. On this basis, the low-temperature Raman peak of dithionite-reduced FeMo-co sample observed by Zhang et al. (1988) at 448 cm^{-1}, which is about 18 cm^{-1} lower than that of the μ_3-S_2 in compound 2, may tentatively be assigned to the S–S stretching frequency of μ_3-S_2 since the redox potential of the dithionite-reduced FeMo-co with valence states of the transition-metal ions known to be $3Fe^{II}$, $3Fe^{III}$, Mo^{IV} (for a review, cf. Orme-Johnson 1985) should be more negative than that of Fe-protein in the oxidized state with 2 ferrous and 2 ferric ions, the same as the average valence of the Fe ions in compound 2. This assignment may be checked by taking the low-temperature Raman spectra of dye-oxidized or electrochemically oxidized FeMo-co samples. Other laser Raman peaks observed at room temperature by Zhang et al. (1988) for the same sample of FeMo-co is shown in Fig. 3a, among them are the 428-cm^{-1} and 364-cm^{-1} peaks (which are close to the 432-cm^{-1} and 368-cm^{-1} Raman peaks observed by Leuchenko et al. (1980) for their FeMo-co sample), and weak Raman peak at 453 cm^{-1}; but Zhang et al. did not find any weak Raman peak around 530 cm^{-1} reported by Leuchenko et al. (1980).

4 Mechanisms of Reductive-Hydrogenation of Nitrogenase Substrates

With model X-5, the probable modes of coordination of the various substrates of nitrogenase and inhibitor (CO), and some of the key steps in nitrogenase-catalyzed reactions of these substrates are illustrated in Fig. 3; these differ very little from that illustrated by model X-2 (Tsai et al. 1979; Tsai 1980) and by model X-4 (Cai et al. 1987). For the sake of clarity, in Fig. 3 two Fe ions not taking part in substrate-binding are omitted and the two μ_3-S_2 endo-ligands at the two opposite corners of the Mo-sharing-twin-cubanes are still simply represented each by a single sulfur atom as in model X-2, with the understanding that due to the fluxionality brought about by the S_2 tridentate-ligands, the internuclear distance between each pair of the substrate-binding Fe ions is slightly more flexible, as discussed above, based on the work of Kubas and Vergamini (1976, 1981) on iron-sulfur cubane clusters containing disulfido-ligands.

Figure 3b shows that with sufficiently high CO partial pressure (0.1 atm.), two adjacent sites on Mo^{IV} may be taken up by CO, probably with slight recession of the two pairs of substrate-binding Fe ions from the Mo^{IV}, thus the coordination of CO converts the two halves of a M-cluster into two 3Fe-3S-type hydrogenase active-sites, as pointed out previously (Cai et al. 1987); two hydrido-ligands, $2H$, may be produced on 2 metal–metal-bonded Fe ions through two cycles of ATP-driven coupled electron- and proton-transfers; one of the hydrido-ligands may migrate over and by cis elimination with the other hydrido-ligand to form H_2. For sufficiently high N_2 partial pressure, N_2 may likewise take up the two available adjacent sites on the Mo^{IV}, as shown in Fig. 3a; but ordinarily it may take up only one of the two sites, probably leaving the other site to a hydrido-ligand. The first step of reductive hydrogenation of coordinated substrate molecule, N_2, may proceed via cis-migration of a hydrido-ligand on a substrate-binding $Fe^{II,III}$ ion to the *exo*-N of the coordinated N_2 to form the coordinated intermediate, N_2H (as shown in Fig. 3a-2), which would be a high potential-energy intermediate if coordinated to only one metal ion (Cai et al. 1987), but with subsidiary bonding with a second metal ion (Fe ion) the activation-energy barrier for this particular step can be lowered considerably (in fact, to some value lower than the activation energy of the enzyme turnover in this case, so this step is not the rate-determining step) this is perhaps one of the most important reasons why a cluster-structural active-center is needed here. Partial stabilization of a high-potential-energy intermediate in order to lower the transition-state activation energy is often the key feature in enzyme catalysis, as pointed out by Pauling more than 40 years ago (cf. Schultz et al. 1990), and more recently has been receiving more and more attention in chemical catalysis. Further reductive hydrogenation of NNH may proceed through NNH_2 intermediate coordinated N-end-on to Mo^{IV} and probably also weakly coordinated to an adjacent Fe ion, and then to the formation of NH_3 and a coordinated NH (Fig. 3a) through sequential transfer of 2 more hydrido-ligands, $2H$, i.e., $2e^- + 2H^+$, and finally to the liberation of one more NH_3. In the presence of D_2 or H_2 in the gaseous phase, the reductive hydrogenation of N_2 to NH_3 is inhibited, and 2HD or 2HH are produced, respectively, for every 2 electrons diverted from N_2 reduction (Burgess et al. 1981), this specific catalysis of 2HD formation by N_2 plus nitrogenase has been interpreted to means that certain N_2-reductive-hydrogenation intermediate, most probably certain N_2H_2 coordinated species reacts with D_2 (or H_2) to give 2HD (or 2HH) (Stiefel et al. 1980; Burgess et al. 1981). Since it does not appear that diimide is a nitrogenase substrate, neither is diimide, nor hydrazine an intermediate in the reductive hydrogenation of N_2, it is more likely that it is the well-documented hydrazido-(2-) ligand coordinated on Mo^{IV} that reacts with the D_2 or H_2 molecule (probably, weakly activated by coordination on an adjacent Fe^{II}) to give 2DH or 2HH, as illustrated in Fig. 3a-3. Note that D_2 is not split into 2 deuterido-ligands of the Fe ion, otherwise deuterated ammonia would be formed. Based on the results of their high-pressure experiment that H_2 formation is independent of pD_2 but dependent on pN_2 to the limit that at least about 27% of the electron flux is diverted to the hydrogen-evolution reaction, Guth and Burris (1983) and Jensen & Burris (1985) have more recently proposed a different scheme for the HD formation reaction without going into mechanistic details about how the coordination activation of D_2 on nitrogenase active-center is specifically catalyzed by N_2. Some pressure dependency would be expected for the rate of diffusion of D_2 to the active site if the mole fraction of D_2 should become vanishingly small on extrapolation of pN_2 to very high value. The interpretation for the limiting EAC of N_2 of about 0.75(?) based on the postulate (Thorneley et al. 1981) that N_2 has to displace a coordinated H_2 (or 2 hydrido-ligands?) for its binding to nitrogenase active-center does not seem to have sufficient justification either. According to Pez et al. (1982), N_2 can be readily absorbed by a binuclear Ti^{II} complex to form μ_3-N_2 ligand, which can be readily protonated to give NH_3, and according to Diamamtis & Dubrawski

(1981), N_2 can even displace H_2O ligand in a Ru^{II}-edta complex to form a dinuclear complex of N_2. In the cluster-structural active-center of nitrogenase, N_2 has the smallest K_m value, or apparently the highest binding affinity among all the exogenous substrates, so N_2 should be able to bind directly to the active-center of nitrogenase, perhaps even more easily than cyclopropene and diazirine, which, according to McKenna et al. [1984], at high enough partial pressure seems to be able to completely inhibit the hydrogen-evolution reaction. An alternative interpretation of the apparently limiting EAC value for N_2 may be conceived as follows: if a weakly coordination-activated D_2 molecule should be able to react with the intermediate species H_2NN to form 2HD with dehydrogenation of the latter species, then a hydrido-ligand, H, on a Fe ion, which is also partially linked to the HNN intermediate species, should be able to react likewise with the HNN by heterolytic elimination of H_2 (similar to heterolytic elimination of H_2 from HZnOH) as an alternative reaction in competition with further reductive hydrogenation of HNN via cis-migration of the hydrido-ligand to HNN to form H_2NN hydrazido(2-) intermediate; in other words, heterolytic cis-elimination of H_2 may be in competition with further reductive hydrogenation of HNN via cis-migration of H. Since the rates of both of these two competitive reactions will conceivably be dependent similarly on the first power of pN_2, at higher pN_2 when both adjacent active-sites are practically occupied by N_2, further increase in pN_2 will have no appreciable effect on the NH_3/H_2 ratio, or the EAC of N_2. On the other hand, an increase in $e^- + H^+$ flux by increasing the ratio of Fe-protein/MoFe-protein, resulting in more copious supply of hydrido-ligands H to take up more of the active-sites on the Fe ions, will suppress the weak coordination of D_2 on the Fe ion site, thereby suppressing the formation of 2HD in favour of NH_3 and H_2 formation, as observed experimentally (for reviews. see Orme-Johnson 1985; Smith et al. 1988).

The mechanisms of reductive hydrogenation of other exogenous substrates have also been illustrated previously with the corner-sharing-twin-cubanes model X-2 (Tsai et al. 1979; Tsai 1980; Wan & Tsai 1981), which differs very little from model X-5 as far as the two adjacent substrate-binding sites are concerned. Some of the key intermediates of the proposed reaction-pathways for these substrates are illustrated in this paper in Fig. 3c-1, which are almost self-explanatory, so, more detailed discussion seems unnecessary here, although some minor refinements can still be made. For example, Fig. 3j-2 might not be one of the major reaction intermediates from the reductive-hydrogenation of coordinated diazirine (Fig. 3j-1), for this would lead to the formation of about equimolar amounts of CH_3NH_2 and NH_3, but this product ratio is not in accord with the known experimental data listed in Table 1. It is more probable that sequential addition of 2 H to the μ_3-coordinated diazirine to disrupt one of the two C–N bond may be followed by disruption of the weak N–N bond on the Mo^{IV} (and also linked to two adjacent Fe ions) to form CH_3N and NH fragments coordinated on the two adjacent sites containing the Mo^{IV}; further reductive hydrogenation of CH_3N may lead to the formation of CH_3NH_2, CH_4,

and NH_3 in approximately equimolar ratios, and further reductive hydrogenation of NH will of course give NH_3 as the sole product; so the ratios NH_3:
CH_3NH_2:CH_4 of the total products will be approximately 3:1:1, in accord with
experimental results (Hardy et al. 1979).

Thus this corner-sharing-twin-cubanes cluster-structural model based on
the postulate that each of the two M-clusters can function independently appears
to be able to give satisfactory elucidation of the reaction mechanisms for all
the known reducible substrates of nitrogenase. So it may not be necessary to
assume that it requires two Mo^{IV} (which are probably more than 7 Å apart) or
two FeMo-co to jointly activate one N_2 substrate molecule, especially since
neither diimide, nor hydrazine appears to be a reaction-intermediate from N_2
reductive-hydrogenation; furthermore, it is known that double-end-on (linear)
coordination of N_2 to two transition-metal atoms or ions usually does not lead
predominantly to the formation of NH_3 on protonation, or reductive hydrogenation, without the formation of appreciable amount of hydrazine (e.g.,
according to Schrock et al. (1984), the $(Ta^{III})_2(\mu_2, w, w')$-$N_2$ complex reacts readily
with HCl in ether to give 85–95% hydrazine; and a $(W^{II})_2$-$(\mu$-$N_2)$ reacts with
aqueous HCl to give N_2H_4 quantitatively). In some cases, double-end-on coordination of N_2 does not appear to be an effective mode of activation of N_2 for
protonation or reductive hydrogenation to form ammonia; for example, according
to Bercaw (1980), the binuclear zirconium complex $[(cp')_2Zr(N_2)]_2(\mu$-$^{15}N \equiv {}^{15}N)$
on treatment with 12N HCl gave hydrazine as the major product (86% yield)
with $^{28}N_2H_4$ and $^{30}N_2H_4$ in about 1:1 ratio and concomitant liberation of
$2N_2$ (while $[(cp')_2Zr(CO)]_2(\mu$-$N_2)$ was not protonated to give hydrazine under
similar conditions); it was interpreted that the complex was probably first
protonated to give $(cp')_2ZrCl_2$, N_2, and a mononuclear cis-$Zr^{II}(^{28}N_2H)$ $(^{30}N_2H)$
complex which was further protonation to form hydrazine with concomitant
liberation of one more N_2 (labelled or unlabelled). Thus, in this case the
dinuclearly and linearly coordinated μ-$N_2(N$-$N = 1.182$ Å) was not protonated
(or reductively hydrogenated) in preference to the mononuclearly, linearly
coordinated N_2 (N-$N = 1.115$ Å).

5 Design of Synthetic Methods and Attempted Synthesis of FeMo-co Model Compounds

Synthesis of FeMo-co model compounds or FeMo-co analogs may provide the
best check for the validity of the proposed mechanism of nitrogenase catalysis
and of the proposed model of nitrogenase active-center. This type of synthetic
work is both challenging and exciting, especially since natural FeMo-co, in spite
of its estimated molecular weight of not more than 1000, still defies X-ray crystal-
structural characterization and more precise elemental analysis, so far its Mo,
Fe, and sulfur atomic ratios being given as 1:(6–8):(6–10) (Nelson et al. 1983;
Smith et al. 1985), or 1:(6–8); (4–10) (Smith et al. 1988).

For the attempted synthesis of FeMo-co model compounds or analogs, the synthetic method we have been using since the early 1980s (Xu et al. 1980) is based essentially upon the following strategy: labile-ligand-assisted self-assembling of $(NR_4)_2MoS_4 - (FeCl_2)_2$ linear trinuclear compound with an excess of $FeCl_2$ (with or without the addition of smaller amounts of $FeCl_3$ and NaH), in DMF or other organic solvents, with the aid of some suitably chosen Lewis bases as neutral ligands, or labilizable chelate-ligands. It is known that $(NR_4)_2MoS_4$ plus excess $FeCl_2$ in concentrated DMF solution can proceed in self-assembling only as far as to the formation of the linear trinuclear complex $(NR_4)_2MoS_4(FeCl_2)_2$ first synthesized by Coucouvanis et al. (1980), in which each sulfido-ligand only as a μ_2 endoligand, not a μ_3 sulfido-ligands as in a $MoFe_3S_4$ cluster framework. This may be due to the strong polarizing-effect of Mo^{VI} acting on the highly polarizable sulfido-ligands as shown by the strong charge-transfer electronic absorption peak. If one or two suitably chosen Lewis-basic ligands can be made to coordinate on the Mo^{VI} carrying the 4 sulfido-ligands, forcing the central cationic Mo of the linear trinuclear complex to assume temporarily a penta-coordination square-pyrimidal, or triangular-bipyrimidal structure, or better a square-bipyrimidal structure, and suppressing the polarizing effect of the central cationic Mo acting on the sulfido-ligands, then it might be possible for each of the sulfido-ligands to coordinate one more Fe ion to become a μ_3-S, and if one of the Lewis-basic ligands is also a μ_3-ligand (e.g., a methoxy ligand, which is known to be a μ_3-OCH_3 ligand in the $(CH_3Zn)_6(\mu_3$-$OCH_3)_8Zn$ corner-sharing-twin-cubanes compound first synthes-ized by Eisenhuth & Wazer [1968]), or better a bis-μ_3-ligand (e.g., an ethylene glycolate anion) then monocubanes, or twin-cubanes clusters might be formed, probably through the formation of intermediate clusters each with one μ_3-Cl ligand loosely bound to $3Fe^{II,III}$ ions at one corner of a monocubane, or at each of the two opposite corners of a twin-cubanes. These two μ_3-Cl might subse-quently be replaced by $2\mu_3$-S, or $2\mu_3$-S_2 ligands, through the addition of $HS^- + Et_3N$, or Na_2S_2, or Li_2S_2. The scheme of this strategy is illustrated in Fig. 5.

Following this scheme, Xu et al. (1980) and Zhang et al. (1985) have used $HOCH_2CH_2O^-$, or $^-OCH_2CH_2O^-$, as well as CH_3O^-, as the Lewis base added into the system containing about 0.01-M $(NR_4)_2MoS_4$ and 0.08 to 0.10-M $FeCl_2$ in DMF at room temperature (under this initial condition, more than 95% of the thiomolybdate was in the linear trinuclear complex) and followed the course of the reaction by taking out samples from time to time for electronic absorption measurement, and for catalytic activity and selectivity assay in the reductive hydrogenation of acetylene with KBH_4 by the method given by Shah & Brill (1977) for chemical catalytic activity assay for samples of natural FeMo-co (their observed t.o.n. being 25 to 34 nmol C_2H_2 reduced/mmol Mo.min, & 99% selectivity to C_2H_4). It was found that, at the beginning of the reaction, only the characteristic electronic absorption peaks of the linear trinuclear complex were observed, and the turnover number of acetylene and selectivity to ethylene were, respectively, only about 5 to 8 nmol C_2H_2/nmol

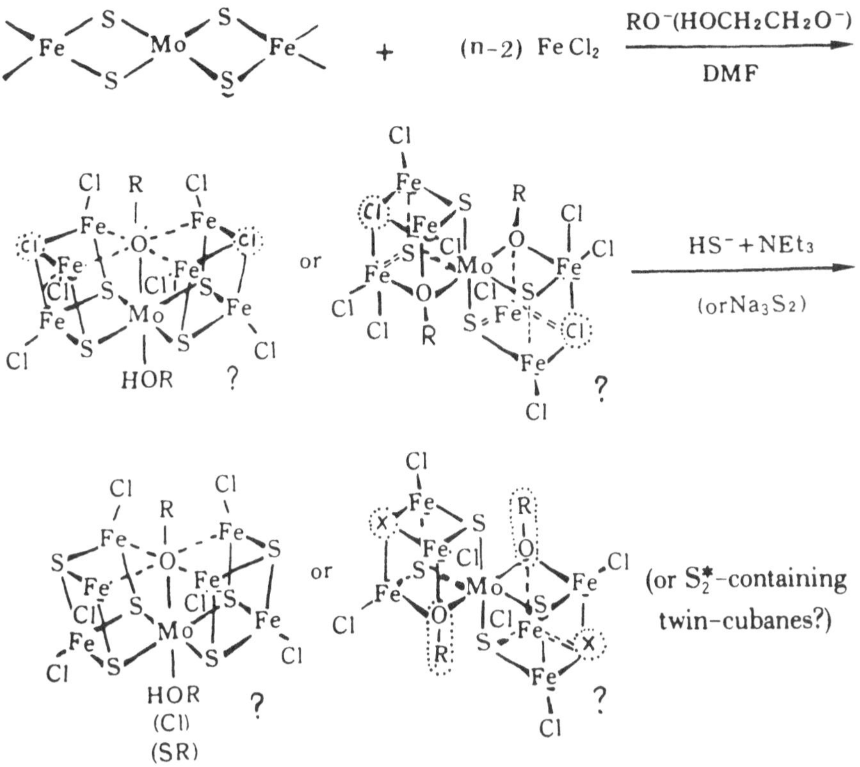

Fig. 5. A scheme of synthetic method for the attempted synthesis of FeMo-co model compounds (model *X-2*, or *X-4?*, on *X-5?*) based on labile-ligand-assisted self-assembly of linear trinuclear $MoS_4(FeCl_2)_2^{2-}$ with more $FeCl_2$ in DMF at $\sim 30\,^\circ C$

Mo.min. & 90 to 91% selectivity to ethylene; and that near the first step of the self-assembling reaction, just before the addition of alkali sulfide, or disulfide, the electronic absorption peaks of the trinuclear complex had almost completely disappeared, indicating the formation of higher dimensional cluster complex with higher extent of electron delocalization and partial reduction of Mo^{VI} to lower valency, as expected; and finally that after the addition of NaSH plus Et_3N and completion of the last step of the synthesis reaction, all of the electronic absorption peaks of the trinuclear complex completely disappeared, and a featureless absorption spectrum with monotonously decreasing absorption amplitude with wave length resulted, very similar to that of natural FeMo-co and quite different from that of the double-cubanes (Wolff et al. 1979, 1980). Some of the samples thus prepared showed very high catalytic activity and selectivity in the reduction of acetylene by KBH_4 (24 nmol C_2H_2/nmol Mo.min. & 94 to 95% selectivity to ethylene) assayed by Shah & Brill's method. Samples of the FeMo-co model compounds (or analogs) thus prepared was obtained in the form of dark-brown, crude crystalline substance, with $Mo:Fe:S:Cl = 1:(6-8):(5-6):(6-8)$, close to that required by the model. But in combination with UW-45, the reconstituted nitrogenase activities of these samples were still low, the best we ever observed with these samples in our own laboratory was 9–18 nmol C_2H_2/nmol. Mo.min. (i.e., ca.3–6% of the t.o.n. of FeMo-co, 300 nmol C_2H_2/nmol Mo.min., as reported by Shah & Brill 1977), and the N_2 reductive-hydrogenation activities about 1.5–3.0 nmol N_2/nmol Mo.min. (Xu et al. 1980; Zeng et al. 1980; Cai et al. 1987). Evidently, the crude crystalline samples of the FeMo-co model compounds thus prepared were still very impure and difficult ot purify; they might contain a high proportion of inactive analogs of FeMo-co, such as those with two labile μ_3-ligands in the trans positions, rather than in the cis positions on the Mo^{IV}; or they might also contain a high proporation of monocubane clusters, especially when a monodentate Lewis base was used as a labile ligand to assist the self assemblying (moreover, a potential μ_3 ligand, such as a methoxy ligand, would first bind the Mo-site like a monodentate ligand, and two of these would very likely take up two *trans* positions, instead of the required *cis* positions on Mo^{IV}. However, it is interesting to note that, in the activity assay of these synthetic samples, the highest reconstituted nitrogenase activity (18 nmol C_2H_2/nmol Mo.min.) was observed with samples (318A; 7012S) synthesized with the use of ethylene glycolate as the labilizable bis-μ_3 chelate-ligand, and with the DMF solution of each of the crude crystalline samples treated with potassium citrate (K_3Cit) for possible *exo*- or *endo*-ligand exchange just before the activity assay; note especially that this final K_3Cit treatment was found to result in a very significant increase (cs. 2–4 fold) in the reconstituted nitrogenase activity (Xu et al. 1980; Zeng et al. 1980); in connection with the recent finding by Madden et al. (1990) and by Collet et al. (1990) that homocitrate is an integral part of FeMo-co, the above findings indicate that the impure samples (318A; 7012S) might actually contain small amounts of the desired FeMo-co model compound, though we were not able to isolate and purify it for structural characterization.

In the course of this investigation, a systematic study has also been made on the effects of a selected variety of Lewis-basic ligands on further self-assembling of the linear trinuclear complex with more $FeCl_2$ to form monocubanes, twin-cubanes, or other higher-dimensional clusters, with the expected higher degree of electron delocalization so that the Mo^{VI} might be automatically reduced by the ferrous ions in the cluster to lower valence state, probably Mo^{IV}, without the addition of any other reducing agent. The course of any self-assembly reaction may be conveniently followed by means of uv-visible electronic absorption spectroscopy, as mentioned above. In this way, Song et al. (1984) and Zhang et al. (1985) have observed that the addition of sufficient amounts of formamide, or N-methylformamide, or triethylamine, or butylamine to the MoS_4^{2-}-$nFeCl_2$ system ($n = 8$ to 10) can slowly bring about almost complete disappearance of the characteristic electronic peaks of the linear trinuclear complex in about two days at room temperature (ca.30 °C). In a subsequent series of experiments, Zhang et al. (1985) have observed that the addition of n-Bu_3P to the system can bring about practically complete disappearance of the electronic absorption peaks of the linear trinuclear cluster in a few hours at room temperature, resulting in a feature-less electronic absorption curve (Fig. 6) similar to that of FeMo-co, or of a monocubane $MoFe_3S_4L_n$ cluster; whereas the addition of Ph_3P shows practically no effect, as anticipated. More recently, Zhou et al. (1990), working with $VS_4^{2-}-nFeCl_2$ system ($n = 3$ to 4), have found that the addition of n-Bu_3P to this system leads to considerably

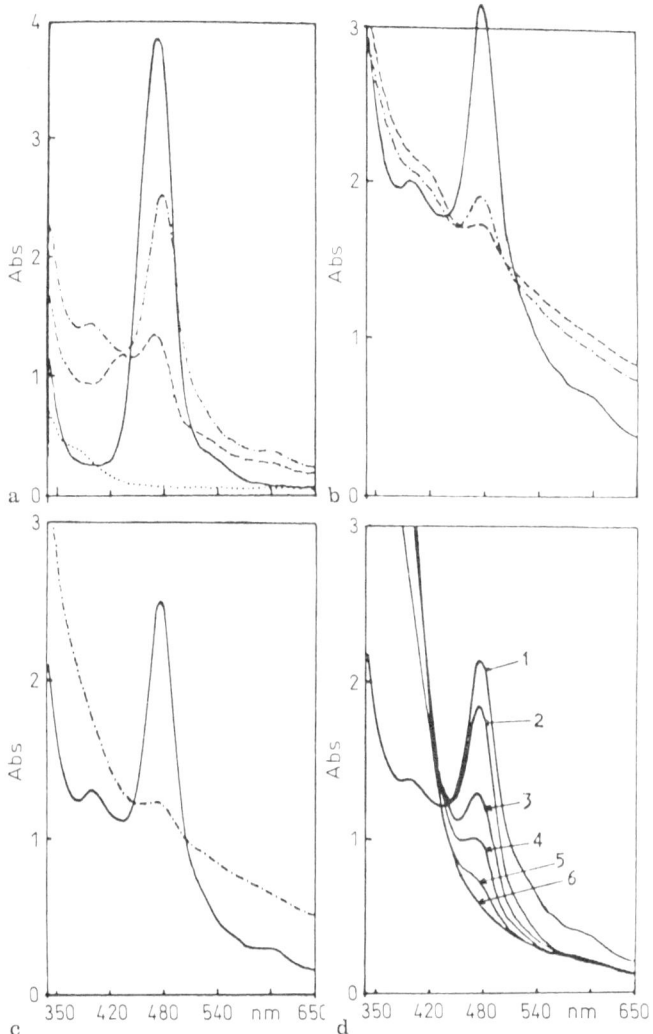

Fig. 6 a–d. Self-assemblying reaction(s) of $(NH_4)_2MoS_4(FeCl_2)_2$ with more $FeCl_2$ in DMF at 25 C induced by a variety of Lewis bases, reaction(s) being followed by changes in electronic absorption with time. **a.** Reference systems: solid line, $(NH_4)_2MoS_4$; dot-dashed line, $[Cl_2Fe(S)_2Mo(S)_2FeCl_2]^{2-}$; dashed line, $[Cl_2Fe(S)_2Mo(S)_2]^{2-}$; dotted line, $FeCl_2$. **b.** $(NH_4)_2MoS_4-8FeCl_2-4Et_3N$ system: solid line, Et_3N just added; dot-dashed line, 24 h later; dashed line, 48 h later. **c.** $(NH_4)_2MoS_4-8FeCl_2-4NMF$ system: solid line, NMF just added; dot-dashed line, 7 days later. **d.** $(NH_4)_2MoS_4-8FeCl_2-2(n$-$Bu_3P)$ system: 1, n-Bu_3P just added; 2–6, taken every 15 minutes

faster change in the electronic absorption spectrum than in the case of the Mo–Fe–S system; in less than 2 hours at room temperature, the change is complete, and the electronic absorption spectrum becomes practically the same as that of the $[VFe_3S_4Cl_3(DMF)_3]^1$ more recently reported by Kovacs & Holm (1986); and the ^{31}P-NMR spectrum of the final system shows that the n-Bu$_3$P is not bound to the vanadium; so it seems that the n-Bu$_3$P ligand only serves to initiate the self-assembly reaction; it is finally displaced from the V^{III} coordination sphere by the DMF solvent molecules present in large excess, probably because the remaining 3 sites on the V^{III} coordination sphere taken together fail to provide enough space to accommodate 2DMF plus a considerably more bulky n-Bu$_3$P ligand.

The reaction pathway of this ligand-assisted self-assembling of the V–3Fe 4S–6L-type monocubane (Fig. 2d) and probably also the Mo–3Fe–4S–6L monocubanes with similar structures might proceed in the following way. Conceivably, by taking up a strong monodentate ligand like R$_3$P, the linear trinuclear complex of $VS_4^{3-}(FeCl_2)_2$ or $MoS_2^-(FeCl_2)_2$, or simply the corresponding dinuclear complexes, $VS_4^{3-}(FeCl_2)$, is forced to assume a triangular bipyramid structure with 3 of the 4S on one side, by only a slight angular readjustment of 2 V–S bonds, i.e., 2S remaining at the equatorial plane and one S moving to one of the 2 apices of the bipyramid, the other apex being taken up by the R$_3$P or Et$_3$N ligand; thus with 3S in conjunction with one of the two Cl-lignads on the Fe^{II}, the anionic skeleton of the monocubane framework is already there, it only needs 2 more FeCl$_2$ to complete the monocubane structure with concurrent readjustment of the valence states of the Mo and Fe through more extensive electron delocalization typical of the 4M–4S cubane clusters. With the decrease in positive valency of the Mo, or V the remaining sulfido-ligand becomes labile, and may be replaced by a neutral solvent ligand, DMF, which is also a fairly strong Lewis base, and 2 more DMF ligands may be taken up by the Mo^{IV} (d^2), or V^{III}(d^2) to complete the octahedral coordination. The loosely-bound μ_3-Cl ligand may be easily displaced by the labilized sulfido-ligand, either as free anion (as S$^=$ in DMF, or as HS$^-$), or a labile ligand on Mo^{IV}, or V^{III} of a second molecule. This reaction pathway may be checked by means of kinetic and spectroscopic methods, and qualitatively by quantum-chemical study.

In view of the strong tendency towards the formation of mono-cubane cluster complexes in the above-mentioned labile-ligand-assisted self-assembly reaction of tetrathiomolybdate-2FeCl$_2$, or tetrathiovanadate-2FeCl$_2$ linear trinuclear complex with excess ferrous chloride in DMF solution, the strategy for successful synthesis of the corner-sharing-twin-cubanes model compounds would seem to be the use of adequate bis-μ_3-labile-chelate ligands which will simultaneously take up two neighboring sites on the central Mo or V ion so as to block the pathway for the formation of mono-cubanes, and at the same time to provide an additional μ_3-ligand (labile, or labilizable) for each half of the corner-sharing twin-cubanes (Figs. 7 and 8), as pointed out in our previous works (Zeng et al. 1980; Xu et al. 1980). A neutral bis-μ_3-chelate ligand would be better than the

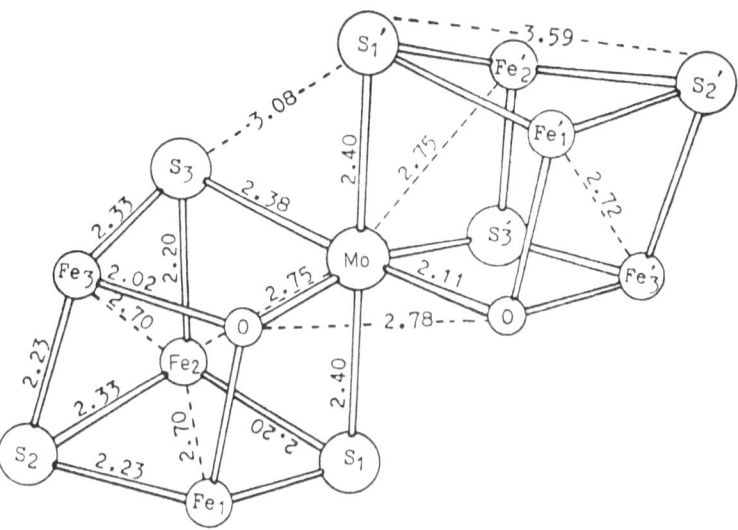

Fig. 7. Corner-sharing-twin-cubanes cluster model *X-2* shown with some structural parameters. (The two *O*-containing labile ligands are 2.78 Å apart, bonding Mo-Fe2 (Mo-Fe2′) 2.75 Å apart, nonbonding. Mo-Fe1, Mo-Fe3 (Mo-Fe1′, Mo-Fe3′), about 3.1 Å apart, nonbonding; S–S (non-bonding) in each half of the twin-cubanes, all about 3.6 Å apart, internuclear distance)

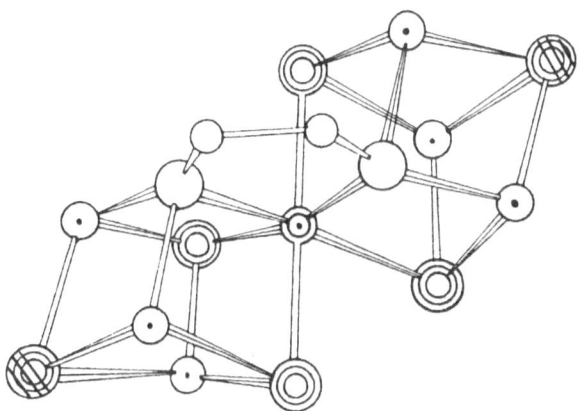

Fig. 8. Model *X-2* (or model *X-5*, with the two *half-shaded double-circles* denoting two μ_3-S$_2$ *endo*-ligands for the sake of clarity) shown with a bis-μ_3-chelate-ligand taking up 2 adjacent μ_3-sites each envolving Mo and 2Fe nuclei

ethylene glycolate dianion, $^-OCH_2CH_2O^-$, previously used by us since this dianion will react with the FeCl$_2$ which is present in large excess. As a first step towards the final goal, it mat be a good idea to try to synthesize first a FeMo-co analog, such as a Mo–6Fe–6S, or a V–6Fe–6S-type of twin- cubanes cluster complex, rather than the probably highly fluxional μ_3-S$_2$-containing twin-cubanes cluster-structural model compound of FeMo-co.

Acknowledgements. This review is based mainly upon the research work on chemical modelling of biological nitrogen fixation carried out in the Catalysis Laboratory of the Chemistry Department in collaboration with the Biochemical Laboratory of the Biology Department of this university. Many of our colleagues and graduate students have made important contributions to this research work, as acknowledged in our previous review published in 1987. We wish to acknowledge our thanks to Professors Xingxian Guo, Aoqing Tang, and Jiaxi Lu for their constant encouragements, supports, and helpful discussions in the course of our work on nitrogen fixation studies.

References

Andrews MA, Knobler CB, Kaesz HD (1979) Crystal & molecular structure of Fe_3 ($NCCH_2CH_2CH_3$) $(CO)_9$, Characterization of a cluster complex with triply bridging nitrile ligand. J Amer Chem Soc 101: 7260

Antonio MR, Teo BK, Orme-Johnson WH, Nelson MJ, Groh SE et al. (1982) Iron EXAFS of the iron-molybdenum cofactor of nitrogenase J. Amer. Chem. Soc. 104: 4703

Bercaw JE (1977) Reduction of molecular nitrogen to hydrazine at titanium and zirconium. In: Newton WE, Postgate JR, Rodriguez-Barrueco C (eds) Recent Development in Nitrogen Fixation, p 25; see also Manriquex JM, Sanner RD, March RE, Bercaw JE (1976) J Amer Chem Soc 98: 3042

Blount JF, Dahl LF, Hoogzand C, Hubel W (1966) Structure of and bonding in an alkyne-non-acarbonyltriiron complex. A new type of iron-acetylene interaction. J Amer Chem Soc 88: 292; cf. also Dahl LF, Smith DL (1962) J Amer Chem Soc 84: 2450

Bulen WA (1976) Nitrogenase from *Av* and reactions affecting mechanistic interpretations. In: WE Newton & CJ Nyman (eds) Proc of the First International Symposium on Nitrogen Fixation, Vol.1, p 177

Burgess BK, Yang S-S, You, C-B, Li J-G, Freisen, GD et al. (1981) In: Gison AH, Newton WE (eds) Current Perspectives in Nitrogen Fixation pp 71

Burgess BK (1984) Structure and activity of nitrogenase— An overview. In: C Veeger & WE Newton (eds) Advances in Nitrogen Fixation Research, pp 103

Cai (Tsai), Q.-R (KR), Zhang, H.-B, Lin G.-D (1987) Cluster catalysis in fixation of nitrogen to ammonia catalyzed by nitrogenase & by iron catalysts. Advances in Science of China—Chemistry, Vol 2, 125

Chen H-B, Lin G-D, Zhang H-T, Tsai K-R (1985) Chemical modelling of ATP-driven electron transport in nitrogenase-catalyzed reactions. J Xiamen Univ (Nat Sci) 24: 448

Collet TA, Hickman AB, Salifoglou T, Wright DA, Orme-Johnson WH (1990) Physical characterization of the molybdenum-iron cofactor of nitrogenase. Final Program 8th Int Congr Nitrogen Fixation (Knoxville, Tenn. 1990), Abstr A-10

Coucouvanis D, Baenziger NC, Simhon ED, Stremple P, Swenson D, Kostikas A et al. (1980) Heterodinuclear di-μ-sulfido bridged dimers containing iron and molybdenum or tungsten. Structures of $(Ph_4P)_2$ $(FeMS_4)$ complexes (M = Mo, W). J Amer Chem Soc 102: 1730

Coucouvanis D, Baenziger NC, Simhon ED, Stremple P, Swenson D, Kostikas A et al. (1980) Synthesis and structural characterization of the $(Ph_4P)_2(Cl_2FeS_2MS_2FeCl_2)$ complexes (M = Mo, W). First example of a doubly bridging MoS_4 unit and its possible relevance as a structural feature in the nitrogenase active site. J Amer Chem Soc 102: 1732

Cramer SP, Gillum WO, Hodgson KO, Mortenson LE, Stiefel EI, Chisnell JR, Brill WJ, Shah VK (1978) The molybdenum site in nitrogenase. A comparative study of MoFe-proteins and the FeMo-cofactor by X-ray absorption spectroscopy. J Amer Chem Soc 100: 3814

Diamantis AA, Dubrawski JV (1981) Preparation and structure of ethylene diaminotetra-acetate complexes of ruthenium (II) with dinitrogen, carbon monoxide, and other π-acceptor ligands. Inorg Chem 20: 1142

Eisenhuth WH, Van Wazer JR (1968) Equilibrium in the methanolysis of dimethylzinc. J Amer Chem Soc 90: 5397

Flank AM, Weininger M, Mortrnson LE & Cramer SP (1986) Single-crystal EXAFS of nitrogenase. J Amer Chem Soc 108: 1049

Guth JH, Burris RH (1983) Inhibition of nitrogenase-catalyzed NH_3 formation by H_2. Biochemistry 22: 5111

Haight GP, Jr (1987) Hydrolysis of phosphate esters and anhydrides: role of metal ions. Coord Chem Rev 79: 293–319. See also Crease IC, Haight GP, Peachey R, Robinson WT, Sargeson AM (1984) Rapid cleavage of chelated pyrophosphate using metal ion complexes. J Chem Soc Chem Commun, pp 1568

Hardy RWF, Burns RC, Parshall GW (1971) Biochemistry of Nitrogen Fixation. Advan Chemistry Series 100: 219

Hardy RWF (1979) Reducible substrates of nitrogenase. In: RWF Hardy, F Bottomley, RC Burns (eds), A Treatise on Nitrogen Fixation, pp 515

Hoch GE, Schneider KC, Burris RH (1960) Hydrogen evolution and exchange, and conversion of N_2O to N_2 by soybean root nodules. Biochim Biophys Acta 37: 273

Hoffman BM, Roberts JE, Orme-Johnson WH (1982) Molybdenum-95 and proton ENDOR spectroscopy of the nitrogenase MoFe-protein. J Amer Chem Soc 104: 860

Jackson EK, Parshall GW, Hardy RWF (1968) Hydrogen reactions of nitrogenase. Formation of molecule HD by nitrogenase & by an inorganic model. J Biol Chem 243: 4932

Jensen BB, Burris RH (1985) Effect of high pN_2 and high pD_2 on NH_3 production, H_2 evolution, and HD formation by nitrogenases. Biochemistry 24: 1141

Jonas K, Brauer DJ, Kruger C, Roberts PJ, Tsay Y.-H (1976) "Side-on" dinitrogen transition metal complexes. The molecular structure of $\{C_6H_5[Na \cdot O(C_2H_5)_2]_2[(C_6H_5)_2Ni]_2N_2NaLi_6(OC_2H_5)_4O(C_2H_5)_2\}_2$. J Amer Chem Soc 98: 74

Kovacs JA, Holm RH (1986) Assembly of vanadium-iron-sulfur cubane clusters from mononuclear and linear trinuclear reactants. J Amer Chem Soc 108: 340

Kruger C, Y-H Tsay (1973) Molecular structure of a π-dinitrogen-nickel-lithium complex. Angew Chem Int Ed 12: 998

Kubas GJ, Vergamini PJ, Eastman MP, Prater KB (1976) Electrochemistry of the ion-sulfer cluster compound $[(n-C_5H_5)_2Fe_2S_2(SC_2H_5)_2]$. Synthesis & properties of the oxidation products. J Organomet Chem 117: 71

Kubas GJ, Vergamini PJ (1981) Synthesis, characterization, reactions of iron-sulfur clusters containing the S_2 ligand: $(Cp_2Fe_2(S_2)(SR)_2)^{0.1+}$, $(Cp_4Fe_4S_5)^{0.1+2+}$, $(Cp_4Fe_4S_6)$. Inorg Chem 20: 2667

Leuchenko LA (1980) Spectroscopic investigation of FeMo-co, *coenzyme A* as one of the probable components of an active site of nitrogenase. BBRC 96: 1384

Likhtenstein GI (1988) Structure & molecular dynamics of metalloenzymes studied by physical label methods. J Mol Catal 47: 129

Lu J-X (1973) Structural-chemical aspects in biological nitrogen fixation process—on the coordination activation and reductive hydrogenation of N_2 in nitrogen fixation. In: *Progress in Chemical Modelling of Biological Nitrogen Fixation, 2nd Compilation (from the 1973 Symposium papers)*, pp 1–53. Science Press, China, 1976; cf. also Nitrogen Fixation Research Group of the Fujian Inst of Structure of Matter, Kexue Tongbao (1975) 20: 540

Lu J-X (1980) Composite "string bag" cluster model for the active centre of nitrogenase. In: WE Newton, WH Orme-Johnson (eds), *Nitrogen Fixation* Vol 1, pp 343

Madden M, Shah V, Kindon N, Burke S, Ludden, P (1990) Dinitrogen reduction by dinitrogenase activated in vitro is dependent upon which diastereomer of 1-fluorohomocitrate is incorporated into the FeMo-co. Final Program, 8th Int Congr Nitrogen Fixation (Knoxville, Tenn 1990), Abstr A-05

McKenna CE, McKenna M-C, Huang CW (1976) Low stereoselectivity in methylacetylene and cyclopropene reactions by nitrogenase. Proc Natl Acad Sci USA 76: 4773

McKenna CE, McKenna M-C, Higa M (1976) Chemical probes of nitrogenase. I Cyclopropene. Nitrogenase-catalyzed reduction to propene and cyclopropane. J Amer Chem Soc 98: 4657

McKenna CE, Huang C-W Jones JB, McKenna M-C, Makajima T, Nguyen HT (1980) Cyclopropenes: New chemical probes of nitrogenase active site interactions. In: WE Newton, WH Orme-Johnson (eds), *Nitrogen Fixation* Vol. 1, pp 223

McKenna CE, Eran H, Nakajima T, Osumi A (1981) Active site probes for nitrogenase. In: AH Gibson, WE Newton (eds) *Current Perspectives in Nitrogen Fixation*, p 358

Mortenson LE, Thorneley RNF (1979) Structure & function of nitrogenase. Ann Rev Biochem 48: 387

Nelson WJ, Levy MA, Orme-Johnson WH (1983) Metal and sulfur composition of iron-molybdenum cofactor of nitrogenase. Proc Natl Acad Sci USA 80: 147

Newton WE, Bulen WA, Hadfield KL, Stiefel EI, Watt GD (1977) HD formation as a probe for intermediates in N_2 reduction. In: WE Newton, JR Postgate, C Rodriguez-Sarrueco (eds), Recent Development in Nitrogen Fixation, pp 119

Newton WE, Burgess BK, Cummings SC, Lough S, McDonald JW, Rubinson JF, Conradison SD, Hodgson KO (1984) Structure aspects and reactivity of the FeMo-co from nitrogenase. In: C Veeger, WE Newton (eds), *Advan Nitrogen Fixation Res.*, pp 160

Nitrogen Fixation Research Group, Kirin Univ (1973) "Chemical Modelling of Biological Nitrogen Fixation". Science Press, China

Nitrogen Fixation Research Group, Chemistry Department, Kirin Univ (1974) Theory of chemical bonding in dinitrogen complexes. Scientia Sinica 17: 192

Nitrogen Fixation Research Group, Chemistry Department, Xiamen Univ (1974) On the mechanism of nitrogenase catalysis and structure of the active centre. J Xiamen Univ Nat Sci 13: (1), 111

Nitrogen Fixation Research Group, Chemistry Department, Xiamen Univ (1976) Aspects of the structure of nitrogenase active-centre coordination catalysis in chemical modelling of biological nitrogen fixation. In: Progress in Chemical Modelling of Biological Nitrogen Fixation in China, 2nd Compilation (from 1973 Symp papers), pp 163–209 Science Press, China, 1976

Nitrogen Fixation Research Group, Chemistry Department, Xiamen Univ. (1976) A model of nitrogenase active-centre and mechanism of nitrogenase catalysis. Scientia Sinica (Engl. Ed.) 34: 460

Nitrogen Fixation Research Group & Inst of Physical Chemistry, Xiamen Univ (1982) Fixation of dinitrogen to ammonia via enzymatic, non-anzymatic catalysis. J Xiamen Univ (Nat Sci) 21: 424

Ooi, B-L, Sykes A.G. (1989) Solution properties and reactivity of the aqua ion of Mo^{IV}_3 incomplete cuboidal Mo/S cluster $(Mo_3S_4(H_2O)_9)^{4+}$. Inorg Chem 28: 3799

Orme-Johnson WH, Munck E (1980) On the prosthetic groups of nitrogenase. In: MP Coughlan (ed.), Molybdenum & Molybdenum Containing Enzymes, pp. 427

Orme-Johnson WH, Lindahl P, Meade J, Warren W, Nelson M, Groh S, Orme-Johnson NR, Munck E et al. (1981) Nitrogenase: prothetic groups and their reactivities. In: AH Gibson & WE Newton (eds) Current Perspectives in Nitrogen Fixation, pp 79

Orme-Johnson WH (1985) Molecular basis of biological nitrogen fixation. Ann Rev Biophys Chem 4: 419

Pez GP, Apgar P, Crissey (1982) Reactivity of $[\mu-(n^1:n^5-C_5H_4)]$ $(n-C_5H_5)_3Ti_2$ with dinitrogen. Structure of a titanium complex with a triply-coordinated N_2 ligand. J Amer Chem Soc 104: 482

Schrock RR, Blum L, Theopold KH (1984) Hydrazido(1-), 1,2-hydrazido(2-), and hydrazido(4-) complexes of tungsten (VI). In: C Veeger, WE Newton (eds), Advances in Nitrogen Fixation Research, pp. 75

Schultz PG, Lerner RA, Benkovic SJ (1990) Catalytic antibodies. C & EN News 68: (22), 26

Shah VK, Brill WJ (1977) Isolation of an iron-molybdenum cofactor from nitrogenase. Proc Natl Acad Sci USA 74: 3249

Shibahara T, Yamamoto T, Kanadani H, Kuroya H (1987) Double-cubane type, molybdenum-sulfur cluster aquo ion, $[(H_2O)_9Mo_3S_4MoS_4Mo_3(H_2O)_9]^{8+}$. J Amer Chem Soc 109: 3495;

Shibabara T (1988) Proc 26th Intern Coord Chem Conf Portugal, 1988 paper A27 (cf Ooi & Sykes, 1989); Akashi H, Shibahara T (1889) Novel cubane-type molybdenum-tin cluster complexes. $(H_2O)_9Mo_3S_4SnS_4Mo_3(H_2O)_9)^{8+}$ and $Mo_3SnS_4(aq)^{6+}$. Inorg Chem 28: 2906

Smith BE, Lowe DJ, Bray RC (1973) Studies by electron paramagnetic resonance on the catalytic mechanism of nitrogenase of Kp. Biochem J 135: 331

Smith BE, Bishop PE, Dixon RA, Eady RR, Filler WA, Lowe DJ, Richards AJM, Thomson AJ, Thorneley RNF, Postagate JR (1985) The FeMo-co of nitrogenase. In: HJ Evans, PJ Bottomley, WE Newton (eds), Nitrogen Fixation Research Progress, pp 597

Smith BE, Buck M, Eady RR, Lowe DJ, Throneley RNF, Ashby G, Deistung J, Eldridge M, Fisher K, Gormal C, Ioannids I, Kent H, Arber J, Flood A, Garner CD, Hasnain S, Miller R (1988) Recent studies on the structure and function of molybdenum nitrogenase. In: Bothe/de Bruijn/WE Newton (eds.), Nitrogen Fixation: Hundred Years after Gustav Fischer, pp 91

Song Y (1984) Effects of labile ligands on self-assembly of tetra-thiomolybdate-2FeCl$_2$ with excess FeCl$_2$. MS thesis. Xiamen Univ

Stephens P, McKenna CE, Smith BE, Nguyen HT, McKenna M-C, Thompson AJ, Delvin F, Jones BJ (1979) CD and MCD of nitrogenase proteins. Proc Natl Acad Sci USA 76: 2585

Stiefel EI, Burgess BK, Wheland S, Newton WE, Corbin JL, Watt GD (1980) $A v$ biochemistry: H_2 $(D_2)/N_2$ relationships of nitrogenase and some aspects of iron metabolism. In: WE Newton, WH Orme-Johnson (eds), Nitrogen Fixation, Vol I, pp 211

Swamy KCK, Day RO, Holmes RR (1988) A new structural form of tin as a double cube. A heptanuclear tin-sulfur cluster. J Amer Chem Soc 110: 7543

Tang Aoqing (1982; 1983) Plenary presentations given at the 1st & 2nd National Conf on Chemical Modelling of Biological Nitrogen Fixation; cf Nitrogen Fixation Research Group, Kirin University (1974; 1976)

Thomas MG, Mutterties EL, Day RO, Day VW (1976) Metal clusters in catalysis. 5. Four electron n²-ligand bonding in clusters and catalytic intermediates. J Amer Chem Soc 98: 4645

Thorneley RNF, Lowe DJ (1983) Nitrogenase of Kp. Biochem J 215: 543

Thorneley RNF, Bergstrom NHJ, Eady RR, Lowe DJ (1989) Vanadium nitrogenase of Azotobacter chroococcum. Biochem J 257: 789

Tsai KR (1965) Catalysis by coordination activation. Chinese Univ J Sci (Chemistry & Chem Eng Ed) 1965: 486

Tsai KR, Lin S-T, Wan H-L (1979) Evolution of a model of nitrogenase active-centre and mechanism of nitrogenase catalysis. J Xiamen Univ (Nat Sci) 18: 30

Tsai KR (1980) Development of a model of nitrogenase active centre and mechanism of nitrogenase catalysis. In: WE Newton, WH Orme-Johnson (eds), Nitrogen Fixation, Vol I, pp 373

Tsai KR, Wan H-L (1981) Coordination catalysis by transition-metal complexes. In: M Tsutsui, Y Ishi, Huang Y-Z (eds.), Fund Res Organomet Chem pp 1

Tsai KR et al. (1990) Coordination catalysis in chemical modelling of biological nitrogen fixation-development of a model of nitrogenase active-centre and mechanism of nitrogenase catalysis. Submitted for publication

Walker M, Mortenson LE (1973) Oxidation reduction properties of nitrogenase from Cp W5. BBRC 54: 669

Wang Z-C, Watt GD (1984) H₂-uptake activity of the MoFe-protein component of Av nitrogenase. Proc Natl Acad Sci USA 81: 376

Wan H-L, Tsai KR (1981) Chemical modelling of biological nitrogen fixation. XIV EHMO study of probable modes of μ₃(n²) coordination of some nitrogenase substrates (cyclopropene etc.) & reductive hydrogenation intermediates of N₂. J Xiamen Univ (Nat Sci) 20: 62

Xu Z-W, Yan C-Z, Ding M-T, Zhang P-X, Lin S-T, Xu L-S, Tsai KR (1980) Chemical modelling of biological nitrogen fixation. VI. Synthesis and catalytic activities of FeMo-co modelling compounds. J Xiamen Unvi (Nat Sci) 19: 34

You C-B, Song W, Zeng D, Tsai KR (1990) ATP binding to Fe-protein and ATP-driven electron transfer in nitrogen fixation. In: Nitrogen Fixation Research in China

Zeng D, Xu L-S, Chen J-F, Zhang F-Z, Hung H-Q, Xu Z-W, Lin G-D, Zhang M-H (1980) The biological activities of the modelling compounds of FeMo-co nitrogenase. J Xiamen Univ (Nat Sci) 19: (4) 78

Zhang F-Z, Xu L-S, Chen J-F, Huang H-Q, Zeng D, Liao Y-Y, Yuan L-W (1988) Some characteristics of laser Raman and electronic absorption spectra of FeMo-co of nitrogenase. J Xiamen Unvi (Nat Sci) 27: 572

Zhang H-T et al. (1985) Unpublished results; cf Zhang H-T et al. (1990)

Zhang H-T, Lin G-D, Song Y, Tsai KR (1990) Spectroscopic studies on labile-ligand-assisted self-assembly of tetrathiomolybdate-2FeCl₂ linear trinuclear cluster with excess FeCl₂ by various Lewis-basic ligands. Submitted for publication

Zhou Z-H, Lin G-D, Zhang H-T, KR Tsai (1990) Spectroscopic study on labile-ligand-assisted self-assembly of tetrathiovanadate-2FeCl₂ linear trinuclear cluster with 2FeCl₂ and catalytic activity of thiovanadate-FeCl₂ system in the reduction of acetylene by KBH₄. Submitted for publication

Zumft WG, Mortenson LE (1975) The nitrogen-fixing complex of bacteria. Biochim Biophys Acta 416: 1

Zumft WG (1978) Isolation of thiomolybdate compounds from the molybdenum-iron protein of Cp nitrogenase Eur J Biochem 91: 345

CHAPTER 5
ATP Binding to Nitrogenase and ATP-Driven Electron Transfer in Nitrogen Fixation[1]

Chong-Biao You[2], Wei Song, Ding Zeng, and K.R. Tsai[2]

[1] This work is supported by the National Science Foundation of the P.R. China, and has also been supported by grants from the National Science and Technology Commission administered through Academia Sinica and through the National Education Commission.
[2] To either of whom correspondence should be addressed.

1 Introduction

ATP-driven electron transfer in nitrogenase catalysis is an important example
of coupled electron- and energy-transfers in certain biological processes, including
the ATP-dependent dark reaction in photosynthesis. It is known that nitrogenase-
catalyzed reduction of H^+, N_2, or other substrates is coupled to the hydrolysis
of ATP, which acts in some unclarified way as "electron activator" in the
transfer of an electron from the reduced Fe-protein to the dithionite-reduced
MoFe-protein, a tetrameric ($\alpha_2\beta_2$) metallo-protein with a molecular weight of
about 220 kDa, containing 2 Mo, about 30 Fe, and about equal number of
acid-labile S, and carrying the substrate-binding sites. The complexation of
MgATP and MgADP with nitrogenase, or more specifically with the Fe-protein,
and the mechanism of the ATP-mediated "electron activation" have been subject
to extensive investigation; many important progress has been made, and many
excellent reviews on these subjects have been published (e.g., Zumft and Mortenson
1975; Mortenson and Thorneley 1979; Burgess 1984; Orme-Johnson 1985; Smith
et al. 1988.). This article will mainly review the research work on these subjects
done by Chinese investigators since the mid 1970s. Biochemical and biophysical
studies of MgATP and MgADP bindings to Fe-protein will first be reviewed,
followed by model studies on adenine nucleotide complexation with synthetic
analogs of the 4Fe–4S center of Fe-protein. A discussion of the probable location
of the MgATP-binding sites is presented, and a mechanism of ATP-driven
electron transfer in nitrogenase catalysis is proposed.

2 Complexation of MgATP and MgADP
with Nitrogenase Fe-Protein and Effects
on the Physicochemical Properties of the Metallo-Protein

2.1 Composition and Optical Spectrum of Nitrogenase Fe-Protein

Nitrogenase Fe-protein from a variety of species is a dimer (γ_2) of two identical
subunits with dimeric molecular weight of about 57 to 73 kDa (Nelson et al.
1982), or around 60 kDa (Burgess 1984). The Fe-proteins from *Azotobacter
vinelandii*(Av2) and blue-green algae *Anabaena cylindrica*(An2) have been
purified by You et al. (1978) and by Lin et al. (1984), using a simple, modified
procedure (You et al. 1983a) and specific apparatus designed by Dai et al. (1985).
The molecular weights were found to be 64 kDa and 62 kDa Av2 and An2,
respectively. The physicochemical properties of these two Fe-proteins, such as
amino acid compositions, UV-visible spectra, etc. have also been examined
(Nitrogen Fixation Group, IAAE 1976; Lin et al. 1984). These results have been
confirmed by the complete amino acid sequences. The amino acid sequences of

the Fe-proteins from Av2, *Anabaena* 7120, *C. pasteurianum* (Cp2), *R. japonicum, R. meliloti, R. trifolii,* and *Parasponia Rhizobium* have also been reported (Howard et al. 1985; Zeng 1987). The Fe-proteins share 68% identical residues (based upon Av2 amino acid sequence of 289 residues). Five cysteine residues are located in highly conserved sequences (residues 38, 85, 97, 132, and 184), and two of them from each subunit may be ligands for the 4Fe–4S center (Howard et al. 1985; You 1987; Zeng 1987).

The Av2 has been determined to contain 4 (3.89) iron atoms (Nitrogen Fixation Group, IAAE 1976) and 4 labile sulfur atoms per dimer (Nelson et al. 1982). A 4Fe–4S cluster was extruded and indentified by comparison of optical spectra with those of synthetic 4Fe–4S cluster complexes (Que et al. 1975; You et al. 1979). Since there are 4Fe per dimer, the cluster has been presumed to bridge the two subunits (Howard et al. 1985).

The optical spectrum of the reduced Fe-protein is featureless, while that of the reversibly oxidized protein has a broad shoulder near 420 to 450 nm (Nitrogen Fixation Group, IAAE 1976). This pattern has also been found in the purified nitrogenase complex from Av (Song and You 1987).

2.2 Binding of MgATP to Fe-Protein

Although ATP is known to be indispensable for nitrogenase catalysis, relatively little is known about the role of MgATP. Some information about this has been obtained from different binding experiments. With column gel filtration, You et al. (1978) demonstrated that Av2, but not Av1, bound MgATP (Table 1). This was first reported by Bui and Mortenson (1968) and latter by Tso and Burris (1973), using a rapid gel-equilibration method. But some conflicting results have been reported. Biggins and Kelly (1970) found that both Kp1 and Kp2 bound MgATP and that inactivation of Kp2 by oxygen did not affect this binding. Moreover, a reinvestigation of MgATP binding to Cp2 indicated 5 ± 2 binding sites (Mortenson and Thorneley 1979), and more recently, Cordewener et al. (1983) found 1.3 ATP-binding sites for Av2.

Table 1. Comparative ATP-binding experiments for nitrogenase component proteins from *A. vinelandii*

Nitrogenase component protein added to the control*	μM	ATP bound μM
Control*	0	0
Fe-protein (dithionite-reduced)	78	135.9
Fe-protein (oxygen-inactivated)	21	83.4
MoFe-protein (dithionite-reduced)	9.3	0

* Control: 10 μM Mg^{2+}, 35 μM ATP in 0.05 M Tris-HCl buffer, pH 8.0

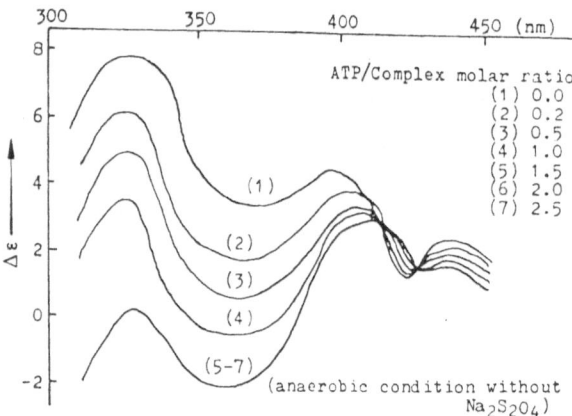

Fig. 1. Effects of MgATP concentrations on CD spectra of Av nitrogenase complex

Indirect methods used to study the binding of MgATP, as well as MgADP, with the Fe-protein are EPR spectroscopy, circular dichroism, and fluorescence titrations (Cordewener et al. 1985). Zumft et al. (1973) concluded from EPR data that Cp2 bound two molecules of MgATP. McKenna et al. (1984) used circular dichroism to study the interactions of MgATP and MgADP with the Fe-protein and found that both Av2 and Kp2 had a high affinity for two molecules of MgADP, and probably also for two molecules of MgATP. You et al. (1983b, 1984) studied the binding of MgATP and MgADP to Av2 with the aid of fluorescence probes, fluorescein mercuric acetate and fluorescamine. Their data indicated the presence of two binding sites for MgATP and at least two for MgATP on Av2. Zeng and Burris (1986) used the reaction betweenthe Fe-protein and bathophenanthroline disulfonate as a probe for the interactions of the Fe-protein with MgATP, their results indicated two binding sites on Av2 and on Kp2. Song and You (1987) found from the CD spectra and fluorescence depolarization of Av nitrogenase complex that the absorbances increased with increasing MgATP/complex molar ratio, and that both the ΔCD and Δp appeared to reach saturation values when the MgATP/complex molar ratio was equal to 2, suggesting 2 binding sites for MgATP on nitrogenase complex. The results also suggested that in the enzyme turnover, the possibility of complexation of dithionite-reduced MoFe-protein with reduced Fe-protein followed by MgATP binding to the enzyme complex may also be considered, though in the absence of MgATP and in the presence of dithionite, complexation of the two enzyme-components is known to be very weak. It is interesting to note that, under anaerobic condition but in the absence of additional amount of dithionite, the 1:1 Av1–Av2 complex can be maintained practically undissociated, with characteristic UV-visible absorption spectrum and CD spectrum different from that of the individual components, indicating some change in the microenvironments of the 4Fe–4S cluster attending the complexation of the two components. Figure 1 shows the effect of MgATP concentration on the CD spectrum of the enzyme complex (Jiang 1981; Song and You 1987). Quantitative

Table 2. MgATP (t) or MgADP (d) binding to Fe-proteins: number of adenylate-binding sites and stoichiometric (macroscopic) dissociation constants

Species (Redox state)	No. of t or d-binding sites (Expt. method)	Dissociation constants mM	Reference
Av2 (red.)	2t (column gel filtration)	0.013–0.016	You et al. (1978)
Av2 (red.)	2t (BPS chelation)	$K_1 = 0.22$ $K_2 = 0.43$	Hageman et al. (1980)
Av2 (red.)	2t (BPS chelation)	$K_1 = 0.14$ $K_2 = 0.45$	Zeng et al. (1986)
Av2 (red.)	2t (flow dialysis)	$K_1 = 0.22$ $K_2 = 1.71 \pm 0.50$	Cordewener et al. (1985)
Av2 (ox.)	2t (flow dialysis)	$K_1 = 0.049 \pm 0.016$ $K_2 = 0.18 \pm 0.05$	Cordewener et al. (1985)
Av2 (red.)	2d (flow dialysis)	$K_1 = 0.091 \pm 0.021$ $K_2 = 0.044 \pm 0.009$	Cordewener et al. (1985)
Av2 (ox.)	2d (flow dialysis)	$K_1 = 0.024 \pm 0.015$ $K_2 = 0.039 \pm 0.022$	Cordewener et al. (1985)
Kp2 (red.)	2t (BPS chelation)	$K_1 = 0.048$ $K_2 = 0.217$	Zeng et al. (1986)
Cp2 (red.)	2t (BPS chelation)	0.085	Ljones et al. (1978)

results concerning the binding of MgATP to nitrogenase Fe-protein from three different species are listed in Table 2, together with the dissociation constants.

2.3 Effects of MgATP and MgADP Bindings on Redox Properties of Fe-Protein

Nitrogenase Fe-protein can be reversibly oxidized with redox dyes, or with MoFe-protein plus MgATP, to an EPR-silent state (Orme-Johnson 1985). The reduced Fe-protein is regarded as a one-electron donor (Mortenson and Thorneley 1979). This view is supported by potentiometric titration of isolated Fe-protein Av2, where the mid-point potential has been found to be -256 ± 20 mV (You and Gao 1987), or -270 mV (Zeng 1987), and the addition of MgATP to Fe-protein resulted in a lowering of the mid-point potential to -390 ± 20 mV, the data fitting very closely to a curve for $n = 1$ (You and Gao 1987) (Fig. 2). Watt (1985) found $Em = -310$ mV for Av2.

2.4 Accessibility of the Iron of the Fe-Protein after MgATP Binding

MgATP is specific for increasing the accessibility of the iron of Fe-protein to ferrous-iron chelators, 2,2′-dipyridyl or bathophenanthroline disulfonate (BPS), (Walker et al. 1973; You et al. 1979; Ljones et al. 1978; Zeng et al. 1986). Fe-protein with high activity is resistant to chelation by Fe(II)-chelators in the

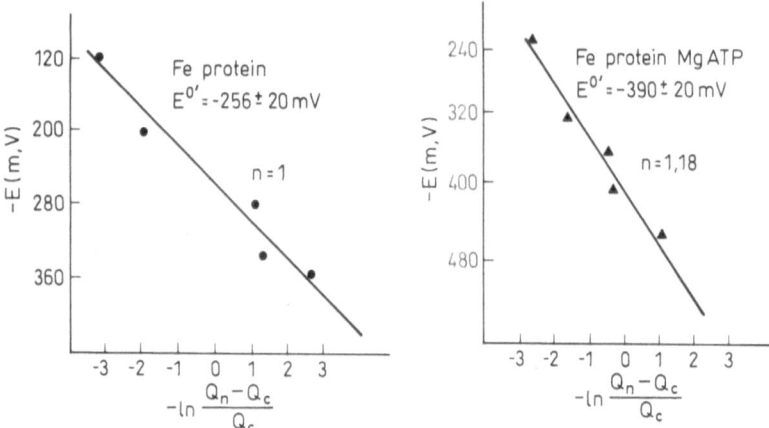

Fig. 2. Midpoint redox potentials of Fe-protein and Fe-protein-MgATP by using coulometric assay at fixed potential.

Q_c: electrocapillary maximum (ecm) of sample after equilibration in corresponding voltage (E); Q_o: total ecm of sample; Eo': potential at $(Q_o - Q_c)/Q_o = 1$. Temp. 25 °C, pH 8.0

absence of MgATP, and little iron is complexed even after 20. The addition of MgATP dramatically increases the rate of reaction between the chelating agent and Fe-protein; more than 75% of the iron reacts in the first hour, and over 97% within 20. No effect was seen with Mg^{2+} or ATP alone (Walker et al. 1973; You et al. 1979). A Hill plot (Fig. 3) of $\log[v/(v_{max} - v)]$ as a function of log(MgATP conc'n) gave a straight line with a slope of about 2.0 Iron in

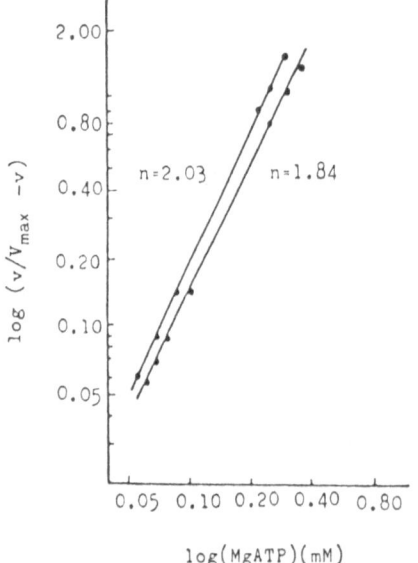

Fig. 3. Hill plot of $\log[v/(V_{max} - v)]$ as a function of log(MgATP) (mM).
1. Reaction time: 60 min
2. Reaction time: 30 min

oxygen-damaged Fe-protein was found to react immediately with 2,2'-dipyridyl, suggesting the disruption of the thiolate ligands by oxygen.

2.5 Effect of MgATP Binding on Sulfhydryl-Group Reactivity

Studies on the reactivity of the amino acid residues of Fe-protein as affected by MgATP binding to the Fe-protein have been made. Iodoacetate was employed to determine to what extent the carboxymethylation of the amino acid residues would vary in the presence or absence of MgATP. We found that 3.84 of the 12 sulfhydryl groups of native Fe-protein dimer were alkylated during a short-term (5–10 min) incubation with indoacetate. The extent of carboxymethylation decreased with increasing amount of MgATP until the ratio of MgATP to Fe-protein became to 2:1. When the Fe-protein was preincubated with this saturated amount of MgATP, only 1.78 sulfhydryl groups were alkylated in a short-term incubation, indicating that 2 sulfhydryl groups per dimer were protected or blockaded by the binding of MgATP to Fe-protein (Fig. 4). These results also suggested the possibility of competition between MgATP and iodoacetate for the same cysteine residues on Fe-protein. These results have also been confirmed by using the sulfhydryl agent PCMB. When the native Fe-protein was preincubated with MgATP for 5–10 min and then incubated with PCMB, or vice versa, same amount of cysteine residues were alkylated by iodoacetate as shown in Table 3 (You et al. 1978, 1979).

Note that Hausinger & Howard (1985) recently made an extensive study of the effects of MgATP and MgADP bindings on the alkylation of the cysteinyl residues of Av2 with iodo[2-^{14}C] acetic or acetamide. They identified the labeled residues from the known sequence data and found that cysteine residues 97 and 132 from each subunit were several times more rapidly labeled in the

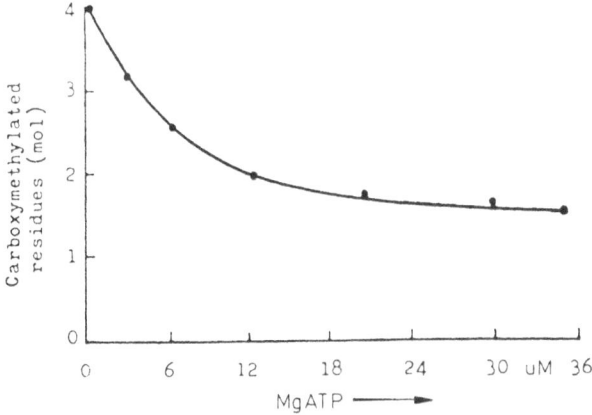

Fig. 4. Effect of MgATP concentration on number of cysteine residues of Av2 carboxymethylated during 5 min incubation with IAA

Table 3. Effect of MgATP or/and PCMB on carboxymethylation of cysteine residues of the Fe-protein Av2 by iodoacetamide

Fe-protein and treatments	Number of carbo-xymethylated Cys per mole Av2 in 5-min treatment
Fe-protein, native	4.11
Fe-protein + MgATP*	1.79
Fe-protein + PCMB**	1.68
Fe-protein + PCMB** (preincubated for 5 min) then + MgATP*	1.73
Fe-protein + MgATP* (preincubated for 5 min) then + PCMB**	1.75

* 10 µM Mg^{2+}, 30 µM ATP.   ** PCMB: Fe-protein = 1:1 (M/M)

presence of MgATP and 2,2′-dipyridyl, indicating that these might be *exo*-ligands for the 4Fe–4S center; they also observed that cysteine residues 85 from each subunit appeared to be protected to some extent (ca. 30%) by MgATP, but not by MgADP, indicating that Cys-85 might be associated with a MgATP-binding site. However, by incubating the Av2 with labeled iodoacetate and MgATP at the same time for a period of up to 30 min, they found little change in the overall level of alkylation of Cys-5, 38, 97, 132, 151, and 184 of each subunit.

2.6 Thermodynamic Parameters of MgATP and MgADP Binding to Av2

The standard enthalpies (ΔH^0) for MgATP binding and for MgADP binding to Av2 have been measured by means of batch microcalorimetry and found to be −4.78 kcal/mol and −5.45 kcal/mol, respectively (You et al. 1984). Their corresponding values of standard free energy changes (ΔG^0) have been estimated to be −6.66 kcal/mol and −6.89 kcal/mol from the dissociation constants (ca. 13.3 µM and 8.9 µM, respectively) of the Av2-adeninenucleotide complexes determined by means of column gel filtration technique (You et al. 1978). From these two sets of ΔH^0 and ΔG^0 values, the corresponding values of ΔS^0 have been estimated to be about +6.3 and +4.8 entropy units, respectively. However, the observed magnitudes of ΔH^0 and the dissociation constants might both be too small since the concentrations of MgATP and of MgADP used might not be high enough for nearly complete bindings in view of the more recent work by Cordewener et al. (1985). Re-evaluation of these important thermodynamic parameters seems desirable.

2.7 Effect of MgATP Binding on the Extent of Deactivation of the Fe-Protein by PCMB

Some insight into the role of ATP in the functioning nitrogenase complex was gained when it was found that the conformation of the Fe-protein changed when it bound MgATP (Zumft et al. 1973). The sigmoidal kinetics, the UV-visible and infrared spectra (You et al. 1978) and the sedimentation behavior of Fe-protein all undergo some changes when Fe-protein binds MgATP. These observations together with the information gained from other biochemical and biophysical methods, such as EPR spectroscopy, CD spectroscopy (Stephens et al. 1981), and physical label techniques (Likhtenstein 1988, and relevant literature cited therein) provide some basis for postulating about the modes of MgATP and MgADP bindings to nitrogenase Fe-protein and the location of MgATP-binding sites relative to the 4Fe–4S center of Fe-protein. However, proposals from different investigations are still at variance; this point will be discussed in more details later.

Here it may be pertinent to mention some information obtained from the studies on the activities of nitrogenase with Fe-protein treated in various ways with MgATP and PCMB (You et al. 1979). The results are shown in Table 4. When Fe-protein was preincubated with PCMB in 1:1 molar ratio for 5 min, and then incubated with MgATP in excess, the preparation was completely inactivated. However, when the order of treatments was reversed, the Fe-protein still retained some activity, indicating some protection of the nitrogenase-activity-related sulfhydryl group(s) or cysteine residue(s) by MgATP, which implied that the bulky MgATP might be bound near the active site of Fe-protein, thus shielding or blockading some of the activity-related sulfhydryl groups against the access of the strong sulfhydryl reagent PCMB.

Table 4. Effect of MgATP or/and PCMB on N_2-reduction activity of Fe-protein in nitrogenase-activity assay*

Fe-protein (total protein 0.578 mg/1.1 ml) and treatments	$^{15}N_2$ fixed %	specific activity (nm N2 reduced/min/mg protein)
Fe-protein (native) + MoFe-protein, 2:1(M/M)	5.78	262.6
Fe-protein (control expt.)	0	0
MoFe-protein (control expt.)	0	0
Fe-protein + PCMB** + MoFe-protein	0	0
Fe-protein + PCMB** (preincubated for 5 min.) then + MgATP + Mofe-protein	0	0
Fe-protein + MgATP (preincubated for 5 min.) then + PCMB** + MoFe-protein	1.32	47.1

* Reaction mixture: 5 µmol $MgCl_2$, 5 µmol ATP, 0.5 mg creatine kinase, 50 µmol creatine phosphate, 60 µmol tris-HCl buffer (pH 7.4), 30 µmol sodium dithionite, in a total volume of 1.1 ml. Fe-protein: MoFe-protein = 2:1 (M/M). Gas phase: 1 atm $^{15}N_2$ (99.36%)
** PCMB: Fe-protein = 1:1 (M/M); after incubation for 5 min, MoFe-protein was added

2.8 Interaction Between MoFe-Protein and Fe-Protein in the Presence or Absence of MgATP

Interaction between the two metallo-protein components in the presence or absence of MgATP has been studied by means of polarographic analysis according to Brdicka reaction for determining the amounts of -SH groups and disulfide bonds with fluorescamine as a probe (You et al. 1979, 1983b). Samples of Av1 and Av2 used in the experiments were purified the by known methods to specific activities of about 1600 and 1400, respectively. The results showed that the polarographic wave heights of the mixtures of two metallo-protein components, either in the presence or absence of MgATP, were somewhat lower than the sum of the polarographic wave heights of the individual components before mixing, indicating some interaction between the two components even in the absence of MgATP, and that some of the sulfhydryl groups were possibly blocked by the formation of the enzyme complex, no appreciable difference in effect being observed with varying amounts of MgATP. More significant decrease in the quenching of FMA and in the number of amino groups reactive with fluorescamine was observed after the two metallo-protein components were mixed (You et al. 1983b). Note that the interaction of the two metallo-protein components in the absence of MgATP was first observed by Orme-Johnson (1972) from EPR experiment.

The unique fast-reacting cysteine residues of Av1 and Av2 have been labeled selectively with fluorescent probes 1,5-BrAEDANS and 5-IAF, which render a particularly effective donor-acceptor pair in fluorescence energy-transfer studies. Steady-state fluorescence measurements have been carried out to demonstrate the presence of energy transfer between the bound donor Av2-AEDANS or Av2-MgATP-AEDANS, and acceptor Av1-AF chromophores in Av1–Av2 complex and Av1–Av2-MgATP complex, the minimum and maximum distances between the fluorophores of Av1 and Av2 have been estimated to be about 35 Å and 58 Å, while the corresponding minimum and maximum distance in Av1–Av2-MgATP complex appear to be somewhat larger (38 Å and 63 Å) (Gao and You 1987). The effect of MgATP-binding and that of MgADP-binding to Fe-protein (Av2) on increasing in Av2 the number of amino groups reactive with fluorescamine appear to be practically the same, from the two series of comparative experiments with varying proportions of adenylates to Av2 (You et al. 1983), indicating that the conformational changes caused by MgATP-binding and by MgADP-binding might be quite similar. The shielding of some of these amino groups reactive with fluorescamine due to complex formation between Av2-MgATP or Av2-MgADP with Av1 is in line with the proposal that the dimeric Fe-protein (with or without bound adenylates) might insert itself into a central pocket of the tetrameric Av1 to form a dense structure with practically no change in the particle size of the MoFe-protein (cf. Mortenson 1987).

2.9 Effect of MgATP Binding on Electron Transfer from Flavodoxin (Fld) to Fe-Protein

Azotobacter vinelandii flavodoxin (Fld) which can serve as electron donor for nitrogenase was isolated, purified, and characterized. The absorption spectra of purified Fld showed peaks at 450 and 600 nm in the semi-reduced state (Fld_{sr}), and at 375 and 455 nm in the oxidized state (Fld_{ox}), but gave no peak in the reduced state (Fld_r). The mid-point redox potentials (Em) of Fld_{ox}/Fld_{sr} and Fld_{sr}/Fld_r measured by coulometry at fixed potentials and calculated from Nernst's Equation were found to be $-280\,mV$ and $-500\,mV$, respectively. The mid-point potentials of four redox couples are calculated and given below:

$$Fld_r + Fe\text{-protein}_{ox} \overset{+e}{\underset{-e}{\rightleftharpoons}} Fld_{sr} + Fe\text{-protein}_r \qquad\qquad 244\,mV$$

$$Fld_r + Fe\text{-protein}_{ox} - (MgATP)_n \overset{+e}{\underset{e}{\rightleftharpoons}} Fld_{sr} + Fe\text{-protein}_r - (MgATP)_n \qquad 110\,mV$$

$$Fld_{sr} + Fe\text{-protein}_{ox} \overset{+e}{\underset{-e}{\rightleftharpoons}} Fld_{ox} + Fe\text{-protein}_r \qquad\qquad 24\,mV$$

$$Fld_{sr} + Fe\text{-protein}_{ox} - (MgATP)_n \overset{+e}{\underset{e}{\rightleftharpoons}} Fld_{ox} + Fe\text{-protein}_r - (MgATP)_n \qquad -110\,mV$$

Thus Fld_r can transfer electron to Fe-protein in the presence or absence of MgATP, and this has been confirmed by acetylene reduction activity; but Fld_{sr} cannot, in the presence of MgATP bound to Fe-protein$_{ox}$. In other words, the results suggest that electron transfer from Fld_r to Fe-protein is easier when MgATP is absent (You and Gao 1987).

2.10 Effects of the Ratio of Nitrogenase Components and of Operating Conditions on the ATP/2e-Ratio

It has been known that (as reviewed by Zumft and Mortenson 1975), in nitrogenase-catalyzed reduction of H^+, N_2 or other substrates, at room temperature and neutral pH, and with sufficient supplies of reductant (to be denoted by R; e.g., $S_2O_4^{2-}$ and in the in vitro experiments) and MgATP (to be denoted by t), about 2 to 2.5 molecules of t are hydrolyzed into MgADP (to be denoted by d) and inorganic phosphate (to be denoted by Pi, which at pH 7.4 is an approximately equimolar mixture of HPO_4^{2-} and $H_2PO_4^-$) for each electron transferred, indicating the coupling of electron transfer to the thermodynamically spontaneous ATP-hydrolysis. However, the $Pi/2e^-$ ratio appears to be varied over a wide range with experimental conditions; for example, at very low Fe-protein/MoFe-protein ratio, the steady-state $Pi/2e^-$ ratio may exceed 20, thus the ATP hydrolysis may not necessarily always be coupled to the electron transfer; while with Fe-protein/MoFe-protein ratio equal to 2:1 and with d/t

ratio in the range of 0.3 to 0.5, $Pi/2e^-$ ratio as low as 2 has been reported by one laboratory (Mortenson & Upchurch 1981).

A value of $Pi/2e^-$ lower than 4 would imply that, under the specific reaction conditions, electron back-flow was negligible, and that the reduced Fe-protein dimer bound with one t and one d (to be denoted by $[2_s^{td}]$, where the subscript s signifies that the species has characteristic EPR signals) might also be able to transfer an electron to the dithionite-reduced MoFe-protein (to be denoted by $[1_s]$), converting the latter into the fully reduced and EPR-silent state, $[1_0]$, with concomitant hydrolysis of one ATP, a possibility which has been considered before (Nitrogen Fixation Research Group, Xiamen University 1982; Cai (Tsai) et al. 1987).

Working in Orme-Johnson's laboratory, using $Av2/Av1 = 10/1, 8/1$, or $2/1$, and with the addition of a fixed amount of creatine phosphate and carefully adjusted small amounts of creatine kinase to regenerate the ATP just about as fast as it was hydrolyzed by nitrogenase catalysis, so as to maintain the selected values of d/t ratio approximately constant for about 20 min after the presteady-state period, and indirectly evaluating the amount of ATP hydrolyzed from the amount of creatine liberated corrected for $\mp \Delta d$, or $\pm \Delta t$ in a steady-state interval of time, Wan (1987) verified that there seemed to be a minimum $Pi/2e^-$ ratio when d/t ratio was about 0.3 to 0.5, as observed by Mortenson and Upchurch (1981); but the minimum $Pi/2e^-$ ratio was found to approach 4, rather than 2. However, there were some differences in the experimental conditions. Appropriate reaction conditions for exploring the possibility of $Pi/2e^-$ ratio lower than 4 has been suggested (NiFRG, Xiamen Univ. 1982; Wan 1987).

It has now been generally agreed that Fe-protein in the reduced or oxidized state can sequentially and reversibly bind $2t$ or $2d$ to form $[2_s^{tt}]$, $[2_o^{tt}]$, or $[2_s^{dd}]$, $[2_o^{dd}]$, by our notations, and that MoFe-protein in the dithionite-reduced state shows very little or no affinity for binding t (or d?). More recently, however, Miller and Eady (1988) observed tight binding interactions of dye-oxidized MoFe-protein and VFe-protein with adenine nucleotides, t and d, from non-equilibrium binding studies. Although the results obtained with dye-oxidized nitrogenase components may not be relevant to the steady-state enzyme-turnover, they may reflect some significance in the pre-steady-state burst of ATP hydrolysis, and obviously also in the ATPase activity of dye-oxidized nitrogenase recently reported by Wassink et al. (1988), who observed that dye-oxidized nitrogenase displayed a pre-steady-state ATPase activity indistinguishable from that of the reduced enzyme. As to be discussed later, these new findings seem to be imply that, though ATP hydrolysis does not occur readily when t binds to Fe-protein, it may occur readily when $[2_s^{tt}]$, $[2_o^{tt}]$, or even $[2_s^{td}]$, $[2_o^{td}]$, further complexes with MoFe-protein, or VFe-protein, regardless of the redox states of the enzyme components, or the presence or absence of electron transfer.

3 Model Studies on Adenine Nucleotides Complexation with Synthetic 4Fe–4S Cluster Complexes and a Proposed Mechanism of ATP-Driven Electron Transfer

3.1 Chemical Modelling of ATP-Driven Electron Transfer in Nitrogenase Catalysis

In the early stage of our studies on this topic, we started with the assumptions that (1) two molecules of MgATP might coordinate directly to the 4Fe–4S center of a reduced dimeric Fe-protein molecule, thereby increasing the ligand field acting on the 4Fe–4S to drive the transfer of an electorn from the Fe-protein to the semi-reduced (i.e., dithionite-reduced) MoFe-protein in the enzyme complex, resulting in the formation of the fully-reduced MoFe-protein and the oxidized Fe-protein with concomitant hydrolysis of an ATP molecule; (2) the huge Fe_nS_n system of the MoFe-protein might serve as an electron reservoir or condenser in some sort of electronic contacts with the Fe-protein on the one hand and with the Mo- and Fe-containing active-center on the other; (3) a second step of ATP-driven electron transfer might be necessary with the remaining MgATP molecule shifting to coordinate to the Mo- and Fe-containing active-center to drive the electron over to the coordinated substrate molecule with concomitant hydrolysis of the second ATP; and (4) excessive consumption of ATP beyond $ATP/2e^- = 4$ might be due to the electron back-flow from the fully reduced MoFe-protein to an oxidized Fe-protein (Nitrogen Fixation Research Group, Xiamen Univ. 1974, 1976, 1982; Tsai 1980). Though the assumption of 2-step ATP hydrolysis was later modified to one step in view of the published work of Hageman et al. (1980), the idea that MgATP coordination to Fe-protein might provide the driving force necessary for the electron transfer remains essentially unchanged, based upon the following simple reasoning: the 4Fe–4S center with 4 thiolate *exo*-ligands of Fe-protein in the reduced or oxidized state is coordinationally unsaturated, the total number of electron in the coordination spheres of the 4 Fe being formally 67, or 66 for the reduced, or the oxidized state, respectively, if each electron-pair bonding 2 Fe is counted twice (Cai et al. 1987); and should be able to coordinate $2t$ or $2d$ sequentially, phosphate and polyphosphate ions being known to have high affinity for coordinating ferric and ferrous ions; the resulting increase in ligand field should be able to produce enough negative shift of redox potential to drive the electron transfer from the redox-couple $[4Fe–4S]^+/[4Fe–4S]^{++}$ of the Fe-protein to the redox core-couple of the P clusters supported to be four $[4Fe–4S]^0/[4Fe–4S]^+$, which normally would be at least 200 mV more negative than the former redox core-couple with only the 4 *exo*-thiolate ligands; finally, complexing of a triphosphate molecule to VO^{3+} or Mn^{3+} is known to greatly enhance the rate of hydrolysis of the triphosphate anhydride-linkages (Haight et al. 1980, 1987), this

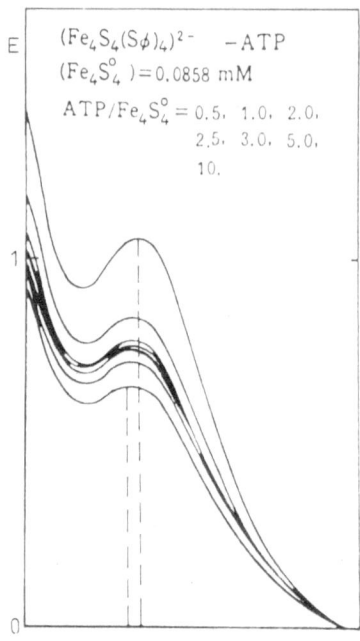

Fig. 5. Suppression of electronic absorption peak (at 458 nm) of $[Fe_4S_4(SPh)_4]^{2-}$ in DMF-H_2O by varying proportions of ATP at pH 7.0

provides a ready explanation for the ATP hydrolysis concomitant with MgATP coordination and electron transfer in nitrogenase catalysis.

In order to test this reasoning, a number of chemical modelling experiments was carried out with synthetic 4Fe–4S cluster complexes, $(NR_4)_2Fe_4S_4(SPh)_4$ (R = Et, n-Bu, or Me), prepared according to the known methods (Holm et al. 1973; Christou and Garner 1979) and purified ATP, ADP, and AMP (m) (Chen et al. 1985; Wu et al. 1988).

Chen et al. (1985) observed that the electronic absorption peak at 458 nm of $[Fe_4S_4(SPh)_4]^{2-}$ in DMF-H_2O (3:1 v/v) was depressed by the addition of t (or ATP) to an extent of about 50% when the concentrations of the cluster complex and ATP were, respectively, 0.098 mM and 0.98 mM. Suppression of the 458-nm absorption peak by the addition of similarly varying amounts of ADP or Pi was appreciably smaller, but much larger than that produced by the addition of AMP. (For the effect of t, see Fig. 5). They also studied the effects of ATP on the cyclovoltametry and polarography of $[Fe_4S_4(SPh)_4]^{2-}$, and found that the addition of ATP to the cluster complex in DMF-H_2O (85:15 v/v) at pH 8.0 and room temperature suppressed the polarographic half-wave at -1.00 V and gave a new half-wave at -1.49 V (Fig. 6); i.e., complexation of the cubane-like cluster anion with ATP appeared to shift the redox potential of the former to more negative value by -0.49V. This was found to promote electron transfer from the cluster anion to indigo carmine or methylene blue in DMF-H_2O at neutral pH and room temperature; the extents of rate-enhancement of the redox reaction produced by the addition of Pi and

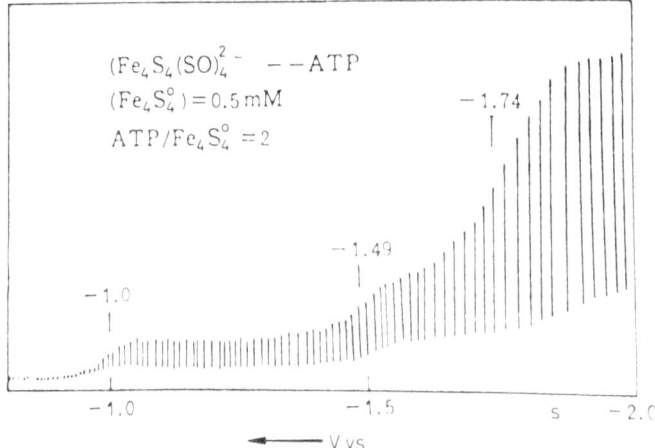

Fig. 6. Shift of polarographic half-wave of $[Fe_4S_4(SPh)_4]^{2-/3-}$ in DMF–H_2O (85:15 v/v) due to addition of ATP, or MgATP at pH 7.5

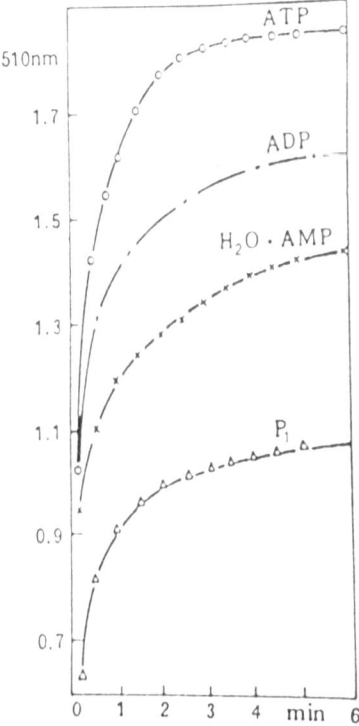

Fig. 7. Promotion of redox reaction between $[Fe_4S_4\cdot(SPh)_4]^{2-}$ and indigo carmine by ATP (ADP, pi, AMP) in DMF–H_2O (4:1 v/v, Tris-HCl pH 7.0)

ADP were appreciably smaller, and that produced by the addition of AMP was very small (Fig. 7). These results indicate that the extents of complexation of $[Fe_4S_4(SPh)_4]^{2-}$ with the three adenine nucleotides decrease in the order ATP > ADP ≫ AMP, apparently in the order of increasing steric hindrance due to increasing proximity of the bulky adenyl group with the *exo*-thiolate ligands in passing from ATP to ADP and to AMP if these nucleotides all coordinate through the terminal phosphate groups (i.e., the α-, β-, and γ-phosphate groups, respectively) to the cluster anion in DMF-H_2O at neutral pH. Using extraction and silver dithiazon test for the detection of any free HSPh, Wu et al. (1988) and Chen et al. (1985) demonstrated that complexation of the cluster anion with ATP (ADP, or AMP, with which the complexation is apparently almost ineffective) did not lead to the displacement of any of the *exo*-thiolate ligands under the experimental conditions.

Wu et al. (1988) found that treatment of ATP with $[Fe_4S_4(SPh)_4]^{2-}$ in DMf-H_2O (4:1 v/v) caused the ^{31}P-NMR peaks of the α-, β-, and γ-PO_3 of ATP to shift downfield by about 8.2 ppm, 7.9 ppm, and 10 ppm, respectively, with the γ-^{31}P-NMR peak almost completely broadened and buried in the noise,

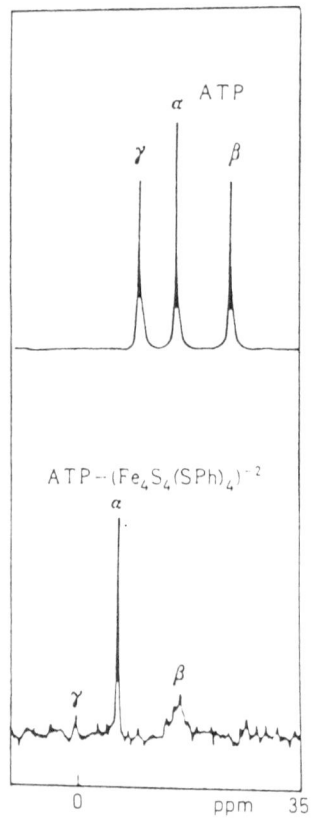

Fig. 8. Downfield shift of ^{31}P-NMR peaks of ATP in DMF-D_2O (4:1 v/v, pH 7.0); Na_2HPO_4 in D_2O as external standard

and the β-^{31}P-NMR peak considerably broadened, while the α-^{31}P-NMR peak remaining narrow and sharp (Fig. 8). The experiment was done in the following way: A mixture of 4.0 ml 15-mM$(Et_4N)_2Fe_4S_4(SPh)_4$ in DMF and 1.0 ml 150-mM ATP in D_2O (adjusted to pH 7.0 with dilute NaOH in D_2O) was shaken and allowed to stand in thoroughly deoxygenated argon atmosphere for 20 min, then centrifuged to remove the excess ATP which remained largely undisolved in the DMF-D_2O (4:1 v/v) medium. The ^{31}P-NMR Spectrum of the supernatant liquid was taken and compared with that of 20 mM ATP in D_2O with and without an equivalent amount of HSC_2H_4OH; Na_2HPO_4 in D_2O was used as external standard and the ^{31}P-NMR peaks of the α-, β-, and γ-PO_3 of ATP were labelled according to the literature (Mortenson and Upchurch 1981). The downfield shiftings of the three ^{31}P-NMR peaks thus observed together with the large broadening of the γ-PO_3 ^{31}P-NMR peak and the virtually unbroadening of the α-PO_3 ^{31}P-NMR peak appear to be quite similar to the downfield shifts of the α-, β-, and γ-PO_3 ^{31}P-NMR peaks of MgATP (by about 8.7 ppm, 9.0 ppm, and 7.7 ppm, respectively), the large broadening and suppression of γ-PO_3 ^{31}P-NMR peak and the virtually unbroadened α-PO_3 ^{31}P-NMR peak caused by the addition of the reduced Fe-protein to MgATP in D_2O, as observed by Mortenson and Upchurch (1981); and this argues strongly that the downfield shiftings and broadenings of the γ- and β-^{31}P-NMR peaks caused by the addition of reduced Fe-protein to MgATP may also be due to interactions of the ^{31}P nuclear spin with the highly delocalized frontier d-electrons of the 4Fe–4S center (with at least two co-existing spin states, cf. Orme-Johnson 1985) of the reduced Fe-protein. The very large broadening and apparent suppression of the γ-PO_3 ^{31}P-NMR peak caused by the addition of the cluster anion $[Fe_4S_4(SPh)_4]^{2-}$ to ATP in DMF-D_2O (4:1 v/v) suggests once again that the coordination of ATP in this model system might be predominantly through the $\gamma^- PO_3$ only.

The effect of ATP (ADP, AMP, or Pi) on the rate of disruption of $[Fe_4S_4(SPh)_4]^{2-}$ by phenanthroline(phen) in DMF-H_2O was also investigated (Wu et al. 1988). It was found that, in DMF-H_2O (3:2 v/v, Tris HCl 25 mM, pH 7.5) and at room temperature, the presence of ATP (1.0 mM) actually accelerated, rather than retarded, the disruption of $[Fe_4S_4(SPh)_4]^{2-}$ (0.10 mM) by phen (about 1.7 mM, introduced as 10-mM DMF solution), as shown by the relative rates of development of the electronic absorption at 510 nm (Fig. 9). Under similar conditions, ADP also accelerated the disruption, though to a smaller extent; AMP showed practically no effect; while Pi appeared to retard the disruption of the cluster anion by phen. In the case of acceleration by ATP, the disruption of the cluster anion by phen appeared to be completed in less than 10 min. A plausible explantation of this acceleration is as follows: The complexation of one or two molecules of ATP (or ADP) with the cluster anion is fast, sequential, and reversible; there may be considerable proportion of 1:1 complex in dynamic equilibrium with other species. Since ATP has much higher affinity for ferric iron than for ferrous iron (similar to the high affinity of phosphate or pyrophosphate for ferric iron), it will tend to delocalize the ferric irons in the cluster anion, making the ferrous irons (the 2 Fe^{II}

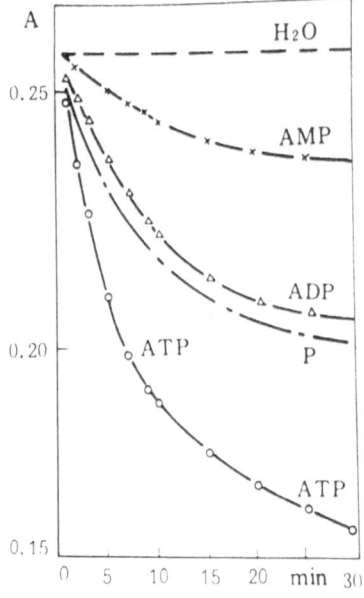

Fig. 9. Sensitization of $[Fe_4S_4(SPh)_4]^2$ cluster disruption by phen due to addition of ATP, or ADP in DMF–H$_2$O (2:1 v/v, Tris-HCl pH 7.5). (Et$_4$N$_2$) $[Fe_4S_4(SPh)_4]$ 125 μM, ATP 1.2 mM. (phen: 0.3 ml of 10-mM phen)

uncomplexed by ATP) more susceptible to chelation by phen, or 2,2'-dipyridyl. However, the effect of ATP (or ADP) appeared to vary with the composition of the solvent media and with the redox state of the cluster anion $[Fe_4S_4(SPh)_4]^{n-}$. For DMF-H$_2$O system with high proportion of DMF, the effect of ATP (or ADP) on the rate disruption of $[Fe_4S_4(SPh)_4]^-$ by phen became one of retarding; and in the case of $[Fe_4S_4(SPh)_4]^-$ in DMF-H$_2$O (3:2 v/v, pH 7.5), ATP (ADP) also appeared to retard the chelation by phen (YH Wu, unpublished results). Further investigation is desirable.

The extent of ATP hydrolysis in relation to electron transfer from the cluster anion $[Fe_4S_4(SPh)_4]^{2-}$ to redox dyes or hydrogen peroxide was also investigated (Wu et al. 1988). No promotion of ATP hydrolyze attending the oxidation of the cluster anion by indigo carmine or methylene blue was detected by determining the Pi liberated by means of a modification of Baginski's (1967) molybdenum blue method (after removal of methylene blue by extraction as chloroform-soluble M.B.$^+$ ClO$_4^-$ ion-pair) and comparing with a reference sample (blank) containing the same amounts of ATP and the cluster anion, but without the redox dye. Small extents of ATP hydrolysis were observed, however, by oxidizing the cluster anion $[Fe_4S_4(SPh)_4]^{2-}$ with hydrogen peroxide in the presence of ATP in DMF-H$_2$O (3:2 v/v) and the extent of ATP hydrolysis was found to decrease with increasing proportion of DMF in the mixed solvents. Thus with DMF-H$_2$O (3:2 v/v), containing 30 μmol of (Et$_4$N)$_2$Fe$_4$S$_4$(SPh)$_4$, 30 μmol of ATP, and 60 μmol of H$_2$O$_2$ in 3.0 ml of solution, the amount of Pi liberated after correction for blank (1.26 μmol Pi liberated in reference sample containing no H$_2$O$_2$) was 2.44 μmol, i.e., ca. 8% of the total amount of ATP;

while with the same amounts of the cluster complex, ATP, and H_2O_2 as above, but with DMF-H_2O (4:1 v/v), only about 4% of the total amount of ATP (30 μmol) was hydrolyzed after correction for blank. Before the determination of Pi in each case by the modified molybdenum blue method, the cluster anions were removed as voluminous precipitate of $(Bu_4N)_2Fe_4S_4(SPh)_4$ by the addition of 0.30 ml of 1.2-M n-Bu_4NI in methanol and 10 ml of water. For detailed procedures, see Wu et al. (1988). The very small extent of ATP hydrolysis thus observed once again indicates that, in DMF-H_2O, neutral pH and room temperature, ATP might coordinate to the cluster anion predominantly through the terminal γ-PO_3 only, rather than through both the γ-PO_3 and the β-PO_3 by bridging chelation, which might render the anhydride linkage between them more susceptible to hydrolysis. The slight increase in the extent of ATP hydrolysis (which was quite reproducibly observed, as reported by Wu et al. (1988) in the presence of H_2O_2, as compared with blank, indicates that the cluster anion in higher oxidation state, presumably $[Fe_4S_4(SPh)_4]^-$, most probably has higher affinity for ATP (or other polyphosphates), so there might be a slightly higher tendency for ATP to coordinate to this cluster anion via the above-metioned bridging-chelation mode even in the presence of the mixed-solvent DMF-H_2O. Note that Haight (1987) and his coworkers have made a systematic investigation on the effects of metal ions on the hydrolysis of the normally sturdy P-O-P anhydride-linkages of ATP and other polyphosphates; their results demonstrate that chelation, or bridging-chelation, of ATP or an inorganic triphosphate to certain multivalent cations (e.g., VO^{III}, or Co^{III}) together with simultaneous binding of the terminal γ-PO_3 to a metal ion (e.g., another Co^{III} carrying an OH ligand, or a hydrated Mg^{2+} in the case of certain enzyme systems quoted by Haight 1987) can greatly enhance the rate of hydrolysis of the anhydride linkage even in the absence of redox reaction.

3.2 The Probable Location of the ATP-Binding Sites on Fe-Protein and the Probable Modes of the ATP-Binding Before and After Formation of the Enzyme Complex

Though it is generally agreed that the 4Fe–4S center of Fe-protein is bound symmetrically between the two identical subunits through exothiolate ligands (Cys-residues 97 and 132 from each subunit, according to Hausinger and Howard 1983), the probable location of the MgATP-binding sites is still a matter of dispute. Some investigators hold the view that these sites may be directly adjacent to the 4Fe–4S center at distances suitable for chemical interaction (Ljones Burris 1978; Likhtenstein 1988), while some other investigators maintain that these sites are quite remote from the 4Fe–4S center (Mortenson, Thornely 1979). More recently, Mortenson (1987) commented that recent results obtained by means of Spin Echo and Linear Field Analysis techniques indicate no difference in the ready accessibility of $H_2O(D_2O)$ to the Fe-protein 4Fe–4S center, no detectable interaction between the ^{31}P-NMR of MgATP and the electronic spins of the 4Fe–4S center after MgATP binding; thus it was concluded that the MgATP-binding sites must be more than 5Å away from the 4Fe–4S center, and that the shift in the mid-point potential of Fe-protein due to nucleotides binding may result from the effects other than polarization of the 4Fe–4S center. He also suggested substantial structural rearrangement of the Fe-protein attending the MgATP binding during which the 4Fe–4S center might become exposed, and that in the enzyme complex the two MgATP molecules might be bound close to the MoFe-center while the Fe-protein 4Fe–4S center might be close to the P clusters so that an ATP-hydrolyzing unit might be formed which might include the MgATP-binding sites of both components and at least two molecules of H_2O carried into the enzyme complex as water of hydration of the 2 MgATP, and that coupled electron- and proton-transfers might occur concomitant with the ATP hydrolysis. Furthermore, the specific sensitizing effect of MgATP for the disruption of the 4Fe–4S center of Fe-protein by Fe^{II} chelators and the inhibition of this sensitizing effect by competitive MgADP binding have been regarded by some investigaters as a strong argument for the dramatic conformational change by MgATP binding at sites remote from the Fe-protein 4Fe–4S center. However, all information derived from the quite similar effects produced by MgATP binding and by MgADP binding to Fe-protein on the changes in CD spectra, EPR spectra, redox potentials, in the extents of FMA fluorescence quenching and chemical modification by iodoacetate, as mentioned above, indicate close interaction between the Fe-protein 4Fe–4S center and the adenine nucleotides, as well as no radical difference in conformational changes caused by MgATP binding and by MgADP binding. Thus a rationalization of interpretations based on experimental facts from all sources is desirable.

In fact, enough informations are now available to allow such rationalized interpretations to be made regarding the probable location and modes of MgATP- and MgADP-bindings to the Fe-protein before and after formation of the enzyme complex.

Consider first the effects of 2t-binding and 2d-binding on the accessibility of the Fe-protein 4Fe–4S center to Fe^{II} chelators and other molecules. If 2t or 2d bind symmetrically to 2 equivalent sites at some distances from 2 opposite faces of the 4Fe–4S cubane-like cluster with 2 opposite faces held symmetrically between 2 identical subunits of the Fe-protein. In the case of 2t binding with the formation of $[2_s^{tt}]$, the gap between the two subunits may be widened to such an extent that, the remaining two opposite faces of the cubane-like cluster may become accessible to the Fe^{II} chelators, and while the Fe^{II}-chelator molecules are approaching the two faces, one or both of the 2t in $[2_s^{tt}]$ may be dissociating and moving away from the other two faces to make more room for the attacking Fe^{II}-chelator, since the binding of 2t is supposed to be sequential, fast, and reversible; while in the case of 2d binding with the formation of $[2_s^{dd}]$, the gap between the two subunits may not be wide enough to accommodate the Fe^{II}-chelator, especially the more bulky and less pliable BPS molecules. So MgADP may act as a competitive inhibitor to the specific sensitization effect of MgATP. However, with smaller reagent molecules, such as iodoacetamide and fluorescamine, the gap between the two subunits of the Fe-protein bound with $2d$ may also be wide enough to accommodate these reagent molecules; so the effect of MgADP appears to be quite similar to that of MgATP in the fluorescence-quenching experiment, as well as in the sulfhydryl alkylation experiment with the exception of certain-SH groups (Cys residue 85 of each subunit, according to Hausinger & Howard 1983), which appear to be partially protected by MgATP. With small molecules like H_2O, D_2O, the gap between the two subunits of Fe-protein without t or d bound may be already wide enough for the approaching of these molecules to the 4Fe–4S center.

The downfield shifts of the $\alpha^{-31}P$, $\beta^{-31}P$, and $\gamma^{-31}P$-NMR peaks in the order of 9 ppm when t (1 mM) was mixed with $[2_s]$ (1 mM) as observed by Mortenson and Upchurch (1981), the similar downfield shifts of the three ^{31}P-NMR peaks of ATP when it was treated with the synthetic cluster complex $[Et_4N]_2Fe_4S_4(SPh)_4$ (12 mM in DMF-D_2O, 4:1 v/v) and the absence of any effect HSC_2H_4OH on the NMR spectrum of ATP in D_2O, as observed by Wu et al. (1988), suggest that the interaction between MgATP and Fe-protein may be mainly a dipole-charge one in nature, rather than chemical bonding between the nucleotide and the sulfphydryl groups of the Fe-protein. Both the γ-P=O (phosphonyl) and the β-P=O dipoles of t in $[2_s^t]$, or of each t in $[2_s^{tt}]$ may interact with the effective core charge of $[4Fe-4S]^+$, resulting in the lowering of electron density around each ^{31}P. In the case of $[2_s^{tt}]$, these phosphonyl dipoles may be symmetrically located so that a near 4 symmetry axis results passing through the center of the 4Fe–4S cluster, application of linear electric field may produce practically no net effect on the dipole-charge interaction energy. Suppose each of the γ-P=O and β-P=O is so located and oriented that the phosphonyl oxygen is about 3.5–4.0 Å away from the nearest $Fe^{II,III}$ of the 4Fe–4S center, with the P=O — Fe angle of about 120°; then the center of each P=O dipole would be about 5.0–5.5 Å away from the center of the 4Fe–4S cluster, and the orientation of the P=O dipole would make an angle of about 110° with the line joining the two centers; the ^{31}P nucleus of the P=O would be about 5.5–6.0 Å away from the center of the 4Fe–4S cluster, just beyond the detection limit of the Spin Echo technique. The dipole-charge interaction is an interaction of much longer range. With the effective core-charge(q) of the $[4Fe-4S]^+$ of $[2_s^{tt}]$ taken to be about $+0.6$ and the effective dipole moment (μ) of each of the 4P=O dipoles (regarded as point dipoles) taken to be about 2.0 debyes, the diople-charge interaction energy ($4.q.\mu.\cos\theta/r^2$) is readily estimated to be about -0.32 to -0.40 erg/molecule, or -0.20 to -0.24 eV, or -4.5 to -5.6 kcal/mol. This is the estimated *net* electrostatic interaction corresponding to the simplified model and the roughly estimated effective core-charge and P=O dipole moment. With this model, the potential energy of an electron in the frontier orbital of the 4Fe–4S cluster would be raised by about $+0.24$ to $+0.30$ eV due to the interaction with the 4P=O dipoles, as estimated from the same formulas; this would correspond to a shift of about

-240 to $-300\,mV$ in the "intrinsic" redox potential or ionization potential of the 4Fe–4S center of the Fe-protein due to the binding of $2t$. Some correction may have to be made for any slight readjustment of the bond angle between each exo thiolate ligand and an adjoining edge of the 4Fe–4S cubane-like cluster. With only one t bound, i.e., in the case of $[2'_s]$, the magnitudes of dipole-charge interaction energy and the negative shift in intrinsic redox potential would be about half of the above estimated values. It is known that classical point-change-point-dipole models some times work fairly well for the long-range electrostatic interactions.

By binding with $2t$ or $2d$, the macroscopic midpoint potential (Em) of Av2 is lowered from about $-310\,mV$ to about $-450\,mV$, respectively (Watt 1985); that of Kp2 from about $-200\,mV$ to about $-320\,mV$ or $-350\,mV$, respectively (Thorneley & Deistung 1988), indicating that Fe-protein in the oxidized state, $[2_o]$, on account of its larger core-charge $[4Fe–4S]^{2+}$, binds $2t$ or $2d$ much more tightly than Fe-protein in the reduced state, $[2_s]$, with smaller core-charge $[4Fe–4S]^+$; thus in the presence of sufficient t or d, the concentration ratio $[2_s]/[2_o]$ between the reduced state and the oxidized state of Fe-protein in the Nernst Equation is changed considerably (ca. 100 fold). But even in the presence of t and d, the physiological electron carrier (e.g., an appropriate ferredoxin or flavodoxin) will have no difficulty in transferring an electron to the nucleotide-bound Fe-protein. The real challenge to the enzyme is to promote the transfer of an electron from the Fe-protein operating between the core states $[4Fe–4S]^+/[4Fe–4S]^{2+}$ to the P-clusters (presumably operating between the core states $[4Fe–4S]^0/[4Fe–4S]^+$, cf. Orme-Johnson 1985) with Em at least $-250\,mV$ more negative than that of the Fe-protein without the bound nucleotides, and to the M-cluster(s) to convert it from the semi-reduced state, $[1_s]$, to the fully reduced state, $[1_o]$ and in a subsequent cycle to the substrate-reducing "super-reduced state", a task which neither the dithionite, nor even a strongly reducing viologen with Em as low as $-640\,mv$ is able to accomplish.

Although dye-oxidized MoFe-protein shows some affinity for MgATP, under in-vitro operating conditions in the presence of dithionite, the MoFe-protein is either in the fully reduced state, $[1_o]$, or in the semi-reduced state, $[1_s]$, which is known to show parctically no tendency to bind MgATP. Let alone this experimental fact, a shift of $2t$ in the enzyme complex to bind the P-clusters or the M-cluster(s) would be working in the wrong direction, for it is an electron in the Fe-protein $[2''_s]$ which needs and additional driving force to drive it over to the MoFe-protein. Obviously, there should be some change in the mode of binding of the MgATP attending the formation of the enzyme complex, and a probable mode of MgATP binding in *the enzyme complex* should provide and adequate explanation both for the increase in the driving force for the protein–protein or protein–protein-substrate electron- and proton-transfers and for the cocomitant ATP hydrolysis. Now, according to the mechanistic model of metal-ions promoted hydrolysis of ATP (or other polyphosphates) depicted by Haight (1987), it requires a metal ion to bridge the γ- and

β-PO$_3$ and another hydrated metal ion to bind the terminal PO$_3$ while an OH-
or H$_2$O ligand on this hydrated metal ion is in the right position for nucleophilic
attach on the terminal PO$_3$ to promote the hydrolysis of the anhydride linkage
between the γ- and β-PO$_3$. So it is postulated that, with the formation of the
enzyme complex, say, [1$_s$] [2$_s''$], some conformational change of the Fe-protein
may take place, which allow the steric constraint imposed upon the bulky
adenosine groups of the 2t to relax a little so that the 4P=O dipoles can now
move closer to the Fe-protein 4Fe–4S center by about 1.3 to 1.8 A, and the
resulting dipole field acting on the electron becomes considerably larger (roughly
by a factor of about (5.0 Å)^2/(3.7 Å)^2; i.e., about double the original value). The
molecular, or intrinsic redox potential of [2$_s''$] should now be quite negative
enough to drive the protein–protein electron-transfer. Since the 4P=O oxygen
are now directly coordinated to the 4Fe^{II,III}, with the length of each Fe–O
dative-bond of about 2.2 Å, each of the 2t now becomes a bridging chelate-ligand,
spanning 2Fe^{II,III} about 2.7 Å apart. The precise mode of binding of the hydrated
Mg^{2+} to the ATP of each coordinated t is not known, but the β-PO$_3$ most
probably takes part in the binding (since after the ATP hydrolysis, the resulting
MgADP is bound to the physiologically oxidized Fe-protein and is not easily
released until the latter is replenished with an electron). The γ-phosphate group
is probably also bound to the hydrated Mg^{2+} ion, making this a bridging
cation, the H$_2$O ligand on the hydrated Mg^{2+} may not be in the right position
for the nucleophilic (S$_N$2) attack at the γ-phosphate group. According to
Mortenson (1987), the iron sites of an isolated Fe-protein dimeric molecule are
readily accessible to H$_2$O (D$_2$O) molecules in the presence or absence of bound

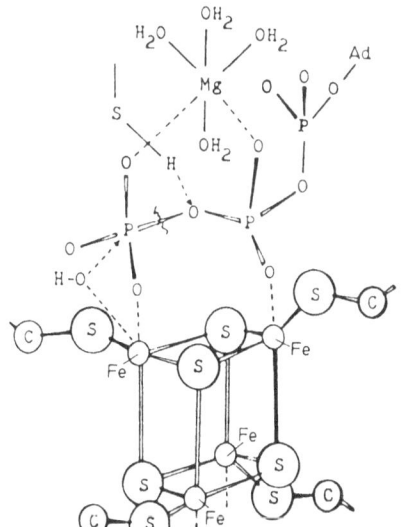

Fig. 10. Probable modes of coordination of
MgATP and MgADP on the 4Fe–4S center of
Fe-protein *in nitrogenase complex*. **a** MgATP as
a bridging bidentate ligand susceptible to nucleophilic S$_N$2 attack at the Γ-phosphate by OH or
H$_2$O ligand (on Mg^{2+}) and concerted protonic
attack at the anhydride oxygen. **b** MgADP as a
bridging bidentate ligand through the β-PO$_3$H
alone

adenylates, So it may be assumed that, when the γ-P–O is moving closer to the iron site for direct coordination during the formation of the enzyme complex, a loosely bound H_2O ligand may be displaced from the iron site, and on its way for nucleophilic attack at the γ-PO_3, this H_2O ligand may transfer a proton to the protein micro-environment and become an OH-ligand (Fig. 10a); concertedly, the anhydride oxygen may pick up a protonic hydrogen from the protein miroenvironment (probably from Cys-85?), thus leading to ready hydrolysis of the anhydride linkage of MgATP, regardless of whether there is electron transfer or no. The protonic hydrogen thus attached to the β-PO_3 may attack the γ-$HOPO_3$–Mg^{2+}, resulting in the formation of $(\gamma HOPO_3H)^-$, (as a *Pi*) and the binding of the Mg^{2+} entirely to the β-PO_3. If there is an electron-transfer, a proton-transfer may conceivably be induced to go along with it; the proton may come from one the 2*Pi* produced from the ATP hydrolysis. In the enzyme complex $[1_s][2_s^{u}]$, with the β-PO_3 of each *t* directly coordinated to the Fe-protein 4Fe–4S cluster, it may not be possible for the γ-PO_3 to be coordinated directly to Mo- or Fe-sites of the Mo-protein M-clusters because of steric hindrance. that the Mo-site plays no part in the direct coordination of the γ-PO_3 is also evident from the fact that the nitrogenase-catalyzed hydrolysis of ATP is not inhibited by CO, which is known to be a potent inhibitor of the Mo-site. As mentioned above, MoFe-protein in the dithionite-reduced state shows practically no affinity for binding MgATP; furthermore, there is no experimental evidence indicating that the Fe-sites of the M-clusters is accessible to H_2O molecules, which would be required for the nucleophilic attack at the γ-PO_3 if the Fe-sites of the M-clusters should take part in promoting the ATP hydrolysis by direct coordination of the γ-PO_3 of *t*. The reason why complexation of the MgATP-bound Fe-protein with the M-cluster-deficient UW-45 does not lead to the promotion of ATP hydrolysis may be due to the difference in conformation between UW-45 and a normal MoFe-protein molecule.

3.3 Mechanism of ATP-Driven Electron (and Proton) Transfer in Nitrogenase Catalysis

A mechanism of 2-step ATP-driven electron transfer (coupled with proton transfer and with 1-step ATP hydrolysis) has been proposed (Cai et al. 1987; We et al. 1988). With the discussion given in the foregoing section, we can now gain a deeper insight into the mechanism. The Fe-protein can bind 2*t* at two equivalent sites and have these 2*t* poised ready for further approach to the 4Fe–4S center, but before the Fe-protein enters into the formation of the enzyme complex, the 2*t* may be held, by steric constraint, each at a comfortable distance from the 4Fe–4S center, with the γ-P=O and β-P=O oxygen at about 3.5 4.0 Å from the nearest $Fe^{II,III}$, so that, in this precursory mode of nucleotide-binding, premature ATP hydrolysis will not occur, and the magnitude of the nagative dipole-field produced by the 4P=O dipoles will not be too large to allow rapid electron transfer from the physiological electron-carrier to the Fe-protein. After complexation with the MoFe-protein (the normal MoFe-protein, not the M-cluster-deficient-type, UW-45) to from $[1_s][2_s^{tt}]$, further conformational change may take place in the Fe-protein with some relaxation of the steric constraint, allowing each of the $2t$ to move closer to the 4Fe–4S center by about 1.3–1.8 Å to coordinate directly as a bridging chelate-ligand. The dipole field is now sufficiently negative to drive the protein–protein, or protein–protein-substrate electron-transfer, with concomitant hydrolysis of $2t$ into $2d$ plus $2Pi$, which may be trapped temporarily inside the enzyme complex, or may diffuse out by virtue of the smaller size as $HPO_4^{2-} + H_2PO_4^-$, leaving behind a proton for coupled transfer with the electron. So the overall reaction of the ATP-driven coupled electron- and proton-transfers is as follows (Cai et al. 1987):

$$[Fe_4S_4(cys)_4]^{3-} + 2MgATP^{2-} + 2H_2O \rightleftharpoons [Fe_4S_4(cys)_4]^{2-}$$
$$+ 2MgADP^- + HPO_4^{2-} + H_2PO_4^- + e^- + H^+$$

Following the discussion in the preceding section, a probable mode of coordination of $2t$ in the enzyme complex $[1_s][2_s^{tt}]$ is shown in Fig. 10a, which is a slight modification of the proposed previously (Cai et al. 1987; Wu et al. 1988), the hydrated Mg^{2+} being now depicted as being bound to the terminal γ-phosphate and the β-phosphate group (instead of the β- and α-phosphate groups), and an OH- or H_2O ligand on a Fe-site may be in position for nucleophilic (S_N2) attack at the terminal phosphate group (Fig. 10a). The ligand dipole field produced by the $2t$ ligands should now be sufficiently negative to drive the protein–protein electron (and proton) transfer in view of the results of the model experiment (Chen et al. 1986; Wu et al. 1988) which showed that the addition of ATP, or MgATP, to $(Bu_4N)_2Fe_4S_4(SPh)_4$ in $DMF-H_2O$ (85:15 v/v, pH 8.0) shifted the polarographaphic half-wave potential ($E_m(3-/2-)$) by about -0.49 V, from -1.00 V to -1.49 V. Each of the $2d$ in the *enzyme complex* $[1_o][2_o^{dd}]$ might coordinate also as a bridging chelate-ligand (or a monodentate ligand) through the β-Po_3 only since the α-PO_3 will not be able to coordinate because of steric hindrance as shown by model experiments (Wu et al. 1988); or more probably the hydrated Mg^{2+} might be bound to the β-PO_3 only (not shown in Fig. 10), which might coordinate through the β-$P=O$ as a monodentate ligand to a Fe^{III} of the Fe-protein 4Fe–4S center; so no hydrolysis of the anhydride bond of d will occur. The ligand field produced by the $2d$ ligands may conceivably be less negative than that produced by $2t$ ligands in the *enzyme complex*, but will still be sufficiently negative to prevent the electron backflow from $[1_o]$ to $[2_o^{dd}]$ in the enzyme complex.

After dissociation from the enzyme complex, $[1_o]$ will not readily reduce substrate (H^+), nor will it readily disproportionate, according to Orme-Johnson and Münck (1980). So a second step of ATP-driven electron-transfer appears to be necessary for the reduction of the substrate. This may be actuated in the following way: After the ATP-driven protein–protein electron-transfer with concomitant ATP hydrolysis and proton transfer, the $[2_0^{dd}]$ (with the entrapped $2Pi$ if they fail to diffuse out fast enough) is obliged to dissociate from the enzyme complex $[1_o][2_0^{dd}]$ for the release of the $2d$ (and the entrapped $2Pi$) and replenishment with electron and MgATP, which may take place in either order as shown explicitly in Fig. 11, which otherwise is essentially the same as that reported previously (Cai et al. 1987; Wu et al. 1988), except that here ATP hydrolysis is supposed to take place also whenever $[2_2^{tt}]$, or $[2_0^{td}]$ complexes with $[1_o]$ (or with $[1_s]$, not shown). After replenishing with electron and MgATP, the intrinsic redox potential of $[2_2^{tt}]$ (and very probably $[2_0^{td}]$ also) may become (by complexing with $[1_o]$) sufficiently negative to drive the protein–protein–substrate electron-transfer with concomitant ATP hydrolysis and proton transfer; in this step of ATP-driven electron-transfer, the MoFe-protein in the fully reduced state $[1_o]$ might be further reduced to a transcient, super-reduced state competent for immediate relay of an electron to the coordinated substrate species. After this the enzyme complex $[1_o][2_0^{dd}]$ is obliged to dissociate again for the Fe-protein to be replenished with electron and MgATP, and for the release of any entrapped Pi. Note that, after the dissociation of the enzyme complex, the two nucleotide-binding sites of the Fe-protein may be constrained to move back to their original positions.

If the enzyme system is initially in the dithionite-reduced state $[1_s]:[2_s]$ with abundant supplies of reductant and t, excess $[2_s]$, and large t/d ratio, then the reaction will proceed along the presteady-state steps $(0) \rightarrow (1) \rightarrow (2)$ (Fig. 11), with ATP-driven electron- and proton-transfers from $[2_s^{tt}]$ to $[1_s]$ coupled with the hydrolysis of 2 ATP for each electron transferred, but no H_2 evolution. After this presteady-state burst of ATP hydrolysis, the reaction will proceed predominantly along the cyclic pathway $(3) \rightarrow (5a)$ $(4a) \rightarrow (4a') \rightarrow (6) \rightarrow (7) \rightarrow (3)$, or alternatively, $(3) \rightarrow (4b) \rightarrow (4b') \rightarrow (4b'') \rightarrow (5b) \rightarrow (6)(7) \rightarrow (3)$, with $Pi/2c^- = 4$. If the enzyme solution is not too dilute, the step (3), the dissociation of the enzyme complex $[1_o][2_0^{dd}]$, will be the rate-determining step (rds), and the enzyme system will exist predominantly in the EPR silent state, as observed by Smith et al. [1973]. However, there will be some chance that $[2_0^d]$ (or $[2_o]$, not shown) may complex with $[1_o]$, and when that happens, the electron backflow (ebf) will take place because of the conceivably higher intrinsic redox potential (i.e., higher electron affinity) of $[2_0^d]$ (or $[2_o]$) in the enzyme complex $[1_o][2_0^d]$ than that of the fully reduced MoFe-protein $[1_o]$; thus a "futile cycle" involving reductant-independent ATP hydrolysis will result if there is sufficient supply of t for the system to proceed along the cyclic pathway: $(8)(9) \rightarrow (10) \rightarrow (11) \rightarrow (11') \rightarrow (1) \rightarrow (2) \rightarrow (3) \rightarrow (4b) \rightarrow (8)$ (Fig. 11) So the overall $Pi/2e^-$ ratio will be somewhat higher than 4 under the steady-state enzyme-turnover in argon with abundant supplies of reductant and large t/d ratio. Note that Mensink and Haaker (1988) have observed that the MgADP tightly bound by Av2(ox) is released only if MgATP is present.

If the t/d ratio is not large, then the possibility that ATP-driven electron- and proton-transfers may occur in $[1_s][2_0^{td}]$ and in $[1_o][2_0^{td}]$ with concomitant hydrolysis of one t only in each case, must be taken into consideration. If the

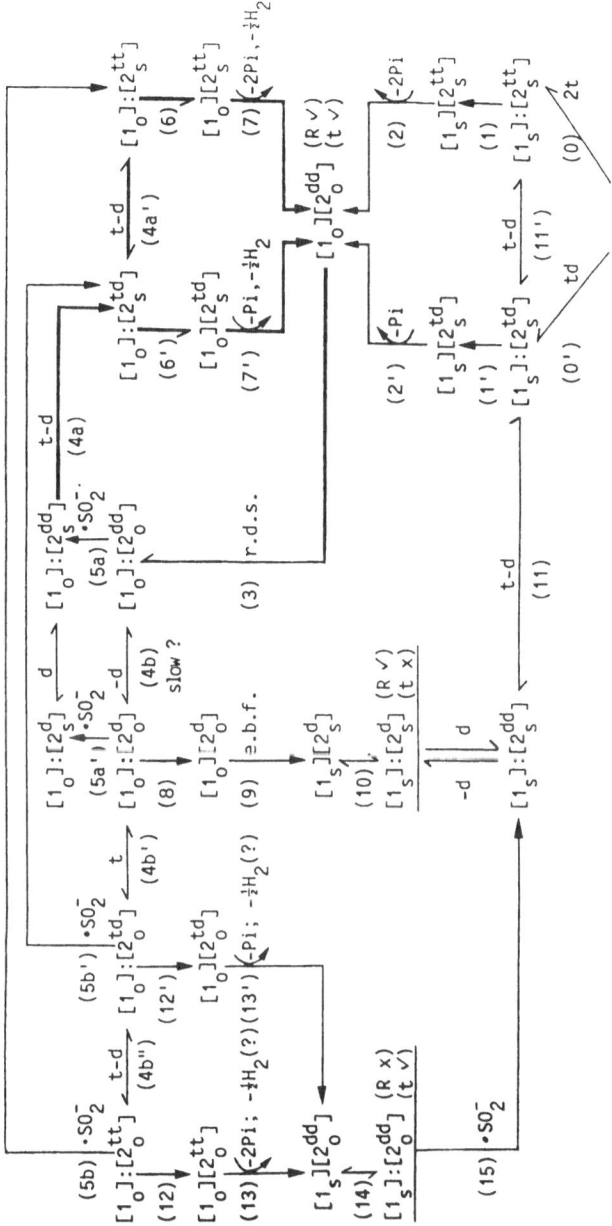

Fig. 11. Diagram showing N_2ase-catalyzed reaction-pathways for a mechanism of ATP-driven electron-transfer and reductant-independent ATP hydrolysis. (Rv)(tv), (Rv)(tx), and (Rx)(tv) denote, respectively, sufficient supplies of reductant and ATP, sufficient supply of reductant but not ATP, and vice versa; the states of the enzyme system underlined being the major states of the system observed under the corresponding conditions denoted

reaction proceeds along the cyclic pathway: $(3) \to (5a) \to (4a) \to (6') \to (7') \to (3)$, or alternatively, $(3) \to (4b) \to (4b') \to (5b') \to (6') \to (7') \to (3)$, the $Pi2e^-$ ratio would be 2. With this and other pathways mentioned above, including the reductant-independent, "futile" ATP-hydrolysis cyclic pathway, all taken into consideration, the overall $Pi/2e^-$ ratio might fall between 2 and 4, as reported by one laboratory (Mortenson and Upchurch 1981).

Two secondary reaction pathways: $(12) \to (13) \to (14) \to (15) \to (11) \to (11') \to (1) \to (2) \to (3) \to (4b) \to (4b') \to (4b'') \to (12)$, and $(12') \to (13') \to (14) \to (15) \to (11) \to (1') \to (2') \to (3) \to (4b) \to (4b') \to (12')$ may not be significant if the reductant is in large excess, but must also be taken into consideration when the reductant is in short supply. Here complexation of $[2_o^{tt}]$, or $[2_o^{td}]$ with $[1_o]$ (or with $[1_s]$, not shown) besides resulting in ATP hydrolysis as discussed above, is also supposed to promote the transfer of an electron from $[1_o]$ to the coordinated substrate species since the intrinsic redox potential of $[2_o^{tt}]$, or $[2_o^{td}]$ might still be more negative than the fully reduced MoFe-protein, $[1_o]$, which alone does not reduce substrate readily. This promotion of electron transfer may be compared with the observation of Ledwith and Schulz (1975), who found that, if a solution of $[Mo_2O_2(cys)_2]^{2-}$ is electrochemically reduced to the Mo^{III} stage and the potential cut off, then C_2H_2 introduced afterward is not reduced; and that the reapplication of potential leads to the immediate C_2H_2 reduction (for similar remarks, cf. Stiefel 1977; Tsai 1980).

If the reductant is exhausted, while the supply of ATP is still sufficient, then it is easy to see from Fig. 11 that steps $(5a)$, $(5a')$, $(5b)$, $(5b')$, and (15) can no longer proceed; so the enzyme system with sufficient supply of ATP will proceed to change along the reductant-independent ATP-hydrolysis cyclic pathway in cycles, at the same time part of the enzyme system will also proceed to change along the two secondary pathways as far as step (14); finally the whole system will end up in the state $[1_s]:[2_o^{dd}]$, exhibiting the characteristic EPR signals of the semi-reduced (i.e., dithionite-reduced) MoFe-protein, as observed by Smith et al. (1973) and by Walker and Mortenson (1973). On the other hand, with sufficient supply of reductant (R $\sqrt{}$), but insufficient supply of t (t x) (Fig. 11), steps $(4a)$, $(4a')$, $(4b')$ $(4b'')$, (11), and $(11')$ can no longer proceed; most part of the enzyme system will proceed to change along $(3) \to (5a)$ to reach the state $[1_o]:[2_s^{dd}]$, which may combine to form the enzyme complex $[1_o][2_s^{dd}]$ (not shown in Fig. 11), in which the intrinsic redox potential of $[2_s^{dd}]$ is most probably somewhat more negative than that of $[1_o]$, and so may promote the transfer of an electron from $[1_o]$ to the substrate, leading to the formation of $[1_s][2_s^{dd}]$; so the system will finally end up in the state $[1_s]:[2_s^{dd}]$. Parts of the enzyme system already in the states $[1_o]:[2_o^{d}]$, $[1_o]:[2_o^{td}]$, and $[1_o]:[2_o^{tt}]$ will also proceed to change untill they also all end up in the state $[1_s]:[2_s^{dd}]$, showing the characteristic EPR singals of both components, as observed by Walker and Mortenson (1973).

Finally, it can also be seen from the diagram (Fig. 11) that the appropriate experimental conditions for examining the possibility of having $Pi/2e^-$ ratio ≤ 4 and ≥ 2 may be as follows: (a) abundant supply of reductant ($S_2O_4^{2-}$) in high concentration, say in the 10 to 20 mM range, so as to decrease as much as possible the chances of having the enzyme system in the states $[1_o]:[2_o^d]$, $[1_o]:[2_o^{td}]$, and $[1_o]:[2_o^{tt}]$ to proceed to from enzyme complexes and to change along the electron-backflow pathway and the two secondary pathways: (b) for fast replenishment of $[2_o^{dd}]$ with electron, it will be more promising to work with C_p or K_p nitrogenase, rather than with Av nitrogenase since the midpoint potential of Av2 is about -110 mV more negative than that of Cp2 or Kp2; (c) abundant supply of concentrated ATP, say in the 2–6 mM range, and an optimum d/t ratio, probably in the 0.3–0.5 range as in the in-vivo nitrogen fixation conditions (and also with a small excess of Fe-protein over the 1:2 ratio); this will insure that $[2_s^{td}]$ predominates over $[2_s^{tt}]$ and $[2_s^t]$, as well as $[2_o^t]$. Note that these conditions were essentially met with in Mortenson and Upchurch's (1981) experiment. So it may be fruitful to confirm their important results by different methods. An overall $Pi/2e^-$ ratio ≥ 4 is an extravagant waste of bio-energy, and NATURE should be able to do better than that. The in-vivo d/t ratio of 0.3 to 0.5, the very slow dissociation of d from $[2_o^{dd}]$ until it is replenished with electron, the negative cooperativity in $2t$-binding to form $[2_s^{tt}]$ vs. the positive cooperativity in the binding of t by $[2_s^d]$ to from $[2_s^{td}]$ according to Cordewener et al. (1985), and the abundant supply of appropriate electron-carrier (e.g., Fld_r) with sufficiently negative redox potential, all seem to be contrived by the enzyme for the purpose of suppressing the $Pi/2e^-$ ratio. Incidentally, if $[2_s^d]$, $[2_s^{dd}]$, $[2_s]$, $[2_t^{st}]$, $[2_t^{tt}]$, and $[2_t^t]$ exist in dynamic equilibrium with the adenine nucleotides in large excess, then their concentration ratios can be readily estimated from the known stoichiometric dissociation constants and the concentrations of the adenine nucleotides.

4 List of Notations

Av1, Cp1, Kp1 and Av2, Cp2, Kp2 for MoFe-proteins (component 1) and Fe-proteins (component 2) from *Azotobacter vinelandii, Clostridium pasteurianum*, and *Klebsiella pneumoniae*; t, d, m, and Pi for MgATP, MgADP, MgAMP, and inorganic phosphate; $[1_s]$, $[1_o]$ for semi-reduced, EPR-active (signified by the subscript "s") MoFe-protein and the fully-reduced, EDR-silent (signified by the subscript "o") MoFe-protein; $[2_s]$, $[2_o]$ for Fe-protein in the reduced, EPR-active state, and in the oxidized, EPR-silent state; $[2_s^{tt}]$, $[2_s^t]$, $[2_s^{dd}]$, $[2_s^d]$, $[2_s^{td}]$, and $[2_o^{tt}]$, $[2_o^t]$, $[2_o^{dd}]$, $[2_o^d]$, and $[2_o^{td}]$, for the Fe-protein $[2_s]$ and $[2_o]$ bound with the adenine nucleotides indicated by the superscripts; R for the reductant $(S_2O_4^{2-})$; $(R\surd)$, $(t\surd)$ signifying sufficient supplies of reductant and ATP.

Acknowledgements. This review is based mainly on nitrogen fixation research done in collaboration with many of our colleagues and graduate students, to whom our thanks are due, especially to Dr. H.B. Chen, for his help in many ways in making this manuscript ready. We are very grateful to Professors Xingxian Guo, Aoqing Tang, and Jiaxi Lu for their organizing a national project for nitrogen fixation studies, without this project and their constant supports and encouragements for more than a decade, all these researches could not have been accomplished.

References

Averill BA, Herskovitz T, Holm RH, Ibers JA (1973) Synthetic analogs of the active sites of iron-sulfur proteins II. Synthesis and structure of the tetra [mercapto-μ_3-sulfido-iron] clusters $[Fe_4S_4(SR)_4]^{2-}$. J Amer Chem Soc 95: 3523

Baginski ES, Foa PP, Zak B (1967) Microdetermination of inorganic phosphate, phospholipids, and total phosphate in biologic materials. Clin Chim Acta 13: 326–332 (Eng) 15: 155

Biggens DR, Kelley M (1970) Interaction of nitrogenase from *K. pneumoniae* with ATP or cyanide. Biochim Biophys Acta 205: 288

Bui PT, Mortenson LE (1968) Mechanism of enzymic reduction of N_2: the binding of ATP and cyanide to the N_2-reducing system. Proc Natl Acad Sci USA 61: 1021

Burgess BK (1984) Structure and reactivity of nitrogenase—An overview. In: Veeger C, Newton WE (eds) Advances in Nitrogen Fixation Research, pp 103. Nijhoff, The Hague

Cai (Tsai) Q (KR), Zhang HB, Lin GD (1987) Cluster catalysis in the fixation of nitrogen to ammonia catalyzed by nitrogenase and by iron catalysts. Advan. Sci. Chian (chemistry) 2: 125

Chen HB, Zhang HT, Lin GD, Tsai KR (1985) Chemical modelling of ATP-driven electron transfer in nitrogenase-catalyzed reactions I. Studies of $[Fe_4S_4(SR)_4]^-$ complexation with ATP using polarographic and electronic absorption spectroscopic methods. J Xiamen Univ (Nat Sci) 24: 448

Christou G, Garner CD (1979) A convenient synthesis of tetrakis [thiolato-μ_3-sulfido-iron]2-clusters. JCS (Dalton) pp 1093

Cordewener J, Haaker H, Veeger C (1983) Binding of ATP to the nitrogen proteins from *A. vinelandii*. Eur J Biochem 132: 47

Cordewener J, Haaker H, Van Ewijk P, Veeger C (1985) Properties of the MgATP and MgADP binding sites on the Fe-protein from A vinelandii. Ibid 148: 499

Cordewener J, Asbroek A, Wassink H, Eady R, Haaker H, Veeger C Ibid 162: 265

Dai LF, He JR, Du DX, Lin HM, Xin WS (1985) purification of nitrogenase components from *Anabaena cylindrica*, Plant Physiol Commun (1): 54

Gao M, You C-B (1987) Energy transfer within nitrogen-fixing chain in *A. vinelandii*. Acta Phytophysiol. Sinica 13: 302

Haaker H, Braaksm A, Cordewener J, Klugkist J, Wassink H, Grande H, Eady RR, Veeger C (1984) Iron-sulfide content and ATP binding properties of nitrogenase component II from *Azotobacter vinelandii*. In: Veeger C, Newton WE (eds) Advances in Nitrogen Fixation Research, pp 123

Hegemann RV, Orme-Johnson WH, Burris RH (1980) Role of MgATP in the hydrogen evolution reaction catalyzed by nitrogenase from *Azotobacter vinelandii*. Biochemistry 19: 2333

Haight GP, Jr (1987) Hydrolysis of phosphate esters and anhydrides: role of metal ions. Coord Chem Rev 79: 293–319. See also Crease IC, Haight GP, Peachey R, Robinson WT, Sargeson AM (1984) Rapid cleavage of chelated pyrophosphate using metal ion complexes. JCS Chem Commun pp 1568

Hausinger RP, Howard JB (1983) Thiol reactivity of the nitrogenase Fe-protein from *Azotobacter vinelandii*. J Biol Chem 258: 13486

Hill CL, Renaud J, RH, Mortenson LE (1977) synthetic analogs of the active sites of iron-sulfur proteins. 15. Comparative polarographic potentials of the $[Fe_4S_4(SR)_4]^{2-,3-}$ and Cp2 redox couples. J Amer Chem Soc 99: 2549

Howard JB, Diets TL, Anderson GL, Maroney M, Que L, Hausinger RP (1985) Mechanism and structure of nitrogenase Fe-protein. In: PW Ludden, JE Burris (eds) Nitrogen fixation and CO_2 metabolism, pp 153

Jiang YM (1981) Role of ATP and catalytic mechanism of nitrogenase. Advan Biochem Biophys China 6: 22

Ledwith DA, Schulz FA (1975) Catalytic electrochemical reduction of acetylene in the presence of molybdenum-cysteine complex. J Amer Chem Soc 97: 6591

Likhtenstein GI (1988) Structure and molecular dynamics of metalloenzymes studied by physical label methods. J Mol Catal 47: 129

Lin HM et al. (1984) Studies on nitrogenase of blue-green algae. In: C Veeger, WE Newton (eds.), p 156

Ljones T, Burris RH (1978) Nitrogenase: the reaction between the Fe-protein and bathophenanthroline disulfonate as a probe for interactions with MgATP. Biochemistry 17: 1866

McKenna CE, Stephens PJ, Eran H, Luo G-M, Zhang F-X, Ding M-T, Nguyen HT (1984) Substrate interactions with nitrogenase and its Fe-Mo cofactor: chemical and spectroscopic investigations. In: C Veeger, WE Newton (eds) Advances in Nitrogen Fixation Research, pp 115

Mensink RE, Haaker H (1988) Interactions of MgATP and MgADP with the iron protein. In: Bothe/de Bruijn/Newton (eds.), Nitrogen Fixation: Hundred years After Gustsv Fischer. Stuttgart, New York

Miller RW, Eady RR (1988) Tight binding interactions of molybdenum and vanadium nitrogenase proteins with adenine nucleotides. Ibid p 90

Mortenson LE, Thorneley RNF (1979) Structure and function of nitrogenase. Ann Rev Biochem 48: 387

Mortenson LE, Upchurch RG (1981) Effect of adenylates on electron flow and efficiency of nitrogenase. In: AH Gibson, WE Newton (eds) Current Perspectives in Nitrogen Fixation, pp 75

Mortenson LE (1987) ATP and nitrogen fixation. In: Ullrich et al. (eds) Inorganic Nitrogen Metabolism, pp 165, Springer-Verlag, Berlin Heidelberg

Nelson MJ, Lindahl PA, Orme-Johnson WH (1982) Bioinorganic chemistry of nitrogenase. In: GL Eichborn, LG Marzilli (eds), Advan Inorg Biochem 4: 1

Nitrogen Fixation Group, IAAE, CAAS (1976) Purification and some properties of iron-protein from A. vinelandii. Acta Mircobiol Sinica 16: 126

Nitrogen Fixation Research Group, Xiamen Univ (1974) On the mechanism of nitrogenase catalysis and structure of the active center. J Xiamen Univ (Nat Sci) 13: (1) 111

Nitrogen Fixation Research Group, Xiamen Univ (1976) A model of nitrogenase active center and the mechanism of nitrogenase catalysis. Scientia Sinica (Engl) 19: 460

Nitrogen Fixation Research Group and Institute of Phys Chem Xiamen Unvi (1982) Fixation of nitrogen to ammonia via enzymic and nonenzymic catalysis. J Xiamen Univ (Nat Sci) 21: 424

Orme-Johnson WH, Hamilton WD, Jones TL, Tso M-YW, Burris RH, Shah VK, Brill WJ (1972) Electron paramagnetic resonance of nitrogenase and nitrogenase components from Cp W5 and Av OP Proc Nat Acad Sci USA 69: 3142

Orme-Johnson WH, Munck E (1980) On the prosthetic groups of nitrogenase. In: MP Coughlan (ed.), Molybdenum and Molybdenum-containing Enzymes, pp 427 Pergamon Press

Orme-Johnson WH (1985) Molecular basis of biological nitrogen fixation. Ann Rev Biophys Chem 4: 419

Que J, Holm RH, Mortenson LE (1975) Extrusion of Fe_2S_2 and $Fe_4S_4{}^*$ cores from the active sites of ferredoxin proteins. J Amer Chem Soc 97: 46

Smith BE, Lowe DJ, Bray RC (1973) Studies by ESR on the catalytic mechanism of nitrogenase of K. pneumoniae. Biochem J 135: 331

Smith BE et al. (1988) Recent studies on the structure and function of molybdenum nitrogenase. In: Bothe/de Bruijn/WE Newton (eds), Nitrogen Fixation: Hundred Years After Gustav Fischer, pp 91

Song W, You C-B (1987) Isolation and purification and some characters of nitrogenase complex from A. vinelandii. Acta Phytophsiol Sinica 3: 35

Stiefel EI (1977) Mechanisms of nitrogen fixation. In: WE Newton, JR Postgate, Rodrigaez-Bareueco C (eds.), Recent Development in Nitrogen Fixation, pp 69–108 Academic Press

Thorneley RNF, Deistung J (1988) Electron transfer studies involving flavodoxin and a natural redox partner, the iron protein of nitrogenase. Biochem J 253: 587

Tsai KR (1980) Development of a model of nitrogenase active center and mechanism of nitrogenase catalysis. In: WE Newton, WH Orme-Johnson (eds.), Nitrogen Fixation, Vol I, pp 373

Tso MW, Burris RH (1973) The binding of ATP and ADP by nitrogenase components from C. pasteurianum. Biochim Biophys Acta 309: 263

Walker M, Mortenson LE (1973) Oxidation reduction properties of nitrogenase from C. paseurianum W5. Biochim Biophys Res Commun 54: 669

Wan H-L (1987) The effects of ATP/ADP and Fe-protein/MoFe-protein ratios on Pi 2e⁻ and minimum Pi/2e⁻ ratios. J Xiamen Univ (Nat Sci) 26: 73

Wassink H, Cordewener J, Haaker H (1988) The ATPase activity of nitrogenase. In Bothe de Bruija/WE Newton (eds), Nitrogen Fixation: Hundred Years After Gustav Fischer, p 139

Watt GD (1985) Redox properties of nitrogenase component proteins from A. vinelandii. In HJ Evans, PJ Bottomley, WE Newton (eds.), Nitrogen Fixation Progress, pp 585

Wu Y-H, Chen H-B, Lin G-D, Yu X-X, Zhang H-T, Wan H-L, Tsai KR (1988) Cluster-complex mediated electron-transfer and ATP hydrolysis. Pure and Appl Chem Vol 60, No. 8, pp 1291

You C, Li J, Song W, Li S (1978) Some properties of iron protein of nitrogenase from A. vinelandii (II). Acta Phystophysiol. Sinica 4: 123-131; You C, Li J, Li X (1979) Some properties of iron protein of nitrogenase from A. vinelandii (III). Ibid. 5: 215

You C, Wang H, Gao M, Ping S (1983a) A method for purification of nitrogenase protein components from A vinelandii. Plant Physiol Commun (3): 51

You C, Wang H, Ping S, Gao M (1983b) Fe-protein of nitrogenase from A. vinelandii (IV). Acta Phytophysiol Sincia 9: 403 (1984) Fe-protein of nitrogenase from A. vinelandii (V). Ibid. 10: 73

You C (1987) The structure and function of nitrogenase. In: C You, Y-M Jiang, HY Song (eds.), Biological Nitrogen Fixation, pp 20-45, Academic Press, China

You C, Gao M (1987) Electron transport within nitrogen fixation chain involving flavodoxin. Acta Phyt physiol. Sinica 13: 174

Zeng D, Burris RH (1986) Studies on the binding of adenylate compounds to the nitrogenase component. J Xiamen Unvi (Nat Sci) 25: 207

Zeng D (1987) The Biology of Nitrogen Fixation, pp 312-334. Xiamen Univ Press, Xiamen, China

Zumft WG, Mortenson LE (1975) Nitrogen fixing complex of bacteria Biochim Biophys Acta 416: 1

CHAPTER 6

Chemical Modelling of the Active Site of Molybdenum–Iron Protein. Synergism of MoFe$_3$S$_4$ Cubane-Like Unit from Physical and Chemical Evidence

Bei-Sheng Kang, Qiu-Tian Liu, and Jia-Xi Lu

The Nitrogen Fixation
and its Research in China
Editor Guo-fan Hong
© Springer-Verlag Berlin Heidelberg 1992

1 Introduction

Investigations on molybdenum–iron–sulfur cluster compounds have been a hot subject since the isolation of FeMoco from MoFepr of nitrogenases (Shah 1977). Both spectroscopic (Conradson 1985, 1987; Hedman 1988) and chemical analysis (Burgess 1984) results have implicated a mixed molybdenum–iron cluster nature with sulfur and oxygen ligand atoms of composition $MoFe_{6-9}S_{4-9}$ for the active site of FeMoco. Many structural models (Lu 1975, 1980; Cai 1982; Cramer 1978; Teo 1979; Christou 1982; Orme-Johnson 1983) having mostly $MoFe_3S_{3-4}$ as a basic structural unit have been proposed, of which the first one that was suggested by Lu in 1973 was named Fuzhou model. It has a string-bag $MoFe_3S_3$ composition with a defected cubane-like skeleton, where the open site is for substrate coordination. Mo-EXAFS (Conradson 1987) and other studies on

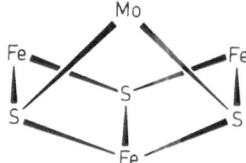

MoFepr or FeMoco have revealed a distorted octahedral Fuzhou model environment for molybdenum atom with $2 \sim 3$ oxygen (and/or nitrogen) atoms at $2.10 \sim 2.12$ Å, $3 \sim 4$ sulfur atoms at 2.37 Å, and $3 \sim 4$ iron atoms at $2.68 \sim 2.70$ Å. Chemically, the $MoFe_3S_4$ cubane-like unit composed of these atoms is one of the most stable arrangements known. Compounds containing this basic structural fragment have many of the characteristics and properties of FeMoco. Although the exact molecular structure of FeMoco remains unknown, cluster compounds with $MoFe_3S_4$ cubane-like unit chelated by thiolato ligands have been regarded as the closest chemical models and widely studied for their chemical and physical properties, and their relevance to the activity of nitrogen fixation is proposed.

Tetrathiomolybdate has often been used as "synthon" in building up the $MoFe_3S_4$ skeleton, as it is both a good molybdenum source and an effective chelating agent furnishing inorganic sulfur atoms. For all the $MoFe_3S_4$-containing clusters shown in the schematic view below, the monocubane series Aa with $dtcR_2$ ligands (dtc = dithiocarbamate) and the dicubane series Ca with arylthiolato ligands that we have prepared and studied in this laboratory, oxidation states n for the $[MoFe_3S_4]^{n+}$ core of 3 (Kang 1986, 1987, 1988; Cai 1984, 1985), 4 (Liu 1986), 5 (Liu 1986, 1987, 1988, 1990), and 6 (Liu 1988) have all been obtained, allowing a detailed understanding of the properties of these $MoFe_3S_4$ clusters be possible. From investigation of the crystal structures, ^{1}H NMR (Liu 1987) and ESR spectra, Mössbauer effects (Kang 1988), magnetic and electrochemical properties, and reactions (Kang 1989) with acyl chlorides, synergism (Kang 1989; Liu 1988) of the $MoFe_3S_4$ core has been proposed as

	R	L(L')	n
a1	Me		0
a2	Et		0
a3	$C_4H_8=R_2$		0
a4	$C_5H_{10}=R_2$		0
a5	Et		1
a6	Bz		0
b1	Me(x = 1)		
b2	$C_4H_8=R_2(x = 2)$		

$[MoFe_3S_4(dtcR_2)_5]^{n-}$ Aa
$[Mo_xFe_{4-x}S_4(dtcR_2)_6]$ Ab

	R	L(L')	n
1	C_6H_4Cl-p	SR ($^iPr_2cat$)	3
2	C_6H_4Cl-p	N_3 (Cl_4cat)	3
3	Et	SR (dmpe)	1

$[MoFe_3S_4(SR)_3LL']^{n-}$ B

	R	L(L')	n
a1	Ph	SR	4
a2	o-tolyl	SR	4
a3	m-tolyl	SR	4
a4	p-tolyl	SR	4
b1	Et	SR	3
b2	Bz	SR	4

$[Mo_2Fe_7S_8(SR)_{12}]^{n-}$ C

	R	L(L')	n
1	Ph	Cl	3
2	Et	Cl	3
3	m-tolyl	Cl	3
4	Ph	SR	3
5	Et	SR	3
6	Ph	SR	5

$[Mo_2Fe_6S_8(SR)_3L_6]^{n-}$ Da

	L(L')
1	SBu^t
2	SPh

$[Mo_2Fe_6S_8(OMe)_3L_6]^{3-}$ Db

Schematic view of compounds containing $MoFe_3S_4$ cubane-like core.

the key function for these compounds and that the catalytic activity in native nitrogenases may also be related to a similar effect of its Mo–Fe–S(O) core.

2 Evidence from Preparation of the Compounds

Whether by one-pot reaction of simple inorganic salts in the presence of chelating agent as in the preparation of series A compounds (reaction 1) or by employing a multi-nuclear iron-thiolate $[Fe_4(SR)_{10}]^{2-}$ as in series Ca compounds (reaction 2), clusters with $MoFe_3S_4$ cubane-like core(s) have always been obtained under these reaction conditions as the most stable ones (Kang 1987):

$$MoS_4^{2-} + FeCl_2 + R_2dtc^- \xrightarrow{DMF} [MoFe_3S_4(dtcR_2)_{5\sim6}]^{0,1-} \qquad (1)$$

$$A$$

$$MoS_4^{2-} + [Fe_4(SR)_{10}]^{2-} \xrightarrow{MeCN} [Mo_2Fe_7S_8(SR)_{12}]^{4-} \qquad (2)$$

$$Ca$$

Besides, some of these cluster compounds (Aa2, Aa3, Aa4) can also be obtained by reaction of linear di- or trinuclear molybdenum–iron–sulfur compounds with dtcR$_2$ ligands, substantiating the general stability of the cubane core (Liu 1988, 1989, 1990). Table 1 has summarized some of the reactions for the preparation

Table 1. Summary of synthetic reactions for $[MoFe_3S_4]^{n+}$ cluster compounds

Reagent	Medium	Compound	n
MoS_4^{2-}, $FeCl_3$, RS^-	ROH	Da4, Da5	3
		Cb1, Cb2	3
MoS_4^{2-}, $FeCl_2$, R_2dtc	DMF	Aa	4, 5
	CH_2Cl_2	Ab	6
MoS_4^{2-}, $[Fe(SR)_4]^{2-}$	MeCN	Da4	3
MoS_4^{2-}, $[Fe_4(SR)_{10}]^{2-}$	MeCN	Ca	3
$MoO_2S_2^{2-}$, $[Fe_4(SR)_{10}]^{2-}$	MeCN	Ca1	3
$Fe(DMF)_6[(FeCl_2)_2MoS_4]$, R_2dtc^-	DMF	Aa2, Aa3, Aa4	5
$Fe(DMF)_6[(FeCl_2)_2MoS_4]$, PhS^-	DMF	Da4	3
$Fe(DMF)_6[(FeCl_2)MoS_4]$, R_2dtc^-	DMF	Aa4	5
$[Fe(SR)_4]^{2-}$, $(FeCl_2)MoS_4]^{2-}$	MeCN	Da4	3
$[(PhS)_2FeS_2FeS_2MoS_2]^{3-}$, $[Fe(SR)_4]^{2-}$	MeCN	Da4	3
$[(FeCl_2)MoS_4]^{2-}$, $[Fe_2S_2Cl_4]^{2-}$, R_2dtc^-	MeCN	Aa4	5
$[(FeCl_2)MoO_2S_2]^{2-}$, $[Fe(SR)_4]^{2-}$	MeOH	Da4	3

of $MoFe_3S_4$ cluster compounds. In spite of the variation in precursors and reaction mediums, there is a great tendency for the synthetic reactions to form cluster compounds with $MoFe_3S_4$ core as an entity, even when sulfur-deficient reactant such as $MoO_2S_2^{2-}$ or $[(FeCl)_2MoO_2S_2]^{2-}$ was employed.

Another interesting feature of the above reactions is that $[MoFe_3S_4]^{n+}$ core oxidation states n of 4, 5, and 6 were obtained by using 1,1-bidentate ligand $dtcR_2$ with formation of monocubane cluster compounds $[MoFe_3S_4(dtcR_2)_{5\sim6}]^{0.1-}$ (A), which is higher than that in dicubane cluster compounds $[Mo_2Fe_7S_8(SR)_{12}]^{4-}$ (Ca, $n = 3$) where monodentate thiolato ligands were employed. The reason rests in the fact that ligand R_2dtc^- exists in two resonance forms I and II:

$$R_2N^+ = C\underset{S^-}{\overset{S^-}{\big\langle}} \leftrightarrow R_2N-C\underset{S^-}{\overset{S}{\big\langle}}$$

$$\text{I} \qquad\qquad \text{II}$$

where form I is the predominant structure (vide infra) in clusters A and the two electron-rich sulfur atoms favor the stabilization of the $MoFe_3S_4$ core in high oxidation states. A similar argument has been reported for $M(dtcR_2)_n$ (Kanatzidis 1985). These multiple oxidation levels from $3+$ to $6+$ for the $MoFe_3S_4$ cluster core are in themselves enough to lead us to propose that these compounds are models for the active site of FeMoco either for their redox characters or for their magnetic and spectroscopic properties.

3 Evidence from Reactions with Acyl Chlorides

Acyl chlorides have often been employed in the preparation of terminal chloride-substituted cubane-like compounds (Wolff 1980; Christou 1980; Palermo 1982, 1983; Johnson 1978; Wong 1978) with the skeletons of the clusters remaining intact and the bridging thiolato groups preserved. While the dicubane cluster Ca1 or Ca3 was allowed to react with acetyl chloride, an unexpected extrusion of the central $Fe(SR)_3$ unit was observed to give dicubane clusters Da1 and Da3 (Kang 1989) as shown in reaction 3:

$$[Mo_2Fe_7S_8(SR)_{12}]^{4-} \xrightarrow{\text{MeCOCl}} [Mo_2Fe_6S_8(SR)_3Cl_6]^{3-} \qquad (3)$$

$$\text{Ca1 or Ca3} \qquad\qquad\qquad \text{Da1 or Da3}$$

In this way, the two types of dicubane clusters Ca and Da are correlated by chemical conversion for the first time because of the great stability of the $[MoFe_3S_4]^{3+}$ core. In contrast, the monocubane cluster Aa with $[MoFe_3S_4]^{5+}$ are less stable under similar reaction conditions and broke down to the mononuclear compound $[Fe(dtcR_2)_2Cl]$ (Kang 1989) during reaction with acyl chloride. This phenomenon implies that the resting state of native MoFepr may have more of the iron atoms (a total of ~22) in reduced states.

4 Evidence from Structural Parameters

Crystallographic data for the two series of $MoFe_3S_4$ cubane clusters Aa and Ca are listed in Table 2. Selected structural data for Aa and Ca are in Tables 3 and 4, respectively, and those reported for D and B in Tables 5 and 6, respectively, for comparison. Representative crystal structures of Aa1, Ab1, and anions of Ca3, and Da1 are shown in Figs. 1–4, respectively. The following structural features have been observed.

1. The atomic distances and bond angles for the $MoFe_3S_4$ core in the same oxidation state do not vary much from compound to compound with different ligands, while significant variations are observed among compounds with altered core oxidation states. Generally speaking, metal to ligand distances are shorter for higher oxidation state of metal atom in the same environment (same types of metal atoms, ligands and coordination numbers). For example, Mo–Sc (c for core) and Fe–Sc distances decrease in compounds approximately

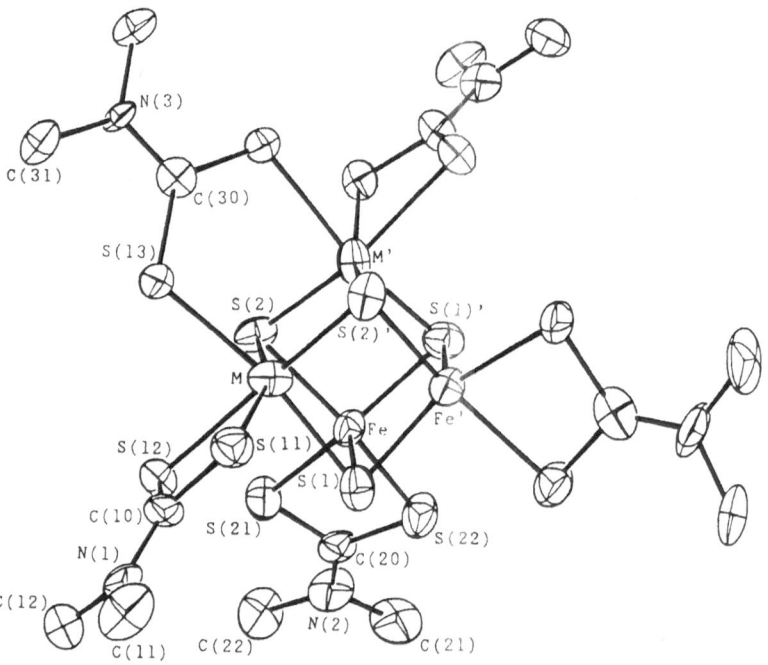

Fig. 1. Structure of Aa1

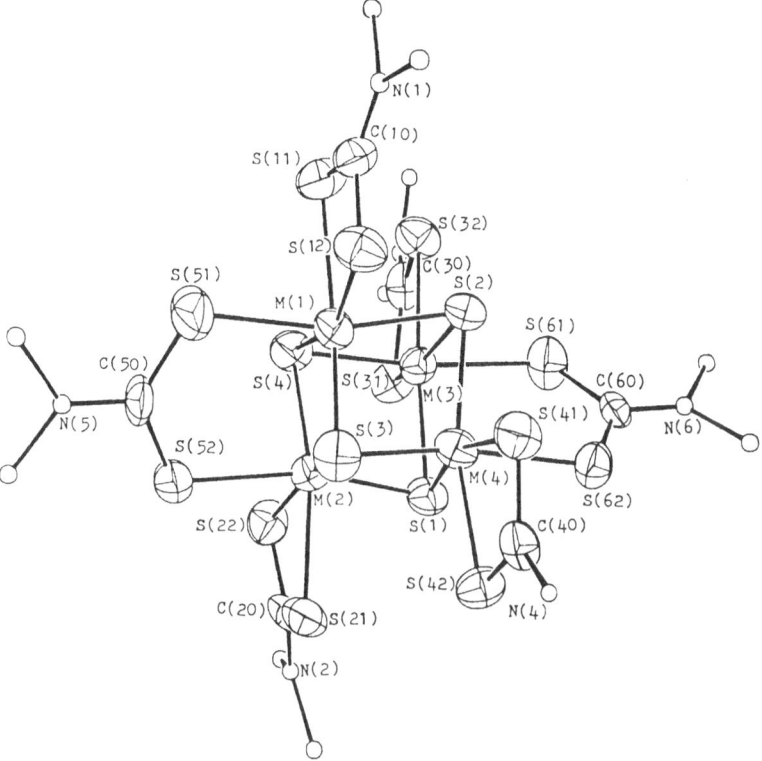

Fig 2. Structure of Ab1

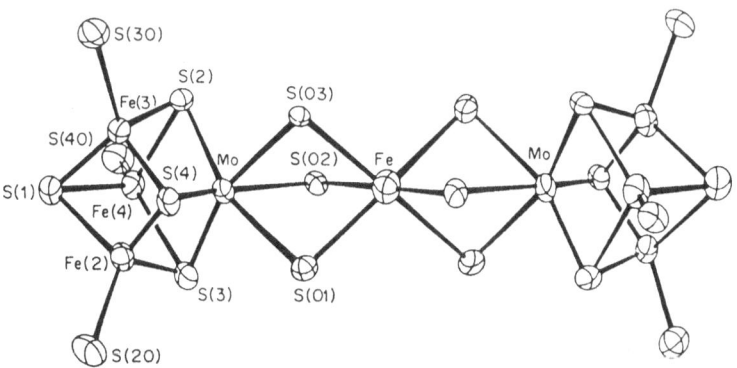

Fig. 3. Structure of Ca3

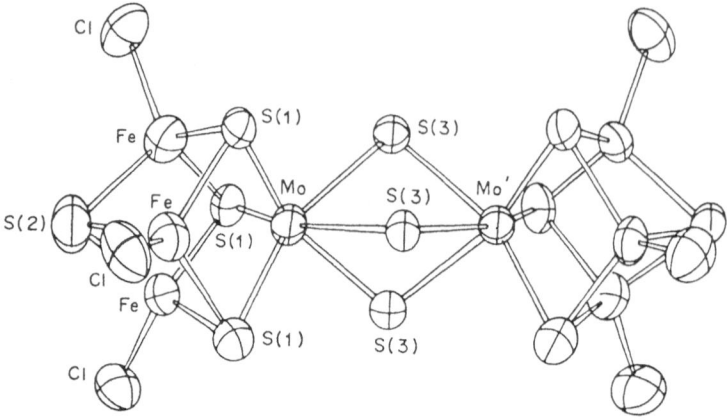

Fig. 4. Structure of Da3

in the order of B > Aa5 > Aa1–Aa4, when the core oxidation state increases from 3+ to 5+.

2. The addition of an electron to a $MoFe_3S_4$ core will cause an overall change of structural parameters as the electron is delocalized among the $MoFe_3S_4$ core atoms and therefore the Fe–S bonds in cluster Aa5 are longer than their counterparts in cluster Aa2. A similar comparison of corresponding bond lengths in $MoFe_3S_4$ cores of Aa and B also indicates the delocalization of the extra electron in clusters B. Therefore, the delocalized distribution of the extra electron causes changes of the whole $MoFe_3S_4$ core rather than localized changes of certain individual metal atom. Such synergism of cluster atoms implies a great stability of the $MoFe_3S_4$ core which would function in its entity as an electron-donor or acceptor in redox processes.

3. Bond distances in compounds Aa1–Aa4 ranging from 2.68 to 2.70 Å for Mo–Fe atoms bridged by $dtcR_2$ simulate the Mo–Fe distances in FeMoco (Conradson 1987). This result strongly suggests that the metal atoms in MoFepr may be chelated by bidentate or polydentate bioligands which bridge the metal atoms to form biomolecules so that the Mo–Fe distances are shortened.

4. The pseudo-cubane core $MoFe_3S_4$ has two intercalating tetrahedra $MoFe_3$ (abbreviated M_4) and S_4 with the tetrahedron of S_4 enclosing that of $MoFe_3$ in a distorted fashion as shown in Figs. 1–4 and by the data in Tables 7 and 8. A distortion of the $[MoFe_3S_4]^{3+}$ core of this kind with S_4/M_4 ratio within the range 2.34 ~ 2.40 has always been observed for cubane cluster compounds whether one or two cubane cores is contained in the molecule as long as the oxidation state of the $MoFe_3S_4$ core is kept constant. When the oxidation level of $MoFe_3S_4$ becomes higher, the ratio of S_4/M_4 lowers drastically as shown in Table 8. These data show that the abstraction of electron(s) from the $MoFe_3S_4$ core affects not only the volume of M_4 but also that of S_4 by

Table 2. Summary of crystal data, intensity collection, and structure refinement parameters for compounds containing the $MoFe_3S_4$ unit

Data	Aa1[a]	Aa2[b]	Aa3[c]	Aa4[d]	Aa5[e]	Ab1[f]	Ab2[g]	Ca1[h]	Ca3[i]	Da3[j]
M_r	1162.68	1174.08	1233.11	1277.83	1304.32	1196.2.	1896.87	2670.3	2982.8	1756.6
a, Å	20.34(1)	15.711(3)	12.544(2)	19.222(7)	22.897(3)	12.060(3)	12.063(2)	12.775(4)	18.022(2)	16.827(3)
b, Å	19.978(2)	16.637(4)	20.478(4)	27.853(4)	12.399(2)	12.064(4)	16.034(2)	13.076(3)	18.375(2)	16.827(3)
c, Å	10.357(3)	21.842(3)	19.582(7)	19.456(3)	20.928(4)	19.595(5)	19.690(4)	20.576(4)	22.254(3)	15.95(2)
α, °		86.93(1)				83.10(2)	90.18(1)	80.00(2)		
β, °		70.84(1)	102.82(2)		97.15(1)	84.93(2)	101.52(2)	81.39(2)		
γ, °		82.61(2)				66.59(2)	100.00(1)	61.51(2)		
V, Å³	4207.6	5347.7	4905	10416	5895	2594	3672.2	2966	6969	3912
Z	4	4	4	8	4	2	2	1	2	2
d_c, g/cm³	1.70	1.458	1.636	1.630	1.47	1.53	1.72	1.494	1.424	1.491
cry. syst	orthorh.	triclinic	monoclinic	orthorh.	monoclinic	triclinic	triclinic	triclinic	monoclinic	hexagonal
space group	Pbcn	P1̄	C2/c	Pbca	$P2_1/c$	P1̄	P1̄	P1̄	$P2_1/n$	$P6_3/m$
scan range 2θ, °	1–26	1–23	1–25	1–22.5	1–24	1–25	1–25	1–22	1–23	1–25
Unique data ($I > 3\sigma(I)$)	1415	7495	2179	2966	6725	5183	5033	4031	5173	1296
R	0.083	0.082	0.057	0.097	0.068	0.069	0.079	0.077	0.064	0.049

[a] $Aa1 \cdot 2CH_2Cl_2$. [b] $Aa2 \cdot CH_3CN$. [c] $Aa3 \cdot 2C_2H_5CN$. [d] $Aa4 \cdot CH_2Cl_2$. [e] $(Et_4N)Aa5 \cdot CH_3CN$. [f] $Ab1 \cdot 2CH_3CN$. [g] $Ab2 \cdot 3.5C_2H_2Cl_4$ with $Mo_2Fe_2S_4$ core. [h] $(Et_4N)_4Ca1$. [i] $(Et_4N)_4Ca3 \cdot 2THF$. [j] $(Et_4N)_3Da3$

Table 3. Selected structural data for cluster compounds Aa

Bond distance (Å) or bond angle (deg)	Aa1	Aa2	Aa3	Aa4	Aa5
$[MoFe_3S_4]^{n+}$	5	5	5	5	4
Mo–Fe[a]	2.694	2.707	2.686	2.696	2.624
Mo–Fe'[b]	2.86	2.75[c]	2.85	2.86	2.87
		(3.00)			
Fe'–Fe'[b]	2.71	2.696	2.697	2.701	2.736
Mo–Sc[d]	2.25[c]	2.28	2.25[c]	2.24[c]	2.27[c]
Fe(Fe')–Sc	2.23[f]	2.22	2.23[f]	2.23[f]	2.27[f]
Sc–Mo–Sc	92–106[c]	101.8	94–106[c]	93–106[c]	99–108[c]
Sc–Fe(Fe')–Sc	94–104[f]	88–107	96–104[f]	95–104[f]	98–108[f]
Fe(Fe')–Sc–Fe'	78.1[f]	75–88	78.3[f]	78.8[f]	76.3[f]
Mo–Sc–Fe(Fe')	78.0[e,f]	74.6	78.0[e,f]	78.5[e,f]	75.5[e,f]
Mo–St[d]	2.46[c]	2.51	2.47[e]	2.47[c]	2.47[c]
Fe(Fe')–St	2.31[f]	2.34	2.32[f]	2.31[f]	2.52[f]

[a] Mo and Fe stand for 6-coordinate atoms. [b] $M = \frac{1}{2}(Mo + Fe)$ and is 6-coordinate, Fe' is 5-coordinate. [c] Mean value for Mo–Fe' (Fe–Fe'). [d] c denotes core, t denotes terminal, same in the following Tables. [e] Mean value includes partial Fe component in Mo. [f] Mean value includes partial Mo component in Fe

Table 4. Selected structural data for cluster compounds C

Bond distance (Å) or bond angle (deg)	Ca1	Ca3	Cb1	Cb2
$MoFe_3S_4$ core				
Mo–Fe	2.716	2.724	2.730	2.735
Fe–Fe	2.692	2.697	2.712	2.696
Mo–Sc	2.354	2.343	2.357	2.356
Fe–Sc	2.262	2.255	2.264	2.262
Sc–Mo–Sc	102.9	102.6	102.7	102.5
Sc–Fe–Sc	109.1	108.4	108.9	108.6
Sc–Fe–Sc'	104.0	104.0	103.8	104.3
Fe–Sc–Fe	73.2	73.5	73.6	73.2
Fe–Sc'–Fe	73.1	72.6	72.4	72.6
Mo–Sc–Fe	72.1	72.6	72.4	72.6
$Mo(\mu-SR)_3Fe(\mu-SR)_3Mo$ bridge				
Mo–Feb	3.594	3.617	3.319	3.462
Mo–Sb	2.567	2.564	2.559	2.567
Feb–Sb	2.537	2.624	2.309	2.529
Mo–Sb–Feb	89.6	90.1	85.8	85.6
Sc–Mo–Sb'	89.2	89.4	89.9	88.4
Sb–Fe–Sb'	76.4	75.7	83.5	79.6
Fe–St	2.264	2.262	2.256	2.252
Mo...Mo'	7.188	7.234	6.638	6.924

Table 5. Selected structural data for cluster compounds D

Bond distance (Å) or bond angle (deg)	Da3	Da	Da5	Da6	Db
[MoFe$_3$S$_4$] core					
Mo–Fe	2.741	2.725	2.723	2.76	2.734
Fe–Fe	2.707	2.691	2.687	2.70	2.695
Mo–Sc[a]	2.330	2.337	2.351	2.37	2.354
Fe–Sc	2.268	2.249	2.260	2.28	2.278
Sc–Mo–Sc	102.65	102.7	102.47	102.0	102.8
Sc–Fe–Sc	107.01	105.5	108.86	108.5	108.4
Sc–Fe–Sc'	104.19		104.17	104.7	103.9
Fe–Sc–Fe	73.49	73.4	73.17	72.8	72.9
Fe–Sc'–Fe	73.05		72.47	71.6	
Mo–Sc–Fe	73.28	72.5	72.46	72.9	72.5
Mo(μ-SR)$_3$Mo bridge					
Mo–Sb[a]	2.583	2.588	2.567	2.62	2.112[b]
S—S	3.058	c	3.111	3.11	2.447[b]
Fe–L	2.190	2.259	2.232	2.30	2.243
Mo–Sb–Mo	93.76	92.23	91.20	93.5	95.8
Sb–Mo–Sb	72.58	73.77	74.59	72.8	71.0
Sb–Mo–Sc	93.63	94.9	90.36	c	92.9
	87.39	85.7	89.18		89.3
Mo...Mo'	3.770	3.685	3.668	3.813	3.13

[a] c denotes core atom linked to Mo atom, b denotes bridge, c' denotes core atom not linked to Mo. [b] For (μ-OMe)$_3$ bridges and Mo–O or O–O distance. [c] Not reported

Table 6. Selected structural data for cluster compounds B

Bond distance (Å) or bond angle (deg)	B1	B2	B3
Mo–Fe	2.76	2.75	2.74, 2.72
Fe–Fe	2.69	2.69	2.71, 2.73
Mo–Sc	2.38	2.37	2.38, 2.37
Fe–Sc	2.27	2.27	2.27, 2.28
Sc–Mo–Sc	101.6	102.1	101.5, 102.4
Sc–Fe–Sc	106.0	106.4	107.6, 107.8
Fe–Sc–Fe	72.3	72.6	73.5, 73.3
Mo–Sc–Fe	72.9	72.6	72.1, 71.9
Mo–St	2.60	—	2.49, 2.57
Fe–St	2.27	2.27	2.25, 2.25

shortening of M–S bonds and lengthening of some of the M–M bonds just as shown in Table 3. Such deformation often predicts decreased stability for cubane core of high oxidation state (vide infra) due to the weakening of the M–M bonds. However, synergism in the MoFe$_3$S$_4$ core is playing an important role in keeping the volume of M$_4$S$_4$ nearly constant throughout all the different oxidation levels of MoFe$_3$S$_4$ core from 2+ to 6+.

Table 7. Comparison of $[MoFe_3S_4]^{3+}$ core volumes of dicubane cluster compounds C and D

Compound	V, Å³ S₄	V, Å³ M₄	S₄/M₄	V, Å³ S₄M₄
Cb1	5.630	2.375	2.37	9.51
Cb2	5.640	2.360	2.39	9.47
Ca1	5.605	2.331	2.40	9.37
Ca3	5.558	2.346	2.37	9.39
Da4	5.482	2.342	2.34	9.33
Da3	5.529	2.379	2.32	9.46
Db1	5.557	2.365	2.35	9.44
Db2	5.625	2.355	2.39	9.45
Da5	5.592	2.333	2.40	9.37

Table 8. Comparison of core volumes for cubane clusters with $[MoFe_3S_4]^{n+}$ unit

Compound	n	V, Å³ S₄	V, Å³ M₄	S₄/M₄	V, Å³ S₄M₄
Da6	2	5.758	2.378	2.42	9.59
B1	3	5.704	2.379	2.40	9.56
Aa5	4	5.183	2.581	2.01	9.80
Aa2	5	4.731	2.589	1.83	9.56
Ab1	6	4.689	2.640	1.78	9.61

5 Evidence for Physicochemical Properties

5.1 ¹H-NMR Spectroscopy

Relatively simple spectra with two sets of signals assigned to α-H of fragments FedtcR₂ and ModtcR₂, respectively, were recorded for Aa1–Aa3 in DMSO-d₆. The data listed in Table 9 showed that the three iron atoms in a $[MoFe_3S_4]^{5+}$ core are equivalent in DMSO solution despite the fact that different coordination environments have been observed by X-ray diffraction (vide supra) of their

Table 9. ¹H-NMR data (ppm) of MoFe₃S₄(R₂dtc)₅ (Aa) in DMSo-d₆ solution at room temperature

Compound	R₂dtcFe, ppm α-H	β-H	R₂dtcMo, ppm α-H	β-H
Aa1	45.7		9.4	
Aa2	32.8		7.9	
Aa3	39.5	2.1	4.7	2.1

single crystals. The possible mechanism as depicted in reaction 4 is that the $Fe-S_t$ bond of the bridging $dtcR_2$ ligand on the six-coordinate iron atom cleaves in solution to afford the equivalence of the three iron atoms.

The proton isotropic shifts of cluster series Ca are listed in Table 10 and their spectra depicted in Fig. 5. The data showed that the six terminal ligands on iron atoms in the two cubane units are equivalent as well as the six ligands bridging the cubane unit and the central Fe^{II} ion, since only one set of signals is observed for each type of protons. Due to the fact that the HOMO of ligating sulfur atoms and the LUMO of the paramagnetic iron atoms are very close to each other and that the atomic orbitals of the substituted R group on the sulfur atom is from a large π-system, delocalization of the unpaired electrons occurs easily over the whole molecular fragment. Such is also the origin of contact interaction which governs the proton isotropic shifts of series Ca cluster compounds. By comparison of ^{1}H-NMR data in this way, structures of Ca2 and Ca4 without single crystal crystallographic data can be deduced from the consideration of synergism of the same cubane core.

5.2 Magnetic Susceptibility

Theoretically, if the magnetic character of a compound is caused by electron spin only, the calculated magnetic moment μ for high temperature approximation

Table 10. ^{1}H-NMR isotropic shifts (ppm) of $(Et_4N)_4[Mo_2Fe_7S_8(SR)_{12}]$ (Ca)

| Compound | $(H/Ho)_{iso}$[a] (ppm) Bridge | | | Terminal | | |
	p-H(CH$_3$)	m-H(CH$_3$)	o-H(CH$_3$)	p-H(CH$_3$)	m-H(CH$_3$)	o-H(CH$_3$)
Ca1	31.0	−14.2	b	11.0	−6.5	b
Ca2	30.6	−16.3	(−25.4)	10.1	−5.9	(−12.3)
Ca3	30.4	−14.2	b	10.1	−6.9	b
		(9.4)			(2.9)	
Ca4	(−31.2)	−13.9	b	(−10.9)	−5.4	b

[a] Determined on Varian FT 80A NMR spectrometer. [b] Undetectable

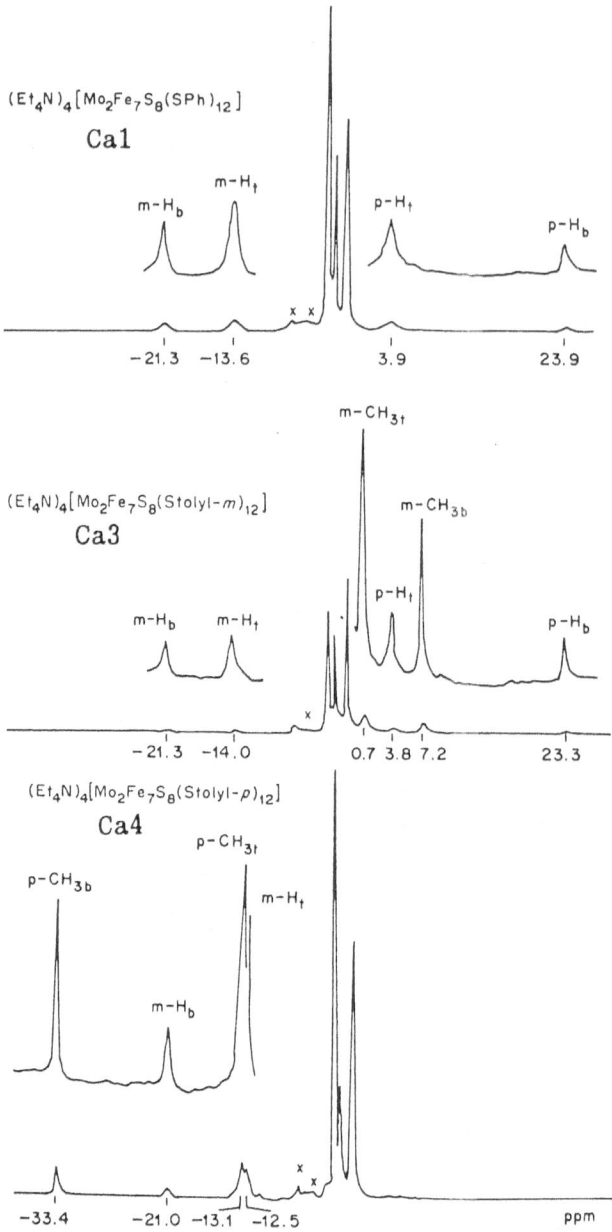

Fig. 5. ¹H-NMR spectra of Ca1, Ca3, and Ca4 in DMSO-d₆ at 298 K. *Ortho*-proton resonances (not shown) occur upfield and are broad

can be obtained from equation $\mu = g\beta[S(S + 1)]^{1/2}$. When the structural fragments (or magnetic centers) are magnetically independent of each other, the total magnetic moment μ_t will be $\mu_t = \sqrt{\sum_i \mu_i^2}$ for i's isolated fragments. The good agreement of the experimental value $(\mu_t)_{obs}$ of 7.3 ~ 7.7 μ_B (Kang 1989) for the compounds Ca with the calculated value $(\mu_t)_{cal}$ of 7.35 μ_B strongly implies that the three magnetic fragments—two $[MoFe_3S_4]^{3+}$ units and a Fe^{2+} bridge—are relatively independent of one another and the spins are 3/2 and 2, respectively. The integrity of $MoFe_3S_4$ core is once again shown by the constancy of the measured magnetic moments.

5.3 Mössbauer Effect

The Mössbauer spectra for Aa2 and Aa5 at 77 K are shown in Fig. 6 and their parameters and those for some related compounds are listed in Table 11. Two sets of data with similar isomer shifts (IS) of 0.37 mm/s but distinctly different quadrupole splittings (QS) of 1.35 and 0.27 mm/s were observed with the latter values corresponding well with those for $Fe_4S_4(dtcR_2)_4$ (5-coordinate iron) and

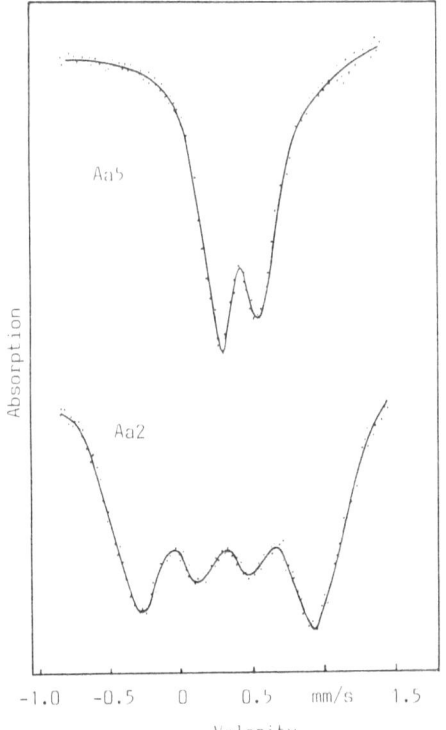

Fig. 6. ^{57}Fe Mössbauer spectra of Aa2 and Aa5 at 77 K

Table 11. Isomer shifts[a] (mm/s) and quadrupole splitting (mm/s) of R_2dtcFe in Aa and related compounds

Compound	IS 289 K	QS	IS 77 K	QS	Relative intensity
$(R_2dtc)_3$Fe	0.39	0.29	0.47	0.55	—
$Fe_4S_4(R_2dtc)_4$	0.34	1.03	0.41	1.42	—
Aa1	0.32	1.32	0.36	1.36	2
	0.32	0.20	0.35	0.19	1
Aa2	0.31	1.21	0.37	1.36	2
	0.29	0.34	0.36	0.28	1
Aa3	0.32	1.20	0.39	1.32	2
	0.31	0.31	0.38	0.33	1
Aa5	0.42	1.11	0.51	1.18	2
	0.43	0.88	0.54	0.86	1

[a] Refer to α-Fe metal at room temperature

Table 12. ^{57}Fe Mössbauer spectral parameters of cluster compounds C

Compound	T K	I.S.[a] mm/s	Q.S. mm/s	Relative intensity
	293	0.31	0.68	5.2
Ca2		0.34	1.04	2.2
		0.90	1.72	1.0
	77	0.41	0.79	3.4
Ca2		0.45	1.32	1.9
		0.96	2.10	1.0
	77	0.41	0.78	5.8
Cb1		0.44	1.17	
		0.43	2.18	1.0
	77	0.42	0.96	5.8
Cb2		0.97	1.90	1.0

[a] Relative to α-Fe at room temperature

Fe$(dtcR_2)_3$ (six-coordinate iron), respectively, indicating the different coordination environments for Fe^{III} atoms in compounds Aa1, Aa2, and Aa3. In comparison with cluster compound Aa2, compound Aa5 with one additional electron for the anion caused an overall change of parameters which is another aspect of synergism of the MoFe$_3$S$_4$ core.

Mössbauer parameters at room temperature and 77 K for compound Ca3 together with those of related compounds from literature (Wolff 1980) are collected in Table 12. The spectrum at 77 K is depicted in Fig. 7. Both the IS and QS of the compounds in the table indicate the presence of two types of Fe atoms in the MoFe$_3$S$_4$ unit: iron(III) and iron(II). Compared to compound Cb1, the extra electron in Ca3 is located at the bridge Fe atom as the parameters of only one iron atom differ drastically between the two compounds. In contrast, the conversion of compound Da4 and Da6 by addition of one electron each to

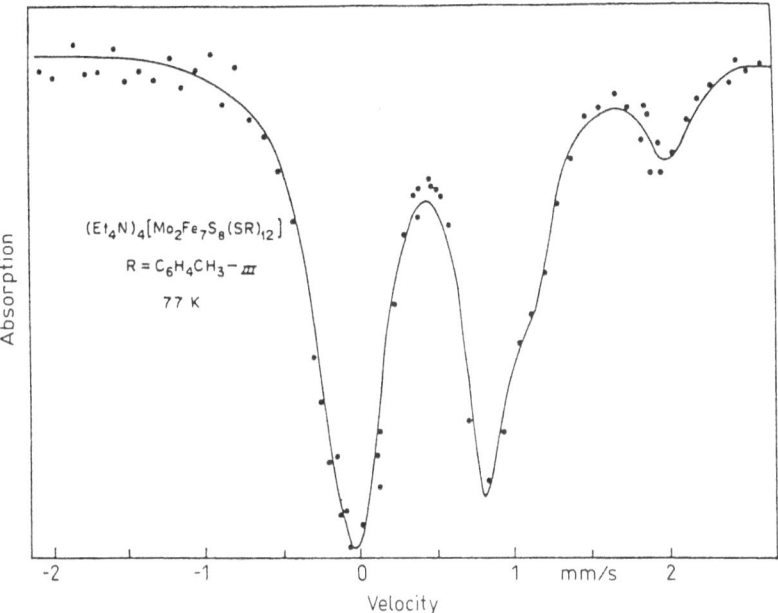

Fig. 7. ^{57}Fe Mössbauer spectra of Ca3 at 77 K

the two cubane units cause a total change of Mössbauer parameters (Christou 1978, 1982) due to synergism of a cluster.

5.4 ESR Spectroscopy

The ESR spectra of Aa3, Aa4, and Aa5 at 77 K are shown in Fig. 8. Assuming a model with antiferromagnetic coupling of metal atoms for Aa3 and Aa4 with $[MoFe_3S_4]^{5+}$, where the three iron atoms are in oxidation state 3^+, a ground-

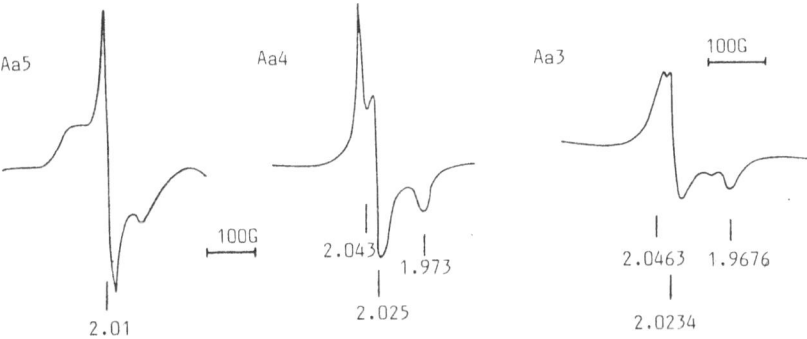

Fig. 8. ESR spectra of Aa3, Aa4, and Aa5

state spin $S = 1/2$ is obtained. In accordance with this argument, the spectrum of Aa5 with $[MoFe_3S_4]^{4+}$ core gave a spin of $S = 1$. Compared to the single $MoFe_3S_4$ cubane clusters with $3+$ core $(S = 3/2)$ (Christou 1982; Armstrong 1982) and $2+$ core $(S = 2)$ (Christou 1982; Mascharak 1983]), it is noted that when the ground-state spins of the clusters decrease from 2, 3/2, 1, 1/2 to 0, the core oxidation states n increase from 2, 3, 4, 5 to 6 in that order. Such orderliness may be relevant to the function of MoFepr which also displays varied electron spin states during nitrogen fixation.

5.5 Cyclic Voltammetry

All the monocubane cluster compounds Aa with same oxidation levels display similar electrochemical behavior as depicted in the cyclic voltammagrams in Fig. 9 and the redox potentials in Table 13. Two quasi-reversible redox couples and one irreversible reduction wave were observed in the range $0.4\,V \sim -1.3\,V$, each corresponds to a one-electron transfer process associated to the $MoFe_3S_4$ core as in reaction 5:

$$[MoFe_3S_4]^{6+} \overset{E_1}{\rightleftharpoons} [MoFe_3S_4]^{5+} \overset{E_2}{\rightleftharpoons} [MoFe_3S_4]^{4+} \overset{E_3}{\longrightarrow} [MoFe_3S_4]^{3+} \quad (5)$$

The irreversible reduction of $4+$ core to $3+$ core implied the destruction of the latter when obtained during electrolysis in the presence of $dtcR_2$ ligand, which has been shown to stabilize cubane cores in high oxidation states.

The redox potentials of compounds Ca are shown in Table 14 and the $E_{p,c}$ values of some related compounds in Table 15. Figure 10 depict the cyclic voltammograms of compounds Ca which showed two redox pairs in the range of $0 \sim -2.0\,V$, each corresponds to a quasi-reversible redox couple for one $[MoFe_3S_4]^{3+}$ core. The redox process is shown in reaction 6 where (ox) is abbreviation for $[MoFe_3S_4(SR)_3]^0$ and (red) for $[MoFe_3S_4(SR)_3]^-$:

$$[(ox)Fe(SR)_6(ox)]^{4-} \overset{E_1}{\rightleftharpoons} [(red)Fe(SR)_6(ox)]^{5-} \overset{E_2}{\rightleftharpoons} [(red)Fe(SR)_6(red)]^{6-} \quad (6)$$

Table 13. Cyclic voltammogram data for $MoFe_3S_4(R_2dtc)_5$ (Aa) in CH_2Cl_2 at room temperature vs. SCE

Compound	$[MoFe_3S_4]^{6+/5+}$			$[MoFe_3S_4]^{5+,4+}$			$[MoFe_3S_4]^{4+,3+}$
	E_1, V $E_{p,c}$	$E_{p,a}$	$E_{\frac{1}{2}}$	E_2, V $E_{p,c}$	$E_{p,a}$	$E_{\frac{1}{2}}$	E_3, V $E_{p,c}$
Aa1	0.13	0.28	0.21	-0.54	-0.37	-0.46	-0.96
Aa2	0.04	0.22	0.13	-0.70	-0.47	-0.59	-0.15
Aa3	0.10	0.22	0.16	-0.55	-0.40	-0.48	-0.97
Aa4	0.10	0.24	0.17	-0.62	-0.45	-0.54	-1.0

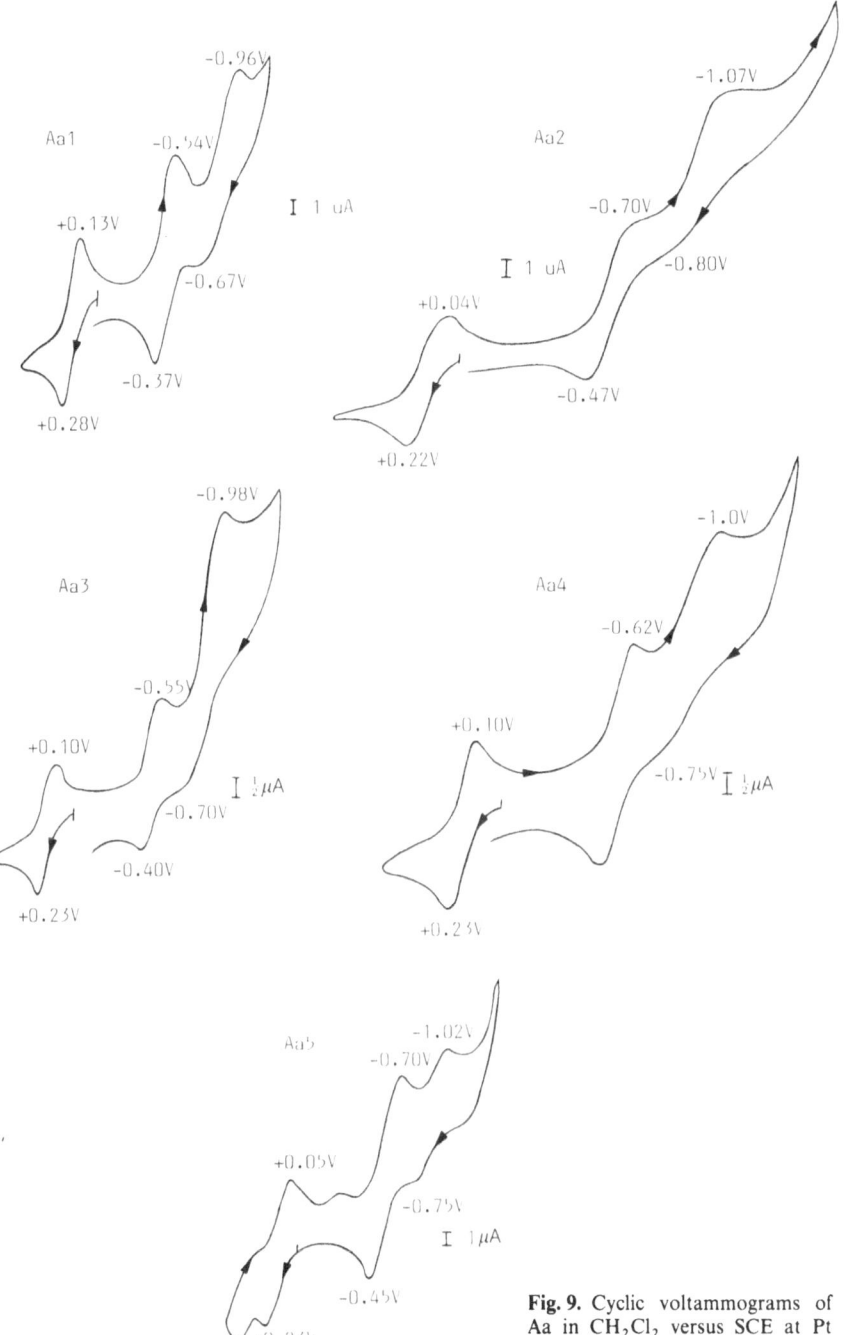

Fig. 9. Cyclic voltammograms of Aa in CH_2Cl_2 versus SCE at Pt electrode

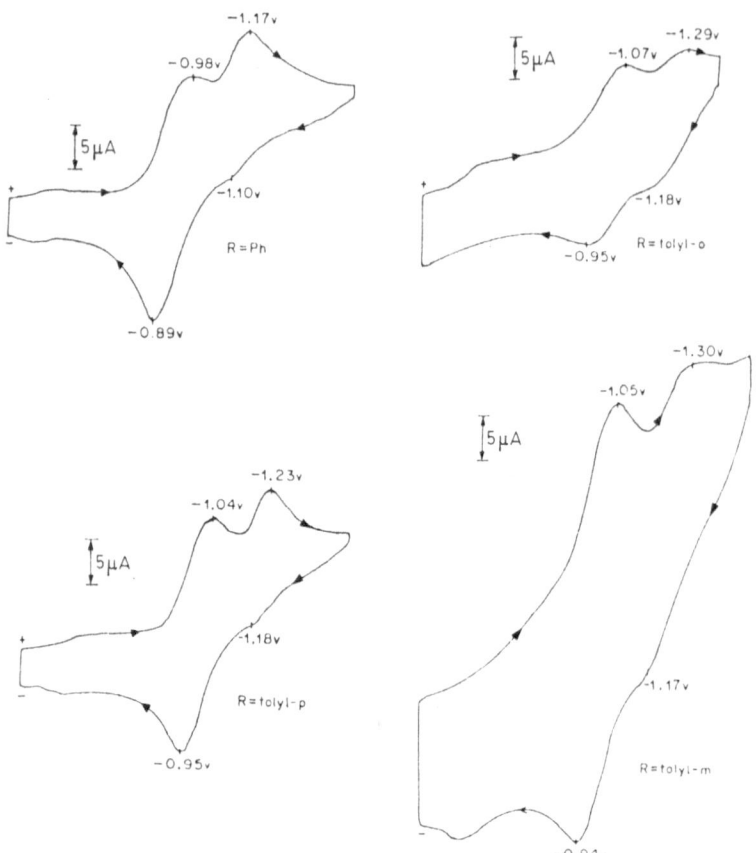

Fig. 10. Cyclic voltammograms of Ca in MeCN versus SCE at Pt electrode

Table 14. Cyclic voltammogram data for $(Et_4N)_4[Mo_2Fe_7S_8(SR)_{12}]^a$(Ca)

Compound[b]	$[Mo_2Fe_7S_8(SR)_{12}]^{4-,5-}$ E_1				$[Mo_2Fe_7S_8(SR)_{12}]^{5-,6-}$ E_2			
	$E_{p,c},V$	$E_{p,a},V$	$E_{1/2},V$	E_{p},mV	$E_{p,c},V$	$F_{p,a},V$	$F_{1/2},V$	E_{p},mV
Ca1	-0.98	-0.89	-0.94	90	-1.17	-1.10	-1.14	70
Ca2	-1.07	-0.95	-1.01	120	-1.29	-1.18	-1.24	110
Ca3	-1.05	-0.94	-1.00	110	-1.30	-1.17	-1.24	130
Ca4	-1.04	-0.95	-1.00	90	-1.23	-1.18	-1.21	50

[a] Pt working and auxillary electrodes, $n\text{-}Bu_4NBF_4$ supporting electrolyte, SCE reference electrode, scan rate 100 mV/s. [b] In MeCN solution at $2\text{-}6 \times 10^{-3}$ M

Table 15. CV data of some dicubane cluster compounds with $[MoFe_3S_4]$ units

Compound	$E_{p.c^a}$, V		E^c, mV
	$[os]-[red]^b$	$[red]-[red]$	
Cb1	−1.67	−1.72	50
Ca1	−0.98	−1.17	190
Ca2	−1.07	−1.29	220
Ca3	−1.05	−1.30	250
Ca4	−1.04	−1.23	190
Da5	−1.56	−1.76	200
Da4	−1.24	−1.44	200
Db2	−1.31	−1.51	200
Da3	−0.84	−1.06	220
Da2	−1.11	−1.31	200

^{a} [ox] denotes $[MeFe_3S_4]^{3+}$ whether in compound C or D, [red] denotes $[MoFe_3S_4]^{2+}$. ^{b} E is the difference of oxidation-reduction potential between ([ox]-[red]) and ([red]-[red])

The value of $|E_{1p,c} - E_{2p,c}| \geq 190\,mV$ for these compounds is comparable to the corresponding value for compounds D ($\sim 200\,mV$) but much larger than for Cb1 ($\sim 50\,mV$), as the latter is bridged by an iron (III) atom in oxidized state which effects strongly and differently the electronic structure of the molecule. This result is an indication that the effect of a reduced cubane unit $[MoFe_3S_4]^{2+}$ upon the second $[MoFe_3S_4]^{3+}$ core is very similar in the two series of compounds Ca and D. This is a further evidence for the relative independence of the three structural fragments in the molecules of compounds Ca and for the strong integrity of a cubane-like core.

6 Final Remarks

Based on analyses of the structural parameters, ^1H NMR, ESR, Mössbauer effects, magnetic susceptibilities, and cyclic voltammetric data of the two series of compounds A and Ca, and with comparison to known compounds B and D, and from the results of chloride substitution reactions, a property of the $MoFe_3S_4$ cubane unit was observed in common: it always functions in an integrity. Namely, synergism among the constituted atoms of a $MoFe_3S_4$ unit is an important factor in the behaviors of these cluster compounds. The addition or abstraction of an electron from it causes changes of the physicochemical properties and structural parameters of the whole cubane unit or even the total cluster rather than local changes of certain metal atom. It may essential that the synergetic effect of the Mo–Fe–S cluster skeleton functions to activate the dinitrogen molecule and assists in its reduction to ammonia for the active center of nitrogenase. To study the synergism of these cluster compounds will definitely be beneficial'to understanding the mechanism of nitrogen fixation of native enzymes and helpful for the chemical simulation of the biological system.

7 References

Armstrong WH, Mascharak PK, Holm RH (1982) Demonstration of the existence of single cubane-type $MoFe_3S_4$ clusters with $S = 3/2$ ground states: preparation, structure, and properties. Inorg Chem 21:1699

Armstrong WH, Mascharak PK, Holm RH (1982) Double bridged double cubanes containing MFe_3S_4 clusters (M = Mo, W). Synthesis, structure, and conversion to spin-quartet single clusters in solution. J Am Chem Soc 104:4373

Burgess BK (1984) Structure and reactivity of nitrogenase—An overview. In: Veeger C, Newton WE (eds) Advances in nitrogen fixation research, Nijhoff, Pudoc, p 103 and references cited therein

Cai JH, Chen CN (1984) Crystal structure of $(Et_4N)_3[Mo_2Fe_6S_8(SPh)_9]$. Jiegou Huaxue 3:33

Cai JH, Kang BS (1984) Crystal structure of $(Et_4N)_4[Mo_2Fe_7S_8(SC_6H_5)_{12}]$. Jiegou Huaxue 3:143

Cai JH, Kang BS (1985) Synthesis and crystal structure of the tri-(μ-methoxy)-bridged double cubane cluster complex $(Et_4N)_3[Mo_2Fe_6S_8(\mu$-OMe$)_3(SPh)_6]$. Jiegou Huaxue 4:82

Cai QR (1982) Fixation of dinitrogen to ammonia via enzymic and non-enzymic catalysis. Acta Sci Natur Univ Amoiensis 21:424

Christou G, Hagen KB, Holm RH (1982) Synthesis, structure and properties of $[Co_8S_6(SPh)_8]^{4-}$ containing an octanuclear Co_8S_6 rhombic dodecahedron related to that of cobalt pentlandite. J Am Chem Soc 104:1744

Christou G, Garner CD (1980) Synthesis and proton magnetic resonance properties of Fe_3MS_4 cubane-like cluster dimers. J Chem Soc Dalton Trans: 2354

Christou G, Garner CD (1980) Ligand substitution reactions of iron-molybdenum sulphur cubane-like cluster dimers; selective halide incorporation. J Chem Soc Chem Comm 613

Christou G, Garner CD, Miller RM (1980) Mössbauer and electrochemical studies on Fe_3MoS_4 and Fe_3WS_4 cubane-like cluster dimers. J Chem Soc Dalton Trans 2363

Christou G, Garner CD, Mabbs FE, King TJ (1978) Crystal structure of $(Bu^n_4N)_4[\{(PhSFe)_3MoS_4\}_2 \cdot (SPh)_3]$: an Fe_3MoS_4 cubic cluster dimer. J Chem Soc Chem Comm 740

Christou G, Mascharak PK, Armstrong WH, Papaefthymiou GC, Frankel RB, Holm RH (1982) Electron-transfer series of $MoFe_3S_4$ double-cubane clusters: electronic properties of components and the structure of $(Et_4N)_5[Mo_2Fe_6S_8(SPh)_9]$. J Am Chem Soc 104:2820

Conradson SD, Burgess BK, Newton WE, Hodgson KO, McDonald JW, Rubinson JF, Gheller SF, Mortenson LE, Adams MWW, Mascharak PK, Armstrong WA, Holm RH (1985) Structural insights from the Mo K-edge X-ray absorption near edge structure of the iron-molybdenum protein of nitrogenase and its iron-molybdenum cofactor by comparison with synthetic Fe–Mo–S clusters. J Am Chem Soc 107:7935

Conradson SD, Burgess BK, Newton WE, Mortenson LE, Hodgson KO (1987) Structural studies of the molybdenum site in the MoFe protein and its FeMo cofactor by EXAFS. J Am Chem Soc 109:7507

Gramer SP, Hodgson KD, Gillum WO, Mortenson LE (1978) The molybdenum site of nitrogenase. Preliminary structural evidence from x-ray absorption spectroscopy. J Am Chem Soc 100:3398

Hedman B, Frank P, Gheller SF, Roe AL, Newton WE, Hodgson KO (1988) New structural insights into the iron–molybdenum cofactor from Azotobacter vinelandii nitrogenase through sulfur K and molybdenum L X-ray absorption edge studies. J Am Chem Soc 110:3798

Holm RH (1981) Metal clusters in biology: quest for a synthetic representation of the catalytic site of nitrogenase. Chem Soc Rev 10:455 and references cited therein

Huang LR, Lin SH (1984) Crystal and molecular structure of $(Et_4N)_3[Mo_2Fe_6S_8(OMe)_3(SBu^i)_6]$. Jiegou Huaxue 3:25

Johnson RW, Holm RH (1978) Reaction chemistry of the iron-sulfur protein site analogues $[Fe_4S_4(SR)_4]^{2-}$. Sequential thiolate ligand substitution reactions with electrophiles. J Am Chem Soc 1978, 100:5338

Kanatzidis MG, Coucouvanis D, Simopoulos A, Kostikas A, Papaefthymiou V (1985) Synthesis, structural characterization and electronic properties of the Ph_4P^+ salts of the mixed terminal ligand cubanes $Fe_4S_4(dtcEt_2)_nX_{4-n})^{2-}$. Two different modes of ligation on the $[Fe_4S_4]^{2+}$ core. J Am Chem Soc 107: 4925

Kang BS, Liu QT, Huang LR, Wu DX, Cai JH, He LJ, Liu HQ, Lu JX (1987) A summary of studies on Mo–Fe–S cluster compounds with $MoFe_3S_4$ unit. Jiegou Huaxue 6: 7

Kang BS, Zhuang BT, Liu QT, Lu JX (1987) Molybdenum–iron–sulfur cluster compounds and chemical modelling of the active center of nitrogenase, in Huang YZ, Qian YL (eds.) Advances in Organometallic Chemistry, Chinese Chem Indus Press, Beijing, p 143

Kang BS, Liu QT, Huang LR, Wu DX, Cai JH, He LJ, Yang Y, Liu HQ, Lu JX (1986) Studies on Mo–Fe–S cluster compounds with $MoFe_3S_4$ unit. Abstracts, 4th Japan–China–US Symposium on Organometallic Chemistry and Catalysis, Tsukuba, Japan, p 128

Kang BS, Cai JH, Chen CN, Lu JX (1986) Synthesis and structural studies on Mo–Fe–S cluster compounds I. Synthesis, crystal and molecular structure of $(Et_4N)_4[Mo_2Fe_7S_8(SPh)_{12}]$. Acta Chimica Sinica 44: 781–786 (Chi); 209 (Eng)

Kang BS, Huang LR, Cai JH, Yang Y, Lu JX (1987) Synthesis and structural studies of Mo–Fe–S cluster compounds. II. Synthesis of di-pseudo-cubane with $Fe(SR)_6$ bridge and structure of $(Et_4N)_4[Mo_2Fe_7S_8(SC_6H_4CH_3-m)_{12}]·2THF$. Acta Chimica Sinica 45: 1152

Kang BS, Cai JH, Wu DX, Liu QT, Weng LH, LU JX (1989) Synthesis and structural studies of Mo–Fe–S cluster compounds V. Transformation of cluster skeletons of compounds with $[MoFe_3S_4]$ unit and crystal structure of $(Et_4N)_3[Mo_2Fe_6S_8(SC_6H_4CH_3-m)_3Cl_6]$. Acta Chimica Sinica 47: 744

Kang BS, Liu HQ, Cai JH, Huang LR, Liu QT, Wu DX, Weng LH, Lu JX (1989) Synergism in molybdenum iron sulphur cluster compounds. Synthesis, structural, magnetic, cyclic voltammetric evidence and reaction of bicubane clusters containing $[MoFe_3S_4]$ unit. Transition Metal Chem 14: 427

Kang BS, Liu HQ, Wang F, Guo Z, Chen WZ, Lu CZ, Wang XP, Lu JX (1988) Synthesis and structural studies of Mo–Fe–S cluster compounds VI. Cyclic voltammetry of dicubane-like $[MoFe_3S_4]$ compounds and the Mössbauer spectrum of $(Et_4N)_4[Mo_2Fe_7S_8(SC_6H_4CH_3-m)_{12}]$. Chinese J Appl Chem 5: 67

Kang BS, Wu DX, Liu HQ (1989) Synthesis and structural studies of Mo–Fe–S cluster compounds. IV. Reaction of dicubane-like $[MoFe_3S_4]$ compounds with acyl chlorides. Acta Chimica Sinica 47: 64

Liu HQ, Wu DX, Kang BS (1987) Synthesis and structural studies on Mo–Fe–S cluster compounds III. Magnetic characterization of double cubane cluster compounds with $Fe(SR)_6$ bridge. Chinese J Magnetic Resonance 4: 13

Liu HQ, Kang BS, Cai JH, Huang LR, Wu DX, Wang F, Guo Z, Cong AZ, Lu JX (1988) Synthesis and structural studies of Mo–Fe–S cluster compounds VII. Spectroscopic study of the $[MoFe_3S_4]$ cubane unit in $(Et_4N)_4[Mo_2Fe_7S_8(SR)_{12}]$ and related compounds. Jiegou Huaxue 7: 171

Liu QT, Huang LR, Kang BS, Liu CW, Wang LL, Lu JX (1986) Investigation of single $MoFe_3S_4$ cubane clusters I. First successful synthesis by spontaneous self-assembly reaction and structure of $(Et_4N)[MoFe_3S_4(Et_2NCS_2)_5]·CH_3CN$. Acta Chimica Sinica 44: 343 (Chi); 107 (Eng).

Liu QT, Huang LR, Kang BS, Yang Y, Lu JX (1987) Synthesis and structure of single $MoFe_3S_4$ cubane cluster compound $[MoFe_3S_4(C_5H_{10}NCS_2)_5]·CH_2Cl_2$. Kexue Tongbao 32: 898 (Eng).

Liu QT, Huang LR, Kang BS, Yang Y, Lu JX (1987) Investigation of single $MoFe_3S_4$ cubane-like clusters II. Synthesis and structure of $[MoFe_3S_4(C_5H_{10}NCS_2)_5]$. Acta Chimica Sinica 45: 133

Liu QT, Huang LR, Yang Y, Lu JX (1988) Investigation of single $MoFe_3S_4$ cubane-like cluster compounds III. The conversion of linear trinuclear cluster into single cubane cluster and the structure of $[MoFe_3S_4(Et_2NCS_2)_5]·MeCN$. Acta Chimica Sinica 46: 1

Liu QT, Lei XJ, Huang LR, Chen WZ, Zhao K, Liu HQ, Lu JX (1990) Studies on the synthesis, structure and electrochemistry of $MoFe_3S_4$ and WFe_3S_4 cubane-like cluster compounds. Science in China (Ser B) 561

Liu QT, Huang LR, Liu HQ, Lei XJ, Wu DX, Kang BS, Lu JX (1990) Structural chemistry of molybdenum–iron–sulfur cluster compounds with single cubane $[MoFe_3S_4]^{n+}$ (n = 4, 5, 6) core and crystal structure of $[MoFe_3S_4(Me_2dtc)_5]·2CH_2Cl_2$. Inorg Chem 29: 4131

Liu QT, Huang LR, Yang Y, Lu JX (1988) Investigation of single $MoFe_3S_4$ cubane-like cluster compounds IV. Synthesis and structure of a novel doubly-bridged single cubane-like cluster compound $MoFe_3S_4(\mu-Me_2NCS_2)_2(Me_2NCS_2)_4·2MeCN$. Acta Chimica Sinica 46: 1075 (Chi); 286 (Eng)

Liu QT, Kang BS, Chen CN, Huang LR, Cai JH, Zhuang BT, Lu JX (1989) Study on the syntheses of cluster compounds containing [MoFe$_3$S$_4$] cubane-like unit. Scientia Sinica (Ser B) 32: 392 (Eng)

Liu QT, Huang LR, Lu JX (1990) Single Mo$_n$Fe$_{4-n}$S$_4$ (n = 0, 1, 2) cubane-like clusters: Preparation from combination of dinuclear complexes and the structure of [Mo$_2$Fe$_2$S$_4$(C$_4$H$_8$dtc)$_6$]·3.5C$_2$H$_2$Cl$_4$. Science in China (Ser B) 787

Lu JX (1975) Primary model for the active site of nitrogenase–and structural requirements for the coordination and activation of dinitrogen molecule. Kexue Tongbao 20: 540

Mascharak PK, Armstrong WH, Mizobe Y, Holm RH (1983) Single cubane-type MFe$_3$S$_4$ clusters (M = Mo, W): synthesis and properties of oxidized and reduced forms and the structure of (Et$_4$N)$_3$[MoFe$_3$S$_4$(SC$_6$H$_4$Cl-p)$_4$(3, 6-(C$_3$H$_5$)$_2$C$_6$H$_2$O$_2$)]. J Am Chem Soc 105: 475

Mascharak PK, Papaefthymiou GC, Armstrong WH, Foner S, Frankel RB, Holm RH (1983) Electronic properties of single- and double-MoFe$_3$S$_4$ cubane-type clusters. Inorg Chem 22: 2851

Orme-Johnson WH (1983) Molybdenum and iron centers in nitrogenase. In: Abstract for 186th ACS Annual Meeting, Washington D.C., 1983, no. 301

Palermo RE, Holm RH (1983) Reactivity properties of single MoFe$_3$S$_5$ cubane-type clusters: synthesis, structure, and ligand substitution reactions of [MoFe$_3$S$_4$Cl$_3$(3, 6-(C$_3$H$_5$)$_2$C$_6$H$_2$O$_2$)(THF)]$^{2-}$. J Am Chem Soc 105: 4310

Palermo RE, Power PP, Holm RH (1982) Ligand substitution properties of the MFe$_3$S$_4$ double-cubane cluster complexes [Mo$_2$Fe$_6$S$_8$(SR)$_9$]$^{3-}$ and [M$_2$Fe$_7$S$_8$(SR)$_{12}$]$^{3-}$ (M = Mo, W). Inorg Chem 21: 173

Palermo RE, Singh R, Bashkin JK, Holm RH (1984) Molybdenum atom ligand substitution reactions of MoFe$_3$S$_4$ cubane-type clusters: synthesis and structures of clusters containing Mo-bonded pseudosubstrates of nitrogenase. J Am Chem Soc 106: 2600

Shah VK, Brill WJ (1977) Isolation of an FeMoco from nitrogenase. Proc Nat Acad Sci USA 74: 3249

Teo BK, Averill BA (1979) A new cluster model for the iron-molybdenum cofactor of nitrogenase. BBRC 88: 1454

Wolff TE, Berg JM, Hodgson KO, Frankel RB and Holm RH (1979) Synthetic approaches to the, molybdenum-iron-sulfur "double-cubane" cluster complexes [Mo$_2$Fe$_6$S$_8$(SEt)$_9$]$^{3-}$ and [Mo$_2$Fe$_6$·S$_7$(SEt)$_8$]$^{3-}$. J Am Chem Soc 101: 4140

Wolff TE, Berg JM, Power PP, Hodgson KO and Holm RH (1980) Structural characterization of the iron-bridged "double-cubane" cluster complexes [Mo$_2$Fe$_7$S$_8$(SC$_2$H$_5$)$_{12}$]$^{3-}$ and [M$_2$Fe$_7$S$_8$·(SCH$_2$C$_6$H$_5$)$_{12}$]$^{4-}$ containing MFe$_3$S$_4$ cores. Inorg Chem 19: 430

Wolff TE, Power PP, Frankel RB, Holm RH (1980) Synthesis and electronic and redox properties of "double-cubane" cluster complexes containing MoFe$_3$S$_4$ and WFe$_3$S$_4$ cores. J Am Chem Soc 102: 4694

Wong GB, Bobrik MA, Holm RH (1978) Inorganic derivatives of iron sulfide thiolate dimers and tetramers: synthesis and properties of the halide series [Fe$_2$S$_2$X$_4$]$^{2-}$ and [Fe$_4$S$_4$X$_4$]$^{2-}$. Inorg Chem 17: 578

Zhang YP, Bashkin JK, Holm RH (1987) Phosphine cleavage of iron(III)-bridged double cubanes: A new route to MoFe$_3$S$_4$ single cubanes. Inorg Chem 26: 694

CHAPTER 7

Synthesis, Structure and Properties of some Mo–Fe–S, Mo–S Model Compounds for the Mo–Fe Center and the First Coordination Sphere of the Mo–Atom in Nitrogenase

Bo-Tao Zhuang, Liang-Ren Huang, Xiao-Ten Zhao, Pen-Cheng Chen, An-Jian Lan, and Jia-Xi Lu

The Nitrogen Fixation
and its Research in China
Editor Guo-fan Hong
© Springer-Verlag Berlin Heidelberg 1992

1 Introduction

Nitrogenase is a two-component redox enzyme that catalyzes the reduction of dinitrogen to ammonia. The structure and function of Mo in the MoFe-protein of this enzyme has been the subject of a number of investigations including the use of EPR, Mössbauer (Muenck et al. 1975; Zimmermann et al. 1978; Huynh et al. 1979) and X-ray absorption spectroscopy (XAS) which includes the extended X-ray absorption fine structure (EXAFS) and X-ray absorption near edge structure (XANES) (Cramer et al. 1978, 1978a; Conradson et al. 1987, 1985). The recent results of these investigations on the MoFe-protein and its FeMo-co have revealed that the active center responsible for catalyzing the reduction of nitrogen to ammonia in nitrogenase involves a iron–molybdenum–sulfur aggregate with the ratio of Fe:Mo:S of 1:6–8:4–9 (Shah et al. 1977; Yang et al. 1982; Nelson et al. 1983; Burgess et al. 1983) and the environment of the molybdenum atom of FeMo-co in and after extrusion from the MoFe-protein matrix contains two to three oxygen atoms at $2.10\,\text{Å}$, three to four sulfur atoms at $2.37\,\text{Å}$ and three to four iron atoms at $2.70\,\text{Å}$. (Conradson et al. 1987, 1985).

Although the single crystal structure of FeMo-co or MoFe-protein has not been determined successfully so that the nature of the full Mo–Fe–S site in nitrogenase has been unclear, this limited structural information in addition to the results from biological investigation (Kennedy et al. 1981; Roberts et al. 1981; Mortenson et al. 1979; Burgess et al. 1984; Hawkes et al. 1984) has elicited the suggestion of various structural models for defining the Fe–Mo–S aggregate, for example, Fe_3MoS_4-cubane-like cluster (I), butterfly type MoS_4Fe_2 (II), Fuzhou string-bag model (III), Xiamen model (IV) and so on (Cramer et al. 1978, 1978a; Lu 1975, 1980; Tsai 1976, 1980, 1983; Teo et al. 1979; Christon et al. 1982; Orme-Johnson 1983). The investigations on the chemistry of Mo–Fe–S cluster compounds and Mo–S complexes in order to attain a synthetic representation of the FeMo-co and Mo-site of nitrogenase have received increasing attention in recent literature (Holm 1981; Coucouvanis 1981).

A great many of Fe–Mo–S cluster compounds including cubane-like clusters of the stoichiometry $[Mo_2Fe_5S_8(\mu_2-L)_3(L')_6]^{3-,5-}$ ($L = RS^-$ or RO^-, $L' = RS^-$ or ArO^-), $[Mo_2Fe_7S_8(SR)_{12}]^{4-}$, $[MoFe_3S_4(cat)(SR)_3L]^{2-,3-}$ (cat = substituted catecholate, L = any of a variety of ligands) and $[MoFe_3S_4(dtc)_5]^{0,1-}$, (dtc = dithiocarbamate) (Holm et al. 1981; Garner et al. 1983; Liu 1988) and "linear" type of $[S_2MoS_2FeL_2]^{2-}$ ($L = ArS^-$, ArO^- or OAc^-), $[Cl_2FeS_2MoS_2FeCl_2]^{2-}$ and $[S_2MoS_2FeS_2Fe^-(SAr)_2]^{3-}$ (Coucouvanis et al. 1979, 1980, 1983; Tieckelmann et al. 1980; Silvis et al. 1981; Teo et al. 1982; Zhuang et al. 1983; Tieckelmann et al. 1980a; Coucouvanis et al. 1980; Mueller et al. 1980) have been synthesized and studied as modelling compounds. Only a few other classes of Fe–Mo–S compounds (Bose et al. 1986, 1989; Eldredge et al. 1988) and Mo–S–O complexes which reflect the XAS data of the first coordination sphere of the Mo atom in FeMo-co (Christou et al. 1980; Wolff et al. 1981) have been reported. Our efforts to approach the structure and functioning of the active center of nitrogenase are to attempt to produce the synthetic model compounds of the FeMo center and the first coordination sphere environment of the Mo atom of nitrogenase by reaction of low-valence molybdenum sources which differ from the high-valence molybdenum source, MoS_4^{2-}, most other research groups have used. We have synthesized some new classes of Fe–Mo–S and Mo–S–O cluster compounds which possess the Mo–Fe, Mo–S, and Mo–O distances comparable to the EXAFS data of nitrogenase and simulate the first coordination sphere environment of the Mo atom in nitrogenase or exhibit the detectable UW45 constitution activity to reduce dinitrogen to ammonia. Herein, we summerize the synthesis, structure and properties of those compounds and discuss the implication derived from the results.

2 A New Class of Molybdenum–Iron–Sulfur Cluster Compounds with μ_3-S, Low-Valence Molybdenum Atom and Iron–Molybdenum Bond Distances Relevant to Nitrogenase. $[MFe_2S_2(CO)_8(S_2CNEt_2)]^-$ (M = M, W)

Reaction of one equiv. of metal(0) dithiocarbamato carbonyl $[M(CO)_4-(S_2CNEt_2)]^-$ (M = Mo (Zhuang et al. 1988, 1989), W (Note 1)) with a solution of $Li_2[Fe_2S_2(CO)_6]$ (Seyferth et al. 1982) (generated in situ from addition of $LiEt_3BH$ in THF to a solution of $Fe_2S_2(CO)_6$ in MeOH at $-78\,°C$) in a mixed solvent of THF, MeOH and MeCN at room temperature for 24 h, affords a dark-red crystalline product $[Et_4N][MFe_2(\mu_3\text{-}S)_2(CO)_8(S_2CNEt_2)]$ ($M = Mo(V), W(VI)$). The yields are 42% and 36% for V and VI, respectively.

The X-ray crystal structures of V and VI (Figs. 1, 2) (Note 2) reveal that V and VI are new classes of Mo(W)–Fe–S clusters which contain neither cubane-like MoFe$_3$S$_4$ nor linear Mo(μ_2-S)$_2$Fe unit but a new type of Mo(μ_3-S)$_2$Fe$_2$ core with the Mo atom in a low oxidation state. As shown in Fig. 1, the anion of V has only Cl symmetry and is best viewed as consisting of a distorted tetragonal pyramid MoS$_2$Fe$_2$ core (with Mo–Fe bond distance of 2.76 Å, Mo–S of 2.4 Å and a pseudo-planar Fe$_2$S$_2$ unit with Fe–S bond length of 2.25 Å, Fe···Fe distance of 3.395 Å and dihedral angle of 162.33° between two Fe$_2$S$_2$ planes) in which the Mo atoms are coordinated by additional bidentate dithio-carbamate (dtc) ligand S$_2$CNEt$_2$ and two terminal carbonyls, and each Fe atom contains three terminal CO ligands. The geometry around the Mo atom is a trigonal prism containing S(1)S(3)C(8) and S(2)S(4)C(7) triangles and both Fe atoms are in five coordination environment. There is a mirror plane, which contains the two μ_3-S, Mo, C(10) and N atoms, in the anion of V if the two ethyls of the dithiocarbamate ligands are neglected. Mo–S (from dtc) and Mo–C bond distance are 2.52 Å and 1.96 Å, respectively, which are comparable to that observed in $[Mo(CO)_4dtc]^-$ (Zhuang et al. 1988) and other Mo(0)–(SR) complexes (Zhuang et al. 1986, 1989a, 1989b; Huang et al. 1989).

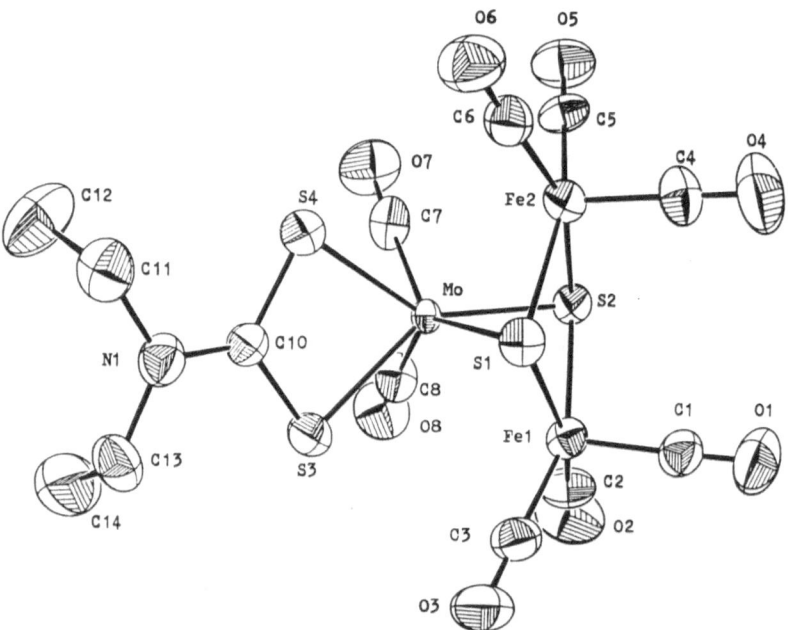

Fig. 1. Molecular structure of the amion of V

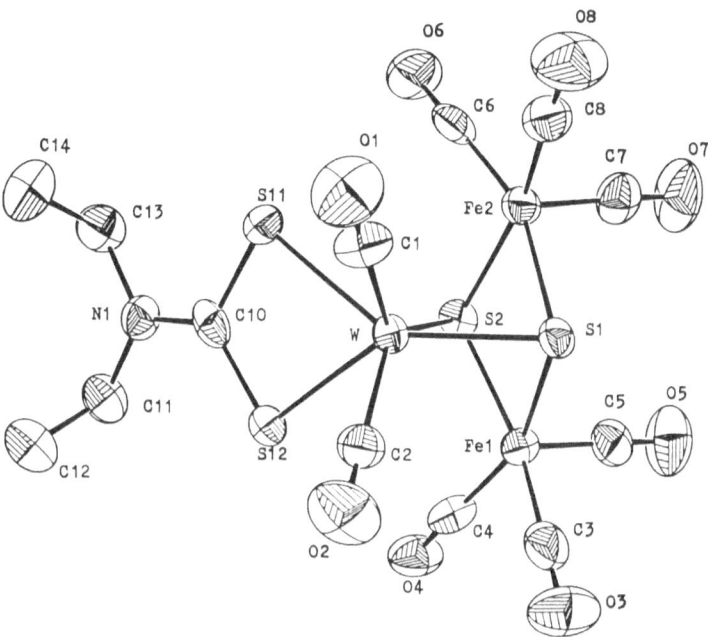

Fig. 2. Molecular structure of the anion of *VI*

The ^{57}Fe Mössbauer spectra (Note 3) of V at 77 K consists of an overlapping quadrupole doublet isomer shift $\delta = 0.008$ mms^{-1} and the quadrupole splitting $\Delta Eq = 0.965$ mms^{-1} which are typical of low spin Fe^{2+} $(S = 0)$ (Gibb 1976; Bancroft 1973). The fact that the magnetic susceptibility measurement shows that V is diamagnetic also supports this judgement. The cyclic Voltametry of V in MeCN indicates a reversible 1e$^-$ oxidation at -0.05 V vs SCE and a quasi-reversible oxidation at $+0.25$ V vs SCE (Note 4).

From the structure and properties, several notable features of cluster V could be pointed out as following: (a) Cluster V is a new class of Fe–Mo–S cluster containing neither known cubane-like Fe$_3$MoS$_4$ nor linear FeS$_2$Mo unit but a Mo(μ_3-S)$_2$Fe$_2$ core with Mo atom in low oxidation state. In terms of the Mo–S (from dtc) and Mo–C bond distance similar to that observed in the Mo(O)–SR complexes and the ^{57}Fe mossbauer spectra parameters indicating low-spin Fe^{+2} it leaves not much room for doubt that the Mo atom in V is in zero oxidation state. (b) V possesses two Mo–Fe bond distances of ca. 2.7 Å, which approximate to the case for FeMo-co, for which EXAFS measurements have indicated the presence of 3 ± 1 iron atoms at ca. 2.68 ~ 2.70 Å from Mo (Conradson et al. 1987, 1985). In addition to the similar Mo–S distance, V thus provides a new synthetic model for understanding the bonding mode of Mo, Fe and S atoms in Fe–Mo center of nitrogenase. (c) The configuration of Fe$_2$S$_2$(CO)$_6$ unit in V has the change in cleavage of Fe–Fe bent bond and increase of the oxidation state of Fe atoms in contrast with the high-valence Mo complexes containing $Fe_2S_2(CO)_6$ unit (for example, $[MoOFe_5S_6(CO)_{12}]^{2-}$ (Bose et al. 1986) and $Fe_2(CO)_6S(\mu_2\text{-}S)_2Mo(NNMePh)_2\text{-}(PPh_3)$ (Dilworth et al. 1986) in which the $Fe_2S_2(CO)_6$ unit remains the configuration similar to that of the free $Fe_2S_2(CO)_6{}^{2-}$ ion (Seyferth et al. 1982). It is obvious that the configuration change of $Fe_2S_2(CO)_6$ unit in addition to the creation of Fe–Mo bond is caused by entrance of the low-valence of Mo atom. As a matter of fact, a similar case was observed for Co-analog (Eremenko et al. 1989). This is also for meeting the need to obey the $9n$-L skeletal bonding electron pairs rule (Tang et al. 1983) (50 electrons and L = 2). Importantly, the fact that introduction of low-valence Mo atom results in configuration change of $Fe_2S_2(CO)_6$ unit implies that the use of low valence Mo source in synthesis is capable of producing a variety of Fe–Mo–S cluster provided that the $Fe_2S_2(CO)_6{}^{2-}$ is used as Fe-sources. (d) The dtc and carbonyls of Mo atom in V will be able to be replaced by another donating metal moieties such as Fe–S moiety according to the reactivity of $[Mo(CO)_4(dtc)]^-$ (Zhuang et al. 1988, 1989). So, there is a possibility that further reaction of cluster V with certain Fe–S compound leads to a Fe–Mo–S cluster compounds with high ratio of Fe/Mo which approximate to the core stoichiometry distribution of FeMo-co.

3 Iron–Molybdenum–Sulfur Compound with Clathrate Crystal Structure: G-Series of Modelling Compound with some Ability to Reduce Dinitrogen to Ammonia

An attempted synthesis of modelling compounds for the cubane-like "Fuzhou string-bag model" (Lu 1975, 1980) of the active center of nitrogenase in biological nitrogen fixation in our laboratory in 1979, has led to the successful isolation of a so-called "G" (Note 5) series of modelling compounds, one of which exhibits somewhat ability to reduce dinitrogen to ammonia under standard assaying conditions.

1 equiv. of $(NH_4)_2MoS_4$ and 4 equiv. of $FeCl_3$ were reacted with 11 equiv. of Grignard reagent EtMgBr in diethylether-THF under argon atmosphere at room temperature for 10 h resulting in a black precipitate, and the fractional crystallization of the resulting black precipitate from DMF or DMF–THF affords the "G" series of crystalline Fe–Mo–S compounds including GA (containing ca. 10% Mo), GB (containing ca. 2.5% Mo) and GC (containing ca. 0.4% Mo) (Lu 1980a; Huang et al. 1979, 1981). On the basis of elemental analysis, GA and GB are assigned the chemical formula $(Mg\cdot6DMF)_6[Mo_6Fe_8S_{25}Cl_{22}]$ and $(Mg\cdot6DMF)_4[MoFe_{10}S_7Cl_8Br_{10}]$, respectively. GC is roughly consistent with the Fe_2S_2-type compound contaminated with a little Mo impurity. Interestingly enough, crystal GB with a moderate Mo content has shown the ability to reduce C_2H_2 and N_2 to C_2H_4 and NH_3, respectively (confirmed beyond any doubt

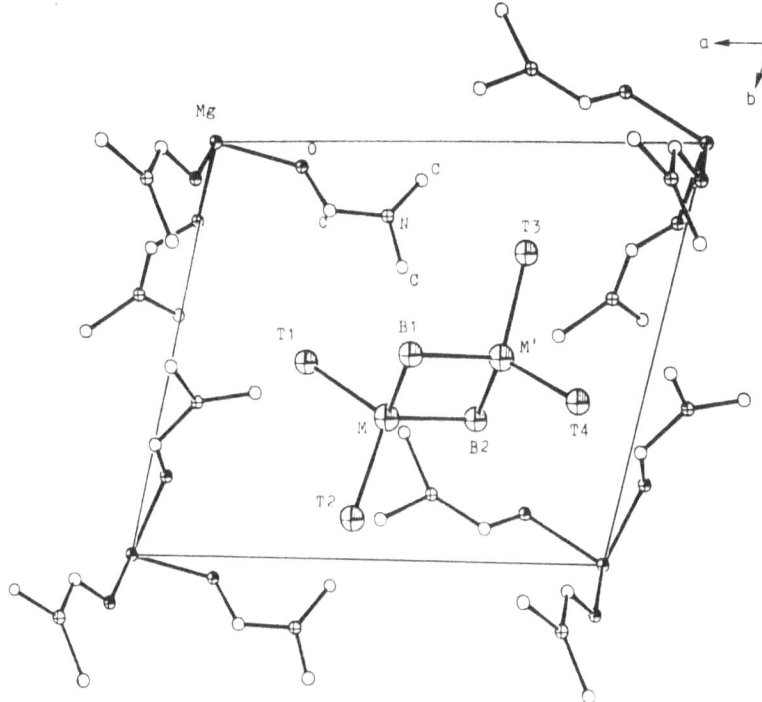

Fig. 3. Prejection of unit cell (001)

by using isotopically labelled N_2), with good selectivity, either in the presence of KBH_4 or in the pseudo-enzyme-catalyzed reaction on combination with UW45 mutant strain plus an ATP generating system (Huang et al. 1981).

The structures of the series G crystals crystallized from DMF are even more informative (Huang et al. 1983, 1983a). All these crystals, irrespective of GA, GB or GC are found to crystallize in the same triclinic unit cell with the following parameters: a = 11.27 Å, b = 9.47 Å, c = 9.25 Å, $\alpha = 87.2°$, $\beta = 71.2°$, $\gamma = 74.9°$; space group P1; Z = 1 (referred to Mg·6DMF). Upon careful analysis of their X-ray diffraction data, it turned out that each of these G series of crystals possesses a clathrate structure with rather simple dianions of the following variable constitution but essentially identical size and configuration:

$$\begin{smallmatrix} T \\ \\ T \end{smallmatrix}\!\!>\!M\!\!<\!\!\begin{smallmatrix} X \\ \\ X \end{smallmatrix}\!\!>\!M'\!\!<\!\!\begin{smallmatrix} T' \\ \\ T' \end{smallmatrix},\ (M, M' = Fe, Fe\ or\ Fe, Mo;\ X, T, T' = S, Cl)$$

enclosed in a triclinic unit cell cage, the corners of which are occupied by the dicationic $[Mg(DMF)_6]^{2+}$ (Fig. 3). Thus, the GA crystal can be reformulated as $(Mg·6DMF)\,[(MoFeS_4Cl_2)_{6/7} - (Fe_2Cl_6)_{1/7}$; in other words, there is imbedded in the dicationic $(Mg(DMF)_6)^{2+}$ cage a cluster dianion $[MoFeS_4Cl_2]^{2-}$ (Tieckelmann et al. 1980) intermixed with some (probably one-sixth as much)

$[Fe_2Cl_6]^{2-}$, and GC is obviously consistent with the formula $[Mg(DMF)_6] \cdot [Fe_2S_2Cl_4]$ containing $[Fe_2S_2Cl_4]^{2-}$ dianion (Wong et al. 1978). For the GB crystal it is plausible to infer that there are enclosed in the $[Mg(DMF)_6]^{2+}$ cage at least three different types of dianions, which could be formed under the reaction conditions, with molecular sizes comparable to and molecular shapes similar to $[MoFeS_4Cl_2]^{2-}$ as follow:

(X = Cl, Br)	(X = Cl, Br)	(X = Cl, Br)
VII	*VIII*	*IX*

Taking the structure into account, the fact that neither GA nor GC but GB exhibits the ability to reduce N_2 to NH_3 seems to be easy to understand. In view of the formation of black Roussinate monoanion $[Fe_4S_3(NO)_7]^-$ (X) (Johansson et al. 1958) by the "dimeric" condensation of two red Roussinate dianions $[Fe_2S_2(NO)_4]^{2-}$ (XI) under the action of either a weak acid or a sulfide ion precipitant of moderate

X	*XI*

strength, thus facilitating the removal of one nitrosyl ligand and one labile sulfur atom, it is reasonable to speculate that some of those dianions *VII* and *VIII*, in particular, will be able to undergo a "dimeric" condensation (in a manner somewhat similar to the case for the Roussinate) under standard assaying conditions to form an active species (probably a polynuclear cluster with cubane-like 'string-bag' Fe–Mo–S core like Fuzhou model (Lu 1975, 1980) exhibiting the reducing ability for N_2 (to form NH_3).

It is interesting to note that the GB crystal with only several dinuclear clusters containing sulfido-bridges is able to exhibit nitrogen fixation reduction activity under assay conditions. This implies that the model compound for the active center of nitrogenase will be able to be synthesized by the reaction of some reactive fragments such as the dinuclear compound containing sulfur bridges under certain reaction conditions and has led to our "Unit construction" approach to the rational synthesis of transition-metal cubane-like clusters by the use of reactive fragments as building blocks (Lu et al. 1989).

4 New Mo–S–O Compounds Containing the O_3MoS_3 Structural Unit to Reflect the XAS Data of the First Coordination Sphere of the Mo Atom in FeMo-co of N_2-ase: Mono-, Di-, and Trinuclear Mo Complexes $[Mo(S, O-C_6H_4-1,2)_3]^-$, $[Mo_2(CO)_3(S, O-C_6H_4-1,2)_3]_2^-$, and $[Mo_3(CO)_7(S, O-C_6H_4-1,2)_3]_2^-$

The reaction of molybdenum hexacarbonyl with o-hydroxyl-phenthiolate and air-treated o-hydroxylphenthiolate in acetonitrile under nitrogen atmosphere affords the dinuclear product $[Et_4N]_2[Mo_2(CO)_3(S, O-C_6H_4-1, 2)_3]$ (XIII) at ~45 °C and the trinuclear compound $[Et_4N]_2[Mo_3(CO)_7(S, O-C_6H_4-1,2)_3]$ (XIII) at ~60 °C (Zhuang et al. 1989c). The mononuclear species $[Et_4N]$-$[Mo(S, O-C_6H_4-1,2)_3]$ (XII) could be isolated when the resulting reaction solution is treated by air in the presence of $[Fe(SR)_4]^{2-}$.

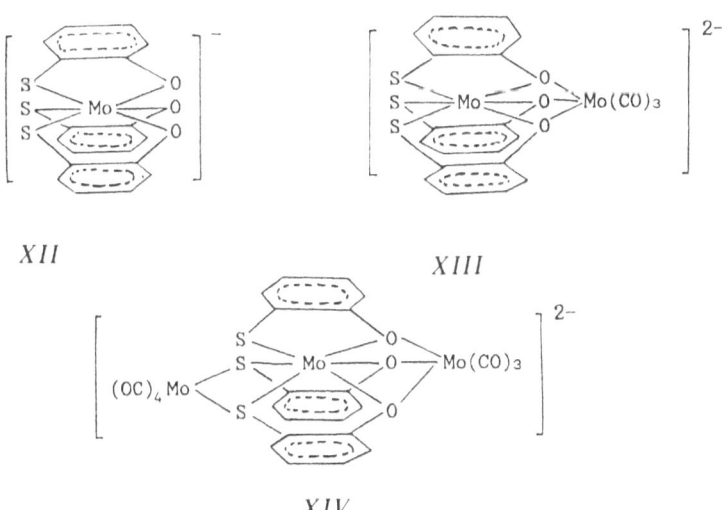

XII XIII

XIV

The X-ray crystal and molecular structures of the anions of XII, XIII and XIV are shown in Figs. 4, 5 and 6, respectively (Note 6). As shown in the figures, the anion of XII possesses a simple mononuclear structure with coordination of three bidentate $O, S-C_6H_6-1,2$ ligands to a Mo^{5+} ion resulting in a distorted

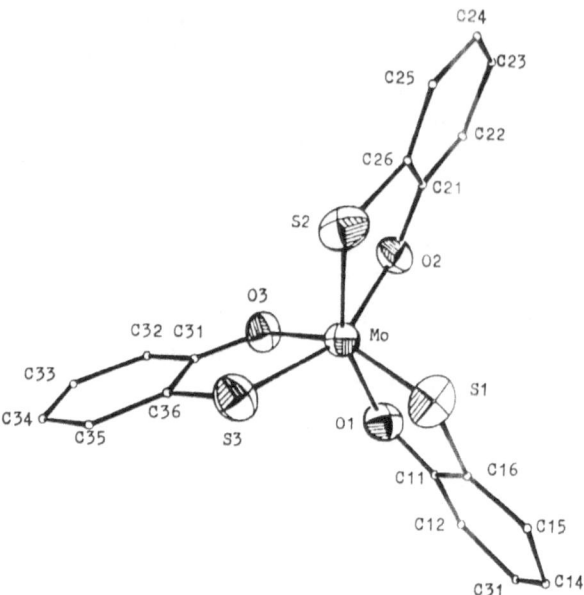

Fig. 4. Molecular structure of the anion of *XII*

trigonal prismatic O_3MoS_3 unit (containing $O(1)O(2)O(3)$ and $S(1)S(2)S(3)$ triangles with a small twist angle) with three Mo–O bond distances of 1.99(1), 1.97(2), 1-99(2) Å and three Mo–S bond lengths of 2.385(5), 2.372(6), 2.377(6) Å. The anion of *XIII* has a pseudo-C_3 axis through the two Molybdenum atoms. The geometry around one Mo atom is a trigonal prism with three S atoms (at 2.314(6), 2.324(5), 2.328(6) Å) and three 0 atoms (at 2.06(2), 2.06(2), 2.06(2) Å) from three bidentate (S, O-C_6H_4-1,2) ligands, and the three 0 atoms, as bridging ligands, are also linked to the other Mo atom which has three terminal carbonyls resulting in a distorted octahedron geometry (containing three Mo–O bond distances of 2.26(2), 2.24(1), 2.26(1) Å and three Mo–C of 1.91(3), 1.88(3), 1.92(2) Å). Thus, the overall coordination geometry of the anion of *XIII* can be described as a trigonal prism and a distorted octahedron sharing a face. The Mo–Mo bond distance is 3.168(2) Å showing the existence of Mo–Mo weak interaction. The different coordination environment around the two Mo atoms obviously indicates the two Mo atoms in *XIII* are in different oxidation states. The anion of *XIV* can be considered to be the structure that consists of *XIII* of which the two sulfur atoms coordinate to an additional $Mo(CO)_4$-moiety resulting in a trinuclear cluster containing three Mo atoms in different coordination environments with two Mo–Mo bond distance of 3.1843(8) and 3.2145(8) Å and MoMoMo angle of 118.93(2). The geometry around Mo(1) (a trigonal prism with three Mo–S distances of 2.372(1), 2.370(1), 2.302(1) Å and three Mo–O bond lengths of 2.070(3), 2.074(3), 2.056(3) Å) and Mo(2) (a distorted octahedron with three Mo–O of 2.270(3), 2.255(3), 2.286(3) Å and three Mo–C

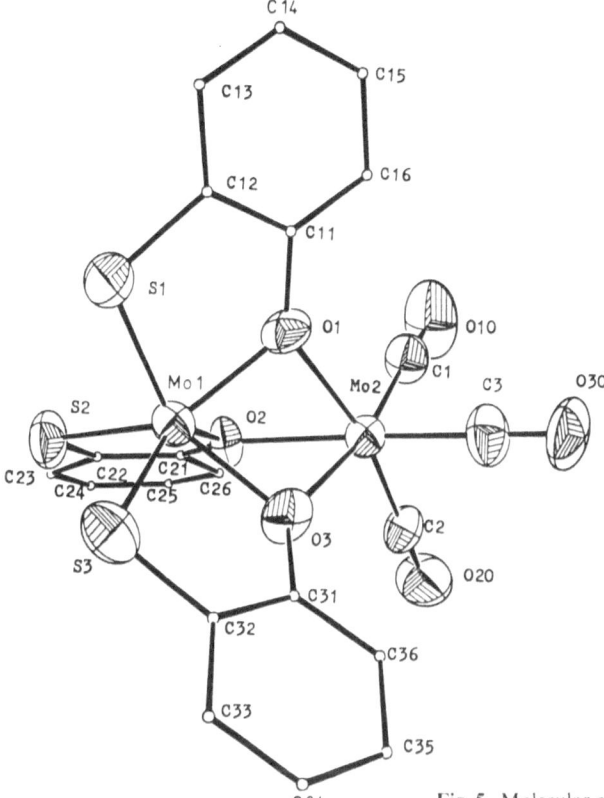

Fig. 5. Molecular structure of the anion of *XIII*

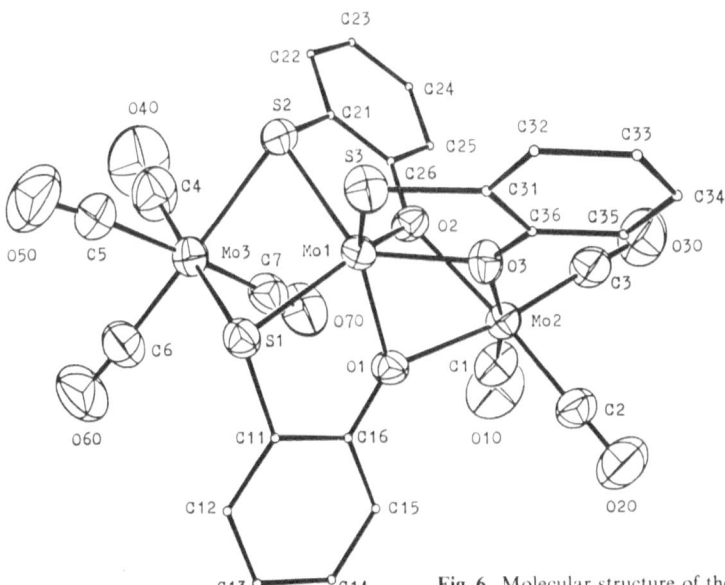

Fig. 6. Molecular structure of the anion of *XIV*

of 1.909(6), 1.898(6), 1.901(6) Å) is almost the same as that of **XIII**. The third molybdenum atom Mo(3) is located in a distorted octahedron with four Mo–C of 1.938(8), 2.031(7), 1.957(7), 2.031(7) Å and two Mo–S of 2.566(1), 2.569(1) Å which are similar to that observed in dinuclear Mo(O)–SR complexes $[Mo_2(CO)_8(SR)_2]^{2-}$ (R = $CH_2COOEt, C_6H_5, C_6H_4OH, C_6H_4CH_2SH$) (Zhuang et al. 1986, 1989a, 1989b; Huang et al. 1989) and the Mo(3)S_2Mo(1) is nonplanar with dihedral angle of 46.23(10)° between Mo(1)S(1)Mo(3) and Mo(1)S(2)Mo(3) planes.

It is interesting that all structures XII, XIII and XIV contain a common structural unit, O_3MoS_3, which possesses trigonal prism configuration and reflects the XAS data of the first coordination sphere of the Mo atom in FeMo-co of nitrogenase (Conradson et al. 1987, 1985) (as shown in Table 1). In addition to $[Mo_2Fe_6S_8(OCH_3)_3(SR)_6]^{3-}$ and $[MoFe_4S_4(C_6H_4-O_2)_3(SEt)_3]^{3-}$ reported by Christou et al. 1980 and Wolff et al. 1981, XII, XIII and XIV are the third classes of model complexes containing O_3MoS_3 unit with the structural data comparable to that derived from XAS study on FeMo-co, but in contrast with $[Mo_2Fe_6S_8(OCH_3)_3(SR)_6]^{3-}$ and $[MoFe_4S_4(C_6H_4-O_2)_3(SEt)_3]^{3-}$ of which the S ligands in the O_3MoS_3 unit are sulfido-ligands, the S-ligands of O_3MoS_3 in XII, XIII and XIV are thiolate ligands.

The XPS (X-ray photoelectron spectra) measurement (Note 7) of XIII and XIV show that XPS of XIII contains two couples of peaks with binding energies at 230.08/232.79, 231.43/234.21 ($3d^{5/2}/3d^{3/2}$) and XPS of XIV consists of three couples of binding energies at 232.0/235.1, 227.5/230.7 and 228.7/232.3 eV ($3d^{5/2}/3d^{3/2}$). This result indicating two and three different valence Mo atoms in XIII and XIV respectively is consistent with the fact that two and three different coordination environments of Mo atoms exist in XIII and XIV respectively. Referring to the binding energies of 228.14/231.34 eV ($3d^{5/2}/3d^{3/2}$) observed in Mo(O)–SR complex $[Et_4N][Mo_2(CO)_8(SPh)_2]$, the oxidation states of the Mo atoms in O_3MoS_3 units of XIII and XIV are almost the same and could be estimated as the value between $+4$ and $+2$. The electrochemistry (CV) of XIII shows a reversible $1e^-$ reduction at -0.83 V vs SCE with two quasi-reversible $1e^-$ oxidation at $+0.28$ and $+0.49$ V vs SCE. In terms of the existence of Mo atoms in widely separated formal oxidation states in XIII it is obvious that the electrochemistry on the two molybdenum centers of XIII is different and, reasonably, the reversible reduction at -0.83 V vs SCE should be derived from the redox that occurs on the high-valence Mo center.

Table 1. Comparison of the structural parameters of MoS_3O_3 unit in X, XI and XII with the XAS data of the as isolated FeMo-co of N_2-ase

	FeMo-co	XII	XIII	XIV
Mo–S (Å)	2.37 ± 0.002	2.38	2.32	2.35
No. of S	3	3	3	3
Mo–O (Å)	2.10 ± 0.02	1.98	2.06	2.07
No. of O	3	3	3	3

Now, from the studies on these Mo–S–O compounds several conclusions could be derived: (i) *XIII* and *XIV* not only simulate the EXAFS data of the first coordination environment of Mo atom in FeMo-co of N_2-ase but also provide a new possibility to interpret the XAS data of FeMo-co. Now that the Mo–S(R) in *XIII* and *XIV* have the same distance as the Mo–S (silfido) in $[MoFe_4S_4(cat)_3(SEt)_3]^{3-}$ and $[Mo_2Fe_6S_8(OCH_3)_3(SR)_6]^{3-}$, it is not impossible that the XAS data of FeMo-co reflects either the sulfido or the thiolate ligand in the active center of nitrogenase. (ii) comparing the Mo–S and Mo–O bond distances of *XI*, *XIII*, *XIV* with that of FeMo-co derived from the EXAFS study (as shown in Table 1) it could be found that the *XIV* is the best one to approximate to FeMo-co. This perhaps means the sulfur and oxygen atoms in nitrogenase are more possible in bridging-SR and bridging-OR forms than in terminal ligand forms and the possible oxidation state of Mo atom in nitrogenase should be lower than Mo^{5+}. (iii) The fact that the S atoms in *XIII* are capable of coordinating to a $Mo(CO)_4$-moiety forming the trinuclear *XIV* implies that it is possible to introduce an iron atom into *XIII* forming Fe–Mo–S–O modelling compounds including the string-bag-model compound which has not been synthesized successfully so far, because, instead of the $Fe(II)L_2$-moiety, the $Mo(CO)_4$-moiety could complex with MoS_4^{2-} or WS_4^{2-} resulting in a series of Mo–Mo(W)–S complexes (Zhuang et al. 1989; Rosenhein et al. 1987) which are similar to the linear Fe–Mo(W)–S modelling compounds (Coucouvanis 1981). (iv) As the structural unit of *XII*, *XIII*, *XIV*, and as a transition-state intermediate in the synthetic reaction, the $[S_3MoO_3]^{2-}$ with trigonal prism configuration has been recognized and obtained in situ. This seems to be available for designing the routes to synthesize the model compounds including the Siamese-twinned dicubane-like (closo- or nido-) cluster compounds with high Fe/Mo ratio in particular.

Acknowledgements. We would like to thank the late Professor Wenkei Huang for his cooperation to synthesize the G-series compounds and acknowledge the grants from the National Natural Science Foundation of China and the Science Foundation of the Chinese Academy of Science in support of this research.

5 References and Notes

Bancroft GM (1973) Oxidation state and electronic configuration of iron in minerals. In: Mössbauer Spectroscopy—An Introduction for Inorganic Chemists and Geochemists, McGraw-Hill, Maidenhead, Berkshire, England, p 156

Bose KS, Lamberty PF, Kovaes JE, Sinn E, Averill BA (1986) Synthesis of a new class of Mo–Fe–S clusters containing the MoS_2Fe_2 unit. Polyhedron 5: 393

Bose KS, Chmielewski SA, Eldredge PA, Sinn E, Averill BA (1989) New Mo–Fe–S clusters via oxidation decarbonylation reaction: The $[MoFe_5S_6(CO)_6L_3]^{n-}$ (L = PEt$_3$, $n = 0$; L = I, $n = 2$) capped cubanes. J Am Chem Soc 111: 8953

Burgess BK, Newton WE (1983) Iron–molybdenum cofactor and its complementary protein from mutant organisms. In: Mueller A, Newton WE (eds) Nitrogen fixation: The chemical-biochemical-genetic-interface Plenum, New York, p 83

Burgess BK (1984) Structure and reactivity of nitrogenase. In: Veeger C, Newton WE (eds) Advances in nitrogen fixation research Martinus Nijhoff, The Hague, p 103

Christou G, Garner CD (1980) Synthesis and proton magnetic resonance properties of Fe_3MS_4. (M = Mo or W) cubane-like cluster dimers. J Chem Soc Dalton Trans 2354

Christon G, Hagen KS, Holm RH (1982) Synthesis, structure and properties of $[Co_8S_6(SC_6H_5)_8]^{4-}$ containing an octanuclear Co_8S_6 rhombic dodecahedron related to that of cobalt pentlandite. J Am Chem Soc 104: 1744

Conradson SD, Burgess BK, Newton WE, Hodgson KO, Mcdonald JW, Rubinson JF, Gheller SF, Mortenson LE, Adams MW, Maschanak PK, Armstrong WA, Holm RH (1985) Structure insights from the molybdenum K-edge X-ray absorption near edge structure of the iron–molybdenum protein of nitrogenase and its iron-molybdenum cofactor by comparison with synthetic iron–molybdenum–sulfur clusters. J Am Chem Soc 107: 7935

Conradson SD, Burgess BK, Newton WE, Mortenson LE, Hodgson KO (1987) Structural studies of the molybdenum site in the MoFe protein and its FeMo cofactor by EXAFS. J Am Chem Soc 109: 7507

Coucouvanis D, Simhon ED, Sweson D, Baenziger NC (1979) "X-ray crystal structure of bis(tetraethylammonium)di-μ-thio-[bis-(phenylthio)-ferrate(III) dithiomolybdate(V)], $[Et_4N]_2$ $[(phS)_2FeMoS_4]$: a dinuclear complex with the $FeMoS_2$ core. J Chem Soc Chem Commun 361

Coucouvanis D, Beanziger NC, Simhon ED, Stremple P, Swenson D, Simopoulos A, Kostikas A, Petrouleas V, Papaefthyniou V (1980) Synthesis and structural characterization of the $(Ph_4P)_2$ $[Cl_2FeS_2MS_2FeCl_2]$ complexes (M = Mo, W). First example of a doubly bridging MoS_4 unit and its possible feature in the nitrogenase active site. J Am Chem Soc 102: 1732

Coucouvanis D (1981) Fe–M–S complexes derived from MS_4^{2-} anions (M = Mo, W) and their possible relevance as analogues for structural features in the Mo site of nitroenase. Acc Chem Res 14: 201

Coucouvanis D, Stremple P, Simhon ED, Swenson D, Beanziger NC, Draganjac M, Chan LT, Simopoulos A, Papaefthymiou V, Kostikas A, Petrouleas V (1983) Dinuclear iron-molybdenum-sulfur complexes containing the FeS_2Mo core. Syntheses, ground-state electronic structures and crystal and molecular structures of the $[(C_6H_5S)_2FeS_2MoS_2]$, $[(C_2H_5)_4N]_2[(C_6H_5S)_2FeS_2WS_2]$ and $[(C_6H_5)_4P]_2[(S_5)-FeS_2MS_2]$ (Mo = Mo, W) complexes. Inorg Chem 22: 293

Cramer SP, Hodgson KO, Gillum WO, Mortenson LE (1978) The molybdenum site of nitrogenase. Preliminary spectroscopy. J Am Chem Soc 100: 3398

Cramer SP, Gillum WO, Hodgson KO, Mortenson LE, Stiefel EI, Chisnell JR, Brill WJ, Shah VK (1978a) The molybdenum site of nitrogenase. 2. A comparative study of molybdenum cofactor by X-ray absorption spectroscopy. J Am Chem Soc 100: 3814

Dilworth JR, Morton S (1986) Iron–molybdenum and iron–tungsten sulphido clusters containing hydrazido(2-) ligands. J Organometallic Chemistry 314: C25

Eldredge PA, Bryan RF, Sinn E, Averill BA (1988) The $[MoFe_6S_6(CO)_{16}]^{2-}$ ion: A new model for the FeMo-cofactor of nitrogenase. J Am Chem Soc 110: 5573

Eremenko IL, Pasynskii AA, Katugin AS, Zalmanovitch VR, Orazsakhatov B, Sleptsova SA, Nekhaev AI, Kaverin VV, Ellert OG, Novotorsev VM, Yanovsky AI, Shklover VE, Struchkov YT (1989) Antiferromagnetic complexes with metal-metal bonds. XVIII. The influence of electronic factors on geometry of heterometallospirane clusters. Molecular structure and magnetic properties of $Cp_2Cr_2(\mu-SCMe_3)(\mu_3-S)_2M(\mu_3-S)_2Fe_2(CO)_6$ (M = Fe, Rh) and $(MeC_5H_4)_2Cr_2(\mu-SCMe_3)(\mu_3-S)_2Co(\mu_3-S)_2Fe_2(CO)_6$. J Organometallic Chem 365: 325

Garner CD, Acost SR, Christou G, Collison D, Mabbs FE, Petrouleas V, Pickett CJ (1983) Iron–molybdenum–sulfur clusters. In: Mueller A, Newton WE (eds.) Nitrogen fixation: The chemical-biochemical-genetic interface, Plenum Press, New York, p 245

Gibb TC (1976) Formal oxidation state in Principles of Moessbauer Spectroscopy. Chapman and Hall, London, p 74

Hawkes TR, McLean PA, Smith BE (1984) Nitrogenase from NifV mutants of Klebsiella pneumoniae contains an altered form of the iron-molybdenum cofactor. Biochem J 217: 317

Holm RH (1981) Metal cluster in biology: quest for a synthetic representation of the catalytic site of nitrogenase. Chem Soc Rev 10: 455

Huang LR, Lu JX (1983) Structural chemistry of G series modelling compound for the active center of nitrogenase—Crystal and molecular structure of $[Mg \cdot 6DMF][MoFeS_4Cl_2]$. Science Sinica(B), 3: 193

Huang LR, Lu JX (1983a) Structural chemistry of G series modelling compound for the active center of nitrogenase—Crystal and molecular structure of $[Mg \cdot 6DMSO][MoFeS_4Cl_2]$. Science Sinica(B) 3: 199

Huang LR, Zhuang B, Lu JX (1989) Crystal structure of $[Et_4N]_2[Mo_2(CO)_8(SCH_2C_6H_4-CH_2SH)_2]$. Jiegou Huaxue 8: 19

Huang WK, Zhao XT, Tan Z, Lin CZ, Jiang, FL, Cui YX, Lu JX (1979) Synthesis onto model compound of simulating the active center of nitrogen fixation in nitrogenase. Lanzhou Daxue Xuebao 4: 160

Huang WK, Tan Z, Cui YX, Zhao XT, Ling CZ, Jiang FL, Huang LR, Lu JX (1981) Attempted synthesis of 'series G' model compounds for the active center of nitrogenase. In: Gibson AH, Newton WE (eds.) Current perspective in nitrogen fixation Australia Acad Sci, Canbera, p 346

Huynh BH, Muenck E, Orme–Johnson WH (1979) Moessbauer studies of the cofactor centers of nitrogenase. Biochim Biophys Acta 576: 192

Johansson G, Lipscomb WN (1958) The structure of Roussin's black salt, $CsFe_4S_3(NO)_7 \cdot H_2O$. Acta Cryst 11: 594

Kennedy C, Cannon F, Cannon M, Dixon R, Hill S, Jensen S, Kumar S, McLean P, Merrick M, Robsem R, Postgate J (1981) "Recent advances in the genetics and regulation of nitrogen fixation" In: Gibson AH, Newton WE (eds) Current perspectives in nitrogen fixation. Elsevier/North Holland, New York, p 146

Liu QT, Kang BS, Chen CN, Huang LR, Cai JH, Zhuang B, Lu JX (1988) Studies on syntheses of $MoFe_3S_4$ cubane-like clusters. Science Sinica(B) 9: 920

Lu JX (Nitrogen Fixation Res Group, Fujian Inst of Res Struct of Matter, Chinese Acad Sci) (1975) A preliminary model for the active site of catalytic nitrogen fixation in nitrogenase—Also on structural criteria for the activation of dinitrogen molecule via coordination. Kexue Tongbao 20: 540

Lu JX (1980) Composite 'String bag' cluster model for the active center of nitrogenase. In: Newton WE, Orme-Johnson WH (eds) Nitrogen fixation, Vol 1. Free-living systems and chemical models. University Park Press, Baltimore, USA, p 343

Lu JX (Nitrogen Fixation Res Group, Fujian Inst Res Struct of Matter, Chinese Acad Sci) (1980a) Attempt to synthesis of 'Di-string-bag' modelling compound for the active center of nitrogenase. Kexue Tongbao 25: 191

Lu JX, Zhuang B (1989) A 'unit construction' approach to the rational syntheses of transition metal cubane-like clusters by the use of reaction fragments as building blocks. Jiegou Huaxue 8: 233

Mortenson LE, Thorneley RNF (1979) Structure and function of nitrogenase. Annu Rev Biochem 48: 387

Mueller A, Sarkar S, Dommrose AM, Filgueira R (1980) Indentification of a doubly bridging thiomolybdate(VI) ligand between iron centers by the resonance Raman effect and a simple preparation of $[(C_6H_5)_4P]_2[Cl_2FeS_2MoS_2FeCl_2]$. Z Naturforsch 35B: 1592

Muenck E, Rhodes H, Orme-Johnson WH, Davis IC, Brill WJ, Shah VK (1975) Nitrogenase. VIII. Moessbauer and EPR spectroscopy. Molybdenum-iron protein component form Azotobacter vinelandii. Biochim Biophys Acta 400: 32

Nelson MJ, Levy MA, Orme-Johnson WH (1983) Metal and sulfur composition of iron-molybdenum cofactor of nitrogenase. Proc Natl Acad Sci USA 80: 147

Orme-Johnson WH (1983) Molybdenum and iron centers in nitrogenase. 186th ACS annual Meeting, Washington, DC, Aug No 301

Rosenhein LD, McDonald JW (1987) Synthesis and characterization of $[(CO)_4MoS_2MS_2]^{2-}$ and $[(CO)_4MoS_2MS_2Mo(CO)_4]^{2-}$ ions (M = Mo, W) species containing group VI(6^+) metal in widely separated formal oxidation states. Inorg Chem 26: 3414

Seyferth D, Henderson RS, Song LC (1982) Chemistry of μ-dithio-bis(tricarbonyliron), a mimic of inorganic disulfides. 1. Formation of di-μ-thiolate-bis(tricarbonyliron) dianion. Organometallics 1: 125

Shah VK, Brill WJ (1977) Isolation of an iron–molybdenum cofactor from nitrogenase. Proc Natl Acad Sci USA 74: 3249

Silvis HC, Averill BA (1981) A synthetic molybdenum–iron–sulfur cluster with phenoxide terminal ligands. Inorg Chim Acta 46: L57

Tang AC et al. (1983) The study on the structural rule of metal cluster compounds. Kexue Tongbao 28: 25

Teo BK, Averill BA (1979) A new cluster model for the iron-molybdenum cofactor of nitrogenase. Biochem Biophys Res Commun 88: 1454

Teo BK, Antonio MR, Tieckelmann RH, Silvis HC, Averill BA (1982) New EXAFS models for the iron sites of the iron molybdenum cofactor of nitrogenase: The $[Cp-CH_3C_6H_4S)_2FeS_2Fes_2MoS_2]^{3-}$ trianion and the $[(C_6H_5O)_2FeS_2MoS_2]^{2-}$ dianion. J Am Chem Soc 104: 6126

Tieckelmann RH, Silvis HC, Kent TA, Huynh BH, Waszczak JV, Teo BK, Averill BA (1980) Synthetic molybdenum–iron–sulfur clusters. Preparation, structures and properties of the $[S_2MoS_2Fe(SC_6H_5)_2]^{2-}$ and $[S_2MoS_2FeCl_2]^{2-}$ ions. J Am Chem Soc 102: 5550

Tieckelmann RH, Averill BA (1980a) Preparation and properties of the bis(phenylmercapto)iron(III)-di-μ-sulfidoiron(II)-di-μ-sulfido-disulfidomolybdate(VI) ion. $[(PhS)_2FeS_2FeS_2MoS_2]^{3-}$. Inorg Chim Acta 46: L35

Tsai KR (Nitrogen Fixation Group, Xiamen Univ) (1976) A model of nitrogenase active center and mechanism of nitrogenase catalysis. Science Sinica 19: 479

Tsai KR (1980) Development of a model nitrogenase active center and mechanism of nitrogenase catalysis. In: Newton WE, Orme-Johnson WH (eds) Nitrogen fixation Vol 1. Free-living systems and chemical models. University park Press, Baltimore, USA, p 373

Tsai KR, Zhong HF, Lin GD, Wu MG, Han GB, Yang HH, Lai WJ, Liao DW (1983) Unified elucidation of nitrogenase catalyzed H_2-evolution reactions with edge-sharing twin-cubanes model and further studies on synthesis of FeMo-co modelling compounds. 186th ACS Annual Meeting, Washington DC, Aug No 95

Wolff TE, Berg JM, Holm RH (1981) Synthesis, structure and properties of the cluster complex $[MoFe_4S_4(SC_2H_5)_3(C_6H_4O_2)_3]^{3-}$, containing a single cubane-type molybdenum–iron–sulfur $(MoFe_3S_4)$ core. Inorg Chem 20: 174

Wong GB, Bobrik MA, Holm RH (1978) Inorganic derivatives of iron sulfide thiolate dimers and tetramers: Synthesis and properties of the halide series $[Fe_2S_2X_4]^2$ and $[Fe_4S_4X_4]^2$ (X = chlorine, bromine, iodine). Inorg Chem 17: 578

Yang SS, Pan WH, Friesen GD, Buwgess BK, Corbin JL, Stiefel EI, Newton WE (1982) Iron-molybdenum cofactor from nitrogenase. J Biol Chem 257: 8042

Zhuang B, Mcdonald JW, Newton WE (1983) Synthesis of the $[AcO)_2FeS_2MS_2]^{2-}$ (M = Mo,W) ions and observation of a new bonding mode for complexes of $[MoO_2S_2]^{2-}$. Inorg Chim Acta 77: L221

Zhuang B, Huang LR, He LJ, Chen WZ, Yang Y, Lu JX (1986) Studies on the synthesis, structure, IR spectra and reactivity of a new dinuclear Mo(O) complex $[Et_4N]_2[Mo_2(CO)_8(SCH_2CO_2Et)_2]$. Acta Chim Sin 4: 294

Zhuang B, Huang LR, He LJ, Yang Y, Lu JX (1988) "Synthesis, structure, reactivity and electro-chemistry of a new molybdenum(O) dithiocarbomato complex $[Et_4N][Mo(CO)_4(S_2CNEt_2)]$. Inorg Chim Acta 145: 225

Zhuang B, Yu PH, Huang LR, Lu JX (1989) Reactivity of metal(O) dithiocarbamato carbonyl complex $[M(CO)_4(S_2CNEt_2)]^-$ (M = Mo,W): A new synthesis and structure of a mixed-metal Mo–W–S complex $[Et_4N]_2[(CO)_4MoS_2WS_2]$. Inorg Chim Acta 162: 121

Zhuang B, Huang LR, He LJ, Yang Y, Lu JX (1989a) Structure of $[Et_4N]_2[Mo_2(CO)_8(\mu\text{-}SPh)_2]$ and structural chemistry of dinuclear molybdenum(0,I) carbonyl complexes containing thiolato-bridges. Inorg Chim Acta 157: 85

Zhuang B, Huang LR, Lu JX (1989b) Crystal and molecular structure of a dinuclear molybdenum(O) carbonyl complex with thiolato-bridges, $[Et_4N]_2[Mo_2(CO)_8(SC_6H_4OH)_2]$. Jiegou Huaxue 8: 103

Zhuang B, Huang LR, Hong GX, Lu JX (1989c) Synthesis, structure and electrochemistry of new di- and trinuclear Mo–S–O complexes containing O_3MoS_3 unit and their relevance to Mo-environment of FeMo-co in nitrogenase. XXVII Int Conf Coordination Chemistry, Broadbeach, Queensland, Australia, July, Abstracts M75

Zimmermann R, Muenck E, Brill WJ, Shah VK, Henzl MT, Rewlings J, Orme-Johnson WH (1978) Nitrogenase. X. Moessbauer and EPR studies on reversibly oxidized molybdenum–iron protein from Azotobacter vinelandii OP. Nature of the iron centers. Biochim Biophys Acta 537: 185

Note 1. Zhuang B, Yu PH, Huang LR, Unpublished data, $[Et_4N][W(CO)_4(S_2CNEt_2)]$ crystallizes in the orthorhombic, space group $Pca2_1$ with a = 19.516(2), b = 8.147(1), c = 14.338(1) Å. V = 2279.8 Å^3, Z = 4, R_1 = 0.046, R_2 = 0.053. W–S bond distances are 2.591(3) and 2.636(4) Å

Note 2. V crystallizes in the triclinic, space group P-1 with a = 9.895(3), b = 11.526(2), c = 14.082(3) Å, α = 85.87(2), β = 88.50(2), γ = 69.61(2) , V = 1502.3 Å^3, Z = 2; R_1 = 0.049, R_2 = 0.051. I I crystallizes in the triclinic space group P-1 with a = 9.898(2), b = 11.564(2), c = 14.071(3) Å. α = 85.79(2), β = 88.64(2), γ = 69.66(1) , V = 1505.9 Å^3, Z = 2, R_1 = 0.044, R_2 = 0.046

Note 3. FH-1918 model Moessbauer spectrometer with computer was used for the measurement and the source was 20 m Ci $^{57}Co(Pd)$

Note 4. Cyclic voltammetry measurements were carried out with a three electrode cell using 0.1 M Bu4NBF$_4$ as the supporting electrolyte and MeCN as the solvent. The working electrode was a glassy carbon disk, solutions were blanketed with N$_2$ and the potentiostate was a CV-IB from BAS (Bioanalytical Systems)

Note 5. "G" for Gansu, China. The G series of crystal compounds were isolated by our research group in cooperation with Prof. Huang WK, Lanzhou University, Lanzhou, Gansu

Note 6. *XII* Crystallizes in the monoclinic, space group P2$_1$/c with a = 13.128(8), b = 11.031(4), c = 19.665(10) Å, β = 108.03 °, V = 2707.4 Å^3, V = 2707.4 Å^3, Z = 4, d = 1.478 g/cm^3, μ = 7.2 cm^{-1}, R$_1$ = 0.069, R$_2$ = 0.071. *XIII*. C$_3$H$_7$OH crystallizes in the monoclinic, space group P2$_1$/c with a = 13.310(2), b = 19.572(3), c = 17.580(2) Å, β = 103.78(1)°, V = 4447.9 Å^3, Z = 4, R$_1$ = 0.072, R$_2$ = 0.082. *XIV* crystallizes in the triclinic, space group P-1 with a = 13.801(2), b = 14.163(3), c = 12.157(2) Å, α = 93.18(2), β = 98.01(3), γ = 83.32(2)°, V = 2335.3 Å^3, Z = 2, μ = 9.6 cm^{-1}, R$_1$ = 0.046, R$_2$ = 0.059

Note 7. XPS were measured on a NP-3G model X-ray Photoelectron Spectrometer from SHENYANG Scientific Instrument Factory of Chinese Acad Sci. The instrumental deviation is less than 0.05 eV. The MgKα X-ray line at 1253.6 eV was used for excitation and the X-ray power supply was run at 15 kV and 10 mA. All samples were run as powder dusted onto gallium in a nickel trough supported by a Mo-holder and under vacuum

CHAPTER 8

Systematic Investigations on the Bonding Property of the M–Fe–S (M = Mo, V, and W) Cluster Compounds and Novel Assumption on the Active Center Models of Nitrogenase

Chun-Wan Liu and Jia-Xi Lu

The Nitrogen Fixation
and its Research in China
Editor Guo-fan Hong
© Springer-Verlag Berlin Heidelberg 1992

1 Introduction

At present, three kinds of nitrogenases are known, i.e. the molybdenum-containing nitrogenase (Mo nitrogenase), the vanadium-containing nitrogenase (V nitrogenase), and a third nitrogen fixation system (nitrogenase 3). The last one has more complicated components, and its activities are lower than those of the former, and the nitrogen fixation mechanism of nitrogenase 3 has not yet become clear (Chisnell et al. 1988). Similar to the Mo nitrogenase, V nitrogenase requires MgATP and a source of low-potential electrons to reduce substrates such as H^+, N_2, and C_2H_2 (Eady et al. 1987). The kinetic and spectroscopic data suggest the same sequence of electron-transfer reactions and a similar overall mechanism (Eady et al. 1987; Dilworth et al. 1988; Hales et al. 1986, a, b). However, obvious differences in substrate specificity have been observed, i.e., Mo nitrogenase reduces C_2H_2 to C_2H_4 whereas V nitrogenase reduces C_2H_2 to C_2H_4 and C_2H_6 (Dilworth 1988). As the assay temperature was lowered from 30 to 5 °C, nitrogen remained an effective substrate for V nitrogenase, but not for Mo nitrogenase (Miller et al. 1988).

The experimental results of ESR, Mössbauer, and EXAFS indicate that there exist the Mo–Fe–S and V–Fe–S clusters in the FeMo and FeV cofactors, respectively (Ciurli et al. 1989a). At present, it is difficult to determine the crystal structures of the FeMo-co and FeV-co by the diffraction method. Therefore, the study of the model compounds of the clusters in FeMo-co and FeV-co is very important in our understanding of the biological functions of the nitrogenases. In this aspect, theoretical investigations of the electronic structure and bonding properties of the model cluster compounds and rationalization of the experimental data of EXAFS, Mössbauer, ESR, MCD, etc., provide important data for the understanding and chemical simulation of biological nitrogen fixation.

2 Bonding Properties of M–Fe–S Clusters (M = Mo, V)

The experimental results of ESR, Mössbauer, and EXAFS reported by Huynh et al. (1979) and Cramer et al. (1978a) provided evidence that there exists a Mo–Fe–S cluster in FeMo-co. It is already clear that the Fe_4S_4 cluster plays the role of electron storage and transfer in the iron protein and ferredoxin. Therefore, investigations on the electronic structure and the bonding properties of the Fe–S and Mo–Fe–S cluster compounds are theoretically significant.

The electronic structure of a series of Mo–Fe–S and Fe–S cluster compounds from dinuclear to nonanuclear listed in Table 1 have been investigated in this laboratory by using the extended Hückel (EH) and EH–MAD–SCC methods (Liu et al. 1986a). It was shown that the extensive electron delocalization is a general feature of the Fe–S cluster compounds. It appeared usually as the larger

Table 1. Mo–Fe–S and Fe–S cluster compounds studied by Liu et al. (1986a)

Type of cluster	Clusters	Method
Dinuclear	$Fe_2S_2X_4^{2-}$ (X = SR, Cl, NO) $Mo_2S_2X_2(SH)_4^{2-}$ (X = S, O) $MoFeS_4X_2^{2-}$ (X = Cl, NO)	MAD–SCC
Linear trinuclear	$Fe_3S_4(SR)_4^{3-}$, $Mo_2FeS_8^{3-}$, $MoFe_2S_6(SR)_2^{3-}$, $MoFe_2S_4Cl_4^{2-}$	
Cyclic trinuclear	$Fe_3S(S_2)_3^{2-}$, $Mo_3S_4(Cp)_3^+$, $Mo_3S_{13}^{2-}$, $Mo_3S_7Cl_7^{3-}$, $Mo_3S_2Cl_9^{3-}$	
Cubane-like tetranuclear	$Fe_4S_4X_4^{2-}$ (X = SR, Cl) $Fe_4S_4(SR)_4^{3-}$, $Fe_4S_4(NO)_4$, $[Mo_2O(NR)S_2(S_2P(OH)_2)_2]_n$ (n = 1, 2) $MoFe_3S_4Cl_3O_3H_4^{3-}$	
"String-bag" tetranuclear	$Fe_4S_3(NO)_7^{1-}$	
"String-bag" hexanuclear	$Fe_6S_9(SR)_2^{4-}$	
Cubane-like tetranuclear	$MoFe_3S_4(dtc)_5^{1-}$ $MoFe_3S_4(S_2CNC_5H_{10})_5$	EHMO
Double-cubane octanuclear	$Mo_2Fe_6S_8(SR)_9^{3-}$	
Double-cubane nonanuclear	$Mo_2Fe_7S_8(SR)_{12}^{3-}$	

contribution of the p electrons on the bridging sulfur atoms to the frontier molecular orbitals and plays an important role in stabilizing the cluster systems.

Some regularities concerning the strength of chemical bonds in the Mo–Fe–S cluster compounds have been obtained by analyzing the Mulliken bond orders P_{A-B} between atoms A and B:

$$P_{Mo-L} > P_{Mo-S_b} > P_{Fe-S_b} > P_{Fe-L} > P_{M-M'}.$$

The order of P_{Fe-L} and P_{Fe-S_b} is reversed when L = NO, while the $P_{M-M'}$ remains the smallest. Among the M–M bonds there is:

$$P_{Mo-Mo} > P_{Mo-Fe} > P_{Fe-Fe}.$$

For the cubane cluster $MoFe_3S_4$, we have:

$$P_{Mo-S} > P_{Fe-S} > P_{Mo-Fe} > P_{Fe-Fe}.$$

Although the M–M bond plays an important role in electron delocalization, the chief contribution to the formation of a stable Mo–Fe–S cluster is found to arise from the bonding interaction between the inorganic sulfur and metal atoms, since the former is an electron donor, possessing the capability of strong electron delocalization.

The MO analysis of the rhomboidal cluster core MS_2M:

$$M \diamond_{S}^{S} M$$

shows that among the six low-lying bonding orbitals, there are four orbitals and two orbitals for the dinuclear clusters $Fe_2S_2Cl_4^{2-}$ and $MoFeS_4L_2^{2-}$ ($L = NO$, Cl), respectively. The same feature also appears in the energy-level diagram of trinuclear clusters.

In the energy-level diagram of the cubane-like tetranuclear cluster Fe_4S_4 $(SH)_4^{2-}$, twelve bonding orbitals are found in the Fe–S* band instead of the six in the dinuclear cluster $Fe_2S_2(SH)_4^{2-}$, and both the frontier orbital band and the Fe–S* (S* hereafter denotes the bridging sulfur) band were broadened. This reveals that in the process of spontaneous assembly reactions, each rhomboidal cluster FeS_2Fe provides two orbitals (b_{1g} and b_{3u}), so that between Fe and S* atoms of these two rhomboidal clusters there are four "new" bonds in the direction perpendicular to the rhomb plane to be formed. Thus the tetranuclear cluster $Fe_4S_4(SH)_4^{2-}$ can be considered as the combination of two dinuclear cluster fragments. The same condition takes place also in the energy-level diagrams of $Mo_2S_2O_2(SH)_4^{2-}$ and $[Mo_2O(NH)S_2(S_2P(OH)_2]_2$.

The energy-level diagram of $Fe_6S_9(SH)_2^{4-}$ is very similar to that of $Fe_4S_4(SH)_4^{2-}$, but with its levels arranged more closely. So it can be considered that a polynuclear cluster consists of the basic rhomboidal connected directly by μ_4-S, μ_3-S, and μ_2-S.

Even if the spontaneous assembly mechanism in the formation of $MoFe_3S_4(S_2CH)_5^n$ ($n = 0, -1$) (see Fig. 1) is not yet explicit at the present time, we can see by analyzing the values of bond orders listed in Table 2 that the M–M and M–S bonds between the two fragments, MoS_2Fe and FeS_2Fe, are weaker than those within each fragment. Therefore, it is easy to imagine that in the spontaneous assembly process, after formation of the above two fragments,

Table 2. Bond orders of $MoFe_3S_4(S_2CH)_5^n$

n	P_{Mo-Fe}	$P_{Mo-Fe'}$	$P_{Fe-Fe'}$	$P_{Fe'-Fe'}$	P_{Mo-S_b}	$P_{Mo-S_b'}$
0	0.103	0.042	−0.031	0.042	0.635	0.578
−1	0.112	0.044	−0.061	0.071	0.640	0.538

	P_{Fe-S_b}	$P_{Fe-S_b'}$	$P_{Fe'-S_b}$	$P_{Fe'-S_b'}$	$P_{Fe'-S_b'}$
0	0.420	0.432	0.529	0.471	0.417
−1	0.412	0.431	0.491	0.465	0.339

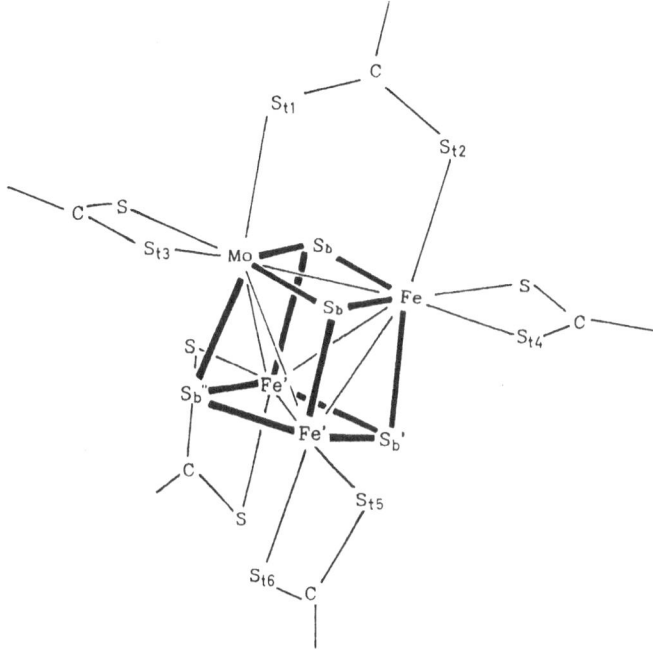

Fig. 1. The structure of $[MoFe_3S_4(S_2CH)_5]^n$ ($n = 0, -1$)

the inorganic sulfur atoms in each fragment donate electrons to the metal atom in another fragment, consequently the tetranuclear cluster is formed from these two fragments.

The previous analysis shows that the Mo–Fe–S clusters contain a group of six low-lying bonding molecular orbitals corresponding to the heterometallic rhomboidal MS_2M' so in the Mo–Fe–S clusters there exists indeed a basic structural unit, that is rhomboidal core. With the strong bonding interaction between the inorganic sulfur and the metal atoms in this unit taken into account, it is suggested that in the process of spontaneous assembly the electron can be transported from the inorganic sulfur to the metal atom, and new M–S bonds will be formed between the "cluster fragments". Thus, it can be seen that the bridging effect of the inorganic sulfur plays an important role in the spontaneous assembly reaction.

A strong π interaction between Fe and NO has been found in the clusters containing terminal nitro ligands, such as $Fe_2S_2(NO)_4^{2-}$, $Fe_4S_4(NO)_4$, Fe_4S_3, $(NO)_7^{1-}$, and also found between Mo and ligand L, as in $Mo_2S_2L_2(SH)_4^{2-}$. Moreover, in this type of clusters, the M–M bond is rather strong.

For the Mo–Fe–S clusters, the difference in magnetic, electronic, and reaction properties, dependent on the terminal ligands, is reflected in the difference of the HOMO–LUMO energy gaps.

In terms of the strength of the interaction between metal atom M and terminal L, the Mo–Fe–S cluster compounds can be classified into "strong terminal cluster" (STC) and "weak terminal cluster" (WTC). The HOMO–LUMO gaps ΔE and relative Mulliken bond order are listed in Table 3. In the case of STC, it has $P_{M-L} > P_{M-S_b}$. The STC is usually diamagnetic and its magnetic susceptibility does not change with temperature (Coucouvanis 1981). The electron delocalization increases on account of a strong π interaction between the metal atoms and ligands, which leads to a large HOMO–LUMO energy gap. Compared with STC, the WTC is usually paramagnetic (Mascharak et al. 1983). The energy level of the frontier orbitals is so closely spaced that the variation of magnetic susceptibility with temperature is relatively large. This is the reason why this type of clusters is a good electron transferor and easy to undergo redox reactions. The strength of bonding in this type of clusters has $P_{M-L} \approx P_{M-S_b}$.

The structure of the clusters in question formed through spontaneous assembly reactions is closely dependent on the kind of ligands in the cluster fragment. For the STC, such as $Fe_2S_2(NO)_4^{2-}$, the terminal ligand NO is not easy to be broken away from the cluster during the reaction because of the strong M–L bond. In certain conditions, however, the M–L bond could be weakened. A good illustration is the spontaneous assembly reaction of the red Roussinate under weak acidification. In this reaction, one of the NO ligands loses from the $Fe_2S_2(NO)_4^{2-}$, and then the red Roussinate is converted into the black Roussinate which possesses a nido "string-bag" structure, but it is very unlikely to lose more NO ligands simultaneously to form a closo cubane-like structure $Fe_4S_4(NO)_4$, even though the latter is also a stable cluster.

Table 3. Effect of terminal ligands

Cluster	Weak terminal cluster (WTC) ΔE (eV)	P_{M-M}	P_{M-L} (P_{M-S_b})	Cluster	Strong terminal cluster (STC) ΔE (eV)	P_{M-M}	P_{M-L} (P_{M-S_b})
$Fe_2S_2(SH)^{2-}$	0.39	0.106	0.511 (0.660)	$Fe_2S_2(NO)_4^{2-}$	2.32	0.145	1.060 (0.590)
$Fe_2S_2Cl_4^{2-}$	0.15	0.092	0.479 (0.479)	$MoFeS_4(NO)_2^{2-}$	1.20	0.095	1.040 (0.782)
				$MoFeS_4Cl_2^{2-}$	0.79	0.126	1.081 (0.469)
				$Mo_2S_4(SH)_4^{2-}$	1.35	0.164	1.130 (0.641)
				$Mo_2S_2O_2(SH)_4^{2-}$	1.42	0.160	0.971 (0.649)
$Fe_4S_4(SH)_4^{2-}$	0.6	0.113	0.599 (0.505)	$Fe_4S_4(NO)_4$	1.34	0.127	1.077 (0.526)
$Fe_4S_4Cl_4^{2-}$	0.5	0.095	0.572 (0.516)	$Fe_4S_3(NO)_7^{1-}$	1.23	0.130	1.122 (0.538)

Table 4. Combination forms of the cluster fragments

Cluster fragments in combination	Conformation of combined cluster	Illustration
STC + STC	nido "string-bag"	$Fe_2S_2(NO)_4^{2-} - Fe_4S_3(NO)_7^{1-}$
WTC + WTC	closo cubane-like	$Fe_2S_2(SR)_4^{2-} - Fe_4S_4(SR)_4^{2-}$
STC + WTC	nido "string-bag"	

Now we can come to the conclusion that only the nido "string-bag" cluster could be formed by using STC as a cluster fragment in spontaneous assembly reactions. If a WTC is used as cluster fragments, then a closo cubane-like cluster can be obtained, so as the formation of $Fe_4S_4(SH)_4^{2-}$ from $Fe_2S_2(SH)_4^{2-}$. Therefore, we can infer that in spontaneous assembly reactions, forming cubane-like clusters, such as $MoFe_3S_4(dtc)_5^{1-}$ and $MoFe_3S_4(S_2CNC_5H_{10})_5$, the WTC fragments is probable to be intermediate in the process. From this argument we can make a further suggestion that if a STC and a WTC were taken at the same time as cluster fragments, then it would be favorable to form a nido "string-bag" cluster. The possible combinations of cluster fragments are summarized in Table 4.

In order to study further the similarity and difference of the electronic structure and bonding property of a variety of M–Fe–S cluster compounds, some linear trinuclear and cubane-like M–Fe–S (M = Mo, W, and V) cluster compounds have been investigated on the level of EHT and MAD–SCC methods.

3 Linear Clusters $[Cl_2FeS_2MS_2FeCl_2]^{n-}$ (M = V, Mo, and W)

The idealized geometries of the three linear clusters, $[Cl_2FeS_2MS_2FeCl_2]^{n-}$ (M = V, n = 3; M = Mo, W, n = 2), are of D_{2d} symmetry (Coucouvanis et al. 1984; Do et al. 1983). As shown in Fig. 2, the molecular orbital level diagrams of these clusters are quite similar to each other, and the frontier orbitals have the same bonding character that results mainly from the M–M interaction (see Fig. 3). As a result of cancellation of the bonding and antibonding interactions, in effect only the δ bonding interaction remains, therefore the M–M interaction is very weak. Examining the bonding interactions in the three clusters (see Table 5) we have the orderings of Mulliken bond orders of $P_{M-S} > P_{Fe-S} > P_{Fe-Cl} > P_{M-Fe}$ and $P_{W-Fe} > P_{Mo-Fe} > P_{V-Fe}$, indicative of the fact that the M–S bonds are the strongest. From Fig. 2 we can see that the M/Fe–S bonds are corresponding to the low-lying MOs, and similar to the cases of the Mo–Fe–S cluster (Liu et al. 1986a). There are four σ bonds and two π bonds corresponding to the rhomboidal MS_2Fe while M = V, W; so we can conclude

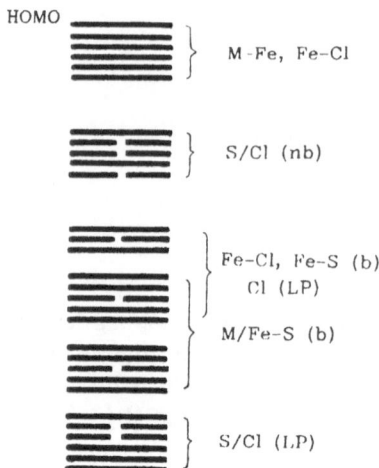

HOMO ——————

} M–Fe, Fe–Cl

} S/Cl (nb)

} Fe–Cl, Fe–S (b)
  Cl (LP)

} M/Fe–S (b)

} S/Cl (LP)

Fig. 2. The characteristics of energy level diagram of trinuclear clusters $[Cl_2FeS_2MS_2FeCl_2]^n$ (D_{2d}) ($M = V$, $n = 3$; $M = Mo$, W, $n = 2$)

that the rhomboidal MS_2Fe is a basic structural unit that builds up the M–Fe–S clusters.

The MO analysis also indicates that in the case of $M = V$ there is no distinct separation between the levels contributed predominantly from the M–S bond and the levels from the Fe–Cl bond, while the opposite is true for the case of $M = W$, i.e., these two types of orbital levels are well-separated, and the case of $M = Mo$ is something in between. This implies that the electron delocalization of these clusters has the ordering of $V > Mo > W$, i.e., the electron transfer ability of the W-containing cluster is weak, which would be unfavorable in nitrogen fixation.

The linear trinuclear cluster can be considered as the one that consists of two $FeCl_2$ and one MS_4^{n-} fragments, and the interaction between these fragments can be treated by using the procedure of fragment orbital analysis. The results show that the width of the d-band in the energy level diagram is in the ordering of $V > Mo > W$. The order of the strength of fragment interaction between the fragments MS_4^{n-} and $Cl_2Fe....FeCl_2$ is $VS_4^{3-} > MoS_4^{2-} > WS_4^{2-}$. This order is in accord with the order of Mulliken bond order between the fragments and the order of the energy gap $\Delta E(LUMO-HOMO)$ of the parent cluster listed in Table 6.

Table 5. Mulliken bond order of $[Cl_2FeS_2MS_2FeCl_2]^{n-}$

Cluster	P_{M-Fe}	P_{M-S}	P_{Fe-S}	P_{Fe-Cl}
$M = V$ $(n = 3)$	0.089	0.837	0.540	0.451
$M = Mo$ $(n = 2)$	0.155	0.867	0.496	0.481
$M = W$ $(n = 2)$	0.165	0.959	0.453	0.504

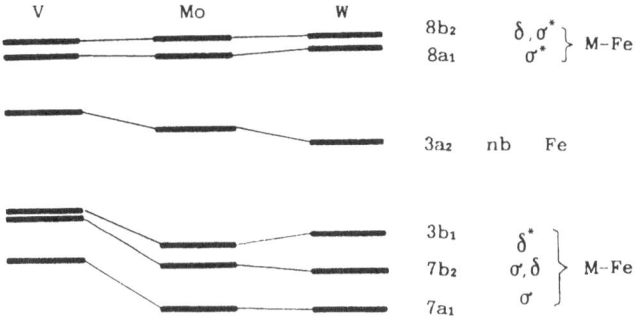

Fig. 3. The characteristics of frontier orbitals nearby HOMO ($8b_2$) of $[Cl_2FeS_2MS_2FeCl_2]^{n-}$

Table 6. The Mulliken bond order and energy gap of clusters $[Cl_2FeS_2MS_2FeCl_2]^{n-}$ and the fragments

Cluster	Fragment	$Cl_2Fe–FeCl_2$	ΔE (LUMO—HOMO) (eV)
M = V $(n = 3)$	$[S_2VS_2]^{3-}$	2.291	0.26
M = Mo $(n = 2)$	$[S_2MoS_2]^{2-}$	2.264	0.04
M = W $(n = 2)$	$[S_2WS_2]^{2-}$	2.112	0.008

4 The Cubane-Like M–Fe–S Clusters (M = V, Mo, and W)

Currently the cubane-like M–Fe–S (M = Mo, V) clusters are presumably the best structural analogs for the active center of the nitrogenases. Up to now, many single and double cubane-like M–Fe–S cluster compounds have been synthesized. A wide variety of physicochemical information about the properties of the analogous clusters has been provided by Mössbauer spectra, ESR spectra and other physical measurements. The comparative properties of the cubane-like clusters containing the isoelectronic cores $[VFe_3S_4]^{2+}$, $[MoFe_3S_4]^{3+}$, and $[WFe_3S_4]^{3+}$, were studied experimentally by Christou et al. (1979, 1980) and Carney et al. (1987). We have examined the electronic structure and bonding characteristics of five cubane-like M–Fe–S (M = V, Mo, and W) clusters (Table 7) at the EHMO level. Generally, the geometries of each cubane in the double-cubane clusters are similar to those of the single-cubane clusters with the same cluster core. Consequently, it would be qualitatively reasonable to examine the bonding properties of the double-cubane clusters from investigating the electronic structure of the single-cubane clusters. The cores of this type clusters are characterized by two interpenetrating tetrahedra: a smaller one consisting of four metal atoms and a larger one consisting of four sulfur atoms. Each Sulfur atom triply bridges three metal atoms. The fragment $Fe_3S_4L_3$ (L = SR, Cl) is the common unit of these clusters. The M–M bonding is readily

Table 7. Cubane-like M–Fe–S (M = V, Mo, W) cluster compounds studied by using EH method

NO.	Cluster	Symmetry	Reference
1	$[VFe_3S_4Cl_3(OCH_2)_3]^{1-}$	C_{3v}	46
2	$[MoFe_3S_4(SH)_6]^{3-}$	C_{3i}	16
3	$[WFe_3S_4(SH)_6]^{3-}$	C_{3v}	91
4	$[MoFe_3S_4(SH)_3(OH)_3]^{3-}$	C_{3v}	10
5	$[WFe_3S_4(SH)_3(OH)_3]^{3-}$	C_{3v}	9

identified from the fragment orbital interaction by dividing the system into two fragments $[Fe_3S_4(SR)_3]^{3-}$ and $[ML_3]^{2+}$. Similar to the linear M–Fe–S clusters, a common feature of bonding in the cubane-like M–Fe–S clusters is the extensive electron delocalization and the stronger M–S* and Fe–S* interactions. The rhomboidal M$\overset{S}{\underset{S}{<>}}$M can be taken as the basic structural unit that builds up the clusters.

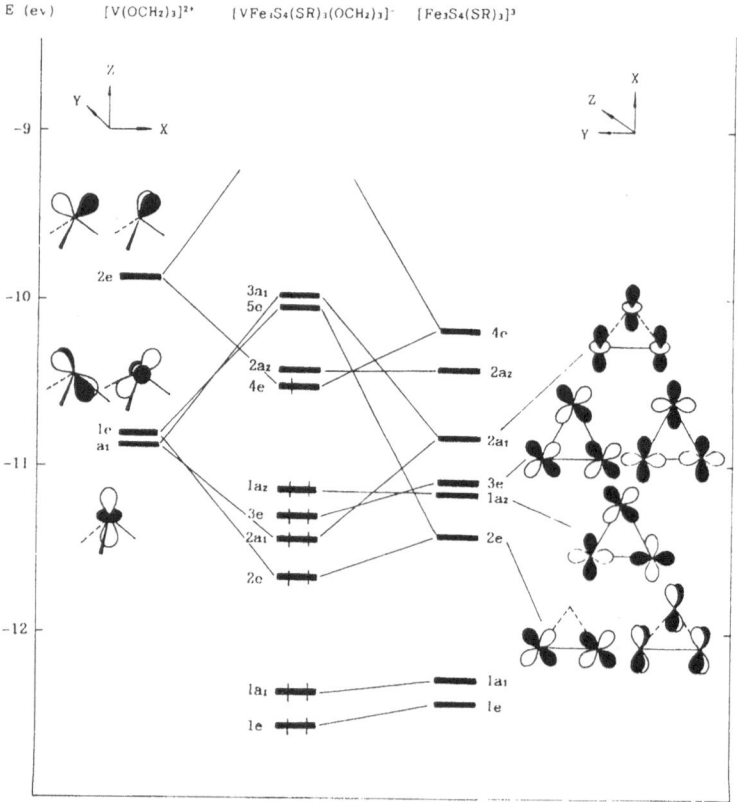

Fig. 4. Energy level correlation diagrams of the frontier orbitals in $[VFe_3S_4(SR)_3(OCH_2)_3]^-$ constructed from fragments $[Fe_3S_4(SR)_3]^{3-}$ and $[V(OCH_2)_3]^{2+}$

The orbital correlation diagrams of the interaction between two fragments $[Fe_3S_4(SR)_3]^{3-}$ (I) and $[V(OCH_2)_3]^{2+}$ (II) is presented in 1 in Fig. 4. It is illustrated that three orbitals, $1e$ and $1a_1$ of I do not take part in the fragment interaction. In the formation of $3e$ and $1a_2$ of 1, and $1a_2$ and $3e$ of I are not affected by the d orbitals of II. The bonding orbitals $2e$ and $2a_1$ and unoccupied antibonding orbitals $5e$ and $3a_1$ of 1 are formed from the interaction between the orbitals $2e$, $2a_1$ of I and a_1, $1e$ of II.

The orbital correlation diagrams of the frontier orbitals of clusters $1-3$ are illustrated in Fig. 5. The Mulliken bond orders (P) of the M–M bonds and energy gaps ΔE(LUMO–HOMO) of $1-3$ are given in Table 8. It is shown that the energy levels of these clusters have similar structures and characters, and the obvious regulations exist in the variation of Mulliken bond order with bond lengths.

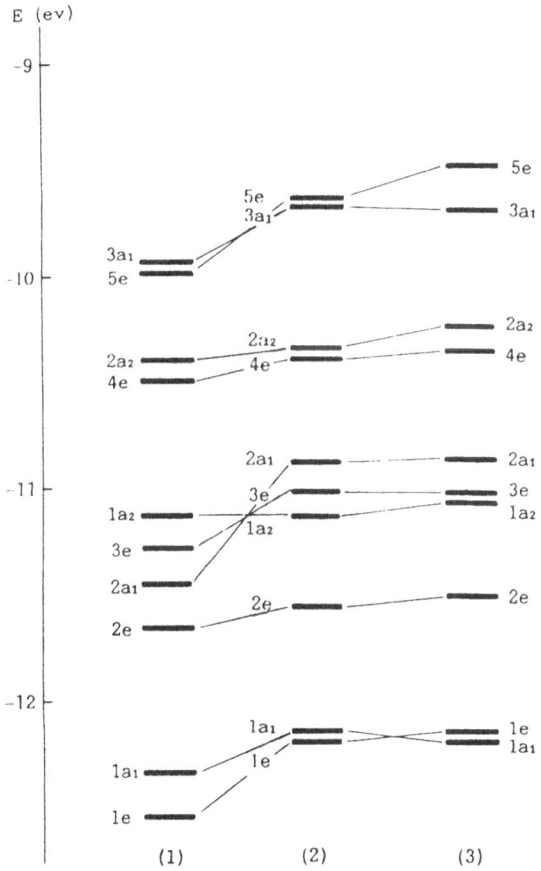

Fig. 5. Energy level correlation diagrams of the frontier orbitals of three cubane-like clusters with the core $[MFe_3S_4]$ {M = V (1), Mo (2), W (3)}

Table 8. Mulliken bond order (P) and the M–M length (R) of three cubane-like clusters

Cluster	1	2	3
P(M–Fe)	0.091	0.115	0.123
R(M–Fe) (Å)	2.777	2.714	2.710
P(Fe–Fe)	0.063	0.053	0.066
R(Fe–Fe) (Å)	2.707	2.746	2.693
ΔE(LUMO–HOMO) (eV)	0.093	0.086	0.129

Synthesizing the above regularities, we can conclude that both the electron delocalization and the energy gap ΔE(LUMO–HOMO) of the vanadium cluster are larger than that of the molybdenum cluster, meaning that the former is more stable than the latter, therefore the reactivity of the V cluster with substrates of nitrogenase is lower than that for the Mo cluster. This is consistent with the experimental fact that the activity of FeVco is lower than that of FeMoco. As mentioned above, the electron-transfer ability of the W cluster is quite weak, and the largest ΔE(LUMO–HOMO) suggests the lowest activity of the tungsten cluster. For this reason, the similarities of the chemistry of Mo and W do not extend to these enzymes. In other words, the WFe cluster would be unlikely to serve as the active center in nitrogenase.

The clusters 4 and 5 can be obtained from substitution of the terminal group $(SR)_3$ bonded to Mo and W by $(OH)_3$ in 2 and 3, respectively. The bonding properties of the clusters before and after the substitution of the thiolato ligands are listed in Table 9. From the data of Table 9, we can see that after the ligand

Table 9. Comparison between the bonding properties of 2–5

Cluster compounds			2	4	3	5
Percent M atomic population of frontier orbitals (%)	LUMO $(2a_2)$	Fe	63.68	68.74	63.20	68.70
	HOMO $(4e)$	M	8.05	9.60	5.80	7.19
		Fe	64.50	65.56	65.26	66.44
Net charge of the core		M	1.823	2.425	1.979	2.507
		Fe	−1.784	−1.579	−1.906	−1.673
		S*	−0.421	−0.547	−0.444	−0.583
Net charge of terminal atoms		S_t/O	−0.657	−1.170	−0.667	−1.151
Mulliken bond order		M–Fe	0.115	0.127	0.123	0.137
		Fe–Fe	0.053	0.054	0.066	0.067
		M–S*	0.472	0.505	0.513	0.543
		Fe–S_t	0.542	0.544	0.542	0.543
		Fe–S*	0.472	0.470	0.462	0.460
		M–S_t/O	0.452	0.301	0.450	0.319
Energy gap ΔE (LUMO–HOMO) (eV)			0.086	0.008	0.129	0.034

substitution, the percentage populations of the Mo and W atoms in the HOMO increase by 19% and 24%, respectively, while both positive charges of the metal atoms and negative charges of the bridging terminal sulfur atoms bonded to the M atoms increase by about 30%. Upon substitution, the M–Fe and M–S* interactions are strengthened but the M–S$_t$ bonds weakened. This can be rationalized by the fact that the electronegativity of the oxygen atom is larger than for the sulfur atom, and therefore the electron donation ability of oxygen is weaker than for sulfur.

The above results show that the existence of an oxygen atom in these cluster systems can increase the metal population in the frontier orbitals. As it was suggested earlier that the active sites of the cluster are the metal centers, the substitution of thiolato ligands by oxo ligands may be favorable in the activity of nitrogenases. This conclusion is further supported by the decrease of the energy gaps $\Delta E(\text{LUMO–HOMO})$ upon substitution. It is thus indicated that the oxygen atoms may play a role in the active center of nitrogenase.

Comparing the electronic structure of the W clusters with that of Mo clusters of the same type (see Table 9), we will come to the same conclusion that the W cluster has lower activity than the Mo cluster because of the smaller percent M atomic populations in frontier orbitals and the larger energy gaps $\Delta E(\text{LUMO–HOMO})$ of W clusters.

5 Structural Rules of the Cubane-Like Transition-Metal Clusters

In the theoretical investigations on chemical simulation of biological nitrogen fixation, a fascinating subject has been the structure-property regularity of the transition-metal clusters. The understanding of the relation between the topological geometries and various properties (magnetism, electronic structure, redox property, etc.) can shed light on the origin of the biological functions of the transition-metal clusters in nitrogenases. Thus, this study will give valuable information for the research of nitrogen fixation on the atomic or molecular levels.

So far, several structural rules of cluster compounds have been established, such as the Wade's "N + 1" rule (Wade 1976), the Polyhedral Skeletal Electron Pair Theory (Mingos 1983, 1985), the Topological Electron Counting Theory (Teo 1985), the rule of bonding capability of transition-metal clusters (Lauher 1978, 1979), the "9N − L" rule (Tang et al. 1983), and the "$nxc\pi$" structural formalism (Xu 1982). These rules set up the relationship between the number of framework electrons (i.e. valent electrons) and the structure of metal polyhedron and are applicable to the electron-rich transition-metal cluster systems, but not very adequate for the electron-deficient metal clusters.

The "9N − L" rule has the following formulae:

$$\text{VBO} + \text{VNBO} = 9N - L \tag{1}$$

where VBO is the number of valent bonding orbitals, VNBO the number of valent nonbonding orbitals in the cluster, N the number of metal atoms in the cluster, and L the number of bonds between the metal–metal atoms. This rule has been extended to the "chain" and clusters containing the basic structural unit, MoS_2Fe, and cubane-like clusters containing the core $MoFe_3S_4$. Moreover, a relationship between the spin property and the structure of Mo–Fe–S clusters has been proposed (Tang et al. 1986):

$$(9N - L) - VE/2 - S = Nu \qquad (2)$$

where VE is the number of valent electrons in the cluster, i.e., the sum of number of valent electrons of the metal atoms and the covalent electrons of ligand atoms, S spin, Nu the number of unoccupied valent bonding orbitals.

The bonding rule in Fe–S clusters was proposed on the basis of systematical studies of the different structural factors and bridging sulfur ligands on the Fe–Fe bonding interactions in the Fe–S cluster compounds (Lin et al. 1987). The average oxidation state (x) of Fe in the Fe–S clusters can be expressed by following equation:

$$x = 3 - 3(N_{\mu_3-S} + 3N_{\mu_4-S})/(N_{Fe}^2 + N_{\mu_2-S} + N_L) \qquad (3)$$

where N_{μ_n-S} and N_{Fe} are the numbers of $\mu_n - S$ and Fe atoms, respectively, N_L the number of terminal ligands on the Fe atom. The results obtained by using Eq. (3) are quite exact for most Fe–S clusters.

In order to explore the structural rules applicable to the electron-deficient transition-metal clusters, we have studied the electron counting of the "chain" and cubane-like Mo–Fe–S clusters (Liu et al. 1988a).

For the clusters shown in Table 10, the bond orders of the terminal Mo–S_b bonds vary from 0.93 to 0.98, significantly larger than those of the M–S_b bonds which are about 0.5 for Fe–S (terminal and bridging) and Fe–Cl. On the basis of the above characteristic and in view of the traditional valence bond theory, the following convention is proposed: the Mo–S_b bond is a σ single bond, the Mo–S_t bond a $\sigma + \pi$ double bond, and the Fe–L (Fe–S and Fe–Cl) bonds are σ single bonds. The numbers of the skeleton electrons in 6–9 are 26, 39, 39, and 36, respectively (see Table 10). According to the multiplicity of the M–L bonds, if the pair of electrons in the Mo–Fe bond is counted separately, i.e., one electron

Table 10. Number of skeleton electrons N_S of Mo–Fe–S "chain" clusters

No.	Cluster compounds	N_S
6	$[S_2MoS_2FeL_2]^2$ (L = SR, Cl, OR)	26
7	$[S_2MoS_2FeS_2Fe(SPh)_2]^{3-}$	39
8	$[S_2MoS_2FeS_2MoS_2]^{3-}$	39
9	$[Cl_2FeS_2MoS_2FeCl_2]^{2-}$	36

is counted as if it should belong to each metal atom, there are 13 electrons around the Fe atom: eight electrons from the four Fe–L bonds, one electron from the Mo–Fe bond and four d electrons. There are also 13 electrons around the Mo atom in 6–8, which have the MoS_4 terminal units: eight electrons from the Mo–S σ bonds, four electrons from the two Mo–S π bonds, and one electron from the Mo–Fe bond. However, there are only ten electrons around the Mo atom in 9, because it contains only four single Mo–S σ bonds and two electrons from the Mo–Fe bonds. In view of this counting, we are led to conclude that the "chain" Mo–Fe–S clusters containing a MoS_4 terminal unit satisfies the "13-electron counting" rule.

The above analysis is also applicable to the bonding of analogous W–Fe–S compounds, e.g., $[S_2WS_2FeS_2WS_2]^{2-}$ (Müller et al. 1984). The total number of skeleton electrons of these two clusters, being 78 and 52, respectively, satisfy the "13-electron counting" rule. This counting rule is also applicable to those Mo–Fe–S compounds, in which the Mo=S double bonds are substituted $Mo{\overset{\textstyle S}{\underset{\textstyle S}{<}}}$ or Mo = O bonds, e.g., $[Cl_2FeS_2MoO(S_2)]^{2-}$ (Müller et al. 1983), $[(S_4)SMoS_2MoS(S_4)]^{3-}$ (Dahlstrom et al. 1981), as well as other clusters.

The framework of the cubane-like Mo–Fe–S cluster is composed of MoL_6 (L = SR, etc.) FeS_4 fragments bridged by sulfur atoms. The bonding properties of the cluster compounds of this type can be readily treated by using a fragment orbital interaction scheme. The total number of skeleton electrons in a single-cubane cluster is 55. Excluding the 36 electrons of the M–L σ bonds, the remaining 19 d electrons will fill the occupied MOs among the frontier orbitals of the MoL_6 and FeS_4 fragments.

The double-cubane Mo–Fe–S cluster is composed of two single Mo–Fe–S cubanes joined through the two bridging units Mo–SR–Fe or joined at the Mo sites either through $Mo(\mu\text{-SR})_3Mo$ or through $Mo(\mu_2\text{-SR})_2(\mu_2\text{-S})Mo$. Since there is no possibility of M–M bonding between the two single cubanes, the total number of skeleton electrons is 110, a sum of two single cubanes. The 55 electron counting can be used to analyze the spin state and oxidation state of metal atoms in single- and double-cubane clusters; the results have been confirmed by Mössbauer data (Holm 1981, Christou et al. 1982). For another cubane-type cluster $[MoFe_4S_4(SC_2H_5)_3(C_6H_4O_2)_3]^{3-}$ the result of the analysis of Fe oxidation states is in accord with Mössbauer results (Wolff et al. 1981).

6 Structural Rule of Cubane-Type Clusters

At present, the Mo–Fe–S clusters form only a small group in the synthetical cubane-type clusters. Due to the similarity in the geometries and ligand coordinations, a more common structural rule is expected. A systematical study on the bonding and electron counts was reported recently by Harris (1989). In terms of the type and coordination environments of metal atoms, we have classified the cubane-type clusters into four classes and investigated the electronic structures of some compounds selected from each class. The following characteristics listed in Table 11 for each class are obtained on the basis of MO analysis. The MO energy level schemes of these clusters are illustrated qualitatively in Fig. 6. The levels above the dotted line (see Fig. 6) are so high lying above the M–M antibonding orbitals that these levels can not accept the electrons donated from the ligands. These levels are called the high-lying nonbonding orbitals (HLNO). The MO analysis indicates that the width of the nonbonding Mo levels of the M–M interaction depends on the chemical properties of the ligands L, the coordination number, and strength of the ligand field. When all the four metal atoms are six coordinated, as the clusters of class IV, all the eight nonbonding orbitals are HLNO so that there are no nonbonding orbitals that are occupied. When one of the four metal atoms is four coordinated and the rest are six coordinated, there are two occupied nonbonding orbitals like the clusters of class III. The number of the occupied nonbonding orbitals will increase, if the number of four-coordinated metal atoms increases. When all the four metal atoms are four coordinated, there are eight occupied nonbonding orbitals. In the systems to which Eq. (1) is applicable, there are no other unoccupied orbitals except the antibonding orbitals. The numbers of antibonding orbitals and bonding orbitals as well as the number of edges of the polyhedron are equal to one another, for example, the cluster $Cp_3Cr_3CoS_4(CO)$. However, for the electron-deficient clusters, such as $MoFe_3S_4(SR)_6^{3-}$ and $Cp_3Cr_3FeS_4(OR)$, where bonding or, nonbonding unoccupied orbitals exist except the antibonding unoccupied orbitals, Eq. (1) is not applicable.

Table 11. Four classes of cubane-like cluster compounds

Class	Cluster compounds	Coordination number		Number of skeleton electrons	Orbital occupations*
		M	M'		
I	$L_3MM'_3S_4L'_3$ (M = V, Mo, W, Re; M' = Fe)	6	4	55	$(nb)^7(b)^{12}(ab)^0$
II	$L_6M_2M'_2S_4L'_2$ (M = V, Mo; M' = Fe, Co, Ni)	6	4	58, 59, 60	$(nb)^8(b)^{10-12}(ab)^{0-2}$
III	$L_9M_3M'S_4L'$ (M = Cr, Mo, W; M' = Fe, Co, Ni, Cu)	6	4	58, 60	$(nb)^{2-4}(b)^{12}(ab)^0$
IV	$L_{12}M_3M'S_4$ (M = Cr; M' = V, Nb)	6	6	59	$(nb)^0(b)^{11}(ab)^0$

* nb–nonbonding, b–bonding, ab–antibonding.

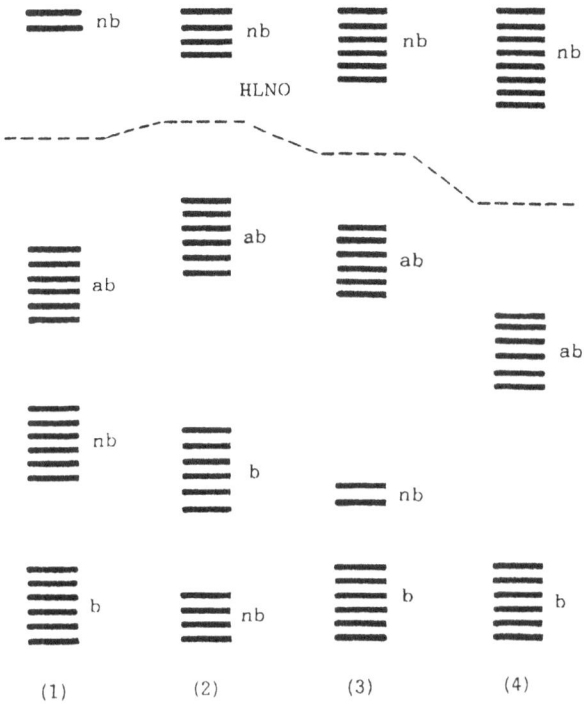

HLNO

(1)   (2)   (3)   (4)

Fig. 6. Qualitative MO energy level schemes for cubane-like cluster compounds of classes I–IV

Summing up our results, two concepts are introduced, that is, "the number of antibonding unoccupied orbitals" and "the number of other unoccupied orbitals" (including nonbonding and bonding unoccupied orbitals), denoted by "A" and "O", respectively. For single-occupied orbitals, the values of A and O are taken to be 1/2. Then Eq. (1) can be rewritten as follows:

$$BMO + NBMO = 9N - A - O \tag{4}$$

or

$$VE = 2(BMO + NBMO) = 2(9N - A - O) \tag{5}$$

The electron counting, electronic configuration of M–M interaction orbitals and the values of A and O obtained from molecular orbital diagrams in Fig. 6 are listed in Table 12. It can be seen that the numbers of skeleton electrons obtained by substituting the values of A and O into Eqs. (4) and (5) agree exactly with the results obtained by the conventional count of assigning electrons to ligands. As can be seen from Table 12, the "$9N - A - O$" rule and "$9N - L$" rule are essentially identical, while $A = L$ and $O = 0$, for clusters such as

Table 12. Electron counting of the cubane-like clusters

Cluster	Number of skeleton electrons (VE)	Number of metal electrons	Orbital occupations	A	O	Ref.
$L_3VFe_3S_4L'_3$	55	19	$(nb)^7(b)^{12}(ab)^0$	6	2.5	46
$L_3MoFe_3S_4L'_3$	55	19	$(nb)^7(b)^{12}(ab)^0$	6	2.5	91
$L_3WFe_3S_4L'_3$	55	19	$(nb)^7(b)^{12}(ab)^0$	6	2.5	91
$L_3ReFe_3S_4L'_3$	55	19	$(nb)^7(b)^{12}(ab)^0$	6	2.5	13
$L_6V_2Fe_2S_4L'_2$	58	18	$(nb)^8(b)^{10}(ab)^0$	6	1	77
$L_6V_2Co_2S_4L'_2$	60	20	$(nb)^8(b)^{11}(ab)^1$	5.5	0.5	78
$L_6V_2Ni_2S_4L'_2$	62	22	$(nb)^8(b)^{12}(ab)^2$	5	0	78
$L_6Mo_2Co_2S_4L'_2$	60	20	$(nb)^8(b)^{12}(ab)^0$	6	0	34
$L_6Mo_2Ni_2S_4L'_2$	62	22	$(nb)^8(b)^{12}(ab)^2$	5	0	22
$L_9Cr_3FeS_4L'$	58	14	$(nb)^2(b)^{10}(ab)^0$	6	1	29
$L_9Mo_3FeS_4L'$	58	14	$(nb)^2(b)^{10}(ab)^0$	6	1	23
$L_9Cr_3CoS_4L'$	60	16	$(nb)^4(b)^{12}(ab)^0$	6	0	74
$L_9Mo_3NiS_4L'$	60	16	$(nb)^4(b)^{12}(ab)^0$	6	0	84
$L_9Mo_3CuS_4L'$	60	16	$(nb)^4(b)^{12}(ab)^0$	6	0	57
$L_9W_3CuS_4L'$	60	16	$(nb)^4(b)^{12}(ab)^0$	6	0	58
$L_{12}Cr_3VS_4$	59	11	$(nb)^0(b)^{11}(ab)^0$	6	0.5	75
$L_{12}Cr_3NbS_4$	59	11	$(nb)^0(b)^{11}(ab)^0$	6	0.5	75

Note: nb–nonbonding, b–bonding, ab–antibonding, A–the number of Antibonding unoccupied orbitals, O–the number of other unoccupied orbitals

Table 13. Electron counting of the homometallic cubane-like clusters

Cluster	Number of metal electrons	Orbital occupations	Result of formula (4)			Ref.
			A	O	VE	
$Mo_4S_4(CN)_{12}$	12	$(nb)^0(b)^{12}(ab)^0$	6	0	60	68
$Mo_4S_4(NH_3)_{12}^{4+}$	12	$(nb)^0(b)^{12}(ab)^0$	6	0	60	82
$Mo_4S_4(Cp)_4$	12	$(nb)^0(b)^{12}(ab)^0$	6	0	60	4
$Cr_4S_4(Cp)_4$	12	$(nb)^0(b)^{12}(ab)^0$	6	0	60	76
$Mo_4S_4(Cp)_4^+$	11	$(nb)^0(b)^{11}(ab)^0$	6	0.5	59	4
$Mo_4S_4(edta)_2^{3-}$	11	$(nb)^0(b)^{11}(ab)^0$	6	0.5	59	81
$Mo_4S_4(Cp)_4^{2+}$	10	$(nb)^0(b)^{10}(ab)^0$	6	1	58	4
$Mo_4S_4(NCS)_{12}^{6-}$	10	$(nb)^0(b)^{10}(ab)^0$	6	1	58	17
$V_4S_4(Cp)_4$	8	$(nb)^0(b)^8(ab)^0$	6	2	56	22
$V_4S_4(Cp)_4^+$	7	$(nb)^0(b)^7(ab)^0$	6	2.5	55	22
$Ti_4S_4(Cp)_4$	4	$(nb)^0(b)^4(ab)^0$	6	4	52	22
$Mo_4S_4(NO)_4(CN)_8^{8-}$	20	$(nb)^0(b)^{12}(ab)^8$	2	0	68	69
$Fe_4S_4(Cp)_4$	20	$(nb)^0(b)^{12}(ab)^8$	2	0	68	80
$Fe_4S_4(NO)_4$	28	$(nb)^{16}(b)^{12}(ab)^0$	6	0	60	31
$Fe_4S_4X_4^{2-}$ (X = Cl, Br, I)	22	$(nb)^{16}(b)^6(ab)^0$	6	3	54	5
$Fe_4S_4(SR)_4^{2-}$	22	$(nb)^{16}(b)^6(ab)^0$	6	3	54	3

Note: nb–nonbonding, b–bonding, ab–antibonding, A–the number of antibonding unoccupied orbitals, O–the number of other unoccupied orbitals

$L_6Mo_2Co_2S_4L_2$, and $L_9Cr_3CoS_4L'$, etc. Therefore, the "9N – A – O" rule is an extension of the "9N – L" rule to the cubane-type transition-metal cluster compounds, and set up an unequivocal relationship between the number of skeleton electrons, coordination of metal atoms, and structures of cubane-like clusters. By using this rule, the characteristics of the geometry and the M–M interaction can be explained and distinguished satisfactorily. Moreover, the "9N – A – O" rule is also applicable to the homometallic cubane-like clusters shown in Table 13.

7 New Assumption on the Active Center Models of Nitrogenase

Quantum-chemical simulation on the coordination activation of nitrogenase substrates has its own particular advantage, because it is much more operative and easily controlled, avoiding the difficulties in synthesis of the chemical species under study. On the basis of summarizing the structural and physicochemical properties of FeMo-co and FeV-co obtained by experiments, to simulate quantum-chemically the functions of the active center of nitrogenase through model compounds is significant in our understanding of the functions and properties of the nitrogenase active center.

During the past 20 years, a variety of multinuclear Mo–Fe–S and V–Fe–S cluster compounds including μ_n-X ligands ($n = 2, 3$; $X = S$, O, Cl) have been synthesized through spontaneous self-assembly reactions. In regard of the mechanism of the spontaneous self-assembly reaction, one of the authors (JX Lu) proposed the "reactive fragment hypothesis": a complicated cluster might be assembled spontaneously from some building blocks, the cluster fragments (Sino-American Joint Symposium on Nitrogen Fixation, University of Wisconsin, Madison, 1982). Analogous compounds of the active center of nitrogenase, possessing nitrogen-fixing activity, might be combined through clusters fragments. From this, the "unit construction approach to the rational syntheses of transition-metal cubane-like clusters by using of reactive fragments as building blocks" was further proposed (Lu et al. 1989).

After the proposal of the Fuzhou model (the "string-bag cluster model") of the active center of nitrogenase (Lu 1975), some new compounds with geometries similar to the model, e.g. $[Fe_3S(S_2C_{10}H_{12})_3]^{2-}$ (Henkel et al. 1981), $[Fe_6S_9(SCH_3C_6H_5)_2]^{4-}$ (Henkel et al. 1982), and $[MoFe_7S_6(SR)_7L_3]^{2-}$ (Christou et al. 1982) were reported. The monoanion of the black Roussinate $[Fe_4S_3(NO)_7]^-$ (called black salt, for short) is the only synthetical model compound with a geometry ideally similar to that of the Fuzhou model-I. It has been known that in weakly acidic conditions and even in the aqueous solution of CO_2, the black salt can be synthesized spontaneously from dianion of the red Roussinate $[Fe_2S_2(NO)_4]^{2-}$ (called red salt). Thus, from the viewpoint of the "reactive fragment hypothesis" the red salt can be considered to be the "cluster fragment" of black salt. The electronic structure of the black salt and red salt have been investigated by using the closed-shell CNDO/2 (S.D scheme) method (Liu et al. 1987). By analysis of the bonding properties on the basis of Mulliken populations, charge densities, MO energies, and MO characters, the following results have been revealed. Strong electron delocalization has been found in these two clusters. In the process of electronic transfer from the red salt to the black salt, the sulfur atoms are an electron donor. The main contribution to the M–M and M–S_b bonds is the interaction between the s, p, and d_{z^2} orbitals on the metal atoms, and the s and p_z orbitals on the sulfur atoms. The contribution of the π interaction between the d orbitals on the metal atoms to the bonding of the cluster framework is quite small. The Fe–NO bonds are stronger than the Fe–S_b bonds in both salts. So it was proposed that in weakly acidic conditions, one bridging sulfur atom would be broken off from one red salt, which is favorable to produce a nido geometry. The results of the CNDO/2 MO calculations for the clusters $[Fe_2S_2(SH)_4]^{2-}$ and $[Fe_4S_4(SH)_4]^{2-}$ were presented and compared with those of the red salt and black salt. Similar features of the electronic structure in clusters containing iron and bridging sulfur with local tetrahedral symmetry are:

(1) The direct M–M interaction is formed by the contribution of the metal σ-bonds;
(2) The metal d-π bonding and antibonding interactions simultaneously exist in the high-lying occupied orbitals. Therefore there is no net M–M d-π bonding contribution;
(3) The occupied orbitals contributed predominantly from the bridging sulfur lone pair in character are not frontier orbitals, and are responsible for relatively stable M–(μ_3-S) bonding.

The different features are: the skeleton electrons in a cluster containing terminal ligand SR are more delocalized than in that containing terminal ligand NO, and the bonding interaction of iron atom with terminal SR in a binuclear complex is weaker than that of iron atom with terminal NO, consequently the terminal SR ligands are easier to be broken off than the bridging sulfur ligands when the binuclear complex $[Fe_2S_2(SH)_4]^{2-}$ is spontaneously converted into $[FeS_4(SH)_4]^{2-}$ and the product possesses a closo cubane-like structure (Cao et al. 1986). A further quantum-chemical study on the mechanism of the "spontaneous self-assembly" reaction of the formation of the black salt monoanion via dimeric condensation of red salt dianion confirmed the above conclusions. The results were compared with similar calculations carried out for the dimeric condensation reaction in the formation of the closo cubane-like 4Fe–4S dianion $[Fe_4S_4(SR)_4]^{2-}$ from the 2Fe–2S dianion $[Fe_2S_2(SR)_4]^{2-}$. Some suggestions for attempted syntheses of nido "string-bag" Fuzhou Model-I modelling cluster compounds are proposed (Liu et al. 1986b, 1989).

The coordination activation effects of the red salt, black salt, $MoFeS_4(NO)_2^{2-}$, and "string-bag" model on the nitrogenase substrates (C_2H_2, N_2, NNH^+, and NCH) have been simulated quantum-chemically by using the EHMO calculations.

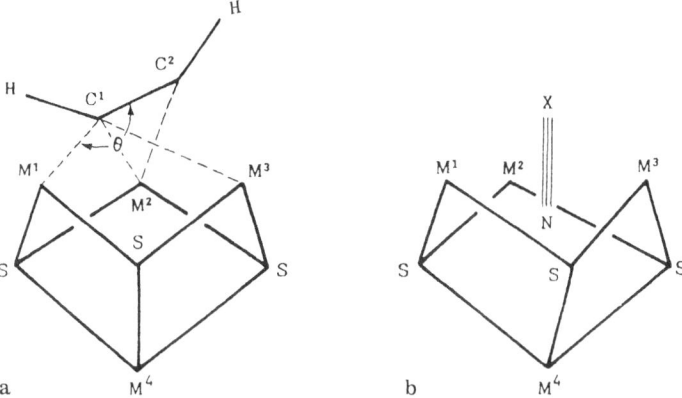

Fig. 7a. The coordination of C_2H_2 to the string-bag model in "cannon mount mode"; **b** the coordination of N–X (X = N, CH, NH) to the string-bag model in "diving mode"

Considering the variation of the total energy of the systems under consideration and the Mulliken populations of the multibonds of substrates, the optimal activated conformation for the system formed from dinuclear clusters and C_2H_2 is $d = 1.2$ Å, where d is the distance between the C≡C bond and MS_2M' plane. For the system formed from the black salt and C_2H_2 in the "cannon mount mode" (Fig. 7a) is $\theta = 140°$, where θ is the $M-C^1-C^2$ angle. In this respect, Fe is more effective than Mo for weakening the C≡C bond. In the system formed from the "string-bag" model and N_2, NNH^+ or NCH in the "diving mode" (Fig. 7b), the multiple bonds of the substrates are weakened substantially. In the "diving mode", the electron density of the nitrogen atom outside the mouth of the bag increases, thus favorable for electrophilic attack. The protonation along the orientation of the N–N axis on the N_2 is more advantageous for activating the N≡N bond. (Liu et al. 1988b).

8 Experimental Evidence for the Active Center Model of Nitrogenase

The important experimental evidence for the hypothetical active center models of nitrogenase are: (1) the proportion of M and S atoms and the local coordination information around Mo measured by using the EXAFS technique, and (2) the electronic and magnetic properties from the data of ESR, Mössbauer, magnetic susceptibility, MCD, etc. These experimental results reflect the chief properties of the active center of nitrogenase from different viewpoints (see Table 14). In Table 14, the M cluster and P cluster are two major types of components existing in MoFe protein (Münck et al. 1975; Zimmermann et al.

Table 14. The chief properties of the FeMo-co and FeV-co

Properties	FeMo-co	FeV-co	Ref.
Proportion of atoms M:Fe:S*	1:6–8:4–10	1:6:5	73, 86, 94
ESR, Mössbauer and MCD data	$S = 3/2$	$S = 5/2, 3/2, 1/2$	72, 95
	$g = 4.6, 3.2, 2.0$	$g = 5.6, 4.35, 3.6, 2.0$	28, 43, 44, 66
		$g = 5.8, 5.4, 4.34, 2.04, 1.93$	67
	positive zero-field splitting	negative zero-field splitting	72
Coordination number of M	6	6	
EXAFS information	Mo–Fe 2.68–2.73 Å	V–Fe 2.69 Å	1, 2
	Mo–S 2.36–2.37 Å	V–S 2.32 Å	14, 15
	Mo–O 2.10–2.12 Å	V–O 2.14 Å	20, 21
Coordinate environment of M	$MoFe_3S_3O_3$	$VFe_3S_3O_3$	14, 15, 28
Summary	M cluster character for reduced form	Only similar to the M cluster character for reduced form	66
	P cluster character for oxidized form	P cluster character for oxidized form	

1978). The M cluster is a Mo–Fe–S cluster for the ground state $S = 3/2$, and the P cluster is a [4Fe–4S] cluster that exhibits diamagnetism in reduced form and $S = 5/2$ in oxidized form (Smith et al. 1982; Johnson et al. 1985). As was studied by Rawlings et al. (1978), the $S = 3/2$ for FeMoco is the result of the spin-coupling between six Fe atoms around the Mo atoms. Due to the difference between the environments of each Fe atom, the magnetic hyperfine coupling constants of three equivalent Fe atoms (called B type) are positive and those of the rest three Fe atoms (A type) are negative. It was shown that spin $S = 3/2$ is the result of high spin-coupling between Fe atoms of two types (Huynh et al. 1980). With the exception of data on bond lengths, further information can be provided by EXAFS. Flank et al. (1986) compared the single-crystal EXAFS spectra of nitrogenase with that of single crystals of the model compounds $(PH_4P_2)[Cl_2FeS_2-MoS_2FeCl_2]$ and $(Et_4N)_3[Fe_6Mo_2S_8(SEt)_9]$, and concluded that the experimental data were not well simulated by the cluster with a linear arrangement of Fe–Mo–Fe atoms, whereas trinuclear clusters with a Fe–Mo–Fe angle between 50° and 130° gave a satisfactory agreement. The tetrahedral $MoFe_3$ and the square-pyramidal $MoFe_4$ clusters also gave satisfactory simulations of the orientation dependence. Therefore the linear trinuclear Fe–Mo–Fe clusters are unlikely to be adequate model compounds for the active center of the MoFe protein. Arber et al. (1987) compared both the EXAFS data of K-edge of V atom in the VFe protein (Acl*) and that of the V–Fe–S cluster $[Me_4N][VFe_3S_4Cl_3(DMF)_3]$ with X-ray data (Kovacs et al. 1986) which indicated that the agreement is quite well except for the V–Fe distances (2.69 Å) which is slightly shorter than that in the V–Fe–S cluster (2.73 Å). This fact shows that the V-containing cofactor in Acl* is similar to the FeMoco, in which the active sites of nitrogenase are contained. The EXAFS data of the V-containing nitrogenase in Avl' (George et al. 1988) supports the conclusion of Arber (1987). It is noted that the V–S and V–Fe distances are slightly longer than the Mo–S ad Mo–Fe distances, respectively. This slight difference led to the discrepancy in the geometries of the M–Fe–S (M=Mo, V) cluster compounds and thereby in the reactivity and mechanism with respect to the reduction of substrates (Smith et al. 1988; Dilworth et al. 1987).

9 New Hypothetical Active Center Models of Nitrogenase

Up to now, more than ten models of the active center of nitrogenase have been proposed from different viewpoints as reported by Gibson et al. (1988). Considering the above-mentioned experimental results and the bonding rules of the Mo–Fe–S clusters (Liu et al. 1988a), we proposed four new models (Liu et al. 1990) on the basis of the "string-bag models" (Lu 1975):

9.1 [FeS$_2$(SR)$_2$]$_3$ Combined with B-String Bag

In model *1*, a B-string bag (Liu et al. 1988b) was combined with three FeS$_2$(SR)$_2$ groups, resulting in cluster [MoFe$_6$S$_9$(SR)$_6$L$_3$]$^{3-}$ as shown in Fig. 8. The Mo

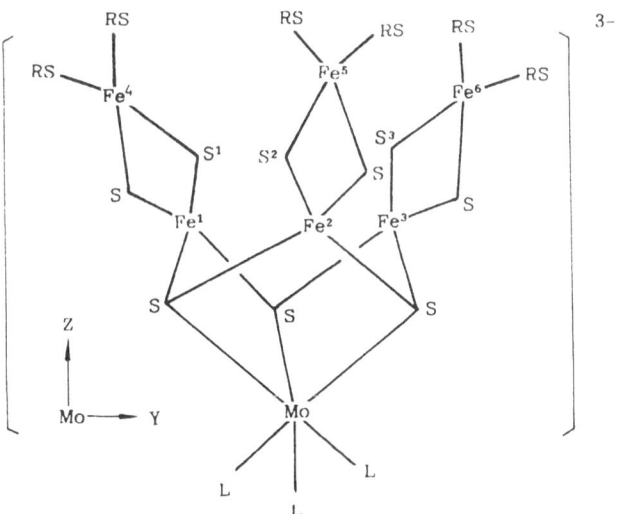

Fig. 8. [MoFe$_6$S$_9$(SR)$_6$L$_3$]$^{3-}$ model (model *1*)

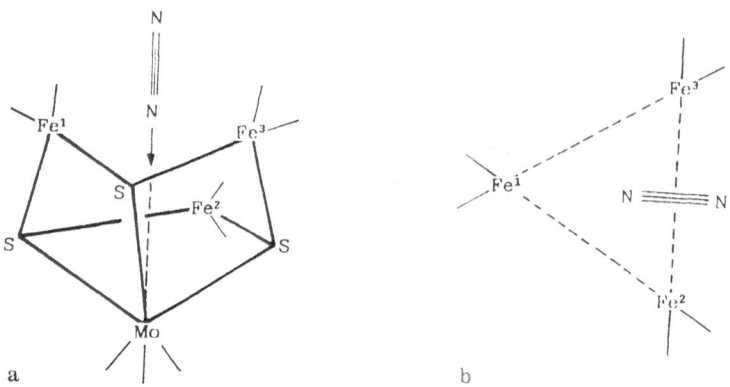

Fig. 9. The coordination of N_2 molecule to the active site of model *1* in "diving mode" **a** and in "laying mode" **b**

atom is six coordinated. The spin-coupling between metal atoms corresponds to the overlap of three Mo–Fe–Fe fragments, and there is high spin coupling ($S = 1/2$) between Fe(III) and Fe(II) (Liu et al. 1988a; Mascharak et al. 1981), so that we have $S = 3/2$, consistent with the spin state of FeMoco.

The active site of model *1* is near the "mouth" consisting of three Fe atoms. The coordination activation effects of model *1* plus N_2 molecule have been simulated quantum-chemically by using EHMO calculation to study the system formed from model *1* plus N_2 in the "diving mode" (**a**) and "laying mode" (**b**), as shown in Fig. 9. The variation of total energy and Mulliken bond order in the process of the N_2 attack to the active site of model *1* show that the optimal activation distances between N and Mo atoms are 2.6 and 2.1 Å for (**a**) and (**b**) models, respectively.

9.2 Cubane-String Bag Complex Models

In Table 14, the EXAFS data reveal that the local structures of both FeMoco and FeVco are similar to that of the cubane-like cluster. The P cluster is a good electron storer and transferer. According to the properties listed in Table 14 and from the viewpoint of multinuclear activation, the other three models *2–4* were proposed (see Fig. 10). Each of these models consists of a cubane and a string-bag structure, and is called the "cubane-string-bag complex models".

The difference in the reduction activity of FeMoco and FeVco is caused by the metal atoms M (M = Mo or V) in each model. The model *2* and model *3* are the clusters containing B- and H-string-bag (Liu et al. 1988b), respectively, with a cubane-like structure. In fact, each model only exhibits some principal properties of the FeMco (M = Mo, V), although consideration was given to all

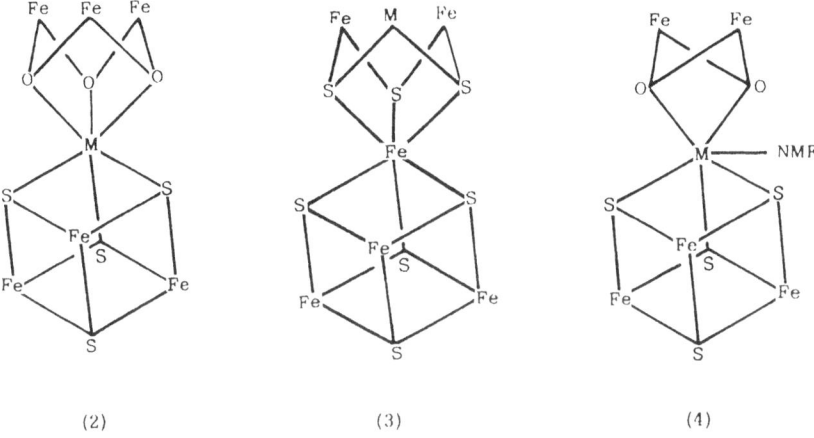

Fig. 10. Cubane-string-bag complex models (M = Mo, V): 2 cubane and B-string-bag complex model; 3 cubane and H-string-bag complex model; 4 cubane and $(\mu_3\text{-O})_2\text{Fe}_2$ complex model

properties of the FeMoco. For example, in model *2*, the requirement of multi-nuclear activation is well satisfied, but the coordination of M atom does not agree with that in the MoFe protein. In model *3*, the biological function in electron transfer of 4Fe–4S cubane is taken into consideration, and the number of ligands on M should be chosen properly so as to satisfy the saturated coordination of the M atom.

The coordinations of N_2 to models *2–4* in the "diving mode" have been simulated theoretically by using the EHMO method. For model *2*, an optimal activation conformation was formed with the N atom at the mouth of the bag at a distance of 0.6 Å. In the case M = Mo, model *2* has a better to weaken the $N{\equiv}N$ bond than the case of M = V, consistent with the experimental results. A similar result was also obtained for model *3*.

When the N_2 molecule approaches to the mouth of the bag in model *4*, the N_2 coordinates onto two Fe atoms in a side-on way, after the N_2 is put into the mouth of the bag and has undergone the interaction of the metal M in a head-on way, therefore the $N{\equiv}N$ bond would be further activated, and under the attack of proton, the $N{\equiv}N$ bond would be broken down and the NH_3 would be formed. Comparing the variation of Mulliken bond order of N_2 for M = Mo and V, we can see that the activation of N_2 by model *4* is in agreement with the experimental observations.

Acknowledgements. The authors are grateful to Mr. Huang Tang, Xing-ru Lin and Jun Li for many constructive suggestions and helpful discussions, and for their assistance in preparation of the manuscript.

10 References

1 Arber JM, Dobson BR, Eady RR, Stevens P, Hasnian S, Garner CD, Smith BE (1987) Vanadium K-edge X-ray absorption spectrum of the VFe protein of the vanadium nitrogenase of Azotobacter chroococcum. Nature (London) 325: 372

2 Arber JM, Flood AC, Garner CD, Gormal CA, Hasnain SS, Smith BE (1988) Iron K-edge X-ray absorption spectroscopy of the iron–molybdenum cofactor of nitrogenase from Klebsiella pneumoniae. Biochem J 252: 421

3 Averill BA, Herskovitz T, Holm RH, Ibers JA (1973) Synthetic analogues of the active sites of iron– sulfur proteins. II. Synthesis and structure of the tetra[mercapto-μ_3-sulfido–iron] clusters, $[Fe_4S_4(SR)_4]^{2-}$. J Am Chem Soc 95: 3523

4 Bandy JA, Davies CE, Green JC, Green MLH, Prout K, Rodgers DPS (1983) Synthesis, crystal structures, and bonding of the molybdenum cubane compounds $[Mo(\mu\text{-}C_5H_4Pr^i)\,(\mu_3\text{-}S)]_4^{n+}$, where $n = 0$, 1, and 2. J Chem Soc Chem Commun 1395

5 Bobrik MA, Hodgson KO, Holm RH (1977) Inorganic derivatives of iron–sulfide–thiolato dimers and tetramers. Structures of tetrachloro-μ-disulfido-diferrate(III) and tetrakis(chloro-μ_3-sulfido– iron) dianions. Inorg Chem 16: 1851

6 Cao HZ, Liu CW, Lu JX (1986) The electronic structure of $[Fe_2S_2(SH)_4]^{2-}$ and $[Fe_4S_4(SH)_4]^{4-}$. Acta Chim Sinica 44: 1197

7 Carney MJ, Kovacs JA, Zhang TP, Papaefthymiou GC, Spartalian K, Frankel RB, Holm RH (1987) Comparative electronic properties of V–Fe–S and Mo–Fe–S clusters containing isoelectronic cubane-type $[VFe_3S_4]^{2+}$ and $[MoFe_3S_4]^{3+}$ core. Inorg Chem 26: 719

8 Chisnell JR, Premakumar R, Bishop PE (1988) Purification of a second alternative nitrogenase from a nif HDK deletion strain of Azotobacter vinelandii. J Bacteriol 170: 27

9 Christou G, Garnar CD, King TJ, Johnson CE, Rush JD (1979) Isolation and Characterization by X-ray crystallography and Mössbauer Measurements of $[NEt_4]_3[Fe_6W_2S_8(SPh)_6(OMe)_3]$, and iron–tungsten–sulphur cubic cluster dimer. J Chem Soc Chem Commun 503

10 Christou G, Garnar CD (1980) Synthesis and proton magnetic resonance properties of Fe_3MS_4 (M = Mo or W) cubane-like cluster dimers. J Chem Soc Dalton Trans 2354

11 Christou G, Mascharak PK, Armstrong WH, Papaefthymiou GC, Frankel RB, Holm RH (1982) Electron-transfer series of $MoFe_3S_4$ double-cubane clusters electronic properties of components and the structure of $[(C_2H_5)_4N]_5[Mo_2Fe_6S_8(SC_6H_5)_9]$. J Am Chem Soc 104: 2820

12 Ciurli S, Holm RH (1989a) Insertion of $[VFe_3S_4]^{2+}$ and $[MoFe_3S_4]^{3+}$ cores into a semirigid tritiolate caritand ligand: regiospecific reactions at a vanadium site similar to that in nitrogenase. Inorg Chem 28: 1685

13 Ciurli S, Carney MJ, Holm RH, Papaefthymiou GC (1989b) Stability range of heterometal cubane-type clusters MFe_3S_4: Assembly of double-cubane clusters with the $ReFe_3S_4$ core. Inorg Chem 28: 2696

14 Conradson SD, Burgess BK, Newton WE, Hodgson KO, McDonald JW, Rubinson JF, Gheller SF, Mortenson LE, Adams MWW, Mascharak PK, Armstrong WH, Holm RH (1985) Structural insights from the molybdenum K-edge X-ray absorption near edge structure of the iron–molybdenum protein of nitrogenase and its iron–molybdenum cofactor by comparison with synthetic Fe–Mo–S clusters. J Am Chem Soc 107: 7935

15 Conradson SD, Burgess BK, Newton WE, Mortenson LE, Hodgson KO (1987) Structural studies of the molybdenum site in the MoFe protein and its FeMo cofactor by EXAFS. J Am Chem Soc 109: 7507

16 Cook MR, Karplus M (1985) Electronic structure of the $MoFe_3S_4(SH)_6^{3-}$ ion: a broken-symmetry metal–sulfur cluster. J Chem Phys 83: 6344

17 Cotton FA, Diebold MP, Diri Z, Llusar R, Schwotzer W (1985) The cuboidal $Mo_3S_4^{6+}$ aqua ion and its derivatives. J Am Chem Soc 107: 6735

18 Coucouvanis D (1981) Fe–Mo–S complexes derived from MS_4^{2-} anions (M = Mo, W) and their possible relevance as analogies for structural features in the Mo site of nitrogenase. Acc Chem Res 14: 201

19 Coucouvanis D, Simhon ED, Stremple P, Ryan M, Swenson D, Baenziger NC, Simopoulos A, Papaefthymiou V, Kostikas A, Petrouleas V (1984) Trinuclear Fe–M–S complexes containing a linear Fe–M–Fe array and a bridging S_2MS_2 unit. Electronic structures and crystal and molecular structures of the $[(C_6H_5)_4P]_2[Cl_2FeS_2MS_2FeCl_2]$ (M = Mo, W) complexes. Inorg Chem 23: 741

20 Cramer SP, Hodgson KO, Gillum WO, Mortenson LE (1978a) The molybdenum site of nitrogenase. Preliminary structural evidence from X-ray absorption spectroscopy. J Am Chem Soc 100: 3398

21 Cramer SP, Gillum WO, Hodgson KO, Mortenson LE, Stiefel EI, Chisnell JR, Brill WJ, Shah VK (1978b) The molybdenum site of nitrogenase. 2. A comparative study of Mo–Fe proteins and the iron–molybdenum cofactor by X-ray absorption spectroscopy. J Am Chem Soc 100: 3814

22 Curtis MD, Williams PD, Butler WM (1988) Preparation, structures and electrochemistry of tetranuclear sulfido clusters $Cp_2M_2M'_2S_{2-4}$ (M = MO, W; M' = Fe, Co, Ni). Inorg Chem 27: 2853

23 Dahlstrom PL, Kumar S, Zubieta J (1981) Synthesis and X-ray structural characterization of $[Me_4N]_3\{[SCH_2CH_2S)MoS_3]_2Fe\}$, the first example of a heteronuclear trimer containing molybdenum in a squarepyramidal geometry. J Chem Soc, Chem Commun 411

24 Darkwa J, Lockemeyer JR, Boyd PDW, Rauchfuss TB, Rheingold AL (1988) Synthetic, structural, and theoretical studies on the electron-deficient cubanes $(RC_5H_4)_4Ti_4S_4$, $(RC_5H_4)_4V_4S_4$, and $[RC_5H_4)_4V_4S_4]^+$. J Am Chem Soc 110: 141

25 Dilworth MJ, Eady RR, Robson RL, Miller RW (1987) Ethane formation from acetylene as a potential test for vanadium nitrogenase in vivo. Nature 327: 167

26 Dilworth MJ, Eady RR, Eldridge ME (1988) The vanadium nitrogenase of Azotobacter chroococcum. Biochem J 249: 745

27 Do Y, Simhon ED, Holm RH (1983) Derivatives of tetrathiovanadate(V): synthesis of the linear heterometallic $Fe(\mu_2\text{-}S)_2V(\mu_2\text{-}S)_2Fe$ core and the structures of $[VS_4]^{3-}$. J Am Chem Soc 105: 6731

28 Eady RR, Sobson RL, Richardson TH, Miller RW, Hawkins M (1987) The vanadium nitrogenase of Azotobacter chroococcum, purification, and properties of the VFe protein. Biochem J 244: 197

29 Eremenko IL, Pasynskii AA, Orazsakhatov B, Ellert OG, Novotortsev VM, Kalinnikov VT, Porai-Koshits MA, Antsyshkina AS, Dikareva LM, Ostrikova VN (1983) Interaction of heteronuclear chromium-containing clusters with Carboxylic acids. Molecular structure of the paramagnetic tetrahedral cluster $Cp_3Cr_3(\mu_3\text{-}S)_4Fe(OOCCMe_3)$. Inorg Chim Acta 73: 225

30 Flank AM, Weininger M, Mortenson LE, Cramer SP (1986) Single-crystal EXAFS of nitrogenase. J Am Chem Soc 108: 1049

31 Gall RS, Chu CT-W, Dahl LF (1974) Preparation, structure and bonding of two cubane-like iron–nitrosyl complexes, $Fe_4(NO)_4(\mu_3\text{-}S)_4$ and $Fe_4(NO)_4(\mu_3\text{-}S)_2(\mu_3\text{-}NC(CH_3)_3)_2$; Stereochemical consequences of bridging ligand substitution on a completely bonding tetrametal cluster unit and of different terminal ligands on the cubane-like Fe_4S_4 core. J Am Chem Soc 96: 4019

32 George GN, Coyle CL, Hales BJ, Cramer SP (1988) X-ray absorption of Azotobacter vinelandii vanadium nitrogenase. J Am Chem Soc 110: 4057

33 Gibson CP, Dahl F (1988) Synthesis and characterization of $(\eta^5\text{-}C_5Me_5)_2Mo_2Fe_2(CO)_9(\mu_2\text{-}CO)$ $(\eta^2,\mu_4\text{-}CO)$, a 62-electron butterfly molybdenum–iron cluster containing a π-bound four-electron-donating carbonyl ligand: a possible structural model for the active dinitrogen molybdenum–iron site in nitrogenase. Organometallics 7: 535

34 Habert TR, Cohen SA, Stiefel EI (1985) Construction of heterometallic "thicubanes" from $M_2S_2(\mu\text{-}S)_2$ core complexes: Synthesis of $Co_2M_2S_4(S_2CNEt_2)_2(CH_3CN)_2(CO)_2$ (M = Mo, W) and structure of the $Co_2Mo_2(\mu_3\text{-}S)_4$ cluster. Organometallics 4: 1689

35 Hales BJ, Langasch DJ, case EE (1986a) Isolation and characterization of a second nitrogenase Fe-protein from Azotobacter-vinelandii. J Biol Chem, 261: 15301

36 Hales BJ, Case EE, Morningstar JE, Dzeda MF, Mauterer LA (1986b) Isolation of a New vanadium-containing nitrogenase from Azotobacter vinelandii. Biochemistry 25: 7251

37 Harris S (1989) Structure, bonding and electron counts in cubane-type clusters having M_4S_4, $M_2M'_2S_4$ and MM'_3S_4 core. Polyhedron 8: 2843

38 Henkel G, Tremel W, Krebs B (1981) $[Fe_3S(S_2C_{10}H_{12})_3]^{2-}$: the first synthetic trinuclear iron–sulfur cluster compound. Angew Chem Int Ed Engl 20: 1033

39 Henkel G, Strasdeit H, Krebs B (1982) $[Fe_6S_9(SCH_2C_6H_5)_2]^{4-}$: a hexanuclear iron–sulfur cluster anion containing the square-pyramidal $[(\mu_4\text{-}S)Fe_4]$ unit. Angew Chem Int Ed Engl 21: 201

40 Holm RH (1981) Metal clusters in biology: quest for a synthetic representation of the catalytic site of nitrogenase. Chem Soc Rev 10: 455

41 Huynh BH, Müenck E, Orme-Johnson WH (1979) Nitrogenase XI. Mössbauer studies on the cofactor centers of the MoFe protein from Azotobacter vinelandii op. Biochim Biophys Acta, Protein Structure 576: 192

42 Huynh BH, Henzl MT, Christner JA, Zimmermann R, Orme-Johnson WH, Müenck E (1980) Nitrogenase XII. Mössbauer studies of the MoFe protein from Clostridium Pasteurianum W5. Biochim Biophys Acta, Protein Structure 623: 124

43 Johnson MK, Thomson AJ, Robinson AE, Smith BE (1981) Protein structure characterization of the paramagnetic centres of the molybdenum–iron protein of nitrogenase from klebsiella pneumoniae using low temperature magnetic circular dichroism spectroscopy. Biochim Biophys Acta 671: 61

44 Johnson MK, Morningstar JE, Bennett DE (1985) Magnetic circular dichroism studies of succinate dehydrogenase. Evidence for [2Fe–2S], [3Fe–xS], and [4Fe–4S] centers in reconstitutively active enzyme. J Biol Chem 260: 7368

45 Kovacs JA, Holm RH (1986) Assembly of vanadium–iron–sulfur cubane clusters from mononuclear and linear trinuclear reactants. J Am Chem Soc 108: 340

46 Kovacs JA, Holm RH (1987) Structural chemistry of vanadium–iron–sulfur clusters containing the cubane-type $[VFe_3S_4]^{2+}$ core. Inorg Chem 26: 711

47 Lauher JW (1978) The bonding capabilities of transition metal clusters. J Am Chem Soc 100: 5305

48 Lauher JW (1979) Bonding capabilities of transition metal clusters. 2. Relationship to bulk metals. J Am Chem Soc 101: 2604

49 Lin Z-Y, Liu C-W (1987) Fe–Fe interaction and electronic structure rules in the Fe–S cluster compounds. Acta Chim Sinica 45: 535

50 Liu C-W, Hua J-M, Chen Z-D, Lin Z-Y, Lu J-X (1986a) The bonding properties of Mo–Fe S clusters. Int J Quant Chem 29: 701

51 Liu C-W, Cao H-Z, Lu J-X (1986b) Quantum chemistry simulation of the proton effect in a spontaneous assembly reaction. Acta Chim Sinica 44: 1191

52 Liu C-W, Cao H-Z, Lu J-X, Zheng S-J, Liu R-Z (1987) The electronic structures of red Roussinate and black Roussinate. Acta Chim Sinica 45: 1

53 Liu C-W, Lin Z-Y, Lu J-X (1988a) The bonding rules of Mo–Fe–S cluster compounds. J Mol Struct (Theochem) 180: 189

54 Liu C-W, Hua J-M, Lu J-X (1988b) Quantum chemical simulation on the coordination activation of nitrogenase substrates. Acta Chim Sinica 46: 315

55 Liu C-W, Cao H-Z, Lu J-X (1989) A quantum-chemical study of the mechanism of the "spontaneous self-assembly" reaction in the formation of the black Roussinate monoanion via dimeric condensation of red Roussinate dianion. J Mol Struct (Theochem) 183: 1

56 Liu C-W, Lin Z-Y, Yang X-F (1990) New assumption on the active centre models of nitrogenase. Chem J Chinese Univ Spec 198

57 Lu S-F, Zhu N-Y, Wu X-T, Wu Q-J, Lu J-X (1989a) The Synthesis and Crystal structures of three Mo–Cu–S cubane-like clusters: $[Mo_3CuS_4]$ $[S_2P(OCH_2CH_3)_2]_3 \cdot I \cdot (\mu_2\text{-OOCCH}_3) \cdot H_2O$. $[Mo_3CuS_4][S_2P(OCH_2CH_3)_2]_3 \cdot I \cdot (\mu_2\text{-OOCCH}_3) \cdot (CH_3)_2SO$ and $[Mo_3CuS_4][S_2P(OCH_2CH_3)_2]_3 \cdot I \cdot (\mu_2\text{-OOCC}_6H_5) \cdot C_5H_5N$. J Mol Struct 197: 15

58 Lu S-F, Zhu N-Y, Wu X-T, Wu Q-J, Lu J-X (1989b) The synthesis and crystal structure of a novel cubane-like tungsten copper sulfur $[W_3CuS_4] \cdot [S_2P(OC_2H_5)_2]_3 \cdot I \cdot \mu_2\text{-}CH_3COO \cdot CH_3CN$. J Mol Struct 197: 33

59 Lu J-X (Nitrogen Fixation Research Group, Fujian Institute of Research on the Structure of Matter, Chinese Academy of Sciences) (1975) A preliminary model for the active site of catalytic nitrogen fixation in nitrogenase—Also notes on structural criteria for the activation of dinitrogen molecules via coordination. Kexue Tongbao, 20: 540

60 Lu J-X, Zhuang B (1989) A unit construction approach to the rational syntheses of transition metal cubane-like clusters by the use of reactive fragments as building blocks. Jiegou Huaxue (J Struct Chem) 8: 233

61 Mascharak PK, Papaefthymiou GC, Frankel RB, Holm RH (1981) Evidence for localized Fe(III)/Fe(II) oxidation state configuration as an intrinsic property of $[Fe_2S_2(SR)_4^{3-}]$ clusters. J Am Chem Soc 103: 6110

62 Mascharak PK, Armstrong WH, Mizobe Y, Holm RH (1983) Single cubane-type $MoFe_3S_4$ clusters (M = Mo, W): synthesis and properties of oxidized and reduced forms and the structure of $(Et_4N)_3[MoFe_3S_4(S\text{-}p\text{-}C_6H_4Cl)_4(3,6\text{-}(C_3H_5)_2C_6H_2O_2)]$. J Am Chem Soc 105: 475

63 Miller RW, Eady RR (1988) Molybdenum and vanadium nitrogenases of Azotobacter chroococcum. Biochem J 256: 429

64 Mingos DMP (1983) Polyhedral skeletal electron pair approach a generalized principle for condensed polyhedra. J Chem Soc, Chem Commun 706

65 Mingos DMP (1985) Interrelationships between the topological electron-counting theory and the polyhedral skeletal electron pair theory. Inorg Chem 24: 114

66 Morningstar JE, Johnson MK, Case EE, Hales BJ (1987a) Characterization of the metal clusters in the nitrogenase Mo–Fe and V–Fe proteins of Azotobacter vinelandii using MCD spectroscopy. Biochemistry 26: 1795

67 Morningstar JE, Hales BJ (1987b) Electron paramagnetic resonance study of the vanadium–iron protein of nitrogenase from Azotobacter vinelandii. J Am. Chem Soc 109: 6854

68 Müller A, Eltzner W, Bögge H, Jostes R (1982a) $[Mo_4^{III}S_4(CN)_{12}]^{8-}$, a cluster with high negative charge and cubane-like Mo_4S_4-moiety—on the significance of cyanothiomolybdates for the proebiotic evolution. Angew Chem Int Ed Engl 21: 795

69 Müller A, Eltzner W, Clegg W, Sheldrick GM (1982b) Formation of metal–metal bonds and conversion of the metal aggregate $\{Mo_4(S_2)_4(S'_2)_2\}$ by atom-transfer and redox reactions at nonequivalent ligands; $[Mo_4S_4(NO)_4(CN)_8]^{8-}$ and anion with a central cubane-like unit. Angew Chem Int Ed Engl 21: 536

70 Müller A, Sarkar S, Bögge H, Jostes R, Trautwein A, Lauer U (1983) $[Cl_2FeS_2MoO(S_2)]^{2-}$, a novel bimetallic complex with unusual electronic structure and a substituted tetrachalcogeno-metalate as "ligand". Angew Chem Int Ed Engl 22: 561

71 Müller A, Hellmann W, Römer C, Römer M, Bögge H, Jostes R, Schimanski U (1984) New homo- and heteronuclear tetrathiometallo complexes, specific to the Fe^{II}/WS_4^{2-} system: the novel tetranuclear $[S_2WS_2FeS_2FeS_2WS_2]^{4-}$ complex with linear metal atom array. Inorg Chim Acta 83: L75

72 Münck E, Rhodes H, Orme-Johnson WH, Davis LC, Brill WJ, Shah VK (1975) Nitrogenase VIII. Mössbauer and EPR spectroscopy. The MoFe protein component from Azotobacter vinelandii op Biochimica et Biophysica Acta, 400: 32

73 Nelson MJ, Levy MA, Orme-Johnson WH (1983) Metal and sulfur composition of iron–molybdenum cofactor of nitrogenase. Proc Natl Acad Sci USA 80: 147

74 Pasynskii AA, Eremenko IL, Orazsakhatov B, Kalinnikov VT, Aleksandrov GG, Struchkov YuT (1981a) Antiferromagnetic complexes with metal–metal bonds. VI. Transformation of the antiferromagnetic metalacycle $(Cp_2Cr_2SCMe_3)$ $(\mu_3$-S$)_2Co(CO)_2$ into the diamagnetic metallo-tetrahedron $Cp_3Cr_3(\mu_3$-S$)_4Co(CO)$. J Organomet Chem 214: 367

75 Pasynskii AA, Eremenko IL, Orazsakhatov B, Kalinnikov VT, Aleksandrov GG, Struchkov YuT (1981b) Antiferromagnetic complexes with metal–metal bonds. VII. Synthesis and molecular structure of $(CpCrSCMe_3)_2SMn(CO)_2Cp$ and formation of the tetranuclear clusters $Cp_4Cr_3MS_4$ (Cr, V, Nb). J Organomet Chem 216: 211

76 Pasynskii AA, Eremenko IL, Rakitin YuV, Novotortsev VM, Ellert OG, Kalinnikov VT, Shklover VE, Struchkev YuT, Lindeman SV, Kurbanov TK, Gasanov GS (1983) Anti-ferromagnetic complexes with metal–metal bonds. IX. Synthesis and molecular structures of methylcyclopentadienylchromium (III) sulfide diamagnetic tetramer and the antiferromagnetic copper (II) bromide adduct of the tetranuclear cluster $(MeC_5H_4)_4Cr_4(\mu_3$-O$)(\mu_3$-S$)_3$. J Organomet Chem 248: 309

77 Rauchfuss TB, Weatherill TD, Wilson SR, Zebrowski JB (1983) Stepwise assembly of heterometallic M_4S_4 clusters. The structure of $(MeCp)_2V_2Fe_2(NO)_2S_4$: A 58e cubane. J Am Chem Soc 105: 6508

78 Rauchfuss TB, Gammon SD, Weatherill TD, Wilson SR (1988) Localized structural effects in the heterometallic thiocubanes $(MeCp)_2V_2M_2S_4(NO)_2$ where $M_2 = Fe_2$, Co_2, and Ni_2. New J Chem 12: 373

79 Rawlings J, Shah VK, Chisnell JR, Brill WJ, Zimmermann R, Münck E, Orme-Johnson WH (1978) Novel metal cluster in the iron–molybdenum cofactor of nitrogenase. Spectroscopic evidence. J Biol Chem 253: 1001

80 Schunn RA, Fritchie Jr CJ, Prewitt CT (1966) Syntheses of some cyclopentadienyl transition metal sulfides and the crystal structures of $(C_5H_5FeS)_4$. Inorg Chem 5: 892

81 Shibahara T, Karoya H, Matsumoto K, Ooi S (1984) A novel cubane-type Mo_4S_4 cluster. J Am Chem Soc 106: 789

82 Shibahara T, Kawano E, Okano M, Nishi M, Kyroya H (1986a) Cubane-type cluster $[Mo_4S_4(NH_3)_{12}]Cl_4 \cdot 7H_2O$. Chem Lett 827

83 Shibahara T, Akashi H, Kuroya H (1986b) Cubane-type $Mo_3FeS_4^{4+}$ aqua ion and X-ray structure of $[Mo_3FeS_4(NH_3)_9(H_2O)]Cl_4$. J Am Chem Soc 108: 1342

84 Shibahara T, Kuroya HJ (1988) Preparation and X-ray structure of cubane-type mixed metal aqua ion, $[Mo_3NiS_4(H_2O)_{10}]^{4+}$. Coord Chem 18: 233

85 Smith JP, Emptage MH, Orme-Johnson WH (1982) Magnetic susceptibility studies of native and thioneine-oxidized molybdenum–iron protein from Azotobacter vinelandii nitrogenase. J Biol Chem 257: 2310

86 Smith BE, Eady RR, Lowe DJ, Gormal G (1988) The vanadium-iron protein of vanadium nitrogenase from Azotobacter chroococcum contains an iron–vanadium cofactor. Biochem J 250: 299

87 Tang A-Q, Li Q-S (1983) A study on the structure rule of metal cluster compounds. Kexue Tongbao 28: 25

88 Tang A-Q, Li Q-S, Sun J-Z (1986) The structure rule of Mo Fe -S cluster compounds. Acta Chim Sinica 44: 1217

89 Teo BK (1985) Molecular orbital justification of topological electron-counting theory. Inorg Chem 24: 1627

90 Wade K (1976) Structural and bonding patterns in cluster chemistry. Adv Inorg Chem Radiochem 18: 1

91 Wolff TE, Berg JM, Power PP, Hodgson KO, Holm RH (1980) Structural characterization of the iron-bridged "double-cubane" cluster complexes $[Mo_2Fe_7S_8(SC_2H_5)_{12}]^{3-}$ and $[M_2Fe_7S_8\cdot(SCH_2C_6H_5)_{12}]^{4-}$ (M = Mo, W) containing MFe_3S_4 cores. Inorg Chem 19: 430

92 Wolff TE, Berg JM, Holm RH (1981) Synthesis structure, and properties of the cluster complex $[MoFe_4S_4(SC_2H_5)_3(C_6H_4O_2)_3]^{3-}$ containing a single cubane-type $MoFe_3S_4$ core. Inorg Chem 21: 174

93 Xu G-X (1982) Structural rules of cluster compounds and related molecules (I). Chem J Chinese Univ (Spec.) 3: 114

94 Yang SS, Pan WH, Friesen GD, Burgess BK, Corbin JL, Stiefel EI, Newton WE (1982) Iron–molybdenum cofactor from nitrogenase. Modified extraction methods as probes for composition. J Biol Chem 257: 8042

95 Zimmermann R, Münck E, Brill WJ, Shah VK, Henzl MT, Rawlings J, Orme-Johnson WH (1978) Nitrogenase X: Mössbauer and EPR studies on reversibly oxidized MoFe protein from Azotobacter vinelandii op, nature of the iron centers. Biochim Biophys Acta, Protein Structure 537: 185

Section II:
Biochemistry and Molecular Genetics of Nitrogen Fixation

CHAPTER 9
Identification and DNA Sequence Analysis
of the *fixX* Gene of *R. leguminosarum* bv. *Viciae*

G.F. Hong, F.D. Ni, Q.S. Ma, and J.A. Downie

Nitrogen-fixing bacteria conduct a reaction which, in chemical terms, is most unusual. The dinitrogen (N_2) molecule is normally very unreactive, partly due to the very stable triple bond which links the two atoms in the structure $N\equiv N$. Therefore catalysts at elevated pressures and temperatures as in the Haber–Bosch process, are normally required to make N_2 reactive. However, nitrogen-fixing organisms reduce N_2 at ordinary temperatures and pressures. The enzyme nitrogenase is responsible for this process, and the way in which it functions is a fascinating problem which has only partly been resolved.

Nitrogenase consists of two subunits; component I (P1) which contains both iron and molybdenum and component II (P2) which contains iron. In addition to the nitrogenase complex, other factors are required for the electron transfers involved in the reduction of N_2 to NH_3. Electron donors (such as reduced ferredoxin or flavodoxin) transfer electrons to P2. Individual electrons are then transferred from P2 to P1 of nitrogenase in separate steps, each of which is accompanied by the hydrolysis of two ATP molecules. Following several rounds of such electron transfers (probably eight) the products of the reaction ($2NH_3$ and H_2) are released.

In the process of determining the DNA sequence of the *fix* region of the symbiotic plasmid (pRL1JI) of *Rhizobium leguminosarum* biovar a short open reading frame was identified. This was identical in length (194 base pairs) with those *fixX* genes identified in other *Rhizobium* species and was located 11 bp downstream of the *fixC* gene. The *fixX* gene identified is strongly homologous to the other *fixX* genes identified in *rhizobia* and this strong conservation was also found in the predicted protein sequences. The motif, Cys**Cys**Cys***Cys found in this sequence is recognised as being found in proteins that form Fe-S clusters and similar motifs have been found in several known ferredexins.

It was found that the GC rich inverted repeat of GCGACCGCACGGTCGC is located in the middle of *fixX*. Conventional DNA sequencing gels could not resolve the compression caused by this repeat (see below), and, to our knowledge, this structure does not exist in any of the other *fixX* genes so far identified in *Rhizobium* species, it's function remains unknown.

The Nitrogen Fixation
and its Research in China
Editor Guo-fan Hong
ⓒ Springer-Verlag Berlin Heidelberg 1992

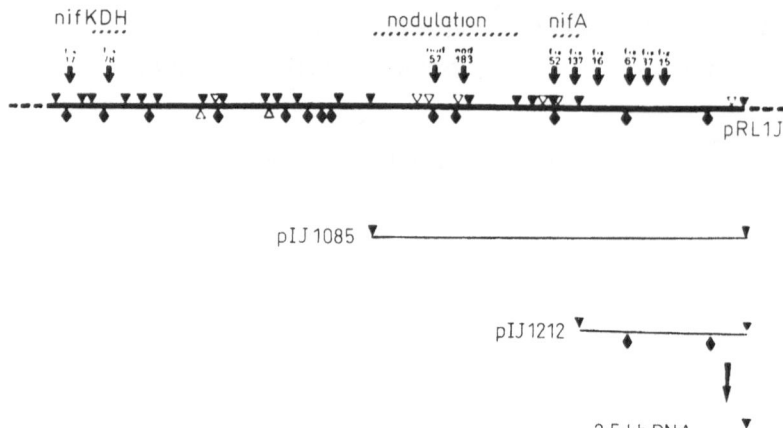

Fig. 1

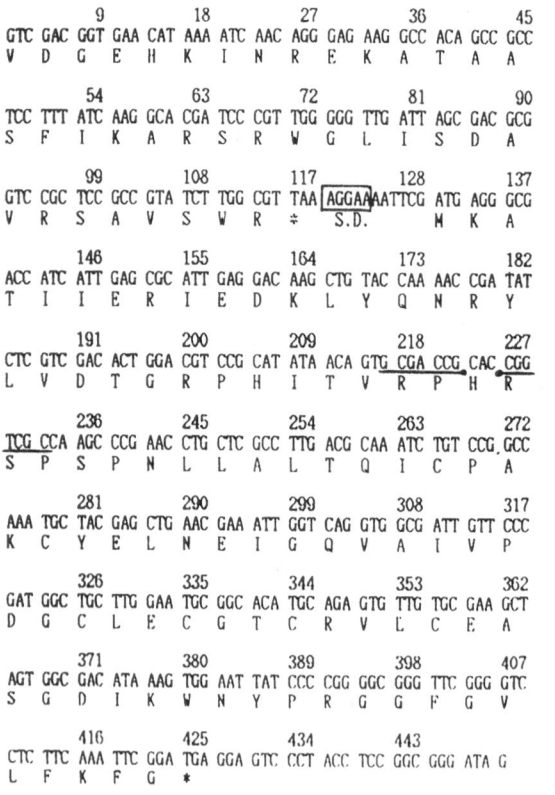

```
            9          18          27          36          45
GTC GAC GGT GAA CAT AAA ATC AAC AGG GAG AAG GCC ACA GCC GCC
 V   D   G   E   H   K   I   N   R   E   K   A   T   A   A

           54          63          72          81          90
TCC TTT ATC AAG GCA CGA TCC CGT TGG GGG TTG ATT AGC GAC GCG
 S   F   I   K   A   R   S   R   W   G   L   I   S   D   A

           99         108         117         128         137
GTC CGC TCC GCC GTA TCT TGG CGT TAA AGGAAAATTCG ATG AGG GCG
 V   R   S   A   V   S   W   R   ÷     S.D.       M   K   A

          146         155         164         173         182
ACC ATC ATT GAG CGC ATT GAG GAC AAG CTG TAC CAA AAC CGA TAT
 T   I   I   E   R   I   E   D   K   L   Y   Q   N   R   Y

          191         200         209         218         227
CTC GTC GAC ACT GGA CGT CCG CAT ATA ACA GTG CGA CCG CAC CGG
 L   V   D   T   G   R   P   H   I   T   V   R   P   H   R

          236         245         254         263         272
TCG CCA AGC CCG AAC CTG CTC GCC TTG ACG CAA ATC TGT CCG GCC
 S   P   S   P   N   L   L   A   L   T   Q   I   C   P   A

          281         290         299         308         317
AAA TGC TAC GAG CTG AAC GAA ATT GGT CAG GTG GCG ATT GTT CCC
 K   C   Y   E   L   N   E   I   G   Q   V   A   I   V   P

          326         335         344         353         362
GAT GGC TGC TTG GAA TGC GGC ACA TGC AGA GTG TTG TGC GAA GCT
 D   G   C   L   E   C   G   T   C   R   V   L   C   E   A

          371         380         389         398         407
AGT GGC GAC ATA AAG TGG AAT TAT CCC CGG GGC GGG TTC GGG GTC
 S   G   D   I   K   W   N   Y   P   R   G   G   F   G   V

          416         425         434         443
CTC TTC AAA TTC GGA TGA GGA GTC CCT ACC TCC GGC GGG ATA G
 L   F   K   F   G   *
```

Fig. 2

Figure 1 shows the physical map and the Tn5 mutations of a region on pRL1JI, a native symbiotic plasmid of *R. Leguminosarum* bv. *viciae*. The *fixX* gene was located in a 2.5 kb DNA region, which was derived from pIJ1212. pIJ1212 was derived from pIJ1085, which covered the whole region of nodulation and *fix* genes on pRL1JI.

Figure 2 shows a 450 bp DNA sequence. The *fixX* gene starts at position 129 and stops at 425. Computer analysis shows that the sequence from 1–117 bp includes part of the *fixC* gene which terminates at position 117. The potential ribosome binding site AGGAA preceding *fixX* open reading frame is within the 11 bp intergenic region. The inverted repeat described above is located to the middle of the gene (from 215 to 231).

Figure 3 is a comparison of the amino acid sequences of the *fixX* genes from different *Rhizobium* species, showing the striking homology that has been

```
        10        20        30        40        50
1. MKATIIERIEDKLYQNRYLVDTGRPHITVRPHRSPSPNLLALTQICPAKCYELNEIGQVA
2. MKTAIAERIEDKLYQNRYLVDAGRPHITVRPHRSPSLNLLALTRVCPAKCYELNETGQVE
3. MKTAIAERIEDKLYQNRYLVDAGRPHITVRPHRSPSLNLLALTRVCPAKCYELNETGQVE
4. MNVEPSVRVEDKLYYNRYLVDAGHPHVRVRAHKTPSPQLITLLKACPARCYELNDNGQVE
5. MKAIVRRVEDKLYQNRYLVDPGRPHISVRKHLFPTPNLIALTQVCPAKCYQLNDRRQVI
6. MKATTIERIEDKLYQNRYLVDTRRPHITVRPHRSPSPSLLALTQICPAKCYEVNEIGQVA
```

```
        70        80        90
1. IVPDGCLECGTCRVLCEASGDIKWNYPRGGFGVLFKFG
2. VTADGCMECGTCRVLCEANGDVECSYPRGGFGVLFKFG
3. VIADGCMECGTCRVLCEANGDVEWSYPRGGFGVLFKSG
4. VTVDGCIECGTCRVIAEPTGDIEWSHPRGGYGVLFKFG
5. IVSDGCLECGTCNVLCGPDGDIEVTYPRGGFGVLFKFG
6. IVSDGCLECGTCRVLAEASGDIKWNYPRGGFGVLFKFG
```

* * *   *   Fig. 3

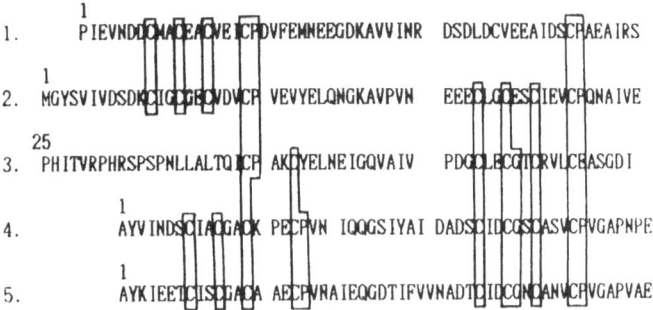

* * *   *   * * *   *   Fig. 4

observed. 1. *R. leguminosarum* 248. 2. *R. meliloti* 1021 (Earl et al. 1987). 3. *R. mililoti* 41 (Dusha et al. 1987). 4. *Bradyrhizobium japonicum* (Iismaa et al. 1987). 5. *R. trifolii* (Gubler et al. 1989). 6. *R. leguminosarum* 3588 (Gropnger et al. 1987).

Figure 4 shows a comparison of the amino acid sequence of the *fixX* gene with those of ferredoxins of several bacteria (Dusha et al. 1987). 1. *Desulfovibrio gigas*, 2. *D. desulfuricans* Norway 3. *R. leguminosarum* 248, 4. *Peptococcus aerogenes*, 5. *Rhodospirillum rubrum*, * marks the position of cystein residues.

The *fixX* gene was completely sequenced on both strands. It was found that no conventional sequencing gels could resolve the compression caused by the inverted repeat mentioned above, but it was perfectly resolved when the sequence reaction was initiated by the heat-stable *Bst* polymerase sequencing system (Ye et al. 1987) followed by the urea/formamide gel analysis (7). In Fig. 2 can be seen the inverted repeat which was underlined.

References

Bankier AT et al. (1987) Random cloning and sequencing by the M13 dideoxymeleotide chain termination method. In: Ray Wu (ed) Methods in enzymology Vol 155, p 51 93

Dusha I et al. (1987) Rhzobium meliloti inser element ISRm2 and its use for identification of the *fixX* gene. J Bacteriol 169: 1403

Earl CD et al. (1987) Genetic and structural analysis of the Rhizobium meliloti fixA. fixB. fixC and fixX genes. J Bacteriol 169: 1404

Gropnger P et al. (1987) Organization and partial sequence of a DNA region of the Rhizobium leguminosarum symbiotic plasmid pRL6JI containing the genes fix ABC, nif A, nif B and a novel open reading frame. Nucl Acid Res 15: 31

Gubler M et al. (1989) The Bradyrhizobium japonicum fix BCX operon: identification of *fixX* and of a 5′ mRNA region affecting the level of the *fix BCX* transcript. Mol Microbiol 3: 141

Iismaa SE et al. (1987) A gene upstream of the Rhizobium trifolii nifA gene encordes a ferredoxin-like protein. Nucl Acid Res 15: 3180

Ye SY, Hong GF (1987) Heat-stable DNA polymerase I large fragment resolves hairpin structure in DNA sequencing. Scientia Sinica 30: 503

CHAPTER 10
Multiple Binding Sites in the Complex Between the *nodA-D* Intergenic Region and the Product of the *nodD* Gene

G.F. Hong, H.M. Cao, F.D. Ni, and Y.Y. Lu

The following findings have been made:

(1) NodD (*nodD* gene product) specifically binds to the *nodA-D* DNA, which spans the intergenic region of *nodD* and *nodA* genes (Hong et al. 1987). (The *nodA-D* DNA sequence is available at the end of this paper).

(2) NodD binds to the *nodA-D* DNA, resulting in as many as seven complexes, which have different velocities when they are subject to electrophoresis on the native polyacrylamide gel.

(3) Naringenin—a plant-produced signal molecule—destabilizes the complexes formed between NodD and the *nodA-D* DNA.

In *Rhizobium leguminosarum* bv *viciae*, there are at least 13 nodulation (*nod*) genes which are responsible for the forming of nodules on the roots of legume. One of those genes called *nodD*, is responsible for the regulation of all the rest *nod* genes as shown below.

induction: requires flavonoid

nodO   T   N   M   L   E   F   D   A   B   C   I   J

(1) NodD specifically binds to the *nodA–D* DNA

Figure 1 is a demonstration of NodD-dependent retardation of the nodA-D DNA. The end-labeled *nodA-D* fragment is exposed to extracts from strain 8401/pKT230 (lane 2) or from 8401/pIJ1518 (lane 3) which have been grown in the absence of inducer. In lane 1 and 4 are controls in which the fragment is not exposed to any extract. The solid arrow indicates the position of the NodD-dependent retarded band and the open arrow the native fragment. Between these two fragments is another band (dotted arrow) which is not present in the controls but which is not dependent on NodD.

It has also been proved that the mutant in *nodD*, which fails to specifically bind to the *nodA-D* DNA, is unable to exercise its regulation (Hong et al. 1987).

8401 is a strain of *Rhizobium leguminosarum* bv. *phaseoli* cured of its symbiotic plasmid, Str (Lamb et al. 1985). pKT230 is a wide host range vector, Kan^R, Str^R

The Nitrogen Fixation
and its Research in China
Editor: Guo-fan Hong
© Springer-Verlag Berlin Heidelberg 1992

(Bagdasarian et al. 1981). pIJ1518 is the cloned DNA fragment containing *nodD* in pKT230, Kan^R (Rossen et al. 1985).

(2) NodD binds to the *nodA-D* DNA forming as many as seven complexes.

NodD is purified to a great extent through an affinity column prepared in this lab. By using this purified NodD, as many as seven complexes are observed, which appear as discrete bands when they are subject to electrophoresis on the native polyacrylamide gel. Figure 2 shows the evidence (from left to right): lane 1 is the control into which only the ^{32}P-labeled *nodA-D* DNA is loaded. From lane 2 throughout the last, the amount of NodD is increasing with the amount of the labelled *nodA-D* DNA unchanged. Instead of a single band as seen in Fig. 1, as many as seven bands are observed in lane 8, 9 and 10 in Fig. 3. Starting from lane 5 through to the last lane, the labelled *nodA-D* DNA disappears, indicating that all free DNA has bound to NodD.

Based on the observations mentioned above we propose a THREE SITES conception, which is explained in the following diagram. As has been reported the protein, which binds to a fragment of DNA at its different locations, generates complexes, which would be seen as discrete bands in the native polyacrylamide

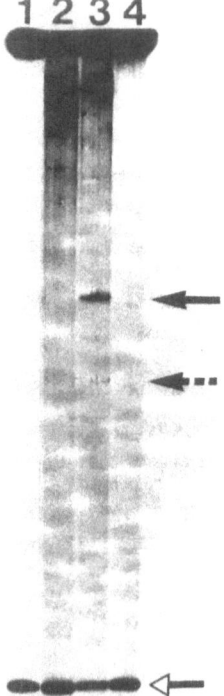

Fig. 1

gel, because they have different conformations (Hen-ming Wu et al. 1984). In the diagram, ▽ stands for NodD; —, for NodA-D DNA; V for binding site.

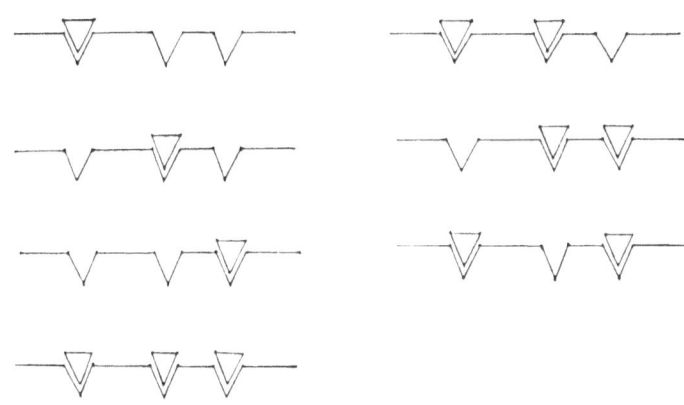

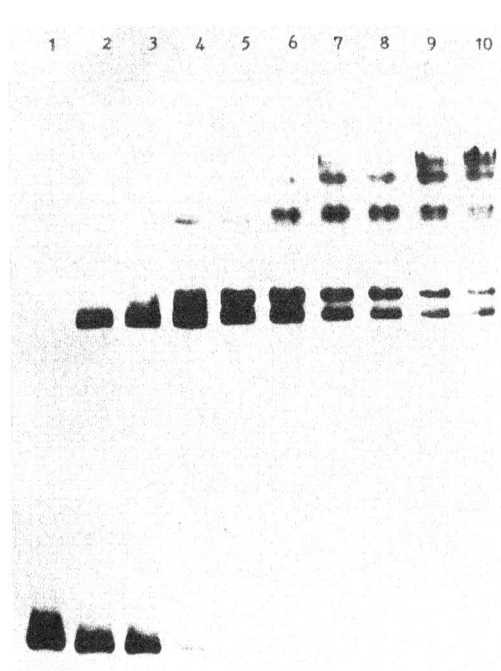

Fig. 2

(3) Naringenin–a plant produced signal-molecule– destabilizes the complexes formed between NodD and the *nodA-D* DNA.

The transcription of *nod* genes requires flavonoids (Firmin et al. 1986). Flavonoids are known to be manufactured in and excreted from the host plant, and naringenin belongs to the flavonoids.

It has not yet been determined whether naringenin enters the Rhizobium cell and interacts with the cell components for its functioning, or it just simply acts upon its receptor (if it is present) on the cell membrane to launch a cascade of reactions, which are required for the transcription of *nod* genes. The observations made in this lab show that naringenin destabilizes to a great extent the complexes formed between NodD and the *nodA-D* DNA, which implies that naringenin may enter the cell and be involved in gene regulation through inter-molecular interactions.

Figure 3 shows that naringenin destabilizes the complexes. Lane 1 is a control, into which only the labeled *nodA-D* is loaded. Lane 2 through 4 are the complexes formed between NodD and the *nodA-D* DNA. Lane 2 contains no naringenin, lane 3 and lane 4 contain naringenin at 0.1 μg/μl, and 1 μg/μl, respectively. It can be seen in lane 4 that one of the retarded bands drastically reduces in its

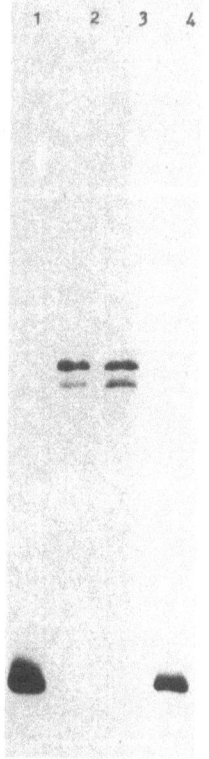

Fig. 3

intensity, showing that naringenin destabilizes the complex, which evidences the in vitro direct interaction between naringenin and NodD and/or *nodA-D* DNA.

nodA

```
CAA  GTT  GTT  TCG  TCG  CCC  AAC  CAC  GAT  CAG  AGC
TGC  GCG  ATA  TTG  TTG  CGC  ATA  CTT  GTT  GGA  TTG
CCG  GTT  AGG  CAA  TCG  AGC  TAA  GAG  GAG  ATG  TAA
GAG  TCT  CAA  ATC  GGG  CCC  CTG  CCC  GGC  GCT  TCG
TTT  TTT  AAG  TTC  CTG  GTC  CTG  ATA  TTG  ATC  AAG
TTC  CGG  GTT  TTC  TAT  AAG  TGA  TCC  GAA  AGA  TTG
GTA  AAA  TTG  ATT  GTT  TGG  ATG  GCA  ATC  ATC  TAT
GGA  ATG  GAT  ATT  CAA  CCC  ATG  CGT  TTT  AAA  GGC
CTA  GAT  CTA  AAC  TTT  CTT  GTA  GCA  TCT
```

nodD

nod A-D DNA sequence. The directions of *nodA* and *nodD* transcription are shown by open arrows at both ends of the sequence. The *nod* boxes are boxed. The inverted sequence is indicated by solid arrows.

Acknowledgements. We would like to thank Drs. A. Johnston and A. Downie for their kind offering strains and cloned DNA, and for their very useful discussion.

References

Bagdasarian M et al. (1981) Gene 16: 237
Firmin JL et al. (1986) Nature 324: 90
Hen-ming Wu, Crothers DM (1984) Nature 308: 509
Hong GF et al. (1987) Nucl Acid Res 15: 9677
Lamb JW et al. (1985) Gene 34: 235
Rossen L et al. (1985) EMBO J 4: 3369

CHAPTER 11
Nitrogenase and FeMo-Cofactor

Yong-Qi Lin and Fei Wang

The Nitrogen Fixation
and its Research in China
Editor Guo-fan Hong
© Springer-Verlag Berlin Heidelberg 1992

1 Introduction

Since biological nitrogen fixation is the major source of ammonia in the natural world, scientists from many nations pay more and more attention to its research. In 1966, it was reported that nitrogenase had two protein components and both of them were essential for nitrogenase activity. Since then great progress concerning this enzyme has been made at the molecular level. In this article, we will review our study on nitrogenase and the FeMo-cofactor.

2 Purification and Properties of Nitrogenase of *Azotobacter vinelandii*

2.1 Purification of Nitrogenase

Method I (Nitrogen Fix. Group, 1974)

The bacterial strain we chose was *Azotobacter vinelandii* Shenyang 230. The cell culture of A.v. 230 was ultrasonicated in an ice bath, after centrifugation the cell-free extract was obtained. The extract was heated to 60 °C in a bath for 10 min and then centrifuged. To the supernatant was added protamine sulfate in a quantity of 7% of the total contained proteins in order to remove nucleic acids by centrifugation. Again protamine sulfate, 19% of total proteins, was added to precipitate nitrogenase. The sediment of nitrogenase and protamine sulfate complex was stirred for 1–2 h in 0.025 mol/l tris-HCl buffer (ph 7.2) containing cellulose phosphate, the crude nitrogenase solution was prepared after centrifugation.

The crude nitrogenase solution was loaded onto DEAE-cellulose columns and eluted with 0.025 mol/l tris-HCl buffer (pH 7.2) to remove unbound proteins;

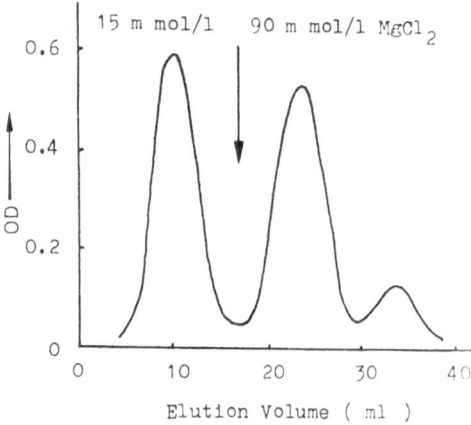

Fig. 1. The elution profile of nitrogenase on DEAE-cellulose column (2.2 × 6.5 cm)

Table 1. Activities recovered in each phase of purification

Sample	Total volume (ml)	Total activity (nmol C_2H_4/min)	Total protein (mg)	(mg/ml)	Specific activity
Crude extract	63	4794	1594	25.2	3.0
Heated at 60 °C	63	7680	1072	17.0	7.2
Precipitated with 10–19% protamine	21	2196	251	12.0	9.2
Fraction of MoFe protein	10	2738	31.6	3.2	86.6
Fraction of Fe protein	12	3282	18.5	1.5	177.6

then eluted with 15 mmol/l $MgCl_2$ followed by 90 mmol/l $MgCl_2$ both in the original buffer. MoFe-protein and Fe-protein fractions were collected, respectively. In Table 1, the activities recovered in each phase of purification are listed. Figure 1 shows the elution profile from the DEAE-cellulose column. Any individual fraction of either MoFe-protein or Fe-protein did not display nitrogenase activity. The activity only occurred when they were mixed. DISC-electrophoretograms indicate that MoFe protein was contained in a single protein band. All the procedures were carried out under protection with pure H_2, the solutions were reduced with $Na_2S_2O_4$.

Method II Preparation of MoFe Protein Crystal
(7th Lab. Inst. Botany 1973)

Azotobacter vinelandii Shenyang 230 was grown in Nichola's culture medium for 18 h. 200 g of cell pellet was suspended and diluted 1:1 in 0.025 mol/l tris-HCl buffer (pH 7.2), the suspension was ultrasonicated and centrifuged, the supernatant was collected. After adjusting to pH 7.2 and protecting with argon gas, the cell-free extract was heated at 60 °C for 5 min in a water bath. Following centrifugation, the supernatant was put on DEAE-cellulose DE-52 column (2.5 × 20 cm). The column was washed with 0.025 mol/l tris-HCl buffer (pH 7.2) containing 0.2 mg/ml $Na_2S_2O_4$, and then eluted with 0.15 mol/l NaCl in the buffer as just indicated.

The MoFe-protein fraction was eluted with 0.25 mol/l NaCl in the above buffer.

The MoFe-protein fraction was concentrated by ultrafiltration in argon gas. When the protein concentration in the solution reached 50 mg/ml, it was diluted 4-fold with 0.025 mol/l tris-HCl buffer (pH 7.2) containing 0.2 mg/ml $Na_2S_2O_4$. The ultrafiltration continued until needle-form crystals formed. DISC-electrophoresis showed that the crystal was homogeneous.

2.2 Properties of MoFe-Protein of Nitrogenase: The Molecular Weight of MoFe-Protein and its Content of Metal Atoms

The molecular weight of MoFe-protein of nitrogenase from *Azotobacter vinelandii* was 225,000 Da as determined by gel filtration on Sephadex G-200 column (Fig. 2) (Nitrogen Fix. Group 1977).

Each molecule of crystallized MoFe-protein contained 1.7 atoms of Mo and 23 atoms of Fe. Each molecule of recrystallized MoFe-protein contained 2 atoms of Mo and 32 atoms of Fe (Guo 1982).

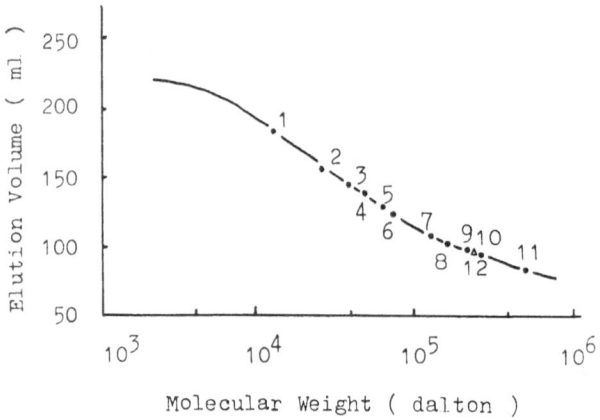

Fig. 2. Determination of the molecular weight of MoFe-protein by Sephadex G-200 column chromatography. 1. cytochrome C, 2. trypsin, 3. pepsin, 4. ovalbumin, 5. catalase, 6. human serum albumin, 7. catalase (dimer), 8. aldolase, 9. γ-globulin, 10. catalase (tetramer), 11. ferritin, 12. MoFe-protein

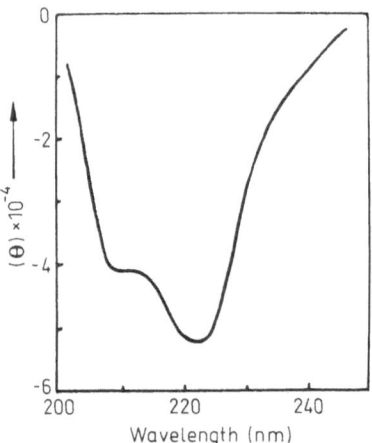

Wavelength (nm)   **Fig. 3.** CD spectrum of native MoFe-protein

2.3 Spectrum Study of the MoFe-Protein

Figure 3 was the circular dichroism spectrum of native MoFe-protein. Calculations based on $[\theta H]222$ showed that there was 38.7% α-helices in MoFe-protein.

3 The Multiple Forms of MoFe-Protein Polymer of Nitrogenase of *Azotobacter vinelandii*

(Lin 1987, 1989(1); Wang 1990)

3.1 Preparation of MoFe-Protein Octamer of Nitrogenase

The MoFe-protein prepared via DEAE-cellulose as described above was further submitted to Sephadex G-200 column chromatography; an elution fraction molecular weight of about 400,000 Da was collected and concentrated. After preparative polyacrylamide gel electrophoresis of the concentrated fraction, the octamer of MoFe-protein was isolated.

3.2 Properties of the Octameric MoFe-Protein

The properties and molecular compositions of the octamer and tetramer and their subunits were studied, and listed in Tables 2 and 3, revealing a close relation between them. It was concluded that the octamer corresponds to the dimer of tetramer with a structure of $\alpha_4\beta_4$. The octamer has a lower specific activity than the tetramer. The octamer and tetramer seem to be different active forms of MoFe-protein.

Table 2. Properties of the two MoFe-proteins

Properties	Tetramer	Octamer
Specific activity (nmol C_2H_4/min/mg pr)	200	50
Molecular weight (Da)	225,000	445,000
Molecular weight of subunits (Da)	62,000 52,000	62,000 52,000
Percentage of metal atoms in MoFe-protein (1 mol. protein:Mo:Fe)	1:1.6:17	1:4.4:40
Percentage of metal atoms in FeMo-co (Mo:Fe)	1:7.9	1:9.9

Table 3. Amino acid compositions of the two MoFe-proteins

Amino acids	Number of AA per molecule of MoFe-protein		
	Tetramer	Octamer	
Asx	199	199	199
Thr	91	93	92
Ser	111	92	107
Glx	228	211	200
Gly	222	163	165
Ala	180	120	135
Cys	28	26	27
Val	187	129	138
Met	55	61	69
Ile	116	108	107
Leu	152	143	152
Tyr	54	70	63
Phe	100	99	82
Lys	142	130	142
His	45	50	44
Arg	86	86	86

3.3 The Biological Significance of the Two Polymeric MoFe-Proteins

The formation of octamer seems to be related to certain cell growth phases. In the early log-phase, the tetramer is the major form, and the octamer becomes the major form both in the late log-phase and in the plateaued phase. The change in forms of MoFe-protein is coincident with that of nitrogenase activity, as can be seen in Fig. 4. When the tetramer was the major form of MoFe-protein,

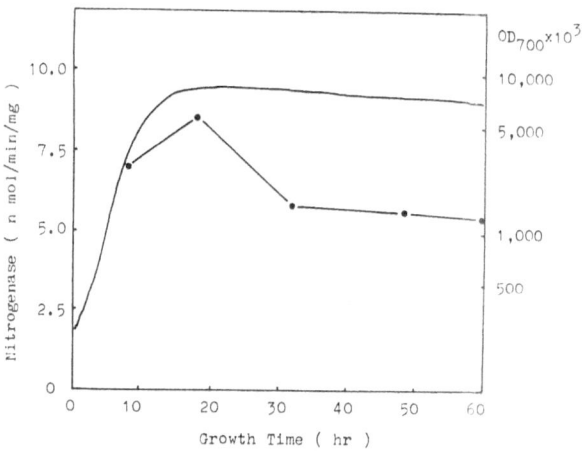

Fig. 4. Changes of nitrogenase activity in cell growth phases, nitrogenase activity (—·—), growth curve (——)

the nitrogenase activity was relatively high; with the octamer being the major form, the activity decreased to a lower level. We suggest that this interchange may serve as a mediator to regulate the nitrogenase activity to meet the physiological requirements in the cell.

It was observed that the octamer formation in vivo can be accelerated by adding ammonia to the culture medium. By adding ammonia one hour after the middle log-phase, the octamer increased 2-fold. It has been known that ammonia inhibits the nitrogenase activity in vivo, so the conversion from tetramer to octamer agreed with the decrease of nitrogenase activity.

4 Chemical Modification of MoFe-Protein of Nitrogenase of *Azotobacter vinelandii*
(Guo 1982; Zhou 1987)

Two kinds of metal actomic clusters were found in the MoFe-protein of nitrogenase, $[Mo, Fe, S^{2-}]$ cluster (FeMo-co) and $[Fe, S^{2-}]$ cluster. The $[Mo, Fe, S^{2-}]$ cluster serves as a core which is combined with dinitrogen molecules in the enzyme molecule. Chemical modifications of the side-chain groups on the MoFe-protein were employed to investigate the characters of the groups binding the metal atomic clusters and the nature of the chemical bonds between the clusters and their binding groups.

4.1 Modification of Sulfhydryl Groups of the MoFe-Protein Molecule

The modifying reagent NEM (*N*-ethylmaleimide) modifies sulfhydryl groups of proteins. When MoFe-protein was modified by NEM, it was discovered that

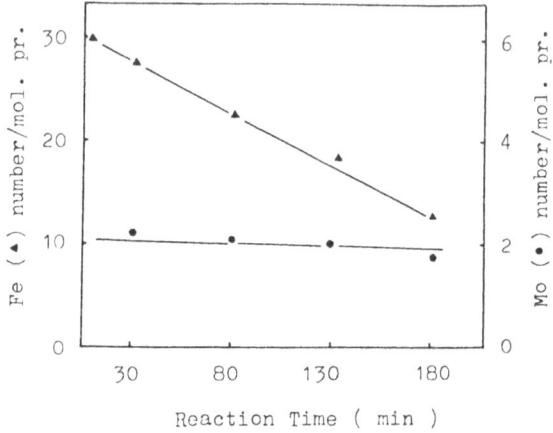

Fig. 5. The changes of contents of Mo (●) and Fe (▲) atoms in MoFe-protein modified with NEM

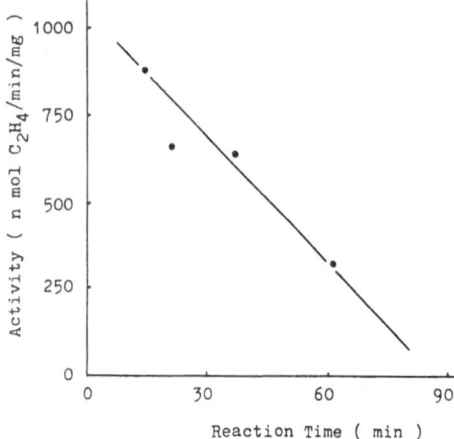

Fig. 6. Activity changes of MoFe-protein in the course of the modification reaction with TNBS

the more–SH groups were modified, the lower the enzyme activity became, and the modification led to the loss of the [Fe, S^{2-}] cluster from the protein molecule with the subsequent reduction in the content of Fe atoms, but the content of Mo atoms remained unchanged (see Fig. 5). This means that the [Fe, S^{2-}] cluster is bound to the protein molecule by sulfhydryl groups, and is essential for the nitrogenase activity.

4.2 Modification of Amino Groups of the MoFe-Protein Molecule

As the amino groups in the MoFe-protein molecule were modified by TNBS (2,4,6-trinitrobenzensulfonic acid), the activity of MoFe-protein decreased rapidly (see Fig. 6). The content of Mo atoms decreased sharply, but not down to zero, whereas the content of Fe atoms almost remained unchanged (see Table 4), indicating that the amino groups in the MoFe-protein played an important role in conjunction between FeMo-co and MoFe-protein. The modification of amino groups did not cause remarkable changes of the protein comformation.

Table 4. Changes in the contents of Fe atoms, Mo atoms and TNP–NH$_2$ groups in modification reactions with TNBS

Reaction time (min)	TNP–NH$_2$ group (nmol/nmol pr.)	Fe atoms (nmol/nmol pr.)	Mo atoms (nmol/nmol pr.)
0	0	32.0	2.0
5	0.98	30.0	2.1
20	1.40	26.5	1.5
60	2.40	23.0	1.0

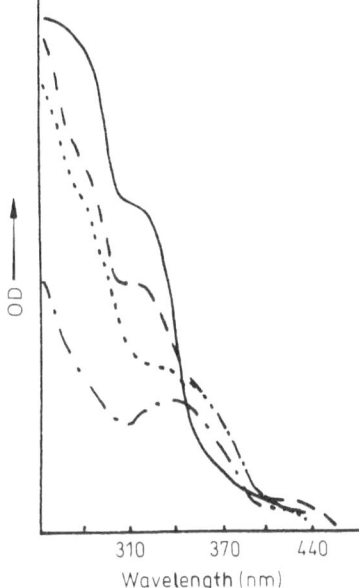

OD

310   370   440
Wavelength (nm)

Fig. 7. Ultraviolet spectra of the MoFe-protein subjected to modification for different lengths of time. 0 min (———), 1 min (————), 2 min (·········) and 5 min (—·—·—·—). The concentration of protein was 0.24 mg/ml

4.3 Modification of Carboxyl Groups of the MoFe-Protein

Woodward Reagent K modifies carboxyl groups of the MoFe-protein. As the reagent reacts with the MoFe-protein, the nitrogenase activity drops rapidly with remarkable changes in its absorption spectra (Fig. 7). The contents of both Mo and Fe atoms in the modified MoFe-protein remained unchanged. So it seems that the carboxyl groups of MoFe-protein are crucial for its conformation, but are not important in conjunction with metal atomic clusters.

5 Preparation, Properties and Structure of the FeMo-Cofactor of the MoFe-Protein of *Azotobacter vinelandii*

Method I (Nitrogen Fix. Group 1978; Lin 1981):

Crystallized MoFe-protein was treated anaerobically with 0.4 mol/l tartaric acid for 10 min in the presence of $Na_2S_2O_4$, then adjusted to pH 4.5 with 6 mol/l NaOH followed by centrifugation to remove the supernatant. The sediment was extracted with NMF. This NMF extract containing FeMo-co was run through a cation exchange resin column to remove NMF, the fraction of FeMo-co was collected, and it was homogeneous as shown on the polyacrylamide gel after electrophoresis.

Table 5. Activities of the reconstituted complex of FeMo-co, WFe-protein and Fe-protein of wild-type *Azotobacter vinelandii*

Source of FeMo-co	Treated with tartaric acid		Treated by method of V.K. Shah (1977)	
	Supernatant	Extract of NMF	Supernatant	Extract of NMF
Specific activity of FeMo-co, (nmol C_2H_4/nmol Mo/min)	31	166	0	106

The molecular weight of FeMo-co was 1500 to 2000 Da, the ratio of Mo:Fe:S = 1:8:8. Table 5 lists the activities of FeMo-co in reconstitution with WFe-protein from non-nitrogen-fixing wild-type *Azotobacter vinelandii*.

Method II (Zhou 1984; Lin 1981):

MoFe-protein was treated with 1% thiophenol for 10 min, and then adjusted to pH 4.5 with 0.5 mol/l HCl, followed by centrifugation. The sediment was washed with water (pH 4.5), and extracted with diethamine. The diethamine extract was concentrated to 1/5 of its original volume under vacuum at room temperature. The product was extracted with ether. The lower layer was extracted with methanol, and then centrifuged; then FeMo-co was obtained.

The advantage of this method is that the [Fe–S] cluster can be removed satisfactorily from the MoFe-protein by chelating with thiophenol, and the purity of prepared FeMo-co is very good. Diethamine is efficient to extract FeMo-co, and allows for easy concentration of the extract (see Table 6).

The Mo:Fe ratio of FeMo-co prepared by this method was 1:7 8 which was similar to that reported by others, who used the NMF method. Reconstituted with cell-free extract of *A.v.* mutant UW45, the activity of this FeMo-co was 175 nmol C_2H_4/nmol Mo/min.

The FeMo-co extracted with diethamine was treated by ether and extracted with methanol; a new [Mo, Fe, S^{2-}] cluster was obtained, with a ratio Mo:Fe of 1:2. The new cluster, after being reconstituted with cell-free extract of *A.v.* mutant UW45, had an activity of 163 nmol C_2H_4/nmol Mo/min. Based on the

Table 6. Mo recovery in FeMo-co preparation

	Total	Thiophenol supernatant	Diethamine extract	Recovery
Mo (μg)	17.3	3.5	15.3	88%

analysis of the results, we suggest a core structure model of Femo-co, as shown below:

6 Purification of Apo-MoFe-Protein of *Azotobacter vinelandii* and it's Reconstitution with Native or Synthetic FeMo-Cofactor

(Cui 1982; Zhou 1982)

6.1 Purification of Apo-MoFe-Protein

150 g cell pellet of *Azotobacter vinelandii* mutant UW45 was suspended in 1:1 (w/v) volume of 0.025 mol/l tris-HCl buffer (pH 7.4), ultrasonicated for 45 min, and centrifuged to collect the cell-free extract. The cell-free extract was purified by DEAE-cellulose DE 11 column (measured 4 × 22 cm). Before loading, the extract was reduced by 0.025 mol/l tris-HCl buffer (pH 7.4) containing 0.2 mg/ml $Na_2S_2O_4$. The column was washed with 0.05 mol/l NaCl in original buffer, and then was eluted with 0.2 mol/l NaCl in original buffer, and a brown fraction was collected, which was the crude apo-MoFe-protein.

Further purification was performed by using preparative PAGE. The brown-color band in the gel was cut off under anaerobic conditions, and apo-MoFe-protein at electrophoresis purity and homogeneity was obtained (Table 7).

6.2 Properties of Apo-MoFe Protein and it's Reconstitution with FeMo-co

The molecular weight of purified apo-MoFe-protein was 115,000 as determined by gel filtration on Sephadex G-200 column. SDS-PAGE analysis showed that apo-MoFe-protein consists of two kinds of subunits with molecular weights of 61,000 and 54,000 Da, which is comparable with MoFe-protein. The amino acid composition of apo-MoFe-protein basically agreed with that of MoFe-protein. 5 atoms of Fe and 5 atoms of S^{2-} were detected in each apo-MoFe-protein molecule, but no Mo could be found, this seems to suggest a $[Fe_4S_4]$ cluster structure.

Table 7. Purification of apo-MoFe-protein of *A. vinelandii* mutant UW45

Step	Volume (ml)	protein (mg/ml)	Total activity (nmol/min)	Specific activity (nmol/min/mg)	Recovery (%)
Cell-free extract	150	29.1	1118.6	0.256	100
0.2 mol/l NaCl eluted fraction	30	8.7	455.7	1.746	40.8
After preparative PAGE	30	1.2	205.5	5.850	18.4

Purified apo-MoFe-protein was reconstituted with FeMo-co, the resultant activity was 53.5 nmol C_2H_4/nmol Mo/min. The mobility of reconstituted apo-MoFe-protein was close to that of the native MoFe-protein in PAGE; the contents of Mo and Fe atoms increased. It seems that the polypeptide chains of apo-MoFe-protein and native MoFe-protein are identical, but their contents of metal atoms and polymeric forms are different. For it's reconstitution with FeMo-co, the apo-MoFe-protein behaved like the native MoFe-protein.

7 Chemical Models of FeMo-co and Their Reconstitution with Apo-MoFe-Protein (Lin 1990)

According to EXAFS data of Mo atoms in MoFe protein (Cramer 1978) FeMo-co possibly includes a 1Mo, 2Fe, 5S-cluster structure. We synthesized a model complex (Liu 1982), 1Mo, 2Fe, 6S cluster, with $MoCl_4(CH_3CN)_2$, CH_3Na, $FeCl_3$, NaHS and PhSH (Zhou 1982). EPR, IR, UV and Mössbauer spectra of the complex were measured. End-bound Mo–S and bridge-bound Mo–S were detected by IR spectrum analysis. The proposed structure of this model is illustrated below:

This model complex displays an acetylene reduction activity of 11.1 nmol C_2H_4/nmol Mo/min., with KBH_4 as reduction reagent. The selectivity for ethylene was 74.8%. Reconstituted with apo-MoFe-protein from *A.v.* mutant UW 45, the acetylene reduction activity of this model complex showed 38.1 nmol

C_2H_4/nmol Mo/min. of the acetylene reduction activity equivalent to 9–10% of the native Femo-co.

We also synthesized another model complex crystal (Zhang 1985), $[(C_2H_5)_4N]_3 \cdot \{[(SCH_2CH_2S)MoS_3]_2Fe\}$, with $[(C_2H_5)_4N]_2MoS_4$, $FeCl_3$ and $HSCH_2CH_2SH$. Structure analyses showed that the model crystal includes the following structure component:

This model had on acetylene reduction activity of 6.87 nmol C_2H_4/nmol Mo/min, with KBH_4 as reaction reagent, the selectivity for ethylene was 97.2%. After its reconstitution with apo-MoFe-protein from *A. vinelandii* mutant UW 45, the model complex displayed a weak activity of 7.0 nmol C_2H_4/nmol Mo/min.

The two above models had similar component structure such as bridge-bound Mo–S and similar ability to reconstitute with apo-MoFe-protein. Because the former model displays a higher activity after reconstitution, it is thought that the structure of $[MoFe_2S_4]^{2-}$ might be more in agreement with the native FeMo-co than that in the former model.

8 Mechanism of Protection of Nitrogenase Against Oxygen by Enzyme Systems

Nitrogenase in nitrogen-fixing bacteria, whether aerobic or anaerobic, is very susceptive to oxygen. Gallon (1981) proposed that the in-vivo protection of nitrogenase against oxygen involved evasion of oxygen, physical screening, respiration to consume oxygen, comformational protection and control of biosynthesis of nitrogenase, etc.

We found that peroxidase (POD), catalase (CAT) and superoxide dismutase (SOD), which take part in oxygen and oxygen-metabolite metabolism, were correlated with nitrogenase (Ma 1987). Figure 8 and Table 8 show that the changes in POD, CAT and SOD activities during cell growth coincide to a certain degree with that of nitrogenase activity, with the maximum activity of SOD, POD and CAT arriving at 30 h, and nitrogenase at 18 h. The reason for the difference might be that CAT, POD and SOD were induced at 18 h to clean the harmful metabolites.

We also investigated the changes of isozymograms of CAT, POD and SOD in the course of cell growth. The results (see Fig. 9) indicate that these isoenzymes might play a role in regulating enzyme activities (Lin 1989(2)).

POD was purified from *A. vinelandii* and employed to investigate the protection effect on nitrogenase in vitro. The cell-free extract of *A.v.* was heated

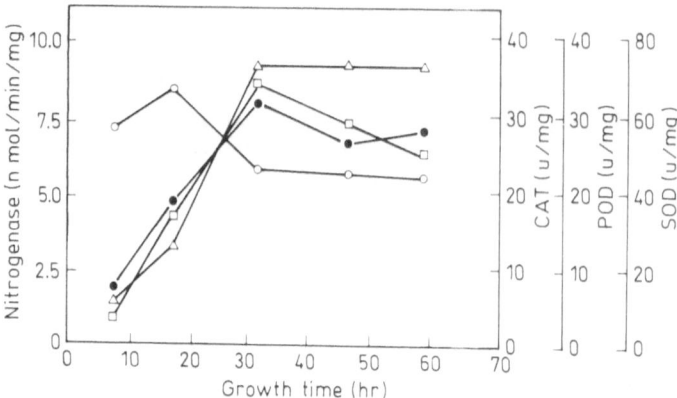

Fig. 8. Activity changes of CAT (△), POD (●), SOD (□) and nitrogenase (○) in cell growth at different times

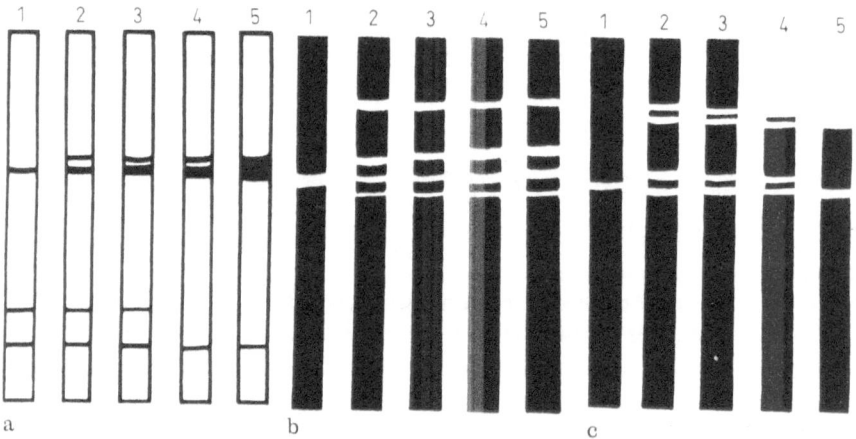

Fig. 9 a–c. Representative isozymograms of POD (**a**), CAT (**b**) and SOD (**c**) of *Azotobacter vinelandii* in cell growth. 1, 2, 3, 4 and 5 mean 8, 18, 32, 48 and 60 age, respectively

Table 8. Comparison of the activities of POD, CAT, SOD and nitrogenase from *Azotobacter vinelandii* at different growth time

| | Activity (units/mg pr. or nmol C_2H_4/min./mg pr.) | | | | |
	at 8	18	32	48	60 h age
CAT	5.0	12.1	36.1	37.4	36.5
POD	2.3	16.9	32.3	27.4	28.4
SOD	12.7	38.0	70.5	58.4	50.7
Nitrogenase	7.2	8.6	5.9	5.7	5.5

Table 9. Protection of nitrogenase by POD against H_2O_2 and O_2 destruction

System	E* + Buffer	E* + POD	E* + Buffer	E* + POD	E* + Buffer
Treatment	Blank	10 μl 5% H_2O_2	10 μl 5% H_2O_2	O_2 (20%) 2 min	O_2 (20%) 2 min
Activity of nitrogenase	100%	90%	59%	65.5%	38.4%

E*: heated cell-free extract

at 50 °C for 5 min under anaerobic conditions to inactivate POD, whereas the activities of nitrogenase and SOD were not affected. After adding purified POD to the heated extract, the protection of nitrogenase by POD against H_2O_2 and O_2 destruction was observed. From Table 9, the system containing heated extract and purified POD preserved 90% of the original activity after adding H_2O_2 to the system; but the system without POD had only 59% of the original activity. The heated extract with purified POD was exposed to air for 2 min, 65.5% of the original activity remained, but the extract without POD, only 38% activity was found. These results verify the protection of nitrogenase by POD against H_2O_2 and O_2 destruction.

We tend to believe that *Azotobacter vinelandii* should have an oxygen-protective system which may include enzymes such as peroxidase, catalase and superoxide dismutase.

9 References

Cramer SP, Gillum WO, Hodgson KO, Mortenson LE, Stiefel EI, Chisnell JK, Brill WJ, Shan VK (1978) The molybdenum site of nitrogenase. 2. A comparative study of Mo–Fe proteins and the iron-molybdenum cofactor by X-ray absorption spectroscopy. J Am Chem Soc 100(12): 3814

Cui TJ, Zhou H, Lin YQ, Yang SC, Tao WS (1982) Synthesis and property study of a model compound of active center of nitrogenase. Chem J of Chinese Universities 3(4): 561

Gallon JR (1981) Trends in Biochem Sci 6: 19

Guo LL, Lin YQ, Tao WS (1982) Studies on the function of sulfhydryl and amino group in MoFe-protein of nitrogenase by chemical modification. Prog Biochem Biophys (China) 6: 26

7th Lab, Inst of Botany, Academia Sinica (1973) Purification and crystallization of Mo–Fe protein, a component of nitrogenase from *Azotobacter vinelandii*. Acta Botanica Sinica 15: 281

Lin YQ, Zheng YH, Li SC (1990) Reconstitution of apoprotein with defective cofactor and the simulated compound of active cofactor of nitrogenase. Acta Scientiarum Naturalium Universitatis Jilinensis 3: 118

Lin YQ, Wang F, Ma B, Chen H, Du JG (1989, 1) Preparation and properties of the molybdenum iron protein octamer of nitrogenase from *Azotobacter vinelandii*. Acta Scientiarum Naturalium Universitatis Jilinensis 1: 113

Lin YQ, Ma B, Du JG, Wang F, Chen H (1989, 2) Relation of nitrogenase with peroxidase, catalase and superoxide dismutase in *Azotobacter vinelandii*. Acta Microbiol Sinica 29(6): 439

Lin YQ, Ma B, Lou DB (1987) The different polymeric molybdenum iron proteins of nitrogenase in *Azotobacter vinelandii* Acta Microbiol Sinica 27(4): 384

Lin YQ, Yang SC, Zhao Y, Zhou H, Cui TJ, Dong QF, Li QS, Zhu ZH, Luo GM, Tao WS (1981) In: Current Perspectives in Nitrogen Fixation. Australian Acad Sci, Canberra, p 355

Liu XS, Xu JQ, Niu SY, Li SQ, Sun CT, Chen YY, Lin YQ, Yang SC, Zhao Y (1982) Synthesis and property study of a model compound of active center of nitrogenase. Chem J of Chinese Universities 3(4): 555

Ma B, Lin YQ (1987) Studies on the isoenzymes of peroxidase and superoxide dismutase in *Azotobacter vinelandii*. Acta Scientiarum Naturalium Universitatis Jilinensis 3: 123

Nitrogen Fixation Study Group, Dept of Chem, Jilin University (1978) The isolation and activity reproduction of the FeMo-co from the nitrogenase Mo–Fe protein in *Azotobacter vinelandii* 230 Acta Scientiarum Naturalium Universitatis Jilinensis 4: 73

Nitrogen Fixation Study Group, Dept of Chem, Jilin University (1977) Acta Scientiarum Naturalium Universitatis Jilinensis 3: 82

Nitrogen Fixation Study Group, Dept of Chem, Jilin University (1974) Studies on the purification and properties of nitrogenase from *Azotobacter vinelandii*. Kexue Tong-Bao (China) 19: 564

Shan VK, Brill WJ (1977) Isolation of an iron-molybdenum cofactor from nitrogenase. Proc Natl Acad Sci USA 74(8): 3249

Wang F, Lin YQ (1990) Studies on the octameric and tetrameric molybdenum-iron protein of nitrogenase from *Azotobacter vinelandii*. Chinese Biochem J 6: 367

Zhang ZG, Wang FS, Lin YQ, Fan TG (1985) Synthesis, structure and properties of a model compound of the "J" system chemistry of the nitrogenase active center. Chem J Chinese Universities 6(6): 544

Zhou H, Zhao ZZ, Fu XQ, Lin YQ (1987) Modification of carboxyl groups in MoFe protein. Prog Biochem Biophys (China) 6: 48

Zhou H, Song WX, Ding YX, Lin YQ (1984) Isolation and characteristics of the iron-molybdenum cofactor with new procedure. Acta Scientiarum Naturalium Universitatis Jilinensis 4: 92

Zhou H, Cui TJ, Lin YQ, Yang SC, Tao WS (1982) Separation and properties of the UW45 FeMo-co-deficient MoFe protein. Acta Scientiarum Naturalium Universitatis Jilinensis 4: 81

CHAPTER 12
Chemical Modification of the Active Center of Nitrogenase

X.P. Dai, J.S. Yang, and C.G. Zhang

The Nitrogen Fixation
and its Research in China
Editor: Guo-fan Hong
© Springer-Verlag Berlin Heidelberg 1992

1 Introduction

The MoFe-protein of nitrogenase consists of two atom clusters (Kurtz et al. 1979, Mortenson et al. 1979). The first is the atom cluster of Fe_4S_4, four of which are contained in each MoFe-protein (M_w 22 kDa). The second is FeMo–co (Shah et al. 1978), two of which are contained in each MoFe-protein; it consists of one Mo, seven to eight Fe and six S^{2-} (or four S^{2-}). The first cluster has the function of electron transfer and the second serves as a chelating and reducing center for the substrate (Shah et al. 1978).

The understanding of how the two clusters connect with the peptide chain is of great significance to elucidate the mechanism of the nitrogen fixation process. The connection of the Fe_4S_4 cluster with the protein is by means of association of Fe with the -S- of the sulfhydryl peptides of the protein, but the exact site of this cluster on the protein is still unknown (Lundell et al. 1981). There is little knowledge about the connection between FeMo–co and the peptide.

Based on chemical modifications of nitrogenase, Sadkov et al. (1979) showed that the Fe-protein consisted of two sulfhydryls. The amino groups of MoFe-protein, when modified with o-phthalyl aldehyde, resulted in complete loss of activity. Guo et al. (1982) studied the way in which the two clusters connect with the protein and its relationship with function. Shah et al. (1986) modified sulfhydryl, phenolic, hydroxyl, imidazolyl and amino groups of the MoFe-protein with p-chloromercuric benzoate (PCMB), dithiobis-2-nitrobenzoic acid (DTNB), trinitrobenzene sulfonic acid (TNBS), diazo-1-H-tetrazole (DHT) and diethyl pyrocarbonate (ETF), respectively. Zhou et al. (1987) modified the carboxyl of the MoFe-protein with Woodward agent K.

2 Modification of the Sulfhydryl Group

Guo modified the sulfhydryl groups with NEM, and observed the effect of modification time on the activity of the MoFe-protein. The activity of the modified MoFe-protein decreased slowly (Table 1), which implied that the protein may not have any highly reactive sulfhydryls. However, as the modifying time increased, the Fe–S cluster starts to appear as seen on anaerobic Sephadex G-15 (Fig. 1). The analysis of the modified protein showed that the decrease in activity and the decrease in Fe content concurred with the Mo content remaining unchanged (Fig. 2, Table 1).

Modified MoFe-protein showed no peak around 420 nm in the circular dichroism spectrum (Fig. 3). Since this peak is a characteristic of the Fe–S cluster, this phenomenon proves that the modification of the MoFe-protein with NEM results in the disconnection of the Fe–S cluster from the protein, thus decreasing its activity. Each molecule of the modified MoFe-protein still

Table 1. Changes in Mo, Fe content and activity of the MoFe-protein following its modification with NEM

Modification time (min)	Specific activity nMC H/ min mg protein	nMFe/nM protein	nMMo/nM protein
Natural	1757	32.0	2.0
35	1318	27.3	2.4
80	1157	17.0	2.2
130	573	19.0	2.2
180	507	12.0	1.4

contains twelve to fourteen Fe atoms and two Mo atoms, which is equivalent to two units of FeMo–co. Thus the modification does not result in the disconnection of FeMo–co. The amino acid analysis of MoFe-protein showed (Table 2) that S-succinylcysteine was increasing with modification, with lysine and histidine unchanged, which showed that the modification is highly specific. Therefore it is very likely that the connection of the Fe–S cluster with the MoFe-protein is through cysteine.

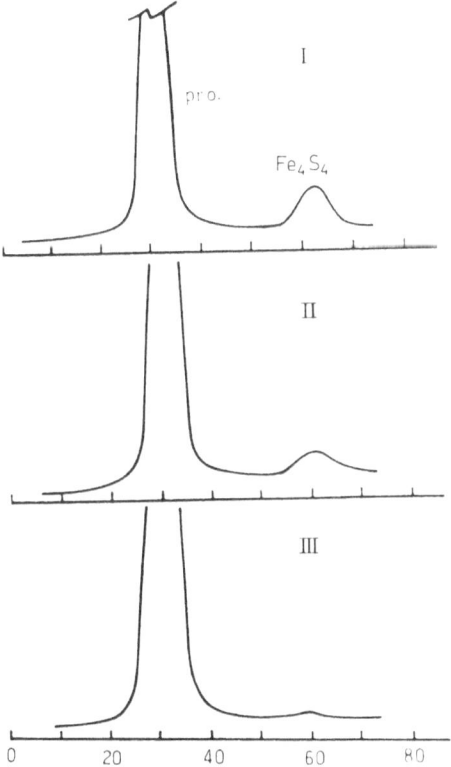

Fig. 1. Elution curve of modified MoFe-protein on anaerobic Sephadex G-15 column (1.7 × 63 cm). The peak at 60 ml is that of Fe_4S_4. Modification time: I, 180 min; II, 80 min; III, not modified

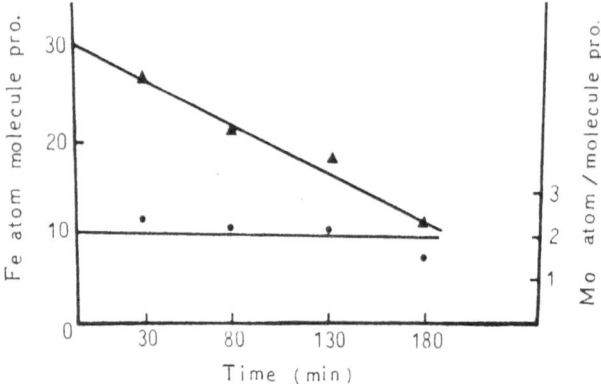

Fig. 2. Relationship between the Mo, Fe contents and the modification time. MoFe-protein was modified with NEM. ●, Mo; ▲, Fe

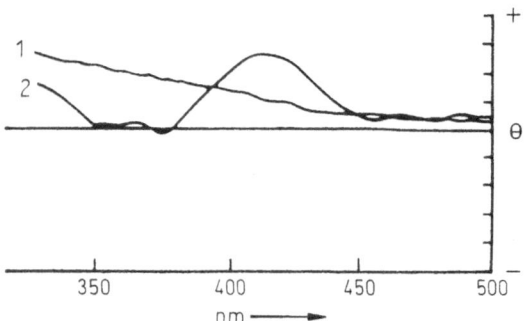

Fig. 3. Circular dichroism spectrum of natural MoFe-protein and MoFe-protein modified with NEM. *1.* Baseline and modified; *2.* not modified

Table 2. Changes in the contents of *S*-succinyl Cysteine, histidine and lysine following modification of MoFe-protein with different amounts of NEM

nM NEM/nM protein	nM *S*-succinyl Cys/nM protein	nM His/nM protein	nM Lys/nM protein
Natural	0	63.5	166
3.5	2	63.0	172
10.5	4	64.0	168
14.0	4	64.0	168
21.0	8	63.0	168

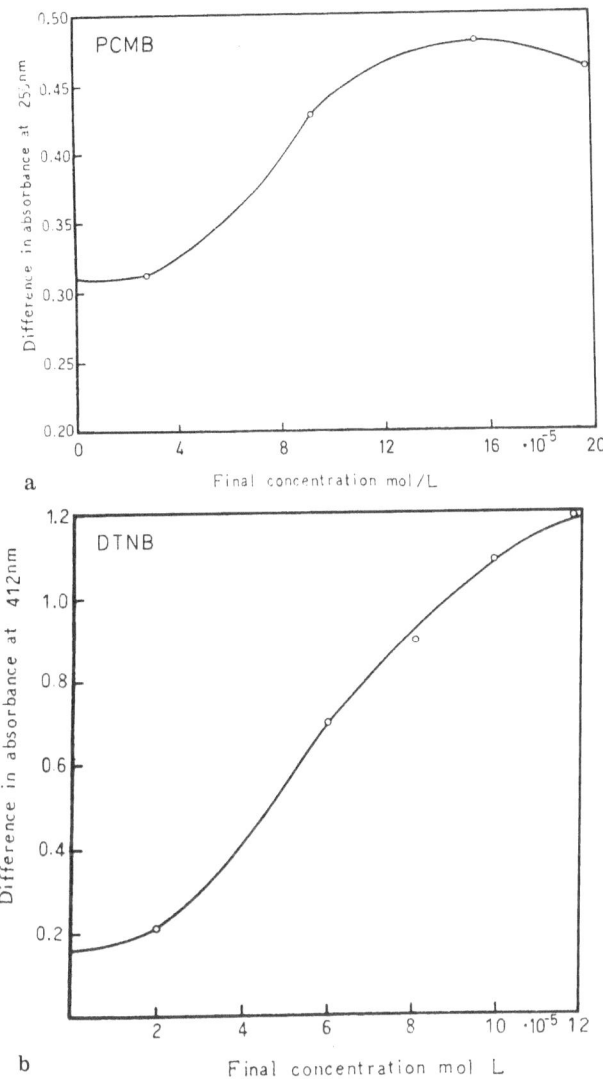

Fig. 4a. Change in absorbance of MoFe-protein titrated with various PCMB concentrations. **b** Change in absorbance of MoFe-protein titrated with various DTNB concentrations

Modified MoFe-protein did not show any significant change in its electrophoresis behavior, molecular weight, and helicity, indicating that the modification did not cause a significant change in its structure. The change in activity could not be attributed to the disconnection of the Fe–S cluster. Shan et al. (1986) modified the sulfhydryl with PCMB and DTNB. Modified MoFe-protein showed its characteristic absorbance at 250 nm and 412 nm, which was in agreement with earlier data (Benesch et al. 1962; Ikura et al. 1976). The optical density of the modified MoFe-protein was enhanced when the concentration of PCMB and DTNB was increased (Fig. 4).

3 Modification of the Amino Group of the MoFe-Protein

TNBS was used for modification of amino groups. This modification caused greater loss of activity than in the case of modification by sulfhydryl (Fig. 5).

When the modified MoFe-protein was analysed on Sephadex G-15, FeMo–co appeared (Fig. 6). The analysis showed that the reduction in activity was accompanied by a decrease in Mo and Fe content (Table 3). The reduction in Mo content reflected the disconnection of one unit of FeMo–co.

The absorbance of the modified protein at 345 nm increased with modification time, which showed that amino groups of the protein were modified. Extinction coefficients revealed that modification of two amino groups caused the disconnection of one FeMo–co. Since FeMo–co is the active center for the nitrogen reduction molecule, the modification of amino groups must cause the loss of activity.

Shan et al. (1986) observed that, following the modification of MoFe-protein with TNBS, the resultant compound showed a new peak at 422 nm, in addition to the normal peak at 367 nm. This might be due to the reaction of TNBS with sodium dithionite or sodium sulfite. The absorbance of MoFe-protein modified with TNBS was significantly increased with an increase of TNBS (Fig. 7).

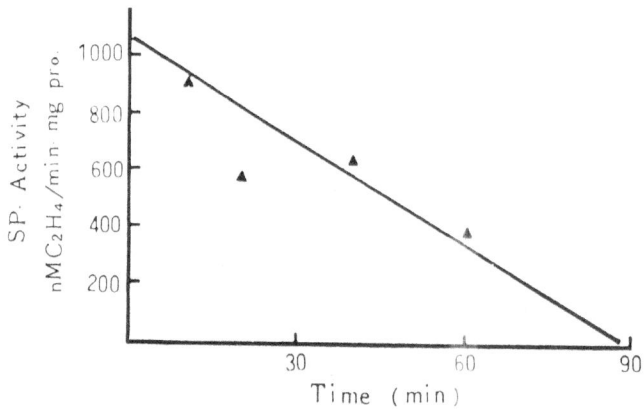

Fig. 5. Correlation between activity and the modification time. MoFe-protein was modified with TNBS

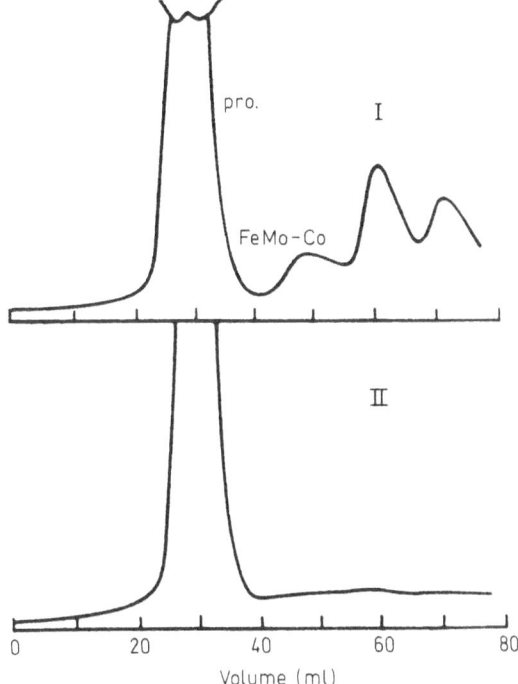

Fig. 6. Elution profile on Sephadex G-15 column. MoFe-protein modified with TNBS. The peak at 50 ml is that of FeMo-co; at 60 ml is TNP-Lys. *I.* Modified; *II.* not modified

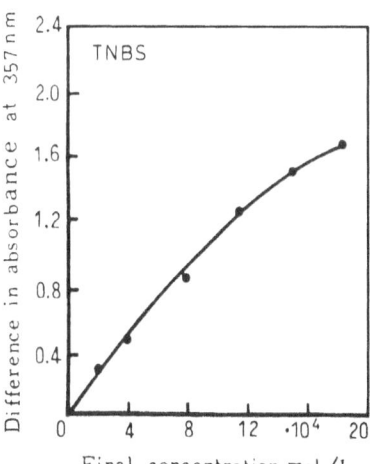

Fig. 7. Change in absorbance of MoFe-protein titrated with various TNBS concentrations

Table 3. The relationship of modified time with the contents of TNP-amino group, Fe and Mo, following modification of MoFe-protein with TNBS

Modification time (min)	nM TNP-amino group/nM protein	nM Fe/nM protein	nM Mo/nM protein
Natural	0	32.0	2.0
5	0.98	30.0	2.1
20	1.40	26.5	1.5
60	2.40	23.0	1.0

4 Modification of the Carboxyl Group of the MoFe-Protein

Zhou et al. (1987) studied the modification of the carboxyl group of MoFe-protein with an excess amount of Woodward K agent, and found that within two min, the activity decreased down to 4% of the original (Table 4). MoFe-protein modified for different lengths of time was separated on Sephadex G-15 and then examined for electrophoretic mobility and ultraviolet spectra were recorded (Fig. 8, Table 5). The data showed that the R_f value of the modified protein was enhanced, when its modification time increased, because MoFe-protein would be more negatively charged by the modifier. The data also showed that the peak at 300 nm was widened, the maximum value was moved to 340 nm. The absorbance ratio of 280 to 250, 280 to 240 could be taken as a way to measure the modifying potential of Woodward K agent (Philip 1971). As the modification continued, the ratios of both 280 to 250 and 280 to 340 became less (Table 6).

Based on the analysis of the ethylamine formation, we found that 96% of the activity was lost after two min of modification where two to five carboxyls per MoFe-protein molecule were modified. MoFe-protein is a tetramer of $\alpha_2\beta_2$ with a molecular weight of 22.5 kDa. Among its nearly 2000 amino acid residues, the total number of glutamic acid, glutamine, aspartic acid, and asparagine are 400. The fact that the modification of only two to five carboxyls leads to 96%

Table 4. The activity loss in MoFe-protein following its modification with Woodward K agent

| | Modification time (min) | | | |
	0	1	2	4
Peak height of C_2H_4 (cm)	8.9	6.3	0.35	0.7
Peak height of C_2H_4 (cm)	3.8	4.6	6.9	8.5
C_2H_4/C_2H_2	2.3	1.4	0.1	0.1
Activity C_2H_4 mmol	22.2	16.4	0.9	0.9
Residual activity %	100	73.8	4.1	4.1

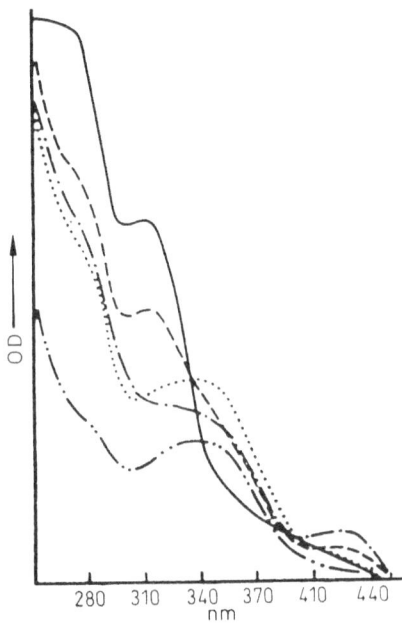

Fig. 8. UV spectra of MoFe-protein modified during different modification times: ———, 0 min, protein concentration 0.23 mg/ml; – – – –, 1 min, protein concentration 0.24 mg/ml; ·······, 2 min, protein concentration 0.24 mg/ml; —··—, 5 min, protein concentration 0.24 mg/ml

Table 5. R_f value* of the modified MoFe-protein

Modification time (min)	0	1	2	4	5
R_f value	0.32	0.327	0.319	0.364	0.377

* polypropylene gel (7.5%) electrophoresis

Table 6. The absorbance ratio changes for the MoFe-protein modified for different lengths of time

Time (min)	Protein concentration	280 nm/340 nm	280 nm/250 nm
0	0.25	3.27	
1	0.24	2.1	0.71
2	0.24	2.0	0.65
3	0.26	1.6	0.64
5	0.24	1.4	0.57

loss in activity predicated a very close relationship between these carboxyls and the active center. In order to know whether the carboxyls are involved in the connection with the Fe–S and FeMo–co clusters or not, we determined the contents of Mo and Fe following the modification of the MoFe-protein. The results showed that there was no change in the MoFe content per single MoFe-protein, indicating that the carboxyls were not the main groups to connect these two clusters.

5 Modification of the Phenolic Hydroxyl and Imidazolyl Group of the MoFe-Protein

Shan et al. (1986) modified the phenolic hydroxyl and imidazolyl groups of the MoFe-protein with DHT and ETF, and determined behaviors of the resulting compounds. The light absorption patterns of the absorption peaks following

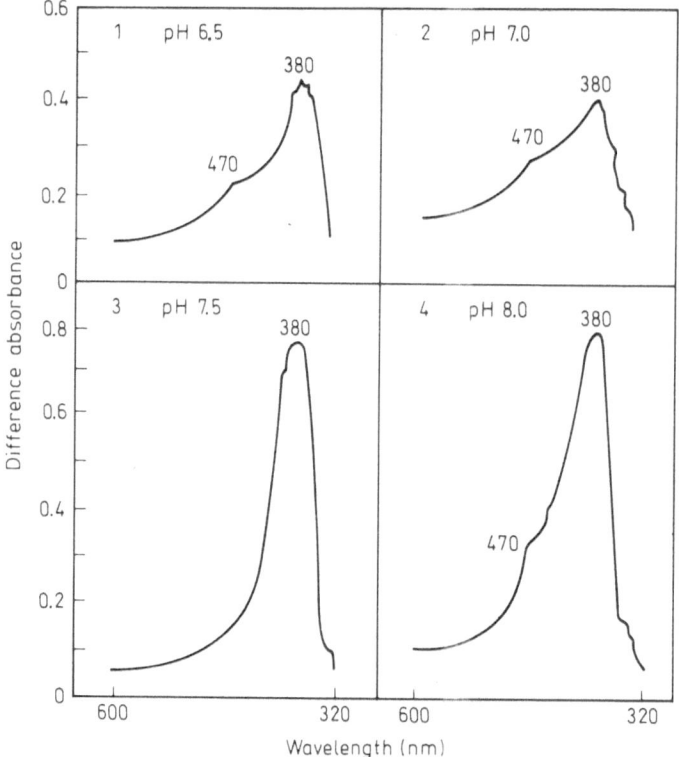

Fig. 9. The variations in the spectral pattern of MoFe-protein modified with DHT at different pHs. Total volume before modification: 2.0 ml; containing: HEPES buffer was used in the modification. Reactions 1, 2, 3 contained 1.08 mg of crystalline MoFe-protein, 4 contained 0.54 mg protein. The total reaction volume for each reaction was 2.6 ml. Concentration of DHT: 22.2×10^{-4} mol/l. Final concentration of sodium dithionite: 0.06 mg/ml; light path of cuvette: 1 cm; temperature: 22 °C

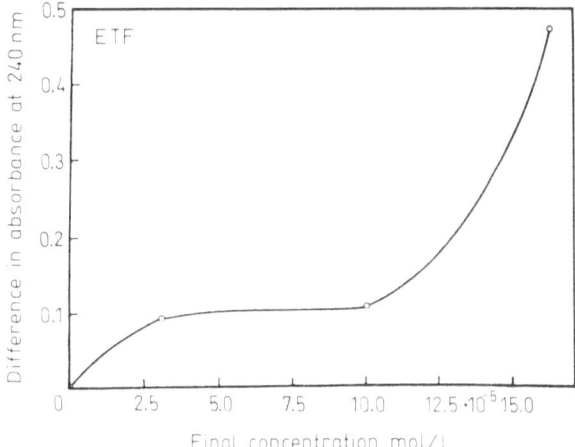

Fig. 10. Change in absorbance of MoFe-protein titrated with various ETF concentrations

the modification with ETF were identical with those reported in the literature. The modification with DHT was usually in sodium carbonate buffer at pH 8.8. It was reported that when tyrosine was modified, the absorption peaks would be at 550 or 380 nm; if histidine is modified, the peaks would be at 470 or 360 nm (Horinish et al. 1964). But the modification carried out at pH 8.8 could lead to severe denaturation of the MoFe-protein (Shan et al. 1985). Therefore, modification was carried out at a wide pH range (pH 6.5, 7.0, 7.5 and 8.0) in HEPES buffer. Changes in the absorption spectra here examined following the modification of MoFe-protein with DHT. The spectra showed an absorption peak at 380 nm and an acromion at 470 nm. As the PH value was lowered, the acromion changed significantly with no sign of change in the absorption peak. It seemed that tyrosine as well as histidine was modified with DHT (Fig. 9).

The modification with ETF was made in HEPES buffer at pH 6.5 to avoid the harmful effect on ETF stability (Berger, 1975, Miles, 1977). Consineau reported (1976) that buffer at higher than pH 7 usually prevented ETF from modifying histidine. HEPES buffer at pH 6.5 would not lead to the denaturation of MoFe-protein. MoFe-protein modified with ETF showed a significant increase in absorbance when ETF was increased (Fig. 10).

6 Activity of the MoFe-Protein Following Modification

The activity of MoFe-protein usually decreases as the concentration of modifier increases. It has been known that the different concentrations of sodium dithionite in the modification system have different effects on the activity of the

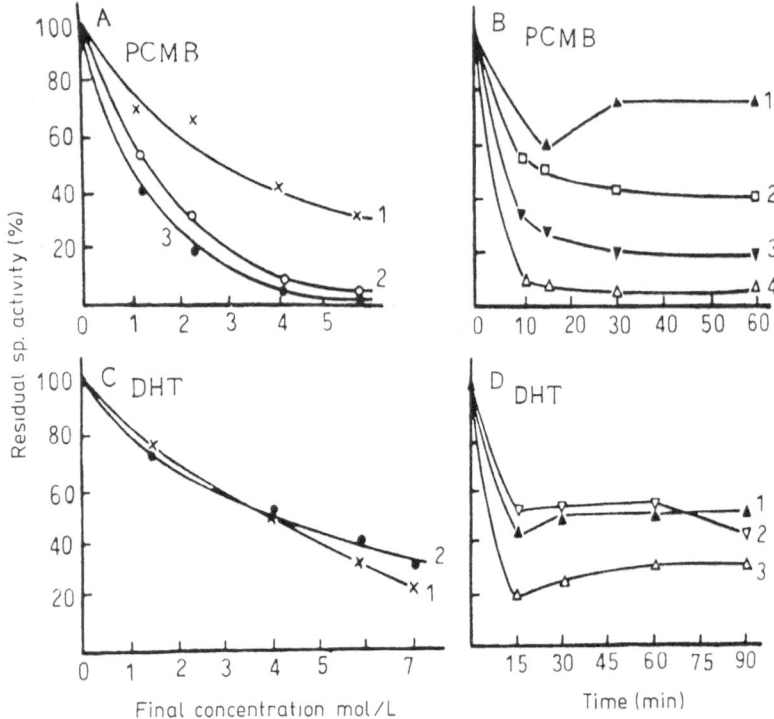

Fig. 11 A–D. Activities of modified MoFe-protein at different concentrations and reaction time of PCMB and DHT. 0.223 mg of crystalline MoFe-protein was disolved in 100 μl of Tris buffer pH 7.2, various amounts of 2.5×10^{-3} mol/l PCMB were added. 2.08 mg of UW45-Fe protein was recombined for activity determination. Temperature of modification: 18 °C.

A. Specific activity C_2H_4 nmol mg protein $\times$ min^{-1} of 1756 is taken as 100%. *1*. $Na_2S_2O_4$ 0.27 mg/ml. allowed to react for 30 min. *2*. $Na_2S_2O_4$ 0.06 mg/ml, allowed to react for 10 min. *3*. $Na_2S_2O_4$ 0.06 mg/ml, allowed to react for 30 min.

B. Specific activity C_2H_4 nmol mg^{-1} protein $\times$ min^{-1} of 900 is taken as 100%. *1*. PCBM 2.27×10^{-4} mol/l, $Na_2S_2O_4$ 0.27 mg/ml. *2*. PCMB 1.19×10^{-4} mol/l, $Na_2S_2O_4$ 0.06 mg/ml. *3*. PCMB 2.27×10^{-4} mol/l, $Na_2S_2O_4$ 0.06 mg/ml. *4*. PCMB 4.17×10^{-4} mol/l, $Na_2S_2O_4$ 0.06 mg/ml.
0.305 mg of crystalline MoFe-protein was disolved in 100 μl of HEPES buffer pH 8.0 various amounts of 3×10^{-2} mol/l DHT were added. 1.38 mg of UW45-Fe protein was recombined for activity determination.

C. Specific activity C_2H_4 nmol mg^{-1} protein $\times$ min^{-1} of 1337 is taken as 100%. *1*. $Na_2S_2O_4$ 0.2 mg/ml, allowed to react for 30 min. *2*. $Na_2S_2O_4$ 0.6 mg/ml, allowed to react for 30 min.

D. Specific activity C_2H_4 nmol mg^{-1} protein $\times$ min^{-1} is taken as 100%. *1*. DHT 3.91×10^{-4} mol/l, $Na_2S_2O_4$ 0.2 mg/ml. *2*. DHT 3.91×10^{-4} mol/l, $Na_2S_2O_4$ 0.06 mg/ml. *3*. DHT 6.92×10^{-4} mol/l, $Na_2S_2O_4$ 0.2 mg/ml

Table 7. Activity changes in modified MoFe-proteins before and after gel filtration

Reagents	Treatments	Sp. activity C_2H_4 nmolmg pro^{-1}.min^{-1}	Changes in activity (%)
Not M	Before GF	1272.3	100
	After GF	964.3	75.8
PCMB	Not M before GF (control)	468.0	—
	M before GF	364.0	100
	M after GF	231.5	63.6
DHT	Not M before GF (control)	1609.0	—
	M before GF	921.0	100
	M after GF	539.6	58.6
DTNB	Not M before GF (control)	502.8	—
	M before GF	369.6	100
	M after GF	283.0	76.6

M: modified;
GF: gel filtration.
Not modified: 0.5 ml (2.75 mg) of crystalline MoFe-protein was passed through the column. 2.23 mg of 230-Fe-protein was used for recombination.
PCMB: 0.5 ml (3.6 mg) of crystalline MoFe-protein modified with 0.2 ml of 5×10^{-3} mol/l PCMB was passed through the column. 2.85 mg of UW45-Fe-protein was used for recombination.
DHT: 0.2 ml (3.05 mg) of crystalline MoFe-protein was in 0.025 mol/l HEPES buffer, and modified with 0.3 ml of 3×10^{-3} mol/l DHT. It was then passed through the column. 1.38 mg of UW45-Fe-protein was used for recombination.
DNTB: 1.0 ml (7.2 mg) of crystalline MoFe-protein was modified with 100 μl of 0.01 mol/l DTNB. It was then passed through the column. 2.85 mg of UW45-Fe-protein was used for recombination

modified protein. For instance, the decrease in the activity of MoFe-protein following the modification with PCMB could be significantly inhibited by sodium dithionite, but it showed no effect on modification with DHT (Fig. 11). The remaining modifier was removed by gel filtration on Sephadex G-25. The activities of modified MoFe-protein before and after gel filtration were compared. It was shown that part of the activity of the MoFe-protein would be lost after the filtration (Table 7). The reason for this could be attributed to the interference of sodium dithionite with the modifiers. Although gel filtration could reduce sodium dithionite to 1/5 of its concentration, it is still high enough to lead to an activity decrease after gel filtration.

In summary, when modifying the MoFe-protein by various modifiers, the following factors should be considered. Firstly, before modification the environment in which the MoFe-protein exists should not cause denaturation of the MoFe-protein. Too acidic or basic buffers should be avoided, because they would denature the MoFe-protein. Secondly, the compounds that interfer with modifiers should be replaced, so that the modification can proceed properly. It was noted that tris-buffer and sodium dithionite would prohibit or moderate the effect of some modifiers, and that gel filtration would reduce the activity of MoFe-protein. Preadding some terminators is essential for slowing down the modifying reaction and a reliable anaerobic environment is important for protecting the activity of the MoFe-protein.

7 References

Benesch R, Benesch RE (1962) Determination of -SH group in proteins. In: Glick D (ed) Methods in biochemical analysis. Vol 10. Interscience, Wiley, New York, London p 43

Berger SL (1975) Diethyl pyrocarbonate: an examination of its properties in buffer solution with a new assay technique. Anal Biochem 67: 428

Consineau J, Meighen E (1976) Chemical modification of bacterial luciferase with ethoxy formic anhydride—Evidence for an essential histidyl residue. Biochemistry 15: 4992

Guo LL, Lin YQ, Tao WX (1982) Studies on the function of sulfhydryl and amino group in MoFe-protein of nitrogenase by chemical modification. Prog Biochem Biophys 6: 26

Horinishi H, Hachimori Y, Kurikara K, Shibara K (1964) States of amino acid residues in proteins. 3. Histidine residues in insulin, iysozyme, albumin and proteinase as determined with a new reagent of diazo-1-H-tetrazole. Biochem Biophys Acta 86: 477

Ikura K, Saaski R, Narita H, Sugimoto E, Chiba H (1976) Multifunctional enzyme, bisphosphoglyceromutase, 2,3-bisphosphoglycerate phosphatase/phosphoglyceromutase from human erthrocytes—Evidence for a common active site. Eur J Biochem 66: 515

Kurtz DM, McMillan RS, Burgess BK, Mortenson LE, Holm RH (1979) Identification of iron–sulfur centers in the iron–molybdenum protein of nitrogenase. Proc Natl Acad Sci USA 76(10): 4986

Lundell DJ, Howard JB (1981) Isolation an sequences of the cysteinyl tryptic peptides from the MoFe-protein of Azotobacter vinelandii nitrogenase. J Biol Chem 256(12): 6385

Miles EW (1977) Modification of histidyl residue in protein by diethyl pyrocarbonate. In: Hirs CH, Timas SN (eds) Methods in enzymology. Vol 47. Enzyme Structure E. Academic Press, New York, p 431

Mortenson LE, Thomely RNF (1979) Structure and function of nitrogenase. Ann Rev Biochem 48: 387

Philip HP (1971) Modification of carboxyl group in bovine carboxypeptidase A. I. Inactivation of the enzyme of N-ethyl-5-phenylisoxazolium-3'-sulfone (Woodward's reagent K) Biochemistry 10: 3163

Research Group on Chemical Nitrogen Fixation (1978) Isolation and activity of iron–molybdenum cofactor (FeMo–co) in MoFe-protein of nitrogenase of Azotobacter vinelandii 230. Acta Sci Nat Univ Jilinensis (4): 73

Sadkov AP, Kotelnikov AI, Gvozdov RI (1977) Investigation of free sulphydryl groups of nitrogenase. Proc USSR Acad Sci, Biol Series 4: 610

Shan VK, Brill WJ (1977) Isolation of an iron–molybdenum cofactor from nitrogenase. Proc Natl Acad Sci USA 74: 3249

Shan VK, Chisenll JK, Brill WJ (1978) Acetylene reduction by the iron -molybdenum cofactor from nitrogenase. Biochem Biophys Res Comm 81: 232

Shan WZ, Yang JS, Zhang CG, Lu AG, Jin SY (1986) Several methods of chemical modification of nitrogenase MoFe-protein from Azotobacter vinelandii. Acta Phytophys Sinica 12(3): 201

Shan WZ, Zhang CG, Yang JS, Jin SY (1985) Effects of pH and temperature on substrate reduction by nitrogenase. Acta Phytophys Sinica 11(3): 293

Zou H, Zao AZ, Fu XQ, Lin YQ (1987) Modification of carboxyl groups in MoFe-protein. Prog Bichem Biophys 6: 48

CHAPTER 13
Genetic Improvement of *Rhizobium* Strains

Larissa A. Sharypova, Inge-M. Pretorius-Güth,
Boris V. Simarov, and Alfred Pühler

The Nitrogen Fixation
and its Research in China
Editor: Guo-fan Hong
© Springer-Verlag Berlin Heidelberg 1992

1 Introduction

Bacteria of the family *Rhizobiaceae* have been studied for nearly 50 years. The main feature which attracted scientists was their ability to form nitrogen-fixing nodules on roots of leguminous plants. Until recently, all root nodule bacteria were considered to be members of the genus *Rhizobium*. Now they are reclassified into five genera: *Rhizobium, Bradyrhizobium, Azorhizobium, Sinorhizobium,* and *Photorhizobium*. The current knowledge of the symbiotic functions of nodule bacteria is based principally on the study of *Rhizobium* and *Bradyrhizobium*. Consequently, this review will deal mostly with the results obtained from these two species. When considering questions common to all, nodule bacteria will be referred to as *Rhizobium* or rhizobia.

From the beginning, the *Rhizobium* selection programs which were undertaken to increase the effectiveness of the legume-*Rhizobium* symbiosis, developed in parallel with basic research which attempted to ascertain the genetic factors responsible for symbiotic nitrogen fixation. This resulted in a constant appreciation of *Rhizobium* selection methods and research approaches. Recent advances in *Rhizobium* genetics made by the application of molecular techniques, gave rise to, on the one hand, new questions, e.g. the deliberate release of genetically ingeneered organisms into environment. On the other hand, these new techniques promise the revolutionary improvement of *Rhizobium* strains. In this context, we will compare the potentials which classical and molecular genetics hold for the construction of *Rhizobium* strains with beneficial symbiotic properties.

2 Mutagenesis and Genetic Transfer in *Rhizobium*

Rhizobia were found to be susceptible to a wide range of chemical and physical mutagens. This allowed the isolation to be made of different kinds of mutants: morphological, auxotrophic, and antibiotic-resistant mutants, as well as mutants with altered carbohydrate and nitrogen metabolism (Beringer et al. 1980). By screening host plants inoculated with mutagenized rhizobia, mutants with altered symbiotic properties were identified. These included mutants which had completely lost their symbiotic functions, e.g. mutants unable to form nodules (Nod$^-$) and mutants inducing nodules but unable to fix nitrogen (Fix$^-$). In addition to this, mutants with quantitatively altered functions, e.g. altered nodulation (determined by the rate of nodule formation), altered nitrogenase activity (assessed by the rate of acetylene reduction), and mutants with altered symbiotic effectiveness (evaluated by the dry matter production of the macrosymbiont), were found.

In addition to the mutation analysis, different methods of genetic transfer were developed for *Rhizobium*. The genetic transformation of rhizobia was first reported by Krasilnikov (1941), who demonstrated the transfer of host specificity between different biovars of *Rhizobium leguminosarum*. In contrast to *Escherichia coli*, it is rather difficult to obtain competent *Rhizobium* cells for transformation. A reason for this could be the complex structure of the *Rhizobium* cell wall or the presence of large amounts of exopolysaccharides. Indeed, rough and slimeless mutants, deficient in some cell wall components, appeared to be better recipients than wild-type strains (Balassa, Gabor 1961; Laptev, Simarov 1981). Despite its low efficiency, transformation was the first method used for the genetic mapping in rhizobia. Moreover, the first genes found to determine the effectiveness of the symbiosis, were located on the *Rhizobium meliloti* chromosome by the use of genetic transformation (Laptev and Simarov 1981). It was shown that mutations which decreased the nitrogenase activity and yields of inoculated plants, were located in the vicinity of *str*, *rif*, *thi* and *exo* loci. Different genetic transformation procedures have been tested, but to date, this method remains an unreliable one for introducing plasmid DNA into *Rhizobium* cells (Dunican, Tierney 1973; Kiss, Kalman 1982; O'Gara, Dunican 1973; Selvaraj, Iyer 1981).

Transduction was found to be a much more efficient method of genetic transfer in *Rhizobium*, than transformation. The first bacteriophage able to transduce genetic markers of *Rhizobium*, was isolated from a lysogenic *R. meliloti* strain by Kowalski (1967). By now, many transducing phages have been identified (particularly for *R. meliloti*) and used extensively for linkage mapping in *Rhizobium*. Almost all rhizobiophages transfer host DNA by generalized transduction (Buchanan-Wollaston 1979; Casadesus, Olivares 1979; Finan et al. 1984). It is noteworthy that during the infection of host cells, the phage enzymes induce the specific depolymerization of *Rhizobium* exopolysaccharides and the release of oligosaccharides. Since oligosaccharides are believed to play an important role in the symbiotic interaction between *Rhizobium* and leguminous plants (Hollingsworth 1984), rhizobiophages could also be used for characterizing the *Rhizobium* cell surface components which are involved in nodule formation.

The treatment of *Rhizobium* cells with different lytic agents results in the formation of sphaeroplasts, which differ from the protoplasts of Gram-positive bacteria in that they retain remnants of undigested cell walls (Filatova et al. 1982). Such sphaeroplasts are capable of cell fusion, especially when stimulated by PEG (Bazhenova and Simarov 1983). The fusion products are unstable hybrids (heterozygotes) which expel the excess DNA during the subsequent cell divisions, to return to the haploid state. It was found that a proportion of the hybrids give rise to stable parental and recombinant phenotypes, whereas the rest maintain the unstable merozygotic state for many generations. The analysis of the hybrid segregation allows the genetic linkage or the allelic relationship between mutations of interest to be determined.

The most convenient and efficient method of transfering genetic material in *Rhizobium* is conjugation. The Pl group plasmids of *Pseudomonas aeruginosa* (RP4, RK2, R68 and related plasmids) lead to a high frequency of conjugation and genetic exchange in *Rhizobium*. By the use of these plasmids, circular genetic maps were constructed for a number of strains of *R. meliloti* and of *R. leguminosarum* (Beringer et al. 1984). Conjugation has opened numerous ways of analysing *Rhizobium* plasmids and through this, has contributed greatly to the elucidation of the role of plasmids in the determination of symbiosis. Of particular interest is the development of mobilizing systems (see Sect. 4.1) which overcame the nontransmissibility of *Rhizobium* plasmids.

There is a number of good reviews describing the application of mutagenesis and genetic transfer in the analysis of *Rhizobium* genomes, which emphasis the control of the symbiotic functions (Beringer et al. 1980; Kondorosi 1986; Long 1989). The next section discusses the application of classical genetics in the selection of improved *Rhizobium* strains. As is the case in higher organisms, *Rhizobium* breeding programmes take advantage of both the mutagenic and recombination variability.

3 Improvement of *Rhizobium* Strains Through Classical Genetics

3.1 Use of Mutagenesis for the Isolation of Strains with Increased Nitrogenase Activity and Symbiotic Effectiveness

In contrast to basic research where the attention is focused on mutants which show a complete loss of the symbiotic functions (Nod$^-$, Inf$^-$ and Fix$^-$), applied research deals with mutants bearing mutations which affect the quantitative characters of the symbiosis only. The results of several laboratories evidence that induced random mutagenesis may either impair or enhance the symbiotic properties (Feodorov, Simarov 1987; Maier, Brill 1978; Shuklay et al. 1989; Williams 1981). The main obstacle is to establish reliable criteria which allow mutations (inherited properties) and phenotypic variations to be distinguished from each other. Since leguminous plants are characterized by a high inter-cultivar variability with respect to their response to bacterial inoculation (Gibson 1962; Mytton, Rys 1985; Nicholas, Haydock 1971; Nutman 1954; Scot 1983; Tan 1981), an assessment of the symbiotic properties of *Rhizobium* requires well-defined test systems and several repetitions to minimize experimental errors. Since mutation occurs rarely, each bacterial clone obtained, needs to be checked on several plants. These plant tests are extremely time consuming. To increase the efficiency of selection, attempts were made to find those properties of free-living rhizobia responsible for a high symbiotic performance. Properties considered to be of relevance, include the overproduction of exopolysaccharides, glycine auxotrophy, resistance to antibiotics, increased activity of nitrogen-assimilating enzymes and increased production of plant hormones precursors (Simarov and Provorov 1990). None proved to be an indicator of a high symbio-

Table 1. Frequency of Fix^{++}, Eff^{++} and Fix$^-$ mutants in rhizobia

Rhizobium species	Mutagen	Frequency of mutants Fix^{++}, Eff^{++}	Fix$^-$	Reference
B. japonicum	NTG	8×10^{-5}	n.d.	Maier and Brill 1978
B. sp. cowpea	NTG	1×10^{-2}	2×10^{-2}	Williams 1981
R. meliloti	UV	1×10^{-2}	1×10^{-2}	Feodorov and Simarov 1987

NTG: N-methyl-N'-nitro-N-nitrosoguanidine
UV: Ultraviolet irradiation
n.d.: not determined

tic potential. Therefore, all mutagenised survivors were tested directly on the host plants. Rhizobia were selected either for the increased nitrogenase activity (Fix^{++}), or for the increased symbiotic effectiveness (Eff^{++}). The frequency of Fix^{++} and Eff^{++} mutants was found to be comparable to the frequency of Fix$^-$ mutants and ranged from 10^{-4} to 10^{-2} (Table 1).

Since mutants which were selected for increased C_2H_2 reduction activity, did not necessarily improve the plant dry mass and nitrogen accumulation, there is evidently, no direct relationship between the ability to increase the C_2H_2 reduction and the host plant yield. There are two possible explanations for this phenomenon:

1) the acetylene reduction rate is an indirect estimation of nitrogenase activity (Minchin et al. 1985);
2) the increased nitrogenase activity may lead to an irrecoverable energy expenditure and concomitant loss of reducing equivalents, which would result in an imbalance in the metabolic exchange which occurs between the symbiotic partners.

Simarov and co-workers (1989) showed that those strains which were isolated as a result of a two-stage selection procedure, were the most suited for agricultural application. The first stage selection was for the enhancement of nitrogenase activity and the second stage, for an increase in the plant dry mass. The second stage of selection usually led to a slight decrease in nitrogenase activity, nonetheless, this new level was higher than in the wild-type strain.

3.2 Use of Recombination Variability for the Isolation of Strains with Improved Symbiotic Properties

The strategy of using recombinant variability was based on the assumption that the progeny of a cross would be heterogeneous with respect to non-selectable genetic characters, since each recombinant would arise from an independent recombination event. Parental pairs were chosen such that the resulting recombinants would be prototrophs showing a Nod$^+$ Fix$^+$ phenotype.

Table 2. Frequency of mutants and recombinants with improved symbiotic properties (Simarov et al. 1989)

Method	No. of clones Tested	Frequency of clones (%) Fix^{++}	Eff^{++}
UV-mutagenesis	900	0.9	0.3
Transformation	231	8.2	2.2
Transduction	107	21.5	3.7
Conjugation	63	11.1	4.8
Sphaeroplast fusion	79	15.2	1.3

Heterosis, a powerful means of plant breeding, can not be applied to *Rhizobium*, since this is a haploid organism. Nevertheless, the methods of genetic transfer were found to be more efficient for selecting improved strains than mutagenesis. This is shown in Table 2, which summarizes the results of extensive investigations of related *R. meliloti* strains which originated from *R. meliloti* CXM1.

The analysis of recombinants showed that Fix^{++} and Eff$^+$ phenotypes were expressed independently. Regardless of the method of isolation, the frequency of Fix^{++} isolates was significantly higher than that of Eff$^+$ isolates (Table 2). Moreover, the increases in nitrogenase activity were more pronounced than those of symbiotic effectiveness. The former exceeded the latter by a factor of at least three. One of the main problems which arises when working with randomly isolated mutants and recombinants, is the absence of genetic markers which permit the improved symbiotic properties to be controlled and maintained. Therefore, the mutants made available for agricultural application, must be tested in several greenhouse and field trials, to ensure the stability of their beneficial properties.

3.3 *Rhizobium* Plasmids and Their Role in Determining Symbiotic Effectiveness

3.3.1 Rhizobium Plasmids

A remarkable feature of nodule bacteria is the presence of large indigenous plasmids. These plasmids and their functions were studied more thoroughly in fast-growing rhizobia, e.g. *R. meliloti* and *R. leguminosarum*, which nodulate temperate legumes. In each strain, the so-called sym-plasmid, carrying determinants for nodulation and nitrogen fixation was identified (reviewed by Prakash and Atherly 1986). In the different *R. leguminosarum* biovars, the sym-plasmids range from 160 to 880 kb in size. In *R. meliloti*, the sym-plasmid is one of the two megaplasmids (more than 1000 kb in size) which are presented in all *R. meliloti* strains tested to date (Burkardt and Burkardt 1984; Burkardt et al. 1987; Hynes et al. 1986).

Much more diversity in the genome organization is found amongst fast growing rhizobia which nodulate tropical legumes. For example, in *Sinorhizobium fredii*, which shows similarities to *R. meliloti* and *R. leguminosarum*, the symbiotic functions are encoded by plasmids, whereas in *Rhizobium loti* and *Azorhizobium caulinodans* ORS571, the symbiotic functions are chromosomal. Slow-growing strains, belonging to the species of *Bradyrhizobium japonicum* and *R. loti*, also carry plasmids, but these do not encode the nodulation and nitrogen fixation functions. The sym-plasmids of *R. meliloti* and *R. leguminosarum*, in addition to controlling nodulation and nitrogen fixation, may control hydrogen uptake, melanin production, the biosynthesis and catabolism of rhizopines, the release of, and resistance to bacteriocins, and exopolysaccharide production (Prakash and Atherly 1986; Murphy et al. 1987).

For a long time, the non-sym-plasmids were denoted as cryptic, because their functions were obscure. It was proposed that these plasmids were responsible for the adaptation of rhizobia to environmental inhabitats. Indeed, recently it was shown that non-sym-plasmids are involved in determining the tolerance to acidity and salinity, the ability to catabolise plant exudates, as well as the production and resistance to bacteriocins. Furthermore, these plasmids may also carry genes which modulate the rate of nodulation, and may carry additional copies of *nod*- and *nif*-genes (Harrison et al. 1988).

The *R. meliloti* non-sym-plasmids are represented by so-called medium size plasmids (ranging from 140 to almost 1000 kb) and by the megaplasmid-2 (Huguet et al. 1983; Bromfield et al. 1987). In general, the medium size plasmids encode functions similar to those of the *R. leguminosarum* non-sym-plasmids, whereas megaplasmid-2 is a plasmid very specific to *R. meliloti*. Megaplasmid-2 could be considered a mini-chromosome, since it is very large and seems to be an indispensable part of the *R. meliloti* genome. All atempts to cure the strain of megaplasmid-2 or megaplasmid-1, resulted in deletion formation (Banfalvi et al. 1981; Toro and Olivares 1986a). To date, the following functions are known to be encoded by the megaplasmid-2: exopolysaccharides synthesis (*exo* genes), C_4-dicarboxylate transport (*dct* genes) and thiamine biosynthesis (*thi* genes) (Finan et al. 1986; Hynes et al. 1986; Watson et al. 1988). Although megaplasmid-2 is considered to be a non-symbiotic plasmid, it plays an important role in the symbiosis: mutations in *exo* genes arrest the nodule invasion by rhizobia (Müller et al. 1988) and mutations in the *dct* genes prevent the expression of nitrogenase activity in bacteroids (Watson et al. 1988).

3.3.2 Contribution of sym-Plasmids to the Symbiotic Effectiveness of *Rhizobium*

To elucidate the role of the sym-plasmids in determining the symbiotic effectiveness, specific crosses were made to introduce the various sym-plasmids into new genetic backgrounds. These experiments were performed with *R. leguminosarum* strains, since their sym-plasmids can be transferred either due to their self-transmissibility or via cointegrate formation with transmissible plasmids. The work in this field was pioneered by De Jong and associates (1981). They transferred the sym-plasmids from three *R. leguminosarum* strains which differed significantly with respect to their symbiotic effectiveness, to a deletion derivative of *R. leguminosarum* strain 300 which lacked the plasmid-linked *nod* and *nif* determinants. It was found that the transferred sym-plasmids did restore the *Nod*⁺ and *Fix*⁺ phenotypes, but did not confer the symbiotic effectiveness of the donors to the recipients. All recombinants showed the same symbiotic effectiveness as the recipients, with respect to acetylene reduction, plant biomass production and nitrogen accumulation. When the sym-plasmids were introduced into the wild-type strain 300, the resulting recombinants showed the same, or decreased symbiotic effectiveness as compared to the recipient. From these results, it can be concluded that the presence of two sym-plasmids in one strain, does not result in enhanced symbiotic effectiveness, instead, the plasmid functions interfered with each other. In this connection, it is interesting to mention the results of Harrison and co-workers (1988), who examined 31 *R. leguminosarum* bv *trifolii* strains and found that the presence of multiple sym-plasmids, and multiple sym-genes were characteristic of strains which displayed the lowest effectiveness.

The only convincing example of strain improvement by plasmid transfer was demonstrated by the introduction of pIJ1008 (a cointegrate of the non-transmissible sym-plasmid pRL6JI and a transmissible non-sym-plasmid pVW5JI) into the *R. leguminosarum* strain 300 (De Jong et al. 1982). The resulting recombinant strain was superior to both parents, as evaluated by the nitrogen accumulation in the host plants. The authors proposed that the increased effectiveness was connected to the expression of the Hup-determinants encoded by pIJ1008. Subsequent investigations however, have shown that genetic determinants, other than the *hup*-genes accounted for the superior symbiotic effectiveness (Sørensen and Wyndaele 1986).

Sørensen and Wyndaele also transferred the sym-plasmids between the *R. leguminosarum* strains MA1 (Hup⁺) and MC1 (Hup⁻). Plants inoculated with MC1 showed a 60% increase in nitrogen content when compared to plants inoculated with MA1. The introduction of pSymMA1 into MC1 or its plasmidless derivative, did not change the symbiotic effectiveness of the recipient. The reciprocal plasmid transfer, the introduction of pSymMC1 into MA1, however, gave rise to recombinants with a higher effectiveness than that of the recipient, and lower than that of the donor. These results evidence that the high symbiotic potential of strain MC1 was determined by the chromosome and that the Hup⁻ pSymMC1 plasmid conferred advantages to the recipient over the Hup⁺ pSymMA1 plasmid. Scøt and co-workers (1986) also constructed a set of *R. leguminosarum* strains, combining sym-plasmids with different genomic backgrounds. The analysis of the symbiotic properties of these strains revealed that the genomic background of *R. leguminosarum* played a larger role than the sym-plasmids tested, in determining the carbon costs of nitrogen fixation, the host plant biomass and the onset of nitrogenase activity. Furthermore, these results demonstrated that no foreign plasmid could confer a symbiotic effectiveness higher than that of the resident plasmid.

In summary, various conclusions can be drawn. Although the sym-plasmids play a crucial role in establishing the symbiosis, the quantitative traits of the symbiosis depend mostly on chromosomal determinants. The symbiotic potential of a recipient strain can be improved only if it is inferior to that of the donor. To date, the contribution of sym-plasmids to the symbiotic potential of *Rhizobium* strains has been studied in *R. leguminosarum* only. However, in *S. fredii* and especially in *R. meliloti* which carries unusually large sym-plasmids, the situation could be different.

3.3.3 Contribution of Non-sym-Plasmids to the Symbiotic Effectiveness of *Rhizobium*

The non-sym-plasmids seem more likely to be involved in determining the competitiveness, rather than the effectiveness of a strain. Since competitiveness is also of great importance for agricultural applications, a few works dealing with this property of *Rhizobium* strains should be mentioned.

In *R. meliloti* RG4, mutations and deletions in the medium-size plasmid pRmeGR4b, decreased the nodulation rate and survival of the strain in the soil (Toro and Olivares 1986b; Sanjuan and Olivares 1988). Curing the plasmids of *R. leguminosarum* VF39 revealed that 3 of the 6 plasmids (especially the largest non-sym-plasmid) were indispensable to the strain's competitive ability (Hynes 1990). There is a little evidence implicating the involvement of the non-sym-plasmids in the determination of the symbiotic effectiveness.

An interesting example of the improvement of *R. meliloti* by the introduction of non-sym-plasmids, was shown by Strobel and colleagues (1985). They transferred the Ri plasmid of *Agrobacterium rhizogenes* TR105 to the *R. meliloti* strain BL116. *Agrobacterium* belongs to the family *Rhizobiaceae* and is apparently more related to *R. meliloti* than to *Bradyrhizobium* or *Azorhizobium* (Gibbins and Gregory 1972; Hooykaas et al. 1982). *Agrobacterium* plasmids can be stably maintained in *Rhizobium* cells. In addition, various plasmids of both species belong to the same incompatibility group. Hence, it can be assumed that plasmids normally found in *Agrobacterium* may be found in natural *Rhizobium* strains also. In *Agrobacterium*, the Ri plasmid was found to be responsible for the hairy root phenotype. *R. meliloti* transconjugants which carried the Ri plasmid, enhanced the nodulation and as a consequence, the nitrogen accumulation in the host plant.

Pankhurst and co-workers (1986) found that in *R. loti*, which bear *nod-* and *nif-*genes on the chromosome, the presence of indigenous plasmids effects the symbiotic effectiveness deleteriously. They demonstrated that by curing the single large plasmid pRlo2037 from *R. loti* NZP2037, the symbiotic effectiveness and competitiveness were enhanced. Further examination of 35 *R. loti* strains (including 25 field isolates) revealed that two did not contain a large plasmid. Both these strains were characterized by high symbiotitic effectiveness which became reduced when plasmid pRlo2037 was introduced.

Thus, the non-sym-plasmids may have either a positive or a negative effect on the symbiotic effectiveness of *Rhizobium*. Clearly, more research is required to determine the implications of plasmid transfer or elimination for the improvement of *Rhizobium* strains. Progress in this field could be achieved by the use of replicon labeling, mobilization and selective elimination techniques of *Rhizobium* plasmids. In conclusion, the improvement of *Rhizobium* by means of classical genetics, relies on the selection of mutants and recombinants which have beneficial symbiotic properties. This can be accomplished in two ways:

1) by random mutation or random recombination of chromosomal determinants;
2) by plasmid transfer which lead to the reassortment of plasmid and chromosomal DNA.

Both approaches deal with formal genetic factors without inquiring about the molecular mechanisms of the effective *Rhizobium*-Legume symbiosis. For the rational improvement of *Rhizobium* strains, strategies and methodology of molecular genetics are required.

4 Molecular Genetics of *Rhizobium*

4.1 Construction of Versatile Plasmid Vectors

Since conventional *E. coli* phage and plasmid vectors can not be applied to *Rhizobium*, the development of *Rhizobium* molecular genetics began with the construction of new vectors. Thus, instead of phage lambda, which was used to deliver transposons into cells of enteric bacteria, plasmid vector pJB4JI was constructed for transposon mutagenesis in *Rhizobium* (Beringer et al. 1978). This vector is a derivative of the promiscuous P1 group plasmid pPH1JI which carries the phage Mu genome and the Tn5 element. Phage Mu DNA abolishes the replication of P-plasmids in *Rhizobiaceae*. The vector therefore, can be used as a suicide vector. After conjugal transfer from *E. coli* to *Rhizobium*, the plasmid is eliminated from recipient cell, with the exception of transposon Tn5 which can be rescued, since it is able to transpose into the *Rhizobium* genome. Since Tn5 inserts randomly into the *Rhizobium* DNA, a wide range of Tn5 induced mutants can be obtained by this procedure. The use of Tn5 insertion mutagenesis is advantageous in that it results in target genes being marked by genetic and physical means. Since most symbiotic genes are not expressed under free living conditions, Tn5-mutagenesis is of particular importance in analysis of these genes.

A DNA cloning system involving pRK2, which was developed for rhizobia and related Gram-negative bacteria by Ditta et al. (1980), made use also of the specific properties of P1 plasmids. As compared to commonly used cloning vectors, pRK2 was too large (56 kb) and contained too many restriction sites. To reduce the size of the plasmid and maintain only single restriction sites suitable for cloning, DNA regions which were non-essential for replication, were deleted. The *trans*-acting transfer functions were separated onto the helper plasmid pRK2013. The resulting vector plasmid pRK290 (20 kb in size) can be transferred in the presence of pRK2013 by conjugation, between *E. coli* and *Rhizobium*, and is maintained stably in both hosts. This property of pRK290 allowed *Rhizobium* gene banks to be kept in *E. coli*, and the expression of cloned DNA to be studied in the *Rhizobium* background.

The idea of a binary vector systems was developed further by Simon and co-workers (1983). They designed a family of so-called pSUP vectors which have become indispensable tools in the genetic analysis of *Rhizobium*. These vectors are derivatives of the well known *E. coli* vectors pBR325, pACYC177 and pACYC184, to which a Mob-site and the *cis*-acting recognition site for the conjugal transfer of P1-plasmids, were added. Since the replication functions of the pSUP vectors are expressed in *E. coli* only, these vectors can not be maintained autonomously in non-enteric hosts. The transfer functions absent from the pSUP plasmids are complemented in trans by the helper plasmid RP4 or one of its derivatives carrying suitable antibiotic resistance markers. To avoid the cotransfer of the vector and the helper plasmid, the latter was integrated into the *E. coli* chromosome, resulting in a donor mobilizing strain S17-1 and its derivatives, SM10 and S68-1.

The pSUP vectors have features in common with the suicide and cloning vectors. In addition, they are able to integrate into the *Rhizobium* genome via both homologous and non-homologous recombination. Their great versatility has made the pSUP vectors suitable for numerous purposes, such as insertional mutagenesis, the marking and mobilization of non-transmissible replicons, the cloning and site directed mutagenesis of DNA. The functions of the pSUP vectors could be extended or modified by adding particular genes and genetic elements. For example, pSUP vectors carrying the *cos*-fragment of the phage lambda, display the properties of cosmids. Being small in size, these vectors are able to incorporate foreign DNA of up to 42 kb in length. The pSUP vectors carrying the *sac*-genes of *Bacillus subtilis* were shown to be suitable vehicles for the selection and cloning of *Rhizobium* indigenous IS-elements, for eliminating plasmids, as well as for the homogenotization of DNA fragments (Hynes et al. 1989).

4.2 Construction of Tn*5* Derivatives

In addition to vector systems, new Tn*5* derivatives should be mentioned. Analogous to Tn*5*, the Tn*5* derivatives continue to play an important role in the analysis of the symbiotic functions of *Rhizobium*. The construction of Tn*5*-Mob which bears the same cis-acting transfer site as the pSUP vectors made it possible to tag and mobilize non-transmissible *Rhizobium* plasmids and to study their functions in different genetic backgrounds. A Tn5 derivative, carrying the promoterless *lacZ*, *ntp*II or *luc* genes, encoding β-galactosidase, neomycin phosphotransferase and luciferase, respectively, can be used to generate *in vivo* transcriptional fusions. These permit the expression of promoter sequences situated upstream of the inserted transposon, to be monitored (Simon et al. 1989). Tn5 carrying the modified *E. coli phoA* gene which encodes alkaline phosphatase, was used as a probe for protein export signals (Manoil and Beckwith 1985). Therefore, Tn5-*pho*A has proved very helpful in the identification and study of transmembrane and secreted proteins. Taking into consideration that *Rhizobium*-Legume interactions occur at the surfaces of plant and bacterial cells, Tn5-*pho*A random mutagenesis is an ideal method for the molecular genetic analysis of cell surface components involved in the symbiosis (Long et al. 1988). Moreover, site-directed mutagenesis using Tn5-*pho*A and Tn5-*lac*Z concurrently, could facilitate the detailed analysis of the topological structure of the membrane protein (Manoil et al. 1990).

4.3 Identification and Cloning of Symbiotic Genes

As mentioned above, most of the symbiotic genes are silent under free living conditions. Consequently, the approaches to identify and clone symbiotic genes rely mainly on DNA manipulation and gene targeting. These approaches are as follows:

1) Screening of *Rhizobium* Gene Libraries with Appropriate DNA and RNA Hybridization Probes.

The study of *Rhizobium* genetics benefited greatly from the knowledge of the *Klebsiella pneumoniae nif* genes. By using hybridization probes carrying the *K. pneumoniae nif* genes, numerous homologous genes were found and characterized in rhizobia, including *nif* HDK, *nif* A and *nif* B (Rossen et al. 1984; Ruvkun and Ausubel 1980; Szeto et al. 1984). Since large number of symbiotic genes are derepressed *in planta* only, bacteroid mRNA can be used for their identification.

2) Tn5-labeling of Symbiotic genes.

Random Tn5-mutagenesis has allowed the identification of genes with diverse symbiotic functions, including the *nod*, *nif*, *fix*, and *exo* genes, to be made. The use of the Tn5-*pho*A element enabled genes which encode membrane-bound symbiotic functions to be identified (Long et al. 1988).

3) Complementation of Mutations in Symbiotic Genes by a Gene Bank of a Wild-type Strain.

The potential of this approach was first demonstrated by Long and co-workers (1982). A gene bank of a wild-type *R. meliloti* strain was introduced "en masse" into a *Nod⁻* mutant. Transconjugants carrying *nod* gene fragments were isolated by the use of a host plant which provided a high selective pressure in favour of the Nod^+ phenotype.

4) Site-directed Mutagenesis and Sequencing of Regions Adjacent to Known Symbiotic Genes.

The majority of symbiotic genes was found to be organized in gene clusters (Martinez et al. 1990). For this reason, genomic clusters, known to control symbiotic functions, were saturated with insertions of transposons and antibiotic resistance cartridges. The phenotype of the resulting mutants enabled additional symbiotic genes to be identified. These were analysed in greater detail by sequencing.

5) Induction Approach for Identifying Symbiotic Genes.

It is well established that derepression of bacterial symbiotic genes is caused by host plant root exudates. The technique of random transposon mutagenesis allows the construction of fusions between genomic promoter sequences and a reporter gene, e.g. *lacZ*. By exposing these fusions to root exudates, it is possible to identify symbiotically regulated genes (Olson et al. 1985).

The approaches mentioned have led to the identification of numerous symbiosis genes in *Rhizobium*. Some of these genes are essential for symbiosis only, e.g. the *nif* and *fix* genes determine the synthesis of an active nitrogenase complex, while the *nod* genes are responsible for the recognition of the host plant and for the induction of nodule formation. Other genes function, both in the free-living and in the symbiotic state, e.g. the *exo* genes determine the exo-polysaccharide synthesis and the penetration of rhizobia via infection threads into the nodules; the *dct* genes govern the transport of C_4-dicarboxylates into both the vegetative and the symbiotic forms of rhizobia. The question arises as to how agriculturally important symbiotic genes could be identified and cloned. Generally, there are two approaches. The first is the traditional genetic approach which is based on the isolation of mutants with altered symbiotic effectiveness and the subsequent analysis of genes involved. The second is the analytical approach employing the genetic, biochemical and physiological knowledge of factors which contribute to a high symbiotic potential. The results of Tn5-labeling of genes responsible for the symbiotic effectiveness in *R. meliloti* illustrate the first approach. The second approach will be discussed in the following section of this review.

4.4 Tn5-Labeling of Genes Responsible for Symbiotic Effectiveness

The first Tn5-mutants with enhanced symbiotic effectiveness (Eff^{++}) were isolated in *R. meliloti* 41, by Plazinski (1981). The mutants exceeded the wild type with respect to host plant biomass production and nitrogen accumulation. To confirm the linkage between the symbiotic phenotype and the Tn5 insertion and also to determine its position on the chromosome, one of the mutants was mated with a set of *R. meliloti* strains which carried defined genetic markers. However, neither the kanamycin resistance (Km^r) gene, nor the Eff^{++} mutation was found to be linked to any of the markers distributed over the chromosome (Plazinski 1981). Therefore, the initial two questions remained unanswered:

1) whether the mutation which increased effectiveness, was caused by the insertion of Tn5,
2) where in the *R. meliloti* genome, such a mutation was located.

In order to analyse these questions in more detail, Sharypova et al. (1987) carried out Tn5 mutagenesis in *R. meliloti* CXM1-105 and CXM1-188 with the use of pSUP2021 and pSUP5011. Nineteen mutants which showed enhanced symbiotic effectiveness, were isolated. In aseptic plant tests, these mutants demonstrated a superior ability to increase the alfalfa shoot mass, but retained the same acetylene reduction activity as the wild-type strain. When the alfalfa plants were grown in non-sterile soil in a greenhouse, the mutants also demonstrated an increased symbiotic effectiveness, as measured by the shoot dry mass and the nitrogen accumulation.

By the use of transduction, the Tn5 encoded Km^r phenotype was transferred from the Eff^{++} mutants to the parental strain. Although the recombinants varied in their symbiotic effectiveness, the majority of them inherited the Eff^{++} phenotype of the donors. Transduction experiments were conducted to ascertain whether the Tn5 insertions were responsible for the enhanced effectiveness. These results demonstrated that, by introducing the Tn5-marked Eff^{++} determinants, the symbiotic potential of a strain could be improved. Whether this transduction approach will find an application in the construction of *R. meliloti* strains remains to be seen, since the phenotypic expression of a gene which is known to control a polygenic trait such as symbiotic effectiveness, would depend greatly on the genetic background. Southern blot analysis revealed that 11 of 19 mutants contained Tn5 inserted in one of the megaplasmids (Sharypova et al. 1990). To distinguish between megaplasmid-1 and megaplasmid-2, the Tn5-induced mutants were conjugated with the *R. meliloti* deletion mutant ZB121 (Banfalvi et al. 1981) and with the plasmidless *Agrobacterium tumefaciens* UBAPF2. This allowed four Tn5 insertions in megaplasmid-1 and five in megaplasmid-2, to be located (Sharypova et al. 1990). Both megaplasmids are evidently involved in the control of symbiotic effectiveness. The three most effective mutants were chosen for further analysis. Total DNA of the mutants was digested with *Eco*RI and cloned into the pSUP202 vector. Clones containing the *R. meliloti* DNA fragments with Tn5 insertions were identified by the expression of kanamycin resistance. The resulting plasmids are currently being investigated.

5 Improvement of *Rhizobium* Strains Through Molecular Genetics

In this section of the review we will discuss the present knowledge of hydrogen recycling, nitrogen fixation and dicarboxylate transport systems and its application to the construction of improved *Rhizobium* strains.

5.1 The Hydrogen Recycling Capacity as a Beneficial Symbiotic Property

All nodulated legumes are known to produce hydrogen as a by-product of the nitrogen fixation reaction. According to some estimations, 25% or more of the nitrogenase electron flux is expended for the evolution of H_2 (Evans et al. 1987). Consequenty, it is believed that the recycling of hydrogen, determined by the *hup* genes, could be beneficial for symbiotic nitrogen fixation. Convincing results were obtained for *B. japonicum*, but not for *R. leguminosarum* nor *R. meliloti*. The lack of positive effects could be attributed to the insufficient H_2 recycling capacity and/or to the ineffective coupling of H_2 oxidation to ATP synthesis (Evans et al. 1987). The *hup* genes were studied in detail, in *B. japonicum* and *R. leguminosarum* (Lambert et al. 1985; Leyva et al. 1990). They were cloned in vector plasmids and can be transferred to any strain of interest. This will not be discussed in detail, since the idea of improving *Rhizobium* strains by the introduction of the *hup* genes has already attracted much attention.

5.2 Regulation of Nitrogen Fixation Genes and Optimization of Their Expression

The nitrogen fixation genes in rhizobia are under the positive control of the *nif*A and *ntr*A genes. The *nif*A gene encodes the transcriptional activator protein, Nif A, while *ntr*A encodes the alternative sigma factor, sigma 54, of the RNA-polymerase. There are several lines of evidence suggesting that modifications of both genes could enhance the symbiotic nitrogen fixation (Cannon et al. 1988; Williams et al. 1990), or at least enhance the expression of specific *nif* genes (Merrick and Coppard 1989; Krey et al. 1990).

Nif A belongs to a family of bacterial regulatory proteins which activate transcription from promoters which require the alternative sigma factor NtrA. This family also includes NtrC, the activator protein for nitrogen assimilation genes, and DctD, the activator protein of the structural gene *dct*A which is responsible for C_4-dicarboxylate transport (Ronson et al. 1987b). All these regulatory proteins share two conservative domains: the Central Domain (CD), and the DNA-binding Domain (DBD) which resides at the C-terminal end of the proteins (Drummond et al. 1986; Ronson et al. 1987b). It is presumed that the CD interacts with the NtrA-RNA polymerase holoenzyme to stimulate the transcription from NtrA-dependent promoters, while the DBD interacts with the DNA at the Upstream Activator Sequence (UAS) of the target promoter (Buikema et al. 1985; Drummond et al. 1986; Morret et al. 1988; Ronson et al. 1987b).

The N-terminal part of *R. meliloti* and *B. japonicum* NifA shows no similarity to other transcriptional regulators which are activated by specific sensor proteins, e.g. NtrB modulates NtrC and DctB modulates DctD. Moreover, recently it was found that the NifA of a certain *R. leguminosarum* strain was devoid of the N-terminal domain (Iisma and Watson 1989) indicating that the N-terminal part of the protein is dispensable for its function. In fact, deletion of the N-terminal part of the protein did not abolish its ability to activate the transcription of *nif* genes (Beynon et al. 1988; Huala and Ausubel 1989). Beynon and co-workers (1988) demonstrated that such a truncated protein could be even more active than the wild-type one.

In *K. pneumoniae*, the NifA activity is regulated by NifL. In the absence of NifL, the NifA protein is insensitive even to high concentrations of oxygen. In contrast to this, the *Rhizobium* NifA protein loses its activity drastically as the oxygen partial pressure increases (Fischer and Hennecke 1987; Fischer et al. 1988; Huala and Ausubel 1989). In order to analyse this oxygen sensitivity of the *R. meliloti* NifA protein, Krey and co-workers (1990) employed in vitro hydroxylamine mutagenesis to isolate a series of oxygen tolerant *nif*A mutants. Some mutations resulted in a several-fold enhancement of the NifA activator function, both under aerobic and microaerobic conditions.

The expression of the *R. meliloti nif*A gene is known to be under the control of a regulatory protein couple FixL–FixJ (David et al. 1988). Under oxygen limited conditions, both in the symbiotic and microaerobic state, the transmembrane protein FixL activates FixJ, which in turn induces the transcription of *nif*A. The manner in which FixJ affects the transcription of *nif*A remains obscure, since the *R. meliloti nif*A upstream sequence has neither similarity to the NtrA-dependent promoters, nor to the other types of known promoters (Buikema et al. 1985). To enhance the *nif*A expression in *R. meliloti*, a constitutive promoter was fused to the *nif*A coding sequence and the fusion was integrated into the inositol locus of the *R. meliloti* genome (Williams et al. 1990). The resulting strain was characterized by an enhanced expression of the *nif*HDK promoter. When tested in a greenhouse, this strain showed a 3–7% increase in the host plant biomass over the wild-type strain. By integrating a second copy of the modified *nif*A gene into another region of the *R. meliloti* genome, a derivative strain was constructed, which gave a 7–15% increase in alfalfa biomass, when measured against the wild type. This increase was statistically significant. The results presented evidence that the increased expression of the nitrogen fixation genes could be achieved by the alteration of NifA activator function and by the modulation of the *nif*A gene expression.

The *ntr*A gene encodes an alternative sigma factor that modifies the specificity of the RNA-polymerase. The NtrA-RNA-polymerase holoenzyme is able to start transcription from NtrA-dependent promoters in the presence of specific activator proteins only, e.g. the nitrogen assimilation genes require NtrC, the nitrogen fixation genes, NifA, and the dicarboxylic acid transport, DctD (Kustu et al. 1989). The *ntr*A gene was identified in six different species of bacteria, including *R. meliloti* and the broad host-range *Rhizobium* sp. NGR234 (Ronson et al. 1987a; Stanley et al. 1989). Mutations in the *ntr*A gene of both *Rhizobium* species abolished nitrogen fixation, C_4-dicarboxylates transport and nitrate assimilation (Ronson et al. 1987a). In addition, a *ntr*A mutant of *Rhizobium* NGR234 showed delayed nodulation (Stanley et al. 1989).

Recently, Merrick and Coppard (1989) identified and analysed two open reading frames (ORF95 and ORF162) situated downstream of the *K. pneumoniae* *ntr*A gene. Mutations in both genes significantly increased the transcription from *ntr*A-dependent promoters, namely those of *nif* H, *nif* L and *gln*A (promoter 2). The acetylen reduction activity in both mutants was enhanced by 40–50% under fully derepressed conditions (without ammonia) and by more than 20-fold under repressed conditions (in the presence of ammonia in the medium). It was suggested that the ORF95 and ORF162 gene products modulate the activity of the NtrA-RNA-polymerase holoenzyme. The open reading frame ORF95, was found to have homologs in *K. pneumoniae*, *Azotobacter vinelandii*, *Pseudomonas putida* and *R. meliloti*; all located downstream of *ntr*A (Merrick and Coppard 1989). This conservation gives rise to the possibility that mutations in the *Rhizobium* ORF95 homologs could enhance the symbiotic nitrogen fixation capacity of the strain.

5.3 Genetic Control of C_4-Dicarboxylate Transport in Relation to Nitrogen Fixation

The nitrogen fixation process requires a great deal of energy and reducing power, which are supplied to the bacteroids by a host plant photosynthate. The molecular form in which photosynthate enters the bacteroids, is not absolutely clear. The respiration and nitrogen fixation of isolated *R. meliloti* bacteriods were shown to be supported by C_4-dicarboxylic acids (dCA): succinate, malate and fumarate (Miller et al. 1988). Moreover, *R. leguminosarum* and *R. meliloti* mutants defective in the dCA transport and metabolism, were found to induce ineffective (Fix⁻) nodules (Engelke et al. 1987; Finan et al. 1983; Ronson et al. 1981). Hence, the dCA are essential for nitrogen fixation. More research is needed however, to reveal whether additional compounds enter bacteroids and how the nitrogen and carbon exchange occurs through the peribacteroid membrane.

While recognising that our understanding of the biochemical cooperation between the plant and bacterial partners is limited, we can nevertheless, outline the possibilities of rationalized improvement of the symbiosis, based on the knowledge of the dCA transport and its genetic determination. The transport of the dCA into rhizobia is determined by three genes; *dct*A, *dct*B and *dct*D, the sequences and organization of which, are well conserved in *R. meliloti* and

R. leguminosarum (Bolton et al. 1986; Engelke et al. 1987; Ronson et al. 1984; Yarosh et al. 1988). The *dct* genes comprise two operons, *dct*A and *dct*BD, which are transcribed divergently. The protein product of the *dct*A gene, DctA, is a transmembrane transport component for dCA. DctB and DctD represent a regulatory protein couple, in which DctB is presumed to be a sensor and DctD a transcriptional activator for the *dct*A gene (Ronson et al. 1987b).

By analysing mutants affected in the dCA transport, Engelke and colleagues (1987) found that the rate of dCA transport into bacteroids correlated well with the symbiotic nitrogen fixation rate. It was proposed that by increasing the dCA transport, the nitrogen fixation rate would be enhanced. To increase the rate of dCA uptake, Birkenhead and co-workers (1988) introduced additional copies of the *R. meliloti dct* genes into *B. japonicum*. Under microaerobic conditions which permit nitrogen fixation in *B. japonicum*, the modified strain was characterised by a higher rate of acetylene reduction than the wild type. Hence, the study of the dCA transport in *Rhizobium* and its role in symbiotic nitrogen fixation, revealed that a coordination between the genetic regulation and metabolic exchange exists between the symbiotic partners. It is expected that this knowledge will open new possibilities for the rational improvement of the *Rhizobium*-legume interaction.

6 Conclusions

Nodule bacteria are agriculturally important organisms. They are widely used as inoculants for leguminous forage crops in different parts of the world. To date, traditional selection methods were utilized to improve the symbiotic performance of *Rhizobium* strains and their leguminous hosts. In this review, the prospects for genetic improvement of *Rhizobium* were considered from the viewpoints of classical and molecular genetics.

Classical genetics allowed genetic factors which determine symbiotic effectiveness to be identified. Mutagenesis and different methods of genetic hybridization were shown to increase the potential of genetic variation in *Rhizobium*. This resulted in the isolation of strains with improved symbiotic performance. Molecular genetics allowed genes to be identified which determine the symbiotic properties of *Rhizobium*, and facilitated the study of the regulation of their expression. By constructing defined mutants which were affected in particular symbiotic functions, details of the bacteroid metabolism were elucidated. This rapidly accumulating knowledge is crucial to the understanding and optimization of the symbiotic nitrogen fixation process.

When considering the problem of improving the effectiveness of *Rhizobium* strains, it is interesting to compare the nature of genes which are altered by random mutations and by recombinant DNA techniques. Random mutations are known to cause the inactivation of gene functions. Therefore, the genes affected should be responsible for the negative control of symbiotic effectiveness. In contrast to this, recombinant DNA techniques permit the targeted alteration of a gene function. Since the majority of symbiotic genes have been found to be controlled by positive regulators, current efforts are directed at modifying genes encoding such regulatory proteins. Structural genes are also considered to be subjects for genetic manipulations, but a better knowledge of the biochemistry of the symbiotic nitrogen fixation process is required to improve the symbiotic effectiveness by altering a single enzymatic activity. To summarize, since different genes are modified by random mutation and by genetic engineering, the approaches complement each other in the construction of *Rhizobium* strains with improved symbiotic performance.

7 References

Balassa R, Gabor M (1961) Mikrobiologia (Microbiology, in Russian) 30:457
Banfalvi Z, Sakanyan V, Koncz C, Kiss A, Dusha I, Kondorosi A (1981) Molec Gen Genet 184: 318
Bazhenova OV, Simarov BV (1983) Dokl Akad Nauk SSSR (Proc Acad Sci USSR, in Russian) 269: 958
Beringer JE, Beynon JL, Buchanan-Wollaston AV, Johnston AWB (1978) Nature 276:633
Beringer JE, Brewin NJ, Johnston AWB (1980) Heredity 45:161
Beringer JE, Johnston AWB, Kondorosi A (1984) In: O'Brien SJ (ed) Genetic maps. Cold Spring Harbor Laboratory, p 202
Beynon JL, Williams MK, Cannon FC (1988) EMBO J 7:7
Birkenhead K, Manian SS, O'Gara F (1988) J Bacteriol 170:184
Bolton E, Higginson B, Harrington A, O'Gara F (1986) Arch Microbiol 144:142
Bromfield ESP, Thurman NP, Whitwill ST, Barran LR (1987) J Gen Microbiol 133:3457
Buchanan-Wollaston V (1979) J Gen Microbiol 112:135
Buikema WJ, Szeto WW, Lemley PV, Orme-Johnson WH, Ausubel FM (1985) Nucl Acids Res 13:4539
Burkardt B, Burkardt HJ (1984) J Mol Biol 175:213
Burkardt B, Schillik D, Pühler A (1987) Plasmid 17:13
Cannon FC, Beynon J, Hankinson T, Kwiatkowski R, Legocki RP, Ratcliffe H, Ronson C, Szeto W, Williams M (1988) In: Bothe H, de Bruijn FJ, Newton WE (eds) Nitrogen fixation: Hundred years after. Gustav Fischer, Stuttgart, New York, p 735
Casadesus J, Olivares J (1979) J Molec Gen Genet 174:203
David M, Daveran M-L, Batut J, Dedieu A, Domerque O, Ghai J, Hertig C, Boistard P, Khan D (1988) Cell 54:671
De Jong TM, Brewin NJ, Johnston AWB, Phillips DA (1982) J Gen Microbiol 128:1829
De Jong TM, Brewin NJ, Phillips DA (1981) J Gen Microbiol 124:1
Ditta G, Stanfield S, Corbin D, Helinski DR (1980) Proc Natl Acad Sci USA 77:7343
Drummond M, Whitty P, Wooton J (1986) EMBO J 5:441
Dunican LK, Tierney AB (1973) Molec Gen Genet 126:187
Engelke T, Jagadish MN, Pühler A (1987) J Gen Microbiol 133:3019
Evans HJ, Harker AR, Papen H, Russel SA, Hanus FJ, Zuber M (1987) Ann Rev Microbiol 41: 335
Feodorov SN, Simarov BV (1987) Sel Biol (Agr Biol, in Russian) 9:44
Filatova OV, Simarov BV, Kuchko VV (1982) Izv AN SSSR, Ser Biol (News of the Acad Sci USSR, in Russian) 1:140
Finan TM, Hartweig E, Le Mieux K, Bergman K, Walker GC, Signer ER (1984) J Bacteriol 159: 120
Finan TM, Kunkel B, De Vos GF, Signer ER (1986) J Bateriol 167:66

284 Larissa A. Sharypova et al.

Finan TM, Wood JM, Jordan DC (1983) J Bacteriol 154: 1403
Fischer H-M, Bruderer T, Hennecke H (1988) Nucl Acids Res 16: 2207
Fischer H-M, Hennecke H (1987) Molec Gen Genet 209: 621
Gibbins AM, Gregory KF (1972) J Bacteriol 111: 129
Gibson AH (1962) Aust J Agric Res 13: 388
Harrison SP, Jones DC, Schümmann PHD, Forster JM, Young JPW (1988) J Gen Microbiol 134:
    2721
Hollingsworth RJ (1984) J Bacteriol 160: 510
Hooykaas PJJ, Peerbolte R, Regensburg-Tuink AJG, de Vries P, Schilperoort RA (1982) Molec
    Gen Genet 188: 12
Huala E, Ausubel FM (1989) J Bacteriol 171: 3354
Huguet T, Rosenberg C, Casse-Delbart F, De Lajudie P, Jouanin L, Batut J, Boistard P, Julliot
    J-S, Dénarié J (1983) In: Pühler A (ed) Molec genetics of the bacteria-plant interaction Springer-
    Verlag, Berlin, Heidelberg p 35
Hynes MF (1990) In: 8th Int Congr Nitrogen Fixation, Knoxville, May 20–26, B-33
Hynes MF, Simon R, Müller P, Niehaus K, Labes M, Pühler A (1986) Mol Gen Genet 202: 356
Hynes MF, Quandt J, O'Connel M, Pühler A (1989) Gene 78: 111
Iisma SE, Watson JM (1989) Molec Microbiol. 3: 943
Kiss G, Kalman Z (1982) J Bacteriol 150: 465
Kondorosi A (1986) In: Broughton WJ, Pühler A (eds) Nitrogen fixation. Clarendon Press, Oxford.
    4: 245
Kowalski M (1967) Acta Microbiol Polonica 16: 7
Krasilnikov NA (1941) Dokl Akad Nauk SSSR (Proc Acad Sci USSR, in Russian) 31: 90
Krey R, Sharypova LA, Klipp W, Pühler A (1990) In: 8th Int Congr Nitrogen Fixation, Knoxville.
    May 20–26, E-10
Kustu S, Santero E, Keener J, Popham D, Weiss D (1989) Microbiol. Rev. 53: 367
Lambert GR, Cantrell MA, Hanus FJ, Russell SA, Haddad KR, Evans HJ (1985) Proc Natl Acad
    Sci USA 82: 3232
Laptev GU, Simarov BV (1981) Docl Akad Nauk SSSR (Proc Acad Sci USSR, in Russian) 201.
    207
Long SR (1989) Ann Rev Genet 23: 483
Long SR, Buikema WJ, Ausubel FM (1982) Nature 298: 485
Long SR, McClure S, Walker GC (1988) J Bacteriol 170: 4257
Leyva A, Palacios JM, Murillo J, Ruiz-Argüeso T (1990) J Bacteriol 172: 1647
Maier R, Brill WJ (1978) Science 201: 448
Manoil C, Beckwith J (1985) Proc Natl Acad Sci USA 82: 8129
Manoil C, Mekalanos JJ, Beckwith J (1990) J Bacteriol 172: 515
Martinez E, Romero D, Palacios R (1990) Crit Rev Plant Science 9: 59
Merrick MJ, Coppard JR (1989) Molec Microbiol 3: 1765
Miller RW, McRae DG, Al-Jobore A, Berndt WB (1988) J Cell Biochem 38: 35
Minchin FR, Sheehy JE, Witty JF (1985) In: Evans HJ, Bottomley PJ, Newton WE (eds) Nitrogen
    fixation research progress p 285
Morret E, Cannon W, Buch M (1988) Nucl Acids Res 16: 11469
Müller P, Hynes M, Kapp D, Niehaus K, Pühler A (1988) Mol Gen Genet 211: 17
Murphy PJ, Heycke N, Banfalvi Z, Tate ME, de Bruijn FJ, Kondorosi A, Tempe J, Schell J (1987)
    Proc Natl Acad Sci USA 84: 493
Mytton LR, Rys GJ (1985) Plant and Soil 88: 197
Nicholas DB, Haydock KP (1971) Aust J Agric Res 22: 223
Nutman PS (1954) Heredity 8: 35
O'Gara F, Dunican LK (1973) J Bacteriol 116: 1177
Olson ER, Sadowsky MJ, Verma DPS (1985) Biotechnology 3: 143
Pankhurst CE, MacDonald PE, Reeves JM (1986) J Gen Microbiol 132: 2321
Plazinski J (1981) Microbiol Lett 18: 137
Prakash RK, Atherly AG (1986) Int Rev Cytol 104: 1
Ronson CW, Astwood PM, Downie JA (1984) J Bacteriol 160: 903
Ronson CW, Lyttleton P, Robertson JG (1981) Proc Natl Acad Sci USA 78: 4284
Ronson CW, Nixon BT, Albright LM, Ausubel FM (1987a) J Bacteriol 169: 2424
Ronson CW, Nixon BT, Ausubel FM (1987b) Cell 49: 579
Rossen L, Ma Q-S, Mudd EA, Johnston AWB, Downie JA (1984) Nucl Acids Res 12: 7123
Ruvkun GB, Ausubel FM (1980) Proc Natl Acad Sci USA 77: 191

Sanjuan J, Olivares J (1988) In: Palacios R, Verma DPS (eds) Molecular genetics of plant-microbe interactions. APS Press, Minnesota, p 200

Scøt L (1983) Physiol Plant 59: 585

Scøt L, Hirsch PR, Witty JF (1986) J Appl Bacteriol 61: 239

Selvaraj G, Iyer VN (1981) Gene 15: 279

Sharypova LA, Fomina-Eshchenko JG, Simarov BV (1990) In: 8th Int Congr Nitrogen Fixation, Knoxville, May 20–26, E-04

Sharypova LA, Onichshuk OP, Novikova NI (1987) Genetika (Genetics, in Russian) 23: 2104

Shukla RS, Singh CB, Dubey JN (1989) Theor Appl Genet 78: 433

Simarov BV, Novikova NI, Sharypova LA, Provorov NA, Aronshtam AA, Kuchko VV (1989) In: Vančura V, Kunč F (eds) Interrelationships between microorganisms and plants in soil. Academia Publ House of the CAS, Praha p 45

Simarov BV, Provorov NA (1990) In: Simarov BV (ed) Geneticheskie Osnovy Selekcyi Klubenkovyh Bakterii (Genetic Principles of the Nodule Bacteria Selection, in Russian) Leningrad, VO Agropromizdat p 167

Simon R, Quandt J, Klipp W (1989) Gene 80: 161

Simon R, Priefer U, Pühler A (1983) Biotechnology 1: 784

Sørensen GM, Wyndaele RJ (1986) Gen Microbiol 132: 317

Stanley J, van Slooten J, Dowling ND, Finan T, Broughton WJ (1989) Mol Gen Gent 217: 528

Strobel GA, Lam B, Harrison L, Hess BM, Lam S (1985) J Gen Microbiol 131: 355

Szeto WW, Zimmermam JL, Sundaresan V, Ausubel FM (1984) Cell 36: 1035

Tan GY (1981) Crop Sci 21: 485

Toro N, Olivares J (1986a) J Appl Environ Microbiol 51: 1148

Toro N, Olivares J (1986b) J Mol Gen Genet 202: 331

Watson RJ, Chan YK, Wheatcroft R, Yang AF, Han S (1988) J Bacteriol 170: 927

Williams PM (1981) Plant and Soil 60: 349

Williams MK, Beynon JL, Ronson CW, Cannon FC (1990) In: 8th Int Congr Nitrogen Fixation, Knoxville, May 20–26, E-09

Yarosh OK, Charles TC, Finan TM (1989) Molec Microbiol 3: 813

CHAPTER 14
The Importance of the Rhizobial Cell Surface in Symbiotic Nitrogen Fixation

Andrew W.B. Johnston

The Nitrogen Fixation
and its Research in China
Editor: Guo-fan Hong
© Springer-Verlag Berlin Heidelberg 1992

1 Introduction

As described elsewhere in this book and in recent reviews (Long 1989; Downie, Johnston 1988], the symbiotic interaction between legumes and bacteria that comprise the "rhizobia" is complex and multifaceted, involving coupled differentiation and development in both partners. In this interaction there is intimate contact between both partners throughout the infection process. There is recognition at the very earliest stages when the bacteria first make contact with the root hairs of its host all the way through to the point at which the bacteroids are fixing nitrogen and are in close, perhaps direct, contact with the plant-specified peribacteroid membrane which surrounds them.

A priori then, it should come as no surprise that structures at the cell surfaces of both partners should play an important role in the establishment of nitrogen-fixing nodules. At least in the case of the bacterial symbiont, genetic evidence has amply confirmed this statement. By the isolation of various classes of mutants and the demonstration that these are affected in nodulation and/or nitrogen fixation, it has clearly been demonstrated that several genes whose protein products themselves are at the cell surface or which are exported from the cells are important for nodulation. Further, other genes whose products are involved in the synthesis of molecules that are located at the rhizobial surface play key roles in the establishment of effective nitrogen-fixing nodules on a range of leguminous plants.

This chapter reviews recent studies that have led to these conclusions. The "cell surface" is termed rather loosely, and those cases are considered in which gene products are directly associated with one or other of the bacterial membranes, the periplasm or where the polypeptide is exported into the growth medium. Further, this article deals with those cases where the gene product is required for the synthesis of other molecules which are themselves membrane-associated or exported. In fact there are some examples in which, perhaps not surprisingly, the proteins which are involved in the synthesis or export of (for example) exopolysaccharides, are themselves located in the membrane.

The various identified genes are grouped into several classes as follows. Firstly, evidence is given to the fact that certain of the products of the so-called "nodulation" (nod) genes which specify the early steps in the nodulation process are located at the cell surface. It will also be shown that at least two transport systems which probably depend on the presence of proteins at the surface of Rhizobium are important for nodulation and genetic evidence is presented suggesting that each of the three polysaccharides that are at the cell surface are variously required for the development of normal nitrogen-fixing nodules. These descriptions only consider those cases in which genetic evidence has conclusively shown that specific surface molecules are intimately involved in the symbiotic interaction.

2 Nodulation Genes

In several species of rhizobia, clusters of *nod* genes which determine the ability of the bacteria to participate in the early steps in the infection and which are of key importance in determining the host range of a given strain have been identified. Only a brief mention is made of those cases that are relevant to the topic of the rhizobial cell surface. By direct sub-cellular localisation or by analysis of the deduced polypeptide products of individual nod genes (obtained from DNA sequence determination), several such Nod proteins have been shown directly or have been suggested to be associated with one or other of the bacterial membranes, the periplasm or to be exported from the cells into the media. Those *nod* genes whose products are associated with the cell surface are as follows.

nodC This *nod* gene is absolutely required for nodulation and indeed all the steps of infection. As shown by John et al. (1988, 1989) and by Johnston et al. (1989), its product is present in the bacterial membrane. Indeed, from the model presented by John et al. (1989), NodC is a very unusual protein, its C-terminus being located in the inner membrane and its N-terminus being in the outer, the extreme N-terminus actually protruding from the cells. The combined actions of the *nodA, B* and *C* genes are required for the production of an exported molecule that has the ability to induce root hair curling and deformation, one of the earliest observable steps in the symbiosis.

– *nodD and nod gene regulation.* Transcription of *nod* genes is tightly regulated, most *nod* genes being transcribed only in cells exposed to particular flavonoid inducers. This induction requires the regulatory gene *nodD*, which Schlaman et al. (1989) showed was associated with the inner membrane of Rhizobium. It is perhaps significant that Recourt et al. (1989) obtained evidence that there was accumulation of flavonoid inducers in this compartment of the cells suggesting that there may be some complex interaction between the inner membrane and NodD. However, this is not required for specific binding of NodD to *nod* gene regulatory sequences (*nod* boxes). Hong et al. [1987] showed that this binding could occur in vitro with membrane-free cell extracts.

In *R. leguminosarum* biovar (bv.) *phaseoli* (which nodulates *Phaseolus* beans), there are three copies of *nodD*, one of which is immediately preceded by a gene (*nolE*) that specifies a protein that has an N-terminus similar in sequence to the signal peptide that directs proteins to the bacterial periplasm (Davis and Johnston 1990a, b). By constructing *nodE-phoA* insertions (see below) it was confirmed that *nolE* specifies a protein that is directed into the periplasm. The function of *nolE* is not known but it is tempting to speculate that it might be involved in transport or processing of inducer molecules.

– *nodE*: Spaink et al. (1989) showed that the product of *nodE* of R. *leguminosarum* was present in the inner membrane. This gene is required for normal nodulation and its product is similar in sequence to polyketide synthetases and fatty acid synthetase, which, interestingly, are cytosolic enzymes in other organisms.

– *nodO*: This gene, found in *R. leguminosarum* bv *viciae* (which nodulates peas, vetches and lentils) specifies a protein that is similar in sequence to a family of bacterial polypeptides that are exported into the growth medium (de Maagd et al. 1989; Economou et al. 1990). Although single mutations in *nodO* have very little effect on nodulation, NodO⁻ strains are severely debilitated for nodulation when other nod mutations [e.g. in *nodF*, *E* or *L*] are also present in the strain.

– *nodI*: In the same operon as *nodC*, the *nodI* gene specifies a protein that is similar to a family of bacterial proteins (including *chvB* and *ndvB*-see below) which are involved in the import or export of a variety of low molecular weight compounds, suggesting that this *nod* gene might also be involved in transport (Evans and Downie 1986). If true, though, the identity of the transported molecule has yet to be determined.

– *nodJ*: Downstream of *nodI* is another gene, *nodJ*, (Evans and Downie 1986) whose sequence revealed that its product was very hydrophobic, indicating that it might be associated with one of the bacterial membranes but this has not formally been proven. The functions of *nodI* and *J* are unknown; mutations in these genes have only a minor inhibitory effect on nodulation.

– *nodX*: Primitive peas of variety Afghanistan fail to be nodulated by "conventional" strains of *R. leguminosarum* bv *viciae*. However, a strain of this species, isolated from Turkey can nodulate these peas, this being due to its possession of an "extra" gene, *nodX*, which is downstream of *nodJ* (Gotz et al. 1985). Sequence determination of *nodX* showed that it, too, specifies a very hydrophobic protein (Davis et al. 1988), suggesting that it is located in a membrane; however this remains to be proved and, as with many of the *nod* genes, its precise function in the nodulation process is unknown.

From the observation presented in this section, it is clear that approximately half of the known *nod* genes specify proteins that are associated with the bacterial cell surface. This is useful to know, and will help in the future in their purification and, perhaps in allowing us to assign a function to them.

3 Transport

Given the traffiking of various molecules that must occur between the two partners during the development of nodules, it would not be surprising if bacterial mutants, defective in certain transport systems, would also be affected in their symbiotic capabilities. This has turned out to be the case for at least two systems, namely those involved in the transport of Fe and dicarboxylic acids.

Nadler et al. (1990) isolated a mutant of *R. leguminosarum* bv. *viciae* that over-accumulated protoporphyrin, a precursor of haem and found that although it nodulated peas and that the bacteria within the nodules differentiated into bacteroids, these were completely defective in nitrogen fixation. It was found that the defect in this mutant was due to its inability to take up Fe from the medium and although it was not formally proved that the product of the mutant gene was located in the bacterial membranes of peroplasm, *a priori*, it seems likely that this is the case. The fact that these non-fixing mutant bacteroids probably make only small amounts of haem lends support to the idea that the bacteria provide this moiety of the leghaemoglobin.

The finding that mutants of *Rhizobium* which cannot take up dicarboxylic acids, such as succinate, are Fix⁻ (Ronson et al. 1981) in nodules (although they can form bacteroids) is consistent with the idea that such compounds are important for the supply of carbon skeletons and energy to these forms of the bacteria. One of the genes that is required for the uptake of these molecules [*dctC*] has been shown to be a periplasmically located transporter and one of the genes that regulates the expression of the *dct* genes is located in the inner membrane where it "senses" the presence of succinate in the medium (Jiang et al. 1989).

4 Rhizobial Polysaccharides

Strains of rhizobia make at least three distinct polysaccharides, all of which are important for effective nodulation, though genetic studies have shown that certain of these may be important in the nodulation of some legumes but not others.

4.1 The Acidic Exopolysaccharide (EPS)

Colonies of most strains of rhizobia are mucoid, this being due to the production of copious amounts of a high molecular weight acidic exopolysaccharide. The particular structures of the EPS's of different rhizobial species and genera differ quite markedly, but share the properties of having a sugar backbone with small side chains at regular intervals. By obtaining mutants that fail to make the EPS and testing their symbiotic proficiency, it has been clearly demonstrated that, at least in some cases, this polymer is required for nodulation or nitrogen fixation.

4.1.1 EPS Synthesis in *R. meliloti*

The most detailed studies on the genetics of EPS synthesis have been done on *R. meliloti* whose hosts include alfalfa. *R. meliloti* makes a succinoglycan in which the backbone comprises 3 glucoses and one galactose residue; attached to every fourth sugar of the backbone is a side chain comprising 4 glucoses, the terminal one of which is succinylated. Paradoxically, the strain of *R. meliloti* on which the most comprehensive genetic studies have been conducted is not, in fact, mucoid. However, its EPS fluoresces under UV light when the bacteria are grown in the presence of the optical brightener calcofluor. The colonies fluoresce and make a "halo" which surrounds them. Thus, it is possible to obtain mutants in exopolysaccharide (*exo*) genes which are completely "dark" or which make reduced amounts of EPS ("dim") or in which the halo is not made. As shown by Leigh et al. (1985) and Long et al. (1988a), all mutants that fail to form EPS are aberrant in their nodulation of alfalfa. These mutations define several different complementation groups in a cluster on a very large plasmid; this is not the same plasmid that contains the nitrogen fixation and nodulation genes in this species (Finan et al. 1986; Hynes et al. 1986). The nodules induced by these EPS mutants fail to fix nitrogen, which is not too surprising since they contained no bacteria nor infection threads! This observation is inherently interesting since it shows that elements of the normal infection process can be "uncoupled" and means that bacteria present on the root surface can elicit certain, though aberrant, responses in the plant. Associated with nodule development is the production of nodule specific proteins (nodulins) by the plant. In the case of alfalfa inoculated with *exo* mutants, only two (of the normally found fourteen) such proteins were induced, indicating that nodule development aborted at a very early stage (Leigh et al. 1987). It was noted (Leigh et al. 1988) that mutations in the *exoB* and *exoC* also abolished the production of lipopolysaccharide (LPS), indicating some hitherto unsuspected link between the synthesis/processing of EPS and LPS. Some of the *exo* mutants which were reduced for their production of EPS were able to form nitrogen-fixing nodules but the levels of fixation were much reduced compared to the wild type.

A particularly interesting mutation was in a gene termed *exoH* (Leigh et al. 1987). The colonies of *exoH* mutants fluoresced but did not make the characteristic halo. In such *exoH* mutants the polymer was not succinylated and the lower molecular weight forms of the EPS (responsible for the halo) are not produced. As with other *exo* mutants, these *exoH* strains induced non-fixing, "empty" nodules indicating that the succinyl modification is required for the function of the molecule. Similar observations were also made by Muller et al. (1988).

There is good evidence that several of the *exo* genes specify proteins that themselves are associated with the cell surface. This was deduced from studies using fusions between the *exo* genes and the *phoA* gene of *Escherichia coli* (Long et al. 1988). *phoA* specifies alkaline phosphatase (AP), an enzyme which is active only when present in the bacterial periplasm. AP is targetted to this compartment via its N-terminal signal peptide which is cleaved by a signal peptidase during the passage of the protein through the cytoplasmic membrane. Long et al. (1988) used a defective version of *phoA* which lacks the signal peptide and so does not make active AP unless it is fused, in frame, to an expressed gene whose product itself is transported to the membrane(s) or the periplasm. Thus the sequences in the "target" protein deliver the fusion protein to the periplasm, exposing its AP part to the periplasm where it will be expressed, this being detected by appropriate staining procedures. Using this approach, Long et al.

(1988b) found that *phoA* insertions into several *exo* genes expressed AP. Also, in *R. leguminosarum*, *phoA* fusions into genes involved in EPS synthesis in that species expressed AP activity indicating that in this species too, the protein products of genes that specify the synthesis of EPS are located in the membrane(s) or the periplasm (JW Latchford and AWBJ, unpublished observations].

It is apparent that the production of EPS in rhizobia is regulated; the amounts of the polymer vary considerably depending on the growth media (for example, N limitation can cause copious amounts of EPS to be made).

Further, bacteroids make little or no EPS (Tully and Terry 1985) suggesting some type of control mechanism. Recently, we (AWBJ and JW Latchford, unpublished observations) found, using transcriptional fusions between *exo* (*pss*) genes of *R. leguminosarum* and *uidA*(gus) and histochemical staining, that transcription of at least one *exo* (also known as *pss*-polysaccharide synthesis) gene in the bacteroids was below the limits of detection. Doherty et al. (1988) identified two chromosomally located genes, *exoR* and *exoS*, mutations in which caused an overproduction of the normal EPS by *R. meliloti*. It was confirmed that in such mutants, transcription of the *exo* genes was enhanced, indicating that *exoS* and *exoR* encode repressors of *exo* gene expression. Strikingly, the *exoR* mutants were very poor at fixing nitrogen in alfalfa nodules implying that deregulation of EPS production is detrimental for the infection process.

A remarkable discovery from Walker's laboratory is that the strain of *R. meliloti* in which his genetic studies were conducted has the potential ability to make two exopolysaccharide molecules which bear little resemblance to each other (Glazebrook and Walker 1989). These authors noted that, at low frequency, *exo* mutant strains gave rise to mucoid colonies; significantly, though, the polysaccharide made by these "revertants" did not fluoresce with calcofluor, indicating that they had a different chemical composition to that of the "original" EPS, termed EPS 1. It transpired that these mucoid colonies made a novel EPS (EPS 2) as a result of a mutation in a regulatory gene, *expR*, which presumably repressed transcription of a normally silent set of alternative genes, termed *exp*, that are involved in the synthesis of this new polymer. The "structural" *exp* genes are on the same large plasmid as the *exo* genes (Glazebrook and Walker 1989); *expR* is on the chromosome. Intriguingly, *expR* mutants regain the ability to fix nitrogen in alfalfa nodules; this shows that the defect caused by the loss of EPS1 can be rectified by the structurally unrelated EPS2. This raises some interesting questions concerning the role of EPS in nodulation; the fact that two different molecules can functionally replace each other suggests that the specific structure of the polymer is not very important (though the symbiotic defect of the *exoH* mutants argues against that) or that there are different "receptors" in the plant which have evolved to recognise different polymers in the appropriate *Rhizobium* species.

4.1.2 EPS Production in Other Rhizobia

Studies on EPS production in species other than *R. meliloti* have shown that its presence can also be (though not always) important for nodulation of other legumes.

For example, Sanders et al. (1981) found that certain EPS⁻ mutants of *R. leguminosarum* bv. *viciae* failed to nodulate peas; however, other such mutants appeared to be unaffected in their symbiotic phenotype. The reason for such differences is unknown; presumably it reflects the step in the biochemical pathway towards EPS synthesis which is blocked in individual mutants and perhaps suggests that if the defect is at a late stage, precursors of the mature EPS can have biological activity. In connection with this idea, it may be relevant to note that Djordjevic et al. (1987) found that addition of intact EPS or even the repeat units to the rhizosphere of clover plants inoculated with exo mutants of *R. leguminosarum* bv. trifolii, restored the Nod⁻ defect of these mutants. However we (AWBJ and JW Latchford; unpublished) and GC Walker (personal communication) were unable to rescue the symbiotic defects of *exo* mutants of *R. leguminosarum* bv. *viciae* or *R. meliloti* on their respective hosts by addition of the EPS's of these species to the legume roots; the reasons for these differences are not known.

A striking feature of the behaviour of mutants defective in EPS production is that the symbiotic phenotype can vary depending on the host legume. Thus, Chen et al. (1985) found that non-mucoid mutants of a strain with a wide host-range on tropical legumes were aberrant (inducing tumour-like structures) on certain hosts but were apparently unaffected on other species of legume. Borthakur et al. (1986) isolated *pss* (equivalent to *exo*) mutations in near-isogenic strains of *R. leguminosarum* which differed only in the identity of their sym plasmids. One of the strains [*R. leguminosarum* bv. *viciae*] had a sym plasmid that specified the ability to nodulate peas and the other contained pRP2JI, a sym plasmid that determines the ability of the strain *R. leguminosarum* bv. *phaseoli* to nodulate *Phaseolus* beans. Strikingly, the EPS⁻ mutant of the former was unable to nodulate peas but in *R. leguminosarum* bv. *phaseoli*, nodulation of *Phaseolus* beans by the non-mucoid strain appeared to be normal. Clearly, then, the particular requirements for exopolysaccharide vary in diffferent host-microbe interactions. Similar observations to those of Borthakur et al. (1986) were made by Diebold and Noel (1989).

Although the sym plasmid in *R. leguminosarum* does not appear to be important in the production of EPS (strains cured of such plasmids make normal amounts of the polymer) there is at least one region of a sym plasmid of *R. leguminosarum* bv. *phaseoli* which affect its production. Borthakur et al. (1985) noted that a gene in such a plasmid, close to other "symbiotic" genes, involved in nodulation and nitrogen fixation, caused strains to be non-mucoid and Nod⁻ when cloned in multi-copy plasmids. This gene, termed psi, specifies a small polypeptide (Mr 10 kDa) with an extremely hydrophobic N-terminus suggesting that it may be located in the membrane (Borthakur et al. 1987), a suggestion that has been confirmed by using *psi-phoA* fusions (AWBJ and JW Latchford, unpublished observations). It was further noted that mutants in *psi* caused the bacteria to form non-fixing, empty nodules reminiscent in morphology to those induced by *exo* mutants of *R. meliloti* (Borthakur et al. 1985; JW Latchford, AWBJ, unpublished observations). It was suggested that the role of *psi* was somehow to shut down EPS production in the nodule and that it is normally expressed by bacteroids but not by free living cells; however, by cloning it, its effects would become "uncoupled" and its effects on EPS production would be seen *in vitro* (Borthakur et al. 1985). We have since shown, using *psi-gus* fusions, that *psi* is indeed expressed at high levels by bacteroids but only at very low level in free living bacteria (JW Latchford and AWBJ, unpublished observations). It follows, therefore, that in *psi* mutant strains EPS production might continue during nodule development and this might be detrimental to nitrogen fixation (cf. the defect of *exoR* mutants of *R. meliloti*-see above). A gene similar in function and, in part, in sequence, has recently been identified in another fast-growing, wide host-range *Rhizobium* (Gray et al. 1990).

It was found that the effects of multi-copy *psi* on EPS production and nodulation could be overcome if the strain also contained the cloned *pss* genes indicating that there may be some interaction between *psi* and *pss* (Borthakur et al. 1988). It was found that *psi* did not affect transcription of *pss* nor vice versa (Borthakur et al. 1988). However, sequence analysis of one of the *pss* genes (*pssA*) of *R. leguminosarum* revealed that it too specified a protein with a hydrophobic domain at its N-terminus and the prediction (Borthakur et al. 1988) that it was located in the membrane has since been substantiated using *pss-phoA* fusions (unpublished). It is possible, therefore that the products of the two genes interact in the membrane and can make normal amounts of EPS. However the precise mechanisms involved in the interaction between the products of the *psi* and *pss* genes remain to be determined.

It is clear from the above that EPS is important for the symbiosis, but many questions still remain. Why is it that EPS production is required for the nodulation/nitrogen fixation of some legumes but not others? How is it that very different polymers in *R. meliloti* can "substitute" for each other? Why is it the case that in *R. meliloti*, *exo* mutants induce aberrant nodules, whereas in *R. leguminosarum* bv. *viciae* certain *exo* mutants fail to induce nodules at all? These questions reflect our ignorance of the precise role of the acidic EPS in the nodulation process.

4.2 $\beta[1 \rightarrow 2]$-Glucan

Strains of *Rhizobium* and *Agrobacterium* make an unusual cyclic $\beta(1 \rightarrow 2)$-glucan (York et al. 1980; Zevenhuizen, Scholten-Koerselman 1979) which is present both in the periplasm and which is exported from the bacteria. It appears to have a role in allowing the bacteria to adapt to changes in external osmotic conditions and also in their motility (Dylan et al. 1990a), but of more relevance to the subject of this review is the fact that mutants which fail to make this molecule are also defective in normal nodule induction. It was found that mutations in one of two genes (*ndvA* and *ndvB* {nodule development}) abolish the ability of *R. meliloti* to induce nitrogen-fixing nodules on alfalfa (Dylan et al. 1986). Indeed, these non-fixing nodules contained no bacteria nor bacteroids, showing that as with some mutants defective in EPS production, the block to nodulation may occur at an early stage in the infection process but that the bacteria outside the roots have the capacity to "bypass" part of the block and so elicit morphological changes in the plants. Interestingly, pseudorevertants, isolated from the occasional pink alfalfa nodule following inoculation with a *ndv* mutant strain of *R. meliloti*, though improved for their symbiotic nitrogen fixation ability, were not restored for their ability to make the glucan, suggesting that the loss of this molecule *per se* may not be the cause of the symbiotic defect (Dylan et al. 1990b).

The *ndv* genes have functional and structural homologues in *Agrobacterium tumefaciens*; in fact the way in which the *Rhizobium ndv* genes were first identified exploited this similarity. Mutants of *Agrobacterium* had been isolated which failed to form tumours and it was shown that these mutant genes were not on the Ti plasmid (Douglas et al. 1985). Recombinant plasmids from a gene bank of *R. meliloti* DNA which restored tumourigenicity to thesee mutants were isolated and, following mutagenesis of the cloned *Rhizobium* DNA, *ndv* mutants were isolated (Dyland et al. 1986). Subsequent comparisons of the *ndv* and *chv* genes showed that they were indeed similar to each other. Analysis of the deduced sequence of the *ndvA* and *chvA* gene products showed that they were related to a family of membrane-associated proteins involved in the import or export of a variety of molecules suggesting that *ChvA* is required for the export of the glucans (Stanfield et al. 1988; Cangelosi et al. 1989; de Iannino and Ugalde 1989). This is supported by the observation that *ChvA* mutants still make the $\beta[1 \rightarrow 2]$-glucan but that it is confined to the cytoplasm; in contrast, mutations in *ndvB* or *chvB* do not make the molecule, indicating that these are the structutal genes required for its synthesis.

It was shown that *chv* mutants of *A. tumefaciens* failed to bind tightly to plant cells (Douglas et al. 1982); more recently, Dylan et al. (1990b) found, similarly, that *ndv* mutants of *R. meliloti* were severely impaired in their binding to the roots of alfalfa seedlings.

It is clear from these studies that mutations which abolish the synthesis of the $\beta[1 \rightarrow 2]$-glucan are severely compromised in their ability to induce normal nodules. However, the molecular basis of this property is not clear nor indeed has it been proved that it is the loss of the glucan itself that is responsible rather than one or other of the pleiotropic defects that are associated with *ndv* mutant strains. If it is indeed the $\beta[1 \rightarrow 2]$-glucans that are responsible, it is conceivable that they act as a protectant against plant defence reactions; one speculative thought stems from the fact that naturally occurring $\beta[1 \rightarrow 2]$-glucans are being considered from an industrial point of view as candidates as "chemical carriers" that would substitute for commercial agents such as cyclodextrans; is it possible that during the infection these glucans act to "mop up" (or concentrate) low-molecular-weight signal molecules or defence compounds that are generated by the plant?

It should be noted that these studies have been conducted on one species of *Rhizobium*, namely *R. meliloti* and the symbiotic role[s] of the corresponding glucans in other rhizobia may not be the same as in that system. In fact, it was shown recently (Miller et al. 1990) that the structure of the cyclic oligosaccharides in *Bradyrhizobium* differs from that of *Agrobacterium* and *R. meliloti*.

4.3 Lipopolysaccharides

As is the case with other Gram-negative bacteria, rhizobial cells make a series of complex polymers, lipopolysaccharides (LPS), at their cell surfaces. These molecules possess a lipid moiety, lipid A, which is anchored to the bacterial membrane and which is conserved among rhizobia of a given species e.g. *R. leguminosarum* (Carlson et al. 1988). Attached to this core is the outer *O*-antigen chain which is variable between strains and even within a strain owing to the variable number of sugar residues that are present in the chain (Carlson 1984).

Genetic studies have shown that this polymer, too, is required for normal nodulation and nitrogen fixation, Mutants of *R. leguminosarum* bv.'s *viciae* and *phaseoli* which fail to make LPS are unable to induce normal nodules (Brink et al. 1990; Noel et al. 1986; Priefer 1989). In this species, most mutants that fail to make LPS induce nodules in which the infection aborts at an early stage such that they contain few, if any, bacteria, The genes involved in LPS production in this species appear to be clustered on the chromosome (Cava et al. 1989; Preifer 1989; Brewin, personal communication). Using monoclonal antibodies, it has been found that bacteroids make particular LPS molecules that are not present in free-living bacteria grown under normal growth conditions (VandenBosch et al. 1989) and that this expression of the antigen within the nodule appears to be due either to the low O_2 therein and/or the particular carbon source since Kennenberg and Brewin (1989) showed that the particular form of LPS that was recognised by the monoclonal antibody and antibody and which was present in bacteroids was also produced by free-living bacteria providing that they were grown at low oxygen tension or on carbon sources, such as glucose, which resulted in a reduction of the pH of the culture medium. The differential expression o the epitope that is recognised by the monoclonal antibody identified by VandenBosch et al. (1989) suggested that the expression of the gene(s) responsible for its synthesis are regulated. This supposition was substantiated by recent observations by Wood et al. (1989) who isolated mutants of *R. leguminosarum* which constitutively expressed the LPS that was normally expressed only in the bacteroids or in free-living bacteria grown at low pH or low pO_2. These mutations would appear to be in a gene that normally represses expression of the corresponding *lps* genes.

As was noted, though, for the differences in effects of mutations that abolish EPS production in different rhizobial strains, so it has also been found that, despite the obvious importance of LPS for the induction of nitrogen-fixing nodules by strains of R. *leguminosarum*, Clover et al. (1989) reported that mutants defective in the synthesis of LPS were unaffected with regard to symbiotic nitrogen fixation on alfalfa.

5 Conclusions

From the data provided above, it is clear that various macromolecules on the surface of rhizobial cells are required for nodulation and/or nitrogen fixation. Analysis of mutants defective in the synthesis of such entities have revealed important clues to aid our understanding of the infection process, but the picture is by no means complete. For example, with the exception of *nodD*, the precise biochemical functions of individual *nod* genes has yet to be established although some "hints" are available from analysis of their sequences. Similarly, although surface polysaccharides are clearly important for the formation of nitrogen-fixing nodules, their exact role in the infection process is not clear.

In this article, I have dealt exclusively with the role of the bacterial cell surface in the infection process. What though, of the legume's role in the process?

Although it is known that legumes make nodule-specific proteins, definitive genetic evidence for their role in the symbiosis has, for the most part, been lacking. An exception, though, concerns a recently reported finding by Diaz et al. (1989). For many years it has been proposed in certain quarters that the particular lectins (proteins made by higher organisms which recognise sugar moieties) are important in the nodulation process. Using transgenic technology, Diaz et al. (1989) introduced the gene that specifies the production of pea lectin into clover; remarkably, the transformed clover plants were nodulated by a strain of R. *leguminosarum* bv. *viciae* (which normally nodulates peas but not clover). Thus, the inter-generic transfer of a single plant gene altered the host-range of the host legume. Since it is likely that plant lectins recognise a surface or exported component (as yet unknown) of the rhizobial nodulation mechanism, I close this chapter in citing an exciting example in which genetic studies with the host legume may form the start of a period in which the community will begin to engage in programmes using the powers of molecular genetics to identify factors made by this member of the symbiosis and which are involved in cellular recognition.

6 References

Borthakur D, Barber CE, Lamb JW, Daniels MJ, Downie JA, Johnston AWB (1986) A mutation that blocks exopolysaccharide synthesis prevents nodulation of peas by *Rhizobium leguminosarum* but not of beans by *Rhizobium phaseoli* and is corrected by cloned DNA from the phytopathogen *Xanthomonas*. Mol Gen Genet 203: 320

Borthakur D, Barker RF, Latchford JW, Rossen L, Johnston AWB (1988) Analysis of *pss* genes of *Rhizobium leguminosarum* required for exopolysaccharide production and nodulation of peas: their primary structure and their interaction with *psi* and other nodulation genes. Mol Gen Genet 213: 155

Borthakur D, Downie JA, Johnston AWB, Lamb JW (1985) *psi*, a plasmid-linked gene that inhibits exopolysaccharide and which is required for symbiotic nitrogen fixation. Mol Gen Genet 200: 278

Borthakur D, Johnston AWB (1987) Sequence of *psi*, a gene on the symbiotic plasmid of *Rhizobium phaseoli* which inhibits exopolysaccharide synthesis and nodulation and demonstration that its transcription is inhibited by *psr*, another gene on the symbiotic plasmid. Mol Gen Genet 207: 155

Brink BA, Miller J, Carlson RW, Noel KD (1990) Expression of *Rhizobium leguminosarum* CFN42 genes for lipopolysaccharide in strains derived from different *R. leguminosarum* field isolates. J Bacteriol 172: 548

Cangelosi GA, Martinetti G, Leigh J, Lee CC, Theines C, Nester EW (1989) Role of *Agrobacterium tumefaciens* ChvA protein in export of β-1, 2-glucan. J Bacteriol 171: 1609

Carlson RW (1984) Heterogeneity of *Rhizobium* polysaccharides. J Bacteriol 158: 1012

Carlson RW, Hollingsworth BL, Dazzo FB (1988) A core oligosaccharide component from the lipopolysaccharide of *Rhizobium trifolii* ANU 843. Carbohydr Res 176: 127

Cava JR, Elias PM, Turowski DA, Noel KD (1989) *Rhizobium leguminosarum* CFN42 genetic regions encoding lipopolysaccharide structures essential for complete nodule development on bean plants. J Bacteriol 171: 8

Chen H, Batley M, Redmond J, Rolfe BG (1985) Alteration of the effective nodulation properties of a fast growing broad host range *Rhizobium* due to changes in exopolysaccharide synthesis. J Plant Physiol 120: 331

Clover RH, Kieber J, Signer ER (1989) Lipopolysaccharide mutants of *Rhizobium meliloti* are not defective in symbiosis. J Bacteriol 171: 4054

Davis EO, Johnston AWB (1990a) Analysis of three nodD genes in *Rhizobium leguminosarum* biovar *phaseoli*; nodDl is preceded by *nolE*, a gene whose product is secreted from the cytoplasm. Mol Microbiol 4: 921

Davis EO, Johnston AWB (1990b) Regulatory functions of the three *nodD* genes of *Rhizobium leguminosarum* bv phaseoli. Mol Microbiol 4: 933

Davis EO, Evans IJ, Johnston AWB (1988) Identification of *nodX*, a gene that allows *Rhizobium leguminosarum* biovar *viciae* strain TOM to nodulate afghanistan peas. Mol Gen Genet 212: 531

de Iannino NI, Ugalde RA (1989) Biochemical characterisation of avirulent *Agrobacterium tumefaciens chvA* mutants; synthesis and excretion of β[1 → 2]-glucan. J Bacteriol 171: 2842

de Maagd RA, Wijffelman CA, Pees E, Lugtenberg BJJ, nodO, a new nod gene of the *Rhizobium leguminosarum* biovar *viciae* sym plasmid pRL1jI, encodes a secreted protein. J Bacteriol 170: 4424

Diaz CL, Melchers LS, Hooykaas PJJ, Lugtenberg BJJ, Kijine JW (1989) Root lectin as a determinant of host-plant specificity in the *Rhizobium*-legume symbiosis. Nature 338: 579

Diebold R, Noel KD (1989) *Rhizobium leguminosarum* exopolysaccharide mutants: biochemical and genetic analyses and symbiotic behaviour on the three hosts. J Bacteriol 171: 4821

Djordjevic, SP, Chen H, Batley M, Redmond JW, Rolfe BJ (1987) Nitrogen fixation ability of exopolysaccharide synthesis mutants of *Rhizobium* sp. strain NGR234 and *Rhizobium trifolii* is restored by the addition of homologous polysaccharides. J Bacteriol 169: 53

Doherty D, Leigh JA, Glazebrook J, Walker GC (1988) *Rhizobium meliloti* mutants that overproduce the *R. meliloti* acidic calcofluor-binding exopolysaccharide. J Bacteriol 170: 4249

Douglas, CJ, Halperin W, Nester EW (1982) *Agrobacterium tumefaciens* mutants affected in attachment to plant cells. J Bacteriol 152: 1265

Douglas CJ, Stalonei RJ, Rubin RA, Nester EW (1985) Identification and genetic analysis of an *Agrobacterium tumefaciens* chromosomal virulence region. J Bacteriol 161: 850

Downie JA, Johnston AWB (1988) Nodulation of legumes by *Rhizobium*. Plant Cell and Env 11: 403

Dylan T, Helinski DR Ditta GS (1990a) Hypoosmotic adaptation in *Rhizobium meliloti* requires $\beta[1 \rightarrow 2]$-glucan. J Bacteriol 1400

Dylan T, Ielpi L, Stanfield S, Kashyap L, Douglas C, Yanofsky M, Nester E, Helinski DR, Ditta G (1986) *Rhizobium meliloti* genes required for nodule development are related to chromosomal virulence genes in *Agrobacterium tumefaciens*. Proc Natl Acad Sci USA 83: 4403

Dylan T, Nagpal P, Helinski DR Ditta GS (1990b) Symbiotic pseudorevertants of *Rhizobium meliloti* *ndv* mutants. J Bacteriol 172: 1409

Economou A, Hamilton WDO, Johnston AWB, Downie JA (1990) The *Rhizobium* nodulation gene *nodO* encodes a Ca^{2+}-binding protein that is exported without *N*-terminal cleavage and is homologous to haemolysin and related proteins. EMBO J 9: 349

Evans IJ, Downie JA (1986) The *nodI* gene product of *Rhizobium leguminosarum* is closely related to ATP-binding bacterial transport proteins; nucleotide sequence analysis of *nodI* and *nodJ*. Gene 43: 95

Finan TM, Hartweig E, LeMieux K, Bergman K, Walker GC (1986) A second symbiotic megaplasmid in *Rhizobium meliloti* encodes exopolysaccharide and thiamine synthesis genes. J Bacteriol 167: 66

Gotz R, Evans IJ, Downie JA, Johnston AWB (1985) Identification of the host-range DNA which allows *Rhizobium leguminosarum* strain TOM to nodulate cv. Afghanistan peas Mol Gen Genet 201: 296

Glazebrook J, Walker GC (1989) A novel exopolysaccharide can function in place of the calcofluor-binding exopolysaccharide in nodulation of alfalfa by *Rhizobium meliloti*. Cell 56: 661

Gray JX, Djordjevic MA, Rolfe BG (1990) Two genes that regulate exopolysaccharide production in *Rhizobium* sp. strain NGR234: DNA sequences and resultant phenotypes. J Bacteriol 172: 193

Hong GF, Burn JE Johnston AWB (1987) Evidence that DNA involved in the expression of nodulation (*nod*) genes in *Rhizobium* binds to the product of the regulatory gene *nodD*. Nucl Acid Res 15: 9677

Hynes MF, Simon R, Niehaus K, Labes M, Puhler A (1986) The two megaplasmids of *Rhizobium meliloti* are involved in the effective nodulation of alfalfa. Mol Gen Genet 202: 356

Jiang J, Gu B, Albright LM, Nixon BT (1989) Conservation between coding and regulatory elements of *Rhizobium meliloti* and *Rhizobium leguminosarum dct* genes. J Bacteriol 171: 5244

John M, Schmidt J, Weineke U, Kondorosi E, Kondorosi A, Schell J (1985) Expression of the nodulation gene *nodC* of *Rhizobium meliloti* in *Escherichia coli*; role of the *nodC* product in nodulation. EMBO J 4: 2425

John M, Schmidt J, Weineke U, Krussman HD, Schell J (1988) Transmembrane orientation and receptor-like structure of the *Rhizobium meliloti* common nodulation protein NodC. EMBO J 7: 583

Johnston D, Roth LE, Stacey G (1989) Immunogold localization of the *NodC* and *NodA* proteins J Bacteriol 171: 4583

Kannenberg E, Brewin NJ (1989) Expression of a cell surface antigen from *Rhizobium leguminosarum* 3841 is regulated by oxygen and pH. J Bacteriol 171: 4543

Leigh JA, Lee CC (1988) Characterizaton of polysaccharides of *Rhizobium meliloti exo* mutants that from ineffective nodules. J Bacteriol 170: 3327

Leigh JA, Reed JW, Hanks JF, Hirsch AM, Walker GC (1987) *Rhizobium meliloti* mutants that fail to succinylate their calcofluor-binding exopolysaccharide are defective in nodule invasion. Cell 51: 579

Leigh JA, Signer ER, Walker GC (1985) Exopolysaccharide-deficient mutants of *Rhizobium meliloti* that form ineffective nodules. Proc Natl Acad Sci USA 82: 6231

Long S, Reed JW, Himawan J, Walker GC (1988a) Genetic analysis of a cluster of genes required for synthesis of the calcofluor-binding exopolysaccharide of Rhizobium meliloti. J Bacteriol 170: 4239

Long S, McClune S, Walker GC (1988b) Symbiotic loci of *Rhizobium meliloti* identified by random *TnphoA* mutagenesis. J Bacteriol 170: 4257

Long SR (1989) *Rhizobium*-legume symbiosis; life together in the underground. Cell 56: 203

Miller KJ, Gore RS, Johnston R, Benesi R, Reinhold VN (1990) Cell-associated oligosaccharides of *Bradyrhizobium* spp. J Bacteriol 172: 136

Muller P, Hynes M, Kapp D, Neihaus K, Puhler A (1988) Two classes of *Rhizobium meliloti* infection mutants differ in exopolysaccharide production and in co-inoculation properties with nodulation mutants. Mol Gen Genet 211: 17

Nadler, KD, Johnston AWB, Chen JW, John TR (1990) A *Rhizobium leguminosarum* mutant defective in symbiotic iron acquisition. J Bacteriol 172: 670

Noel KD, Vandenbosch KA, Kulpaca B (1986) Mutations in *Rhizobium phaseoli* that lead to arrested development of infection threads. J Bacteriol 168: 1392

Priefer UB (1989) Genes involved in lipopolysaccharide production and symbiosis are clustered on the chromosome of *Rhizobium leguminosarum* biovar *viciae* VF39. J Bacteriol 171: 6161

Recourt K, van Brussel AAN, Driesen AJM, Lugtenberg BJJ (1988) Accumulationn of a *nod* gene inducer, the flavonoid naringenin, in the cytoplasmic membrane of *Rhizobium leguminosarum* bv *ivciae* is caused by the pH-dependent hydrophobicity of naringenin, J Bacteriol 171: 4370

Ronson CW, Lyttleton P, Robertson JG (1981) C_4 – dicarboxylate transport mutants of *Rhizobium trifolii* form ineffective nodules on *Trifolium repens*. Proc Natl Acad Sci Usa 78: 4284

Sanders R, Raleigh E, Signer ER (1981) Lack of correlation between exopolysaccharide production and nodulation ability in *Rhizobium*. Nature 292: 148

Schlaman HRM, Spaink HP, Okker RJH, Lugtenberg BJJ (1989) Subcellular localization of the *nodD* gene product in *Rhizobium leguminosarum*. J Bacteriol 171: 4686

Spaink HP, Roest H, Vierboom M, Okker RJH, Wijffelman CA, Lugtenberg BJJ (1989) The *Rhizobium nodE* protein as a major determinant of host specifity. In: Lugtenberg BJJ (ed) Signal molecules in plants and plant-microbe interactions. Springer-Verlag, Berlin, p 359

Stanfield SW, Ielpi L, O'Brochta D, Helinksi DR, Ditta GS (1988) The *ndvA* gene product of *Rhizobium meliloti* is required for $\beta[1 \rightarrow 2]$-glucan production and has homology to the ATP-binding export protein HlyB. J Bacteriol 170: 3523

Tully RE, Terry M (1985) Decreased exopolysaccharide synthesis by anaerobic and symbiotic cells of *Bradyrhizobium*. Plant Physiol 79: 445

VandenBosch KA, Brewin NJ, Kannenberg E (1989) Developmental regulation of a *Rhizobium* cell surface antigen during growth of pea root nodules. J Bacteriol 171: 4537

Wood EA, Butcher GW, Brewin NJ, Kannenberg EL (1989) Genetic derepression of a developmentally regulated lipopolysaccharide antigen from *Rhizobium leguminosarum* 3841. J Bacteriol 171: 4549

York WS, McNeil M, Darvill AG, Albbersheim P (1980) Beta-2-linked glucans secreted by fast-growing species of *Rhizobium*. J Bacteriol 142: 243

Zevenhuizen LPTM, Scholten-Koerselman HJ (1979) Surface carbohydrates of *Rhizobium*. 1β-1,2-glucans, Antoine van Leeuwenhoek J Microbiol 45: 165

CHAPTER 15

Rhizobium Meliloti Symbiotic Nitrogen Fixation: Identification of *NIF* and *FIX* Genes; The *NIF FIX* Cascade Regulatory Pathway

Pierre Boistard

The Nitrogen Fixation
and its Research in China
Editor Guo-fan Hong
© Springer-Verlag Berlin Heidelberg 1992

1 Introduction

Rhizobium meliloti fixes nitrogen in symbiotic association with a limited number of legumes among which alfalfa (*Medicago sativa*). During the past 15 years it has been the subject of intense genetic studies with the result that it is now one of the best-known nitrogen-fixing organisms. It has largely benefitted from the recent progress in bacterial genetics and molecular biology and conversely some of the most significant technological advances in bacterial genetics have been achieved using *R. meliloti* (or other *Rhizobium* species) as a model system. Devising new tools for genetic studies of symbiotic nitrogen fixation was necessary because mutant phenotypes cannot be observed easily. Most *Rhizobium* species fix nitrogen only in the symbiotic state. Nitrogen fixation takes place in a specialized plant organ, the nodule which constitutes an adapted ecological niche for the nitrogen-fixing differentiated bacteroids (see the review by Long 1989). Therefore the screening of bacterial mutants deficient for nitrogen fixation needs time-consuming plant inoculation. Transposon mutagenesis was a first significant advance because it allows a positive screening of the mutagenic event (Beringer et al. 1978). Reverse genetics allows to identify the phenotype encoded by a given DNA sequence, for example one which shows homology with an already identified gene of another organism (Ruvkun and Ausubel 1981).

This paper is devoted to a record of the main achievements in the identification and characterization of *R. meliloti* genes needed for symbiotic nitrogen fixation and in the study of their regulation

2 The Use of Reverse Genetics for the Identification of *nifHDK* Genes of *R. meliloti*

Central in the process of nitrogen fixation is the nitrogenase complex which catalyses the reduction of gaseous dinitrogen into ammonia.

The nitrogenase complex is made up of two components. Component one or nitrogenase sensu stricto is a tetramer of two different polypeptides and carries the iron–molybdenum cofactor endowed with the catalytic properties of the enzyme. Component two, a homodimer, is the nitrogenase reductase responsible for the transfer of electrons to component one (for a review see Dixon 1984). Biochemical studies had revealed that both component one and component two were highly conserved proteins, were they isolated from any one of the prokaryotic nitrogen-fixing organism (see for example Ruvkun and Ausubel 1980).

Genetic studies in *Klebsiella pneumoniae*, an enteric bacterium which reduces nitrogen to ammonia in anaerobic cultures deprived of combined nitrogen, led to the identification of the structural genes for the polypeptides of components one and two, *nifDK* and *H*, respectively (Dixon 1984).

Because the gene products were highly conserved, it was hypothesized that the coding sequences were also conserved. Using *nif HD* probes from *K. pneumoniae*, it was possible by Southern hybridization technique to detect homologous sequences in the *R. meliloti* genome (Ruvkun and Ausubel 1980).

This led to the cloning of the restriction fragments hybridizing to the *K. pneumoniae nif* probe.

Reverse genetics, that is, mutagenesis of the cloned sequence in *Escherichia coli* followed by marker exchange between the mutagenized cloned fragment and its wild-type counterpart in *R. meliloti*, was used to obtain transposon Tn5 insertions in the *R. meliloti* sequences homologous to *K. pneumoniae nif* genes. The resulting insertion mutants were unable to fix nitrogen in symbiosis (Fix⁻ phenotype). This was the demonstration that the *R. meliloti* sequences were encoding products needed for symbiotic nitrogen fixation. Amino acid sequences deduced from partial DNA sequencing confirmed that *nif H* and *nif D* genes were present in this region together with a *nif* gene whose DNA hybridized with *K. pneumoniae nif K* (Ruvkun et al. 1982).

3 The *nifHDK* Genes of *R. meliloti* are Part of a Cluster of Genes Required for Symbiotic Nitrogen Fixation

A second working hypothesis which proved to be fruitful was that, like in *K. pneumoniae*, genes involved in nitrogen fixation were clustered in *R. meliloti*.

In *K. pneumoniae* 20 genes required for nitrogen fixation are clustered on the chromosome (Dixon 1984; Arnold et al. 1988). In addition to the *nifHDK* genes which code for the structural components of nitrogenase, several genes are involved in the processing of each component. Among those, *nif E, N* and *B* are required for the synthesis and insertion of the iron-molybdenum cofactor of component one. Others code for electron transporters and *nif A* and *nif L* are regulatory genes. The NifA protein is a transcriptional activator which interacts with the σ^{54}RNA polymerase holoenzyme (reviewed in Gussin et al. 1986). Promoters of genes transcribed by the σ^{54}RNA polymerase are characterized by a $-24-12$ consensus sequence different from the -35 and -10 consensus sequences characteristic of the promoters recognized by σ^{70} (reviewed in Thöny and Hennecke 1989).

In addition to this $-24-12$ sequence, *nif* promoters generally contain upstream activating sequences at about -100 which bind the transcriptional activator NifA (Gussin et al. 1986). The transcriptional activity of NifA is inhibited by the NifL protein in the presence of oxygen or of combined nitrogen (Dixon 1984).

Assuming that similarly to *K. pneumoniae, nif* genes are clustered in *R. meliloti*, several groups have used reverse genetics to identify nitrogen fixation genes in the vicinity of *nifHDK* genes by transposon mutagenesis of DNA

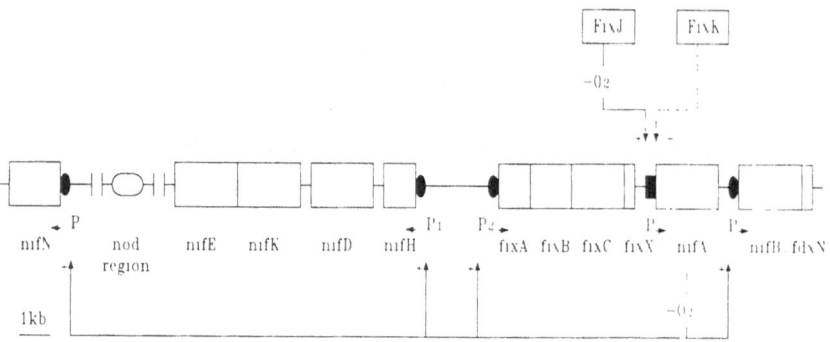

Fig. 1. The *R. meliloti nif* cluster. The *open rectangles* represent the different ORF's. The *filled symbols* represent the promoters upstream of each transcription unit. The *filled rectangle* means that the *nif A* promoter has a structure different from that of the other *nif* promoters. The *small arrows* indicate the direction of transcription. The *dotted line* means that it is not known whether FixK mediated negative regulation of *nif A* expression is by a direct interaction with the *nif A* promoter

carried by overlapping cosmid clones. Amino acid sequence comparisons between the products of these newly identified *R. meliloti* genes and *K. pneumoniae* Nif proteins allowed the identification of *nifE*, *nifN*, *nifA* and *nifB* homologs (Fig. 1).

One remarkable feature of this cluster is that it is interrupted by a *nod* cluster which carries genes needed for bacterial infection and nodule morphogenesis (Long 1989). In addition to genes which code for proteins homologous to *K. pneumoniae* Nif proteins, the *nif* cluster carries *fixABCX* genes whose homologs have been found in *Azotobacter vinelandii*. Fix X is highly homologous to an *Azotobacter* ferredoxin (Earl et al. 1987). In *Bradyrhizobium japonicum* and *Azorhizobium caulinodans* these genes are required for *ex planta* nitrogen fixation (Kaminski et al. 1988; Gubler and Hennecke 1986). However, their precise function has not been established neither has a function been ascribed to the ferredoxin coding *fdxN* gene located downstream of *nifB* except that it is needed for nitrogen fixation (summarized in Masepohl et al. 1989).

As a consequence of the discovery of *fixABCX*, it is convenient to distinguish *R. meliloti nif* genes which are homologous to *K. pneumoniae nif* genes from *fix* genes which are required for nitrogen fixation but show no homology to *K. pneumoniae* genes. However, it is remarkable to note that in spite of these differences concerning the gene products, a striking similarity between the *K. pneumoniae nif* cluster and the *R. meliloti nif* cluster concerns their regulation. In both organisms the various operons which constitute the *nif* cluster require NifA for their activation (Szeto et al. 1984).

4 Most of the Characterized Nitrogen Fixation Genes (*nif* and *fix*) are Located on a Megaplasmid in *R. meliloti*

R. meliloti belongs to the family of Rhizobiaceae which also comprises the genus *Agrobacterium*.

Both agrobacteria and rhizobia are able to establish an intricate relationship with plants. Rhizobia establish a symbiotic relationship, whereas agrobacteria transform plants, which needs a very elaborate system of introduction of genetic material into the host plant. In the 1970s it was demonstrated that a large *Agrobacterium* plasmid, the Ti plasmid (Van Larebeke et al. 1974), was playing an essential role in tumour induction. For this reason several groups examined the presence of large plasmids in various *Rhizobium* species.

Using a method which had proved successful for the isolation of plasmid Ti DNA in *Agrobacterium tumefaciens*, Nuti et al. were the first to demonstrate convincingly the presence of large plasmids in *Rhizobium leguminosarum* (Nuti et al. 1977). Subsequently, they showed that *nif* sequences were present on large plasmid DNA (Nuti et al. 1979). In *R. meliloti*, similar procedures as those used for *R. leguminosarum* led to the detection of plasmids in the range of $90-200 \times 10^6$ M$_w$ (Casse et al. 1979). However, none of these plasmids carried *nif* genes. By using the more sensitive Eckhardt's procedure it was possible to identify a new class of very large plasmids which were present in all the strains tested (Banfalvi et al. 1981; Rosenberg et al. 1981). In fact, two megaplasmids were 1200-1500 kb in size and one of them was shown to carry *nif* genes and *nod* genes. For this reason this megaplasmid was called pSym (for symbiotic plasmid).

Because this plasmid was carrying *nod* genes involved at the early stages of symbiotic interactions as well as *nif* genes which are expressed in the mature nodule we made the hypothesis that pSym could also carry *fix* genes needed for symbiotic nitrogen fixation and not identified so far.

In fact, this reasoning was a generalisation of the working hypothesis that led to the identification of the *nif* cluster.

In order to examine the presence of *fix* genes on the pSym we devised a method for the cloning of large fragments of pSym, up to 300 kb, starting from the *nif* region. These cloned fragments were used for several purposes:

  (i) deletion of the corresponding pSym region;
 (ii) transposon mutagenesis by using reverse genetics on large pSym fragments;
(iii) identification of pSym sequences actively transcribed during symbiosis (David et al. 1987; Renalier et al. 1987).

A combination of these different approaches led to the identification of two new symbiotic clusters on pSym.

One cluster is located about 200 kb downstream of the *nif*H promoter. The central part of the cluster is actively expressed in the symbiotic state. It carries reiterated genes a second functional copy of which is present in a *fix* cluster 40 kb upstream of *nif*H.

For convenience we will call the fix cluster located 200 kb downstream of *nifH* the *fix* cluster VI and the *fix* cluster 40 kb upstream, the *fix* cluster VI' (Renalier et al. 1987).

5 The *fix* Cluster VI Carries Structural Genes Which Code for Products Whose Sequence Predicts a Membrane Location as well as Regulatory Genes

Complete DNA sequence analysis was performed on the whole 13 kb *fix* cluster VI. Twelve open reading frames were predicted from sequence data (Fig. 2). Complementation analysis as well as examination of the gene sequences led to the conclusion that *fixGHIS* belong to one transcription unit. Each of the predicted products contains several putative transmembrane helices. The open reading frames overlap by one base pair which is suggestive of a translational coupling. Therefore, a likely hypothesis is that the products of the four ORF's associate in precise stoichiometric ratios to form a membrane complex with coordinated biochemical functions. Comparison of the FixI protein sequence with data banks showed that FixI is homologous to P-type ATPases which use ATP derived energy to pump various cations across biological membranes. FixG contains two cystein clusters of the type found in bacterial ferredoxins: CysxxCysxxCysxxxCys (Pro). In ferredoxins such motifs are known to coordinate iron–sulfur centers. They are also found in other redox proteins such as fumarate reductase and succinate dehydrogenase (Kahn et al. 1989).

Therefore, it is tempting to speculate that *R. meliloti* bacteriods need to pump a cation in order to be able to fix nitrogen and that a redox reaction would be coupled to this transport process.

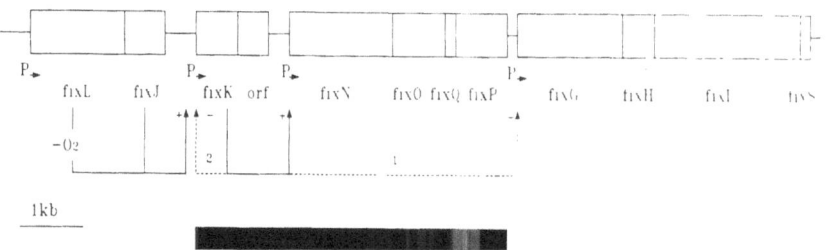

Fig. 2. The *R. meliloti fix* VI cluster. The *filled rectangle* represents the reiterated region which is highly expressed in the endosymbiotic state. P represents the promoters upstream of each transcription unit. The *interrupted arrow 1* means that the hypothetical regulatory role of FixK on *fixGHIS* expression is only inferred from the presence of an anaerobox in the region upstream of *fixG* (see text). The *interrupted arrow 2* means that the negative autoregulation of *fixK* is not necessarily mediated by a direct interaction between FixK and *pfixK*.

fixNOQP belong to the reiterated part of the cluster. Demonstration that they are *fix* genes was made possible by recombining Tn5 insertions in the downstream copy of the *fix*VI cluster with deletions of the whole upstream *fix*VI' region. They constitute one transcription unit which is actively transcribed inside the bacteroid. Protein sequence comparisons revealed a significant homology between FixP and cytochromes c. This fact together with the induction of *fixN* expression in microaerobic cultures (see below) could indicate that *fixNOQP* code for components of an electron transport chain which allows an active respiration of bacteroids in the microaerobic environment created by the plant nodule (J. Batut, M.L. Daveran and D. Kahn, manuscript in preparation).

The Reiterated Part of the *fix* Cluster VI was Shown to Carry a Regulatory Gene, *fixK*

The same technique used to demonstrate that *fixNOQP* genes are needed for nitrogen fixation, that is mutagenesis of both reiterated copies allowed to identify the *fixK* gene (Batut et al. 1989). Nucleotide sequence analysis showed that FixK was homologous to two *E. coli* regulatory proteins, Fnr and Crp. Crp is needed for activation of genes involved in sugar catabolism and its activity depends on the metabolic effector cAMP. FNR is the activator of genes which code for components of anaerobic electron transport chains such as dissimilatory nitrate reductase. Because of these homologies, we decided to look at regulatory properties of FixK. Using a *fixN-lacZ* fusion we could show that *fixK* was necessary for the expression of *fixN*. The discovery of this new regulatory gene *fixK* was consistent with the previous observation that the expression of the reiterated part of the *fix* cluster was independent of *nifA* (David et al. 1987) Therefore it was established that *fix* and *nif* genes of *R. meliloti* belong to two regulons: the *nif* cluster which comprises at least 4 transcription units, *nifHDKE*, *nifN, fixABCX, nifBfdxN* is under the control of *nifA*. The reiterated *fixNOQP* transcription unit is expressed under the control of *fixK* (Batut et al. 1989).

A further level of regulation of *nif* and *fix* gene expression was identified as a result of the discovery that both the *nifA*-dependent regulon and the *fixK*-dependent regulon are subject to a common regulation mediated by the couple of regulatory genes *fixLJ* (David et al. 1988).

Search for sequence homologies led to the conclusion that FixL and FixJ belong to the family of prokaryote two component regulatory systems. These systems are characterized by a conserved mode of signal transduction between a sensor protein and a regulator which most frequently is a transcriptional activator. According to the current model, the sensor protein is a histidine phosphokinase able to autophosphorylate and to transfer a phosphate group to an asparatate residue of the regulator. The phosphorylated form of the regulator is the active form (reviewed in Stock et al. 1989).

6 The Oxygen-Dependent Regulatory Cascade
of *nif* and *fix* Gene Expression in *R. meliloti*

One of the major questions concerning the symbiotic interaction was how the host plant triggers the expression of nitrogen fixation genes.

In the enteric *Klebsiella pneumoniae* the *nif* promoters are activated by the NifA transcriptional activator whose activity is decreased by NifL in the presence of oxygen or of combined nitrogen. The two regulatory genes *nifL* and *nifA* belong to the same operon whose transcription is activated by NtrC in response to a low ratio of glutamine to α-ketoglutarate. The transcriptional activity of NtrC is determined by its degree of phosphorylation by NtrB. At low ratios of glutamine to α-ketoglutarate NtrB phosphorylates NtrC whereas at high ratios NtrB in conjunction with the PII protein dephosphorylates NtrC (see for example the review by Gussin et al. 1986).

In *R. meliloti* an NtrC functional homolog has been identified but an *ntrC* mutant is not affected for symbiotic nitrogen fixation (Gussin et al 1986). Therefore, it was concluded that the expression of nitrogen fixation genes is not regulated by the regulatory pathway which responds to the level of combined nitrogen in the free-living bacteria.

A major breakthrough came from experiments which showed that *nifA* expression can be induced by lowering the oxygen concentration in an *R. meliloti* culture (Ditta et al. 1987).

Shortly after it was shown that the expression of the NifA independent *fixN* gene could also be induced in microaerobic bacterial cultures (David et al. 1988).

Because the expression of the *nifA* regulatory gene and the *nifA* independent *fixN* gene required the regulatory genes *fixLJ* in symbiotic condition, it was of interest to see whether the same requirement was also observed in microaerobic bacterial cultures.

The fact that it was indeed the case suggested that *fixLJ* were responsible for the oxygen dependent expression of *nifA* and *fixN*.

The identification of the respective roles of the FixL and FixJ proteins on the expression of *nifA* and of *fixK*, the activator of the *fixNOQP* operon, was possible by assaying the expression of *nifA* or *fixK-lacZ* fusions in *E. coli* strains carrying plasmids allowing the *fixL* and *fixJ* genes to be expressed under the control of different promoters. It was shown that FixJ is a transcriptional activator whose activity is modulated by FixL in response to oxygen concentration (De Philip et al. 1990). Recently it has been shown that FixL is a hemoprotein which provides a likely mechanism for the sensing of oxygen concentration (Gilles-Gonzales et al. 1991).

The fact that the two same genes *fixL* and *fixJ* were needed for symbiotic expression of *nifA* and *fixK* as well as for their microaerobic induction in bacterial cultures was a strong indication that low oxygen pressure was responsible for symbiotic expression of nitrogen fixation genes.

It has been shown that oxygen pressure is strongly reduced in the invaded part of nodule. The nodule cortical cells and endodermis constitute a strong diffusion barrier which limits oxygen concentration in the inner nodule tissues where bacteroids are thought to consume oxygen actively (Witty et al. 1986).

Therefore the major signal by which the plant triggers the expression of the nitrogen fixation genes is produced as a consequence of the plant developmental program induced by the bacterial infection.

Variation in the oxygen concentration is a major developmental signal for the synthesis of the nitrogen apparatus; this is further indicated by the fact that the *R. meliloti* NifA protein itself is oxygen sensitive (Fischer et al. 1988). This was demonstrated by measuring NifA-activated expression of *nif* promoters in *E. coli* in the presence or absence of oxygen when the *nifA* gene was expressed independently of oxygen concentration. This has been confirmed recently by in vivo experiments in which the binding of NifA to *nif* promoters was studied (Morett et al. 1991).

7 Questions and Prospects

7.1 Functions of Structural *fix* Genes

Until now, no *fix* gene has been assigned a precise function, except for the regulatory genes *fixL, J* and *K*. However *fixABCX* and *fdxN* belong to one regulon activated by NifA, which also activates the *nif* genes. This suggests that those *fix* genes could be directly involved in the functioning or synthesis of the nitrogenase complex. Furthermore, as mentioned above *fixABCX* genes have been identified in *Azorhizobium caulinodans* which is able to fix nitrogen in pure cultures. Mutants in the *fixABCX* operon have a Fix⁻ phenotype in pure culture. In addition to the fact that it is an indication that *fixABCX* play a direct role in nitrogen fixation this should facilitate the identification of their function.

Regulation of the transcription of the *fixGHIS* operon could not be studied because of the low level of expression of these genes in the endosymbiotic state. However, upstream of *fixG* it has been possible to identify an anaerobox present also upstream of the FixK-dependent *fixNOQP* operon (Colonna-Romano et al. 1990). This could mean that *fixGHIS* belong to the same regulon as *fixNOQP* and therefore could participate in the same physiological pathway or in the same adaptative response.

7.2 Role of FixK

In addition to its role as an activator of the *fixNOQP* operon and its possible involvement in the expression of *fixGHIS*, FixK is also a repressor of the transcription of the two regulatory genes *nifA* and *fixK* (Batut et al. 1989). The mechanism by which this repressor affects *nifA* and *fixK* transcription is not clear yet. Is it a direct effect on *nifA* and *fixK* regulatory sequences or could it be an indirect one, through a modulation of the transcription of the *fixLJ* operon?

Both Fnr and Crp proteins activate their target genes in response to a metabolic signal. The question then arises whether FixK needs an effector for its transcriptional activity on target genes and, if this is the case, what is the chemical nature of this effector. In this respect it is interesting to note that FnrN, a FixK homolog, has been identified in *Rhizobium leguminosarum*. When introduced into *R. meliloti, fnrN* is able to promote the microaerobic expression of *fixN* in a *fixJ* mutant. This suggests that the FnrN protein senses oxygen concentration. The FnrN protein has an N-terminal 21 amino acid stretch with several cysteine residues which is absent from FixK. A similar sequence is present in the *E. coli* oxygen sensor Fnr (Colonna-Romano et al. 1990). It is possible that originally FixK was an oxygen-sensitive Fnr-like protein needed for activation of genes involved in the microaerobic life of *R. meliloti*. One can hypothesize that oxygen control was then taken up by the FixLJ pair with the consequence that FixK expression itself and not activity is oxygen controlled.

7.3 Are Signals Other than Oxygen Concentration Involved in the Control of the Expression of *nif* and *fix* Genes?

Oxygen concentration is a developmental signal which correlates *nif* and *fix* gene expression to the nodule morphogenesis. Because nitrogen fixation is an energy-consuming process, it is very likely that its efficiency is tightly regulated according to the nitrogen needs of the plant and to the energy available.

A priori this control could be exerted either at the metabolic level or at the level of gene expression.

Evidence that the nitrogen metabolism of the plant could regulate the efficiency of symbiotic nitrogen fixation comes from experiments where plant glutamine synthetase activity was inhibited by a bacterial toxin. Glutamine synthetase is the first plant enzyme which is involved in the assimilation pathway of the ammonia produced by the endosymbiotic bacteria. The phytopathogenic bacteria *Pseudomonas syringae* pat. *tabaci* produce a toxin which is an inhibitor of glutamine synthetase. When alfalfa is inoculated with a toxin-producing strain, there is a significant increase in the dry weight of *R. meliloti* inoculated plants as well as of nitrogenase activity per nodule (Knight and Langston-Unkefer 1988).

The relationship between the flux of carbon towards the endosymbiotic bacteria and the efficiency of nitrogen fixation is illustrated by the fact that *R. meliloti dctA* mutants deficient for the transport of succinate have a Fix⁻ phenotype (Yarosh et al. 1989 and references therein). Interestingly, the *dct* genes are located on the second *R. meliloti* megaplasmid. This second megaplasmid also carries genes for the synthesis of an exopolysaccharide which is involved in the process of infection of the nodule by the symbiotic bacteria. Mutants impaired in the utilization of other carbon sources are not affected for symbiotic nitrogen fixation, an indication that succinate is a preferred carbon source for the nitrogen-fixing bacteroid. Furthermore, in symbiotic conditions mutants affected in succinate transport show a more severely decresed expression of *nif* genes compared with the expression of constitutive genes. Therefore the energy supply to bacteroids seems to influence nitrogen fixation at the level of gene expression in addition to its direct metabolic role (Birkenhead et al. 1990). The mechanism by which this genetic regulation operates remains to be elucidated.

In conclusion, a lot of information has accumulated concerning the regulatory pathway which correlates the development of the nodule with the building up of the bacterial nitrogen-fixing apparatus. Oxygen concentration appears to be the major developmental signal involved in this coupling between nodule morphogenesis and the expression of bacterial genes. However, much remains to be known about the molecular mechanisms by which this regulatory pathway operates.

Besides this developmental regulation, the physiological coupling between bacterial nitrogen fixation and the C/N host plant metabolism remains largely to be explored.

Acknowledgements. I am grateful to Michel David for helpful criticism of the manuscript and for drawing the figures. I thank Josiane Prévot for her dedicated help in the preparation of the manuscript.

Work in the author's laboratory was supported by the Biotechnology Action Programme of the European Communities.

8 References

Arnold W, Rump A, Klipp W, Priefer U, Pühler A (1988) Nucleotide sequence of a 24, 206 base-pair DNA fragment carrying the entire nitrogen fixation gene cluster of *Klebsiella pneumoniae*. J Mol Biol 203: 715

Banfalvi Z, Sakanyan V, Koncz C, Kiss A, Dusha I, Kondorosi A (1981) Location of nodulation and nitrogenase fixation genes on a high molecular weight plasmid of *Rhizobium meliloti*. Mol Gen Genet 184: 318

Batut J, Daveran-Mingot ML, David M, Jacobs J, Garnerone AM, Kahn D (1989) fixK, a gene homologous with *fnr* and *crp* from *Escherichia coli*, regulates nitrogen fixation genes both positively and negatively in *Rhizobium meliloti*. EMBO J 8: 1279

Beringer JE, Beynon JL, Buchanan-Wollaston AV, Johnston AWB (1978) Transfer of the drug-resistance transposon Tn5 to *Rhizobium*. Nature 276: 633

Birkenhead K, Noonan B, Reville WJ, Boesten B, Manian SS, O'Gara F (1990) Carbon utilization and regulation of nitrogen fixation genes in *Rhizobium meliloti*. Molec Plant-Microbe Interact 3: 167

Casse F, Boucher C, Julliot JS, Michel M, Dénarié J (1979) Identification and characterization of large plasmids in *Rhizobium meliloti* using agarose gel electrophoresis. J Gen Microbiol 113: 229

Colonna-Romano S, Arnold W, Schlüter A, Boistard P, Pühler A, Priefer U (1990) An Fnr-like protein encoded in *Rhizobium leguminosarum* biovar *viciae* shows structural and functional homology to *Rhizobium meliloti* FixK. Mol Gen Genet 223: 138

David M, Domergue O, Pognonec P, Kahn D (1987) Transcription patterns of *Rhizobium meliloti* symbiotic plasmid pSym: identification of *nifA* independent *fix* genes. J Bacteriol 169: 2239

David M, Daveran ML, Batut J, Dedieu A, Domergue O, Ghai J, Hertig C, Boistard, Kahn D (1988) Cascade regulation of *nif* gene expression in *Rhizobium meliloti*. Cell 54: 671

De Philip P, Batut J, Boistard P (1990) *Rhizobium meliloti* FixL is an oxygen sensor and regulates *R. meliloti nifA* and *fixK* genes differently in *Escherichia coli*. J Bacteriol 172: 4255

Ditta G, Virts E, Palomares A, Kim CH (1987) The *nifA* gene of *Rhizobium meliloti* is oxygen regulated. J Bacteriol 169: 3217

Dixon RA (1984) The genetic complexity of nitrogen fixation. J Gen Microbiol 130: 2745

Earl CD, Ronson CW, Ausubel FM (1987) Genetic and structural analysis of the *Rhizobium meliloti fixA, fixB, fixC* and *fixX* genes. J Bacteriol 169: 1127

Fischer HM, Bruderer T, Hennecke H (1988) Essential and non-essential domains in the *Bradyrhizobium japonicum* NifA protein: identification of indispensable cysteine residues potentially involved in redox activity and/or metal binding. Nucl Acids Res 16: 2207

Gilles-Gonzales MA, Ditta GS, Helinski DR (1991) A haemoprotein with kinase activity encoded by the oxygen sensor of *Rhizobium meliloti*. Nature 355: 170

Gubler M, Hennecke H (1986) *fixAB* and *C* genes are essential for symbiotic and free-living microaerobic nitrogen fixation. FEBS Lett 200: 186

Gussin GN, Ronson CW, Ausubel FM (1986) Regulation of nitrogen fixation genes. Ann Rev Genet 20: 567

Kahn D, David M, Domergue O, Daveran ML, Ghai J, Hirsch P, Batut J (1989) *Rhizobium meliloti fixGHI* sequence predicts involvement of a specific cation pump in symbiotic nitrogen fixation. J Bacteriol 171: 929

Kaminski PA, Norel F, Desnoues N, Kush A, Salzans G, Elmerich C (1988) Characterization of the *fixABC* region of *Azorhizobium caulinodans* ORS571 and identification of a new nitrogen fixation gene. Mol Gen Genet 214: 496

Knight TJ, Langston-Unkefer PJ (1988) Enhancement of symbiotic dinitrogen fixation by a toxin-releasing plant pathogen. Science 241: 951

Long SR (1989) Rhizobium-legume nodulation: life together in the underground. Cell 56: 203

Masepohl B, Krey R, Riedel KU, Reiländer H, Jording O, Puchel H, Klipp W, Pühler A (1990) Analysis of the *Rhizobium meliloti* main *nif* and *fix* cluster in Endocytobiology IV. P. Nardon, V. Gianinazzi-Pearson, AM Grenier, L Margulis, DC Smith (eds). INRA Paris 1990 p 63

Morett E, Fischer H-M, Hennecke H (1991) Influence of oxygen on DNA binding, positive control, and stability of the *Bradyrhizobium japonicum* NifA regulatory Protein.

Nuti MP, Ledeboer AM, Lepidi AA, Schilperoort RA (1977) Large plasmids in different *Rhizobium* species. J Gen Microbiol 100: 241

Nuti MP, Lepidi AA, Prakash RK, Schilperoort RA, Cannon FC (1979) Evidence for nitrogen fixation (*nif*) genes on indigenous *Rhizobium* plasmids. Nature 282: 533

Renalier MH, Batut J, Ghai J, Terzaghi B, Ghérardi M, David M, Garnerone AM, Vasse J, Truchet G, Huguet T, Boistard P (1987) A new symbiotic cluster on the pSym megaplasmid of *Rhizobium meliloti* 2011 carries a functional *fix* gene repeat and a *nod* locus. J Bacteriol 169: 2231

Rosenberg C, Boistard P, Dénarié J, Casse-Delbart F (1981) Genes controlling early and late functions in symbiosis are located on a megaplasmid in *Rhizobium meliloti*. Mol Gen Genet 184: 326

Ruvkun GB, Ausubel FM (1980) Interspecies homology of nitrogenase genes. Proc Natl Acad Sci USA 77: 191

Ruvkun GB, Ausubel FM (1981) A general method for site-directed mutagenesis in prokaryotes. Nature 289: 85

Ruvkun GB, Sundaresan V, Ausubel FM (1982) Directed transposon Tn5 mutagenesis and complementation analysis of *Rhizobium meliloti* symbiotic nitrogen fixation genes. Cell 29: 551

Stock JB, Ninfa AJ, Stock AM (1989) Protein phosphorylation and regulation of adaptative responses in bacteria. Microbiol reviews 53: 450

Szeto WW, Zimmermann JL, Sundaresan V, Ausubel FM (1984) A *Rhizobium meliloti* symbiotic regulatory gene. Cell 36: 1035

Thöny B, Hennecke H (1989) The −24/−12 promoter comes of age. FEMS Microbiology reviews 63: 341

Van Larebeke N, Engler G, Holsters M, Van den Elsaker S, Zaenen I, Schilperoort RA, Schell J (1974) Large plasmids in *Agrobacterium tumefaciens* essential for crown gall-inducing ability. Nature 252: 169

Witty JF, Minchin FR, Skot L, Sheehy JE (1986) Nitrogen fixation and oxygen in legume root nodules. Oxford Surv. Plant Mol Cell Biol 3: 275

Yarosh OK, Charles TC, Finan TM (1989) Analysis of C4-dicarboxylate transport genes in *Rhizobium meliloti*. Mol Microbiol 3: 813

CHAPTER 16
The Nodulation of Legumes by Rhizobia

A. Economou and J.A. Downie

The Nitrogen Fixation
and its Research in China
Editor Guo-fan Hong
c Springer-Verlag Berlin Heidelberg 1992

1 Introduction

The *Rhizobium*-legume symbiosis represents a close association of soil Gram-negative bacteria with plants such as clovers, peas and soybeans, all of which are members of the large family of *Leguminosae*. Rhizobia invade the roots (and in some cases the stems) of the host plant and induce the formation of a specialized plant organ, the nodule (a tubular or spherical highly organized structure) in which the bacteria reside. Within this protected environment, the bacteria are sustained by the supply of carbon sources derived from the plant photosynthate, while they reduce nitrogen from the atmosphere, providing the plant with ammonia.

The incorporation of atmospheric dinitrogen into organic material as a result of the *Rhizobium*-legume symbiosis is estimated to account for one third of the total nitrogen intput needed for world agriculture (Gallon and Chaplin 1987). The use of legumes to enhance soil fertility, to provide a high-protein vegetable diet and as a source of forage and firewood is of particular importance to many developing countries. The agronomic importance of this symbiosis has thus justified extensive research which in the course of the past hundred years has led to the accumulation of an immense wealth of information. As a consequence, the interaction of *Rhizobium* with legumes is probably one of the best understood of all plant-microbe interactions (Young and Johnson 1989; Martinez et al. 1990).

Apart from its importance for agriculture, this fascinating association is also one of great inherent biological interest. Research into this interaction, which involves the interwined differentiation of a eukaryote and a prokaryote, requires the study of cell-to-cell recognition and communication, signal transduction, bacterial and plant differentiation as well as plant physiology and organ development. The techniques employed in this area range from studies in microbial ecology to the molecular analysis of nodule-specific promoters in transgenic plants. It is beyond the scope of this chapter to review all aspects of the wide variety of legume nitrogen-fixing symbioses and we will focus on the rhizobial genes required for the nodulation of legumes.

2 Different Rhizobia and Their Host Plants

Initiallly, all bacterial strains that form nodules on leguminous plants, were classified under the genus *Rhizobium*; subsequent studies revealed that rhizobia show a pronounced nodulation specificity for their host-plants. This led to the definition of "cross-inoculation groups" in which the bacteria are classified according to their ability to establish symbioses with a particular group of plants (Table 1). Three genera of rhizobia have now been recognised-*Rhizobium*, *Bradyrhizobium* and *Azorhizobium*. Members of the genus *Rhizobium* (Table 1)

Table 1. Host plants nodulated by different rhizobia

Rhizobial strains	Some of the host plants nodulated
Rhizobium	
R. leguminosarum	
biovar viciae	Pea (Pisum), vetch (Vicia)
biovar trifolii	Clover (Trifolium)
biovar phaseoli	Bean (Phaseolus)
R. meliloti	Alfalfa (Medicago)
R. loti	Trefoil (Lotus)
R. fredii	Soybean (Glycine)
Rhizobium sp. NGR234	Siratro (Marcoptillium)
Bradyrhizobium	
B. japonicum	Soybean (Glycine)
B. sp. (Vigna)	Cowpea (Vigna)
B. sp. (Lupinus)	Lupin (Lupinus)
B. sp. (Arachis)	Peanut (Arachis)
B. sp. (Parasponia)	Parasponia
Azorhizobium	
A. caulinodans	Sesbania rostrata

are relatively fast growing strains each of which nodulates a limited range of (normally temperate) legumes. Previously, the biovars *trifolii*, *phaseoli* and *viciae* of *R. leguminosarum* were called *R. trifolii*, *R. phaseoli* and *R. leguminosarum* respectively, but more recently it has been recognised that they all belong to one species.

The genus *Bradyrhizobium* (Table 1) contains a group of slow-growing rhizobia, which generally nodulate quite a wide range of host plants and one strain of *Bradyrhizobium* can nodulate the only non-legume (*Parasponia*) known to be nodulated by any of the rhizobia. Whereas *Rhizobium* and *Bradyrhizobium* spp. can fix N_2 in legume nodules, they cannot utilise N_2 as sole source of nitrogen during free-living culture. In contrast, the genus *Azorhizobium* is characterised by a strain (*Azorhizobium caulinodans*) that can grow in vitro using N_2. This strain was isolated from stem nodules (cauli = stem) from the tropical legume *Sesbania rostrata*. It is quite likely that as more rhizobia are characterised from a wider range of uncultivated legumes more genera will be defined.

The question of host-specificity is one of central interest in the study of the *Rhizobium*-legume symbiosis, it may hold the key to our understanding of the important evolutionary questions concerning the properties of legumes (and *Parasponia*) which enabled this symbiosis to evolve, as well as those properties that distinguished symbiosis from pathogenicity (Halverson and Stacey 1986; Djordjevic et al. 1987; Keen and Staskawicz 1988; Young and Johnston 1989). Successful nodulation is the result of several recognition events between the two partners and this elaborate recognition system is expressed at many stages during the symbiotic interaction.

3 The Nodulation Process

In essence, the formation of nodules depends on the initiation of cell division to form the new nodule meristem and a mechanism whereby the rhizobial bacteria come into contact with and infect the cells in this meristem. The position and type of nodule development varies greatly from legume to legume. Soybean and pea nodules are typical representatives of two broad classes of nodules called "determinate" and "indeterminate" respectively. Soybean nodules are spherical with a transient peripheral (determinate) meristem and an open vascular system. Pea nodules are cylindrical and have an indeterminate apical meristem and a closed vascular system. In other legumes it appears that nodule initiation can occur at sites of lateral root emergence—this is true of the stem nodules on *Sesbania rostrata* (Dreyfus et al. 1984) and the root nodules on peanuts (Chandler 1978). In soybeans, multiple sites of cell division are initiated at various places in the root-usually only a few cell layers deep within the root cortex, whereas in peas the nodule initiates deep in the cortex close to the pericycle layer of cells.

The different types of nodules (Sprent 1979, 1989) have significant implications with regard to the infection process—i.e. getting the bacteria to the site of the nodule primordia. When the nodule meristem is close to the epidermis the bacteria have the potential to gain access via crack entry. However, rhizobia infecting temperate legumes such as peas must penetrate many cortical cell layers, and in such legumes, infection is dependent upon the formation of an extensive infection thread. Since this type of infection seems to be the best studied and possibly the most highly developed, we will describe this process.

Initially, the sequence of events leading to a nodule probably starts with chemotactic attraction of rhizobia to compounds exuded by the roots of the host plant (Götz et al. 1982; Caetano-Annollés 1988; Bergman et al. 1988; Armitage et al. 1988). The bacteria attach to the root hair surface probably via specific Ca^{2+}-dependent adhesins and cellulose fibrils (Smit et al. 1987, 1989) although such attachment may not be necessary for nodulation to proceed, at least in soybeans (Vesper and Bhuvaneswari 1988).

A diffusible signal (Fig. 2 and see below) from the bacterium causes root hairs to deform and curl into "shepherd's crook" shapes, thus entrapping the bacteria within the curl (Yao and Vincent 1969; Bhuvaneswari et al. 1980; Bhuvaneswari and Sollheim 1985). The bacteria proliferate and penetrate the root-hair probably after a loosening and partial dissolution of the plant cell wall (Robertson et al. 1985; Turgeon and Bauer 1985; Ridge and Rolfe 1986).

As penetration occurs, root hair cell wall growth is reoriented and becomes introvertive, while appositional cell wall growth (mainly fibril material) takes place (Bauer 1981). The resulting hollow ingrowth develops into a tubular tunnel-like structure, the infection thread. Bacteria grow along the growing infection thread and ultrastructural studies show them to be surrounded by a mucigel composed of cell-wall polysaccharides, a plant-derived matrix-glycoprotein and their own extracellular polysaccharides (Bauer 1981; Callaham and Torey 1981; Robertson and Lyttleton 1982; VandenBosch et al. 1989). Coincident with root hair curling, a *Rhizobium*-made diffusible signal initiates cortical cell divisions, several cell layers away from the tip of the infection (Libbenga and Harkes 1973; Newcomb 1976; Truchet et al. 1980; Calvert et al. 1984; Dudley et al. 1987). It is this meristem that subsequently forms the actual body of the nodule.

The infection thread tip, following the root hair cell nucleus in basipetal migration (Lloyd et al. 1987), grows inter- and intra-cellularly towards the newly developed meristem, where it starts branching and ramifying, penetrating individual nodule cells (Turgeon and Bauer 1985; Bassett et al. 1977). The bacteria it carries, are released ("pinched off" as infection droplets) through the unwalled apical tip of the infection thread into the host cytoplasm, by a process that resembles endocytosis (Mellor and Werner 1987). Thus the bacteria are now enveloped by a plant membrane derived from the Golgi and endoplasmic reticulum (Basset et al. 1977; Robertson et al. 1978; Robertson et al. 1985). This membrane, termed the peri-bacteroid membrane (PBM), forms an interface of key importance, essential for the transfer of metabolites between the plant and the bacteria. Furthermore, the PBM may control proper nodule development by allowing only compatible surface interactions to occur (Dilworth and Glenn 1984; Robertson et al. 1985; Bradley et al. 1986). The fate of the PBM-enclosed bacteria varies from legume to legume. In some legumes such as peas, the bacteria increase in size and develop into pleiomorphic, differentiated "bacteroids" each surrounded by the PBM, whereas in other legumes such as soybeans, the bacteroids go through one or two doublings within the PBM. The mature bacteroids are now capable of reducing N_2 and the "fixed" nitrogen, in the form of ammonia, diffuses to the host cytoplasm, where it is assimilated into either amides or ureides. Only about half of the plant cells in the nodule contain bacteria and uninfected cells adjacent to infected host cells have also been proposed to participate in the maintenance of nodule functions and the final nitrogen assimilation (Bergmann et al. 1983).

4 Plant Genes Involved in Nodulation

The complex series of events outlined above, requires controlled, coordinated expression of both bacterial and plant genes. The contribution of legume genes in symbiosis has been studied by classical genetic and molecular genetic approaches. Since it has been recently reviewed in considerable detail (Verma and Delauney 1988; Nap and Bisselling 1990a, b), only a brief overview will be presented here.

Many plant mutants with altered symbiotic phenotypes have been identified (these arising spontaneously or after mutagenic treatment). The phenotypes of these mutants include inability to fix nitrogen, failure to nodulate (blocked at different stages of nodule development; i.e. cortical cell division or root hair curling), "supernodulation", nitrate-tolerant nodulation and have been most widely studied in pea, soybean and alfalfa (Postma et al. 1988; Vance et al. 1988; Carroll and Matthews 1990; Weeden et al. 1990). A precise understanding of the biochemical functions of any of these mutated genes is yet to be established. It is significant to note that several non-nodulation pea mutants are also unable to be infected by mycorrhizae, showing that some legume genes are required for both this plant-fungal symbiosis and the plant-*Rhizobium* symbiosis (Duc et al. 1989).

The molecular approaches used, aimed at identifying specific proteins expressed by the plant as a response to infection and nodule formation (termed "nodulins"; Legocki and Verma 1980) have included in vitro translation of total nodule RNA and identification of protein products on two-dimensional gels, construction of appropriate cDNA libraries and differential screening (Govers et al. 1985; Verma and Delauney 1988; Nap and Bisseling 1990a, b) and the use of monoclonal antibodies (Brewin et al. 1985; Bradley et al. 1988).

Two broad classes of nodulins have been distinguished. "Early" nodulins are the products of plant genes expressed during the infection process and initial steps in the development of nodule structure. Many of the "late" nodulins are involved in establishing the appropriate conditions for nitrogen fixation and assimilation. Those early nodulins that have been characterised in detail appear to have the characteristics of structural proteins since they consist of multiple repeat units probably containing several hydroxyproline residues, typical of proteins located in the plant cell wall (Franssen et al. 1989; Scheres et al. 1990; van de Wiel et al. 1990). In some situations (e.g. with exopolysaccharide deficient mutant rhizobia or spontaneously-nodulating alfalfa) nodules can be induced in which the plant nodule cells are devoid of bacteria (Finan et al. 1985; Hirsch et al. 1989; Truchet et al. 1989). It appears that some early nodulins can be expressed in these empty nodules (Hirsch et al. 1989; Truchet et al. 1989) and it is likely that several of these early nodulins can be induced at a distance by a bacterially synthesised signal molecule.

Several late nodulins have been identified in the mature nodule (Verma and Delauney 1988; Nap and Bisseling 1990a, b; Forde and Cullimore 1989) and many of their functions have been identified. The most abundant nodule protein is leghaemoglobin which is a high-affinity oxygen-binding protein. Oxyleghaemoglobin and deoxyleghaemoglobin are in equilibrium in the nodule cell cytoplasm such that the free oxygen concentration is maintained at about $0.001-0.1\,\mu M$ (Appleby 1984). (This varies from legume to legume depending upon the specific affinities of different leghaemoglobins for oxygen.) Thus, although the free oxygen concentration is low, the potential flux of oxygen is high. These criteria meet the unusual demands of the nitrogen-fixing bacteroids-namely a low free oxygen concentration (since O_2 inactivates nitrogenase) and a high rate of oxygen consumption to fuel the energy-demanding process of reducing N_2 to ammonia. In some legumes there are multiple leghaemoglobin genes (Marcker et al. 1984) and a considerable effort is currently going into understanding the regulation of these genes (Stougaard et al. 1990).

The enzyme which assimilates the bacteroid-made ammonia is the plant glutamine synthetase which forms glutamine from ammonia and glutamate in an ATP-dependent reaction. Glutamine synthetase is a multisubunit enzyme and in *Phaseolus vulgaris* there are nodule and leaf specific forms (Forde and Cullimore 1989). The nodule-specific gene was cloned (Cullimore et al. 1984) and confirmed to be expressed only in nodules (Gebhardt et al. 1986). Homologous genes have been identified from peas (Tingley et al. 1987), alfalfa (Dunn et al. 1988) and soybeans (Sengupta-Gopalan and Pitas 1986) and these are also strongly induced in nodules, although in some cases the genes may be expressed at lower levels in other plant tissues.

In legumes with indeterminate nodules (e.g. peas, alfalfa, clover) most of the glutamine formed by glutamine synthetase is used in an aminotransfer reaction (catalysed by asparagine synthetase) to form asparagine from aspartate and much of the nitrogen translocated from these nodules is moved as aparagine. Two asparagine synthetase genes have been cloned (Tsai and Coruzzi 1990) and both are expressed to high levels in pea nodules, although they were also expressed in leaves and cotyledons as might be expected since asparagine plays an important role in nitrogen transport.

In legumes with determinate nodules (e.g. soybeans), the bulk of the fixed nitrogen is not translocated as asparagine. Instead, the fixed nitrogen enters the amino acid pool in the nodule, is metabolised via the ureide biosynthetic pathway and the uric acid produced is converted to allantoin by the enzyme uricase, one of the most abundant enzymes in soybean nodules. The nodule uricase is a tetramer of a 35,000 M_r subunit that is nodule specific and unrelated to leaf or stem tissue uricase. The gene for the nodule-specific uricase has been cloned (Bergmann et al. 1983) and immunolocalisation experiments have shown the uricase to be located in the peroxisomes of unifected nodule cells (Nguyen et al. 1985; VandenBosch and Newcomb 1986).

A key enzyme in the carbon metabolism of nodules is sucrose synthase, which in nodules can catalyse the degradation of sucrose to UDP-glucose and D-fructose. The sucrose synthase gene has been cloned (Thummler and Verma 1987) and it has been suggested that sucrose synthase is inactivated by haem released from degraded leghaemoglobin formed during nodule senescence. This could have the effect of stopping carbon supply to the bacteroids during nodule senescence (Thummler and Verma 1987).

In addition to these nodulins of defined function, there are several which have been recognised as nodule specific but of unknown function (see Verma and Delauney 1988). Many have been identified via cDNA cloning and some have been recognised using monoclonal and polyclonal antibodies. Monoclonal antibodies have been used in immunolocalisation experiments which have shown that some of the nodulins are localised in the peribacteroid membrane (Bradley et al. 1988), and identified a glycoprotein as a major component of the matrix of the infection thread (VandenBosch et al. 1989). However, the role of these remains to be established. Recent reports of successful *Agrobacterium* mediated transformation of legumes provides a long-awaited tool for their genetic manipulation (Petit et al. 1987; Hinchee et al. 1988; White and Greenwood 1987; Puonti-Kaerlas et al. 1990; Köhler et al. 1987). Diaz et al. (1989) capitalised upon this development and the availability of the cloned pea lectin gene to test whether or not this lectin gene contributed to plant-determined host-specificity. Transgenic clover roots carrying the pea lectin gene were found to be nodulated by *R. leguminosarum* biovar *viciae* which could not nodulate untransformed roots. These results showed that control of host-specificity is partly mediated by the root lectins of legumes presumably interacting with a rhizobial carbohydrate during the formation in infection threads. This finding not only provided the long sought-after proof for the importance of lectins in symbiosis, it also identified unequivocally the first legume protein molecule involved in host-specificity.

5 Rhizobial Genes Involved in Nodulation

Substantial developments have been made in identifying rhizobial genes and gene products contributing to nodulation and N_2-fixation. Several genetic tools were developed and rapidly advanced *Rhizobium* genetic research. These include: systems for gene transfer in *Rhizobium* by conjugation and transduction; wide host-range cloning vectors; transposon mutagenesis; construction of reporter-gene transposon probes to analyse gene expression; and "reverse genetics" (Beringer et al. 1980; Long 1989; Downie and Johnston 1988; Simon 1989; Simon et al. 1989; Martinez et al. 1990; Simon and Priefer 1990). This dramatic progress is reflected upon by the large number of symbiotic genes from various rhizobia, that have been isolated and characterized (Young and Johnson 1989; Applebaum

▶

B. For *R.l.* bv. *viciae*: Rossen et al. (1984); Shearman et al. (1986); Evans and Downie (1986); Surin and Downie (1988); Surin et al. (1990); Economou et al. (1990); De Maagd et al. (1989b); Canter-Cremers et al. (1989).

C. For *R.l.* bv. *viciae* strain TOM: Davis et al. (1988); Hombrecher et al. (1984)

D. For *R.l.* bv. *trifolii* strain ANU843: Schofield and Watson (1986); McIver et al. (1989); Spaink et al. (1989); Surin et al. (1990); Surin and Downie (1988), and see Djordjevic et al (1987)

E. For *R.l.* bv. *phaseoli* strain 8002: Davis and Johnston (1990a).

F. For *R. loti*: B. Scott, pers. comm. and Chua et al. (1985); Ward et al. (1989).

G. For *Rhizobium fredii* (?) strain NGR234: Horvath et al. (1987); Bachem et al. (1985); Bassam et al. (1986). For (other) *R. fredii* strains see Appelbaum et al. (1988); Sadowsky et al. (1988); Ramakrishnan et al. (1986); Russel et al. (1985); Lewin et al. (1990).

H. For *Azorhizobium caulinodans* strain ORS571: Goethals et al. (1989, 1990); Van den Eede et al. (1987).

I. For *Bradyrhizobium japonicum* strain 110: Nieuwkoop et al. (1987); Deshmane and Stacey (1989); Göttfert et al. (1989, 1990a, b); Appelbaum et al. (1987).

J. For *Bradyrhizobium* sp. (*Parasponia*): Scott (1986); Scott and Bender (1990); Marvel et al. (1987)

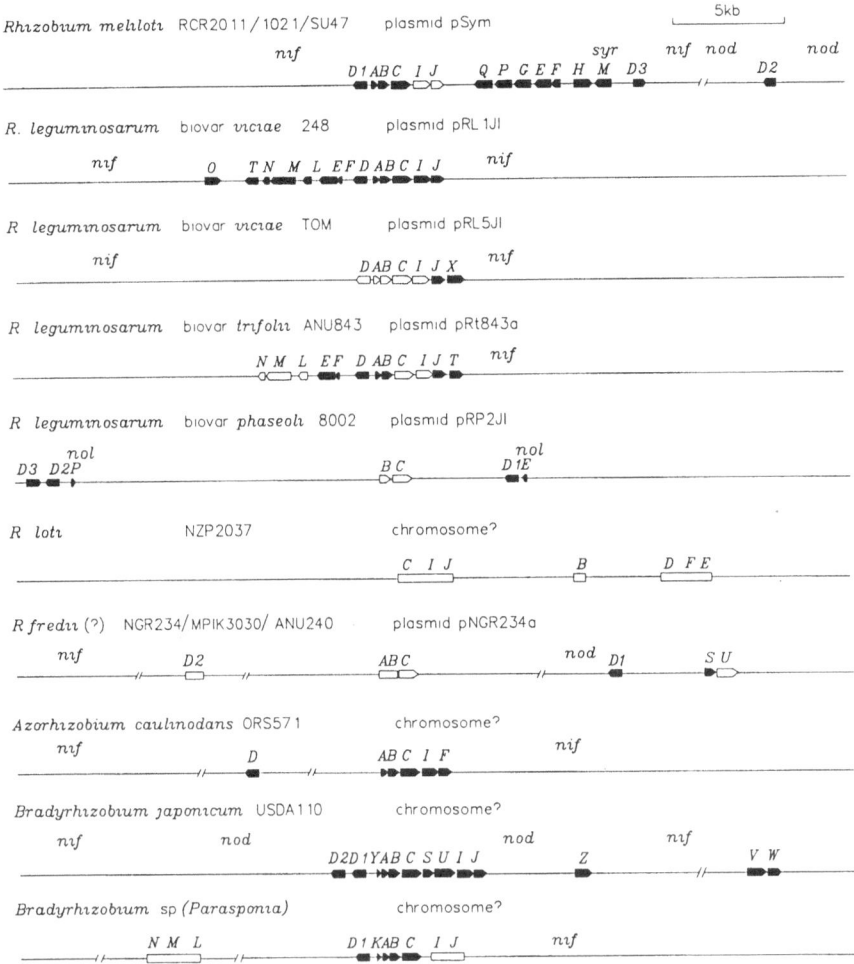

Fig. 1. Genetic map of nodulation (*nod* and *nol*) genes from the genera *Rhizobia, Azorhizobium* and *Bradyrhizobium*. The map was adapted and updated from that originally presented by Young and Johnston (1989). Genes with the same name are homologous as defined by: DNA sequencing analysis (black arrows) and DNA hybridization of genetic complementation experiments (open boxes or when transcriptional orientation is known, open arrows). The genes *nodJ* of R.l. bv. *viciae* strain TOM, *nodC* and *nodI* of R.l. bv. trifolii and *nodC* of the broad host-range *Rhizobium* strain NGR234/MPIK3030 have not been completely sequenced. *nif*, indicates loci containing *nif* and/or *fix* genes; *syrM* (symbiotic regulator), is a recently identified regulatory gene involved in *nod* gene activation in *R. meliloti* (Mulligan and Long 1989; Barnett and Long 1990). Regions marked *nod*, represent loci shown by transposon insertions to be important for host-specific nodulation. The references used for the compilation of this map were as follows:

A. For *R. meliloti* strain RCR2011: Eggelhoff et al. (1985); Jacobs et al. (1985); Cervantes et al. (1989); Schwedock and Long (1989); Debellé and Sharma (1986); Fisher et al. (1987); Renalier et al. (1987); Honma et al. (1990); Sharma and Signer (1990). For the extensively characterized *R. meliloti* strain 41 (not included in this map) see also work by: Gerhold et al. (1989); Göttfert et al. (1986); Török et al. (1984); Horvath et al. (1986); Putnocky et al. (1986). This strain was shown to contain a *nod* operon downstream of *nodD1*, which comprises the genes *nodM, nolFGHI* and *nodN*, transcribed towards *nodD1* (N. Baev and A. Kondorosi, pers. comm.).

(*Caption continued on p. 322*)

1990; Martinez et al. 1990). Because of the relative ease of obtaining bacterial symbiotic mutants (compared to those in plants), such mutants have also been used to study plant proteins involved in differentiation and nodule biogenesis (Mellor et al. 1989).

Three large groups of *Rhizobium* genes essential for the establishment of a successful symbiosis have thus far been identified and include: (a), genes encoding surface carbohydrates, and these genes are described in detail by Johnston in this book; (b), genes involved in carrying out and supporting nitrogen fixation as described by Boistard in this book, and (c), genes involved in host recognition and nodule formation (*nod* genes). The remainder of this chapter will describe some recent work on *nod* genes. Figure 1 summarises the organisation of the *nod* genes in several rhizobia.

In the *R. leguminosarum* biovars and in *R. meliloti*, the *nod* genes are present on large indigenous ("symbiotic") plasmids. On the pSym pRL1JI in *R. l.* bv. *viciae*, 13 *nod* genes have been identified, organised in five operons *nodD*, *nodABCIJ*, *nodFEL*, *nodMNT* and *nodO* (see Fig. 1). With the exception of *nodO*, similar genes (as judged from DNA sequence and hybridisation data) were identified in *R. l.* bv. *trifolii* with the difference that *nodT* is located downstream of *nodJ* (Fig. 1) rather than in the *nodMN* operon. However, despite strong sequence homology, some of these genes such as *nodE* are functionally distinct (Spaink et al. 1989; Surin and Downie 1989).

R. meliloti appears to have more nodulation genes than the *R. leguminosarum* biovars (Fig. 1). The *nodABC* genes are functionally equivalent since mutations in these genes from one strain can be complemented with the genes from another (Fisher et al. 1985). Whilst the *nodFE* genes are present in *R. meliloti* (Fig. 1) it appears that they are somewhat different (allelic) from the *nodFE* genes in the *R. leguminosarum* biovars (Fisher et al. 1985). The *nodHGPQ* genes (Fig. 1) characterised in *R. meliloti* appear not to be present as inducible *nod* genes in the *R. leguminosarum* strains studied.

In general, most rhizobia appear to have more than one *nodD* gene (Fig. 1) with the exception of the *R. leguminosarum* biovars *viciae* and *trifolii*. Recent studies in less well characterised rhizobia revealed that at least the *nodDABC(IJ)* genes are highly conserved among rhizobia although the arrangements of these genes may vary (Fig. 1). Also, individual strains seem to have additional *nod* genes some of which have not as yet been extensively characterised (see Fig. 1 and references cited therein). It is possible that the number of *nod* genes identified may depend on the actual strain of *Rhizobium* species studied.

6 Signal Exchange in the Rhizobial-Legume Symbiosis

It is now apparent that during the infection and nodulation process, specific low molecular weight molecules mediate signal transduction between the rhizobia and the legume roots. Initially legumes secrete flavonoid and/or

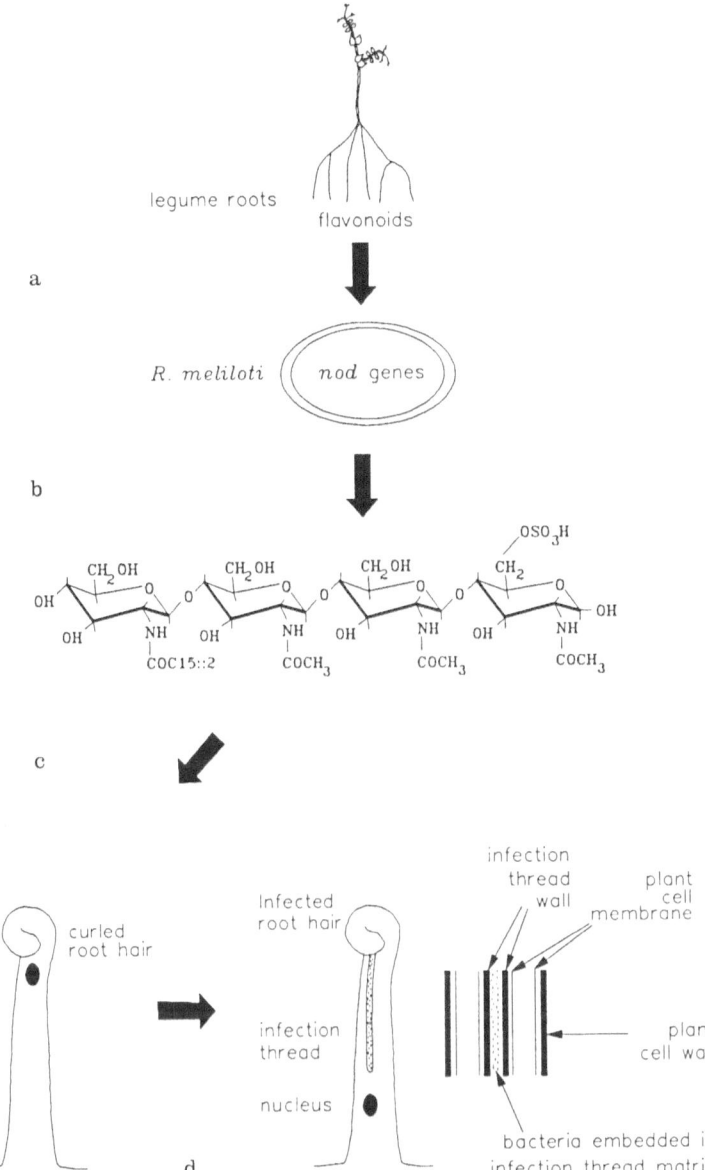

Fig. 2 a–d. Schematic representation of the sequence of events during the early stages of nodule biogenesis. **a.** Flavonoid molecules exuded from legume roots induce rhizobial *nod* gene expression. **b.** Several Nod proteins are involved in the synthesis of a substituted glycolipid molecule, which the bacteria secrete. **c.** The glycolipid molecule interacts with the plant root cells and triggers two processes: curling of root hairs (which entrap rhizobia) and divisions of cortical cells that will form the nodule meristem (not shown). **d.** Rhizobia advance into root hairs through a plant-made infection thread and eventually reach and infect cells of the nodule meristem

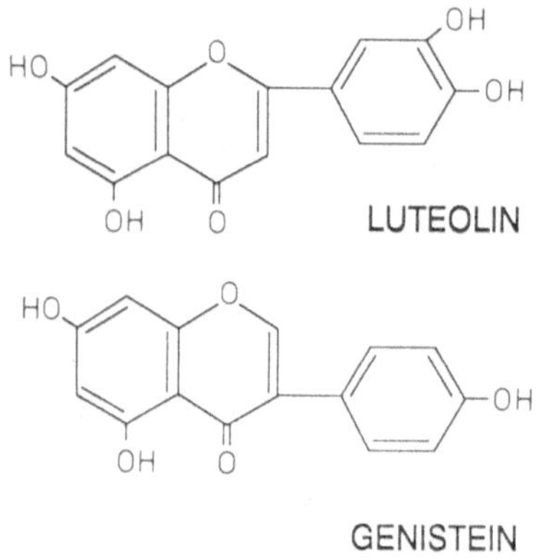

LUTEOLIN

GENISTEIN

Fig. 3. Structures of a flavone (luteolin) and an iso-flavone (genistein) that can induce rhizobial *nod* gene expression. These compounds are derived from plant phenylpropanoid metabolism which also leads to the production of phytoalexins, lignin and flower pigments (Lamb et al. 1989)

isoflavonoid molecules (Figs. 2 and 3) that induce rhizobial *nod* gene expression after interacting with the regulatory *nodD* gene products. Subsequently, several *nod* genes are induced and many of their products are involved in the biosynthesis of low molecular weight signal molecules that initiate a programme of nodule development in the legume root. Different *nod* genes play different roles in this process and the following sections describe the known roles of the *nod* genes and some possible roles for *nod* gene products whose functions have not yet been defined.

6.1 Plant Signals Recognised by Rhizobia

The identification of the first signal in the legume-rhizobia interaction arose from studies on *nod* gene regulation. Since the biochemical role of the *nod* genes was not known, this work was facilitated by the use of transcriptional or translational fusions of the *E. coli lacZ* gene to the rhizobial *nod* genes. In vitro fusions were made with the *nod* genes from *R. meliloti* (Mulligan and Long 1985), *R. l.* bv. *viciae* (Rossen et al. 1985; Spaink et al. 1987a), *B. japonicum* (Kosslak et al. 1987; Banfalvi et al. 1988; Göttfert et al. 1988), *R. l.* bv. *phaseoli* (Davis and Johnston, 1990b) and *Azorhizobium caulinodans* (Goethals et al. 1990). In vivo fusions were made using Mu (*lac*) insertions in the *nod* genes of *R. meliloti* (Horvath et al. 1987), *R. l.* bv. *trifolii* (Innes et al. 1985), *R. fredii* (Olson et al. 1985) and the broad host-range *Rhizobium* strain NGR234 (Bassam et al. 1988). These studies established that in these rhizobia the *nodD* gene is the only constitutively expressed *nod* gene (in defined minimal media), while all the other *nod* genes are transcribed at very low levels in such conditions (if at all). When cells are grown in the presence of legume root exudates, there is induction of *nod* gene transcription; this induction requires the presence of the *nodD* gene which encodes a positively-acting regulatory protein (Mulligan and Long 1985; Rossen et al. 1985; Innes et al. 1985; Shearman et al. 1986).

Legume exudates contain sets of several different flavonoids (Kosslak et al. 1987; Zaat et al. 1988, 1989; Van Brussel et al. 1990; see Fig. 3) and several studies have identified flavonoid molecules which can act as potent inducers of *nod* genes from different rhizobia. Such examples are luteolin and a methoxy chalcone for *R. meliloti* (Peters et al. 1986; Györgypal et al. 1988; Maxwell et al. 1989), naringenin and hesperetin for *R. l.* bv. *viciae* (Firmin et al. 1986; Zaat et al. 1987b); 7,4-dihydroxyflavone (DHF) for *R. l.* bv. *trifolii* (Redmond et al. 1985); naringenin for *A. caulinodans* (Goethals et al. 1990); daidzein, genistein and apigenin for *R. fredii* (Sadowsky et al. 1988); daidzein and genistein for *B. japonicum* (Kosslak et al. 1987; Göttfert et al. 1988); genistein and naringenin for *R. l.* bv. *phaseoli* (Davis and Johnston 1990b).

Interestingly, certain isoflavonoid molecules, such as genistein, which act as inducers of *nod* gene transcription in *B. japonicum* (Kosslak et al. 1987) can antagonize *nod* activation in *R. l.* bv. *viciae* (Firmin et al. 1986). Conversely narigenin, a very potent inducer of *R. l.* bv. *viciae nod* genes, acts as an antagonist of *nod* gene induction in certain *B. japonicum* strains (Kosslak et al. 1990). Another isoflavonoid *nod* gene "anti-inducer" (umbelliferone) has been identified for *R. meliloti* (Peters and Long 1988) and *R. trifolii* (Djordjevic et al. 1987).

Spaink et al. (1987b) and Horvath et al. (1987), showed that the differences in the ability of purified flavonoids to induce *nod* gene expression was determined by the *nodD* genes. It was proposed that the flavonoid molecules may directly interact with the NodD protein, in particular with the non-conserved C-terminal region (Horvath et al. 1987). This proposal was further supported by the work of Burn et al. (1987), who chemically mutagenised the *R. l.* bv. *viciae nodD* gene and isolated a point mutation that caused a substitution of asparagine for an aspartate residue in the C-terminal domain of NodD. This allowed the mutant protein to induce *nod* genes in the absence of any flavonoid. Most significantly, the mutant induced a very high level of *nod* gene expression in the presence of the isoflavonoid genistein which normally acts as an "anti-inducer" (Burn et al. 1987). Subsequently, similar mutations that led to flavonoid-independent activation of *nod* genes were found to affect the C-terminus of the *R. l.* bv. *trifolii* NodD protein (McIver et al. 1989). On the basis of these observations and work on hybrid *nodD* genes (Spaink et al. 1989b) it was concluded that specificity for the flavonoids is determined by the primary structure of the NodD protein.

Thus, rhizobia have the potential to "recognise" different legumes by virtue of a proposed interaction of the flavonoids and the NodD proteins (a direct interaction between NodD and flavonoid has not yet been demonstrated). It has been proposed that the flavonoids partition (by nonactive transport) into the bacterial inner membrane (Recourt et al. 1989) and since Schlaman et al. (1989) have identified membrane-associated NodD, it is possible that a flavonoid-NodD interaction could occur at the bacterial membrane.

Many rhizobia contain multiple alleles of *nodD* (Fig. 1) and it is likely that the different NodD proteins extend the range of legumes that can be nodulated by recognising different flavonoids secreted from the roots of different legume species. For example, *R. meliloti* contains three copies of *nodD*, called *nodD1*, *nodD2* and *nodD3* (Göttfert et al. 1986; Honma and Ausubel 1987; Györgypal et al. 1988; Mulligan and Long 1989; Honma et al. 1990) each of which responds differently to exudate from various host plants.

Moreover, in some cases it is possible to alter the range of legumes nodulated by transferring *nodD* alleles from one strain of *Rhizobium* to another. Thus for example Horvath et al. (1987) showed that a *nodD* gene from a broad host range *Rhizobium* strain could confer upon *R. meliloti* the ability to nodulate siratro, a normal host plant for the broad host range *Rhizobium* (but not normally a host for *R. meliloti*).

The dual role of recognition and regulation mediated by the NodD proteins fits well with the observation that the NodD proteins belong to a family of bacterial transcriptional activators (Shearman et al. 1986; Horvath et al. 1987; Henikoff et al. 1988; Kofoid and Parkinson 1988; Appelbaum et al. 1988; Burn et al. 1989; Goethals et al. 1990; Barnett and Long 1990). These homologous proteins activate transcription of other genes (often encoding enzymes of metabolic pathways), in response to the presence of small molecular weight inducer molecules such as salicylate for the NahR protein, diaminopimelic acid for LysR or β-lactam antibiotics for AmpR (Henikoff et al. 1988; Schell et al. 1989). This group of proteins, termed the LysR family, shares significant overall sequence homology, which is particularly pronounced at their N-termini. This region contains a predicted helix-turn-helix DNA binding motif (Henikoff et al. 1988; Burn et al. 1989), while their C-termini have been proposed to be involved in interactions with their respective inducer molecules (Henikoff et al. 1988). The NodD proteins bind to a highly-conserved sequence (*nod*-box) that precedes the flavonoid-inducible *nod* operons (Rostas et al. 1986; Shearman et al. 1986; Schofield and Watson 1986; Spaink et al. 1987a). This interaction has been demonstrated using NodD-containing cell extracts of *R. l.* bv. *viciae* (Hong et al. 1987) and substantially purified NodD1 or NodD3 proteins from *R. meliloti* (Fisher et al. 1988) that retard the electrophoretic mobility of DNA fragments containing *nod*-box sequences. Using the more refined DNA footprinting technique it was shown that NodD proteins from *R. meliloti* bind to a 55–60 bp sequence corresponding closely to the *nod*-box (Kondorosi et al. 1989; Fisher and Long 1989). The DNA-protein interaction occurs in the absence of flavonoids; Kondorosi et al. (1989) did observe that flavonoids could enhance the binding of the *R. meliloti* NodD to the *nod* box, although this has not been noted by other groups.

The various *nod* operons do not appear to be under the exclusive control of *nodD*. Kondorosi et al. (1989) have identified a repressor of *nod* gene expression, while Dusha et al. (1989) and Wang and Stacey (1990) demonstrated a role for ammonia in the regulation of the *nod* genes of *R. meliloti* and *B. japonicum*. Further, it is apparent the the *nodD*-like gene syrM (symbiotic regulator) can activate the expression of the *R. meliloti nodD3*, which can induce other *nod* genes even in the absence of flavonoids (Mulligan and Long 1989; Barnett and Long 1990). In *B. japonicum* the *nodD1* gene not only activates the expression of other *nod* genes, but also stimulates its own expression in the presence of soybean root exudate (Banfalvi et al. 1988).

Currently, we do not fully understand the mechanism by which the NodD proteins recognise the different flavonoids and activate *nod* gene transcription, although current studies may soon resolve this. However, it is clear that the net result of this recognition is the synthesis by the other *nod* gene products of specific signal molecules that are exported and recognised by the legume, which then initiates nodule morphogenesis (Fig. 2).

6.2 Rhizobia make Host-Specific Signals Recognised by Plants

It was recognised by Bhuvaneswari and Solheim (1985) that when *R. l.* bv. *trifolii* was inoculated onto clover roots, a soluble, low molecular weight ($\approx 1{,}200$) compound was formed which could induce root hair deformation on aseptically grown clover plant. Subsequently, Zaat et al. (1987a) showed that when *R. l.* bv. *viciae* was grown in minimal medium suppplemented with a flavonoid (to induce *nod* gene expression), the growth medium contained a compound that induced root hair deformation on vetch. The formation of this active factor was dependent on *nod* gene expression since its formation was blocked in strains mutated in the *nodAB* or *C* genes. Banfalvi and Kondorosi (1989) showed that when the *nodABC* genes from *R. meliloti* were expressed in *E. coli*, the resultant strain made a root hair curling factor that was found in the growth medium. Lerouge et al. (1990a) purified an active root hair deformation factor from the growth medium supernatant of a culture of *R. meliloti* which expressed the *nod* genes at a high level. The proposed structure of this compound is shown in Figure 2. When added to alfalfa roots this compound can induce root hair curling and nodule formation, even in the absence of rhizobia (Lerouge et al. 1990b).

As described earlier a large number of nodulation genes has been identified in different strains of rhizobia (Fig. 1) and (most of) these *nod* genes are induced by flavonoids or isoflavonoids. The question then arises as to which of these *nod* genes are involved in the formation and/or modification of low molecular weight signal molecules similar to that shown in Fig. 2.

The *R. meliloti nodABC* genes are involved in the biosynthesis of the root-hair curling signal molecule (Lerouge et al. 1990a) possibly by forming the tetraglucosamine backbone (Fig. 2) although John et al. (1985, 1988) proposed an additional role for NodC as a transmembrane receptor-like molecule. The functional homology between the *nodABC* genes of *R. meliloti* and those of other rhizobia would suggest that different rhizobial signal molecules contain a similar polyglucosamine backbone. The *nodM* gene product is strongly homologous to the *E. coli* enzyme glucosamine synthase which makes glucosamine-6-phosphate from fructose-6-phosphate using glutamine as an amino donor (Downie et al. 1990). Although mutations in *nodM* do not block nodulation or root hair curling on vetch (Surin and Downie 1988) this can be understood since it is likely that rhizobia have a second glucosamine synthase activity required as a "housekeeping" gene to form glucosamine when the *nod* genes are not induced. This possibility is supported by DNA hybridization experiments using a *nodM* probe, which revealed a second region of DNA homologous to *nodM* in biovars *viciae* and *trifolii* (Surin and Downie 1988). The *nodM* gene product could confer a competitive advantage possibly by making available an abundant supply of glucosamine which could be used to make the signal molecule(s).

Two other *nod* genes, *nodIJ* are often found in the same operon as *nodABC* (Fig. 1). Mutation in *nodI* or *nodJ* cause a slight delay in nodulation in some strains but not in others (Downie et al. 1985; Djördjevic et al. 1985; Debellé et al. 1986; Nieuwkoop et al. 1987). The predicted protein sequences of NodI and NodJ indicated that they are likely to be involved in membrane transport (Evans and Downie 1986) on the basis that NodI is homologous to a large group of membrane-associated transport proteins (Higgins et al. 1986; Riordan et al. 1989) and was shown by Schlaman et al. (1990) to be membrane-associated, while NodJ is a transmembrane protein (Evans and Downie 1986; Surin et al. 1990). The *nodIJ* gene products may be involved in the export of the root hair curling factor, and the delayed nodulation phenotype caused by mutation in *nodI* or *nodJ* could be explained on the basis that with *nodI* or *nodJ* mutants, sufficient root hair curling factor leaks out (or is released by lysed cells) to allow nodulation to proceed-albeit at a slightly slower rate.

The *nodFE* genes are key determinants of host specificity, a conclusion based on the observations that (a) mutations of these genes can alter the range of host plants nodulated (Djördjevic et al. 1985; Horvath et al. 1986) and (b) transfer of these genes from one *R. leguminosarum* biovar to another is accompanied by a change in specificity of legumes nodulated (Djördjevic et al. 1985; Surin and Downie 1989; Spaink et al. 1989a). This occurs despite the highly conserved amino acid sequences of the NodFE proteins (Spaink et al. 1989a). The biochemical roles of NodF and NodE have not been identified although their predicted amino-acid sequences do give a strong indication as to their possible function. The NodF protein is homologous to acyl carrier proteins (Shearman et al. 1986; Horvath et al. 1986; Debellé and Sharma 1986) while the NodE protein is homologous to β-ketoacyl synthases (condensing enzymes) that are involved in fatty-acid and polyketide biosynthesis (Johnston et al. 1988; Bibb et al. 1989; Shearman et al. 1989). Given that there is a fatty acid substituent on the root hair curling factor identified by Lerouge et al. (1990a) (Fig. 2) it is possible that the *nodFE* genes are involved in determining the type of the fatty acid such that it can confer some degree of host specificity. However, it should be noted that mutation of the *nodFE* genes does not eliminate the ability of the bacteria to induce nodulation or root hair deformation (Downie et al. 1985; Djödjevic et al. 1985; Horvath et al. 1986; Debellé et al. 1986). Therefore it is possible that intermediates in the formation of the signal molecule retain root hair deformation activity or alternatively that in the absence of the *nodFE* genes some other genes can partially subsitute for their function.

The *R. meliloti* genes *nodH* and *nodQ* appear to modify a vetch-specific root hair curling factor into an alfalfa-specific signal (Faucher et al. 1989). This modification is related to the substitution of the sulphate group on the signal molecule (Fig. 2) because mutations in *nodH* and *nodQ* block the attachment of the sulphate group (Roche et al. 1990). This change in the signal molecule results in mutants of *R. meliloti* which have acquired the ability to nodulate vetch but are Nod⁻ on their normal host, alfalfa (Faucher et al. 1988).

The *nodL* gene also plays a role in specificity of nodulation since mutation of *nodL* in biovar *viciae* strongly inhibits nodulation of peas but not of vetch (Surin and Downie 1988). NodL is probably an acetyltransferase based on its homology with other acetyl transferases (Downie 1989), and it is possible that it could be involved in transferring an acetyl group to the signal molecule.

The *nodG* gene from *R. meliloti* plays a role in determining host-specific infection (Horvath et al. 1986; Debellé et al. 1986; Swanson et al. 1987). The *nodG* gene product is homologous to ribitol dehydrogenase (Debellé and Sharma 1986) and other oxidoreductases such as the ketoreductase (Sheldon et al. 1990) that play a role in the reduction of the keto group formed during fatty acid biosynthesis. Therefore it is possible that NodG is involved in the synthesis (or possibly modification?) of the fatty acid on the root hair curling molecule.

It is clear therefore that several *nod* genes have the potential to be involved in the formation of a host-specific signal molecule similar to the one shown in Fig. 2. However, it is also clear that this is not the complete story since there are other *nod* genes that are unlikely to contribute to such a biosynthetic pathway. Thus for example the bv. *viciae nodO* gene encodes a secreted Ca^{2+}-binding protein (de Maagd et al. 1989a, b; Economou et al. 1989, 1990) that may interact directly with plant cell membranes and appears to be found only in strains of bv *viciae*. Downie and Surin (1990) have shown that the *nodO* gene can allow nodulation to proceed even in the absence of the *nodFE* genes.

In *B. japonicum* other genes involved in host-specific nodulation have been identified: mutations in *nodV* or *nodW* almost completely block nodulation on *Vigna* spp. whereas nodulation on soybeans is only delayed by about 3 days (Göttfert et al. 1990a). These genes are likely to be involved in regulation since they are homologous to the large family of bacterial two-component regulatory systems. One of these systems has been found in *Agrobacterium tumefaciens* to recognise plant-made metabolites and thereby induce the *Agrobacterium vir* genes (Stachel and Zambryski 1986; Cangelosi et al. 1990). Therefore by analogy it is possible that in *B. japonicum* there is a second regulatory system that may specifically recognise exudates from some plants and induce a different set of nodulation genes from those induced by *nodD*.

In addition to these genes of predicted functions there are several *nod* genes for which no homologies have been found. Some of these genes have been shown to have clear phenotypic effects—for example, the *nodX* gene found in one strain of biovar *viciae* is required for the ability of that strain to nodulate a variety of peas (cv. Afghanistan) not nodulated by strains lacking *nodX* (Götz et al. 1985; Davis et al. 1988). However, for several of the nodulation genes such as the *nodK*, *nodY* and *nodSU* genes in *Bradyrhizobium* spp. (Fig. 1) little is currently known about their role or function in nodulation. It is possible for example that some *nod* genes are only important during the nodulation of one variety of legume and that legume is not routinely used for laboratory nodulation tests. Alternatively, some of the *nod* genes could play a role in competitiveness, and no phenotype is seen unless that strain is tested in a mixed inoculation test. It is anticipated that future work will lead to an understanding of the biochemical functions of all of the *nod* genes. It is clear that despite the significant progress that has been made in understanding the molecular basis of the signal exchange between rhizobia and their legume hosts, there is still much to understand about this communication.

7 Conclusions

In the development of the rhizobial-legume symbiosis there are several steps at which recognition occurs between the two partners. Initially, the bacteria recognise the type of flavonoids secreted from the legume roots. This recognition is mediated by rhizobial NodD proteins which sense a range of flavonoids and then induce expression of other *nod* genes. This initial step is not an exclusive recognition since exudates from some legumes can induce *nod* genes in some bateria that do not nodulate that legume host.

Some of the induced *nod* gene products are involved in the synthesis of low molecular weight glycolipids that constitute part of the bacterial response to the plant. These signal molecules determine host specificity by subtle alterations in their structure made by *nod* gene products and the resulting molecules are recognised at very low concentrations by legume roots. Although these signal molecules are a prerequisite for root-hair curling, infection and nodule morphogenesis, there are additional characteristics that the bacteria must have for normal infection and nodule development to occur. Thus for example a secreted protein plays a role in nodulation in at least one symbiosis. The bacterial cell surface is also very important and exopolysaccharide is essential for normal infection even though empty nodules can develop in its absence. The bacterial outer membrane lipopolysaccharide also plays a crucial role, especially in the later stages of nodule development.

Now that the chemical structures of different rhizobial signal molecules are being established, the next goal is to understand how the plant recognises these various signals that play a role in nodule initiation and development. Perhaps this may also help us understand some other more general aspects of the cell to cell communication that must occur between different plant cells during development.

Acknowledgements. We are indebted to Andrea Davis for the preparation of the figures. We are grateful to Peter Young for allowing us to use and update his original version of Fig. 1. We also wish to thank Marie-Anne Barny and Elmar Kannenberg for critical comments on the manuscript. Work in the authors' laboratory was supported by the Agricultural and Food Research Council. Additional financial support was provided by the European Community and the John Innes Foundation.

8 References

Applebaum ER (1990) The *Rhizobium/Bradyrhizobium*-legume symbiosis. In: Gresshoff PM (ed) Molecular biology of symbiotic nitrogen fixation, CRC Press, p 131

Applebaum ER, Thompson DV, Idler K, Chartrain N (1988) *Rhizobium japonicum* USDA 191 has two *nodD* genes that differ in primary structure and function. J Bacteriol 170: 12

Applebaum ER, Hennecke H, Lamb JW, Göttfert M (1987) Int Pat Appl PCT/US87/01421

Appleby CA (1984) Leghemoglobin and *Rhizobium* respiration Ann Rev Plant Physiol 35: 443

Armitage JP, Gallagher A, Johnston AWB (1988) Comparison of the chemotactic behavior of *Rhizobium leguminosarum* with and without the nodulation plasmid. Mol Micro 2: 743

Bachem CWB, Kondorosi E, Banfalvi Z, Horvath B, Kondorosi A, Schell J (1985) Identification and cloning of nodulation genes from the wide host range *Rhizobium* strain MPIK3030. Mol Genet 199: 271

Banfalvi Z, Kondorosi A (1989) Production of root hair deformation factors by *Rhizobium meliloti* nodulation genes in *Escherichia coli*; HsnD (*nodH*) is involved in the plant host-specific modification of the NodABC factor. Plant Mol Biol 13: 1

Banfalvi Z, Niewkoop A, Schell M, Besl L, Stacey G (1988) Regulation of *nod* gene expression in *Bradyrhizobium japonicum*. Mol Gen Genet 214: 420

Barnett MJ, Long SR (1990) DNA sequence and translational product of a new nodulation-regulatory locus: SyrM has sequence similarity to NodD proteins. J Bacteriol 172: 3695

Bassam BJ, Djordjevic MA, Redmond JW, Batley M, Rolfe BG (1988) Identification of a *nodD*-dependent locus in the Rhizobium strain NGR234 activated by phenolic factors secreted by soybeans and other legumes. Mol Plant-Microbe Inter 1: 161

Bassam BJ, Rolfe BG, Djordjevic MA (1986) *Macroptilium atropurpureum* (siratro) host specificity genes are inked to a *nodD*-like gene in the broad host range *Rhizobium* strain NGR234. Mol Gen Genet 203: 49

Basset B, Goodman RN, Novacky A (1977) Ultrastructure of soybean nodules. Release of rhizobia from the infection thread. Can J Microbiol 23: 573

Bauer WD (1981) Infection of legumes by rhizobia Ann Rev Plant Physiol 32: 407

Bender GL, Nayudu M, Le Strange KK, Rolfe BG (1988) The *nodD1* gene *Rhizobium* strain BGR234 is a key determinant in the expression of host range to the non-legume *Parasponia*. Mol Plant-Microbe Inter 1: 259

Bergman K, Gulash-Hofee M, Horestadt RE, Larosiliere RC, Ronco PG, SU L (1988) Physiology of behavioural mutants of *Rhizobium meliloti*: evidence for a dual chemotaxis pathway. J Bacteriol 170: 3249

Bergmann H, Preddie E, Verma DPS (1983) Nodulin-35: A subunit of specific uricase (uricase II) localized in uninfected cells of nodules. EMBO J 2: 2333

Beringer JE, Brewin NJ, Johnston AWB (1980) The genetic analysis of *Rhizobium* in relation to symbiotic nitrogen fixation. Heredity 45: 161

Bhuvaneswari TV, Solheim B (1985) Root hair deformation in the white clover/*Rhizobium trifolii* symbiosis. Physiol Plant 63: 25

Bhuvaneswari TV, Turgeon G, Bauer WD (1980) Early stages in the infection of soybean (Glycine max L Merr) by *Rhizobium japonicum*. 1. Localization of infectible root cells. Plant Physiol 66: 1027

Bibb MJ, Biro S, Motamedi H, Collins JF, Hutchinson CR (1989) Analysis of the nucleotide sequence of the *Streptomyces glaucescens tmcI* genes provides information about the enzymology of ployketide antibiotic biosynthesis. EMBO J 8: 2727

Bradley DJ, Wood EA, Larkins AP, Galfre G, Butcher GW, Brewin NJ (1988) Isolation of monoclonal antibodies reacting with peribacteroid membranes and other components of pea root nodules containing *Rhizobium leguminosarum*. Planta 173: 149

Bradley DJ, Butcher GW, Galfre G, Wood EA, Brewin NJ (1986) Physical association between the peribacteroid membrane and lipopolypolysaccharide from the bacteroid outer membrane in *Rhizobium*-infected pea root nodule cells. J Cell Sci 85: 47

Brewin NJ, Robertson JG, Wood EA, Wells B, Larkins AP, Galfre G, Butcher GW (1985) Monoclonal antibodies to antigens in the peribacteroid membrane from *Rhizobium*-induced root nodules of pea cross-react with plasma membranes and Golgi bodies. EMBO J 4: 605

Burn JE, Hamilton WD, Wootton JC, Johnston AWB (1989) Single and multiple mutations affecting properties of the regulatory gene *nodD* of *Rhizobium*. Mol Microbiol 3: 1567

Burn JE, Rossen L, Johnston AWB (1987) Four classes of mutations in the NodD gene of *Rhizobium leguminosarum* and *R. phaseoli*. Genes Dev 1: 456

Caetano-Anolles G, Crist-Estes DK, Bauer WD (1988) Chemotaxis of *Rhizobium meliloti* to the plant flavone luteolin requires functional nodulation genes. J Bacteriol 170: 3164

Callaham DA, Torrey JG (1981) The structural basis for infection of root hairs of *Trifolium repens* by *Rhizobium*. Can J Bot 59: 1647

Calvert HE, Pence MK, Pierce M, Malik NSA, Bauer DW (1984) Anatomical analysis of the development and distribution of *Rhizobium* infections in soybean roots. Can J Bot 62: 2375

Canter-Cremers HCJ, Spaink HP, Wijfjes HM, Pees E, Wijffelman CA, Okker RJH, Lugtenberg BJJ (1989) Additional nodulation genes on the Sym plasmid of *Rhizobium leguminosarum* biovar *viciae*. Plant Mol Biol 13: 163

Cangelosi GA, Ankenbauer RG, Nester EW (1990) Sugars induce the *Agrobacterium* virulence genes through a periplasmic binding protein and a transmembrane signal protein. Proc Natl Acad Sci USA 87: 6708

Carroll BJ, Mathews A (1990) Nitrate inhibition of nodulation in legumes. In: Molecular Biology of Symbiotic Nitrogen Fixation ed. Gresshoff PM CRC Press: pp 159

Cervantes E, Sharma SB, Millet F, Vasse J, Truchet G, Rosenberg C (1989) The *Rhizobium meliloti* host range *nodQ* gene encodes a protein which shares homology with translation elongation and initiation factors. Mol Microbiol 3: 745

Chandler MR (1978) Some observations on infection of *Arachis hypogaea* L. by *Rhizobium*. J Exp Bot 29: 749

Chua K-Y, Pankhurst Ce, Macdonald PE, Hopcroft DH, Jarvis BDW, Scott DB (1985) Isolation and characterization of transposon Tn-5 induced symbiotic mutants of *Rhizobium loti*. J Bacteriol 162: 335

Cullimore JV, Gebhardt C, Saarelainen R, Miflin BJ, Idler KB, Barker D (1984) Glutamine synthetase of *Phaseolus vulgaris* L: Organ specific expression of a multigene family. J Mol Appl Genet 2: 589

Davis EO, Johnston AWB (1990a) Analysis of three nodD genes of *Rhizobium leguminosarum* biovar *phaseoli; nodd1* is preceded by *nolE*, a gene whose product is secreted from the cytoplasm. Mol Microbiol 4: 921

Davis EO, Johnston AWB (1990b) Regulatory functions of the three *nodD* genes of *Rhizobium leguminosarum* biovar *phaseoli*. Mol Microbiol 4: 933

Davis EO, Evans IJ, Johnston AWB (1988) Identification of *nodX*, a gene that allows *Rhizobium leguminosarum* biovar *viciae* strain TOM to nodulate Afghanistan peas. Mol Gen Genet 212: 531

De Maad RA, Spaink HP, Pees E, Mulders IHM, Wijfjes A, Wijffelman CA, Okker RJH, Lugtenberg BJJ (1989a) Localization and symbiotic function of a region on the *Rhizobium leguminosarum* Sym plasmid pRLIJI responsible for a secreted flavonoid-inducible 50kD protein. J Bacteriol 171: 1151

De Maagd RA, Wijfjes AHM, Spaink HP, Ruiz-Sainz JE, Wijffelman CA, Okker RJH, Lugtenberg BJJ (1989b) *nodO*, a new *nod* gene of the *Rhizobium leguminosarum* biovar *viciae* Sym plasmid encodes a secreted protein. J Bacteriol 171: 6764

Debellé F, Rosenberg C, Vasse J, Maillet F, Martinez E, Denarie J, Truchet G (1986) Assignment of symbiotic developmental phenotypes to common and specific nodulation (*nod*) genetic loci of *Rhizobium meliloti*. J Bacteriol 168: 1075

Debellé F, Sharma SR (1986) Nucleotide sequence of *Rhizobium meliloti* RCR 2011 genes involved in host specificity of nodulation. Nucl Acids Res 14: 7453

Deshmane N, Stacey G (1989) Identification of *Bradyrhizobium nod* genes involved in host-specific nodulation. J Bacteriol 171: 3324

Diaz CL, Melchers LS, Hooykaas PJJ, Lugtenberg BJJ, Kijne JW (1989) Root lectin as a determinant of host-plant specificity in the *Rhizobium*-legume symbiosis. Nature 38: 579

Dilworth M, Glenn A (1984) How does a legume nodule work? TIBS 108: 519

Djordjevic MA, Gabriel DW, Rolfe BG (1987) *Rhizobium*—the refined parasite of legumes. Ann Rev Phytopath 25: 145

Djordjevic MA, Redmond JW, Batley M, Rolfe BG (1987) Clovers secrete specific phenolic compounds which either stimulate or repress *nod* gene expression in *Rhizobium trifolii*. EMBO J 6: 1173

Djordjevic MA, Schofield PR, Rolfe BG (1985) Tn5 mutagenesis of *Rhizobium trifolii* host-specific nodulation genes result in mutants with altered host-range ability. Mol Gen Genet 200: 463

Downie JA (1989) The *nodL* gene from *Rhizobium leguminosarum* is homologous to the acetyl transferases encoded by *lacA* and *cysE*. Mol Microbiol 3: 1649

Downie JA, Johnston AWB (1988) Nodulation of legumes by *Rhizobium*. Plant Cell and Environment 11: 403

Downie JA, Surin BP (1990) Either of two *nod* gene loci can complement the nodulation defect of a *nod* deletion mutant *Rhizobium leguminosarum* bv *viciae*. Mol Gen Genet 222: 81

Downie JA, Economou A, Scheu AK, Johnston AWB, Firmin JL, Wilson KE, Cubo MT, Mavridou A, Marie C, Davies A, Surin BP (1990) The *Rhizobium leguminosarum* bv *viciae* NodO protein compensates for the exported signal made by the host-specific nodulation genes. In: Nitrogen Fixation: Achievements and Objective, (eds) Gresshoff PM, Roth LE, Stacey G, Newton W, Chapman and Hall, pp 201–206

Downie JA, Knight CD, Johnston AWB, Rossen L (1985) Identification of genes and gene products involved in nodulation of peas by *Rhizobium leguminosarum*. Mol Gen Genet 198: 255

Dreyfus B, Alazard D, Dommergues YR (1984) Stem-nodulating *Rhizobia*. In: Current Perspectives of Microbial Ecology, (eds) Klug MG, Reddy CE. American Society for Microbiology, Washington, DC: p 161

Duc G, Trouvelot A, Gianinazzi-Pearson V, Gianinazzi S (1989) First report of non-mycorrhizal plant mutants (Myc⁻) obtained in pea (*Pisum sativum* L) and fababean (*Vicia faba* L). Plant Science 60: 215

Dudley ME, Jacobs TW, Long SR (1987) Microscopic studies of cell divisions induced in alfalfa roots by *Rhizobium meliloti*. Planta 171: 289

Dunn MK, Dickstein R, Feinbaum R, Burnett BK, Peterman TK, Thoidis G, Goodman HM, Ausubel FM (1988) Development regulation of nodule-specific genes in a alfalfa root nodules. Mol Plant-Microbe Inter 1: 66

Dusha I, Bakos A, Kondorosi A, de Bruijn FJ, Schell J (1989) The *Rhizobium meliloti* early nodulation genes (*nodABC*) are nitrogen-regulated: isolation of a mutant strain with efficient nodulation capacity on alfalfa in the presence of ammonium. Mol Genet 219: 89

Egelhoff TT, Fisher RF, Jacobs TW, Mulligan JT, Long SR (1985) Nucleotide sequence of *Rhizobium meliloti* 1021 nodulation genes: *nodD* is read divergently from *nodABC*. DNA 4: 241

Evans I, Downie AJ (1986) The *nodI* product of *Rhizobium leguminosarum* is closely related to ATP-binding bacterial transport proteins: nucleotide sequence of the *nodI* and *nodJ* genes. Gene 43: 95

Faucher C, Camut S, Denarie J, Truchet G (1989) The *nodH* and *nodQ* host range genes of *Rhizobium meliloti* behave as avirulence genes in *R. leguminosarum* bv. *viciae* and determine changes in the production of plant-specific extracellular signals. Mol Plant-Microbe Inter 2: 291

Faucher C, Maillet F, Vasse J, Rosenberg C, Van Brussel AAN, Truchet G, Denarie J (1988) Evidence that *Rhizobium meliloti nodH* gene determines host specificity *via* an extracellular signal. J Bacteriol 170: 5489

Finan TM, Hirsch AM, Leigh JA, Johansen E, Kuldav GA, Deegan S, Walker GC, Singer ER (1985) Symbiotic mutants of *Rhizobium meliloti* that uncouple plant from bacterial differentiation. Cell 40: 869

Firmin JL, Wilson KE, Rossen L, Johnston AWB (1986) Flavonoid activation of nodulation genes in *Rhizobium* reversed by other compounds present in plants. Nature 324: 90

Fisher RF, Egelhoff TT, Mulligan JT, Long SR (1988) Specific binding of proteins from *Rhizobium meliloti* cell-free extracts containing NodD to DNA sequences upstream of inducibe nodulation genes. Genes Dev 2: 282

Fisher RF, Long SR (1989) DNA footprint analysis of the transcriptional activator proteins NodD1 and NodD3 on inducible *nod* gene promoters. J Bacteriol 171: 5492

Fisher RF, Swanson JA, Mulligan JI, Long SR (1987) Extended region of nodulation genes in *Rhizobium meliloti* 1021. II Nucleotide sequence, transcription start sites and protein products. Genetics 117: 191

Fisher RF, Tu JK, Long SR (1985) Conserved nodulation genes in *Rhizobium meliloti* and *Rhizobium trifolii*. Applied Environ. Microbiol 9: 1439

Forde BF, Cullimore JV (1989) The molecular biology of glutamine synthetase in higher plants. Oxford Surv Plant Mol Cell Biol 6: 247

Franssen HJ, Thompson DV, Idler K, Kormelink R, Van Kammen A, Bisseling T (1989) Nucleotide sequence of two soybean ENOD2 early nodulin genes encoding Ngm-75. Plant Mol Biol 14: 103

Gallon JR, Chaplin AE (1987) An Introduction to Nitrogen Fixation, Cassell

Gebhardt C, Oliver JE, Forde BG, Sarrelainen R, Miflin BJ (1986) Primary structure and differential expression of glutamine synthetase genes in nodules, roots and leaves of *Phaseolus vulgaris*. EMBO J 5: 1429

Gerhold D, Stacey G, Kondorosi A (1989) Use of a promoter specific probe to identify two loci from *Rhizobium meliloti* nodulation regulon. Plant Mol Biol 12: 181

Goethals K, Gao M, Tomekpe K, van Montagu M, Holsters M (1989) Common *nodABC* genes in *nod* locus 1 of *Azorhizobium caulinodans*: nucleotide sequence and plant-inducible expression Mol Gen Genet 219: 289

Goethals K, Van den Eede G, Van Montagu M, Holsters M (1990) Indentification and characterization of a functional *nodD* gene in *Azorhizobium caulinodans* ORS571. J Bacteriol 172: 2658

Göttfert M, Hitz S, Hennecke H (1990a) Identification of *nodS* and *nodU*, two inducible genes inserted between the *Bradyrhizobium japonicum nodYABC* and *nodIJ* genes. Mol Plant-Microbe Inter. 3: 308

Göttfert M, Grobe P, Henneke H (1990b) Proposed regulatory pathway encoded by the *nodV* and *nodW* genes determinants of host specificity in *Bradyrhizobium japonicum*. Proc Natl Acad Sci USA 87: 2680

Göttfert M, Lamb JW, Gasser R, Semenza J, Hennecke H (1989) Mutational analysis of the *Bradyrhizobium japonicum* common *nod* genes and further *nod* box-linked genomic DNA regions. Mol Gen Genet 215: 407

Göttfert M, Weber J, Hennecke H (1988) Induction of a *nodA-lacZ* fusion in *Bradyrhizobium japonicum* by an isoflavone. J Plant Physiol 132: 394

Göttfert M, Horvath B, Kondorosi E, Putnoky P, Rodriguez-Quinones F, Kondorosi A (1986) At least two *nodD* genes are necessary for efficient nodulation of alfalfa by *Rhizobium meliloti*. J Mol Biol 191: 411

Götz R, Evans IJ, Downie JA, Johnston AWB (1985) Identification of the host-range DNA which allows *Rhizobium leguminosarum* strain TOM to nodulate cv Afghanistan peas. Mol Gen Genet 201: 296

Götz R, Limmer N, Ober K, Schmit R (1982) Motility and chemotaxis in two strains of *Rhizobium* with complex flagella. J Gen Microbiol 128: 789

Govers F, Gloudemans T, Moerman M, van Kammen A, Bisseling T (1985) Expression of plant genes during the development of pea root nodules. EMBO J 4: 861

Györgypal Z, Iyer N, Kondorosi A (1988) Three regulatory *nodD* alleles of diverged flavonoid-specificity are involved in host-dependent nodulation by *Rhizobium meliloti*. Mol Gen Genet 212: 85

Halverson LJ, Stacey G (1986) Signal exchange in plant-microbe interactions. Microb Reviews 50: 193

Henikoff S, Haughn GW, Calvo JM, Wallace JC (1988) A large family of bacterial activator proteins. Proc Nat Acad Sci USA 85: 6602

Higgins CF, Hiles ID, Salmond GPC, Gill DR, Downie JA, Evans IJ, Holland IB, Gray L, Bucke SD, Bell AW, Hermodson MA (1986) A family of related ATP-binding subunits coupled to many distinct biological processes in bacteria. Nature 323: 448

Hinchee MAW, Connor-Ward DV, Newell CA, McDonnell RE, Sato SJ, Gasser CS, Fischoff DA, Re DB, Fraley RT, Horsch RB (1988) Production of transgenic soybean plants using *Agrobacterium*-mediated DNA transfer. Bio/Technol 6: 915

Hirsch AM, Bhuvaneswari TV, Torrey JG, Bisseling T (1989) Early nodulin genes are induced in alfalfa root outgrowths elicited by auxin transport inhibitors. Proc Natl Acad Sci USA 86: 1244

Hombrecher G, Götz R, Dibb NJ, Downie JA, Johnston AWB, Brewin NJ (1984) Cloning and mutagenesis of nodulation genes from *Rhizobium leguminosarum* TOM, a strain with extended host range. Mol Gen Genet 184: 293

Hong G-F, Burn JE, Johnston AWB (1987) Evidence that DNA involved in the expression of nodulation *nod* genes is *Rhizobium*, binds to the product of the regulatory gene *nodD*. Nucl Acids Res 15: 9677

Honma MA, Ausubel FM (1987) *Rhizobium meliloti* has three functional copies of the *nodD* symbiotic regulatory gene. Proc Natl Acad Sci USA 84: 8558

Honma MA, Asomaning M, Ausubel FM (1990) *Rhizobium meliloti nodD* genes mediate host-specific activation of *nodABC*. J Bacteriol 172: 901

Horvath B, Bachem CWB, Schell J, Kondorosi A (1987) Host-specific regulation of nodulation genes in *Rhizobium* is mediated by a plant-signal interacting with the *nodD* gene product. EMBO J 6: 841

Horvath B, Kondorosi E, John M, Schmidt J, Török I, Györgypal Z, Barabas I, Wieneke U, Schell J, Kondorosi A (1986) Organisation, structure and symbiotic function of *Rhizobium meliloti* nodulation genes determining host specificity for alfalfa. Cell 46: 335

Innes RW, Kuempl PL, Plazinski J, Canter-Cremers H, Rolfe BG, Djordjevic MA (1985) Plant factors induce expression of nodulation and host range genes in *Rhizobium trifoli*. Mol Gen Genet 201: 426

Jacobs TW, Egelhoff TT, Long SR (1985) Physical and genetic map of a *Rhizobium meliloti* nodulation gene region and nucleotide sequence of *nodC*. J Bacteriol 162: 469

John M, Schmidt J, Wieneke U, Kondorosi E, Kondorosi A, Schell J (1985) Expression of the nodulation gene *nodC* of *Rhizobium meliloti* in *Escherichia coli*: role of the *nodC* gene product in nodulation. EMBO J 4: 2425

John M, Schmidt J, Wieneke U, Krüssmann HD, Schell J (1988) Transmembrane orientation and receptor-like structure of the *Rhizobium meliloti* common nodulation protein NodC. EMBO J 7: 583

Johnston AWB, Burn JE, Economou A, Davis EO, Hawkins FKL, Bibb MJ (1988) Genetic factors affecting host range in *Rhizobium leguminosarum*. In: Molecular Genetics of Plant–Microbe Interactions, eds Palacios, R and Verma, DPS. APS Press, St Paul, MN: p 378

Keen NT, Staskawicz B (1988) Host range determinants in plant pathogens and symbionts. Ann Rev Microbiol 42: 421

Kofoid EC, Parkinson, JS (1988) Transmitter and receiver modules in bacterial signal proteins. Proc Nat Acad Sci USA 85: 4981

Köhler F, Golz C, Eapen S, Kohn H, Schieder O (1987) Stable transformation of moth bean *Vigna aconitifolia* via direct gene transfer. Plant Cell Rep 6: 313

Kondorosi E, Gyuris J, Schmidt J, John M, Duda E, Hoffmann B, Schell J, Kondorosi A (1989) Positive and negative control of *nod* gene expression in *Rhizobium meliloti* is required for optimal nodulation. EMBO J 8: 1331

Kosslak RM, Bookland R, Barkei J, Paaren HE, Appelbaum ER (1987) Induction of *Bradyrhizobium japonicum* command *nod* genes by isoflavones isolated from *Glycine max*. Proc Nat Acad Sci USA 84: 7428

Kosslak, RM, Joshi RS, Bowen, BA, Paaren HE, Appelbaum ER (1990) Strain-specific inhibition of *nod* gene induction in *Bradyrhizobium japonicum* by flavonoid compounds. Applied and Environ Microbiol 56: 1333

Lamb CT, Lawton MA, Dron M, Dixon RA (1989) Signals and transduction mechanisms for activation of plant defences against microbial attack. Cell 56: 215

Legocki RP, Verma DPS (1980) Identification of "nodule-specific" host proteins (Nodulins) involved in the development of *Rhizobium*-legume symbiosis. Cell 20: 153

Lerouge P, Roche P, Faucher C, Maillet F, Truchet G, Prome JC, Denarie J (1990a) Symbiotic host-specificity of *Rhizobium meliloti* is determined by a sulphated and acylated glucosamine oligosaccharide signal. Nature 344: 781

Lerouge P, Roche P, Prome J-C, Faucher C, Vasse J, Maillet F, Camut S, de Billy F, Barker DG, Denarie J, Truchet G (1990b) *Rhizobium meliloti* nodulation genes specify the production of an alfalfa-specific sulfated lipo-oligosaccharide signal. In: Nitrogen Fixation: Achievements and Objectives, eds Gresshoff PM, Roth LE, Stacey G, Newton W, Chapman & Hall, 177–186

Lewin A, Cervantes E, Wong C-H, Broughton WJ (1990) *nodSU*, two new *nod* genes of the broad host range *Rhizobium* strain NGR234 encode host-specific nodulation of the tropical tree *Leucaena leucocephala*. Mol Plant–Microbe Inter 3: 317

Libbenga, KR, Harkes PAA (1973) Initial proliferation of cortical cells in the formation of root nodules in *Pisum sativum*. Planta 114: 17

Lloyd, CW, Pearce KJ, Rawlins DJ, Ridge RW, Shaw PJ (1987) Endoplasmatic microtubules connect the advancing nucleus to the tip of legume root hairs, byt F-actin is involved in basipetal migration. Cell Motil Cytoskeleton 8: 27

Long, SR (1989a) Life together in the underground. Cell 56: 203

Marcker A, Lund M, Jensen EO, Marcker KA (1984) Transcription of the soybean leghemoglobin gene during nodule development. EMBO J 3: 1691

Martinez E, Romero D, Palacios R (1990) The *Rhizobium* genome. Crit Rev Plant Sci 9: 59

Marvel DJ, Torrey JG, Ausubel FM (1987) *Rhizobium* symbiotic gene required for nodulation of legume and non-legume hosts. Proc Nat Acad Sci USA 84: 1319

Maxwell CA, Hartwig UA, Joseph CM, Phillips DA (1989) A chalcone and two related flavonoids released from alfalfa roots induce *nod* genes of *Rhizobium meliloti*. Plant Physiol 91: 842

McIver J, Djordjevic MA, Weinmanan JJ, Bender GL, Rolfe BG (1989) Extension of host range of *Rhizobium leguminosarum* bv. *trifolii* caused by point mutations in *nodD* that result in alterations in regulatory function and recognition of inducer molecules. Mol Plant-Microbe Inter 2: 97

Mellor RB, Garbers C, Werner D (1989) Peribacteroid membrane nodulin gene induction by *Rhizobium japonicum* mutants. Plant Mol Biol 12: 307

Mellor RB, Werner D (1987) Peribacteroid membrane biogenesis in mature legume root nodules. Symbiosis 3: 75

Mulligan JT, Long SR (1989) A family of activator genes regulates expression of *Rhizobium meliloti* nodulation genes. Genetics 122: 7

Mulligan JT, Long SR (1985) Induction of *Rhizobium meliloti nodC* expression by plant exudate requires *nodD* Proc Nat Acad Sci USA 82: 6609

Nap J-P, Bisseling T (1990a) The roots of nodulins. Physiol Plant 79: 407

Nap J-P, Bisseling T (1990b) Nodulin function and nodulin gene regulation in root nodule development. In: Molecular Biology of Symbiotic Nitrogen Fixation, ed Gresshoff PM CRC Press, p 181

Newcomb W (1976) A correlated light and electron microscopic study of symbiotic growth and differentiation in *Pisum sativum* root nodules. Can J Bot 54: 2163

Nguyen T, Zelechowska M, Foster V, Bergmann H, and Verma DPS (1985) Primary structure of the soybean nodulin-35 gene encoding uricase II localized in the peroxisomes of uninfected cells of nodules. Proc Natl Acad Sci USA 82: 5040

Nieuwkoop AJ, Banfalvi Z, Deshmane N, Gerhold D, Schell MG, Sirotkin KM, Stacey G (1987) A locus encoding host range is linked to the common nodulation genes of *Bradyrhizobium japonicum*. J Bacteriol 169: 2631

Olson ER, Sadowsky MJ, Verma DPS (1985) Identification of genes involved in the *Rhizobium*-legume symbiosis by Mu-*d1* (Kan, *lac*)-generated transcription fusions. Bio/Technol 3: 143

Peters NK, Frost JW, Long SR (1986) A plant flavone, luteolin, induces expression of *Rhizobium meliloti* nodulation genes. Science 233: 377

Peters NK, Long SR (1988) Alfalfa root exudates and compounds which promote or inhibit induction of *Rhizobium meliloti* nodulation genes. Plant Physiol 88: 396

Petit A, Stougaard J, Kuhle A, Marcker KA, Tempe J (1987) Transformation and regeneration of the legume *Lotus corniculatus*: A system for molecular studies of symbiotic nitrogen fixation. Mol Gen Genet 207: 245

Postma, JG, Jacobsen E, Feenstra WJ (1988) Three pea mutants with an altered nodulation studied by genetic analysis and grafting. J Plant Physiol 132: 424

Puonti-Kaerlas J, Eriksson T, Engstrom P (1990) Production of transgenic pea (*Pisum sativum* L.) plants by *Agrobacterium tumefaciens*. Theor Appl Genet 80: 246

Putnoky P, Kondorosi A (1986) Two gene clusters of *Rhisobium meliloti* code for early essential nodulation functions and a third influences nodulation efficiency. J Bacteriol 167: 881

Ramakrishnan N, Prakash RK, Shantharam S, Duteau NM, Atherly AG (1986) Molecular cloning and expression of *Rhizobium fredii* USDA 193 nodulation genes: extension of host range for nodulation. J Bacteriol 168: 1087

Recourt K, van Brussel AAN, Driessen AJM, Lugtenberg BJJ (1989) Accumulation of a *nod* gene inducer, the flavonoid naringenin, in the cytoplasmic membrane of *Rhizobium leguminosarum* biovar *viciae* is caused by the pH-dependent hydrophobicity of naringenin. J Bacteriol 171: 4370

Redmond JW, Batley M, Djordjevic MA, Innes RW, Kuempl P, Rolfe BG (1985) Flavones induce expression of nodulation genes in *Rhizobium*. Nature 323: 632

Renalier M, Batut J, Ghai J, Terzaghi B, Gherardi M, David M, Gavnerone A, Uasse J, Truchet G, Huguet G, Boistard P (1987) A new symbiotic cluster on the pSYM megaplasmid of *Rhixobium meliloti* 2011 carries a functional *fix* gene repeat and a *nod* locus. J Bacteriol 169: 2231

Ridge RW, Rolfe BG (1986) *Rhizobium* sp. degradation of legume root hair cell wall at the site of infection thread origin. Appl Environ Microbiol 50: 717

Riordan JR, Rommens JM, Kerem B-S, Alon N, Rozmahel R, Grzelczak Z, Zielenski J, Lok S, Plavsic N, Chou J-L, Drumm ML, Lannuzzi MC, Collins FS, Tsui L-C (1989) Identification of the cystic fibrosis gene: Cloning and characterization of complementary DNA. Science 245: 1066

Robertson JG, Lyttleton P, Bullivant S, Grayston GF (1978) Membranes in lupin root nodules I. The role of Golgi bodies in the biogenesis of infection threads and peribacteroid membranes. J Cell Sci 30: 129

Robertson JG, Lyttleton P (1982) Coated and smooth vesicles in the biogenesis of cell walls, plasma membranes, infection threads and peribacteroid membranes in root hairs and nodules of white clovers. J Cell Sci 58: 63

Robertson JG, Wells B, Brewin NJ, Wood EA, Knight CD, Downie JA (1985) The legume-*Rhizobium* symbiosis: a cell surface interaction. J Cell Sci Suppl 2: 317

Rolfe B, Gresshoff P (1988) Genetic analysis of legume nodule initiation. Ann Rev Plant Physiol 39: 297

Rossen L, Johnston AWB, Downie JA (1984) DNA sequence of the *Rhizobium leguminosarum* nodulation genes *nodA,B* and *C* required for root hair curling. Nucl Acids Res 12: 9497

Rossen L, Shearman CA, Johnston AWB, Downie JA (1985) The *nodD* gene of *Rhizobium leguminosarum* is autoregulatory and in the presence of plant exudate induces the *nodABC* genes. EMBO J 4: 3369

Rostas K, Kondorosi E, Horvath B, Simoncsits A, Kondorosi A (1986) Conservation of extended promoter regions of nodulation genes in *Rhizobium*. Proc Nat Acad Sci USA 83: 191

Russell P, Schell MG, Nelson KK, Halverson LJ, Sirotkin KM, Stacey G (1985) Isolation and characterisation of the DNA region encoding nodulation functions in *Bradyrhizobium japonicum*. J Bacteriol 164: 1301

Sadowsky MJ, Olson ER, Foster VE, Kosslak RM, Verma DPS (1988) Two host-inducible genes of *Rhizobium fredii* and characterization of the inducing compound. J Bacteriol 170: 171

Schell, MA, Sukordhaman M (1989) Evidence that the transcription activator encoded by the *Pseudomonas putida nahR* genes is evolutionarily related to the transcription activators encoded by the *Rhizobium nodD* genes. J Bacteriol 171: 1952

Scheres B, Van de Wiel C, Zalensky A, Horvath B, Spaink H, Van Eck H, Zwartkruis F, Wolters AM, Gloudemans T, Van Kammen, Bisseling T (1990) The ENOD12 gene product is involved in the infection process during the pea-*Rhizobium* infection. Cell 60: 281

Schlaman HRM, Okker RJH, Lugtenberg BJJ (1990) Subcellular localization of the *Rhizobium leguminosarum nodI* gene product. J Bacteriol 172: 5486

Schlaman HRM, Spaink HP, Okker RJH, Lugtenberg BJJ (1989) Subcellular localization of the *nodD* gene product in *Rhizobium leguminosarum* J Bacteriol 171: 4686

Schofield PR, Watson JM (1986) DNA sequence of *Rhizobium trifolii* nodulation genes reveals a reiterated and potentially regulatory sequence preceding *nodABC* and *nodFE*. Nucl Acids Res 14: 2891

Schwedock J, Long SR (1989) Nucleotide sequence of protein products of two new nodulation genes of *Rhizobium meliloti, nodP* and *nodQ*. Mol Plant–Microbe Inter 9: 181

Scott KF (1986) Conserved nodulation genes from the non-legume symbiont *Bradyrhizobium* sp. (*Parasponia*). Nucl Acids Res 14: 2905

Scott KF, Bender GL (1990) The *Parasponia-Bradyrhizobium* symbiosis. In: Molecular Biology of Symbiotic Nitrogen Fixation, ed Gresshoff PM, CRC Press, p 231

Sengupta-Gopalan, Pitas (1986) Expression of nodule-specific glutamine synthetase genes during nodule development in soybeans. Plant Mol Biol 7: 189

Sharma SB, Signer ER (1990) Temporal and spatial regulation of the symbiotic genes of *Rhizobium meliloti in planta* revealed by transposon Tn5-*gusA*. Genes and Dev 4: 344

Shearman CA, Rossen L, Johnston AWB, Downie JA (1986) The *Rhizobium* gene *nodF* encodes a protein similar to acyl carrier protein and is regulated by *nodD* plus a factor in pea root exudate. EMBO J 5: 647

Sheldon PS, Kekwick RGO, Sidebottom C, Smith CG, Slabas AR (1990) 3-oxoacyl-(acyl-carrier protein) reductase from avocado (*Persea americana*) fruit mesocarp. Biochem 272: 713

Sherman DH, Malpartida F, Bibb MJ, Kieser HM, Hopwood DA (1989) Structure and deduced function of the granaticin producing polyketide synthase gene cluster of *Streptomyces violaceoruber* Tü22. EMBO J 8: 2717

Simon R (1989) In: Promiscuous plasmids of Gram-negative bacteria, (Thomas CM, ed). Academic Press p 207

Simon R, Prieffer UB (1990) Vector technology of relevance to nitrogen fixation research. In: Molecular Biology of Symbiotic Nitrogen Fixation ed Gresshoff PM, CRC Press, p 13

Simon R, Quandt J, Klipp W (1989) New derivatives of transposon Tn5 suitable for mobilization of replicons, generation of operon fusions and induction of genes in Gram-negative bacteria. Gene 80: 161

Smit G, Kijne JW, Lugtenberg BJJ (1987) Involvement of both cellulose fibrils and a Ca^{2+}-dependent adhesin in the attachment of *Rhizobium leguminosarum* to pea root hair tips. J Bacteriol 169: 4294

Smit G, Logman TJJ, Boerrigter METI, Kijne JW, Lugtenberg BJJ (1989) Purification and partial characterization of the *Rhizobium leguminosarum* biovar *viciae* Ca^{2+}-dependent adhesin, which mediates the first step in attachment of cells of the family *Rhizobiaceae* to plant root hair tips. J Bacteriol 171: 4054

Spaink HP, Okker RJH, Wijffelman CA, Pees E, Lugtenberg BJJ (1987a) Promoters in the nodulation region of the *Rhizobium leguminosarum* Sym plasmid pRL1JI. Plant Mol Biol 9: 27

Spaink HP, Weinman J, Djordjevic MA, Wijffelman CA, Okker RJH, Lugtenberg BJJ (1989a) Genetic analysis and cellular localization of the *Rhizobium* host specificity-determining NodE protein. EMBO J 8: 2811

Spaink H, Wijffelman CA, Okker RJH, Lugtenberg BJJ (1989b) Localization of functional regions of the *Rhizobium nodD* product using hybrid *nodD* genes. Plant Mol Biol 12: 59

Spaink HP, Wijffelman CA, Pees E, Okker RJH, Lugtenburg BJJ (1987b) *Rhizobium* nodulation gene *nodD* is a determinant of host specificity. Nature 328: 337

Sprent JI (1979) The Biology of Nitrogen Fixing Organisms. Maidenhead, McGraw-Hill

Sprent JI (1989) Which steps are essential for the formation of functional legume nodules? New Phytol 111: 129

Stachel SE, Zambryski PC (1986) *virA* and *virG* control the plant-induced activation of the T-DNA transfer process of *A. tumefaciens*. Cell 46: 325

Stougaard J, Jorgensen J-E, Christensen T, Kuhle A, Marcker KA (1990) Interdependence and nodule specificity of *cis*-acting regulatory elements in the soybean leghemoglobin lbc_3 and N23 promoters. Mol Gen Genet 222: 353

Surin BP, Downie JA (1988) Characterization of the *Rhizobium leguminosarum* genes *nodLMN* involved in efficient host specific nodulation. Mol Microbiol 2: 173

Surin BP, Downie JA (1989) *Rhizobium leguminosarum* genes required for expression and transfer of host specific nodulation. Plant Mol Biol 12: 19

Surin BP, Watson JM, Hamilton WDO, Economou A, Downie JA (1990) Molecular characterization of the nodulation gene *nodT* from two biovars of *Rhizobium leguminosarum*. Mol Microbiol 4: 245

Swanson JA, Tu JK, Ogawa J, Sanga R, Fisher RF, Long SR (1987) Extended region of nodulation genes in *Rhizobium meliloti* 1021 I. Phenotypes of Tn5 mutants. Genetics 117: 181

Thummler F, Verma DPS (1987) Nodulin100 of soybean is the subunit of sucrose synthetase regulated by the availability of free heme in nodules. J Biol Chem 262: 14730

Tingley SV, Walker EL, Coruzzi GM (1987) Glutamine synthetase genes of pea encode distinct polypeptides which are differentially expressed in leaves, root and nodules. EMBO J 6: 1

Török I, Kondorosi E, Stepkowski T, Posfai J, Kondorosi A (1984) Nucleotide sequence of *Rhizobium meliloti* nodulation genes. Nucl Acids Res 12: 9509

Truchet G, Barker DG, Camut S, de Billy F, Vasse J, Huguet T (1989) Alfalfa nodulation in the absence of *Rhizobium*. Mol Gen Genet 219: 65

Truchet G, Michel M, Dénarié J (1980) Sequential analysis of the organogenesis of lucerne (*Medicago sativa*) root nodules using symbiotically-defective mutants of *Rhizobium meliloti*. Differentiation 16: 163

Tsai FY, Coruzzi GM (1990) Dark-induced and organ-specific expression of two asparagine synthetase genes in *Pisum sativum*. EMBO J 9: 323

Turgeon BG, Bauer WD (1985) Ultrastructure of infection-thread development during the infection of soybean by *Rhizobium japonicum*. Planta 163: 328

Van Brussel AAN, Recourt K, Pees E, Spaink HP, Tak T, Wijffelman CA, Kijne JW, Lugtenberg BJJ (1990) A biovar-specific signal of *Rhizobium leguminosarum* bv. *viciae* induces increased nodulation gene-inducing activity in root exudate of *Vicia sativa* subsp *nigra* J Bacteriol 172: 5394

Van de Wiel C, Scheres B, Franssen H, Van Lierop M-J, Van Lammeren A, Van Kammen A, Bisseling T (1990) The early nodulin transcript ENOD2 is located in the nodule parenchyma (inner cortex) of pea and soybean root nodules. EMBO J 9: 1

Van den Eede G, Dreyfus B, Goethals K, Van Montagu M, Holsters M (1987) Identification and cloning of nodulation genes from the stem-nodulating bacterium ORS571. Mol Gen Genet 206: 291

Vance CP, Egli MA, Griffith SM, Miller SS (1988) Plant regulated aspects of nodulation and N$_2$ fixation. Plant Cell Env 11: 413

VandenBosch KA, Bradley DJ, Knox JP, Perotto S, Butcher GW, Brewin NJ (1989) Common components of the infection thread matrix and the intercellular space identified by immunocytochemical analysis of pea nodules and uninfected roots. EMBO J 8: 335

VandenBosch KA, Newcomb EH (1986) Immunogold localization of nodule-specific uricase in developing soybean root nodules. Planta 167: 425

Verma DPS, Delauney AJ (1988) Root nodule symbiosis: Nodulins and nodulin genes in plant gene research. In: Temporal and Special Regulation of Plant Genes, eds Verma DPS, Goldberg RB). Springer-Verlag, p 169

Vesper SJ, Bhuvaneswari TV (1988) Nodulation of soybean roots by an isolate of *Bradyrhizobium japonicum* with reduced firm attachment capability. Arch Microbiol 150: 15

Wang S-P, Stacey G (1990) Ammonia regulation of *nod* genes in *Bradyrhizobium japonicum*. Mol Gen Genet 223: 329

Ward LJH, Rockman ES, Ball P, Jarvis BDW, Scott DB (1989) Isolation and characterization of a *Rhizobium loti* gene, required for effective nodulation of *Lotus pendunculatus*. Mol Plant–Microbe Inter 2: 224

Weeden NF, Kneen B, Larve TA (1990) Genetic analysis of *sym* genes and other nodule related genes in *Pisum sativum*. In: Nitrogen Fixation: Achievements and Objectives, Gresshoff PM, Roth L E, Stacey G, Newton W eds, pp 323

White DWR, Greenwood D (1987) Transformation of the forage legume *Trifolium repens* L. using binary *Agrobacterium* vectors. Plant Mol Biol 8: 461

Yao PY, Vincent JM (1969) Host specificity in the root hair "curling factor" of *Rhizobium* spp Aust J Biol Sci 22: 413

Young JPW, Johnston AWB (1989) The evolution of specificity in the legume-*Rhizobium* symbiosis. TREE 4: 341

Zaat SAJ, Schripsema J, Wijffelman CA, Van Brussel AAN, Lugtenberg BJJ (1989) Analysis of the major inducer of the *Rhizobium nodA* promoter from *Vicia sativa* root exudate and their activity with different *nodD* genes. Plant Mol Biol 13: 175

Zaat SAJ, Van Brussel AAN, Tak T, Pees E, Lugtenberg BJJ (1987a) Flavonoids induce *Rhizobium leguminosarum* to produce *nodDABC* gene-related factors that cause thick short roots and root hair responses on common vetch. J Bacteriol 169: 3388

Zaat SAJ, Wijffelman CA, Mulders IHM, Van Brussel AAN, Lugtenberg BJJ (1988) Root exudates of various host plants of *Rhizobium legumonosarum* contain different sets of inducers of *Rhizobium* nodulation genes. Plant Physiol 86: 1298

Zaat, SAJ, Wijffelman CA, Spaink HP, Van Brussel AAN, Okker RJH, Lugtenberg BJJ (1987b). Induction of the *nodA* promoter of *Rhizobium laguminosarum* sym plasmid pRL1JI by plant flavonones and flavones. J Bacteriol 169: 198

CHAPTER 17

Nitrogen-Fixing *Enterobacter*: Structure of *nif*-Gene Group, *nif*-Plasmid Spread and Risk Assessment for Releases

Walter Klingmüller

To provide basic knowledge for their exploitation in association with cereals, nitrogen-fixing *Enterobacter agglomerans* strains from the rhizosphere of wheat were studied from two view points: 1) The molecular basis of nitrogen fixation, and 2) The risk of gene spread upon release.

The nitrogen fixation genes were identified on large indigenous plasmids of 100–200 kb. Two of these, *pEA3* and *pEA9*, were characterized by the following methods: preparing cosmid gene libraries, DNA hybridization against known *nif*-gene probes from *Klebsiella*, subcloning with promoter probe vectors, restriction analyses, *lacZ*-transcriptional fusions, and DNA sequencing. Plasmid *pEA3* contains the *nif*-genes in a continous cluster similar in order to *Klebsiella*, where they are on the chromosome. However, in contrast to that organism, the two genes coding for electron transport proteins, *nifF* and *nifJ*, are linked, transcribed together in the order of function of the proteins, and located on one side of the *nif*-gene group. It is suggested that the *nif*-region on these plasmids of *Enterobacter* represents a cornerstone in bacterial *nif*-gene group development.

Plasmid *pEA9*, through labeling with transposons and through mating experiments with cured, plasmid-free *Enterobacter* recipients, has been shown to be self-transmissible at a rate of 10^{-4} per donor for optimized laboratory conditions on LB plates. To assess the risk of uncontrolled (*nif*) gene spread for releases of such *nif*-plasmid containing *Enterobacter* strains, model experiments in soil were carried out in the laboratory, simulating environmental conditions upon release. Plasmid transfer occurred in sterilized soil, but it was not detectable in native soil due to energy limitation. Although addition of rich media (LB) elicited plasmid transfer in native soil, a wheat rhizosphere was ineffective. These data indicate that the potential of nitrogen-fixing *E. agglomerans* for disturbing the native environment upon release are low.

The Nitrogen Fixation
and its Research in China
Editor: Guo-fan Hong
(C) Springer-Verlag Berlin Heidelberg 1992

1 Introduction

During recent years bacterial nitrogen fixation has been studied in many laboratories and with increasing intensity. Not only biochemical methods, but also molecular genetic techniques have been applied to such research.

The best-known nitrogen-fixing bacterium is still *Klebsiella pneumoniae*, which is free living and related to *E. coli*. It was therefore easily amenable to all relevant methods developed for the latter.

The genus *Rhizobium*, also nitrogen-fixing, has been exploited for a long time to improve yields of leguminous plants. In contrast to *Klebsiella*, rhizobia are symbiotic bacteria, i.e. they infect the roots of host plants and initiate nodules in which the bacteria fix nitrogen, the energy for which is provided by plant photosynthesis. In return the bacteria contribute ammonium to the plant. Data on the genetics of nitrogen fixation in rhizobia are now accumulating with remarkable speed.

In countries where nutrition is mainly based on cereals (e.g. maize, wheat and rice), other nitrogen-fixing bacteria have attracted considerable interest. *Azospirilla*, in tropical and subtropical climates, occur in high numbers in the rhizosphere, on the root surface and within the roots of various C_4 and C_3 cereals and grasses (Klingmüller 1988a; Skinner et al. 1989). Recent genetic studies were directed towards the collection of *nif* structural mutants, regulatory mutants, and mutants probably defective in general nitrogen control; in addition, resistance mutants and mutants with alterations in a number of physiological or metabolic properties were obtained (Singh and Klingmüller 1986a; Abdel Salam and Klingmüller 1987; Polsinelli et al. 1991). Such mutants are required for elucidating the molecular basis of plant-bacterial interaction, and for devising strategies for obtaining beneficial effects of inoculations.

In temperate climates other bacteria have to be found with the potential to assist agriculture via biological nitrogen fixation. An ideal bacterium should not only be able to fix nitrogen and to live in close association with plant roots (e.g. wheat) in order to utilize the exudates from the roots as an energy source, but also form a substantial proportion of the root microflora in such climates. We have taken up this problem, identified such bacteria, as *Enterobacter agglomerans* strains, and studied them extensively (Kleeberger et al. 1983; Singh et al. 1983; Klingmüller 1985; Singh and Klingmüller 1986b; Klingmüller 1988b; Singh et al. 1988; Kreutzer et al. 1989; Klingmüller et al. 1989). In the following report, a short review of the earlier results is given, together with the presentation of recent, yet unpublished findings. For later releases, the question of risk assessment is also addressed. By this it is hoped to stimulate further research on this promising group of rhizosphere bacteria, and to specify the problems inherent in bacterial releases for agriculture.

2 Materials and Methods

Bacterial strains
Enterobacter agglomerans 243, 333, 334, 335, 339, all nitrogen-fixing, were isolated from the rhizosphere of wheat and barley (Kleeberger et al. 1983). These strains have large plasmids: Numbers 333 to 339 one each, number 243 two (Singh et al. 1983).

Enterobacter agglomerans 339 (pEA9) nif$^+$, Nal^r, Rp^r and its cured derivative *E. agglomerans* 339, nif$^-$, Sm^r, Rp^r were derived from the above strains.

For the mating experiments the donor plasmid was labelled either by transposon Tn 1725 (Cm^r) or Tn5-Mob (Km^r, Neo^r) (Min 1986; Herterich 1986; Klingmüller et al. 1989). Both transposons are in vitro recombinant, Tn5-Mob contains the Sau 3A-Mob-fragment of RP4.

E. coli HB101 (Boyer and Rouillard-Dussioux 1969). *E. coli* C 600 and *E. coli* C 2110, polA-1, his, rha, Nal^r from the Center for Molecular Genetics, UCSD, San Diego.

Plasmids and Transposons
pMS175-2::Tn1725 Amp^r, Tet^r, Cm^r, Mob$^+$ (Singh and Klingmüller 1986a).
pSUP5011, Amp^r, Cm^r, Tn5 Kan^r::Mob, Col-replicon (Simon 1984).
pRK415, Tet^r, RP4-replicon (Keen et al. 1988).

Growth Media
All bacterial strains were grown in LB medium, including (g per 1 H_2O): tryptone, 10; yeast extract, 5; and NaCl, 5; *Escherichia coli* at 37 °C and *Enterobacter agglomerans* at 30 °C.

Antibiotics
In the selective plates, tetracycline (tet) was applied at 15 µg/ml, kanamycin (km) at 20 µg/ml, nalidixic acid (nal), cycloheximide and nystatin at 50 µg/ml, neomycin (neo), carbenicillin (cb) and rifampicin (rp) at 100 µg/ml, and streptomycin (sm) at 200 µg/ml, final concentration. For example, plates to select for exconjugants against a soil suspension background contained km, neo, sm, cycloheximide and nystatin. Plates to select for recipients from LB-plate matings contained only sm. The spontaneous rate of mutation of the donor to Sm^r was $< 10^{-10}$.

Measuring Nitrogen Fixation
Nitrogen fixation was assessed for bacterial suspensions grown in N-free minimal medium with glucose (NFDM, for *E. coli* strains) or with sucrose (NFSM, for *E. agglomerans* strains) under anaerobic conditions, adding N_2 and using the C_2H_2 reduction assay with minor modifications. For details see Klingmüller (1985).

Soil
Soil was agricultural sandy loam, taken from the experimental farm of Bayreuth University at exactly that position where a first test release is being scheduled

(Klingmüller 1990). It had 2.2% C_{org}; 11.4 C/N; 7.2 meq CEC/100 g; 78% base saturation; pH 5.2; 10.5 mg P_2O_5, 29.4 mg K_2O and 9.8 mg MgO/100 g. It was air dried, to 8.8% remaining water content, sieved and distributed in 50 g aliquots into 250 ml Erlenmeyer flasks. To make samples sterile, they were autoclaved at 121 °C, 1 bar, for 45 min.

Bacterial Matings

In standard matings, donor and recipient cells were grown as overnight cultures in liquid LB medium with appropriate antibiotics. They were then washed, resuspended in saline and mixed 1:1.

For plate matings, 0.4 ml of that mix, plus 1.6 ml LB were spread onto LB-plates. After one night, or 3 days of incubation (risk assessment experiments), the cell suspension was taken up in saline (5 ml), washed once and plated at appropriate dilutions on LB plates supplemented with antibiotics as required (see page 345), to identify donors, recipients and exconjugants.

For matings in soil, a 4-ml volume of cell mix in either saline or LB was added to the 50 g soil in each flask and evenly distributed with a spatula. The flasks were stoppered with cotton and incubated for 3 days. The soil was then resuspended in 50 ml saline, stirred well for 30 min and allowed to settle for 20 min. From the supernatant, aliquots were taken for dilution series and platings made on LB plates, supplemented appropriately as above.

For matings in *non-sterilized soil*, prospective exconjugant colonies coming up were further screened for rifampicin resistance, and in critical cases for the presence of *pEA9* (agarose gel electrophoresis of lysates) and for nif$^+$-phenotype (acetylene reduction assay). Transfer-rates are generally expressed as number of exconjugants per donor cells on the mating plates or in the flasks at the end of the mating period.

Curing of Plasmids

E. agglomerans strains were cured by a heat treatment procedure, i.e. growing and subculturing the cells in nutrient broth at 36.5 °C, for up to 30 passages and assaying their nitrogenase activity every second time. As a control, the strains were subcultured at 30 °C. For details see Singh et al. (1983).

Preparation of Plasmid DNA and Electrophoresis

For rapid detection of plasmid DNA, crude lysates from a small volume of the bacterial culture (1.0 ml) were prepared following either the method described by Kado and Liu (1981) or modified laboratory protocols. Whenever necessary, large quantities of plasmid DNA were purified by cesium chloride-ethidium bromide (CsCl-EtBr) density gradient centrifugation. Plasmids were separated on 0.7% agarose gels and then stained with ethidium bromide. Miniprep DNA for restriction digests was obtained by the alkaline lysis method (Maniatis et al. 1982). For separating restriction fragments, 0.8% agarose gels containing EtBr were used.

Colony Hybridization, southern Transfer and southern Hybridization
These manipulations were done according to Maniatis et al. (1982) and Singh et al. (1988).

3 Results

3.1 Taxonomical Survey

To choose the appropriate bacteria, we first conducted a taxonomical survey screening for bacteria in the innermost rhizosphere of wheat and barley roots isolated from fields in the vicinity of Bayreuth, Germany. The frequencies of bacteria of different groups were registered.

A detailed account of the results obtained has been published (Kleeberger et al. 1983). As can be seen from Table 1, the largest class of bacteria in the innermost rhizosphere of both barley and wheat was *Pseudomonas* (52%). Only very few reports exist in the literature, implying that Pseudomonads may be natural nitrogen-fixers. Therefore, this class of bacteria did not seem suitable for our purposes. However, the next largest class (23%), isolated from the innermost rhizosphere of wheat, was that of *Enterobacteriaceae*, containing the very homogenous group of *Enterobacter agglomerans* as the largest component, amounting to 14%. These bacteria are related to *E. coli* but also to *Klebsiella*. It is therefore not surprising that some of them fix nitrogen. To study this property of *E. agglomerans* in detail seemed of great interest, recalling that these bacteria, as derived from Table 1, and in contrast to *Klebsiella*, were isolated as members of the innermost rhizosphere.

Table 1. Bacteria in rhizosphere of cereals (percent of total). Roots were either carefully washed and surface sterilized by H_2O_2-treatment, or only washed, before homogenizing them and spreading aliquots onto nutrient agar plates to obtain individual colonies. These were then classified

	Barley		Wheat	
	washed	washed and H_2O_2 treated	washed	washed and H_2O_2 treated
	($n = 61$)	($n = 45$)	($n = 28$)	($n = 62$)
Pseudomonas spp.	34	53	46	52
Enterobacteriaceae		16	4	23*
Achromobacter spp.	—	18	—	8
Corynebacterium spp.	61	13	50	15
Bacillus spp.	5	5	—	2

* *E. agglomerans* 14, *E. cloacae* 5, *S. liquefaciens* 4

3.2 Analysis of *nif*-Genes in *Enterobacter agglomerans*

Altogether 50 strains of *E. agglomerans* were isolated, five of which were found to fix nitrogen. We have analyzed these strains on the molecular level, seeking the location, distribution and structural composition of their nitrogen-fixation genes (Singh et al. 1983; Singh and Klingmüller 1985; Singh and Klingmüller 1986b; Singh et al. 1988; Kreutzer et al. 1989). Recalling that the *Klebsiella* *nif*-genes form a cluster of about 25 kb on the chromosome but that certain *Rhizobium* species have *nif*-genes located on plasmids, we decided first to screen the *E. agglomerans* plasmids. We conducted an investigation which lead to the identification of large indigenous plasmids with molecular weights between 73 and 100 MDa (Table 2).

Two of the five nif⁺ strains, viz. nos. 335 and 339, could be cured of their plasmids by heat treatment. Then, not only did they lose the respective plasmid, but also their nitrogen-fixing capability. This was a first indication that at least some (if not all) of the *nif*-genes were localized on these plasmids. The plasmid DNA was separated from the chromosomal DNA in agarose gels, and the *Klebsiella nifKDH* gene region was used as a radioactive probe, which was found to hybridize with a homologous region on one plasmid in each strain, confirming the presence of *nifKDH*

Further investigations focussed on plasmid *pEA3* from strain 333. This plasmid, from more recent investigations, is 110 kb in size (Kreutzer 1986; Singh et al. 1988). By preparing a cosmid gene bank with overlapping fragments of this plasmid, subcloning, restriction mapping and DNA-hybridization against suitable *nif*-gene probes from *Klebsiella*, a physical and genetic map of the plasmid was obtained. The plasmid contains a 23 kb cluster of *nif*-genes which shows homology (by Southern hybridization and heteroduplex analysis) not only to the *KDH* genes, but to nearly all other *nif*-genes of *Klebsiella pneumoniae* M5al. Apart from *nifJ* and *F*, which are located on the right end of the *pEA3* *nif*-gene cluster (near *nifQB*) (Kreutzer et al. 1991), all the other *nif*-genes on *pEA3* are organized in the same manner as in *K. pneumoniae*. Preliminary

Table 2. Designations and molecular weights of plasmids from *nif*-positive *E. agglomerans* strains

E. agglomerans strain	No. of plasmids	Plasmid designation	Molecular weight ($\times 10^6$) + S.E.M.[a]
243	2	pEA1	73 ± 4
		pEA2	100 ± 1
333	1	pEA3	77 ± 2
334	1	pEA4	77 ± 2
335	1	pEA5	100 ± 1
339	1	pEA9	100 ± 1

[a] The molecular weights of plasmids were determined on the basis of relative mobilities in agarose gels using plasmid standards of known molecular weight. There were five molecular weight determinations performed for each plasmid

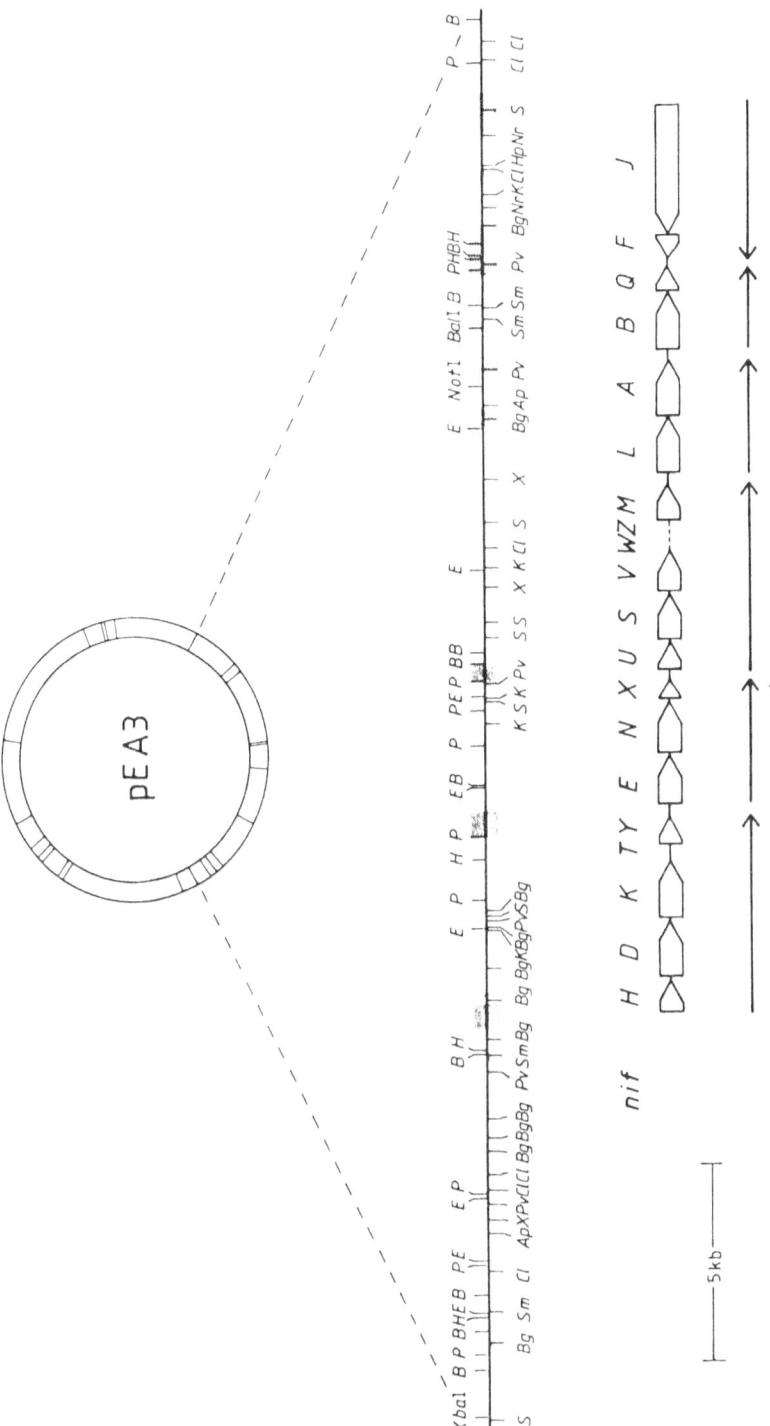

Fig. 1. Organization of *nif*-gene group in *E. agglomerans* 333 (pEA3). *Top*: BamHI restriction map of pEA3. *Center*: Restriction map of insert in cosmid clone peaMS2-2. Those regions sequenced are dashed. *Bottom*: Position and orientation of *nif*-genes and preliminary arrangement in operons. From Kreutzer (1989)

evidence (Steibl 1989; Dippe, personal communication) points to the possibility that the same holds for others of the remaining four *nif*-plasmids from *E. agglomerans*, and in particular for *pEA9*. A *BamHI* restriction map of *pEA3* and a detailed restriction map of the 23 kb *nif*-region on *pEA3* is presented in Fig 1.

To check whether the nif information in this region is complete, i.e. sufficient for nitrogen fixation, we have tried to transfer it as entire functional unit into *E. coli* and to look for nitrogen fixation of such cells. The cosmid vector had a high copy number which very likely would interfere with nitrogen fixation in *E. coli* (Riedel et al. 1983). Therefore, the cosmid clones as such were inappropriate. Instead, low copy number vectors had to be cloned into those constructs and be used in combination with a polA⁻ *E. coli* strain, to block the colE1-origin.

One such construct was #27-1 (Fig. 2). By partially digesting this with *BamHI* and allowing self ligation, construct #49 was obtained, which had only the RK-origin. The former construct, transformed into *E. coli* polA⁻, and the latter, even in *E. coli* HB101, made the recipients *nif*-positive (acetylene-reducing). This demonstrates that the genes we are studying are not only

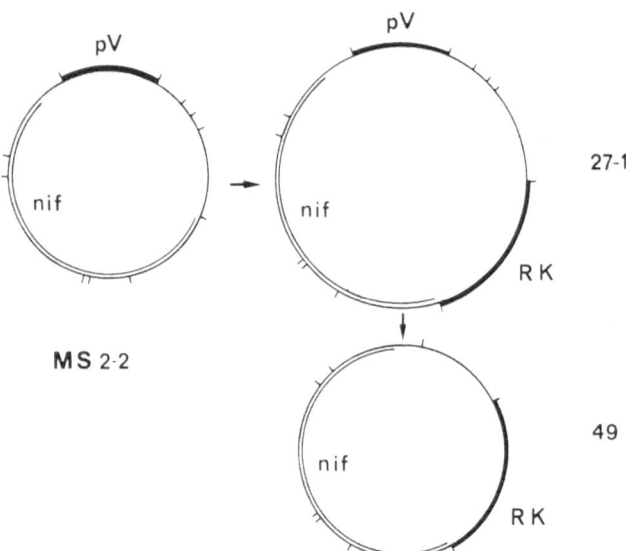

Fig. 2. Nif plasmid construction with broad host range vectors. *Left*: Cosmid peaMS2-2 from a pEA3-*BamHI* gene bank. pV: cosmid pV34, derived from pHC79, with Cb^r gene and *colE1*-replicon; nif: *nif*-region of pEA3; *BamHI* restriction sites are indicated.*Upper right*: One of the hybrid plasmids. #27-1, obtained from the former cosmid, after partially digesting it with *BamHI* and ligating the fragments into the single *BamHI* site of vector plasmid pRK415. The latter carries the tet^r genes. In hybrid plasmid #27-1, pRK415 is closely linked to the pEA *nif*-gene group. *Lower right*: One of the smaller derivatives of plasmid 27-1, viz. #49, obtained after partially digesting 27-1 with *BamHI*, self ligation and selection for Tet^r, Cb^s. In hybrid plasmid #49, cosmid pV34 and some additional material of the former plasmid is lost

DNA-fragments, homologous to *nif*-genes on the molecular level, but are functional *nif*-genes themselves, and that this gene group contains all the genetic information required to fix nitrogen (Klingmüller 1988b).

Recent studies on the nucleotide sequence of electron transport protein genes J and F, and on their transcription, as probed by transcriptional fusions with promoter probe vectors containing the *lacz* gene, have shown that these 2 genes in *E. agglomerans*, in contrast to *Klebsiella*, form one single operon. This operon is located on the right side of the *nif*-gene group beyond *nifQ*, and transcribed from a *nif*-promoter, in the opposite direction of all other *nif*-genes of the group (Fig. 3).

Comparing all N_2-fixing bacterial groups studied so far the *nif*-gene group in *E. agglomerans* hence offers the best-ordered example of *nif*-genes known. It can be considered operationally as an archaetype in *nif*-gene group evolution,

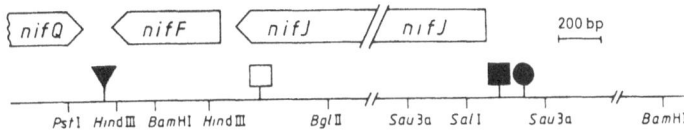

Fig. 3. Restriction map of the right end of the *nif*-gene group in *E. agglomerans* 333. The positions of the genes *nifQ*, *nifF* and *nifJ* as revealed by DNA-sequencing are indicated by *open arrows* above the map. The positions of the *nifJ* promoter and a promoter-like motif are indicated by *filled* and *open boxes*, respectively. The *nifJ* upstream activating sequence is represented by a *filled circle*, and the putative transcriptional terminator by a *filled triangle*. From Kreutzer, Dayananda and Klingmüller (1990)

J F   K D H

E. agglomerans, pEA3, plasmid, well ordered

Rhizobium meliloti, plasmid, complex

Bradyrhizobium japonicum, chromosome, complex

Azotobacter vinelandii, chromosome, additional groups

Azospirillum, chromosome, A gene separate

Anabaena, chromosome, rearrangement cycle

Klebsiella, chromosome, F gene interspersed

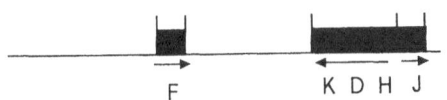

F   K D H J

Fig. 4. A sorting of nitrogen-fixing bacteria, based on the organization of their *nif*-region(s). *Top*: *E. agglomerans*, nif-plasmid pEA3, with (indicated as *blocks*) structural *nif*-genes HDK (*right*) and electron transfer genes J and F (*left*). In between (not marked) the complete set of all other *nif*-genes. The arrows give the direction of transcription. *Bottom*: *Klebsiella*, *nif*-region on chromosome, with genes indicated as above. The electron transfer genes are dispersed. Of them *nifF* in the midst of the (not marked) complete set of all other *nif*-genes. *Center*: Other cases of *nif*-gene arrangement in N_2-fixing bacteria, listed from *Rhizobium* to *Anabaena*, according to plasmid or chromosomal location and to chaotic or ordered arrangement

from which all other *nif*-gene groupings have originated. Alternatively, the *Enterobacter nif*-gene group can be considered as a final stage, into which all other *nif*-gene groupings will ultimately develop. The possible sorting of nitrogen-fixing bacteria according to this principle is exemplified in Fig. 4.

3.3 Transmissibility

To check for transmissibility of these *nif* plasmids to other bacteria by plasmid conjugation, plasmid pEA9 was taken as an example and was marked in *E. agglomerans* 339 by in-vivo transfer of Tn1725 (Cm^r) (Min 1986; Herterich 1986; Klingmüller et al. 1989). Plasmid pEA9 was chosen because the host strain #339 had been amenable to curing. This offered the opportunity to use strains with marked plasmid as donors and the cured strain as the recipient in homologous matings.

For instance, *E. agglomerans* 339T1, Rp^r (pEA9, Cm^r) was mated with the cured but Sm^r or Nal^r derivatives of *E. agglomerans* 339. Exconjugants were selected for Cm^r and Sm^r or Nal^r, and checked further by screening for plasmid content, by Southern hybridization against ^{32}P-labelled Tn1725, and by measuring acetylene reduction.

The rate of plasmid transfer thus obtained was on the order of 10^{-4} to 10^{-5} per recipient cell, depending on mating conditions, titer of donor and recipient cells, and type of donor and recipient strains used. The exconjugants contained the Tn-labelled form of plasmid pEA9 and could reduce acetylene (viz. were nif$^+$) with rates similar to the wild type of the recipients. It should be noted that two other plasmid mutants (labelled in different positions) were ineffective in conjugation experiments. Their tra$^-$ phenotype could be due to the integration of the transposon in a *tra* gene, or to other alterations.

In additional experiments, the host range of transfer was checked, offering other bacterial strains as recipients in matings with 339 T1, e.g. *E. agglomerans* 335, also cured of its nif plasmid. Under the conditions found optimal for the former (homologous) case, transfer rates of 4×10^{-4} per recipient cell were obtained. Again plasmid profiles, Southern hybridization and acetylene reduction assays proved that the exconjugants had obtained the *Tn*-labelled plasmid. Experiments with *E. cloacae* MF10 and *E. coli* HB101 as recipients, however, were negative. This indicates that the host range for transfer of plasmid pEA9 is narrow.

3.4 Risk Assessment for Releases

Since our Enterobacter agglomerans strains offer promise as biofertilizers of crop plants, field tests with such strains, and with derivatives of them improved by in-vitro recombinant methods seem urgently needed. Such trests raise the question of environmental risks. Points to consider are for instance: Would these bacteria-upon release-propagate unpredictably in the field? Would they transmit their genetic material, in particular their nif-plasmids, to other bacteria, and if so, with what consequences? To obtain data on these points, we have carried through the experiments described in the following paragraphs. For detailed accounts see Klingmüller et al. 1990 and Klingmüller 1991.

3.4.1 Comparing Two Different In-Vitro Recombinant Transposon Labels

In a first series of experiments we studied survival and plasmid transfer in homologous matings, as influenced by the plasmid label used, which was either Tn1725 or Tn5-Mob.

The results for washed cells in saline, are presented in *Table 3a* and *b*, respectively. It can be seen that in principle, the two types of matings gave similar results, both in survival data and in plasmid transfer rates. Most important is that although plasmid transfer is possible on LB plates and in sterile soil, it is extremely rare in natural soil.

In all following experiments, the Tn5-Mob labelled plasmid was used exclusively, to be able to check later for any helper effect of RP4-like plasmids from the environment on transfer of plasmids with related Mob-site.

Table 3a. Bacterial survival and plasmid transfer in standard matings of *E. agglomerans* 339 (pEA9::Tn1725) × *E. agglomerans* 339. Cell suspensions in saline, incubation at 22 °C for 3 days; means of 10 independent samples each, and standard error of the mean, where indicated

	LB-plates*	Soil sterile in flasks	Soil non-sterile in flasks
Recipients total,** zero time	$3.5 \pm 0.3 \times 10^8$	$1.7 \pm 0.15 \times 10^9$	$1.7 \pm 0.15 \times 10^9$
Recipients total, end of mating	$2.5 \pm 0.3 \times 10^9$	$9.3 \pm 1.6 \times 10^8$	$1.9 \pm 0.3 \times 10^8$
Donors total, zero time	$4.9 \pm 0.7 \times 10^8$	$2.4 \pm 0.4 \times 10^9$	$2.4 \pm 0.4 \times 10^9$
Donors total, end of mating	$3.1 \pm 0.3 \times 10^{10}$	$4.7 \pm 0.5 \times 10^9$	$1.2 \pm 0.3 \times 10^9$
Exconjugants per donor, end of mating	$3.6x \pm 0.8 \times 10^{-4}$	$2.3x \pm 0.7 \times 10^{-5}$	$0***$

* LB is Luria broth
** For comparisons of LB and soil data, the total cell number per plate or flask, and not the titer per g soil is given
*** Limit of detection $8/10^9$ donors surviving at three days. However, some colonies, verified as transconjugants, occurred in 2 out of 10 samples

Table 3b. Bacterial survival and plasmid transfer in standard matings of *E. agglomerans* 339 (pEA9::Tn5-Mob) × *E. agglomerans* 339. Cell suspensions in saline, incubation at 22 C for 3 days: means of 6 independent samples each, and standard error of the mean, where indicated

	LB-plates	Soil sterile, in flasks	Soil non-sterile, in flasks
Recipients total, zero time	$2.7 \pm 0.3 \times 10^8$	$1.4 \pm 0.14 \times 10^{10}$	$1.4 \pm 0.14 \times 10^{10}$
Recipients total, end of mating	$1.8 \pm 0.2 \times 10^{10}$	$1.4 \pm 0.04 \times 10^{10}$	$9.6 \pm 0.9 \times 10^9$
Donors total, zero time	$3.3 \pm 0.7 \times 10^8$	$1.6 \pm 0.35 \times 10^{10}$	$1.6 \pm 0.35 \times 10^{10}$
Donors total, end of mating	$2.3 \pm 0.2 \times 10^{10}$	$1.2 \pm 0.2 \times 10^{10}$	$9.2 \pm 2.5 \times 10^9$
Exconjugants per donor, end of mating	$3.1 \pm 1.1 \times 10^{-4}$	$2.6 \pm 1.5 \times 10^{-6}$	$0*$

* Limit of detection $2/10^8$ donors surviving at three days; other details as in table 3a

3.4.2 Varying Mating Time, With and Without Energy Source

Other groups (Trevors and Starodub 1987; van Elsas et al. 1987; Krasowsky and Stotzky 1987) have reported on the lack of, or low rates of R-plasmid transfer in non-sterilized soil, but claimed to obtain significant transfer upon supplementing the soil with clay minerals or with nutrients. For plasmid pEA9 we have therefore checked the influence of bentonit, sucrose and Luria Broth in more detail. This last substrate seemed essential, since rich media will probably be used to grow cells if releases are undertaken. Adding bentonit (10%) to unsterile soil and inoculating the mixture with cells suspended in saline did not give any exconjugants. Sucrose, however, was effective and even more so was the addition of Luria Broth (LB) (Fig. 5).

Overnight cell cultures were washed twice and resuspended either in fresh LB or in saline for the inoculation of the soil samples. With the LB (4 ml per 50 g soil, Fig. 5a) high numbers of exconjugants (up to 3×10^{-4} per donor) were obtained, together with drastic cell propagation. Here the highest titers of donors, of recipients and also of exconjugants were obtained after one day. Later, the titer of all three classes of cells declined. In contrast, with the saline cell suspensions (Fig. 5b) no exconjugants were obtained. As reported before, only the donors propagated slightly in the beginning, and again, the titer of donors and recipients declined later on.

3.4.3 Indigenous Bacteria: Increase of Titer with Energy Source

It was expected that the data presented in Sect. 3.4.2 would be related to the titer of indigenous bacteria, which upon the addition of LB to the samples

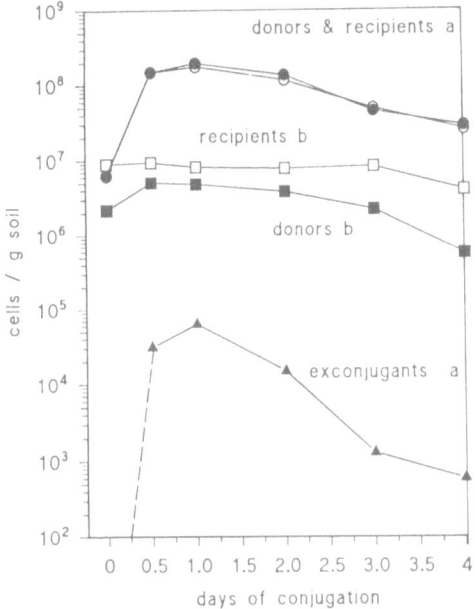

Fig. 5. Survival (donor and recipient cells per g soil) and plasmid pEA9 transfer (exconjugants per g soil) in homologous matings of *E. agglomerans* 339 in natural soil for increasing mating periods with (**a**) and without (**b**) addition of LB. Mating conditions: 22 °C, pH 5.2, humidity 15.5%. The symbols give the means from each 3 replica flasks per point. The number of exconjugants without LB (b) was zero at all times (limit of detection between 2.5/10⁶ and 3.10⁷ surviving donors), that with LB (a) was zero for zero time conjugation

should increase. To determine the amount of this increase, flasks with non-sterilized soil were prepared as before, either with LB or without, inoculated with the recipient strain alone as a reference, or kept uninoculated. Samples were either processed and plated onto appropriate complete or selective plates immediately, or after one day's incubation as before. The results are presented in Table 4.

It can be seen that the titer of indigenous bacteria, *without LB*, is about 1.3×10^5 per gram of soil, and does not increase with incubation (Table 4, line a). There are about 1.1×10^3 indigenous streptomycin-resistant bacteria (Table 4 line b).

Table 4. Titer of indigenous bacteria per g soil, with and without addition of LB, without or after 1 day incubation at 22 °C. Flasks with 50 g non-sterile soil were either left uninfected, or superinfected with 3×10^9 *E. agglomerans* 339. Means of 6 to 12 independent samples each, and 1 × S.E.M. Fungi suppressed by nystatin and cycloheximide

	Zero time		One day incubation	
	−LB	+LB	−LB	+LB
a) Titer of indigenous bacteria	$1.3 \pm 0.14 \times 10^5$	$2.4 \pm 1.1 \times 10^5$	$1.1 \pm 0.08 \times 10^5$	$3.4 \pm 0.4 \times 10^7$
b) Native Sm^r bacteria	$1.1 \pm 0.24 \times 10^3$	$1.6 \pm 0.42 \times 10^3$	$2.3 \pm 0.36 \times 10^3$	$5.0 \pm 0.8 \times 10^4$
c) Titer of super-infected bacteria	$2.4 \pm 0.24 \times 10^7$	$2.1 \pm 0.28 \times 10^7$	$2.7 \pm 0.32 \times 10^7$	$6.0 \pm 0.54 \times 10^8$

After inoculation with *E. agglomerans* 339 Sm^r, about 2.4×10^7 cells per gram soil can be recovered by the plating procedure (Table 4 line c).

If LB is added to the flasks (4 ml per 50 g soil), the titers for zero time incubation are roughly the same as without LB, as expected (Table 4, second column). However, for LB with 1 day's incubation, together with the strong increase in the number of inoculated bacteria (about 25 x), there is an even stronger one in the number of indigenous bacteria (about 250 x). Still, their titer (for culturable, aerobic Gram-negative bacteria) is distinctly lower than that of the inoculated ones (about 1/20), as was desired in these experiments to obtain clear-cut results.

3.4.4 Varying Mating Temperature

Non-sterilized soil was inoculated with the mating mix, in 4 ml LB (complete medium) per 50 g soil sample, to obtain exconjugants at all, and incubated for 3 days at temperatures from 18 to 26 °C. The results are presented in Fig. 6. It can be seen that the titers of donor and recipient are not influenced, but an optimum for plasmid-transfer is around 22 °C.

3.4.5 Varying Mating pH

LB-plates were buffered with $2 \times$ Soerensen buffer (20) to values from pH 5.0 to pH 8.0. then the mating mix was added on top and incubated at 22 °C for 3 days. The mixture was then processed and plated onto selective plates as

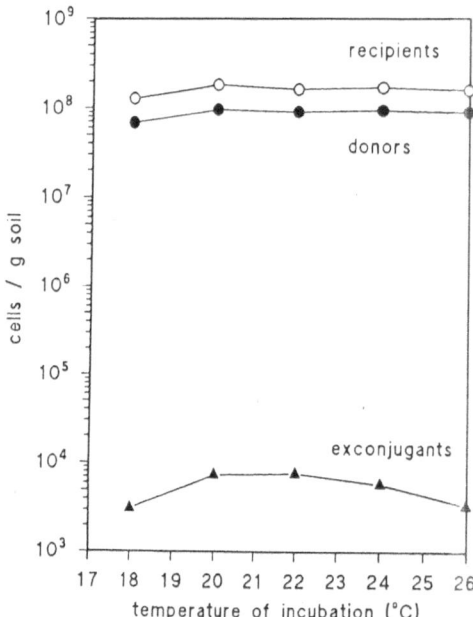

Fig. 6. Survival (donor and recipient cells per g soil) and plasmid pEA9 transfer (exconjugants per g soil) in homologous matings of *E. agglomerans* 339 in natural soil for different temperatures. Mating conditions: pH 5.2, humidity 15.5%, 3 days incubation, with 4 ml LB per 50 g soil. The symbols each give the means from 3 replica flasks

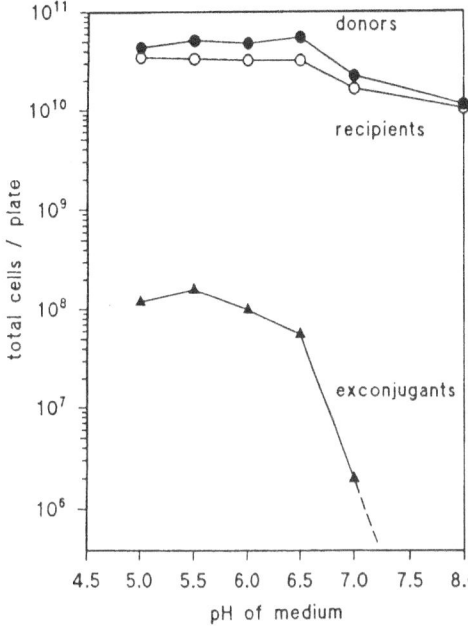

Fig. 7. Survival (total donor and recipient cells per plate) and plasmid pEA9 transfer (total exconjugants per plate) in homologous matting of *E. agglomerans* 339 on LB-plates for different pH. Mating conditions: 8.3 cm plates with LB-agar of graded pH, cell mix (2 ml) on top, 3 days incubation at 22 °C. The symbols each give the means from 2 replica plates

before (Fig. 7). Growth of both donor and recipient was rather independent of pH in the range checked. Plasmid-transfer was optimal around pH 5.5 and zero for pH 8.0.

3.4.6 Varying Moisture Content

Non-sterile soil of graded humidity was incubated. After adding the mating mix, it contained from 10.6 to 24.0% water (-7.6 to -3.8 bar water potential, as measured with a dew point microvoltmeter HR 33T, Wescor Inc., USA). The samples were incubated at standard conditions for 3 days. As above, to obtain plasmid transfer at all, the mating mix contained LB instead of saline. It was found that dry soil (10.6% final rel. humidity) gives distinctly higher numbers of exconjugants (3×10^3 per g soil) than moist soil (Fig. 8). Also in dry soil the number of survivors of both donor and recipient was higher.

3.4.7 Bulk Soil Versus Wheat Rhizosphere

Recently, a stimulating effect of the innermost rhizosphere, possibly due to energy-rich root exudates, on the transfer of plasmid RP4 has been described (van Elas et al. 1988a and b). We have checked this possibility in our system with plasmid pEA9 in prolonged experiments where pregerminated wheat seedlings were put on top of the inoculated non-sterilized soil in the flasks (7 seedlings/flask) and grown for up to 5 weeks. At two weeks the roots were already forming a dense mat in the soil. The soil was then resuspended as before

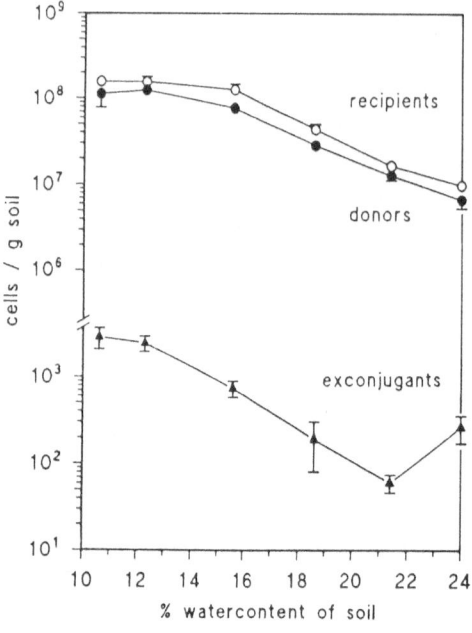

Fig. 8. Survival (donor and recipient cells per g soil) and plasmid pEA9 transfer (exconjugants per g soil) in homologous matings of *E. agglomerans* 339 in natural soil for graded moisture content. Mating conditions: 22 °C, pH 5.2, 3 days incubation with 4 ml LB per 50 g soil. The symbols each give the means from 4 samples (2 independent experiments). The standard error of the means is indicated, where it exceeds the size of the symbols

Table 5. Bacterial survival and plasmid transfer in standard matings of *E. agglomerans* 339 (pEA9) × *E. agglomerans* 339 either with wheat (7 seedlings per flask) or without. Cell suspensions in saline, non-sterilized soil, 14 days incubation, light-temperature cycles of 16 h light, 22 °C and 8 h dark, 14 °C. Means from 8 independent samples each, and standard error of the mean, where indicated

	Bacteria alone	Bacteria plus wheat, free soil
Recipients total, zero time	$1.5 \pm 0.12 \times 10^9$	$1.5 \pm 0.12 \times 10^9$
Recipients total, 14 days	$1.3 \pm 0.2 \times 10^7$	$9.9 \pm 1.3 \times 10^7$
Donors total, zero time	$1.5 \pm 0.15 \times 10^9$	$1.5 \pm 0.15 \times 10^9$
Donors total, 14 days	$4.9 \pm 0.5 \times 10^6$	$7.6 \pm 1.3 \times 10^7$
Exconjugants per donor	0*	0**

* Limit of detection $2/10^6$ surviving donors at 14 days and $8/10^9$ donors at zero time
** Limit of detection $1.6/10^7$ surviving donors at 14 days and $8/10^9$ donors at zero time

(free soil) and samples plated. In addition, the remaining roots were washed carefully, homogenized with quartz sand, and the material (rhizosphere homogenate) plated as before. In Table 5, as an example, data for 14 days incubation, with and without plants, analysing the free soil, are given.

Up to that time, and in contrast to 3 days incubation, there was a drastic decrease in the survival of both donor and recipient cells in the flasks without plants. This decrease was less dramatic if wheat roots were present (shielding effect). There were no transconjugants in the free, non-sterilized soil, neither with nor without plants, and none in the root material, not even after 5 weeks incubation (data not shown). Hence exudates, if present, in contrast to LB or sucrose, are not sufficient to bring transfer-rates up to the level of detection.

3.4.8 Further Experiments

The influence of inoculation density was checked in experiments analogous to Sect. 3.4.1 with sterile and non-sterile soil (Klingmüller et al. 1990, loc. cit. Table 5a and b). Good transconjugation rates were obtained with 10^9 donors and recipients per 50 g soil sample (e.g. 2×10^7 donors and recipient cells per g soil). Higher inoculation densities gave somewhat higher absolute numbers of exconjugants but a distinct decrease in the rate per donor. Lower densities gave fewer absolute numbers, but a slight increase in the rate. For statistical reasons in most experiments reported above, 10^9 donors and recipients per sample were chosen.

The influence of soil packing was checked by varying the compression of inoculated soil before incubation. The soil was non-sterile, and enriched with LB as before. To achieve an equal compression over the volume, tubes containing 10 g soil, inoculated with donor and recipient cells at a 1 to 1 ratio as before, were spun in 40 ml beakers in a swing out rotor at steps ranging from 100 to 2000 g, or kept uncompressed and then incubated for 3 days at 22 °C. Harvesting and plating the cells for the determination of cell propagation and for selection of exconjugants gave highest values for loose and for only slightly compressed soil (100 g), with rates of 3×10^{-6} exconjugants per donor, decreasing to zero for soil compressed with 2000 G. Hence our method of mixing the cells into the soil by stirring with a spatula is favoring exconjugants.

To test whether the possible presence of toxic factors in our non-sterilized soil is the reason for the lack of plasmid transfer in such soil, we have checked soils from 3 other locations in analogous experiments. Again, no transfer was obtained in non-sterile soil (Klingmüller et al. 1990). Such an assumption is therefore very improbable.

Further experiments on the host range of nif-plasmid transfer to other closely related strains confirmed that for optimized conditions on LB plates, the host range is narrow. Transfer was only possible to some *E. agglomerans* strains of the same systematic subgroup as the donor (#I) at low rates, and to some of subgroup II, at even lower rates. No transfer was detected to others, or to *E. cloacae*.

4 Discussion

Bacterial nitrogen fixation is a property with potential to satisfy the nitrogen requirement of agricultural crops. For cereals, attention has to be focussed on strains from their rhizosphere. Such strains include, among others, *Enterobacter agglomerans*.

In the above report, the *genetic basis of nitrogen fixation* in such strains was described. The nif-plasmids they harbour, the *nif*-gene group on them, and the possibilities to transfer this gene group to other soil bacteria were studied.

The location of the *nif*-gene group on plasmids is different from *Klebsiella pneumoniae* and several other *Enterobacteriaceae* where the *nif*-genes are on the chromosome (e.g. *Klebsiella oxytoca*: Wang et al. 1985; *Enterobacter cloacae*: Zhu et al. 1986). It is, however, similar to *Rhizobium leguminosarum* and *meliloti*. Yet, in these rhizobia, the *nif*-genes, although on plasmids, are dispersed in several subclusters, interrupted by *fix* genes, *nod* genes and by other material. This arrangement is different from *E. agglomerans*, where they form one contiguous, well-ordered group of genes, similar to *Klebsiella*. In other words, in *E. agglomerans*, they surpass *Klebsiella* by their particularly well-ordered arrangement: *nif*-genes *J* and *F*, which have similar function but are dispersed in *Klebsiella*, are located in tandem arrangement on one side of the group, transcribed together as one operon.

For environmental considerations, it is remarkable, that some of the *Enterobacter* plasmids, carrying the *nif*-gene group, are self-transmissible. Their spread to other *Enterobacteriaceae* may be a factor in the fluctuation of numbers of nif-fixing bacteria in the rhizosphere of cereals, observed in different locations (Haahtela 1985; Lindberg and Granhall 1984; Väisänen et al. 1985; Jagnow 1984, 1988) and environmental situations.

The compact and well-ordered organization of the *nif*-genes in these *Enterobacter* strains and their positioning on transmissible plasmids can be taken as an indication of a cornerstone in *nif*-gene evolution and dissemination. This system can be envisaged as the one plausible archaetype, from which all other N_2-fixing systems have originated and diverged. Alternatively, it can be envisaged operationally as the ultimate step in the arrangement of these genes towards which all other N_2-fixing bacterial systems are developing. It will be interesting in this context to learn more about the organization of the *nif*-genes in primitive and in highly advanced bacterial groups.

Apart from elucidating further the molecular basis of nitrogen fixation in *E. agglomerans* and its correlation and connection to *nif*-gene group evolution, further work on these strains seems required with emphasis on their application, for instance: Develop methods to increase the absolute number of nif$^+$ *Enterobacter* in the rhizosphere. Develop methods to spread the nif information from nif$^+$ *Enterobacter* to other rhizosphere bacteria that do not yet have it. Find ways to amplify the *nif*-gene group in *Enterobacter* cells to an optimal number. e.g. by providing the gene group in tandem arrangement or by increasing the copy number of the plasmids. Make *nif*-genes derepressed, e.g. by mutation in the *nifAL* operon, so that N_2-reduction does not stop once the bacterium's N-requirement is satisfied. Make cells leaky for NH_4^+ or amino acids synthesized, e.g. by making them transport- or uptake-deficient (amt⁻) so that the plant roots have easier access to the fixed nitrogen. Integrate nif plasmid genes into the recipient's chromosome to avoid uncontrolled spread of the nif information in a release situation. Some of these points are closely connected to the biological safety issue.

Our studies on *risk assessment* document that the conditions in natural soil are strongly opposed to plasmid transfer. This is in line with reports of others (i.e. Trevors et al. 1987), and our own earlier findings (Schilf and Klingmüller 1983; Döhler and Klingmüller 1988). One reason for this could be the action of protozoa (e.g. Casida 1989). However, although they may contribute to the disappearance of the inoculated bacteria, it is unlikely that protozoa would discriminate and preferentially feed on exconjugants. Another, more attractive reason could be the competition of indigenous bacteria with the inoculated ones. However, as documented in Table 4, the titer of culturable indigenous bacteria in our natural soil samples was much less than that of the inoculated bacteria. Further, upon addition of LB, the titer of both groups of bacteria increased, and in particular that of the indigenous group. But these were just the conditions where exconjugants were obtained. This shows that competition is not the cause for low rates of plasmid transfer. It can be concluded that the lack of plasmid transfer in natural soil is primarily due to energy limitation. The occurrence of exconjugants in sterilized soil can then be explained by the assumption that by killing the indigenous microbes, energy containing substrates become available to the inoculated bacteria. The finding that adding LB to the samples elicited high propagation and plasmid transfer raises the question whether LB grown material as inoculant bears the risk of uncontrolled gene spread. However, calculating the total liquid required, if a farmer were to apply the future inoculant as undiluted LB suspension at a rate of 4 ml per 50 g soil, on the basis of an equal distribution of the inoculant in the upper 10 cm layer of the field, it can be shown that per hectare 80,000 l LB would be needed, which – apart from labour – means costs of approximately 200,000.-DM or US $150,000 alone for the substrate. Hence not only cheaper substrates, e.g. molasses, but also smaller amounts of the inoculant, as in routine *Rhizobium* inoculations, would have to be used. Since in our experiments a dilution of as little as 1/100 of the LB cancelled the stimulating effect (Klingmüller et al. 1990), the problem of plasmid transfer induced by LB supplementation would not arise in agricultural practice.

Apart from energy depletion, the possible transfer of nif plasmid pEA9 is restricted by its narrow host range, hence barely significant in nature as a means of gene spread. It is known from other studies that the chances of gene spread by chromosomal transfer, by transduction or by transformation are even smaller than those of plasmid transfer, hence negligible here. Our data, indicating only low chances of gene spread for our bacterial system in natural soil, add to the findings of others with other bacterial systems. Together they tend to throw some doubt upon fanciful speculations that invoke more thrilling scenarios.

Acknowledgements. I am grateful to Mrs. Christine Fentner, Marion Steinlein and Ruth Hösl for excellent technical assistance, and to Prof. R. Aldag, Dipl.-Geoökol. Ch. Hartmann and Dipl.-Biol. I. Simmeth for providing soil data. Dipl. agr. Ch. Rappold helped with some of the experiments. Susan Helms improved the English. Support for this work was provided by grants from the Bundesministerium für Forschung und Technologie, Bonn, and the European Community, Brussels, to W.K.

5 References

Abdel-Salam MS, Klingmüller W (1987) Transposon Tn5 mutagenesis in *Azospirillum lipoferum*: isolation of indole acetic acid mutants. Mol Gen Genet 210: 165

Boyer HW, Rouillard-Dussioux D (1969) Complementation analysis of the restriction and modification of DNA in *Escherichia coli*. J Mol Biol 41: 459

Casida LE Jr (1989) Protozoan response to the addition of bacterial predators and other bacteria to soil. Appl Envir Microbiol 55: 1857

Döhler K, Klingmüller W (1988) Genetic interaction of *Rhizobium leguminosarum* biovar *riceae* with Gram-negative bacteria. In: Klingmüller W (ed) Risk assessment for deliberate releases: The possible impact of genetechnically engineered microorganisms on the environment Springer-Verlag Berlin-Heidelberg New York Tokyo, p 18

van Elsas JD, Govaert JM, van Veen JA (1987) Transfer of plasmid pET-30 between bacilli in soil as influenced by bacterial population dynamics and soil conditions. Soil Biol Biochem 19: 639

van Elsas JD, Trevors JT, Starodub ME (1988a) Plasmid transfer in soil and rhizosphere In: Klingmüller W (ed) Risk assessment for deliberate releases Springer-Verlag Berlin-Heidlberg New York Tokyo, p 89

van Elsas JD, Trevors JT, Starodub ME (1988b) Bacterial conjugation between pseudomonads in the rhizosphere of wheat. FEMS Microbiol Ecol 53: 299

Haahtela K (1985) Nitrogenase activity (acetylene reduction) in root-associated, cold-climate species of *Azospirillum*, *Enterobacter*, *Klebsiella* and *Pseudomonas*, growing at various temperatures. FEMS Microbiol Ecol 31: 211

Herterich S (1986) Untersuchung der Transfereigenschaft des Plasmids pEA9 aus *Enterobacter agglomerans* 339 und Entwicklung eines Transposon-Mutagenesesystems. Diplom-thesis, University of Bayreuth

Jagnow G (1984) Stickstoffbindende Bakterien in der Rhizosphäre I. Bindungsraten im Freiland, Artenvielfalt und Verbreitung. Kali-Briefe (Büntehof) 17: 341

Jagnow G (1988) Enterobacteriaceae in the rhizosphere of wheat, barley and ryegrass: Fractions of nitrogenase-positive strains in field experiments with different doses of nitrogen fertilizer. In: Bothe H, de Brujin FJ, Newton WE (eds) Nitrogen fixation: hundred years after. Gustav Fischer Stuttgart-New York

Kado CI, Liu ST (1981) Rapid procedure for detection and isolation of large and small plasmids. J Bacteriol 145: 1365

Keen NT, Tamaki S, Kobayashi D, Trollinger D (1988) Improved broad-host range plasmids for DNA cloning in Gram-negative bacteria. Gene 70: 191

Kleeberger A, Castorph H, Klingmüller W (1983) The rhizosphere microflora of wheat and barley with special reference to Gram-negative bacteria. Arch Microbiol 136: 306

Klingmüller W (1985) Studies on natural and artificial nitrogen-fixing soil bacteria. In: Sinha U, Klingmüller W (eds) Trends in molecular genetics. Spectrum Publ House. Patna, New Delhi, p 37

Klingmüller W (ed) (1988a) *Azospirillum* IV: genetics, physiology, ecology. Springer-Verlag. Berlin Heidelberg New York Tokyo

Klingmüller W (1988b) Molecular analysis of nitrogen fixation in *Enterobacter*. Proc Int Symp on Plant Biotech. Gyeongsang Natl Univ Chinju, Korea, p 41

Klingmüller W (1991) Plasmid transfer in natural soil: Assessing nitrogen fixing *Enterobacter* for safe release. FEMS Microbiol Ecol 85: 107

Klingmüller W, Herterich S, Min BW (1989) Self-transmissible nif-plasmids in *Enterobacter*. In: Skinner FA, Boddey RM, Fendrik I (eds) Nitrogen fixation with non-legumes. Kluwer, Dordrecht, p 173

Klingmüller W, Dally A, Fentner C, Steinlein M (1990) Plasmid transfer between soil bacteria. In: Fry JC, Day MJ (eds) Bacterial genetics in natural environments. Chapman and Hall, London, p 133

Krasovsky VN, Stotzky G (1987) Conjugation and genetic recombination in *Escherichia coli* in sterile and non-sterile soil. Soil Biol Biochem 19: 631

Kreutzer R (1986) Restriktionskartierung eines nif-plasmids von *Enterobacter agglomerans* und genetische Charakterisierung der von ihm getragenen *nif*-Gene. Diplom-thesis, University of Bayreuth

Kreutzer R, Singh M, Klingmüller W (1989) Identification and characterization of the *nifH* and *nifJ* promoter regions located on the nif-plasmid pEA3 of *Enterobacter agglomerans* 333. Gene 78: 101

Kreutzer R, Steibl HD, Dayananda S, Dippe R, Halda L, Buck M, Klingmüller W (1991) Genetic characterization of nitrogen fixation in *Enterobacter* strains from the rhizosphere of cereals. In: Polsinelli M, Materassi R, Vinzencini M (eds) Nitrogen fixation. Kluwer, Dordrecht, p 25

Lindberg T, Granhall U (1984) Isolation and characterization of dinitrogen-fixing bacteria from the rhizosphere of temperate cereals and forage grasses. Appl Envir Microbiol 48: 683

Maniatis T, Fritsch E, Sambrook J (1982) Molecular cloning. Cold Spring Harbor Laboratory, Cold Spring Harbor, N.Y.

Min BW (1986) Erzeugung von nif⁻-Mutanten des Stammes *Enterobacter agglomerans* 339 durch Transposonmutagenese und Untersuchung der Transfereigenschaft seines Plasmids pEA9. Diplom-thesis, University of Bayreuth

Polsinelli M, Materassi R, Vincenzini M (1991) Nitrogen fixation. Kluwer, Dordrecht

Riedel GE, Brown SE, Ausubel FM (1983) Nitrogen fixation by *Klebsiella pneumoniae* is inhibited by certain multicopy hybrid nif plasmids. J Bacteriol 153: 45

Schilf W, Klingmüller W (1983) Experiments with *Escherichia coli* on the dispersal of plasmids in environmental samples. Recombinant DNA Technical Bulletin, Washington, 6 (3): 101

Simon R (1984) High frequency mobilization of Gram-negative bacterial replicons by the in vitro constructed Tn5-mob transposon. Mol Gen Genet 196: 413

Singh M, Klingmüller W (1985) Nif-plasmids in free-living nitrogen fixing soil bacteria. In: Sinha U, Klingmüller W (eds) Trends in molecular genetics. Spectrum Publ. House, Patna New Delhi, p 15

Singh M, Klingmüller W (1986a) Transposon mutagenesis in Azospirillum *brasilense*: isolation of auxotrophic and nif⁻ mutants and molecular cloning of the mutagenized *nif*DNA. Mol Gen Genet 202: 136

Singh M, Klingmüller W (1986b) Cloning of pEA3, a large plasmid of *Enterobacter agglomerans* containing nitrogenase structural genes. Plant and Soil 90: 235

Singh M, Kleeberger A, Klingmüller W (1983) Location of nitrogen fixation (*nif*) genes on indigenous plasmids of *Enterobacter agglomerans*. Mol Gen Genet 190: 373

Singh M, Kreutzer R, Acker G, Klingmüller W (1988) Localization and physical mapping of a plasmid-borne 23 kb *nif*-gene cluster from *Enterobacter agglomerans* showing homology to the entire *nif*-gene cluster of *Klebsiella pneumoniae* M5al. Plasmid 19: 1

Skinner FA, Boddey RM, Fendrik I (1989) Nitrogen fixation with non-legumes. Kluwer, Dordrecht

Steibl HD (1989) Klonierung des Plasmids pEA9 von *Enterobacter agglomerans* 339 und Analyse der plasmidalen *nif*-Genorganisation. Diplom-thesis, University of Bayreuth

Trevors JT, Starodub ME (1987) R-plasmid transfer in non-sterile agricultural soil. System Appl Microbiol 9: 312

Trevors JT, Barkar T, Bourquin AW (1987) Genetransfer among bacteria in soil and aquatic environments: A Review. Can J Microbiol 33: 191

Väisänen O, Haahtela K, Bask L, Kari K, Salkinoja-Salonen M, Sundman V (1985) Diversity of nif gene location and nitrogen fixation among root-associated *Enterobacter* and *Klebsiella* strains. Arch Microbiol 141: 123

Wang PL, Koh SK, Chung KS, Uozumi T, Beppu T (1985) Cloning and Expression in *E. coli* of the whole *nif* genes of *Klebsiella oxytoca*, a nitrogen fixer in the rhizosphere of rice. Agric Biol Chem 49 (5): 1469

Zhu JB, Li ZG, Wang.LW, Shen SS, Shen SC (1986) Temperature sensitivity of a nifA-like gene in *Enterobacter cloacae*. J Bacteriology 166: 357

CHAPTER 18
Molecular Population Genetics and Evolution of Rhizobia

J.P.W. Young

The Nitrogen Fixation
and its Research in China
Editor· Guo-fan Hong
ⒸSpringer-Verlag Berlin Heidelberg 1992

1 Introduction

Studies of the natural genetic diversity of rhizobia have given us insights into the genetic structure and evolution of bacterial populations, and have also improved our understanding of the nitrogen-fixing symbiosis between legumes and rhizobia, which is important in many agricultural systems worldwide (I use "rhizobia" in the broad sense to cover all legume nodule symbionts). In legume cultivation, the farmer provides the plant partner, which is usually the product of an extensive breeding and selection process, and therefore genetically rather uniform. Sometimes the bacterial partner is added artificially too, in the form of an inoculant containing one or a few strains, but more often there is already a sufficient population of rhizobia living in the soil. Thus the cultivation of legumes is a partnership between a highly bred plant and a "wild" population of bacteria. The farmer has little control over the genetic composition of the indigenous population of rhizobia, and in fact there is abundant evidence that such populations are diverse, even in fields with a long agricultural history. In our laboratory during the last few years, we have tried to describe this diversity in ways that allow us to draw conclusions about the genetic interrelationships among the strains that make up these populations. Most of these studies have been on the species *Rhizobium leguminosarum*, but recently we have also been looking at the wider issue of the evolutionary relationships among rhizobia, in the broad sense, and between rhizobia and other groups of bacteria. In this contribution I shall summarize our progress to date in both these areas.

2 Natural Genetic Diversity in Rhizobia

Genetic diversity has been demonstrated for virtually every character that has been examined in a sufficient sample of rhizobium isolates. Historically, the first "genetic" character to be widely studied was serotype. Stevens (1925) demonstrated antigenic variation within various species of rhizobia, and this variation has been characterized by various methods in so many studies since then that it would be unjust to cite just a few. Isolates differ markedly in resistance to bacteriophage (Barnet 1972; Buchanan-Wollaston 1979), to bacteriocins (Hirsch 1979), and to low levels of antibiotics (Beynon & Josey, 1980; Stein et al. 1982). The pattern of abundant proteins separated by gel electrophoresis shows some variation within species (Noel & Brill 1980; Roberts et al. 1980), while gels stained for specific enzyme activities also show variation (Fottrell & O'Hora 1969; Mytton et al. 1978). At the DNA level, the pattern of fragments seen in a restriction digest of total DNA is variable, as is the pattern of hybridization with specific probes (Hadley et al. 1983). All strains of *R. leguminosarum*, and of most other rhizobial species, carry large plasmids which vary in number and size in different isolates. Strains within a species differ in competitiveness for nodulation (Vincent & Waters 1953), and in their ability to grow and to nodulate at low pH (Jones & Morley 1981; Russell & Jones 1975) and to grow on various carbon sources (Robert et al. 1982). El-Sherbeeny et al. (1977) found that isolates from a single population of *R. leguminosarum* differed significantly in the level of nitrogen fixation and plant growth that they supported when tested with *Vicia faba*.

There is undoubtedly genetic variation underlying all the characters I have just mentioned, but if our aim is to describe the genetic diversity of bacterial populations then it is much easier to interpret the information from some techniques than from others. Differences in effectiveness or environmental tolerance may be of real ecological and agricultural significance, but we cannot know whether they reflect small or large genetic differences without a detailed investigation of each instance. Antigenic phenotypes reflect the underlying genetic variation more directly, but despite their very extensive use in studies of soil populations, there is still no information on the genes responsible for antigenic variation in rhizobia. Similar considerations apply to resistance to bacteriophages and other agents. Gel electrophoresis of total cellular proteins or of restriction digests of total DNA gives "fingerprints" with many bands that can give an indication of overall genomic similarity. These methods have the advantage that they sample many parts of the genome and therefore give a representative picture, but the corresponding disadvantage is that it is not usually possible to say which part of the genome is varying, nor is it easy to quantitate the variation reliably.

Most rhizobia have large plasmids in addition to the main chromosome and, in some species, many genes crucial for the symbiosis are known to be encoded on a plasmid (the Sym plasmid). Because of this, it is important that we are able to distinguish between genetic differences that are chromosomal and those that are due to variation in the plasmids. This is crucial because plasmids, which are potentially transmissible from one cell to another, may have a very different evolutionary history and population structure from the chromosomal genomes with which they coexist.

For all the above reasons, the techniques that have given the most useful information about genetic structure of rhizobial populations are those that look fairly directly at variation in specific genes or small regions of the genome, namely enzyme polymorphism (allozymes) or restriction fragment length polymorphism (RFLP). In our laboratory, and in that of Professor D.G. Jones in Aberystwyth, we have used allozymes as the main tool to look at chromosomally encoded variation, and RFLPs to characterize plasmid variation, specifically variation around the *nod* and *nif* genes on the Sym plasmid. The main conclusions of this work are summarized in the next two sections.

3 Studies of Enzyme Polymorphism

Many amino acid substitutions in enzymes alter the surface charge of the protein, and hence the electrophoretic mobility, without impairing catalytic function. Such variants, called "allozymes", are found at high frequency in natural populations of virtually all organisms, as population geneticists discovered when they surveyed many different species during the 1970s. This universal polymorphism, in which several different variants are often of roughly equal frequency, eradicated the idea that natural populations were made up of a predominant "wild-type" and rarer mutant forms: all species are naturally diverse. This gave rise to considerable debate as to whether allozymic variation was maintained by selection or merely accumulated because it was selectively neutral: in certain cases there is persuasive evidence one way or another, but a general answer has not emerged. Nevertheless, allozyme variation provides a very powerful probe for investigating the genetic structure of natural populations: one can estimate the level of genetic diversity, the extent of recombination, of population subdivision, of migration between populations and gene exchange between species, and so on.

The 1970s saw many such studies on eukaryotes, but there were only a handful of studies on bacteria until the 1980s, when a number of extensive studies were published (see Selander et al. 1986; Young 1989). Most of our own work has been on *R. leguminosarum*, and has been published in a number of papers (Young 1985; Young et al. 1987; Harrison et al. 1987, 1988, 1989). In the majority of our studies we have classified each isolate for the electrophoretic mobility of three enzymes: glucose-6-phosphate dehydrogenase (G6PD), superoxide dismutase (SOD), and β-galactosidase (BGAL). Each distinct allele is coded by a different letter (in alphabetical order of mobility) so that, for example, the enzyme electrophoretic type (ET) FMP has the F allele for G6PD, the M allele for SOD, and the P allele for BGAL. We believe that each of these enzymes is encoded by a chromosomally encoded locus, since strains cured of their plasmids nevertheless retain their ET (Young 1985; B.D. Eardly, personal communication). Although data are presented here for just three polymorphic enzymes, representing three genetic loci, there is good evidence that these are reasonably representative of chromosomal loci in general. Both total protein patterns on denaturing gels (Young 1985) and a survey using twelve polymorphic enzyme systems (B.D. Eardly and J.P.W.Y., unpublished) indicate that the three-locus ETs do correspond to real genetic groupings, although of course there is hidden genetic variation within each ET class.

Table 1 is a compilation of the data from our papers on allozymes: I will try to indicate briefly the major conclusions that can be drawn from them.

(1) *Rhizobium leguminosarum* populations are polymorphic. In terms of the number of alleles per locus and the allelic diversity in each population they are more polymorphic than most eukaryotic species; this is typical of most bacterial species examined so far.

(2) *R. leguminosarum* populations are highly clonal, since there is strong linkage disequilibrium between loci. This can be seen in sample 1 (row 1) of Table 1, for example: the ETs MFF and SSQ are the two most frequent types, and recombination between them might be expected to generate a number of other types such as MFQ, MSQ, SSF, SFF, and so on. Many of these potential "recombinants" are rare or absent, however, implying that chromosomal recombination is certainly restricted. This, too, appears to be the typical situation in bacterial species.

(3) Each plant in a cultivated crop forms nodules with a wide range of rhizobial genotypes, and each nodule on the root is essentially an independent colonization event. The relevant data are presented by Young et al. (1987), and are not reproduced here, though sample 1 of Table 1 was derived by summing the individual plant samples. The detailed data showed that the level of genetic diversity among isolates from a single plant was almost as great as in the field sample as a whole, and that there was almost no correlation in genotype between the isolates from adjacent nodules on one root.

(4) The three biovars of *R. leguminosarum* share a common pool of chromosomal genotypes. When we sampled nodules from peas (*Pisum sativum*) and white clover (*Trifolium repens*) growing in the same field plot (samples 4 and 5) we found that many of the ETs occurred in both samples. Isolates from French beans (*Phaseolus vulgaris*) in the same plot were much less diverse (sample 6), but the predominant ET (MFF) was also the one most frequently isolated from peas and clover. The three biovars (*viciae*, *trifolii*, and *phaseoli*, forming nodules on *Pisum*, *Trifolium*, and *Phaseolus* respectively) are believed to belong to a single species on taxonomic and genetic grounds: there are no consistent distinguishing features other than host range (Jordan 1984), chromosomal genes can be recombined (Johnston & Beringer 1977), and host-range determinants can be exchanged among them (Johnston et al. 1978). The ET data show that to a large extent the biovars behave in the field as a single species with a single (chromosomal) gene pool. Since the host-range determinants are plasmid-borne, we can think of the population as a set of chromosomal types, each of which can carry either a *viciae*, say, or a *trifolii* plasmid. This implies that plasmids are free to move from one chromosomal type to another, but in fact this is an oversimplification, as we shall see later.

(5) The biovar *phaseoli* Sym plasmid is found in a much less diverse range of chromosomal types than the other two biovars (sample 6 compared with 4 and 5). This was initially surprising since, worldwide, *phaseoli* is said to be the most diverse of the biovars (Crow et al. 1981). The low diversity may reflect the fact that biovar *phaseoli* has no native host plants in Europe: *Phaseolus* beans were imported from the Americas and have been grown here for just a few centuries. Although the *phaseoli* isolates from Hainford have a chromosomal background that is typical of "native" strains, they do not appear to have arisen from one of the other biovars by simple mutation

Table 1. Enzyme polymorphism in indigenous *Rhizobium* populations: (a) sample details and literature reference; (b) number of isolates of each distinct three-locus enzyme type (ET; see text)

	Location	Host plant	Sample source	Reference
1	Morley 1, Norfolk	*Pisum sativum* cv. Maro	field nodule	Young et al. 1987
2	Morley 1, Norfolk	*Pisum sativum* JI 1195	soil as inoculum	Young et al. 1987
3	Morley 2, Norfolk	*Pisum sativum* JI 1195	soil as inoculum	Young et al. 1987
4	Hainford, Norfolk	*Pisum sativum* JI 1194	field nodule	Young 1985
5	Hainford, Norfolk	*Trifolium repens* cv. G. Huia	field nodule	Young 1985
6	Hainford, Norfolk	*Phaseolus vulgaris* cv. Fino	field nodule	Young 1985
7	Hainford, Norfolk	*Medicago sativa* cv. Europe	field nodule	Young 1985
8	Colney 1, Norfolk	*Trifolium repens* cv. Nesta	soil as inoculum	Harrison et al. 1987
9	Colney 1, Norfolk	*Trifolium repens* cv. Mikanova	soil as inoculum	Harrison et al. 1987
10	Colney 1, Norfolk	*Trifolium repens* cv. Kersey	soil as inoculum	Harrison et al. 1987
11	Colney 1, Norfolk	*Trifolium repens* cv. S100	soil as inoculum	Harrison et al. 1987
12	Colney 2, Norfolk	*Trifolium repens* cv. Nesta	soil as inoculum at 10^{-2}	Harrison et al. 1987
13	Colney 2, Norfolk	*Trifolium repens* cv. Nesta	soil as inoculum at 10^{-3}	Harrison et al. 1987
14	Colney 2, Norfolk	*Trifolium repens* cv. Nesta	soil as inoculum at 10^{-4}	Harrison et al. 1987
15	Colney 2, Norfolk	*Trifolium repens* cv. Nesta	soil as inoculum at 10^{-5}	Harrison et al. 1987
16	Colney 2, Norfolk	*Trifolium repens* cv. Nesta	soil as inoculum at 10^{-6}	Harrison et al. 1987
17	Tal-y-Bont, Dyfed (pH 4.3)	*Trifolium repens* cv. Nesta	soil as inoculum	Harrison et al. 1989
18	Bronant 1, Dyfed (pH 4.4)	*Trifolium repens* cv. Nesta	soil as inoculum	Harrison et al. 1989
19	Bronant 2, Dyfed (pH 5.9)	*Trifolium repens* cv. Nesta	soil as inoculum	Harrison et al. 1989
20	Pwllpeiran, Dyfed (pH 6.1)	*Trifolium repens* cv. Nesta	soil as inoculum	Harrison et al. 1989
21	Bowstreet, Dyfed (pH 6.2)	*Trifolium repens* cv. Nesta	soil as inoculum	Harrison et al. 1989
22	Frongoch, Dyfed (pH 6.3)	*Trifolium repens* cv. Nesta	soil as inoculum	Harrison et al. 1989
23	Penglais, Dyfed (pH 6.5)	*Trifolium repens* cv. Nesta	soil as inoculum	Harrison et al. 1989
24	Costessey, Norfolk (pH 6.7)	*Trifolium repens* cv. Nesta	soil as inoculum	Harrison et al. 1989
25	Garveston, Norfolk (pH 6.9)	*Trifolium repens* cv. Nesta	soil as inoculum	Harrison et al. 1989
26	Blackawton, Devon (pH 7.5)	*Trifolium repens* cv. Nesta	soil as inoculum	Harrison et al. 1989
27	Colney 3, Norfolk (March)	isolated directly from soil	soil plated on agar medium	M. Wexler (unpubl.)

	F M P	M F F	M F K	M F L	M F M	M F P	M F Q	M F S	M S K	M S L	M S M	M S Q	M S S	S S M	S S Q	S S S	D F PV	B F V	other	total
1	1	50	10	6	1	16	2	9	20	20	2	4	2	45	61					249
2		41	2	2		4	5		19	13	2	3		1	12					104
3		50	1	3		3			15		1	1	1	3	8					86
4	20	20	1		2	2			2		16	3	19	1	1		1		1	89
5		44				1		7			4		35		2		1			94
6		40																		41
7		1															13	14		28
8			2		8	1			13		2	3	2	4	2				2	39
9			4		4				10		4	2	2		6					32
10					5				9		5		2		2					23
11					3				11		2	1	4	1	4				3	29
12			3		9	4			4		7	1	3	2	1				4	38
13			6		11	1			4		5	1	3	2					2	35
14			6		6	1			5		9	1	4	2	2				2	38
15			3		7				5		9		5						1	30
16					2						3								1	6
17					44	1					6	1								51
18					17							1								19
19					5						10	27	8	6	30					86
20			3		6	7			5		8	8		17	40					90
21			10		15				15		8	7	2		1					55
22			21		33	24			3		14									88
23			1		48				4		6		1	5	7					83
24					9	10			5		36		15			18				76
25					10	5			3		34		19	34	3					79
26			5		7						7	1	8	1	5					94
27		3	2								2	1		1	5				2	16

of the host-range genes, since their *nod* genes are extremely diverged from those of biovars *viciae* and *trifolii* (Young and Johnston 1989). Instead, it seems more likely that they are hybrids, in which a "foreign" Sym plasmid has been acquired by a "local" chromosomal type. It will be interesting to establish whether the Sym plasmid is closely related to any of the diverse types that are found among strains from *Phaseolus* nodules in the Americas (Martinez et al. 1985). Recent work on the genetic diversity of isolates from Central and South America has shown that many of them need to be placed in new species, as they are well outside the range of variation found in *R. leguminosarum* biovars *viciae* and *trifolii* (Piñero et al. 1988). I shall return to this point in a later section.

(6) *Rhizobium meliloti* behaves in the field, as in the laboratory, as a separate species from *R. leguminosarum*. The ETs of most strains from *Medicago sativa* are quite different to those from the various host plants of *R. leguminosarum* (sample 7 compared with 4, 5 and 6). There are exceptions, though: one isolate from *M. sativa* had the ET MFF typical of *R. leguminosarum*, and one each from clover and pea had the DF(PV) pattern found in many *R. meliloti* strains. These isolates were found to nodulate only poorly when tested on the host species from which they were isolated, but the genetic nature of their Sym plasmids has not been investigated. Strains with properties intermediate between those of *R. meliloti* and *R. leguminosarum* have also been reported by others (Eardly et al. 1985).

(7) The level of genetic diversity of *R. leguminosarum* is similar at sites in different parts of Britain (apart from very acid soils – see below). Furthermore, samples from different sites contain most of the same ETs (though their relative proportions may vary), and these ETs show the same strong linkage disequilibrium. For example, compare samples 19 to 23 from Wales, 24 and 25 from Norfolk (350 km away), and 26 from Devon.

(8) In very acid soils there is a lower diversity of *Rhizobium* strains able to nodulate clover (as well as a lower density of cells). Samples 17 and 18 are from sites of very low pH, and both are dominated by the ET MFM. By contrast, a much wider spectrum of strains is found in sample 19, which was taken close to 18 but from a site whose pH had been raised by liming, and in sample 21, which was from a site of naturally higher pH that is geographically intermediate between the two acid sites 17 and 18.

(9) Samples obtained by using soil to inoculate test plants under artificial conditions are comparable in genetic diversity to those obtained from nodules on plants growing directly in the field, but there may be significant differences in the ratio of different genotypes. Sample 2 has most of the same ETs as sample 1, but some are in very different proportions (note the deficiency of SSM and SSQ). There could be many reasons for this discrepancy: the very different environmental conditions of a plant grown in vitro on agar rather than in the field is perhaps the most likely influence, but there is also a difference in the host genotype and the population in the field may vary seasonally so that the soil at the time of sampling did

not carry the same population as was present when the field nodules were formed.

(10) Different cultivars of white clover select a similar range of *Rhizobium* ETs when presented with the same soil sample as inoculum (samples 8 to 11). If we are going to draw conclusions about the soil populations of bacteria on the basis of samples collected from nodules on host plants, it is important to assess possible sources of bias in the sampling technique. It appears that, in this instance, host cultivar is not a major factor, although this is not universally true as there are some reported instances of cultivar differences in strain preference (Russell & Jones 1975; Dughri & Bottomley 1984).

(11) Sampling does not seem to be biased by gross differences in competitiveness for nodulation. We examined this by looking at the strains isolated from nodules formed at different dilutions of the soil inoculum: if certain strains form nodules out of proportion to their numerical abundance in the soil, then they should be diluted out at lower densities, allowing more abundant but less competitive strains to form more of the nodules. In fact, the range of ETs observed is similar at all dilutions, right down to the level at which most plants do not receive sufficient cells to nodulate (samples 12 to 16).

(12) Not all rhizobia in the soil have the genes necessary for nodulation. All the samples 1–26 were obtained by isolating bacteria from legume root nodules, and this clearly means that we have only sampled those soil bacteria that have the ability to enter root nodules (and, in most cases, to stimulate their formation). However, symbiotic ability is a plasmid-encoded characteristic in *R. leguminosarum*, and nonsymbiotic derivatives can readily be obtained in the laboratory. The frequency of nonsymbiotic strains in the soil cannot be assessed if we always "filter" our samples by obtaining them from nodules, but unfortunately *Rhizobium* has few distinguishing properties other than symbiotic ability, and no-one has yet devised a selective medium to isolate *Rhizobium* species directly from the soil. It is very important to obtain information on this "hidden" population, though, as it could potentially form a significant part of the gene pool of the species. We have spread soil dilutions on agar plates containing lactose as sole carbon source, and selected colonies that resembled *R. leguminosarum* in appearance (M. Wexler and J.P.W.Y., unpublished). We then checked a number of other characteristics, and determined the ET of those isolates that were still likely candidates. Of 16 such isolates obtained during the winter from an experimental plot (sample 27), 14 had ETs that had previously been found in *R. leguminosarum* from nodules, and the other two had ETs that were clearly related to *R. leguminosarum* patterns. On this basis, we can consider that all these strains are *R. leguminosarum* as far as their chromosomal genotype is concerned. Half of them were able to nodulate *Trifolium repens*, and these also hybridized with a *nifH* gene probe, but the remaining strains did not nodulate test hosts for any of the biovars of *R. leguminosarum* and the majority had no detectable *nifH* DNA. Thus it

seems that only about half the soil population was symbiotically competent. The sample was too small to draw any firm conclusions about differences in ET frequencies between symbiotic and nonsymbiotic strains. Several other groups have also isolated rhizobia directly from soil, and our findings are in general agreement with theirs (Bottomley & Dughri 1989; Jarvis et al. 1989; Soberon-Chavez & Najera 1989).

4 Studies of Sym Plasmid Polymorphism

The host-range determinants of each *R. leguminosarum* strain are encoded on a large plasmid, the Sym plasmid, and the fact that strains of different host range often have closely related chromosomes, as identified by ET, implies that Sym plasmids must sometimes be transferred (unless we postulate that each host-range biovar, or each ET, has arisen independently several times, which is implausible). We wanted to look at this question in more detail, so we investigated the genetic diversity of Sym plasmids in two samples (samples 1 and 4 in Table 1) from field populations of *R. leguminosarum* biovar *viciae* that had previously been characterized for chromosomally-encoded enzyme electrophoretic polymorphism. We used a subset of the strains, chosen to represent abundant chromosomal types. Five overlapping cloned DNA fragments from the Sym plasmid pRL1JI (covering the *nod* genes and some *nif* genes) were used as hybridization probes to identify restriction fragment variation in the homologous genes of the isolates. In addition, a clone of the β-galactosidase gene region (Lac) was used as a probe, providing further data on chromosomal relatedness which confirmed our ET analysis in that each ET had its own characteristic RFLP patterns, and provided additional data since several different patterns were found within some of the ETs.

The full data are published by Young & Wexler (1988), and the main conclusions that can be drawn are as follows.

(1) The Sym region is very polymorphic. The average DNA sequence divergence (estimated from the similarity of the restriction fragment profiles) was twice as great for the Sym region as for the chromosomally-encoded Lac region. Most isolates had so many restriction fragment changes relative to the standard plasmid pRL1JI that it was not possible to deduce their restriction map by simple comparison.

(2) There is little evidence for recombination between distantly related Sym regions. Isolates which are similar when probed with one end of the Sym region tend to be similar at the other end too.

(3) The relationship between plasmid variation and chromosomal variation is strikingly nonrandom. This was unexpected, since we began with the idea that Sym plasmids probably moved fairly freely among the different chromosomal types in the population, as suggested by the ET data on biovars *viciae*

Table 2. Correlation between Sym plasmid polymorphism and chromosomally encoded polymorphism (Young & Wexler 1988)

Chromosomal variants[1] (ET/Lac)	Sym plasmid variants[2]	
	Hainford population	Morley population
FMP/O	A1 A1 A1 A1 A1 A1 C1 C1 C1	A1
MFF/B	E2 E2 E5	A1 A1 E1 E2 E3 E4 E4 G1 O1 O2 Q1 Q2
MFF/P	A1 A1 A2 D1 D2 D2 N1	
SSQ/J		A1 A1 B1
SSQ/M		A1 E4 E5 F1 F2 F2
SSQ/K	H3 H3	
MSM/R	H2	
MSM/L	H1	
MSM/N	J1 K1 K1 K1 K2 K2 K2 K5	K1
MFS/H		H1
MFS/I		K5
MFS/F	K9 K9 K9 K9	
MFS/E		I1 I1 I1 I2 I3
MSS/E	K4 K6 K6 K6 K12 K12 L1 L1 L1 M1	
MSS/D		P1
MFK/C		K8 K8
MFK/Q		K3 K3 K3 K7 K8 K10 K11

[1] Defined by electrophoretic type (ET) and restriction fragment pattern (Lac).
[2] Defined by restriction fragment variation; types with the same letter but different numbers are similar but not identical

and *trifolii.* Table 2 shows the distribution of Sym types across chromosomal types in our samples: it is clear that plasmids are not distributed randomly since, for example, the ET MFF never has plasmids of the K group, even though these are abundant in the populations.

(4) Nevertheless, there is some plasmid transfer between different chromosomal classes. Indistinguishable plasmids are found in quite different chromosomal backgrounds, and very different plasmids occur in isolates with the same chromosomal type. For example, the Sym type A1 was isolated from chromosomal types FMP/O, MFF/B, MFF/P, SSQ/J and SSQ/M. These five chromosomal types also share a number of other Sym types, none of which are found in the isolates which have other ETs. These, in turn, have a different group of plasmids in common. It seems as though the species is subdivided into at least two groups, and plasmids are shared within but not between the groups.

(5) The plasmid/chromosome association is not just a local accident due to chance colonization of a particular field by certain clones. Essentially the same pattern was found in different years at two sites 25 km apart (Table 2).

It seems that there must be hidden "barriers" of some sort that regulate the flow of plasmids through the populations, allowing some transfers to occur successfully but not others. The barriers could be at the stage of plasmid transfer, or plasmid maintenance, or of function. We have initiated laboratory studies to look at plasmid transfer between representatives of the various chromosomal types, but it is too early to draw any conclusions yet. We will not be able to say that we really understand the situation until we have shown that the experimental data on plasmid transfer are consistent with the observed distribution of plasmids in natural populations.

5 Molecular Phylogeny of Rhizobia

There are many reasons why it is important that we gain a clear understanding of the evolutionary relationships among rhizobium species and between rhizobia and other bacteria. I would like to think that the conclusions from our studies of *Rhizobium leguminosarum* are of wider generality, and we need to know the place of this species in the bacterial family tree if we are to make sensible extrapolations to other species. Population genetics is the study of genetic variation and interaction *within* a species, but the vexed question of the boundaries between bacterial species, and the possibility of genetic transfer from one species to another, force us to consider the neighbouring species as a potential part of the population: we need to know which species are closely related, and whether they are in fact separated by a genetic discontinuity. The rhizobia of most legume species have yet to be isolated and described, and when new isolates are obtained it is often difficult and time-consuming to determine whether they are closely related to known species. Furthermore, it has become apparent that rhizobia are not a coherent evolutionary group: some rhizobia are more closely related to nonsymbiotic species, including members of such well-studied groups as the purple photosynthetic bacteria, than they are to other rhizobia. If we are to make sense of the evolutionary history of the symbiosis, we need to assess the genetic relationships among species by methods that do not depend on symbiotic function but, rather, reflect the relatedness of chromosomal genomes. Knowledge of these relationships would also benefit the study of biochemistry, physiology and genetics, since those aspects of the biology of a rhizobium species that are not directly part of symbiotic function will probably have more in common with close nonsymbiotic relatives than with more distant symbiotic species.

For all these reasons, we needed a convenient and general method for assessing the genetic relatedness of rhizobia and their relatives and for reconstructing their evolutionary tree. In recent years the 16S ribosomal RNA (rRNA) molecule has assumed a pivotal role in ascertaining the phylogenetic relationships of bacteria. It is an appropriate tool for many reasons (Woese 1987), not least of which is the rapidly growing wealth of publicly available sequence data. Early studies of 16S rRNA for bacterial phylogeny were based on quantitative hybridization between total DNA and labelled rRNA (De Smedt & De Ley 1977), or on catalogues of RNA oligonucleotides released by T_1 ribonuclease (Woese et al. 1983), but both of these approaches require specialized techniques and fail to extract the full information available. More recently, nucleotide sequences have been obtained, and with the advent of the polymerase chain reaction (PCR) and methods for the direct sequencing of the amplified DNA (Innes et al. 1988; Winship 1989), reliable sequences can be obtained relatively rapidly. There are not yet any published sequences for the 16S genes of rhizobia; our current knowledge of their relationships (discussed in detail by Young 1991) is based on catalogues and hybridization (Woese et al. 1983; Hennecke et al. 1985; Jarvis et al. 1986; Dreyfus et al. 1988).

We have used PCR-based DNA sequencing to obtain data on part of the 16S rRNA genes of a number of strains of rhizobia and related bacteria, including the type strains of all the currently valid rhizobial species. The part we have sequenced corresponds to positions 44 to 337 in the published sequence of the *E. coli* 16S gene (Woese 1987), and varies in length from 260 to 264 bases in different rhizobia. An evolutionary tree can be reconstructed from comparisons of the sequences (Saitou & Nei 1987), and this is shown in Figure 1 (based on unpublished data of H.L. Downer, B.D. Eardly, B.D.W. Jarvis and J.P.W.Y. on the rhizobia and *Rhodopseudomonas*, and on published sequences in the EMBL data bank). The main conclusions are set out below, including some which confirm the results of studies based on other techniques (1–6) and some that are novel (7–8).

(1) *Rhizobium leguminosarum* and *R. meliloti* are closely related, and are on the *Agrobacterium* branch.

(2) *R. loti* and *R. galegae* are more distant members of this branch.

(3) *R. fredii* is very closely related to *R. meliloti*. This is in agreement with the data on rRNA hybridization obtained by Jarvis et al. (1986), but contrasts with the finding by Chen et al. (1988) that the behaviour of *R. fredii* strains was very diverse in a range of biochemical tests. We have examined representatives of the different phenotypic groups identified by Chen et al., and find that they all have exactly the same sequence for the 16S rRNA fragment, and this sequence is also the same in the type strain of *R. meliloti*. It seems that, despite their phenotypic diversity, *R. fredii* strains are closely related in genotype, at least for the 16S genes.

(4) *Bradyrhizobium japonicum* is genetically very diverse, and should probably be divided into several different species. The groups I, Ia and II recognized by DNA–DNA hybridization (Hollis et al. 1981) are also apparent from 16S sequence.

(5) *B. japonicum* is closely related to the purple nonsulphur photosynthetic bacterium *Rhodopseudomonas palustris*, and in fact the diversity within *B. japonicum* is so great that some strains are more distant from the type strain than is *R. palustris*.

(6) *Azorhizobium caulinodans* is a more distant relative of *Bradyrhizobium*.

(7) Although *Rhizobium loti* and *R. galegae* are distant from the *R. leguminosarum/meliloti/fredii* group, they appear to be specifically related to each other.

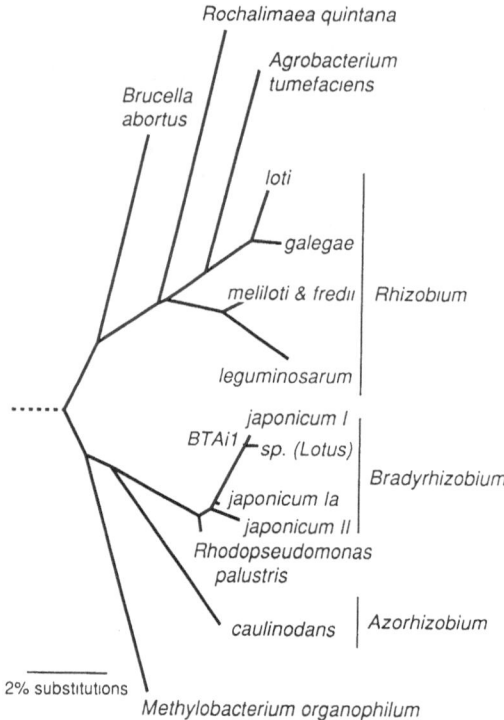

Fig. 1. Phylogenetic tree of the rhizobia and some closely related bacteria, based on sequences of part of the 16S ribosomal RNA gene. The tree was reconstructed by the Neighbor-Joining method from Jukes-Cantor estimates of mutational distance, and rooted using *Erythrobacter longus* and *Escherichia coli* as outgroups

(8) BTAil, a stem-nodulating rhizobium of *Aeschynomene indica* which is also a photosynthetic bacterium (Evans et al. 1990), is a member of the *Bradyrhizobium/ Rhodopseudomonas* cluster. This is a very satisfying result, since this organism provides the missing link between the symbiotic *Bradyrhizobium* and the photosynthetic *Rhodopseudomonas*.

The tree in Fig. 1 includes representatives of all the rhizobial species that are currently validly named, but as more rhizobia are described it is inevitable that new species will be needed. Indeed, we have recently obtained sequences that do not match those of any of the existing species; they belong to rhizobia from various tropical legumes, and I have no doubt that many more species will be named as more isolates are described and it becomes easier to determine their phylogenetic affinities.

6 Conclusions

I have described our studies of genetic diversity within a bacterial species, and our studies of relationships among the genera and species of rhizobia. These two lines of approach converge on the species boundary, and the nature of this boundary is one of the major uncertainties in our current understanding of bacterial evolution (Young 1989). Is the concept of discrete bacterial species merely an artifact invented by bacteriologists in order to describe parts of a genetic continuum, or are bacterial species real biological entities, and if so, what are the barriers to genetic flow that maintain their separate identities? We hope that our studies of rhizobia will not only increase our specific knowledge of this important group of organisms, but will also shed some light on these broader questions.

7 References

Barnet YM (1972) Bacteriophages of *Rhizobium trifolii*. J Gen Virol 15: 1

Beynon JL, Josey DP (1980) Demonstration of heterogeneity in a natural population of *Rhizobium phaseoli* using variation in intrinsic antibiotic resistance. J Gen Microbiol 118: 437

Bottomley PJ, Dughri MH (1989) Population size and distribution of *Rhizobium leguminosarum* bv. *trifolii* in relation to total soil bacteria and soil depth. Appl Envir Microbiol 55: 959

Buchanan-Wollaston AV (1979) Generalized transduction in *Rhizobium leguminosarum*. J Gen Microbiol 112: 135

Chen WX, Yan GH, Li JL (1988) Numerical taxonomy study of fast-growing soybean rhizobia and a proposal that *Rhizobium fredii* be assigned to *Sinorhizobium* gen. nov. Int J Syst Bacteriol 28: 392

Crow VL, Jarvis BDW, Greenwood RM (1981) Deoxyribonucleic acid homologies among acid-producing strains of *Rhizobium*. Int J Syst Bacteriol 31: 152

De Smedt J, De Ley J (1977) Intra- and intergeneric similarities of *Agrobacterium* ribosomal ribonucleic acid cistrons. Int J Syst Bacteriol 27: 222

Dreyfus B, Garcia JL, Gillis M (1988) Characterization of *Azorhizobium caulinodans* gen nov, sp nov, a stem-nodulating nitrogen-fixing bacterium isolated from *Sesbania rostrata*. Int J Syst Bacteriol 38: 89

Dughri MH, Bottomley PJ (1984) Soil acidity and the composition of an indigenous population of *Rhizobium trifolii* in nodules of different cultivars of *Trifolium subterraneum* L. Soil Biol Biochem 16: 405

Eardly BD, Hannaway DB, Bottomley PJ (1985) Characterization of rhizobia from ineffective alfalfa nodules: ability to nodulate bean plants (*Phaseolus vulgaris* (L.) Savi.). Appl Environ Microbiol 50: 1422

El-Sherbeeny MH, Mytton LR, Lawes DA (1977) Symbiotic variability in *Vicia faba*. 1. Genetic variation in the *Rhizobium leguminosarum* population. Euphytica 26, 149

Evans WR, Fleischman DE, Calvert HE, Pyati PV, Alter GM, Subba Rao NS (1990) Bacteriochlorophyll and photosynthetic reaction centers in *Rhizobium* strain BTAi 1. Appl Envir Microbiol 56: 3445

Fottrell PF, O'Hora A (1969) Multiple forms of D(−)3-hydroxybutyrate dehydrogenase in *Rhizobium*. J Gen Microbiol 57: 287

Hadley RG, Eaglesham ARJ, Szalay AA (1983) Conservation of DNA regions adjacent to *nifKDH* homologous sequences in diverse slow-growing *Rhizobium* strains. J Mol Appl Genet 2: 225

Harrison SP, Young JPW, Jones DJ (1987) *Rhizobium* population genetics: effect of clover variety and inoculum dilution on the genetic diversity sampled from natural populations. Plant Soil 103: 147

Harrison SP, Jones DG, Schünmann PHD, Forster JW, Young JPW (1988) Variation in *Rhizobium leguminosarum* biovar *trifolii* Sym plasmids and the association with the effectiveness of nitrogen fixation. J Gen Microbiol 134: 2721

Harrison SP, Jones DG, Young JPW (1989) *Rhizobium* population genetics: genetic variation within and between populations from diverse locations. J Gen Microbiol 135: 1061

Hennecke H, Kaluza K, Thöny B, Fuhrmann M, Ludwig W, Stackebrandt E (1985) Concurrent evolution of nitrogenase genes and 16S rRNA in *Rhizobium* species and other nitrogen fixing bacteria. Arch Microbiol 142: 342

Hirsch PR (1979) Plasmid-determined bacteriocin production by *Rhizobium leguminosarum*. J Gen Microbiol 113: 219

Hollis AB, Kloos WE, Elkan GH (1981) DNA:DNA hybridization studies of *Rhizobium japonicum* and related *Rhizobiaceae*. J Gen Microbiol 123: 215

Innes MA, Myambo KB, Gelfand DH, Brow MAD (1988) DNA sequencing with *Thermus aquaticus* DNA polymerase and direct sequencing of polymerase chain reaction amplified DNA. Proc Natl Acad Sci USA 85: 9436

Jarvis BDW, Gillis M, De Ley J (1986) Intra- and intergeneric similarities between the ribosomal ribonucleic acid cistrons of *Rhizobium* and *Bradyrhizobium* species and some related bacteria. Int J Syst Bacteriol 36: 129

Jarvis BDW, Ward LJH, Slade EA (1989) Expression by soil bacteria of nodulation genes from *Rhizobium leguminosarum* biovar *trifolii*. Appl Envir Microbiol 55: 1426

Johnston AWB, Beringer JE (1977) Chromosomal recombination between *Rhizobium* species. Nature 267: 611

Johnston AWB, Beynon JL, Buchanan-Wollaston AV, Setchell SM, Hirsch PR, Beringer JE (1978) High frequency transfer of nodulating ability between strains and species of *Rhizobium*. Nature 276: 634

Jones DG, Morley SJ (1981) The effect of pH on host plant 'preference' for strains of *Rhizobium trifolii* using fluorescent ELISA for strain identification. Annals Appl Biol 97: 183

Jordan DC (1984) Rhizobiaceae. In: Kreig NR (ed) Bergey's Manual of Systematic Bacteriology, vol 1. Williams & Wilkins, Baltimore, p 234

Martinez E, Pardo MA, Palacios R, Cevallos MA (1985) Reiteration of nitrogen fixation gene sequences and specificity of *Rhizobium* in nodulation and nitrogen fixation in *Phaseolus vulgaris*. J Gen Microbiol 131: 1779

Mytton LR, McAdam NJ, Portlock P (1978) Enzyme polymorphism as an aid to identification of *Rhizobium* strains. Soil Biol Biochem 10: 79

Noel KD, Brill WJ (1980) Diversity and dynamics of indigenous *Rhizobium* populations. Appl Envir Microbiol 40: 931

Piñero D, Martinez E, Selander RK (1988) Genetic diversity and relationships among isolates of *Rhizobium leguminosarum* biovar *phaseoli*. Appl Envir Microbiol 54: 2825

Robert FM, Molina JAE, Schmidt EL (1982) Properties of *Rhizobium leguminosarum* isolated from various regions of Morocco. Ann Microbiol (Inst Pasteur) 133A: 461

Roberts GP, Leps WT, Silver LE, Brill WJ (1980) Use of two-dimensional polyacrylamide gel electrophoresis to identify and classify *Rhizobium* strains. Appl Envir Microbiol 39: 414

Russell PE, Jones DG (1975) Variation in the selection of *Rhizobium trifolii* by varieties of red and white clover. Soil Biol Biochem 7: 15

Saitou N, Nei M (1987) The neighbor-joining method: a new method for reconstructing phylogenetic trees. Mol Biol Evol 4: 406

Selander RK, Caugant DA, Ochman H, Musser JM, Gilmour MN, Whittam TS (1986) Methods of multilocus enzyme electrophoresis for bacterial population genetics and systematics. Appl Envir Microbiol 51: 873

Soberon-Chavez G, Najera R (1989) Isolation from soil of *Rhizobium leguminosarum* lacking symbiotic information. Can J Microbiol 35: 464

Stein M, Bromfield ESP, Dye M (1982) An assessment of a method based on intrinsic antibiotic resistance for identifying *Rhizobium* strains. Ann Appl Biol 101: 261

Stevens JW (1925) A study of various strains of *Bacillus radicicola* from nodules of alfalfa and sweet clover. Soil Sci 20: 45

Vincent JM, Waters LM (1953) The influence of the host on competition amongst clover root nodule bacteria. J Gen Microbiol 9: 357

Winship PR (1989) An improved method for directly sequencing PCR amplified material using dimethyl sulphoxide. Nucl Acids Res 17: 1266

Woese CR (1987) Bacterial evolution. Microbiol Rev 51: 221

Woese CR, Stackebrandt E, Weisburg WG, Paster BJ, Madigan MT, Fowler VJ, Hahn CM, Blanz P, Gupta R, Nealson KH, Fox GE (1983) The phylogeny of purple bacteria: the alpha subdivision. Syst Appl Microbiol 5: 315

Young JPW (1985) *Rhizobium* population genetics: enzyme polymorphism in isolates from peas, clover, beans and lucerne grown at the same site. J Gen Microbiol 131: 2399

Young JPW (1989) The population genetics of bacteria. In: Hopwood DA, Chater KF (eds) Genetics of Bacterial Diversity Academic Press, London, p 417

Young JPW (1991) Phylogenetic classification of nitrogen-fixing organisms. In: Stacey G, Burris RH, Evans HJ (eds) Biological Nitrogen Fixation. Chapman & Hall, New York

Young JPW, Demetriou L, Apte RG (1987) *Rhizobium* population genetics: enzyme polymorphism in *Rhizobium leguminosarum* from plants and soil in a pea crop. Appl Envir Microbiol 53: 397

Young JPW, Johnston AWB (1989) The evolution of specificity in the legume-rhizobium symbiosis. Trends Ecol Evol 4: 341

Young JPW, Wexler M (1988) Sym plasmid and chromosomal genotypes are correlated in field populations of *Rhizobium leguminosarum*. J Gen Microbiol 134: 2731

Young JPW, Downer HL, Eardly BD (1991) Phylogeny of the phototrophic rhizobium strain BTAi1 by polymerase chain-reaction based sequencing of a 16S rRNA gene segment. J Bacteriol 173: 2271

CHAPTER 19
Identification of *nif* and *nod* Genes in *Frankia*

Q.S. Ma, Y.H. Cui, L.M. Chen, M. Qin,
Y.L. Wang, and X.L. Bai

The Nitrogen Fixation
and its Research in China
Editor Guo-fan Hong
© Springer-Verlag Berlin Heidelberg 1992

1 Introduction

Frankia is a group of slow-growing, gram-positive, filamentous, spore-forming, branching and vesicle-forming actinomycetes (Lechevalier 1984) that can induce symbiotic nitrogen-fixing root nodules on non-leguminous actinorhizal plants. Actinorhizal plants are perennial dicots and taxonomically diverse, occurring in at least 23 genera belonging to 8 different families (Benson 1988). Nearly all actinorhizal plants (but *Datisca*) are woody shrubs or trees which are important economically in timber and pulp production. Since a majority of them are pioneer colonizers on nitrogen-poor depaurerate soils, they are of great influence ecologically in land reclamation, reforestation and global nitrogen cycles (Silvester 1974; Torrey 1978). The *Frankia*-actinorhizal symbiosis rivals the rhizobium-legume symbiosis in the amount of nitrogen they fix and the efficiency of the nitrogen-fixing process.

There are both similarities and differences between *Frankia*-actinorhizal symbiosis and the *Rhizobium*-legume one. Since structural genes for nitrogenase (*nif* HDK) have been highly conserved among all diazotrophs so far studied (Ruvkun 1980), it is most likely that similar *nif* HDK genes could be found in *Frankia*. It was observed that these two types of symbiotic associations result in effective nitrogen fixing and have a similar process of nodule formation. Rhizobia fix nitrogen only in symbiotic association with their host plants in the family of Leguminosae (a sole exception is *Parasponia*, a nonleguminous tree belonging to *Ulmaceae*), while *Frankia* can form nitrogen-fixing nodules on the wide range of their host plants in the genera.

A number of nodulation genes (*nod*) have been identified in *Rhizobium*. Since *Frankia* shares features common with those of *Rhizobium* in their respective nodule development, it is conceivable to find similar *nod* gene or genes in *Frankia*, the study of which would hopefully help us to better understand the evolution concerning biological symbiosis in general, and contribute to the development of new symbioses.

Research on *Frankia* has been progressing slowly until a *Frankia* isolate from *Comptonia peregrina* was reported for the first time by Callaham et al. in 1978. Since then, many reports on *Frankia* physiology (Tjepkema et al. 1986) and taxonomy (Lechevalier 1984) have been published with the increasing number of new isolates emerging. These results have been recently reviewed (Simonet et al. 1989). However, no suitable systems, i.e. transposon mutagenesis, transformation, endogenous cloning vectors etc. are available for genetic manipulation for *Frankia*; advances in its genetics are rather limited. The usual approaches for studying *Frankia* symbiotic genes have been through DNA–DNA hybridizations with DNA probes isolated from *Rhizobium* and/or *Klebsiella* with known symbiotic functions to identify corresponding genes of interest from the *Frankia* gene library (in *E. coli*) by sequence homology. Normand et al. (1986) and Simont et al. (1989) have recently reviewed the advances in *Frankia* molecular biology and genetics.

China is a country rich in actinorhizal plants, of which more than 40 species (including almost all the important ones) are well growing and distributing on immense areas from the temperate zone through the tropical and subtropical zone. Since the late 1970s, studies such as a botanical survey of the actinorhizal plants, in-vitro isolation of endophytes, determination of growth conditions and developing new taxonomy methods have been carried out in some detail, as described in this book by Ding and Chen et al. These studies served as a base for molecular and genetic research on *Frankia*. The molecular genetics study on *Frankia* in our lab was initiated in late 1986. This paper is to review our research including some unpublished results (Cui et al. 1990; Cui 1990).

Our goal is to identify symbiotic genes in *Frankia*. Multiple *Frankia* gene libraries were constructed, and two cloning strategies were employed for identifying symbiotic genes. Sequences of *nif* HDK was used in DNA–DNA hybridizations for identifying corresponding genes in *Frankia*, based on the parallels in the early stages of nodule development between *Frankia* and *Rhizobium*. Studies on similarities and complementarities of their nodulation genes were also conducted.

2 Construction of *Frankia* Genomic Libraries

In genetic studies it is a feasible way to clone DNA in the *E. coli* system for those organisms whose genetic analysis system has not been developed. Genes of interest can be screened with known DNA sequences as probes and the resultant genes can be amplified in *E. coli* host cells.

In *Rhizobium* genetic research pLAFR1 (Friedman et al. 1982) has been used widely for gene cloning. pLAFR1 is a broad-host-range cosmid vector among many gram-negative bacteria and can be mobilized into G⁻ recipient. Foreign DNA once cloned in pLAFR1/*E. coli*, thus, can be analyzed in vitro and can be mated into proper *Rhizobium* recipients to test possible functions.

For genetic comparison, three typical *Frankia* strains belonging to different compatibility groups were selected and used to construct genomic libraries. They are At4, Cc01 and Hr16, the nodule endophytes of *Alnus tinctoria*, *Casuarina cunninghamiana* and *Hippophae rhamnoids* respectively.

2.1 *Frankia* Total DNA Preparation

High-molecular genomic DNA is the prerequisite for further genetic studies. Early studies on *Frankia* genetics had been hindered at this step by difficulties in lysing the *Frankia* cell wall efficiently, a critical step in the isolation of high-molecular-weight DNA. Although the *Frankia* cell wall is similar to that of other actinomycetes (Lechevalier et al. 1982), it seems to be rather resistant to the usual lytic mixture used for these microorganisms. At first, Normand and co-workers (1983) demonstrated that the use of lysozyme in combination with a drastic extraction method (10% SDS at 90 °C and vortexing) allowed the detection of a small plasmid (< 35 kb) and the extraction of small amounts of DNA. A similar lysis procedure based mainly on the chemical dissolution of the cell wall by hot SDS was used by Dobritsa (1985) and An (1983) for isolation of *Frankia* DNA. Later Simonet et al. (1984) demonstrated that the use of achromopeptidase, an enzyme initially described by Ogawa et al. (1983) for *Streptomyces* protoplast formation, in combination with lysozyme, resulted in a complete digestion of *Frankia* cell walls.

Two procedures were employed routinely for isolation of *Frankia* total DNA in our lab, and both caused an efficient lysis of fresh *Frankia* cells grown in the simple inorganic medium BAP (Murry et al. 1984). One is the same as described by Simonet et al. (1985); and the other was developed in our lab, i.e. using only lysozyme to complete lysis of fresh *Frankia* cells grown in BAP without $MgSO_4$ but with 0.1% glycine.

Using uncut lambda DNA as a molecular weight standard, it appears that the DNA prepared from two *Frankia* strains were qualified, but not for the DNA from Hr16, which was a smear as seen on the gel, which may be the result of stronger nuclease activities of these particular strains (*Elaeagnus*-compatible group), as have been described by Normand et al. (1983).

2.2 *Frankia* Gene Library Construction

The genomic libraries were constructed as basically described by Grosveld et al. (1982). Partially digested, size-fractionated *Frankia* DNA was ligated with *Eco*RI-digested, dephosphorylated vector DNA for 14 h at 12 °C in a total volume of 5 μl using T4 DNA ligase. The DNA was then packaged into phage lambda in vitro and the mixture was used to infect *E. coli* strains ED8767 of late logarithmic stage. Transductant colonies were selected on LB plates containing 15 μg/ml tetracycline (the tetracycline resistance was conferred by vector pLAFR1). Successful construction of genomic library is dependent on the quality of the target DNA, vector DNA and packaging extracts used. Both the target DNA and vector DNA were checked by ligation. Several ratios of target DNA to vector DNA were tried in ligation and packaging for best transfection efficiency. The results showed that in the best condition the average number of colonies occurring per plate was 40–80 (some even as high as 100 or more) for At4 and Cc01, 5 for Hr16 and 0.2–0.4 for the control of self-ligated vector DNA. The efficiency was approximately 4,000–8,000 recombinants per μg of target DNA (the amount of DNA was estimated by agarose gel electrophoresis) for both At4 and Cc01, but only about 250 for Hr16. The results shown above were scaled up and were used for large-scale ligation and packaging. 4,000, 3,600 and 700 recombinant colonies were obtained for At4, Cc01 and Hr16, respectively.

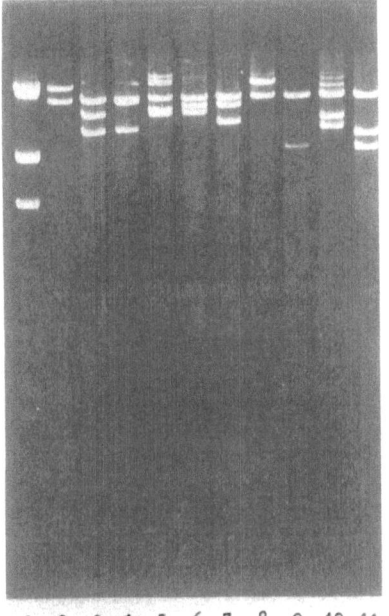

Fig. 1. Electrophoresis of *Eco*R1-digested clone DNA from *Frankia* At4 library.
Lane 1. *Hind*III-digested lambda DNA
Lane 2–11. 10 random selected clones from *Frankia* At4 library

Since the genome size of *Frankia* strains ranges up to 10,000 kb (An et al. 1984) and the size of possible inserts in pLAFR1 is about 20–35 kb, 2,000 clones are required for the genome representation at 99% confidence level. This demand was met for the gene banks of At4 and Cc01, but unfortunately not for Hr16.

Ten Tc^R clones from each gene bank were chosen at random to isolate plasmid DNA using the small-scale isolation procedure. Digestion of these plasmids with *Eco*RI (Fig. 1) demonstrated that the inserts were mostly of large fragments and were different from each other in size. This is due to high G + C content (68%–72%) of *Frankia* strains (An et al. 1983).

3 Cloning of *nif*-like Genes in *Frankia*

In *Rhizobium* genes involved in symbiotic association were grouped into two types: nodulation (Nod) and nitrogen fixation (Fix) genes. The latter can be subdivided into *nif* and *fix* genes, *nif* genes were found to be in homology with the nitrogen fixing genes in *Klebsiella pneumoniae*, a well studied nitrogen-fixing bacterium, in which a cluster of 17 *nif* genes involved in nitrogen fixation has been identified (Robson et al. 1983). Any genes found in *Rhizobium* or other diazotrophes which have the functions equivalent to that of the genes in the cluster are called "*nif*" genes, and the rest which are not equivalent are called "*fix*" genes.

The *nif* structural genes encoding the nitrogenase are highly conserved among diazotrophes, *nif* H codes for the polypeptides of the Fe protein and *nif* DK codes for the alpha and beta subunits of the MoFe protein. Simonet et al. (1989) proved that the three *nif* HDK genes are contiguous in *Frankia* strain Ar13. However, discrepancy to this observation has been reported (Normand et al. 1988). Since a very limited number of *Frankia* strains has been tested so far, it is far from clear how in *Frankia* the *nif* genes as a whole are arranged, not to mention the arrangement of *nif* HDK.

We have constructed three *Frankia* gene libraries, and it is then interesting to clone the *nif* region for further study.

3.1 Hybridizations of *Frankia* Total DNA with *nif* HDK Probe from *K. pneumoniae*

*Bam*HI-digested *Frankia* total DNA was hybridized against a *nif* HDK probe from *K. pneumoniae* (pSA30). The results (Fig. 2) showed that the *nif*-hybridizing fragments were different in size; 3 kb, 1.9 kb and 0.8 kb fragments were detected in the At4 genome, 2.1 kb and 1.7 kb in Hr16 and only a 5.5 kb fragment in all six *Casuarina*-compatible strains (including Cc01) isolated from both *Casuarina* and *Allocasuarina*.

Hennecke et al. (1985) have shown that among the nitrogenase genes *nif* H appears to be the most conserved. The degree of homology is closely related to the relationships determined by other criteria for strain classification. A similar observation was found for the *nif* D gene. This means that the nitrogenase genes are some of the most highly conserved portions of the genome as mentioned by An et al. (1985). Hennecke et al. (1985) have demonstrated that the nitrogenase genes and the 16S rRNA, as a well-known taxonomical criterion employed for determining the phylogenetic position of bacteria, have evolved in parallel. Such results argue against the idea of lateral transfer of genes between nitrogen-fixing bacteria, but for the notion that *nif* genes have evolved in parallel and independently in the bacteria that carry them. These observations suggest that the measurement of *nif* sequence similarity could be used as a taxonomic criterion in addition to the 16S rRNA. However, it is unrealistic to determine *nif* gene sequences routinely from a large number of strains. Recently Normand and his co-workers (1988) have used the *nif* HD region of *Frankia* Ar13 as an homologous probe to detect *nif* RFLPs in a number of *Frankia* strains. Interestingly, the results obtained were well correlated with other molecular biological criteria used in delineating *Frankia* species, such as patterns for isozymes and whole-cell proteins, as well as DNA–DNA homology. This further suggests that *nif* gene sequence similarity, more precisely, the *nif* HDK RFLPs, could be used as a taxonomical criterion within the genus *Frankia*.

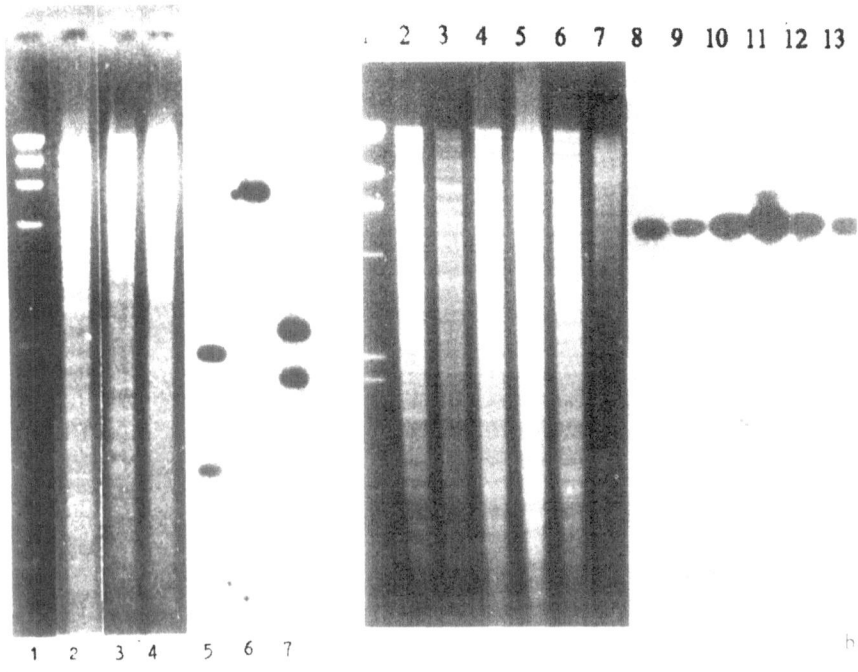

Fig. 2. Hybridization of *Frankia* total DNA against ³²P-labelled pSA30 containing *nif* HDK genes derived from *K. pneumoniae*.
a. Lane 1. *Hind*III-digested lambda DNA
  Lane 2.3.4. At4, Cc01 and Hr16 total DNA digested with *Bam*H1, respectively.
  Lane 5.6.7. Southern blotted lane 2,3,4. hybridized ³²P-labelled pSA30
b. Lane 1. *Hind*III-digested lambda DNA
  Lane 2-7. Total DNAs of *Frankia* spp. Cc01, CG01, CcR01, Ac105, Ce24 and A11I1 digested with *Bam*H1, respectively
  Lane 8-13. Southern blotted lane 2-7 hybridized with ³²P-labelled pSA30
*. *Frankia* spp, Cc01, Cg01, CcR03, Ac105, Ce24 and AllI3 are nodule endophytes of *Casuarina cunninghamiana, C. glaceca, C. cristata, Allocauarina littoralis, C. equisetifolia*, and *Ac. cunninghamiana*, respectively

Compared with those described by Normand et al. (1988), a novel *nif* RFLP was found in both *Alnus*- and *Elaeagnus*-compatible group in this work. This shows that the situation within these two compatible groups may be more complex than previously thought when the number of strains isolated in pure culture increases. It is interesting to note that the *nif* HDK-hybridizing fragments derived from six *Casuarina*-compatible strains isolated from *Casuarina* and *Allocasuarina* were identical, indicating that this group may be more conservative in evolution than the other two groups. Since problems were encountered in the isolation of *Frankia* from nodules of *Casuarina*, little information of it is available (Gauthier et al. 1984). It can be deduced tentatively that the 5.5 kb *Bam*HI *nif*-hybridizing fragment could be used as a marker for identifying "typical" *Casuarina*-type strains.

3.2 Cloning of *nif*HDK Fragment in *Frankia* Cc01

In order to obtain autoradiographs of High resolution and less background, it is essential to use high-homologous probe DNA in DNA-DNA hybridization. Therefore, a *nif* gene clone from *Frankia* as a probe would be much better than any of the *nif* clones from *K. pneumoniae* in studying the *Frankia nif* gene arrangements. As described above, all *nif*HDK-hybridizing sequences are located in 5.5 kb *Bam*HI fragments of the genomes of *Casuarina*-compatible strains, thus making it easy to single out the fragments and clone them.

Frankia Cc01 total DNA was digested completely with *Bam*HI, and the digests were separated on 0.7% agarose gel. The gel region containing 5.5 kb fragment was cut out and was managed to allow it running into the low melting gel (LMP). The DNA isolated from the LMP was ligated with *Bam*HI-digested and dephosphorylated pBR322 DNA. The mixture was used to transform *E. coli* recipient HB101. Two identical *nif* clones (called pCc1GX) were isolated

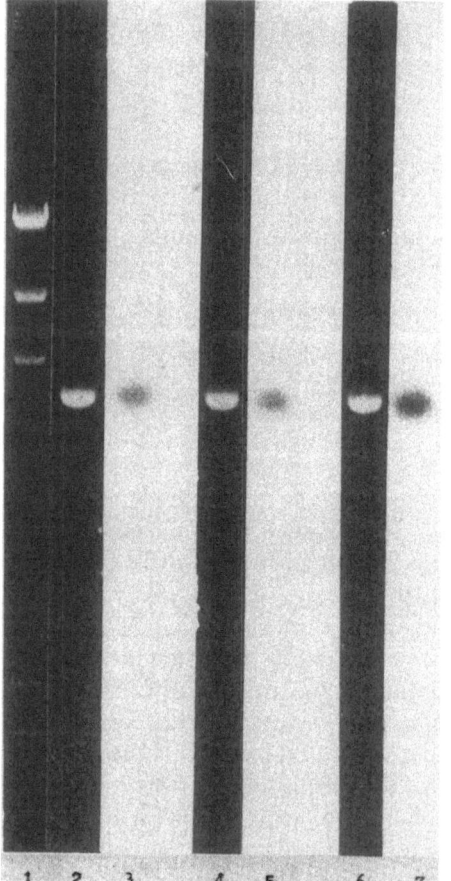

Fig. 3. Hybridizations of pCc1GX against *nif* H. *nif* D and *nif* K genes.
Lane 1. λ DNA/*Hin*dIII marker.
Lane 2,4,6. 5.5 kb *Bam*HI fragment from pCc1GX.
Lane 3,5,7. Southern blotted Lane 2, 4, 6. hybridized with ^{32}P-labelled pPC1201 (*nif* H). pPC1202 (*nif* D) and pPC1203 (*nif* K)

from 192 Sm^{r}Tc^s transformants. Hybridizations of pCc1GX against pPC1201, pPC1202 and pPC1203 (They are *nif* H, *nif* D and *nif* K subclones of pSA30, respectively) (Sibold et al. 1985) showed that pCc1GX harboured all the *Frankia* Cc01 nitrogenase structural genes (Fig. 3).

3.3 Isolation and Characterization of *Frankia nif* Cosmid Clones

The *Frankia* gene libraries were screened by in-situ colony hybridization using the *Frankia* Cc01 *nif* HDK gene fragment contained in pCc1GX as a probe. 89 and 18 strong positives were found from At4 and Cc01 banks, respectively.

Restriction map analysis demonstrated that all of the 89 positive spots from the At4 bank turned out to be identical, which were designated pAt1GX. The insert of pAt1GX is a 24 kb *Eco*RI fragment. When pAt1GX digested with *Bam*HI and hybridized with *nif* HDK probe it turns out that pAt1GX contains the 1.9 kb and most of the 0.8 kb *nif*-hybridizing fragment, which consists of about 0.6 kb *Bam*HI–*Eco*RI fragment, together with a piece of vector (see Fig. 4).

The "positive clones" from Hr16 bank were digested with *Eco*R1 and hybridized with pCc1GX. 13 kb *nif*-hybridizing bands were observed in all of the 17

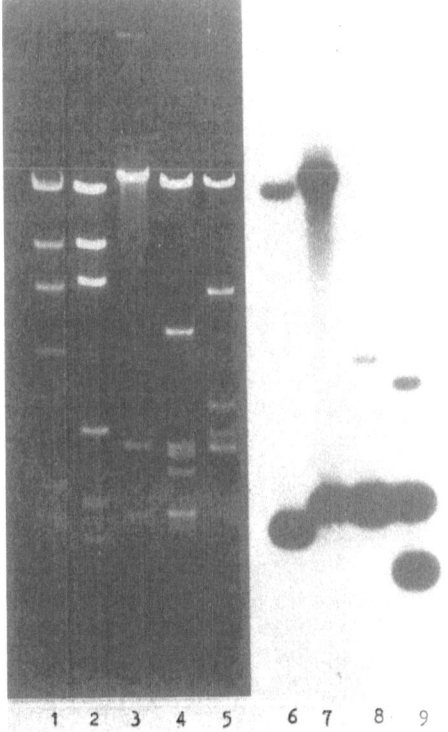

Fig. 4. Hybridization of pAt1Gx, pHr14GX, pHr17GX, and pHr18GX to ^{32}P-labelled pCc1GX.
Lane 1. λ DNA/*Hind*III marker.
Lane 2–5. *Bam*H1 digested pAt1GX, pHr14GX, pHr17GX and pHr18GX, respecively.
Lane 6–9. Southern blotted lane 2–5 hybridized with ^{32}P-labelled pCc1GX

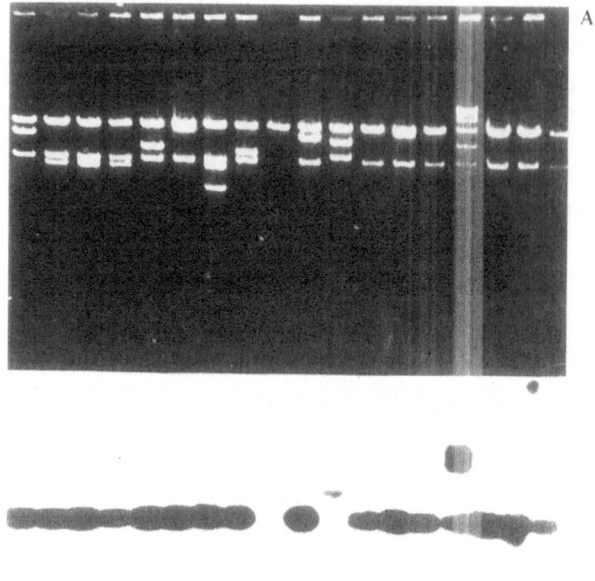

1   2   3   4   5   6   7   8   9   10   11   12   13   14   15   16 17   18   B

Fig. 5. Identification of *nif* clones from Hr16 gene bank.
A. Lane. 1–18. (*upper*) *Eco*R1 digested pHr1GX to pHr18GX, respectively.
B. Lane. 1–18. (*lower*) *Eco*R1 digested pHr1GX to pHr18GX DNA hybridized with ^{32}P-labelled pCc1GX, showing the strong 13 kb band and the weak 17 kb band. Two top bands in Lane 15 are due to partial digestion

clones except pHr11GX, which harbours a 17 kb *nif*-hybridizing band that was a little weaker in intensity than the 13 kb hybridizing bands (Fig. 5). Of the 17 clones three were chosen for electrophoresis analysis. The digestion patterns of these 3 clones were different with a few bands in common. After hybridization with pCc1GX it turns out that pHr18GX contains both 2.1 kb and 1.7 kb *nif*-hybridizing bands, while the other two clones contain only the 2.1 kb band (Fig. 4). These cosmid clones may be overlapped, but the locations of the 2.1 kb and 1.7 kb fragments in pHr18GX were not determined.

Frankia nif gene fragments have been isolated in other labs. Ligon and Nakas (1987) found that on a 49 kb DNA fragment, which was isolated from an *Alnus*-compatible strain FaCl, the *nif* H gene was not present but carried both *nif* D and *nif* K genes only. Their data suggested that the genes for the Fe protein (*nif* H) and MoFe protein (*nif* DK) may be separated by at least 10 kb. This is in agreement with the *nif* gene arrangement in *Bradyrhizobium* species

(Fuhrmann and Hennecke 1982; Kaluza et al. 1983; Scott et al. 1983), where *nif* H is separated from *nif* DK by at least 13 kb and is transcribed as a separate operon. However, this is contrary to the following results reported by Normand et al. (1988) who demonstrated that five out of five *Frankia* strains tested (two *Elaeagnus*-compatible strains, EUN1f and HRN18a; two *Alnus*-compatible strains, AxcN24d and ArI3; and one *Casuarina*-compatible strain, CeD) exhibited close linkage of *nif* D and *nif* H genes. This is similar to the situation in fast-growing *Rhizobium* species (Ruvkun et al. 1982; Corbin et al. 1983; Downie et al. 1983; Schofield et al. 1983; Rolfe and Shine 1984).

In this work, it was demonstrated, through Southern hybridization, that the 1.9 kb *Bam*HI *nif*-hybridizing fragment in pAt1GX and the 2.1 kb *nif*-hybridizing fragment in pHr18GX hybridized strongly with both pPC1201 and pPC1202 (data not shown), suggesting that *nif* H and *nif* DK genes, or at least part of them, are located within the 1.9 kb and 2.1 kb DNA fragment in *Frankia* At4 and Hr16, respectively. These data, together with the previous results that the pSA30-hybridizing fragments were located on a 5.5 kb fragment in six *Casuarina*-compatible strains, suggest that the *nif* H and *nif* DK genes are closely linked with each other in all 8 *Frankia* strains used in this work. It appears that the linkage of *nif* H and *nif* DK in *Frankia* may be a rather common situation. However, the *Frankia* isolates tested so far represent only a small proportion of the actinorhizal endophyte population, it is still too early to draw a general conclusion.

Generally, genes involved in the same function are usually clustered together in prokaryotes. In *Rhizobium*, especially in fast-growing *Rhizobium* species, genes involved in symbiotic nitrogen fixation are closely linked (Long 1989). Recently, Simonet et al. (1989) showed that, in *Frankia* ArI3, the nitrogenase structural genes (*nif* HDK), their regulatory gene (*nif* A), FeMoCo gene (*nif* B) and "early nodulation gene" are located within a 14.4 kb DNA fragment only. As a next step in our work, we will focus on the analysis of the genes flanking *nif* HDK.

4 Cloning of Nodulation Genes in *Frankia*

With the progress in the *nif* gene cloning in *Frankia*, its nodulation genes become the most interesting target for studies on *Frankia* genetics. *Frankia* can form nodules on actinorhizal non-legume plants in a similar way in which *Rhizobium* does. What kind of *nod* genes does *Frankia* have? What mechanisms do they have by which nodules are formed on non-legumes? To what extent do they share homology with *Rhizobium* *nod* genes? To answer these questions, it is necessary, first of all, to identify *Frankia* *nod* genes.

Frankia *nod* gene(s) can be identified simple by DNA–DNA hybridization if they show good homology with *Rhizobium* *nod* genes. If this is not the case, the possible way one can try is to do the functional complementation in

Rhizobium system but not in the *Frankia* system, because it is still not available. The *Rhizobium* nodulation genes of *R. meliloti*, *R. leguminosarum* and *R. trifoli* have been cloned and analyzed in detail (Long et al. 1982; Jacobs et al. 1985; Torok et al. 1987). One group of the genes designated *nod*-DABC, the so-called "common" *nod* genes, were demonstrated to be generally conservative both structurally and functionally within all *Rhizobium* species so far studied.

Because of the similarities in the early stages of nodule development such as bacteria-plant recognition and root hair curling between the *Rhizobium*-legume and *Frankia*-actinorhizal symbiosis, it seems likely that nodulation genes determining these events may be conserved between *Rhizobium* and *Frankia*. Several attempts have been made in this lab to demonstrate the hybridization of the *nod* genes from *R. leguminosarum* with fragments derived from the *Frankia* genome.

4.1 Hybridization Against *Frankia* Total DNA with *Rhizobium* Nodulation Genes

Total DNA from *Frankia* spp. strains At4. Cc01 and Hr16 was digested with *Eco*RI and *Bam*HI, and the digests were hybridized with radioactively labelled pIJ1216 containing *nod*DABC genes of *R. leguminosarum* (Rossen et al. 1985). No hybridization was observed (data not shown). Then the whole colonies of the two genomic libraries from *Frankia* At4 and Cc01 were screened by in-situ colony hybridization with the same probe of pIJ1216. Seven and six colonies showing relatively strong hybridization were obtained from *Frankia* At4 and Cc01 libraries, respectively.

Plasmid DNA was isolated from these 13 clones and digested with *Eco*RI, then transferred to nitrocellulose paper. Hybridization was done with ^{32}P-labelled pIJ1216. No hybridization was found even in the stringency of $2 \times$ SSC but with the vector DNA being hybridized (data not shown). This result was confirmed several times and was in agreement with those by Simonet et al. (1989).

4.2 Cloning of *Frankia* *nod*D-like Sequence by Direct Complementation

8400 is a *R. leguminosarum* bv. *phaseoli* strain cured of its Sym plasmid. When the Sym plasmid pRL1JI of *R. leguminosarum* was introduced into its Str^R derivative 8401 (Lamb et al. 1982), the recipient can form nodules on pea plants. A number of *nod* gene mutants were constructed by Tn5 insertions into the *nod* region of plasmid pRL1JI (Ma 1983). When these mutated pRL1JIs were introduced into 8401, nodulation was abolished (Table 1).

Clones of *Frankia* At4 gene library were pooled in groups, each of which contained 480 clones. The pooled clones were separately transconjugated into

Table 1. Nodulation characteristics of 8401 derivatives

No.	Strain derivative	Nodulation characteristics (on pea)
	8401	−
	8401 (pRL1JI)	+
1	8401 (pRL1JI nodD: : Tn5)	−
2	8401 (pRL1JI nodB: : Tn5)	−
3	8401 (pRL1JI nodC: : Tn5)	−
4	8401 (pRL1JI nodD: : Tn5)/pAt2GX	+
5	8401 (pRL1JI nodB: : Tn5)/pAt2GX	−
6	8401 (pRL1JI nodC: : Tn5)/pAt2GX	−

8401 *nod*D mutant derivative recipient (1) with the helper strain pRK2073. Tc^R transconjugants from each mating were pooled again and inoculated on peas.

Two weeks after inoculation, nodules appeared on pea roots inoculated with 8401 (pRL1JI), and another two weeks later, two nodules appeared on one pea plant's roots inoculated with 8401 *nod*D mutant derivative (1) Tc^R transconjugants, indicating that *nod*D deficiency in the 8401 derivative strain had been complemented.

These two nodules were 1 2 mm in size in pale. Bacteria were isolated from these two nodules. After purification and plasmid isolation a "common" cosmid containing a 24 kb *Eco*RI insert was found, which was called pAt2GX.

pAt2GX (in *E. coli*) was transferred separately into 8401*nod*D, B, C, mutant derivative recipients (1, 2, 3) by triparent mating with the helper strain pRK2073. Tc^R transconjugants containing pAt2GX from each mating were inoculated on pea plants. Three to four weeks after inoculation nodules appeared on pea roots inoculated with 8401 *nod*D mutant derivative (1) Tc^R transconjugants (Fig. 6) (Table 1). These nodules were 0.5 × 2 mm in size with white or pale red color. The plants had less nodules than the control one inoculated with 8401 (pRL1JI). No nodule appeared on pea inoculated with 8401, 8401*nod*D mutant derivative (1), and 8401 *nod*B, C mutant derivative (2, 3) Tc^R transconjugants containing pAt2GX.

The complementation of *Rhizobium nod*D mutant, 8401 (pRL1JI*nod*D: : Tn5) by *Frankia* DNA clone pAt2GX is good evidence showing that *Frankia* carries a *nod*D-like gene whose function is similar in the early nodule development as in *Rhizobium*. It seems unlikely that the complementation was due to the back-mutation of *nod*D mutant since nodulation has not been seen in control tests with the *nod*D mutant as inoculants, although back-mutations of 8401 (pRL1JI*nod*E: : Tn5) were often found because of the leaky property of *nod*E: : Tn5 mutants (data not shown).

The failure to complement 8401 (pRL1JI*nod*B, *nod*C) mutants implied that a different way to induce nodules may exist in *Frankia*. However, since the insert in pAt2GX is only 24 kb in length the absence of *nod*B, C-like genes in pAt2GX can not be ruled out.

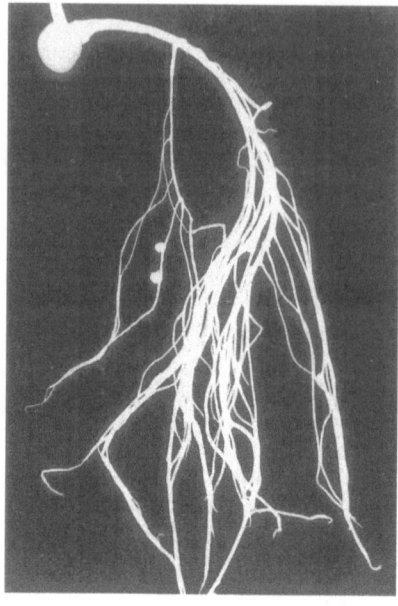

Fig. 6. Nodules formed on pea roots inoculated with 8401 (pRL1JI*nodD*: :Tn5)/pAt2GX

5 Concluding Remarks

The goal of this work was to identify symbiotic genes in *Frankia*. The DNA fragments in homology with *nif* HDK probe (pSA30) have been found in three *Frankia* strains, and these fragments were cloned. Our further studies have shown that *nif* HDK in *Frankia* is contiguous. In *Casuarina*-compatible *Frankia* strains the *nif* HDK gene arrangement is less variable, which can be used as a specific RFLP pattern for strain classification.

Several *nif* HDK-containing cosmid clones have seen isolated. We still do not know what functions of the genes flanking *nif* HDK genes are. It is important to establish a *Frankia* system for genetic analysis. Recently progress has been made in this lab to develop a method to introduce foreign genes into *Frankia* protoplasts by electroporation. Once the whole transformation system in *Frankia* has been completed the flanking sequences of *nif* HDK may be analysed in certain detail and more genes involved in symbiosis can be identified.

The identification of cosmid clone (pAt2GX) containing a *nod*D-like gene is an exciting finding. This identification is confirmed by direct functional complementation, not by DNA sequence homology. Considering the great differences in G + C content in the genomes between *Frankia* and *Rhizobium*, it is conceivable that different codon usage is probable, which may result in poor DNA homology between them. In fact the homology between *K. pneumoniae nif* HDK probe and the *Frankia* genome is not strong. That was why we have used the *Frankia nif* HDK clone in subsequent experiments after the first confirmation of the existence of the *nif* HDK clone in *Frankia*.

In *Rhizobium* the so-called common *nod* genes include *nod*D and *nod*ABC. Now it turns out that *Frankia* displays a *nod*D-like function. What does this mean? Many studies have proved that the *nod*D gene is a key gene involved in the earliest stages of symbiotic recognition between *Rhizobium* and legumes (Spaink et al. 1987). The *nod*D product appears to be a transcriptional activator protein that binds to the promotors of inducible *nod* genes (Hong et al. 1987; Henikoff et al. 1988; Fisher et al. 1988) in a region closely corresponding to the *nod* box (Rostas et al. 1986; Long 1989). The *nod*D gene needs plant-synthesized compounds (flavonoids) for its activity to initiate the early nodulation processes (Mulligan and Long 1985; Innes et al. 1985; Rossen et al. 1985; Shearman et al. 1986; Spaink et al. 1987; Basom et al. 1988; Spaink et al. 1989). These flavonoids include luteolin, 7, 4'-dihydroxyflavone (DHF), daidzein, etc. Therefore, the *nod*D gene is possibly the earliest bacterial gene to initiate the nodulation process.

In *Rhizobium* strain NGR234 (Trinick 1980) which is capable to nodulate a non-legume host *Parasponia*, the *nod*D1 gene can interact with plant extracts from a number of nonhost plants including *Casuarina* to induce transcriptional activation of *nod* genes (Bender et al. 1988; Bassam et al. 1988; Strange et al. 1990). This shows that the *nod*D gene could be less specific. In this sence it is not surprising that *Frankia* has *Rhizobium nod*D function.

Although a *nod*D-like gene has been found in *Frankia*, many questions are still to be answered. For example, what are the other *nod* genes in *Frankia*? Do they function in the same way as the *Rhizobium nod* genes do? Are they arranged in operons and regulated in the same mode as in *Rhizobium*? How are they able to form nodules on actinorhizal plants?

At the moment we have several interesting clones to work with. Studies on these clones will hopefully provide us with more information about *Frankia* symbiosis. We hope that it will be possible in the future to construct super-*Frankia* strains which would show much higher host-endophyte association efficiency that can be used for wood production and environmental protection. And possibly we are also able to extend symbiotic associations to other important crops.

6 References

An CS, Riggsby WS, Mullin BC (1985) Relationships of *Frankia* isolates based on deoxyribonucleic acid homology studies. Int J Syst Bacteriol 35: 140

An CS, Riggsby WS, Mullin BC (1984) DNA relatedness of *Frankia* isolates. In: Veeger C, Newton W (eds) Advances in nitrogen fixation research. Martinus Nijhoff-Junk, The Hague

An CS, Wills JW, Riggsby WS, Mullin BC (1983) Deoxyribonucleic acid base composition of 12 *Frankia* isolated. Can J Bot 61: 2859

Bassam BJ, Djordjevic MA, Redmond JW, Batley M, Rolfe BG (1988) Identification of a *nod*D-dependent locus in the *Rhizobium* strain NGR234 activated by phenolic factors secreted by soybeans and other legumes. Mol Plant-Micobe Interact 1: 161

Bender GL, Nayudu M, Le Strange KK, Rolfe BG (1988) The *nodD1* gene from *Rhizobium* strain NGR234 is a key determinant in the extension of host range to the nonlegume Parasponia. Mol Plant-Microoobe Interact 1: 259

Benson DR (1988) The genus *Frankia*: Actinomycete symbionts of plants. Microbiol Sci 5: 1

Burgess BK (1984) Structure and reactivity of nitrogenase-an overview. In: Veeger C, Newton W (eds) Advances in nitrogen fixation research. Martinus Nijhoff-Junk, The Hague. p 103

Callaham DDel, Tredici P, Torrey JG (1978) Isolation and cultivation in vitro of the actinomycete causing root nodulation in *Comptonia*. Science 199: 899

Corbin D, Barran L, Ditta G (1983) Organization and expression of Rhizobium meliloti nitrogen fixation. Proc Natl Acad Sci 80: 3005

Cui YH, Wang YL, Qin M, Zeng WY, Bai XL, Ding J, Ma QS (1990) Isolation and comparison of *Frankia nif* gene fragments. Science Bulletin (in Chinese) in press

Cui YH (1990) Molecular cloning of symbiotic genes in *Frankia*. PhD thesis, Shenyang Institute of Applied Ecology, Academic Sinica

Dobritsa SV (1985) Restriction analysis of the *Frankia* spp genome FEMS Microbiol Lett 29: 123

Dowine JA, Ma QS, Knight CD, Hombrecher G, Johnston AWB (1983) Cloning of the symbiotic regions of *Rhizobium leguminosarum*: the nodulation genes are between the nitrogenase genes and a *nif*A-like gene. EMBO J 2: 947

Fisher HM, Bruder T, Hennecke H (1988) Essential and non-essential domains in the *Bradyrhizobium japonicum* NifA protein: identification of indispensable cysteine residues potentially involved in redox reactivity and/or metal binding. Nucl Acids Res 16: 2207

Friedman AM, Long SR, Brown SE, Brikema WJ, Ausubel FM (1982) Construction of a broad host range cosmid cloning vector and its use in the genetic analysis of *Rhizobium* mutants. Gene 18: 289

Fuhrman M, Hennecke H (1982) Coding properties of cloned nitrogenase genes from *Rhizobium japonicum*. Mol Gen Genet 187: 419

Ganthier DL, Diem EG, Dommergnes YR (1984) Tropical and subtropical actinorhizal plant. Pesq Agropoc Bras 19: 119

Grosveld FG, Lund T, Murray EJ, Mellor AL, Dahl HHM, Flavell RA (1982) The construction of cosmid libraries which can be used to transform eukaryotic cells. Nucl Acid Res 10: 6715

Henikoff S, Haughn GW, Calvo JM, Wallace JC (1988) A large family of bacterial activator protein. Proc Natl Acad Sci USA 85: 6602

Hennecke H, Kaluza K, Thony B, Fuhrmann M, Ludwig, Wand, Stackebrandt E (1985) Concurrent evolution of nitrogenase genes and 16S rRNA in *Rhizobium* species and other nitrogen fixing bacteria. Arch Microbiol 142: 342

Hennecke N, Fischer HM, Gubler M, Thony B, Anthamatten D, Kullik I, Ebeling S, Fritsche S, Zurcher T (1988) Regulation of *nif* and *fix* genes in *Bradyrhizobium japonicum* occurs by a cascade of two consecutive steps of which the second one is oxygen-sensitive. In: Nitrogen Fixation: Hundred years after Fischer Stuttgart, p 339

Hong GF, Burm JE, Johnston AW (1987) Evidence that DNA involved in the expression of nodulation (*nod*) genes in *Rhizobium* binds to the product of the regulactory genes nodD. Nucl Acid Res 15: 9677

Innes RW, Kuempel PL, Pazindski J, Canter Cremers H, Rolfe BG, Djordjevic MA (1985) Plant factors induce expression of nodulation and host-range genes in *Rhizobium trifoli*. Mol Gen Gent 201: 426

Jacobs TW, Egelhoff TT and Long SR (1985) Physical and gentic map of a *Rhizobium meliloti* nodulation gene region and nucleotide sequence of *nodC*. J Bacteriol 162: 469

Kaluza K, Furhmann M, Hahn M, Regensburger B, Hennecke H (1983) In: *Rhizobium japonicum* the nitrogenase genes *nif*H *nif*KD are separated. J Bacteriol 155: 915

Lamb JW, Hombrecher G, Jonhston AWB (1982) Plasmid-determined nodulation and nitrogen-fixation abilities in *Rhizobium phaseoli*. Mol Gen Gent 186: 449

Lechevalier MP, Horriere F, Lechevalier H (1982) The biology of *Frankia* and related organisms. Develop Indust Microbiol 23: 51

Lechevalier MP (1984) The taxonomy of the genus *Frankia*. Plant Soil 78: 1

Ligon JM, Nakas JP (1987) Isolation and characterization of *Frankia* sp. strain FaC1 genes involved in nitrogen fixation. Appl Environ Microbiol 53: 2321

Long SR, Buikema WJ, Ausubel FM (1982) Cloning of *Rhizobium meliloti* nodulation genes by direct complementation of *nod* mutants. Nature (London) 298: 485

Long SR (1989) *Rhizobium*-legume nodulation: life together in the underground. Cell 56: 203

Ma QS (1983) Molecular genetics of pRL1JI, a symbiotic plasmid of *Rhizobium leguminosarum*. PhD thesis, University of East Anglia, UK

Mulligan JI, Long SR (1985) Induction of *Rhizobium meliloti* nodC expression by plant exudate requires *nod*D. Proc Natl Acad Sci 82: 6609

Murry MA, Fontaine MS, Torrey JG (1984) Growth kinetics and nitrogenase induction in *Frankia* sp. HFP ArI3 grown in batch culture. Plant Soil 78: 61

Normand P, Lalonde M (1986) The genetics of actinorhizal *Frankia*. Plant Soil 90: 429

Normand P, Simonet P, Bardin R (1988) Conservation of *nif* sequence in *Frankia*. Mol Gen Genet 213: 238

Normand P, Simonet P, Butour JL, Rosenberg C, Moiroud A, Lalonde M (1983) Plasmids in *Frankia* sp. J Bacteriol 155: 32

Ogawa H, Imai S, Satoh A, Kojima M (1983) An improved method for the preparation of *Streptomyces* and *Micromonospora* protoplasts. J Antibiotics 36: 184

Roberts GP, Brill WJ (1981) Genetics and regulation of nitrogen fixation. Ann Rev Microbiol 35: 207

Robson R, Kennedy C, Postgate JR (1983) Progress in comparative genetics of nitrogen fixation. Can J Microbiol 29: 954

Rolfe BG, Shine J (1984) *Rhizobium-Leguminosae* symbiosis: the bacterial point of view. In: Verma DPS, Hohn TH (eds) Genes involved in microbe-plant interactions. Springer Verlag, New York, p 95

Rossen L, Shearman CA, Jonston AWB, Downie JA (1985) The nodD gene of *Rhizobium leguminosarum* is autoregulatory and in the presence of plant exudate induces the nodABC genes. EMBO J 4(13A): 3369

Rostas K, Kondorosi E, Horvath B, Simoncsits A, Kondorosi A (1986) Conservation of extended promoter regions of nodulation genes in *Rhizobium*. Proc Natl Acad Sci 83: 1757

Ruvkun GB, Ausubel FM (1980) Interspecies homology of nitrogenase genes. Proc Natl Acad Sci USA 77: 191

Ruvkun GB, Sundaresan V, Ausubel FM (1982) Directed transposon Tn5 mutagenesis and complementation analysis of *Rhizobium meliloti* symbiotic nitrogen fixation genes. Cell 29: 551

Schofield PR, Djordjevic MA, Rolfe BG, Shine J, Watson JM (1983) A molecular linkage map of nitrogenase and nodulation genes in *Rhizobium trifolii*. Mol Gen Genet 192: 459

Scott KF, Rolfe B, Shine J (1983) Nitrogenase structural genes are unlinked in the non-legume symbiont *Parasponia rhizobium* DNA 2: 139

Shearman CA, Rossen L, Johnston AWB, Downie JA (1986) The Rhizobium leguminosarum nodulation gene nodF encodes a polypeptide similar to acyl-carrier protein and is regulated by nodD plus a factor in pea root exudate. EMBO J 5: 647

Sibold L, Pariot D, Bhatnagar L, Henriquet M, Aubet JP (1985) Hybridization of DNA from methanogenic bacteria with nitrogenase structural genes (*nif* HDK). Mol Gen Genet 200: 40

Silvester W (1974) Ecological and economic significance of the nonlegume symbiosis. In: Newton WE, Nyman C (eds) Proc 1st Int Symp Nitrogen Fixation Washington State University Press, Corvallis 2: 489

Simonet P, Capellano A, Navarro E, Bardin R, Moiroud A (1984) An improved method for lysis of *Frankia* with achromopeptidase allows detection of new plasmids. Can J Microbiol 30: 1292

Simonet P, Normand P, Hirsch AM, Akkermans ADL (1989) The genetics of the *Frankia*-actimorhizal symbiosis. In: Molecular biology of symbiotic nitrogen fixation. CRC Press, p 77

Simonet P, Normand P, Moiroud A, Lalonde M (1985) Restriction enzyme digestion pattern of *Frankia* plasmids. Plant Soil 87: 49

Spaink HP, Wijffelman CA, Pees E, Okker RJH, Lutenberg BJJ (1987) *Rhizobium* nodulation gene nodD as a Determinant of host specificity. Nature 328: 337

Spaink HP, Wijffelman CA, Okker RJH, Lugtenberg BJJ (1989) Localization of functional regions of the *Rhizobium* nodD product using hybrid nodD genes. Plant Mol Biol 12: 59

Strange KK, Bender GL, Djordjevic MA, Rolfe BG, Redmond JW (1990) The *Rhizobium* strain NGR234 nodD1 gene product responds to activation by the simple phenolic compounds vanillin and isovanillin present in white seeding extracts. Mol Plant-Microbe Interactions 3: 214

Tjepkema JD, Schwintzer CR, Benson DR (1986) Physiology of actinorhizal nodules. Ann Rev Plant Physiol 37: 209

Torok I, Kondorosi E, Stepkowski T, Posfai J, Kondorosi A (1984) Nucleotide sequence of *Rhizobium meliloti* nodulation genes. Nucl Acids Res 12: 9509

Torrey J (1978) Nitrogen fixation by actinomycete-nodulated angiosperms. Bioscience 28: 586

Trinick MJ (1980) Relationships amongst the fast-growing rhizobia of *Lablab purpureus*, *leucicephala*, *Mimosa* spp., *Acacia farnesiana* and *Sesbania grandiflora* and their affinities with other rhizobial groups. J Appl Bacteriol 49: 39

CHAPTER 20
Generalized Nitrogen Regulation
in the Purple Non-Sulfur Photosynthetic Bacteria

Hong-Yu Song

The Nitrogen Fixation
and its Research in China
Editor Guo-fan Hong
( ) Springer-Verlag Berlin Heidelberg 1992

1 Introduction

A quarter century ago, the researchers in our laboratory had investigated the mechanism of hydrogen evolution in the photosynthetic bacterium *Rhodobacter palustris*, which was later found to be a nitrogenase-catalyzed reaction (Song and Huang 1964). Since the mid-70's, the mechanism of nitrogenase regulation has attracted much attention, and great progress has been made. The members of the photosynthetic bacteria were selected as favourite organisms for the study of the regulation of nitrogenase by many scientists because of their versatile growth types, and the results were highly rewarding.

The non-sulfur purple bacteria (*Rhodospirillaceae*), unlike other diazotrophs, regulate the activity of nitrogenase by a short-term response upon addition of certain nitrogenous compounds. Ammonia, glutamine, asparagine and urea inhibit the nitrogenase activity rapidly and completely. The inhibition is reversible, and on the exhaustion of the combined nitrogen source the activity is fully restored. This regulatory phenomenon has been termed nitrogenase switch-off and switch-on (Zumft and Castillo 1978). The results demonstrated that the inactivation of nitrogenase activity by ammonia or amides was due to a covalent modification of one subunit of the Fe-protein in the enzyme (nitrogenase reductase) (Kanemoto and Ludden 1984). During this process, glutamine synthetase (GS), the central enzyme in nitrogen assimilation is involved. In this paper, we will present the new results concerning the regulation of nitrogenase in photosynthetic bacteria and discuss the role of glutamine synthetase in it.

2 Molecular Properties of GS in *Rhodobacter Sphaeroides*

The molecular characteristics of GS in *R. sphaeroides* is similar to those from *Escherichia coli*. GS from *R. sphaeroides* is a dodecamer with a molecular weight of about 680 kD (Wang and Song 1987a), and is inhibited by the derivatives of glutamine by a covalent modification mechanism (Wang and Song 1987b). Besides, GS from *R. sphaeroides* shows some interesting properties related to nitrogenase regulation. It is known that when there is covalent adenylylation of GS, the biosynthetic activity of GS is expressed by its adenylylation state (number of subunits adenylylated/dodecamer) which in turn is regulated by the glutamine/2-ketoglutarate (GLN/2-KG) ratio. From the experimental results one can estimate that the ratio of GLN/2-KG needed for half of the GS subunits to be adenylylated is more than ten times lower in *R. sphaeroides* than in *E. coli* (Wang and Song 1989a). In other words, the GS system from this photosynthetic bacterium is over ten times more sensitive than that from *E. coli*. Because the regulation of nitrogenase activity by ammonia is mediated by GS, we suggest that this may explain the higher sensitivity of the nitrogenase system in photosynthetic bacteria to ammonia.

Experiments indicated that a binding site for 2-ketoglutarate exists in GS from *R. sphaeroides*. 2-KG inhibits both transferase and biosynthetic activities of GS. 2-KG did not compete with any substrate of GS, the inhibitory effects only emerged when the three substrates of GS were near their saturating concentrations, it being a case of uncompetitive inhibition (Wang et al. 1989). Such inhibition has so far not been reported in bacteria, though the activation of GS by 2-KG has already been demonstrated in mammalian cells (Tate and Meister 1971). It has been suggested that the ratio GLN/2-KG represents the nitrogen status of the cells (Stadtman and Ginsburg 1974), and that when the ammonia concentration in the medium is low, 2-KG can not carry on its function because of lower substrate concentration. When the concentration of ammonia is high enough to be saturated, the inhibitory effect of 2-KG is manifested, which in turn inhibits ammonia assimilation, thus maintaining a balance between nitrogen and carbon metabolism.

Test in our laboratory revealed the operation of a bicyclic cascade control system of GS in *R. sphaeroides* (Wang and Song 1989a), which had been demonstrated only in a few species of bacteria including *E. coli* (Rhee et al. 1985). In vitro experiments on partially purified adenylyltransferase from *R. sphaeroides* showed an activity of P_{II} regulatory protein (Wang and Song, in press).

3 Molecular Characterization of Glutamate Synthase from *R. Sphaeroides*

Glutamate synthase (GOGAT) was purified from *R. sphaeroides* to electrophoretic homogeneity (Zhou and Song 1989a). The enzyme as a monomer has its molecular weight of approximately 138 kDa. The UV-visible spectrum of the enzyme suggests that it be an iron–sulfur flavoprotein. The apparent optimum pH of the enzyme is between 8.3–8.9, and its K_m values for L-glutamine, 2-ketoglutarate and NADPH are 830, 150 and 6 μM, respectively. The products of the reaction (L-glutamate and NADP) and several amino acids (Met, Leu, Asp and Asn) have been proved to inhibit the activity. An analogue of glutamine, 6-diazo-5-oxo-L-norleucine, was found to be a potent inhibitor of the enzyme.

4 Glutamate Synthase Mutant Lacking Nitrogenase Activity in *R. Sphaeroides*

A glutamate auxotroph was isolated from *R. sphaeroides* in our laboratory. Biochemical analysis revealed that this mutant was deficient in GOGAT activity. In comparison with the parent strain the mutant had a low activity of GS, but changed to a much higher level of adenylylation state. The mutant is unable to grow with either nitrogen gas or ammonia as nitrogen sources and is pleiotropic with respect to the utilization of a wide range of nitrogen sources, such as proline, histidine, arginine, serine, adenine, etc. Measurements of intracellular amino acid pool showed that the glutamine content in the mutant was 16-fold higher than in the parent strain. It seems that a highly responsive feedback system was operated for the modulation of the function of nitrogenase in response to the changes in pools of metabolic intermediates; and the intracellular level of glutamine was the key to the feedback modulation of nitrogenase activity (Song et al. 1984).

Mutations in a number of bacteria causing pleiotropic phenotypes, which are deficient in the ability to utilize a wide range of nitrogen sources, have been described. They include the mutations in the gene for GOGAT, referred to as $gltB$ in $E. coli$ and asm in $Klebsiella aerogenes$, $K. pneumoniae$ and $Azospirillum brasilense$ (Pahel et al. 1978; Brenchley et al. 1973; Bani et al. 1973). Similar phenotypes were found in $glnG$ mutants of $E. coli$ and ntr mutants of $K. pneumoniae$ and also in nac mutants of $K. aerogenes$ (Pahel and Tyler 1979, Mereick 1983, Bender et al. 1983). However, all of these mutants except asm ($gltB$) have normal levels of GOGAT activity, suggesting that the mutation described here may occur in the structure gene for GOGAT.

Nitrogenase activity was found in asm mutants of $K. pneumoniae$ (Streicher et al. 1972, Shanmugam 1978). We, at variance with these results, however, were unable to detect the nitrogenase activity in GOGAT mutant of $R. sphaeroides$ which failed to fully derepress GS. In view of these facts, there seem to be two mechanisms: 1) The loss of GOGAT activity enriches the glutamine pool, which, as a result inhibits the activity and the deadenylylation of GS. The low level of unadenylylated GS, in turn, represses nitrogenase synthesis (Song et al. 1984); 2) GOGAT itself affects the regulation of either GS or a nitrogen control factor or that the asm mutations affects a common regulator needed for both GS and GOGAT synthesis. If this is true, then the asm mutations described here not only affect GS activity but also have given rise to a new factor necessary for GS and GOGAT regulation. Further study of asm mutants should provide insight into these mechanisms as well as asm as overall control of nitrogenase synthesis.

5 Nitrogenase Switch-Off by Ammonia

Since the early investigation on the regulation of nitrogenase activity in $R. palustris$ (Zumft and Castillo 1978) and $R. capsulata$ (Song et al. 1979), a great number of papers have been published. The most important information cited from these researches are:

1) The regulation of nitrogenase activity is related to energy metabolism;
2) The inhibition of nitrogenase activity is a result of covalent modification of one subunit of Fe-protein;
3) The inhibitory effect of ammonia is mediated by glutamine synthetase.

Experiments in our laboratory demonstrated that the sensitivity of nitrogenase to ammonia varied with the stage of growth, but a correlation between the sensitivity of nitrogenase and the level of GS was found (Wu et al. 1984). The nitrogenase from the cells with high GS level was more sensitive than from those with low level. The duration of inhibition after the addition of ammonia was also dependent upon the GS level (Sun et al. 1985). Thus, with higher GS level the addition of ammonia caused a more immediate and severe decrease of nitrogenase activity, but the nitrogenase activity was recovered early (Sun et al. 1985). Further more, in cells with low GS level the inactivation of nitrogenase by glutamine was transmitted by deadenylylated GS (Sun et al. 1985). However, the GS in the cells grown photosynthetically was highly adenylylated (Wu et al. 1984), and, according to Stadtman et al. (1974) (Rhee et al. 1985), should be in the inactive state. When the cells in this state were treated with glutamine, the nitrogenase activity was not influenced and the adenylylation state for GS remained unchanged (Still in its highly adenylylated state) while the cells were treated with 2-KG plus phosphate prior to the addition of glutamine. GS in cells were adenylylated and were in active form, the nitrogenase was inhibited by glutamine (Zhou and Song 1989b). It was noteworthy that we failed to detect any inhibitory effect of ammonia in situ.

6 Derepression of Nitrogenase in a Glutamate Auxotroph of *R. Sphaeroides*

Biosynthesis of nitrogenase was found to be repressed by ammonia. The repressive nature was related to the cell's ability to convert ammonia into glutamine and glutamate. There are considerable evidences suggesting that photosynthetic bacteria assimilate ammonia via GS and GOGAT pathways. It is interesting to note that a glutamate auxotroph isolated from *R. sphaeroides* is unable to synthesize nitrogenase (Song et al. 1984), which is different from the same types of mutants isolated from *Rhodospirillum rubrum* and *Rhodobacter capsulata* (Weare 1978; Wall and Gest 1979). These results indicated that GOGAT and GS played a regulatory role in the synthesis of nitrogenase in *R. sphaeroides*. Nitrogenase was induced by the treatment with MSX (L-methionine-DL-sulphoximine) in the GOGAT-deficient mutant of *R. sphaeroides*, which exhibited a Nif⁻ phenotype in RCVB medium supplemented with glutamate (Wang and Song 1988). However, when MSF, an inhibitor of GOGAT, was added to the medium, the expression of nitrogenase in the parent strain was delayed, and the addition of glutamine enhanced the delaying effect. All the results presented above indicate that glutamine level mediates the synthesis of nitrogenase. A further deduction is proposed according to the following experiments. When GOGAT mutant grows in a nitrogen-limited medium (0.5 mM of glutamate), nitrogenase can also be induced by the addition of 2-ketoglutarate with the level of enzyme proportional to the concentration of 2-ketoglutarate

(Wang and Song 1989b). The analysis of glutamine pool indicates that the GOGAT mutant accumulates glutamine. Incubation of the mutant in nitrogen-limited medium lowers the glutamine pool. But at the moment when the nitro-genase is derepressed, the amount of glutamine in the GOGAT mutant will be doubled as compared with that of the wild strain. These results suggest that ratio GLN/2KG rather than the absolute concentration of glutamine, mediates the synthesis of nitrogenase.

7 Conclusions

Our knowledge of the regulation of nitrogenase in photosynthetic bacteria has been greatly enhanced during the past ten years, but problems still remain. We have known that GS is the primary target of ammonia shock, which results in the covalent modification of one subunit of the Fe-protein but we have not known what is in between.

It was proposed that the repressive effect of ammonia on nitrogenase synthesis be transmitted by the bicyclic cascade control system of GS and, more precisely, by its P_{II} regulatory protein (Gussin et al. 1986). This system has also been demonstrated in *R. sphaeroides* (Wang and Song 1989). Experiments in our laboratory suggest that the repressive effect of ammonia on nitrogenase be mediated by the ratio glutamine/2-ketoglutarate. Here a problem arises as to which component senses the signal of GLN/2-KG ratio. Is it the UTase as proposed for *E. coli*? To answer this question, a detailed study of the cascade control system of GS in photosynthetic bacteria would be necessary.

Acknowledgements. I would like to thank the members of my research group at the Shanghai Institute of Plant Physiology – Yongqiang Wu, Jinhua Sun, Xing Wang, Monggan Wu and Changxi Zhu – for their contributions to the research and ideas presented.

8 References

Bani D, Barlerio C, Bazziculupo M, Favilli F, Gallori E, Polsinelli M (1973) Isolation and characteri-
 zation of glutamate synthase mutants of *Azospirillum brasilense*. J Gen Microbiol 119:239
Bender RA, Snyder PM, Bueno R, Quinto M, Magasanik B (1983) Nitrogen regulation system of
 Klebsiella aerogenes: the *nac* gene. J Bacteriol 156:444
Brenchley JE, Prival MJ, Magasanik B (1973) Regulation of the synthesis enzymes responsible for
 glutamate formation in *Klebsiella aerogenes*. J Biol Chem 248:6122
Gussin GN, Ronson CW, Ausubel FM (1986) Regulation of nitrogen fixation genes. Ann Rev Genet
 20:567
Kanemoto RH, Ludden PW (1984) Effect of ammonia, darkness and phenazine mehthosulfate on
 whole-cell nitrogeneses activity and Fe protein modification in *Rhodospirillum rubrum*. J Bacteriol
 158:713

Mereick MJ (1983) Nitrogen control of the *nif* regulation in *Klebsiella pneumoniae*: involvement of the *ntrA* gene and analogies between *ntrC* and *nif* A. EMBO J 2:39

Pahel G, Tyler B (1979) A new *glnA*-linked regulatory gene for glutamine synthetase in *Escherichia coli*. Proc Natl Acad Sci USA 76:4544

Pahel G, Zelenetz AD, Tyler BH (1978) *gltB* and regulation of nitrogen metabolism by glutamine synthetase in *Escherichia coli*. J Bacteriol 133:139

Rhee SG, Chock PB, Stadtman ER (1985) Nucleotidylations involved in the regulation of glutamine synthetase in *Escherichia coli*. In: The Enzymology of Post-Translational Modification of Protein. Freedman RD, Hawkins HC (eds) Academic Press, London. 2:273

Shanmugam KT, O'Gata F, Andersen K, Valentine RC (1978) Biological nitrogen fixation. Ann Rev Plant Physiol 29:263

Song HY, Huang YD (1964) The inorganic hydrogen donor for photosynthetic hydrogen evolution in *Rhodopseudomonas palustris*. In Ann Symp of Biochem of Shanghai Biochem Soc. Part II-1

Song HY, Wu MG, Yu BL, Chem HC, Chen BJ (1979) Physiological control of nitrogen-fixing activity in photosynthetic bacterium *Rhodopseudmonas capsulata*. Acta Phytophysiol Sinica 5:141

Song HY, Wu YQ, Yu BL (1984) Glutamate synthase mutant lacking nitrogenase activity in *Rhodopseudomonas sphaeroides*. Acta Microbiol Sinica 24:149–155

Streicher SL, Gurney EG, Valentine RC (1972) The nitrogen fixation genes. Nature 239:495

Stadtman ER, Ginsburg A (1974) The glutamine synthetase of *Escherichia coli*: Structure and control. In: Boyer PD (ed) The Enzymes. Academic Press, New York, p 755

Sun JH, Wu MG, Song HY (1985) Involvement of glutamine synthetase in short-term regulation of nitrogenase activity by ammonia in *Rhodopseudomonas capsulata*, Acta Phytophysiol Sinica 11:171

Tate SS, Meister A (1971) Regulation of rat liver glutamine synthetase: activation by 2-ketoglutarate and inhibition by glycine, alanine and carbamyl phosphate. Proc Natl Acad Sci USA 68:781

Wall JD, Gest H (1979) Derepression of nitrogenase activity in glutamine auxotrophs of *Rhodopseudomonas capsulata*. J Bacteriol 137:1459

Weare NM (1978) The photoproduction of hydrogen and ammonia fixed from nitrogen by a derepressed mutant of *Rhodospirillum rubrum*. Biochim Biophys Acta 502:486

Wang X, Song HY (1987a) Purification and characterization of glutamine synthetase from the photosynthetic bacterium *Rhodopseudomonas sphaeroides*. Acta Biochim Biophys Sinica 19:172

Wang X, Song HY (1987b) Regulation properties of glutamine synthetase from the photosynthetic bacterium *Rhodopseudomonas sphaeroides*. Acta Biochim Biophys 19:136

Wang X, Song HY (1988) The induction of nitrogenase by MSX in a GOGAT mutant (Nif⁻) of *Rhodopseudomonas sphaeroides*. Acta Phytophysiol Sinica 14:56

Wang X, Song HY (1989a) Allosteric regulation of the state of adenylylation of glutamine synthetase in permeabilized cell preparations of *Rhodopseudomonas sphaerodes*. Science in China (Series B) 32:960

Wang X, Song HY (to be published) Isolation and purification of the glutamine synthetase bicyclic cascade components: adenylyltransferase and Shapiro's regulatory protein, P_{II}. Acta Biochim Biophys Sinica

Wang X, Song HY (1989b) 2-ketoglutarate-dependent induction of nitrogenase from a GOGAT mutant of *Rhodobacter sphaeroides*. Acta Phytophysiol Sinica 15:269

Wang X, Zhu MZ, Song HY (1989) Allosteric regulation of glutamine synthetase from *Rhodobacter sphaeroides* by 2-ketoglutarate. Acta Biochim Biophys 21:401

Wu MG, He YQ, Sun JH, Song HY (1984) Glutamine synthetase in *Rhodopseudomonas capsulata*. Acta Phytophysiol Sinica 10:232

Zhou XL, Song HY (1989a) Purification and characterization of glutamate synthase from *Rhodobacter sphaeroides*. Acta Phytophysiol Sinica 15:313

Zhou XL, Song HY (1989b) Effects of ammonia and glutamine on the regulation of nitrogenase activity in *Rhodobacter sphaeroides*. Acta Phytophysiol Sinica 15:354

Zumft WG, Castillo F (1978) Regulatory properties of the nitrogenase from *Rhodopseudomonas palustris*. Arch Microbiol 117:53

CHAPTER 21
Leghemoglobins and Their Gene Expressions

Y.X. Jing, X.J. Gu, G.S. Li, and X.Q. Shan

The Nitrogen Fixation
and its Research in China
Editor Guo-fan Hong
ⓒ Springer-Verlag Berlin Heidelberg 1992

1 Introduction

All nitrogen-fixing legume root nodules resulting from infection by symbiotic *Rhizobium* contain leghemoglobin. Leghemoglobin facilitates and regulates the flow of oxygen to the nitrogen-fixing bacteroids, resulting in creating the microaerobic environment necessary for the nitrogenase activity (Wittenberg et al. 1972; 1974; Wittenberg 1976). Leghemoglobins from soybean (Appleby et al. 1975; Ellfolk 1960), kidney bean (Lehtovaara et al. 1974), yellow lupine (Broughton et al. 1972; Kuranova et al. 1976), serradella (Broughton et al. 1972), broad bean (Richadson et al. 1975), snake bean (Broughton et al. 1971), *Sesbania* (Bogusz et al. 1987; Gu et al. 1989), which form large round nodules, and from pea (Lehtovaara et al. 1980), clover (Thulborn et al. 1979), alfalfa (Jing et al. 1982, 1983a; Lullien et al. 1987), which form elongated nodules with terminal meristems, have been purified and studied.

Rapid progress has been made in the research on leghemoglobin genes and their expression (Barker et al. 1988; Bojsen et al. 1983; Brison et al. 1982; Jensen et al. 1988; Jing 1985, 1988a; Kiss et al. 1987; Lee et al. 1983; Mauro et al. 1985; Welters et al. 1989), with the main focus on that of soybean (Hyldig-Nielsen et al. 1982; Mauro et al. 1988; Stou-gaard et al. 1986, 1987a, b; Wiborg et al. 1982), *Sesbania* (de Bruijn et al. 1988, 1989; Jensen et al. 1988; Metz et al. 1988, Welters et al. 1989) and alfalfa (Barkea et al. 1988, Jing 1985, 1989a; Kiss et al. 1987).

Some of our research work on leghemoglobins and their gene expression is described below.

2 Leghemoglobins and Their Characteristics

2.1 Leghemoglobins from Alfalfa and *Sesbania*

We isolated and purified the leghemoglobins for the first time from alfalfa (*Medicago sativa* L. Vernal) (Jing et al. 1982, 1983a). There are two major (LbII and LbIII) and three minor (LbI, LbIV and LbV) components of molecular weight 14.5 kDa. In addition, four extra proteins in the in-vitro translation products were immunoprecipitated. These proteins were slightly larger than the five components and were more acidic. Since our procedures for RNA isolation did not allow to remove nRNA from cytoplasmic RNA, and these four proteins were not detected in total nodule cytosol extracts, our preliminary experiments suggested that they may be the in-vitro translation products of the unspliced mRNA precursor (Jing et al. 1982, 1983a). In our further in-vitro translation experiments, eight more proteins were detected on the gel, which were immunoprecipitated by alfalfa antileghemoglobin serum as described by Lullien et al. (1987).

Sesbania cannabina leghemoglobins were also fractionated into five components of molecular weight 14.5 kDa for Lb1, L2, Lb3, and 16.0 kDa for Lb4 and Lb5 (Gu et al. 1989). The fact that there are seven leghemoglobin components in the nodules on both the stem and roots of *Sesbania rostrata* (Bogusz et al. 1986), suggested that the two species belonging to the same genus may have a different number of leghemoglobin genes.

2.2 Immunology of Leghemoglobins from the Nodules of Different Legumes

It is known that soybean leghemoglobin cross-reacts immunologically with the main components of leghemoglobins from serradella. The major components of leghemoglobin from snake bean (*Vigna sinesis*) only partially cross-reacted with antiserum to the soybean leghemoglobins, while no cross-reaction was observed between the lupine and soybean leghemoglobins (Hurrell et al. 1977).

Antiserum prepared against purified alfalfa leghemoglobins cross-reacts strongly with nodule extracts of pea (P. sativum), clover (*T. repens*), *Medicago falcata* and milk vetch (*Astragalus sinicus*) in Ouchterlony double diffusion test, but does not cross-react with nodule extracts of soybean (*G. max*), lupine, (*L. angustifolus*), cowpea (*V. unguiculata* Blackeye), jackbean (*C. ensiformis*), and black locust (*R. pseudoacacia*). Legumes in the latter group form spherical nodules. The above observations seems to be related to the nodule shapes and the S Q value (Jing et al. 1982). We further confirmed that the same pattern of precipitation lines exists for the leghemoglobins from the same morphological nodules, and the other pattern for those from different morphological nodules (Gu et al. 1989; Jing et al. 1983a).

2.3 Leghemoglobin Synthesis and its Adjustment after Dark Treatment of Legumes

There have been reports on leghemoglobin synthesis in nodules after dark treatment of legumes (Dilworth et al. 1976; Roponen 1970). When plants with effective nodules were placed in the dark, degeneration of leghemoglobin occurred fast (Roponen 1970). In contrast to this, there was no leghemoglobin loss in nodules after dark treatment of yellow lupine for up to 4 days (Dilworth et al. 1976). Perhaps, the leghemoglobin in lupine nodules was unusually stable. In senescent nodules the loss of nitrogenase was followed by that of leghemoglobins (Becana et al. 1986; Klucas 1974), but the loss of the two parameters is not always tightly coupled (LaRue et al. 1978).

We observed the leghemoglobin synthesis during rejuvenation after dark treatment of alfalfa (Table 1) (Jing 1988c; Jing et al. 1990b).

The fact that in the dark little radioactivity was observed in nodules probably reflects the lack of biosynthesis. After one day treatment the percentage of

Table 1. Distribution of S-methionine in various components of leghemoglobin

Days in dark before labeling	% of total nodule cpm in Lbs	% Distribution of Lb cpm in components			
		I + II	III	IV	V
0	4.85	33.6	46.1	13.9	6.4
1	6.94	23.7	41.4	17.4	14.5
3	1.71	0.0	55.0	29.8	15.2
6	1.64	0.0	54.9	32.9	12.2

leghemoglobin synthesis increased by 2.1%, and after 3 and 6 day treatment, it decreased by 3.1% and 3.2%, respectively, as compared to that of the control. At the same time, the differential for synthesis of the different leghemoglobin components was observed during the rejuvenation after dark treatment. One day treatment resulted in the decrease in synthesis of components I + II, III, but in the increase in components IV, V. After three and 6 day treatment no synthesis was made for components I + II. On the contrary, the other components, especially, IV, V increase 2-fold. It seems that the differential synthesis of the different components may be related to the regulation of their gene expression, and/or to the translation. The parameters may simply be a reflection of the energy status of nodules and/or of plants.

2.4 Hemoglobins from Non-Legume Nodules

More and more examples indicate that the nodules from non-legumes such as *Parasponia* (Appleby et al. 1983) (which is infected by *Rhizobium*), *Alnus, Myrica, Casuarina, Ceanothus*, and *Elaeagnus* (which are the actinorhizal plants) have hemoglobin (Tjepkema 1983, 1984). After treatment of *Rhizobium*, we have obtained rice nodules with some nitrogenase activity, and the antiserum prepared against the purified *Sesbania cannabina* leghemoglobin was used for the Western blot reaction with extract of rice nodules, and a positive colour reaction was observed (Jing et al. 1990c). We consider, therefore, that the bacteroids in rice nodules may have the possibility to produce hemoglobin and nitrogenase. It was suggested that the leghemoglobin may be renamed as plant hemoglobin (Appleby 1984; Jing 1987).

Leghemoglobin-like DNA sequences were reported to exist in non-legumes (Hattori et al. 1985; Roberts et al. 1985) besides *Parasponia* (Bogusz et al. 1988). Of special interest is the presence of the cross-hybridizing sequence in birch (*Betula alleghaniensis*) which is related to alder (*Alnus*), but not to be nodulated. It seems likely that the globin genes are much more widely distributed in the plant kingdom than has previously been thought. Searching for leghemoglobin-like sequences in a number of plants may have implications in the elucidation of evolutionary processes and become useful to find new hosts into which the nitrogen-fixing system may possibly be introduced.

3 Leghemoglobin Genes and Their Expression

3.1 Populations of Leghemoglobin Messager-RNA

The multicomponents of leghemoglobin are translated from the populations of mRNA in different S values (Jing et al. 1982). When poly(A)-containing RNA was fractionated in a sucrose gradient, three separate populations of mRNA (with approximate S-values of 6, 9 and 26) were active in the in-vitro translation of alfalfa leghemoglobin components (Jing et al. 1983b). The fastest sedimenting population was enriched with mRNA that coded for LbIII, while the slowest sedimenting population was enriched with messages for LbI and LbII. The other population was enriched with messages for LbIV and LbV. Verma et al. (1974) investigated soybean leghemoglobin and did not point out the role of different populations of mRNA in protein synthesis.

3.2 Relationship Between Leghemoglobin Gene Transcription and the Nodule Age

Alfalfa blooms around the 40th day after being planted in a 25 °C chamber with 14 daylight hours. Generally, the activity of nitrogenase is high during the 15–30 days after appearance of nodules in bright red, and then gradually declines after flowering and almost disappears on the 70th day because of the senescence of nodules which become dark red in colour. Poly(A)-containing alfalfa leghemoglobin mRNAs isolated from the nodules at ages of 30, 45, 60 days were in vitro translated into the products, which were immunoprecipitated by alfalfa antileghemoglobin IgGs. The results reveal that the translation rate of leghemoglobin mRNAs in old nodules is different, though not significantly, from that in young nodules. Although the activity of nitrogen fixation reduces in the nodules around the 60th day and the rate of leghemoglobins in transporting oxygen drops, their mRNAs still remain and can be translated into proteins (Jing 1986, 1988b; Jing et al. 1990a).

Marcker et al. (1984) studied the transcription of genes for soybean leghemoglobin cmponents and pointed out the expression sequence of each component gene during the early development of nodules. Our experiment indicated that during a particular time (7-12 days) after activation of all leghemoglobin genes, the mRNAs increased apparently, and even in nodules of 7 weeks old the leghemoglobin genes were still active in transcription (Jing et al. 1990a). Govers et al. (1985) and Barker et al. (1988) have also observed this. Our results also showed that the leghemoglobin genes for major components I, II, III and minor components IV, V were transcribed at high and low levels, respectively. Regardless of the aging of nodules, the mRNAs for most of the leghemoglobin components remainded at a constant level (Jing 1988b; Jing et al. 1990a), which was in agreement with the observation by Marcker

(1984) that all leghemoglobin genes were active for a long period during the lifetime of a nodule, unlike the globin genes of vertebrates which are regulated by developmental mechanisms.

3.3 Translational Regulation of Leghemoglobin Genes

In eukaryotic cells the regulation may be exerted both at the transcriptional and translational level (Metzler 1977). The regulation of gene expression at the transcriptional level is of fundamental importance to both eukaryotes and prokaryotes, but that at the translational level as an additional mechanism influencing the final product of a gene is also of great importance (Holein et al. 1985; Hurell et al. 1976).

a. In-vitro translation inhibition: It is possible that the translation of leghemoglobin mRNA in nodules may depend on the content of leghemoglobins themselves, which, reaching a certain extent, would automatically feedback-inhibit the in vitro translation of the mRNA.

Circular dichroism studies of leghemoglobins from various legumes showed that they have a similar overall polypeptide chain conformation and heme environment and orientation. Immunoprecipitation studies, however, suggested that the antigenic determinants on the surface of leghemoblobins vary considerably (Hamilton et al. 1982; Hurrell et al. 1977). The antisera against alfalfa leghemoglobins could immunocross-react with alfalfa leghemoglobin components and pea leghemoglobins, but not with soybean leghemoglobins. On the contrary, the translation products of pea leghemoglobin mRNAs could be immunoprecipitated by alfalfa antileghemoglobin IgGs, while those of soybean leghemoglobin mRNAs could not (Jing et al. 1983a). This indicates

Table 2. Effect of alfalfa and soybean leghemoglobins on the translation of alfalfa leghemoglobin mRNAs

Cell-free translation system	Counts after immunoprecipitation by alfalfa antileghemoglobin IgG (cpm)*	Alfalfa leghemoglobin synthesis (%)
$\alpha\alpha$**-Lb mRNA	1420 (background)	
$\alpha\alpha$-Lb mRNA	3872	100
$\alpha\alpha$-Lb mRNA +Lb	151	3.8
$\alpha\alpha$-Lb mRNA +LbII	253	6.5
$\alpha\alpha$-Lb mRNA +LbIII	-59	0.0
$\alpha\alpha$-Lb mRNA +Sb-Lb	6085	175

* The counts do not include the background
** $\alpha\alpha$-alfalfa

Table 3. Effect of alfalfa and soybean leghemoglobins on pea leghemoglobin mRNA translation

Cell-free translation system	Counts after immunoprecipitation by alfalfa anti-leghemoglobin IgG (cpm)*	Pea leghemoglobin synthesis (%)
Pea Lb mRNA	1659	100
Pea Lb mRNA+-Lb	0	0
Pea Lb mRNA+sb-Lb	2302	138

* not including the background

Table 4. Effect of hemin, bovine hemoglobin and bovine albumin on alfalfa leghemoglobin mRNA translation

Cell-free translation system	Counts after immunoprecipitation by alfalfa antileghemoglobin IgG (cpm)*	Alfalfa leghemoglobin synthesis (%)
αα**-Lb mRNA	3872	100
αα-Lb mRNA+hemin	4486	115
αα-Lb mRNA+hemoglobin	4713	121
αα-Lb mRNA+BSA	3762	97

* not including the background
** αα-alfalfa

that both alfalfa and pea leghemoglobins have identical, or partially identical antigenic determinants on their surfaces. The results from Tables 2 and 3 seem to explain that the leghemoglobins from different legumes possessing identical antigenic determinants can mutually inhibit their mRNA translations. Alfalfa leghemoglobins not only inhibit the translation of their leghemoglobin mRNAs, but also that of pea leghemoglobon mRNAs (Table 3). However, the alfalfa translation system is not influenced by the soybean leghemoglobins and other proteins such as bovine hemoglobin and bovine serum albumin (Table 4).

b. The role of heme in in-vitro translation of leghemoglobin: There have been many reports about heme affecting the hemoglobin synthesis in reticulocyte lysate (Jensen et al. 1986; Jing 1986; Ochoa et al. 1979; Revel et al. 1978). In this system the heme exerted a great influence upon hemoglobin synthesis. The translation went on only for a few minutes if no hemin was present in the system, but could maintain several hours if hemin was added to it. Several experiments have been done on the mechanism of hemin influencing hemoglobin synthesis (Hamilton et al. 1982; Jenses et al. 1986; Revel et al. 1978). They showed that the translation rate of catalase mRNA in the cell-free system of yeast mutant lacking heme was lowered, but was promoted and enhanced in the system with hemin (Hamilton et al. 1982; Hurrell et al. 1976). Recently Jensen et al. (1986) have reported that in *Saccharomyces cerevisiae*, heme regulates the expression of chimeric antibiotic resistance genes containing the

5' untranslated region of the soybean Lbc3 mRNA, probably by increasing the translational efficiency of the chimeric mRNAs. In our experiments with the alfalfa leghemoglobin mRNA translation system, to which was added hemin, protein synthesis was found to be increased (Table 4; Jing 1986). These results are in agreement with those just mentioned above. If to this system are added soybean leghemoglobin (Tables 2, 3) and bovine hemoglobin (Table 4), both of which contain protoheme IX, the results are the same as those with hemin added (Table 4; Jing 1986), but no effect was found by Verma et al. (1974, 1988) on in-vitro translation of this mRNA in the presence of heme.

Based on the feedback-inhibition and heme effect on the alfalfa leghemoglobin synthesis in the in-vitro translation system, we presume that the inhibition of leghemoglobin mRNA translation may depend upon whether the antigenic determinants on the protein surface are identical or not. The existance of heme and protoheme IX in the translation system may determine whether or not the translation is promoted. If protoheme IX and the proteins having identical antigenic determinants exist simutaneously in the system, the latter may play a leading role in the inhibition of translation, whereas the former may not act. If protoheme IX and the proteins having different antigenic determinants and possessing protoheme IX exist in the translation system, the latter may not inhibit the translation, but the former may function as an agent promoting protein synthesis (Jing 1986). The mechanism by which antigenic determinants of leghemoglobins play the role in their mRNA translation should be further studied.

4 Synthesis and Cloning of Alfalfa Leghemoglobin cDNAs

In order to investigate the chromosomal leghemoglobin gene organization of alfalfa and to contribute to better understanding of the induction and regulation mechanisms of leghemoglobin synthesis in the alfalfa-*Rhizobium meliloti* symbiosis, cDNA probes were constructed. The double-stranded cDNAs were synthesized from poly(A) mRNAs isolated from nodules and inserted into the *Sal*I site of pBR322 using synthetic *Sal*I linker or by adding the homopolymer tail (Jing 1985, 1988a). Four or five groups of cDNA of different molecular weight were observed on the gel. Four leghemoglobins and two putative unspliced mRNA products were synthesized by in-vitro translation (Fig. 1A; Jing et al. 1988a). It was observed that the synthesis of leghemoglobin I, II (Fig. 1B) was arrested by the cDNA prepared from the clone. By the release of the mRNA hybridized to the cDNA clone, the unspliced mRNA products and leghemoglobin V can be synthesized (Fig. 1C). These results confirm that the selected clone contains cDNA, possibly synthesized from either the mRNA for component V, or the mRNA precursor and is useful as a probe for the selection of leghemoglobin genes in the genomic library.

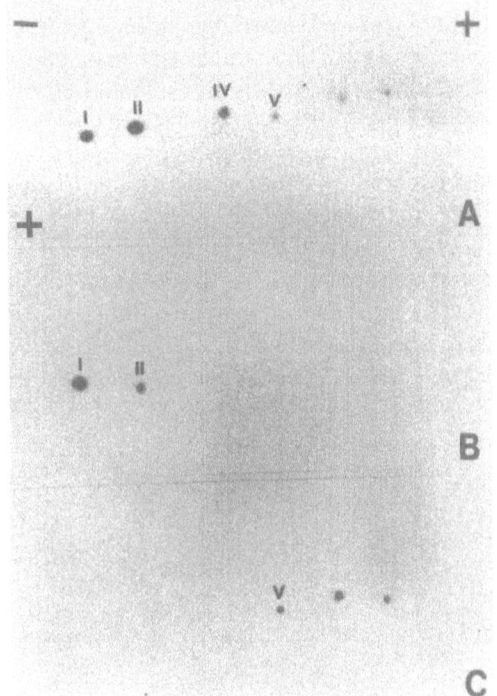

Fig. 1. Fluorographs of S-methionine labeled polypeptides synthesized in wheat grem cell-free system on poly(A)-RNA from alfalfa nodules the polypeptides were separated by 2D-polyacrylamide gel electrophoresis. The dried gel was fluorographed at −80°C for 5h using Kodak XR-film. A: in vitro translation products immunoprecipitated by an anti-alfalfa Lb antiserum; B: hybrid-arrested and immunoprecipitated polypeptides; C: hybrid-release translated polypeptides. The polypeptide I. II. IV and V in A were Lb I. II. IV, respectively

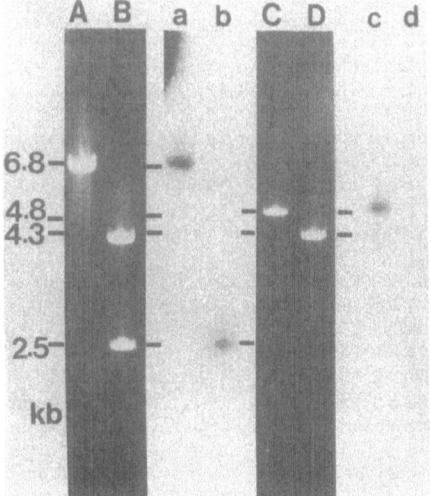

Fig. 2. Southern blot hybridization of alfalfa Lb cDNA to soybean Lb cDNA prepared from p2431 (Truelsen et al. 1979). Approximately 0.5 µg of plasmid DNA was separated by 0.7% agarose gel electrophoresis, blotted onto nitrocellulose paper and hybridized against ^{32}P-nick translated soybean Lb cDNA. The fragment sizes in kilobases (kb) were determined by the use of molecular weight standards from BRL. Track A: PstI digest of the alfalfa Lb cDNA clone (pBR322 inserted with alfalfa Lb cDNA in SalI site of the plasmid); Track B: SalI digest of nonlinearized A (the 4.3 kb band was pBR322 and 2.5 kb band was alfalfa Lb cDNA insert); Track D: PstI linearized PBR322 as a negative control

The size of the cloned cDNA was estimated to be about 2.5 kb (Fig. 2B, b; Jing et al. 1989a). The molecular weight of matured leghemoglobins ranges from 15.0 to 20.0 kDa. Based on the molecular weight of matured leghemoglobins, the size of a full length copy of cDNA of the processed mRNA should be less than 500 bp. Since our cDNA clone is 2.5 kb in size, the insert is likely to be a cDNA clone of unspliced mRNA precursor, because our procedures for RNA isolation did not allow to remove nRNA from cytoplasmic RNA. This is also consistent with previous reports that the leghemoglobin genes of soybean contain three big intervening sequences (Brison et al. 1982; Hyldig-Nielsen et al. 1982) and some higher-molecular-weight proteins in the in-vitro translation products could be immunoprecipitated by the antiserum prepared from purified alfalfa leghemoglobins. Recently, there was a report indicating the existence of two groups of alfalfa leghemoglobin cDNAs which differ significantly in sequence (Barker et al. 1988).

5 References

Appleby CA, Nicola NA, Hurrell JG, Leach SJ (1975) Characterization and improved separation of soybean leghemoglobin. Biochem 14: 4444

Appleby CA, Tjepkema JD, Trinick MJ (1983) Hemoglobin in a nonleguminous plant, parasponia: possible genetic origin and function in nitrogen fixation. Science 220: 951

Appleby CA (1984) Leghemoglobin and Rhizobium respiration. Ann Rev Plant Physiol. 35: 443

Barker DG, Gallusci P, Lullien V, Khan H, Gherard M, Huguet T (1988) Identification of two groups of leghemoglobin genes in alfalfa (*Medicago sativa*) and a study of their expression during root nodule development. Plant Mol Biol 11: 761

Becana M, Gogorcena Y, Aparico-Tejo BM, Sanchez-Diag M (1986) Nitrogen Fixation and leghemoglobin content during vegetative growth of alfalfa. J Plant Physiol 123: 117

Bogusz D, Kortt AA, Appleby CA (1987) Sesbania rostrata root and stem nodule leghemoglobin: Purification and relationship among the seven major components. Arch Biochem Biophys 254: 261

Bogusz D, Appleby CA, Landsmann J, Dennis ES, Trinick MJ, Peacock WJ (1988) Functioning hemoglobin genes in non-nodulating plants. Nature 331: 178

Bojsen K, Abiladsten D, Jensen EO, Paluden K, Marcker KA (1983) The chromosomal arrangement of six soybean leghemoglobin genes. EMBO J 2: 1165

Brison N, Verma DPS (1982) Soybean leghemoglobin gene family: Normal, pseudo, and truncated genes. Proc Natl Acad Sci USA 79: 4055

Broughton WJ, Dilworth MJ (1971) Control of leghemoglobin synthesis in snake bean. Biochem J 125: 1075

Broughton WJ, Dilworth MJ, Godfrey CA (1972) Molecular properties of lupine and serradella leghemoglobin. Biochem J 127: 309

de Bruijn FJ, Jensen EO, Metz BA, Hoffmann HJ, Welters P, Marcker KA, Schell J (1988) Interaction of transacting factors with distinct DNA elements in the promoter region of the *Glycine max* and *Sesbania rostrata* leghemoglobin genes. In: Palacios R, Verma DPS (eds) Molecular genetics of plant-microbe interaction. APS Press, St. Paul. MN USA, pp 333

de Bruijn FJ, Grunenberg GFB, Hoffmann HJ, Metz B, Ratet P, Simons-Schreier A, Szabados L, Welters P, Schell J (1989) Regulation of plant genes specifically induced in nitrogen-fixing nodules: role of *cis*-acting elements and *trans*-acting factors in leghemoglobin gene expression. Plant Mol Biol 13: 319

Dilworth MJ, Covertry DR (1976) Stability of leghemoglobin in yellow lupine nodules. In: Newton WE, Postgate JR, Rodriguez-Barrueco C (eds) Recent developments in nitrogen fixation. Academic Press, London

Ellfolk N (1960) Crystalline leghemoglobin, I. Purification procedure. Acta Chem Scand 14: 609

Govers F, Gloudemans T, Moerman M, van Kammen A, Bisseling T (1985) Expression of plant genes during the development of pea root nodules. EMBO J 4: 861

Gu XJ, Li GS, Jing Y (1989) Leghemoglobin from root nodules of Sesbania cannabina. Chinese J Bot 1: 116

Hamilton B, Hofbauer R, Ruis H (1982) Translational control of catalase synthesis by hemin in the yeast Saccharomyces cerevisiae. Proc Natl Acad Sci USA 79: 7609

Hattori J, Johnson DA (1985) The detection of leghemoglobin-line sequences in legumes and nonlegumes. Plant Mol Biol 4: 285

Holein D, McPherson J, Goh SH, Bornstein P (1981) Regulation of protein synthesis: Translational control by procollagen-drived fragments. Proc Natl Acad Sci USA 78: 6163

Hurrell JGR, Broughton WJ, Dilworth HJ, Minasian E, Leach SJ (1976) Comparative structural and immunochemical properties of leghomoglobins. Eur J Biochem 66: 389

Hurrell JGR, Thulborn AR, Broughton WJ, Dilworth MJ, Leach SJ (1977) Leghemoglobin immunochemistry and phylogenetic relationship. FEBS Lett 84: 244

Hyldig-Nielsen JJ, Jensen EO, Paludan K, Wiborg O, Garett R, Jorgensen P, Marcker KA (1982) The primary structure of two leghemoglobin genes from soybean. Nucl Acid Res 10: 689

Jensen EO, Marcker KA, Villadsen IS (1986) Heme regulates the expression in Saccharomyces cerevisiae of chimeric genes containing 5'-flanking soybean leghemoglobin sequences. EMBO J 5: 843

Jensen EO, Marcker KA, Schell J, de Bruijn FJ (1988) Interaction of a nodule specific, trans-acting factor with distinct DNA elements in the soybean leghemoglobin 1bc3 5'-upstream region. EMBO J 7: 1265

Jing Y, Paau AS, Brill WJ (1982) Leghemoglobins from alfalfa (Medicago sativa L. vernal) root nodules. I. Purification and in vitro synthesis of five leghemoglobin components. Plant Sci Lett 25: 119

Jing Y, Paau AS, Brill WJ (1983a) Methods of isolation and some characteristics of leghemoglobins from alfalfa (Medicago sativa L. vernal) nodules. Acta Botanica Sinica 25: 220

Jing Y, Paau AS, Brill WJ (1983b) mRNAs of leghemoglobins from alfalfa root nodules. Acta Botanica Sinica 25: 431

Jing Y (1985) Synthesis and cloning of cDNA coding for alfalfa leghemoglobin. Acta Botanica Sinica 27: 263

Jing Y (1986) Regulation of alfalfa leghemoglobin synthesis at translational level. Scientia Sinica, Series B, 29: 50

Jing Y (1987) Plant hemoglobin and its gene expression. In: Wu Xiangyu, Huang Zongzhen (eds) Advances in Plant Physiol Biochem. Chinese Academic Press, Beijing, p 141

Jing Y (1988a) Synthesis and cloning of cDNAs coding for alfalfa leghemoglobins. In: Swaminathan MS, Hu H, Shao QQ (eds) Genetic manipulation in crops. CASSELL TYCOOLY, pp 237

Jing Y (1988b) Expression of leghemoglobin genes of alfalfa (Medicago sativa L. vernal) during root nodule development. In: Bothe H, de Bruijn FJ, Newton WE (eds) Nitrogen fixation: Hundred years after Gustav Fischer, Stuttgart, New York, p 636

Jing Y (1988c) Expression of leghemoglobin genes in senescent and dark treatment of root nodules of alfalfa (Medicago sativa L. vernal). In: Book of Second Japan-China Bilateral Symposium on Biophysics, May 16–20, Kyoto, Japan, p 325

Jing Y, Cho MJ, Paay AS, Brill WJ (1989a) Molecular cloning of cDNA synthesized from a putative leghemoglobin mRNA precursor isolated from alfalfa nodules. Chinese J Bot 1: 9

Jing Y, Paau AS, Brill WJ (1990a) Transcription and translation of alfalfa leghemoglobin genes during root nodule development. Science in China 33: 1078

Jing Y, Shan XQ (1990b) Nitrogen fixation and leghemoglobin synthesis during rejuvenation after dark treatment of alfalfa. Chinese J Bot 2: 7

Jing Y, Li GS, Shan XQ, Li JG (1990c) Rice root nodules with acetylene reduction activity. In: Gresshoff PM, Roth E and Stacey G (eds) Nitrogen fixation: achievements and objectives. Chapman and Hall, New York, London p 829

Kiss G, Vegh Z, Vincze E (1987) Nucleotide sequence of a cDNA clone encoding leghemoglobin III from Medicago sativa. Nucl Acid Res 15: 3620

Klucas RV (1974) Studies on soybean nodule senescence. Plant Physiol 54: 612

Kuranova IP, Grebenko AI, Konareva NV, Syromyatnikova IF, Barynin VV (1976) Separation of leghemoglobin from lupine (Lupinus luteus L.) nodules into components. Biokim. 41: 1603

LaRue TA, Child JJ (1978) Leghemoglobin and nitrogenese activity in peas. In: Proc of Steenbock-Kettering Int Symp on Nitrogen Fixation. June 12–16, 1978, University of Wisconsin (Madison), Abstract of Poster and Lecture Session, p 42

Lee JS, Brown GG, Verma DPS (1983) Chromosomal arrangement of leghemoglobin genes in soybean. Nucl Acid Res 11: 5541
Lehtovaara P, Ellfolk N (1974) The primary structure of kidney bean leghemoglobin. FEBS Lett 43: 239
Lehtovaara P, Lappalainen A, Ellfolk N (1980) The amino acid sequence of pea (*Pisum sativum*) leghemoglobin. Biochim Biophys Acta 623: 98
Lullien V, Barker DG, de Lajudie P, Huguet T (1987) Plant gene expression in effective and ineffective root nodules of alfalfa (*Medicago sativa*). Plant Mol Biol 9: 465
Marcker A, Lund M, Jensen EO, Marcker KA (1984) Transcription of the soybean leghemoglobin genes during nodule development. EMBO J 3: 1691
Mauro VP, Nguyen T, Kafinkes P, Verma DPS (1985) Primary structure of the soybean nodulin-23 gene and potential regulation elements in the 5'-flanking regions of nodulin and leghemoglobin genes. Nucl Acid Res 13: 239
Mauro VP, Verma DPS (1988) Transcriptional-activation nuclei from uninfected soybean of a set of genes involved in symbiosis with *Rhizobium*. Mol Plant-Microbe Interaction, 1: 46
Metz B, Welters P, Hoffmann HJ, Jensen EO, Schell J, de Bruijn FJ (1988) Primary structure and promoter analysis of leghemoglobin genes stem-nodulated tropical legume *Sesbania rostrata*: conserved coding sequences, *cis*-elements and *trans*-acting factor. Mol Gen Gent 214: 181
Metzler DE (1977) Enzymes: The protein catalysts of cells. In: Biochemistry: The chemical reaction of living cells. Academic Press, New York, SanFrancisco, London, p 301
Ochoa S, de Haro C (1979) Regulation of protein synthesis in eukaryotes. Ann Rev Biochem 48: 549
Revel M, Groner Y (1978) Post-transcriptional and translational controls of gene expression in eukaryotes. Ann Rev Biochem 47: 1079
Richadson M, Dilworth MJ, Seawen MD (1975) The amino acid sequence of leghemoglobin I from root nodules of broad been (*Vicia faba* L.) FEBS Lett 51: 33
Roberts MP, Jafar S, Mullin BC (1985) Leghemoglobin-like sequences in the DNA of four actinorhizal plants. Plants Mol Biol 5: 333
Roponen J (1970) The effect of darkness on the leghemoglobin content and amino acid levels in the root nodules of pea plants. Plant Physiol 23: 452
Stougaard J, Marcker KA, Otten L, Schell J (1986) Nodule-specific expression of a chimeric soybean leghemoglobin gene in transgenic *Lotus corniculatus*. Nature 321: 669
Stougaard J, Abildsten D, Marcker KA (1987a) The *Agrobacterium rhizogenes* pRiTL-DNA segment as a gene vector system for transformation of plants. Mol Gen Genet 207: 251
Stugaard J, Sandal NN, Gron A, Kuhle A, Karcker KA (1987b) 5'-analysis of the soybean leghemoglobin Lbc3 gene: regulatory element required for promoter activity and organ specificity. EMBO J 6: 3565
Thulborn KR, Minasian E, Leach SJ (1979) Leghemoglobin from *Trifolium subterraneum* purification and characterization. Biochim Biophys Acta 578: 476
Tjepkema JC (1983) Hemoglobin in the nitrogen-fixing root nodules of actinorhizal plants. Can J Bot 61: 2924
Tjepkema JD (1984) Oxygen, hemoglobins, and energy usage in actinorhizal nodules. In: Veeger C, Newton WE (eds) Advances in nitrogen fixation research. Martinus Nijhoff, Boston, p 467
Truelsen E, Gausing K, Jochimsen B, Jorgensen P, Marcker KA (1979) Cloning of soybean leghe- moglobin structural gene sequences synthesized in vitro. Nucl Acid Res 6: 3061
Verma DPS, Nash DT, Schulman HM (1974) Isolation and in vitro translation of soybean leghemoglobin mRNA. Nature 251: 74
Verma DPS, Delauney J (1988) Root nodule symbiosis: nodulin and nodulin genes. In: Verma DPS, Goldberg RB (eds) Plant gene research: temporal and spatial regulation of plant genes. Springer-Verlag, Wien, New York, p 169
Welters P, Metz BA, Schell J, de Bruijn FJ (1989) Nucleotide sequence of the *Sesbania rostrata* leghemoglobin (Srglb3) gene. Nucl Acid Res 17: 1253
Wiborg O, Hyldig-Nielsen JJ, Jensen EO, Paludan K, Marcker KA (1982) The nucleotide sequences of two leghemoglobin genes from soybean. Nucl Acid Res 10: 3487
Wittenberg JB, Appleby CA, Wittenberg BA (1972) The kinetics of the reaction of leghemoglobin with oxygen and carbon monoxide. J Biol Chem 247: 527
Wittenberg JB, Bergersen FJ, Appleby CA, Turner GC (1974) Facilitated oxygen diffusion. J Biol Chem 249: 4057
Wittenberg JB (1976) Facilitation of oxygen diffusion by intracellular leghemoglobin and myoglobin. In: Jobsis FF (ed) Oxygen and physiological functions. Prof Inf Lib, Dallas, Texas, p 228

Section III:
Biology of Nitrogen Fixation

CHAPTER 22
The Biological Nitrogen Fixation Systems Adopted in Rice Paddy Fields in China

C.S. Fan

The Nitrogen Fixation
and its Research in China
Editor Guo-fan Hong
© Springer-Verlag Berlin Heidelberg 1992

1 Introduction

The rice paddy fields in China are mainly distributed in the southern provinces in the subtropical and temperate climatic regions. The Yangtze River Valley is the center of rice plantation which stretches from the east coast to the western mountainous areas of Szechuan and Yunnan provinces. This vast area of more than 20 million hectares of rice paddy fields have gone through several hundreds of years' of agricultural practices of rice-green manure and rice-wheat or rape types of paddy/dryland rotational farming. A type of humus-rich and high N-content rice paddy soil has been developed with high productivity for grains (Xiong and Li 1987). Farmers used to grow a crop of leguminous green manure, milk vetch (*Astragalus*) or vetch (*Vicia*) after rice harvesting in late autumn for the effect of symbiotic nitrogen fixation of *Rhizobium* and to use it as base manure to supply organic matter and nitrogen for the next rice crop (Chiao 1986). During the growing period of rice plants, in addition to the different naturally developed filamentous and unicellular nitrogen-fixing blue-green algae in the stagnant water layer of the rice field, farmers also grow *Azolla* in the rice fields serving as top-dressing green manure to supply nitrogen and other nutrients (Liu 1981). *Azolla* infected with a blue-green alga, *Anabaena azollae*, is a very effective symbiotic nitrogen fixation system on shallow water surface in temperature regions. Thus the opportunities of free-living and symbiotic blue-green algae for nitrogen fixation come into full play in Chinese rice paddy fields.

Table 1. Biological N-fixation systems in paddy fields

Biological N-fixation systems	Main genera	N-fixing ability kg N/ha/yr
Free-living N-fixation systems		
Aerobic N-fixing bacteria	*Azotobacter, Beijerinckia Azomonas, Derxia, Azospirillum, Arthrobacter, Bacillus*	5.0–10.0
Anaerobic N-fixing bacteria	*Clostridium, Desulfotomaculum*	3.5–7.0
Facultative anaerobic N-fixing bacteria	*Acaligenes, Enterobacter, Klebseilla*	1.5–3.0
Photosynthetic N-fixing bacteria	*Rhodospirillum, Rhodopseudomonas*	+
N-fixing blue-green algae	*Anabaena, nostoc, Tolypothrix, Stigonema, Calothrix* etc.	25–50
Symbiotic N-fixation system		
Rhizobium-Astragalus	*Rhizobium astragli*	80–110
Rhizobium-Vicia	*Rhizobium leguminosarum*	60–100
Anabaena-Azolla	*Anabaena azollae*	160–300
Associative N-fixation system		
Azospirillum-rice	*Azospirillum brasilense*	+
Acaligenes-rice	*Acaligenes fecalis*	+
Enterobacter-rice	*Enterobacter cloacae*	+
Azotobacter-Paspalum	*Azotobacter paspali*	+

Besides the customary application of farm manure and green manure to the paddy fields, the ploughing between crops renders the turning over of plant stubbles into the soil supplementing the organic matter as energy source of the free-living N-fixing bacteria in soil to carry on their N-fixing activities both under aerobic and anaerobic conditions. Furthermore, there are also specific N-fixing bacteria living in association with the roots of rice plants and the rotational crops in the winter and spring seasons to supply nutrients directly to the growing plants (Table 1).

The paddy fields have provided a natural ecological environment for various N-fixing organisms, yet the cultivation systems have artificially further promoted the N-fixing effect of the microorganisms. All these, while supplying the nitrogen and other nutrients for the growing rice plants, will also gradually raise the N-content of the soil.

2 Free-Living N-Fixation System in Rice Paddies

The rice paddies in China, being long under the system of annually rotational farming, provide a rather high content of organic matter, are nearly neutral in reaction, with a variety of moisture content, aeration and other ecological environments for the development of various N-fixing microorganisms both in the soil and in the water layer in different seasons. During the summer-autumn seasons, when the rice plants are growing, a rich growth of plankton develops in the surface water layer in which large quantities of different blue-green algal species develop to contribute nitrogen nutrition to rice plants as well as nitrogen to the soil (Huang 1984). At the same time, various kinds of nitrogen-fixing bacteria develop in the water layer, in soil or epiphytically on the slimy sheaths of algal filaments to carry on their nitrogen-fixing activities. They are enriched with the decomposed organic matter of the soil and the photosynthetic products secreted from the growing rice plants and the developing algal flora. In the winter-spring seasons, the fields are planted with wheat, leguminous green manure or other crops, the soil aeration is improved for aerobic activities of soil micro-organisms. The decomposition of organic matter left over by growing crops steps up when the temperatures rise in spring. Available nutrients and energy sources supplied in the soil promote the development of free-living N-fixing bacteria of both aerobic and anaerobic forms in the different soil layers (Chen 1979).

2.1 Nitrogen-Fixing Bacteria

More than 30 species of free-living bacteria belonging to 10 genera have been recorded with the ability to fix nitrogen in rice paddy fields, including *Azotobacter chroocoocum, A. vinelandii, Bacillus polymyxa, Xanthobacter autorophicus,*

Azospirillum lipoferum, species of *Arthrobacter*, *Methylobacter* etc. which are all in aerobic forms. In some fields where the soils are acid at around pH 5.0, *Beijerinckia indica* and *Derxia gummosa* are the characteristic aerobic N-fixing species. Among the anaerobic forms, *Clostridium pasteurianum*, *C. butyricum*, and the species of *Desulfotomaculum* and *Desulfovibrio* are commonly distributed. Some facultative forms such as species of *Klebsiella*, *Enterobacter*, *Acaligenes* are also well developed (Tseng 1987; Sprent 1979). Although the aerobic N-fixing bacteria are not very active when the paddy fields are flooded with water, their development and quantity of nitrogen fixed are promoted when the water in the field is drained off during the ripening stage of rice and the soil aeration is improved by ploughing for winter crops. The anaerobic and facultative forms attend to a relatively large degree of the year round development in the paddy fields since the soil moisture and organic matter content are relatively suitable (Yoshida et al. 1982). The photosynthetic bacteria, *Rhodospirillum rubrum* and *Rhodopseudomonas capsulatus* also attend some degree of development in the water layer of the paddy fields during the growing stage of rice plants. They are anaerobes and can utilize light energy in nitrogen fixation.

Although these bacteria are measured up to a certain intensity of N-fixing activity in pure culture studies, their actual nitrogen-fixing efficiency in the fields

Table 2. Main species of free-living N-fixing bacteria in paddy fields

Bacteria	N-fixing efficiency
Aerobic N-fixing bacteria	
Azotobacter chroococcum	$+++$*
Azotobacter vinelandii	$+++$
Beijerinckia indica	$++$
Derxia gummosa	$++$
Xanthobacter autotrophicus	$++$
Arthrobacter sp	$+$
Bacillus polymyxa	$+$
Azospirillum lipoferum	$+$
Azospirillum brasilense	$+$
Methylococcus capsulatus	$+$
Anaerobic N-fixing bacteria	
Clostridium pasteurianum	$++$
Clostridium butylicum	$+$
Desulfotomaculum orientis	$+$
Facultative anaerobic N-fixing bacteria	
Acaligenes fecalis	$+$
Enterobacter cloacae	$+$
Klebsiella pneumoniae	$+$
Photosynthetic N-fixing bacteria	
Rhodopseudomonas capsulatus	$+$
Rhodopseudomonas globiformis	$+$
Rhodospirillum rubrum	$+$

* The increasing number of "$+$" means the increasing of nitrogen-fixing efficiency

is adversely affected by the seasonal fluctuation of temperature, moisture and energy source supply in the soil, as well as by the application of nitrogen fertilizer during the growing stage of rice plants. It is estimated that the total quantity of nitrogen fixed by free-living bacteria in a hectare of rice field would be 10–20 kg. Recently in China, returning the rice straw to the field has been advocated for supplementing soil organic matter to enlarge the C/N ratio in the soil in order to promote the development of free-living N-fixing bacteria, hence to increase the N-fixing efficiency, and to improve the productivity of the rice paddy fields as a whole (Table 2).

2.2 Nitrogen-Fixing Blue-Green Algae (Cyanobacteria)

More than 40 species of nitrogen-fixing blue-green algae (cyanobacteria) have been found to be growing in Chinese paddy fields. They grow in the water layer above the soil surface and become the most important nitrogen fixers in the growing season of rice plants. Most of them are in filamentous forms with slimy sheath of varying thickness. Species of *Anabaena, Nostoc, Cylindrospermum Oscillatoria, Schizothrix, Anabaenopsis, Calothrix, Gleotrichia, Rivularia* etc. are in simple filament. The species of *Tolypothrix, Scytonema* and *Plectonema* have false branched filaments, while the filaments of the species of *Stigonema* and *Fischerella* are truly branched. There are also some unicellular, coccal forms in aggregates of various shapes such as *Gloeothece, Synechococcus* etc. Among these, species of *Anabaena, Nostoc, Tolypothrix* and *Calothrix* are commonly distributed over the paddy fields in China (Chen 1987). With the exception of the unicellular species and some filamentous forms such as *Oscillatoria, Lyngbya, Phormidium* and *Plectonema*, most of the filamentous species are equipped with specially differentiated thick-walled heterocysts as nitrogen-fixing cells scattered in the filament among vegetative cells or located at the ends of the filament. Those species bearing no heterocysts also show some degree of nitrogen fixation caused by nitrogenase synthesized in their vegetative cells when they are growing in the environment with light but limited oxygen supply (Stewart 1980).

Blue-green algae are mesophilic in nature. It is the shading effect of rice plants in the fields that keeps the temperature and light intensity in the water layer several degrees lower than in the open fields in hot summer days, thus providing a more suitable condition for their development under the rice plants. Research on the nitrogen-fixing blue-green algae naturally developed in the paddy field had shown that the amount of nitrogen fixed during 3 months' growth with the rice plants was calculated as 15–20 kg per hectare, while the corresponding figure in paddy fields in subtropical regions planted with two crops of rice will reach 25–50 kg per hectare. Seeding with a pure culture of *Anabaena* A-686 with a high nitrogen-fixing efficiency, as proved in an experimental rice plot, produced 10 times more algal mass than the unseeded plot in a growing period of 3 months under the rice plants. The amount of nitrogen fixed is estimated at about 100 kg per hectare (Li 1981). The growth of these

Table 3. The nitrogen and organic matter content of the paddy field soil after seeding with blue-green algae culture (Jiangsu Academy of Agricultural Science, 1984)

Treatment	Algal mass kg/mu*	Soil N-content %	Soil organic matter content %
Seeding with *Anabaena*	5000	0.112	1.84
Naturally growing wild blue-green algae (as control)	2000	0.098	1.57

* mu = 1/15 hectare

Table 4. The increase in rice production after seeding with N-fixing blue-green algae in the paddies. (Jiangsu Academy of Agricultural Science, 1984)

Rice varieties	Treatment	Yield kg/mu	Increase in production	
			kg/mu	%
Song-geng 2 (1977)	seeding with blue-green algae	277	28.5	11.0
	Control*	248.5		
Jia-Nong (1978)	seeding with blue-green algae	384	26.5	7.5
	Control*	357.5		
Non-you 2 (1978)	seeding with blue-green algae	517.5	66.5	14.7
	Control*	451.0		
Yuan-feng-zao (1978)	seeding with blue-green algae	359	31.0	9.5
	Control*	328		

* The controls were the paddies without seeding with N-fixing blue-green algae, but with naturally growing wild blue-green algae. mu = 1/15 hectar

algae has shown a beneficial effect on the development of the root system of rice seedlings. The rice seedlings observed are stronger with the dry wt. per 100 seedlings being 35.2% higher than the control. Ammonium (NH_4^+) as a form of nitrogen compound fixed by the blue-green algae during their growing stage in the paddy fields is continuously excreted to the water layer and soil for the nutrition of rice plants. Besides, some growth-promoting substances secreted from the algal filaments are also beneficial to the growth of rice plants. Experimental results from Jiangsu Agricultural Academy show that by seeding N-fixing blue-green algae into experimental rice paddies the production of rice grains will generally be increased 10% over the corresponding controls (Table 4) (Huang 1984).

3 Associative Nitrogen Fixation System in the Rice Rhizosphere

The fibrous root system of rice plants spreads in the soil in a large area. It not only secretes some amount of organic substances during plant growth but also transports air into the soil from above-ground parts of the plant. The rhizosphere of the rice plant is therefore richer in organic nutrition and provides to some extent an aerobic condition for the development of a more flourishing microflora as compared with the soil area outside the rhizosphere. Since the rice seedlings are planted closely, the developing root systems are intermingled among themselves. Therefore a comprehensive rhizospheric microflora of the whole field is formed, which is in action during the rice-growing season. Many kinds of N-fixing bacteria are growing near or on the root surfaces, and some are even living in the intercellular spaces of the root cortex, where the special nutritional environment is favorable to their associative nitrogen fixation. All field tests by the acetylene-reduction method on the rhizospheres of different rice varieties in various paddy fields show obvious nitrogenase activities. The common species of associative N-fixing bacteria isolated are *Acaligenes fecalis, Enterobacter cloacae, Klebsiella pneumoniae, Azospirillum brasilense, Bacillus polymyxa, Xanthobacter autotrophicus* etc. They used to be the natural inhabitants in paddy fields as heterotrophs and do gain a more favorable environment in the rice rhizosphere for growth and nitrogen fixation (Table 5).

Two Gram negative short rods, *Acaligenes fecalis* A15 and *Enterobacter cloacae* E26 have been isolated from the rice roots in Guangdong Province (Qiu et al. 1981, You et al. 1982). They multiply in the slimy layer of the young root surface of rice plants with a significant beneficial effect on the growth of the plants. The nitrogenous compounds including some amino acids secreted through their nitrogen fixation activity are absorbed by rice roots and transported to the above-ground parts to a large extent. This has been proved in experiments by ^{15}N-tracing technique after inoculation of either species onto the rice seedlings. The results also show that the nitrogen fixation activity of *Acaligenes fecalis* A15 is higher than that of *Enterobacter cloacae* E26. It has

Table 5. Nitrogenase activities in rhizospheres of different rice varieties; He Fuheng (1984) Zhejiang Academy of Agricultural Science

Rice varieties	Sampling period	Nitrogenase activities $\mu m\ C_2H_4/g$ root /h
Long-shaped rice Yuan-feng-zao	Booting stage (Spike developing stage)	31.9
Hybrid rice Jun-you 9	Booting stage	20.1
Round-shaped rice Hu-xuan 19	Booting stage	82.9
Glutinous rice Gui-nuo	Booting stage	52.0

Table 6. Nitrogenase activities of root systems of different rice varieties in their different developmental stages

Rice varieties	Nitrogenase activities of root systems nMol C_2H_4/ml (liquid culture)/h		
	Tillering stage	Heading stage	Maturation stage
97A	124.62	265.53	0
6532	382.72	422.90	13.27
97A × 7532	305.20	755.76	1.51
Nong-ken 58	342.32	386.83	2.01
Guang-lu-ai	239.34	562.80	0
IR 34	74.0	797.57	6.03

been observed that the mix-inoculation with both species showed a better effect than the inoculation with single species (You 1984).

The associative nitrogen fixation by *Azospirillum brasilense* on rich plants occurs in most paddy fields either in the subtropic or temperate regions. This organism grows on the rice root surface with its number increasing from the tillering stage to the flowering and milky stages of the rice plant growth, and declining when the plant reaches maturity (Wang 1984, 1985), but with the highest activity of nitrogen fixation in the flowering stage, which is obviously in correspondance with the ample secretion of metabolic substances from the plant in that stage (Table 6).

4 Symbiotic Nitrogen Fixation Systems in Chinese Paddy Fields

4.1 *Rhizobium-Astragalus* Symbiotic Nitrogen Fixation System

A red-flowered legume, *Astragalus sinicus*, has been traditionally cultivated for hundreds years, which has been used as winter-green manure for rice paddy fields in the Yangtze River territory. The cultivation area has been extended northward to the Hui River Valley (Fan et al. 1978) and southward to the subtropic regions of Guangdong and Guangxi provinces in the last 3 decades, to improve the soil fertility of paddy fields for higher yields of rice. The *Rhizobium* in symbiosis with this plant exerts a high efficiency of nitrogen fixation. The nitrogenase activity as measured by the acetylene-reduction method was as high as 2000–3000 µM C_2H_4 per gram of fresh nodule per hour. A crop of green foliage, when ploughed into the soil as base manure for rice planting in late April, can have a yield of 50 000 kg per hectare. The nitrogen content of the green foliage is about 0.35% and the calculated amount of nitrogen fixed by this symbiotic system is 100–120 kg per hectare in a growing season (Table 7) (Chiao 1986).

Table 7. The effect on the soil fertility of growing Astragalus sinicus, in rice paddy fields after using as green manure (Jiangxi Agricultural Research Institute)

Experimentals (on red soil)	Total soil N-content %			Total organic matter content %		
	Before Astragalus	After Ast.	% of increase	Before Astragalus	After Ast.	% of increase
	0.139	0.167	20	2.726	3.06	12.2
II	0.129	0.148	15.5	2.59	2.92	13.0

The green foliage used as green manure is ploughed into the soil one week before the field is flooded with water and prepared for the rice seedling planting. During the more-or-less anaerobic decomposition of the manure materials in the watery soil, nitrogen and other mineral nutrients are gradually released to meet the needs of rice plant growth.

In addition to be ploughed into the soil directly as base manure for early rice, the green foliage may be harvested and composted with rice straw and river mud for anaerobic fermentation. The humified farm manure thus prepared is used to supply nitrogen as well as to improve the physical properties of the paddy soil for the benefit to the late rice (Chiao 1986).

Symbiosis of *Rhizobium astragali* with *Astragalus sinicus* is very specific. The *Rhizobium* exists in the soils only where *Astragalus sinicus* has ever grown. Since the cultivation of *A. sinicus* as green manure of the paddy fields in the Yangtze Valley has been on a traditional rotation scale, root nodules can be observed as soon as the *A. sinicus* seedlings are two weeks old in these fields. The nitrogenase activity of those young nodules can be readily detected by the acetylene-reduction method. However, while the rhizobial cells are multiplying successively in a long period of saprophytic life in the soil, the efficiency of nodule formation and nitrogen fixation will decrease and degenerate as an ineffective strain (Fan and Lou 1980, 1984). Suggestions are made to inoculate the *A. sinicus* seeds with effective strains of *Rhizobium* at the time of sowing to ensure the good nodulation of the *A. sinicus* seedlings, hence the high efficiency of symbiotic nitrogen fixation of the growing *Astragalus* plants. The application of rhizobial inoculum has been the regular practice for cultivation of *Astragalus* in a newly extended area.

The efficiency of nitrogen fixation of *Rhizobium-Astragalus* symbiosis attends the highest peak when the *Astragalus* plants are in full growing stage with only few flowers and well-developed root nodules. It is the proper stage of using as green manure, since the plants have been found to contain a higher percentage of nitrogen after the treatment (Table 8).

Vetch is another legume used as green manure in Chinese paddy fields in the northern regions where the winter temperature is lower than in the Yangtze Valley. There are several species of *Vicia* in cultivation, and the symbiotic *Rhizobium* is inhabiting the soils. A yield of 20 000–30 000 kg of green foliage per hectare can be obtained with a nitrogen content of 0.38% on fresh wt. basis.

Table 8. The nitrogen contents and nodule nitrogenase activities of *Astragalus sinicus* in their different growth stages

Growth stages	N-content % (on dry wt. basis)	Nodule nitrogenase activities μMol C_2H_4/plant/h
Seeding stage (in early spring)	5.83	31.59
Flower bud stage	3.9	196.2
Early flowering stage	3.62	243.7
Flowering stage	3.56	186.23
Pod bearing stage	3.07	17.71

It is a green manure alternate for *Astragalus* for the regional adaptation. In order to strengthen the symbiotic nitrogen fixation research with *Astragalus* on the breeding of high-yield varieties, selection of effective *Rhizobium* strains has been carried on.

4.2 *Anabaena-Azolla* Symbiotic Nitrogen Fixation System

Azolla is a common summer weed on the water surface of ponds, ditches and streams in the lower Yangtze Valley. This plant has a high nitrogen content due to its symbiosis with a nitrogen-fixing blue-green alga, *Anabaena azollae*, and has long been used by Chinese farmers as green manure in rice paddy fields. In the last two decades, *Azolla* was made to propagate in nursery water fields in spring, and then used as propagules by spreading among the rice plants in the fields. This water fern can grow fast on the water surface under the rice plants with a ten-fold increase in green weight after one or two months' growth. The green biomass will be used as top-dressing manure in the tillage stage and early flowering stage of rice plants. Studies have been carried out in the last 30 years on the efficiency of symbiotic nitrogen fixation and its effect on the increase of rice yields with the aim to select more efficient species or varieties of *Azolla* in appropriate symbiosis with *Anabaena* strains suitable for cultivation elsewhere. There have been several selected varieties of an indigenous species of *Azolla imbricata*, and some exotic species such as *Az. filicuoides* and *Az. caroliniana*. A total yield of 40 000–50 000 kg of fresh biomass per hectare with 0.35% nitrogen can be obtained by their fast growing with rice plants in the paddy fields within a period of 6 months. 200–300 kg of nitrogen per hectare can be obtained from the symbiotic nitrogen fixation which serves as top-dressing manure in the paddy fields after turning it over to the soil (Table 9). This green manure decomposes fast due to the high moisture and high temperature, the nitrogen and other minerals and organic nutrients are readily released within 2–3 weeks, which would meet the needs of rice growth. On the other hand, part of the organic matter will be humified, which would improve the physical property of the paddy soil (Liu 1981; Lee 1987).

Table 9. The amount of nitrogen fixed by the *Azolla* growing among the rice plants; Azolla Research Center, Fujian, China (1985)

Azolla species	Biomass kg/ha/year	N-content %/fresh wt.	N-fixed kg/ha/year
Az. imbricata	45 000–50 000	0.35	160–185
Az. filiculoides	60 000–70 000	0.38	230–290

The shading of rice plants creates suitable temperature and light intensity which are favorable for the continuous growth of *Azolla* in the paddy fields. The *Azolla* plants which remain on the surface of the water layer after being used as top-dressing manure will continue to grow and multiply for a next harvest. Due to the differences in *Azolla* species or varieties, the duration of growth with rice plants and field management, the yield of biomass and the amount of nitrogen fixed are different. In the lower Yangtze Valley where rice plants are growing once a year, *Azolla imbricata* will grow only about 3 months with the rice plants, a biomass of 20 000–30 000 kg per hectare can be obtained, while *Az. filiculoides*, a high-yield species, can have a harvesting biomass of 35 000–40 000 kg per hectare. On the north border of subtropic regions, where 2 crops of rice are prevailing and the fields are flooded with water and covered with rice plants successively from April to November, the growing period of *Azolla* in the fields lengthens to 7–8 months, which results in the biomass being 70–100% higher than in the single rice crop area.

The symbiont of *Azolla*, *Anabaena azollae* is a filamentous alga, living in the dorsal cavities of the scale leaves of *Azolla*. The filament consists of egg-shaped vegetative cells scattered among some large-shaped and thick-walled heterocysts. The heterocysts contain nitrogenase and photosystem I and are specialized in the function of nitrogen fixation by using light energy. Since the heterocysts contain no photosystem II, there is no oxygen produced in the photosynthesis to inhibit the nitrogenase activity in the cell. Carbon and mineral nutrients of the heterocysts are supplied from the neighbouring vegetative cells, while the nitrogen fixed in the heterocysts goes as nitrogen nutrition to supply the vegetative cells, thus the *Azolla* plant as well (Fay 1980; Lee 1987).

5 The Bio-Nitrogen Fixation Systems and Their Activities in Different Rice Paddy Fields in China

Rice paddy fields in China are distributed in vast areas of subtropic and temperate regions. Due to the differences in soil properties, climatic conditions, cultivation systems and field management, the microfloras are different to a great extent in their composition and activity. The development of nitrogen fixation systems and their efficiency of N-fixation likewise vary in different fields, hence the soil fertility and the yield of rice are affected considerably. The rice paddy fields in the provinces of Hainan, Guangdong and Guangxi, and the south part of Hunan province in the subtropic region with rather high temperature and more rainfall annually, have 2 crops of rice a year as the prevailing cultivation system. The fields are usually watered more than 10 months in a year for the growth of rice plants. Abundant growth of algae consisting of many efficient species of nitrogen-fixing blue-green forms in the surface water layer is the characteristic microbial flora in these fields. *Nostoc, Calothrix, Tolypothrix, Cylindrospermum, Schizothrix, Anabaena* etc. are the most common members of free-living filamentous blue-green algae responsible mainly for the nitrogen fixation in these fields. A large population of heterotrophic bacteria of either aerobic or facultative forms develops on the slimy sheaths of algal filaments or lives freely in the water layer and the soil surface beneath, on an ample nourishment secreted from the algal filaments and rice plants. *Azotobacter, Beijerinckia, Xanthobacter, Bacillus, Azospirillum, Arthrobacter, Klebsiella, Enterobacter* and *Acaligenes* are active members of the nitrogen-fixing group in this bacterial population, while in the soil beneath the water layer in the paddy fields the *Clostridium* and *Desulfotomaculum* are the dominant anaerobic nitrogen fixers. All these nitrogen-fixing microorganisms are growing well yearly in the paddy fields of this subtropic region with high efficiency of N-fixation under the high soil moisture and high temperature, and in ample supply of soil organic substances as well as the secreted metabolic substances from the growing plants. Besides, the associative nitrogen fixation by *Azospirillum, Klebsiella, Enterobacter, Acaligenes,* etc. exerts a prominent neutritional effect on the rice plant growth.

With the large extent of bio-nitrogen fixation activities in rice paddy fields of the subtropic region, there is enough nitrogen supply for a moderate yield of rice annually even with no application of nitrogen fertilizers. Signh noted that whether or not to supplement nitrogen to the rice paddy fields in India was greatly dependant upon the N-fixation by microorganisms which developed naturally in the fields and of which the blue-green algae played the most important role. The field experiments with no application of nitrogen fertilizers which were carried out in the International Rice Research Institute in the Philippines also illustrated that after a successive harvest of 22 corps of rice there was no decrease of soil N-content in this field (Watanabe et al. 1978).

In the lower Yangtze Valley the center of temperate climatic rice region of China, a rotational cultivation system of rice and winter dryland crop is prevailing. There are one or two crops of rice in summer and autumn seasons and a crop of wheat or legume, as green manure or grain crop in winter and spring. The paddy fields are flooded with water beginning from the early summer to late autumn for rice plant growth. Various N-fixing microorganisms develop well in this period as in those paddy fields in the subtropic region. However, the predominant species and their N-fixing activities vary to some extent due to the variation in climatic conditions and soil properties.

The blue-green algae species of *Anabaena, Nostoc* and *Cylindrospermum* grow predominantly in the surface water layer of the paddy fields, but their development and activities are limited to the rice-growing seasons. It was estimated that the amount of nitrogen provided by naturally developed bio-nitrogen fixing systems in a paddy field could reach 40 kg per hectare (Yamaguchi 1978). However, cultivation of *Azolla* with the rice plants has been a recommended practice to supplement nitrogen to the growing rice plants. The growing of green manure legumes such as *Astragalus sinicus* or vetch in the winter season once every 2–3 years in paddy fields has been considered a major supplement of nitrogen and organic matter to the soil.

The paddy fields located on the lake deposit plains and alluvial areas in the further northern part of the temperate region are available for rice plant growth only in the summer and early autumn months. There is generally no winter crop due to the lower temperature and low soil moisture in the winter season. The activity of the soil microflora is mostly limited to the growing season of the rice plants. Various bio-N-fixing systems enrich the soil with nitrogen, but the efficiency is usually lower than in those fields in the south. Cultivation of *Azolla* is impossible due to the difficulty of keeping this plant over winter to use as "seed" for vegetative production in the summer season, except vetch, which could be used as a winter crop to supplement the nitrogen by symbiotic nitrogen fixation.

Paddy fields for rice cultivation are also established on some saline-alkaline soils of the coastal area, on acid red soil of rainy and hilly areas in the south provinces, and on the muddy humified soil in the mountains flooded by spring water. These are low-productive fields with rice planting only in the summer months, when the fields are irrigated and flooded with water. The bio-nitrogen-fixing activities are influenced primarily by the nature of the soils. In saline-alkaline soil, the aerobic free-living bacterium *Azotobacter chroococcum* is the predominant species and blue-green algae such as *Anabaena* and *Nostoc* as well as the associative N-fixing bacteria are contributing most of the supplementary nitrogen to the soil. In acid soils, *Beijerinckia* instead of *Azotobacter* becomes an important aerobic form to fix nitrogen in their free-living state in the surface water layer. *Clostridium* species are active in soil beneath the surface water layer under the acidic and anaerobic conditions of the soil rich in organic matter. However, the blue-green algae do not grow very well due to the acidic nature of the soil. Their nitrogen-fixing activities therefore are not as important as in alkaline soils. The winter leguminous green manure serves as a main supplementation of nitrogen to these low-productive paddy fields as base manure for rice planting in the next spring.

6 Conclusions

The paddy fields in subtropic southern China are covered with rice plants in a year-round system for 2 crops. During the winter months, the fields are usually kept in water for some water vegetables or the water is drained away for a dryland winter crop. The annual production of these fields may be 12-15 tons of grain per hectare. While on those fields in the Yangtze Valley which is in the temperate zone, a rotational cultivation system of rice-wheat or rice-legume is adopted with an annual yield of 10-12 tons of grain. It is estimated that such a harvest of grain together with straw may bring forth 600-800 kg of nitrogen from the soil per hectare. Besides, a large quantity of phosphorus, potassium and other mineral nutrients are also removed from the soil. Of course, heavy application of nitrogen fertilizer as well as other mineral fertilizers is necessary to mantain the high production. Since different biological N-fixing systems are naturally developed in these paddy fields, the full play of those systems to provide nitrogen as well as other organic nutrient substances through their metabolic activities will be a great investment for the rice plant nutrition and for soil fertility improvement.

The rice cultivation systems and green manuring practices in Chinese paddy fields under different climatic conditions have provided natural habitats for different bio-nitrogen fixation systems in action. With the improvement of the habitat's environments for various N-fixation systems through proper field practices and management, an increase of rice productivity will be expected. These practices have long been experienced by Chinese farmers through their rice cultivation, which is reflected in the recent advances in biological N-fixation research in China.

7 References

Chen HK (1979) Microbiology of rice paddy fields. In: Soil Microbiology (in Chinese). Scientific Technology Press, Shanghai

Chen Y (1987) Nitrogen fixation by blue-green algae (in Chinese) In: You, Chiang, and Song (eds) Biological nitrogen fixation. Science Press

Chiao P (ed) (1986) The green manure of China (in Chinese). Agricultural Press, Beijing

Fan CS, Tsai DT, Lou WJ (1978) Northward extension of *Astragalus Sinicus* in China. J of Nanjing Agricultural University

Fan CS, Lou WJ (1980) Studies on the specificity of *Rhizobium* and *Astragalus* symbiosis (in Chinese). J of Nanjing Agricultural University

Fan CS, Lou WJ (1984) Physiological and biochemical studies on the symbiotic nitrogen fixation of *Rhizobium*-legume (in Chinese). J of Nanjing Agricultural University

Fay P (1980) Nitrogen-fixation in heterotysts. Recent Advances in Biological Nitrogen Fixation, p 121

Huang YS (1984) Nitrogen-fixing blue-green algae (in Chinese). Agric Press, Beijing

Lee CH (1987) *Azolla* in biological nitrogen fixation. In: You, Chiang, Song (eds) Biological nitrogen fixation. (in Chinese). Science Press

Li SH (1981) Studies on the nitrogen-fixing blue-green algae as fertilizer in the rice late crop. Acta Hydrobil Sinica 7: 417

Liu CC (1981) The utilization of *Azolla* in rice production in China. In: Gibson AH, Newton WE (eds) Current perspectives in nitrogen fixation. Aust Acad Science

Qiu YS et al. (1981) Study on the nitrogen-fixing bacteria associated with rice roots (in Chinese). Acta Microbiol Sinica 21(4): 473

Song HY (1987) The nitrogen fixation by photosynthetic bacteria (in Chinese). In: You, Chiang, and Song (eds) Biological nitrogen fixation. Science Press

Sprent JI (1979) The biology of nitrogen-fixing organisms. McGraw-Hill, New York
Stewart WDP (1980) Some aspects on nitrogen fixation of cyanobacteria in rice paddies. In: Nitrogen fixation. Acad Press, London
Tseng D (1987) Biology of nitrogen fixation (in Chinese). Amoy Univ Press
Wang ZF (1984) Studies on the nitrogen-fixing bacteria on rice root surface (in Chinese). Microbiology 9(4): 176
Watanabe I, Cholitkul W (1978) Field studies of bionitrogen fixation in Rice Paddy Soil. Proc Int Symp Nitrogen and Rice, Phillipine
Wang ZF (1985) Relation of H metabolism and nitrogen fixation in *Azospirillum* (in Chinese). Acta Microbiol Sinica 25(1): 54
Xiong Y, Li CK (eds) (1987) The soils of China; (2nd edn in Chinese). Science Press, Beijing
Yamaguchi M (1978) Bio-nitrogen fixation in flooded rice paddies. Proc Int Symp Nitrogen and Rice, Phillippines
Yoshida T, Rinaudo G (1982) Heterotropic nitrogen-fixation in rice paddy soil. In: Microbiology of tropical soils and plant productivity, pp 75
You CB, Qiu YS et al. (1982) Studies on the associative nitrogen fixation of *Acligenes faecalis* A-15, and *Entobacter cloacae* E-26 with rice plant. (in Chinese). J Chin Agr Sci
You CB (1984) Some properties of the nitrogen-fixing associative symbiosis of *Acaligenes faecalis* A-15 with rice plant (in Chinese). Adv Nitrogen-Fixation Research 3
You CB (1987) Associative nitrogen fixation (in Chinese). In: You, Chiang and Song (eds) Biological nitrogen fixation. Science Press

CHAPTER 23
Characteristics, Distribution, Ecology, and Utilization of *Astragalus Sinicus*-Rhizobia Symbiosis

H.K. Chen, F.D. Li, and Y.Z. Cao

The Nitrogen Fixation
and its Research in China
Editor Guo-fan Hong
© Springer-Verlag Berlin Heidelberg 1992

1 Introduction

Astragalus sinicus L. is a symbiotic nitrogen-fixing leguminous green manure crop traditionally grown in rice fields in the lower Yangtze Valley, southeastern China and Japan. In the lower Yangtze Valley, after the rice crop is harvested in September/October, *A. sinicus* seeds are sown to the drained field. The crop is plowed under at full blossoming stage in April/May; after that the field is flooded and prepared for transplanting rice seedlings. A fair yield of the green manure crop amounts to thirty to forty metric tons per hectare of fresh material, including tops and roots, containing approximately four kilograms nitrogen per metric ton, if well nodulated by efficient rhizobia. *A. sinicus* is grown yearly or in alternation with winter wheat, barley or rapeseed.

2 *Rhizobium huakei* (Chen et al. 1990): The *A. sinicus* Rhizobia

Detailed taxonomic studies of the *A. sinicus* rhizobia were carried out only very recently (Chen 1990). A tentative name, *Rhizobium astragali*, was given to the *A. sinicus* rhizobia, which first appeared in a textbook of microbiology in 1959 (Chen 1959) following the then current emphasis on cross-inoculation at the species level. Chan et al. (1988) placed the organism in the genus *Bradyrhizobium*. A new species, *Rhizobium huakui*, was proposed by Chen et al. (1990) for the rhizobia isolated from *A. sinicus* L., based on numerical classification studies comparing two hundred phenotypic features of nine strains of rhizobia isolated from *A. sinicus* with those of forty eight strains of *Rhizobium* spp., *Bradyrhizobium* spp., *Sinorhizobium fredii* and *Agrobacterium* spp., as well as polyacrylamide gel electrophoresis patterns of cell protein, DNA G + C content and DNA–DNA hybridization. All nine isolates from *A. sinicus* are closely related, clustering as

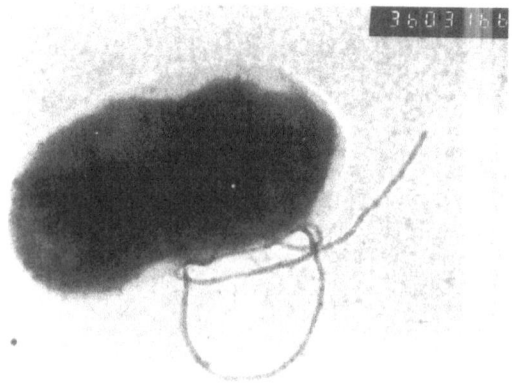

Fig. 1.

one phenon within the genus *Rhizobium*. The morphology and colony characteristics of *R. huakui* are similar to those of other *Rhizobium* spp., except that
it is motile by a single polar or subpolar flagellum (Fig. 1), which is unique in
the genus. *R. huakui* grows well in mannitol yeast extract medium but does not
or very poorly grow in beef extract nutrient medium, similar to most other
Rhizobium spp., Potassium nitrate or urea serves satisfactorily as sole nitrogen
source. One or more vitamins (thiamin, nicotinic amide, etc.) are necessary for
different strains (Zhou and Cao 1981).

A synthetic medium was formulated for large-scale production of liquid
cultures of *R. huakui* and compared favourably against yeast extract-based or
soybean sprout extract-based liquid medium.

3 Host-Rhizobia Specificity of *A. sinicus* Symbiosis

Astragalus is the largest genus in Leguminosae, containing 1500 to 2000 species.
Allen and Allen (1981, pp. 78–80) listed 91 nodulated species. Results of plant-
infection tests in early years suggested that symbiotic associations within the
genus *Astragalus* were exclusively *inter se* (Chen and Shu 1944; Ishizawa 1954);
though there were reports to the contrary (Wilson 1939; Wilson and Chen 1947).
Allen and Allen mentioned in their monograph (1981, p. 77) that the host plant-
rhizobia relationships of 18 species of *Astragalus* were markedly versatile. Not
all *Astragalus* rhizobia nodulated *Astragalus* species tested. Five strains nodulated
2 species of *Phaseolus*, but not *P. lunatus*, 3 strains nodulated *Medicago sativa*
and *Mililotus* sp., and 3 strains nodulated *Vigna* sp. All 18 strains were non-
infective on *Lupinus*, *Glycine*, *Trifolium*, *Pisum*, *Lathyrus* and *Vicia*. Chen and
Shu (1944) tested the nodule-forming ability of rhizobia isolates from nodules
of twenty four species of eighteen genera of Leguminosae on *A. sinicus* host.
All five isolates of *A. sinicus* rhizobia and one isolate from nodules of *Desmodium
heterotyllum* formed nodules on *A. sinicus* host. All the remaining isolates did
not form nodules on *A. sinicus* host, namely, isolates from *Vicia hersuta*, *V. sativa*,
V. sp., *Trifolium pratense*, *T.* sp., *Molilotus indica*, *Medicago hispida*, *M. sativa*,
Soja max, *Cajanus cajan*, *Lespedeza striata*, *Indigofera suffruticosa Phaseolus
aureus*, *P. vulgaris*, *P. acoritifolius*, *Vigna sinensis*, *Crolatoria strata*, *Alysicarpus
vaginalis*, *Atylosia scarabeoides*, *Derris elliptica*, and an unidentified legume.
Ning et al. (1988) found that rhizobia isolated from *A. sinicus* did not nodulate
A. adsurgens and vice versa. Three *A. sinicus* rhizobia isolates tested by Chen
et al. (1990) nodulated *A. adsurgens* and *A. alginosus* but not *A. membranacens*
and *A. mongholicus*; they did nodulate *Vicia villosa*, *Phaseolus vulgaris* and
Sesbania sp., but not *Hedysarum mongolicum*, *Pisum sepium*, *Trifolium repens*,
Melilotus albus, *Lotos corniculatus*, *Glycine max*, *Lupinus* sp. and *Vigna sinensis*.
It is known that there is no clear-cut demarcation dividing host-rhizobia
cross-inoculation groups. All evidence considered, the *A. sinicus* rhizobia
symbiosis may be taken as a monospecific cross-inoculation system, although

interspecific and intergeneric cross-inoculation relations do happen. Observations of extension experts have confirmed this justification. On fields where nodules are naturally formed on *Vicia* spp., *Faba faba*, *Glycine max* or other legumes, *A. sinicus* sown for the first time fail to nodulate if not properly inoculated.

4 The Symbiotic Plasmids

Same as in other fast-growing rhizobia, the nodulation and nitrogenase gene determinants are located on megaplasmids. One to four megaplasmids are present in the *A. sinicus* rhizobia examined. Table 1 presents the number of megaplasmids, phenotypic expression of nodulation and nitrogen fixation of wild and mutant strains of *A. sinicus* rhizobia. Sone strains contain only one megaplasmid and are positive of nodulation and nitrogen fixation. Strain 7653R contains two megaplasmids. The heat-cured mutant, strain 7653R1, has lost one of the wild-type megaplasmids and becomes uninfective to *A. sinicus*. The acridine orange-cured mutant strain 7653R5 has also lost one megaplasmid and behaves likewise. The wild-strain S52 contains also two megaplasmids and

Table 1. Number of megaplasmids and the properties of nodulation and nitrogen fixation of different strains of *A. sinicus* rhizobia

Strain	Number of megaplasmid	Nodulation	Nitrogen fixation	Reference
7653[a]	1	+	+	Zhang et al. 1986
7653R[a]	2	+		Zhou et al. 1987
7653R1[b]	1	−		Zhou et al. 1987
7653R5[c]	1	−		Zhou et al. 1987
76531[d]	4	+	+	Wang et al. 1988
S52[a]	2	+	+	Wang et al. 1986
S52S2[b]	1	+	−	Zhou et al. 1987
HR104[a]	3	+	+	Wang et al. 1988
HR1042[b]	2	+	+	Wang et al. 1988
HR107[a]	3	+	+	Wang et al. 1988
HR112[a]	2	+	+	Wang et al. 1988
Ra31[a]	1	+	+	Wang et al. 1988
Ra1[a]	1	+	+	Wang et al. 1988
Ra74[a]	1	+	+	Wang et al. 1988
Ra81[a]	1	+	+	Wang et al. 1988
CZ74[a]	2	+	+	Wang et al. 1988
SR72[a]	2	+	+	Lin et al. 1986
Ra27[a]	2	+	+	Zhang et al. 1986
HR101[a]	2	+	+	Wang et al. 1988

Notes:
[a] wild type (isolated from nodules of the field host)
[b] heat cured
[c] acridine orange cured
[d] all references except this one are given by members of this laboratory

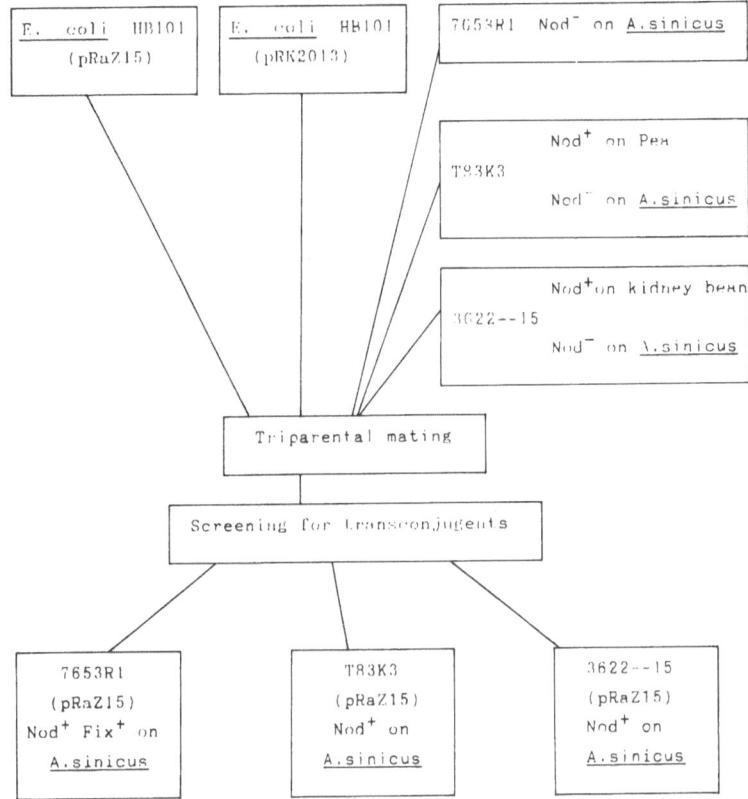

Fig. 2.

is nodulation positive and nitrogen fixation positive. Its heat-cured mutant, strain S52S2, has lost the larger megaplasmid (pRaS52b) and becomes nodulation positive but nitrogen fixation negative. Radioprobing shows that the *nod* genes are located on the smaller megaplasmid (pRaS52a), responsive positively to ^{32}P-labelled pRmSL42 probe containing *nod*DABC, and the *nif* genes are located on the larger megaplasmid (pRaS52b), responsive positively to ^{32}P-labelled pIJ1242 probe containing *nif*KDH (Wei and Li 1989).

Zhang et al. (1989) prepared a gene bank of *A. sinicus* rhizobia strain 7653R. By triparental mating (*E. coli* HB18f gene bank as donor, *E. coli* HB101 (pK2013) as helper and *A. sinicus* rhizobia Nod⁻ strain 7653R1 as receptor) and selection by plant nodulation test, are constructed plasmid pRaZ15 was obtained. Strain 7653R1 (pRaZ15) regains abilities of the nodulation and nitrogen fixation on *A. sinicus*. pRaZ15 was transferred into *Phaseolus vulgaris* rhizobia strain 3622-15 and *Pisum sativum* rhizobia strain T83K3, and the resulting 3622-15 (pRaZ15) and T83K3(pRaZ15) nodulated *A. sinicus* (Fig. 2). Probed by ^{32}P-labelled pRmZ1 containing R.m. *nod*DABC or ^{32}P-labelled pSA30 containing in K.p. *nif* KDH, pRaZ15 responds positively, indicating that the pRaZ15 contains both *nod* and *nif* genes, which is in agreement with the nodulation test.

5 Infection and Nodule Development

The infection and nodule development of *A. sinicus*-rhizobia symbiosis is similar to the *R. leguminosarum* type. Infection of the root hair takes place after the elongation and curling of the root hair stimulated by the presence of rhizobia. Nodule development is visible on the seedling by the naked eye in six or seven days after inoculation of the rhizobia to germinated seeds. Mature nitrogen-fixing nodules are oblong to cylindrical in shape. The dividing meristem lies at the terminal region of the nodule. Reddish infected bacteroidal tissue develops progressively behind the meristem until the nodule stops developing. Two strands of vescular bundles lie between the infected bacteroidal tissue and the cortex surrounding the nodule.

The bacteroids are mostly club-shaped, approximately forty to fifty times larger than the unswelled rod-shaped bacteria present in the infection threads or in in-virtro culture. The bacteroidal host cell contains vast amounts of swelled bacteroids and a small number of rod-shaped bacteria. Microculture of bacteroids and bacteria released from bacteroidal host protoplast (Fig. 3) reveals that only the rod-shaped bacteria multiply (Fig. 4), while the club-shaped bacteroids remain undivided (Cao et al. 1984; zhou et al. 1985). This is common to *A. sinicus* nodules and the nodules of alfalfa, clover and vetch, but not with soybean nodules. The 'bacteroids' in the soybean bacteroidal tissue are rod-shaped, unswelled, and almost all multiply themselves when transferred to the culture medium. Table 2 presents the comparisons of viable counts and direct counts of organisms released from bacteroidal host protoplasts of *A. sinicus*, *Medicago sativa*, *Trifolium repens*, *T. subterraneum*, *Vicia crocca* and *Glycine max* (Cao et al. 1984).

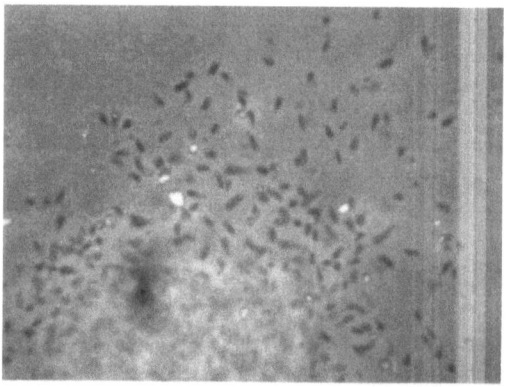

Fig. 3.

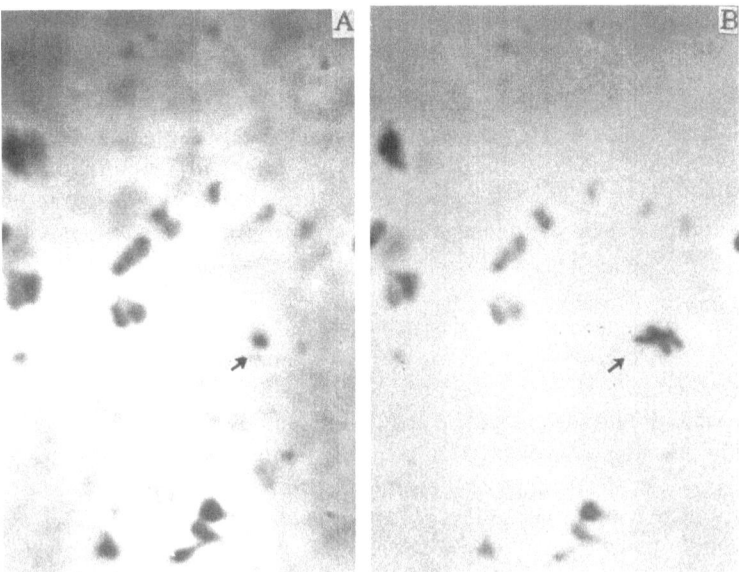

Fig. 4.

Table 2. Viable and total counts of the rhizobia released from bacteroidal host protoplasts from nodules of different legumes (Cao et al. 1984)

Legume	Protoplasts counted		Average number of rhizobia per protoplast	Ratio V:T
A. sinicus	Viable	17	63.9	0.27:100
	total	103	23375.5	
Medicago sativa	Viable	28	26.5	
	Total	15	22755.0	0.11:100
Trifolium repens	Viable	15	17.7	
	Total	11	19509.0	0.09:100
Trifolium subterraneum	Viable	29	19.3	
	Total	15	17813.0	0.11:100
Vicia crocca	Viable	28	29.9	
	Total	15	23973.0	0.12:100
Glycine max	Viable	52	50903	
	Total	216	62781	81.08:100

6 Distribution and Expansion of *A. sinicus*

In the Yangtze Valley and the southern provinces, the autumn-sown legumes, cereals and rapeseeds are grown over winter in properly drained rice fields. Among the legumes, *A. sinicus* and *Vicia crocca* are grown for green-manuring, and *Faba faba* is grown chiefly for seeds. Under the condition of energy stress and short supply of chemical nitrogen fertilizers, nitrogen-fixing green manure crops not only contribute a good deal of nitrogen nutrient for the succeeding rice crop, but also organic matter to the soil. However, the legumes can only grow in properly irrigated and drained rice fields. In the rain-fed rice fields, the fields have to be left fallow in order to receive and conserve water for the next rice crop. In the lowland, without properly developed irrigation and drainage system, the fields are flooded all year round, no upland winter crop could grow. Owing to the progress of construction of reservoirs and irrigation systems in the past forty years, the acreage of the winter legumes and other winter crops were very much expanded in the rice-growing region.

Traditionally there was an interboundary cut roughly at 112 degrees east longitude right across the Hubei and Hunan provinces. *A. sinicus* was grown chiefly in the east up to the sea coast. *Vicia crocca* was grown chiefly to the west and the southwestern provinces. Since the late nineteen fifties, double-cropping rice systems, i.e., an early rice crop followed by a late crop rice in the same year, is extended in the Yangtze Valley. *A. sinicus* fits better than *V. crocca* as green manure crop for the double-cropping rice system. Therefore, *A. sinicus* has taken the place of *V. crocca* in the traditional *V. crocca* region whereever the double-cropping rice system is practised.

For both reasons mentioned above, the acreage of *A. sinicus* expanded significantly in the rice-growing region in the sixties and seventies. The expansion of the acreage of *A. sinicus* in Hubei province between 1957 and 1975 increased approximately three-fold, according to a provincial sensus:

	hectares (10^4)
1957	29
1962	29
1965	59
1970	67
1973	89
1975	106

7 Occurrence and Introduction of *A. sinicus* Rhizobia in Rice Field Soils

In fields where *A. sinicus* has never been sown, the soil is generally devoid of *A. sinicus* rhizobia. Table 3 presents the result obtained by comparing the occurrence of *A. sinicus* in the soil of two rice fields on the same farm at Shizishan,

Table 3. Occurrence of *A. sinicus* rhizobia in two rice field soils at the same farm developed on Yangtze alluvial, Shizishan, Wuhan, 1964 (Cao and Li, unpublished)

Date of sampling	Field I[a]		Field[a]	
	Field condition	Rhizobia[b]	Field condition	Rhizobia[b]
Feb. 2	*A. sinicus*	94 000	Fallow	0
Mar. 31	*A. sinicus* blossoming	29 000	Water-logged	3.6
Jan. 11	Rice planted	30 000	Rice planted	0
Sept. 14	Soil plowed	1 800	Soil plowed	0
Dec. 16	*A. sinicus*	180 000	Rapessed	0

Notes:
[a] field I *A. sinicus* established for years; field II *A. sinicus* never grown before
[b] Number of *A. sinicus* rhizobia per gram dry soil, estimated by the plant nodulation method, see text for brief description

Wuhan, developed on Yangtze alluvial. Estimation of the quantity of *A. sinicus* rhizobia in soil was carried out by a modified Brokwell's nodulation method (Brokwell 1968). Surface-sterilised seeds of *A. sinicus* were seeded to sterilised plant culture tubes. Serial dilutions of soil samples were introduced into the tubes as inoculant. The presence of *A. sinicus* rhizobia in the soil dilutions was indicated by the nodules formed on the seedlings grown for five weeks in the growth chamber. The number of *A. sinicus* rhizobia were estimated in five replicates by the MPN method. In a field where *A. sinicus* had never been grown, no *A. sinicus* rhizobia were detected. In a field where *A. sinicus* was established for years, soil samples taken in different seasons contained thousands to hundred thousands *A. sinicus* rhizobia per gram of dry soil.

For a field where *A. sinicus* was sown for the first time, proper inoculation of *A. sinicus* rhizobia not only ensures successful nodulation and symbiotic nitrogen fixation, but also ensures the integration of the rhizobia as a relatively stable member of the microbial population of the soil. Table 4 presents data of an experiment carried out at Xing'andu, Wuhan. A rice field where *A. sinicus* has never been sown before was divided into two parts. On one part *A. sinicus* was sown without inoculation, and on the other part *A. sinicus* was sown and properly inoculated. At the first sampling, before *A. sinicus* seeds were sown, soil in both parts contained no *A. sinicus* rhizobia, as shown by the plant nodulation method described above. At the second sampling, when the *A. sinicus* had grown for a month, the quantity of soil *A. sinicus* rhizobia in the inoculated parts amounted to 33 thousand per gram of dry soil. In the uninoculated part, the soil contained only 10 *A. sinicus* rhizobia per gram of dry soil, most probably from carried-over seeds. During the entire growth period of *A. sinicus* the quantity of *A. sinicus* rhizobia measured up to hundred times in both parts.

Field surveys carried out by this laboratory and other institutions offered confirming observations. Where *A. sinicus* were sown for the first time, no

Table 4. Occurrence and abundance of *A. sinicus* rhizobia in rice field where *A. sinicus* was sown for the first time with or without artificial inoculation at Xingandu, Wuhan, 1964 (Cao and Li, unpublished)

Date of sampling	Field condition	Rhizobia per gram of dry soil[a]	
		Uninoculated	Inoculated
Sept. 6	Ratoon rice field flooded	0	0
Nov. 18	*A. sinicus*	10	35 000
Jan. 16	*A. sinicus*	4 480	145 000
Mar. 15	*A. sinicus* in full blossom	1 380	163 000
May. 17	Seeds ripe	318 000	10 100 000

[a] Estimated by the plant nodulation method

Table 5. Pot experiment on the nodulation of *A. sinicus* on soil taken from a field where no *A. sinicus* was grown for seven years but was grown yearly before, Xiaolingwei, Nanjing (Huang 1983)

	Uninoculated	Inoculated
Fresh weight (top and roots) per pot (g)	27.15	55.70
Number of nodules per plant	12.2	52.4

Note: Average value of 6 replicated pots, 15 plants per pot

nodule was formed or only a few seedlings were nodulated. Once a population of the *A. sinicus* rhizobia is established in the rice field soil, it would persist for years even if no succeeding crop of *A. sinicus* is grown. Pot experiments were carried out at Xiaolingwei, Nanjing (Huang 1983). Sterilised pots were filled with soil taken from a rice field where *A. sinicus* was grown yearly before 1975 but no *A. sinicus* was grown afterwards. Surface-sterilised seeds of *A. sinicus* were sown to the pots with or without artificial inoculation. After 92 days of culture, an average of 14.2 nodules per plant were formed in the uninoculated pots and 52.4 nodules per plant were formed in the inoculated pots. The result proved that *A. sinicus* rhizobia did exit for seven years as an integrated member of microbial population of the soil (Table 5).

8 Isolation of Efficient Strains

Isolation and selection of *A. sinicus* rhizobia were carried out at this laboratory and other institutions, mostly in the late fifties, sixties and seventies. No centralised culture collection institute was established until the establishment of the China

Table 6. Field performance of *A. sinicus*-rhizobia symbiosis of four strains of *A. sinicus* rhizobia, examined 80 days after sowing, Honghu, Hubei, 1973 (Cao and Li, unpublished)

	Uninoculated control	Inoculated with strain			
		A16	A19	A21	A26
Dry weight per plant (g)	3.3	5.5	4.8	5.5	5.5
Branching	0.2	1.4	1.1	1.4	1.4
Nodules per plant	2.1	19.5	9.6	13.7	17.0
% N content of oven-dry matter of aerial growth	1.5	3.8	3.75	3.34	3.85

Note: 4 replicated plots were laid out for each treatment; 20 plants were taken randomly from each plot. Figures shown are average values.

Committee for Culture Collection of Microorganism (CCCCM) and China Centre for Type Culture Collection (CCTCC) in 1979. Efficient strains of *A. sinicus* rhizobia were obtained and lost. Table 6 presents a representative field experiment related to the performance of strains selected by this laboratory for the purpose of selecting strain used in inoculant preparation.

9 Preparation and Quality of Inoculant

Most inoculant preparations are peat based. Neutral peat is sterilised in an autoclave and packed in bottles or polychloroethylene bags. Liquid cultures of rhizobia are produced in aerated ferment. Aliquotes of liquid culture are added

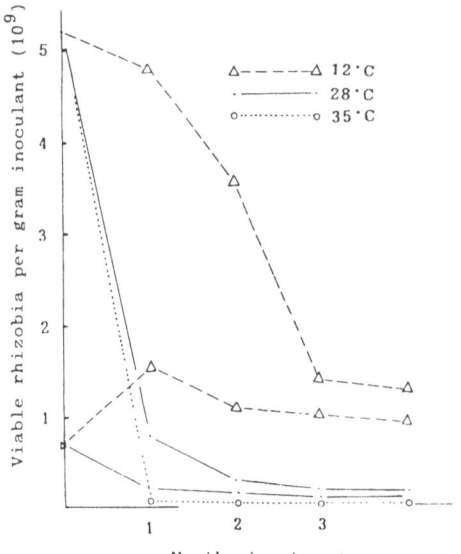

Fig. 5.

to sterilised peat bottles or bags aseptically. The preparations are kept for five days and examined for quality. All unproperly packed, puddled or visibly molded bottles or bags are discarded. Samples are taken from each batch for bacteriological examination. Samples are serially diluted and plated on agar media. A batch is discarded if the samples are contaminated or of very low count.

   A. sinicus is sown in September/October. The inoculants have to be prepared, stored and distributed in summer season. Quality of the preparation is determined by the process of production as well as, or even more by the condition and duration of storage and distribution. At high temperature, the viable counts of rhizobia drop drastically. Figure 5 presents data of changing trends of the viable counts of inoculant preparations during four months of storage at 12 °C, 28 °C and 35 °C. The viable count of inoculant preparation kept at a much higher level at 12 °C than those at higher temperature. At 35 °C, the viable count droped to a level of under one million rhizobia per gram of preparation.

10 Application of Inoculant

In agricultural practice, 40 g peat-based inoculant preparation is mixed with every kg of *A. sinicus* seeds, consisting of approximately 280 000 seeds. Three to four kg seeds are sown to one hectare of drained rice field, either on prepared

Table 7. Sand pot experiments estimating nodulation effectiveness of *A. sinicus* seed-rhizobia mixture (Zhou, Hu, Cao and Li 1965, unpublished)

Inoculant		Nodulation effectiveness						
28 °C, 1 day	A/1.75	10^6	10^5	10^4	10^3	10^2	10^1	10^0
Viable rhizobia	B	4/4	4/4	4/4	4/4	4/4	4/4	—
per gram								
52.4×10^3	C	4.0	3.4	4.1	6.3	4.8	0.7	—
28 °C, 30 days	A/2.90	10^6	10^5	10^4	10^3	10^2	10^1	10^0
Viable rhizobia	B	4/4	4/4	4/4	4/4	4/4	2/4	—
per gram								
10.8×10^8	C	5.7	3.5	4.5	3.4	3.6	0.4	—
12 °C, 90 days	A/0.36		10^5	10^4	10^3	10^2	10^1	10^0
Viable rhizobia	B		4/4	4/4	4/4	4/4	3/4	1/4
per gram								
10.8×10^8	C		3.7	4.6	3.6	4.7	1.9	0.08
25 °C, 90 days	A/0.15		10^5	10^4	10^3	10^2	10^1	10^0
Viable rhizo-	B		4/4	4/4	4/4	4/4	3/4	0/4
bia per gram								
4.5×10^8	C		3.2	3.3	2.4	3.2	2.0	0

Note: A-Number of viable rhizobia per seed inoculated
   B-Pots inoculated in 4 replicates
   C-Nodules per plant, 3 seed hills per pot, 3 seeds per hill

land after single cropping rice, or 1 to 2 weeks before harvest of the second rice crop of the double-cropping system, between rows of rice stand.

Abundant inoculation is emphasised. Sand experiments were carried out to assess the number of viable rhizobia in inoculant preparation required per seed for satisfactory nodulation. In an experiment carried out with inoculant stored at 20 °C for different durations, the minimum number of viable rhizobia per seed required for satisfactory nodulation was 175 and 200 viable rhizobia per seed for inoculant preparations stored for 1 and 30 days, respectively (Table 7). In another experiment, inoculant preparations were stored at different temperature (12 °C and 25 °C) for the duration of 90 days, the minimum number of viable rhizobia per seed for satisfactory nodulation was 36 and 15, respectively (Table 7). No significant improvement in nodulation was observed, even hundred thousand times more viable rhizobia than the above minimum was added.

11 Benefit of Artificial Inoculation to *A. sinicus*-Rhizobia Symbiosis

Many field experiments were carried out by this laboratory and other institutions on the benefit of artificial inoculation of *A. sinicus* rhizobia to the *A. sinicus*-rhizobia symbiosis. Table 8 cites two field plot experiments carried out by this laboratory at Jingmen County, Hubei Province, in 1965, on rice fields where *A. sinicus* was grown for the first time. Statistics of pot and field experiments, carried out by various institutions showed that, similar to the cited experiments artificial inoculation of *A. sinicus* rhizobia was almost invariably necessary for *A. sinicus* grown for the first time in rice field soils. On the other hand, on fields where *A. sinicus* was carried out in farmers' fields, the vegetative growth of *A. sinicus* can only be estimated by pooling samples taken at random on a plot. The standard deviations of mean values of replicated plots so obtained are high. Table 9 presents data of an experiment, in which 28 percent increase of mean value of four replicated plots inoculated by an efficient strain of *A. sinicus* rhizobia over the mean value of uninoculated control plot is statistically not significant

Table 8. Effect of rhizobia inoculation on fresh weight (tops only) of *A. sinicus* grown for first time on fields at Jingmen County, Hubei Province, 1965 (Hu and Li, unpublished)

Experiment	Treatment	MT per hectare[a]	% increment
I	Uninoculated	10.50	
	Inoculated	23.25	179**
II	Uninoculated	21.00	
	Inoculated	37.50	79*

[a] Mean of 4 replicated plots arranged randomly

Table 9. Field experiment on the effect of inoculation of
A. sinicus rhizobia strains on field where *A. sinicus* was
grown successfully in previous years at Qianjiang, Hubei,
1975–1976 (Li et al., unpublished)

Treatment	Fresh wt., MT per hectare[a]	S.D.
Unino. contral	23.92	± 1.995
Ino. st. A16	21.43	± 2.487
Ino. st. A1105	25.99	± 1.429
Ino. st. A1106	30.62	± 2.487
Ino. st. A1107	25.20	± 3.679
Ino. st. A1108	25.90	± 3.506

[a] Mean value of 4 replicated plots for each treatment

Table 10. Survey of the effect of artificial inoculation on *A. sinicus* of 52 fields in 12
counties in Hubei Province, 1966 (Li, Hu and Zhou, unpublished)

Uninoculated		Inoculated	
Fresh wt., MT per hectare	Fields	Fresh wt., MT per hectare	Fields
A. sinicus grown for the first time			
Below 7.5	3		
7.5–15	3	Below 15	3
7.5–22.5	3	15–30	8
		Over 30	10
A. sinicus grown previously			
Below 7.5	1		
7.5–15	3	Below 15	0
7.5–22.5	5	15–30	2
		Over 30	11

at 5 percent probability level. Table 10 presents a field survey of fifty two fields
in twelve counties in Hubei Province for the effect of artificial inoculation of
A. sinicus rhizobia in fields where *A. sinicus* were grown for the first time or
for some years previously. Although factors beside artificial inoculation affect
the yield and nitrogen content of *A. sinicus*, it is justified to conclude that in
terms of practical farming, artificial inoculation is necessary to rice fields new
to *A. sinicus* as well as to fields where *A. sinicus* has been grown previously.

12 Agronomic Considerations

Proper agronomic management is demanded for high yield of the crop and high nitrogen-fixing activity of *A. sinicus* rhizobia symbiosis. *A. sinicus* cultivar, growth season, water-air regime, phosphorus supply and nodulation by efficient rhizobia are the predominant factors.

There are two major types of *A. sinicus* cultivars. The large-leaflet-type cultivars grow more vigorously but blossom two or three weeks later than the small-leaflet-type cultivars. Gross yield and succulence considered, full blossom stage is the most suitable time for plowing under the green matter for manuring. Choice of cultivar type is determined by the time demanded for plowing under the green matter, which, in turn is determined by the time of transplanting rice seedlings.

The *A. sinicus* crop has two active growth periods before and after the severe winter. In the winter, when the daytime temperature falls below 14 °C, there is little or no aerial vegetative growth. The length of the dormant period varies with locality and winter severity of the year.

Water or air stress retards the development of *A. sinicus*. *A. sinicus*-rhizobia symbiosis suffers more than the host plant. In agricultural practice, in the diked rice fields, impeded drainage often causes soil air stress, especially during the wet spring season. In fields containing enough native rhizobia or properly inoculated, soil air stress causes the formation of small nodules containing little bacteroidal tissue which is colourless or only very lightly pink coloured, due to suppressed development of leghemoglobin. Moreover, the life span of the nodules is short and decays early. Shallow ditches dug at the periphery of and across the rice field is practised to improve the soil air regime. On the other hand, water stress is often met during the autumn drought. Light irrigation is practised to relieve water stress. Table 11 presents data of a pot experiment on the effect of the water-air regime on the development of *A. sinicus* and nodulation. Seeds were inoculated and sown to pots containing heavy loam. At the time of sowing, the water content was held at the water-holding capacity of the

Table 11. Pot experiment on the effect of the water-air regime on *A. sinicus* nodulation (Cao, unpublished, 1965)

Water-air regime (% water holding capacity)	Average value per plant		
	Leaves	Plants nodulated (%)	Nodules
100% down to 40%	3.1	0	0.0
100% down, kept at 60%	3.3	69	2.7
100% down, kept at 80%	3.7	97	9.6
Kept at 100%	3.7	80	3.4

Note: Soil (heavy loam) pot experiment (4 replicates), seeded and inoculated, grown for 24 days.

Table 12. Effect of superphosphate and inoculation on the yield of *A. sinicus*

Treatment	Green matter MT per hectare
Uninoculated, P unfertilised	4.58
Inoculated, P unfertilised	6.08
Uninoculated, P fertilised	8.61
Inoculated, P fertilised	12.22

Note: Field experiment carried out by the commune farming team of Macheng, Hubei, 1965. *A. sinicus* grown for the first time. 300 kg of superphosphate applied per hectare. Yield estimated in late March at the stage before blossoming, under the supervision of members of the Laboratory of Soil Microbiology, Huacung Agriculture College

soil. In one treatment, the water content was left to dry naturally down to 40 percent of its water-holding capacity. In other treatments, the water contents were kept at 60, 80 or 100 percent of their water-holding capacity. At the lowest water content level, the plant growth suffered slightly, but no nodule was formed at all. On the other extreme, in pots in which the water content was held at the soil's water-holding capacity, plant growth did not suffer but the number of nodules formed was very much reduced.

Except soils developed on purple shale soils, which are rich in total and available phosphorus, most soils in the Yangtze Valley are developed on weathering materials of granite, schist, tertiary, quarternary or recent deposites, which are poor in total and available phosphorus. *A. sinicus*, similar to other legumes, is very responsive to phosphorus fertiliser. Table 12 presents data of a field experiment carried out by a farming team on their own communal field, at Macheng, Hubei. The combined effect of inoculation and phosphorus fertiliser was much stronger than that when they were treated separately.

13 References

Allen ON, Allen E (1981) The leguminosae. A source book of characteristics, uses and nodulation. University of Wisconsin Press, Madison, USA

Brockwell J (1968) Accuracy of a plant-infection technique for counting populations of *Rhizobium trifolii*. Appl Microbiol 11: 377

Cao YZ, Zhou JC, Chen HK (1984) Differentiation and viability of nodule bacteria in host cells. Scientia Sinica (Series B) 27: 583

Chan CL, Lumpkin TA, Root CS (1988) Characterisation of *Bradyrhizobium* sp. (*Astragalus sinicus* L.) using serological agglutination, intrinsic antibiotic resistance, plasmid visualization and field performance Plant Soil 109: 85

Chen HK, Shu MK (1944) Notes on the root nodule bacteria of *Atragalus sinicus* L. Soil Sci 58: 291

Chen HK (ed) (1959) Microbiology. Higher Education Press, Beijing (in Chinese)

Chen WX, Li GS, Qi YL, Wang ET, Yuan HL, Li JL (1990) *Rhizoboum huakui* sp.nov. isolated from the root nodules of *Atragalus sinicus*. Int J Syst Bacteriol (accepted for publication)

Huang LK (1983) Inoculant efficacy of green manuring rhizobia in the traditionally legume grown area (in Chinese). Jiangsu Agr Sci 1983(1):47

Ishizawa S (1954) studies on the root nodule bacteria of leguminous plants II. part 1. Cross-inoculation test (in Japanese, English summary). J Soil Manure (Japan) 24:297

Lin LF, Gong HY, Zhou LM, Cen YH (1988) Study on the plasmid profiles from fast-growing rhizobia (In Chinese, English summary). Acta Microbiol Sinica 26:271

Ning KZ, Li YF, Huang YL (1988) Nodulation behavior of *Astragalus adsurgens* and *Astragalus sinicus* (in Chinese) Microbiol Magazine 8:56, Shengyang, China

Wang CL, Chen JB (1988) The plasmids pattern and endoantibiotic resistence of *Rhizobium astragali* strains (in Chinese, English summary). J of Huazhong Agr Univ 7:15

Wei H and Li FD (1989) Physical evidence for plasmid-borne symbiotic genes in *Rhizobium astragali* (in Chinese, English summary). J of Huazhong Agr Univ 8:121

Wilson JK (1939) Leguminous plants and their associated organisms. Cornell Univ Agric Exp Sta Mem 221

Wilson JK, Chin JH (1947) Symbiotic studies with isolates from nodules of species of *Astragalus*. Soil Sci 63:119

Zhang ZM, Zhou JC, Xu YC, Chen HK, Li FD, Fan YL (1986) Studies on plasmids of *Rhizobium astragali*. Isolation, purification and enzyme-digestion of plasmids DNA from *Rhizobium astragali* (in Chinese, English summary). J of Huazhong Agr Univ 5:326

Zhang ZM, Chen HK, Li FD, Fan YL (1989) Construction of gene library and isolation of *nod* genes of *Rhizobium astragali*. In: Li Shuxuan (ed) Application of biotechnology in agriculture pp 30–32. Shanghai Scientific and Technical Publ

Zhang ZM, Zhou JC, Chen HK (1989) Identification of *hsn* and *nif*HDK genes on pRaZ15 incorporated to *Rhizobium astragali* (in Chinese, English summary). J of Huazhong Agr Univ 8:285

Zhou JZ, Cao YC (1981) Nutrient requirement of the fast growing *Rhizobium* J of Huazhong Agr Univ 3:44 (in Chinese, English summary)

Zhou JC, Tchen YT, Vincent JM (1985) Reproduction capacity of bacteroids in nodules of *Trifolium repens* L and *Glycine max* (L) Merr Planta 163:473

Zhou JC, Zhang ZM, Huang CJ, Li FD, Chen HK (1987) Studies on *Rhizobium* plasmids II. Tests of plasmid elimination of *R. astragali*. J of Huazhong Agr Univ 6:156 (in Chinese, English summary)

CHAPTER 24
Nitrogen Fixation Associated with Rice Plants

C.B. You*, W. Song, and H.X. Wang

---

* To whom correspondence should be addressed

The Nitrogen Fixation
and its Research in China
Editor. Guo-fan Hong
© Springer-Verlag Berlin Heidelberg 1992

1 Introduction

In China, chemical fertilizer and newly bred high yielding rice varieties as well as hybrid rice are being exploited to meet the demands of a rapidly growing population for more food grains. But, the energy requirement, the price of nitrogen fertilizer and the risk of environmental pollution are increasing with each passing day. In this situation, it is found that biological nitrogen fixation is a most important alternative for nitrogen fertilizer input into cropping systems without substantial loss in yield (Döbereiner and Pedrosa 1987).

Since the rediscovery of *Spirillum lipoferum* in 1974 (Döbereiner and Day 1976) the extension of dinitrogen fixation to cereals and forage grasses remains a major challenge in agricultural research. Many novel plant-associated nitrogen-fixing bacteria have been isolated and identified (Boddey and Döbereiner 1988), and research has been directed towards determining whether these bacteria are able to contribute agronomically significant quantities of nitrogen to their respective hosts. It is impossible, of course, to replace conventional nitrogen fertilization completely with biological nitrogen fixation in growing cereals. But if associative nitrogen fixation could be induced to contribute a part of nitrogen to cereal crops, grain yield might become less dependent on industrially produced nitrogen fertilizer. For the developing countries even a little saving on fertilizer would be of interest economically.

The total area of rice fields in China approximates to 26% of her cultivated land, and the output of rice grain accounts for 45% of the total output of food crops (Xu 1981). Therefore the maintenance of nitrogen fertility in rice paddies is very important (Yoshida and Rinaudo 1982). Fortunately, evidence that flooding favours N_2-fixation in soils has been obtained from long-term fertility tests and N balance studies (Watanabe 1986).

Although the amount of dinitrogen fixed in rice fields is still a subject being debated, estimates made by Burns and Hardy (1975) of 30 Kg N ha^{-1} yr^{-1} may be quite reasonable. Total N fixed by lowland rice fields is calculated at 3.2 million t N per year (Watanabe 1985). These figures have attracted considerable interest among researchers. Our laboratory has been involved in investigating rhizosphere nitrogen fixation in rice since the mid-70s, and, much progress has been made.

In this paper, we shall present the results obtained in relation to physiology, biochemistry and genetics of rice root associated diazotrophs with the emphasis on *Alcaligenes faecalis*.

2 Nitrogen-Fixing Bacteria Associated with Rice Roots and Their Population

Rice is grown in flooded conditions in almost all places in China. Flooding the soil creates aerobic or photic and anaerobic or nonphotic environments and favours the maintenance and multiplication of autotrophs and heterotrophs

(Watanabe 1986). A wide range of N_2-fixing bacteria is found in rice soil: *Pseudomonas, Azotomonas, Azotobacter, Beijerinckia, Desulfovibrio, Azospirillum, Enterobacter*, etc., which suggests that heterotrophic N_2 fixation adapts to widely diverse soil environments (Yoshida and Rinaudo 1982).

In China the great majority of bacteria associated with rice roots and rhizosphere soil belongs to the *Enterobacteriaceae, Pseudomonas* and *Alcaligenes* (Qiu et al. 1980; He et al. 1986; Xu et al. 1987; Jia et al. 1989). *Arthrobacter, Azospirillum, Azotobacter, Bacillus, Derxia, Flavobacterium* etc. have also been reported as nitrogen-fixing inhabitants of rice roots (Wu et al. 1985; He et al. 1896; Jia et al. 1989). Some of them have been identified as new species or strains e.g. *Flavobacterium oryzae* sp.nov.M-sm-1612 (Liu et al. 1980; Huang et al. 1982), *Derxia gummosa* var, *Peritricha* 7954 (Wu et al. 1985). Quite different from other reports (Watanabe 1985; Yoshida and Rinaudo 1982), *Alcaligenes faecalis* is widespread in paddy soil of China, and constitutes the predominant diazotrophic strain isolated from wetland rice root (You and Qiu 1982; He et al. 1986; Jia et al. 1989), although the percentage of Enterobacteriaceae is higher in paddy soil (Jia et al. 1989). This nitrogen-fixing activity of *A. faecalis* has not been reported before (Qiu et al. 1981). *A. faecalis* strain A15 is a Gram-negative organism. It is of rod shape with peritrichous flagella containing no lipid bodies, oxidase positive and litmus milk alkaline. It can use a variety of organic acids such as malic, lactic, succinic, benzoic, etc., but not sugar as its source of carbon. The mole percentage $G + C$ of DNA is 62.95 to 63.93 (Qiu et al. 1981a). Its generation time is 2–3 h when supplied with ammonia in its limited concentration and 12–13 h when grown under diazotrophic conditions (You et al. 1983a)

The number of N_2-fixing bacteria in the rhizosphere and/or on the rice root surface varies with the stages of growth. Nitrogen-fixing activities are lower at the initial tillering stage, and then gradually increase, and reach the peak from heading to the ripening stages (Qian et al. 1981). In the tillering stage, they are 2×10^7 N_2-fixing bacteria per gram fresh soil in rice rhizosphere, and 7×10^6 per gram fresh root on the root surface, and are 2×10^8 and 6×10^8 respectively for the heading stage (Jia et al. 1989).

The magnitude of contribution of diazotrophs to the maintenance of nitrogen fertility of paddy soils might be dependent on environmental conditions and has not been clearly determined yet. However, the nitrogen-fixing activity in rice root core is several hundred times as much as in the rice root (Qian et al. 1981).

3 Physiology and Biochemistry

Physiological and biochemical studies on rice root-associated diazotrophs cover energy requirements, nitrogen metabolism, hydrogen metabolism, oxygen protection, chemotaxis, association with plants, nitrogenase properties etc. Among these *Alcaligenes faecalis* has been studied in great detail in our laboratory (You and Wang 1990).

Environmental factors such as carbon sources, fixed nitrogen, etc., directly or indirectly affect nitrogen fixation in rice rhizosphere. Despite the fact that active nitrogenase is irreversibly demaged by O_2 (Gallon 1981) in flooded paddy soils, however, the reduced soil layer, which exists a few millimeters beneath the soil surface, is an anaerobic environment where Eh is predominantly negative, and dissolved oxygen tension (DOT) is rather low, and oxygen has less effect on nitrogen fixation in paddy soils (Watanabe 1985; Zhang et al. 1990).

3.1 Energy Metabolism

Biological nitrogen fixation is an energy demanding process. In optimum conditions, nitrogenase consumes 4 electron pairs and 16 ATP to reduce N_2 to 2 NH_3 concomitant of H_2 (Postgate 1982). The obligatory evolution of H_2 can lead to a significant waste of reducing equivalents and ATP. Most aerobic diazotrophs, nevertheless, possess H_2-uptake hydrogenase, which is capable of recovering and utilizing a portion of the H_2 evolved (Xu 1987).

3.1.1 Carbon Source

Rice root-associated diazotrophs e.g. *A. faecalis, Enterobacter cloacae* and *Klebsiella* spp. utilize organic acids or carbohydrates as the sole carbon source. The photosynthetic host supplies one or more reduced carbon compounds to the bacteria to meet the energy requirement for nitrogen fixation (Ela et al. 1982; Lin and You 1989). *A. faecalis* and *E. cloacae* are associated with rice plant to fix dinitrogen without extra carbon source, but when extra carbon sources are added, the amount of nitrogen fixed increases significantly (Table 1). This complementary effect was observed in the mixed culture of *A. faecalis* and *E. cloacae* either in pure culture or in association with rice plant, with the nitrogen-fixing rate in mixed culture increasing dramatically (Zhou et al. 1981). This might be due to he excretion of lactate by *E. cloacae* during its growth, and the lactate could serve as a carbon source for *A. faecalis* growth and N_2 fix (You et al. 1981).

3.1.2 Metabolism of Poly-β-Hydroxybutyrate (PHB)

PHB is a source of carbon and energy mobilized under conditions of starvation (Dawes and Senior 1973). *A. faecalis* accumulate a large amount of PHB under N_2 fixing conditions, but less than 2% in the NH_4^+-grown cells at the same DOT (Li et al. 1989). The derepression effect of NH_4^+ on PHB accumulation is probably due to the conservation of the carbon source into nitrogenous compounds instead of PHB (Tal and Okon 1985).

3.1.3 Hydrogen Metabolism and Carbon Dioxide Utilization

A. faecalis is chemolithotroph capable of growing on CO_2 plus H_2 as sole energy sources. Ribulose biphosphate carboxylase and H_2-uptake hydrogenase are present in autotrohphically grown cells of *A. faecalis* (Li et al. 1989).

Table 1. Nitrogen fixation of *A. faecalis* and *E. cloacae* associated with rice

Rice inoculated with bacteria			Culture medium			Rice root			Rice leaf		
C source	Seed[1]	Bacteria[2]	$^{15}N\%$ a.e.	Total N (mg)	N fixed (μg)	^{15}N a.e.	Total N (mg)	Uptak. N Fixed (μg)	$^{15}N\%$ a.e.	Total N (mg)	Uptak. N Fixed (μg)
—	—	A-15	0.907	0.50	50	0.037	2.70	11	0.017	11.15	17
—	—	E-26	0.970	0.70	75	0.048	2.80	15	trace	11.50	trace
—	—	A-15+E-26	1.120	0.74	92	0.014	2.85	36	trace	9.80	trace
—	+	A-15	2.205	1.26	310	0.083	3.65	33	0.015	15.05	25
—	+	E-26	2.377	1.42	377	0.030	3.15	10	0.013	16.50	23
—	+	A-15+E-26	2.907	1.44	467	0.092	4.65	47	0.010	21.40	23
Malate	—	A-15	5.529	0.50	300	1.580	0.71	125	0.056	1.93	12
Sucrose	—	E-26	1.591	1.05	187	0.506	0.58	33	0.073	1.83	14
Malate+Sucrose	—	A-15+F-26	5.053	0.53	290	2.701	0.77	232	0.057	1.87	11
Inocul. E-26 +Sucrose	—	A-15+E-26	4.320	1.12	544	1.764	0.42	82	0.197	1.61	35

[1] Seed: "—" without seed; "+" with seed. [2] A-15: *A. faecalis* A-15; E-26: *E. cloacae* E-26
* Abundance of N induced 8.94%. Seedlings without seed treatment were 36 plants, and with seed treatment were 48 plants.
Culture time: 48 h. Seedlings with carbon source treatment were 8 plants. Culture time: 69 h.

Table 2. H$_2$-supported nitrogenase activity of *A. faecalis* A-15 grown with C source and wiithout C and N source

H$_2$(%)	Reaction time (h)			
	11(−)	12(+)	24(+)	36(+)
0	0.000	0.59	2.37	5.78
5	0.003	0.69	2.67	6.37
15	0.190	0.52	1.88	4.31
25	0.240	0.61	2.35	6.06
35	0.230	0.27	2.52	3.88
45		0.21	1.48	2.50

Reaction temp.: 30 °C; Reaction gas: C$_2$H$_2$ 10%; Nitrogenase activity: C$_2$H$_4$ µmol ml^{-1} culture. (+): with C source; (−): without C and N source.

H$_2$-uptake and evolution by diazotrophs involve three enzymes, a H$_2$-uptake hydrogenase, a reversible hydrogenase and nitrogenase (Xu 1987). H$_2$-uptake hydrogenase are present in a number of rice root-associated diazotrophs, in microaerobe *A. faecalis*, and in anaerobes *E. cloacae, K. planticola, Pseudomonas saccharophila* (Li et al. 1989; Yuan et al. 1990). The H$_2$-uptake hydrogenase activity present in N$_2$-grown cells of bacteria indicated above is sufficient to recycle a portion of hydrogen produced by nitrogenase. H$_2$ evolution was observed by direct inhibition of the hydrogenase activity with CO or C$_2$H$_2$ in *E. cloacae* (You et al. 1981). Hydrogenase activity of *E. cloacae* grown anaerobically with NO$_3^-$ tends to increase significantly at the concentration range of 1–3 mM (You et al. 1981). Nitrous oxide and nitrite serve as electron acceptors for hydrogenase of bacterial cells grown anaerobically on nitrate (Tibelius and Knowles 1984). Ni^{2+} is specifically required for synthesis and/or activity of H$_2$-uptake hydrogenase of *A. brasilense, A. lipoferum, D. qummosa* and *A. chroococcum* (Pedrosa and Yates 1983; Partrige and Yates 1982). However, it is not found in the case of *A. faecalis* (Li et al. 1989).

Molecular hydrogen supports the nitrogen-fixing activity of *A. faecalis* either in the presence or in the absence of carbon source in the medium. However, the nitrogen-fixing rate supported by H$_2$ in the absence of carbon source is higher than in its presence (Table 2) (Li et al. 1989). That carbon starvation is essential for H$_2$-dependent acetylene reduction has been reported previously (Walterbury et al. 1983).

3.1.4 Nitrogen Fixing Efficiency

This is an intergrated index of complex parameters which can be used for evaluation of the nitrogen fixation potential of a diazotroph and of its usefulness as a biofertilizer. Several factors may affect the nitrogen-fixing efficiency of diazotroph. The maximum nitrogen-fixing activity of *A. faecalis* goes up to 1800 nmol. C$_2$H$_4$ formed/ml culture per h, or about 40 mg of N assimilated/g malic acid consumed (You and Qiu 1982). The optimal temperature, pH and pO_2 for growth and nitrogen-fixing activity of *A. faecalis* are 30 °C, 7.0 and 0.016 atm., respectively (You et al. 1983a). It should be noted that the effects of these factors on the level of nitrogen-fixing activity vary with the growth stages of bacteria. The higher the level of nitrogen-fixing activity, the weaker are the effects of these factors (You et al. 1983b).

3.2 Nitrogen Metabolism

Dinitrogen fixation and nitrate assimilation are two fundamental biological processes by which most diazotrophs obtain the nitrogen required for their metabolism. There are no reports of diazotrophs capable of carrying out nitrification (Pedrosa 1988).

3.2.1 Nitrogenase

Nitrogenase is an enzyme which mediated the reduction of dinitrogen to ammonia. Nitrogenase consists of two protein components: a MoFe protein (dinitrogenase) and a Fe protein (dinitrogenase reductase). Both are required for activity and are extremely oxygen sensitive. Nitrogenase from *K. pneumoniae*, *A. vinelandii*, *A. chroococcum*, *C. pasteurianum*, *Azospirillum* spp. etc has been purified and well characterized (You 1987). However, the purified Fe protein from *A. brasilense* and *A. lipoferum* requires activation by an activating enzyme similar to that for the photosynthetic diazotrophs *R. rubrum*, *Rhd. capsulata* etc. (Ludden et al. 1988; Song 1987).

Active nitrogenase complex was purified from *A. vinelandii* (Song and You 1987) and was detected in crude extracts of *E. cloacae* (Raju et al. 1972; You et al. 1981). Component proteins were purified from *A. faecalis*, but only MoFe protein was purified to electrophoretic homogeneity.

Nitrogenase activity of MoFe protein is about 1500 nmol. C_2H_4 formed/mg protein/min. The molecular weight of tetrameric MoFe protein was 226 kDa. It was $\alpha_2\beta_2$-type tetramer with subunits of 57 and 60 kDa. The metal content was 2 g atom Mo, 32 g atom Fe. The amino acid residue composition was 1790, and acidic residues were twice as much as the basic ones in the case of *A. vinelandii*, etc. (You et al. 1979; Song et al. 1989). MoFe protein could be oxidized by thionine with 6 redox equivalents, and the oxidation process was divided into two steps with 4 and 2 equivalents respectively, suggesting it contained 2 kinds of metal clusters: 4 P-clusters and 2 M-clusters (Song et al. 1989; Orme-Johnson 1985). The Fe protein of *A. faecalis* did not require activation by activating enzyme and divalent ion e.g. Mn^{2+} as in the case of *A. brasilense* (You et al. 1983a).

3.2.2 Regulation of Nitrogenase Synthesis

NH_4^+, amino acids, nitrate, nitrite and carbamoyl phosphate are known to repress nitrogenase synthesis (Postgate 1982). High concentration of NH_4^+

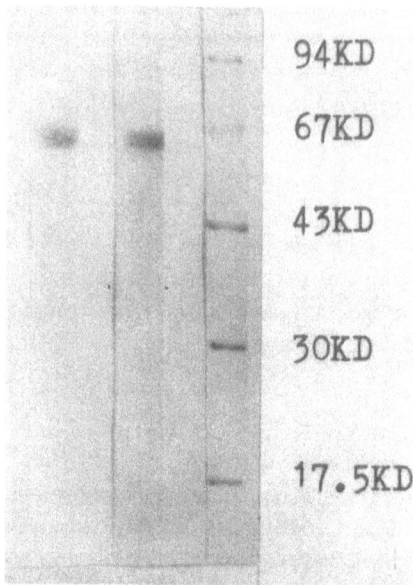

Fig. 1. Photogram of the SDS-PAGE of MoFe protein fron N₂-grown (Afl) and NH₄⁺-grown (Afl) *Alcaligenes faecalis*

effectively represses nitrogenase synthesis, but this possibly does not hold for *A. faecalis*.

By using SDS-PAGE, DEAE-cellulose chromatography, rocket and rocket line immunoelectrophoresis, it was shown that nitrogenase was synthesized by *A. faecalis* in the presence of 30 mM NH_4^+ in the medium, even though nitrogen-fixing activity was absent (You et al. 1988a; Zhang et al. 1989). The MoFe protein from NH_4^+-grown cells of *A. faecalis* also was purified to electrophoretic homogeneity (Fig. 1). The physico-chemical properties such as amino acid composition, molecular weight, redox equivalents, metal contents and UV-visible, fluorescence spectra patterns of MoFe protein from NH_4^+-grown cells were similar to those of N₂-grown cells, but they differed from one another only in circular dichroism (CD) spectra (Fig. 2), suggesting their molecular structures might be different (Song et al. 1989). By aid of antibiotic chloramphenicol, the further biosynthesis of nitrogenase was inhibited in NH_4^+-grown cells. The activity of nitrogenase synthesized during NH_4^+-grown peroid was derepressed when the NH_4^+ was removed from the culture medium (Fig 3) (Zhang et al. 1989).

3.2.3 NH_4^+ Assimilation and Regulation of Nitrogenase Activity

In rice root-associated bacteria such as *K. oxytoca*, *E. cloacae*, *A. faecalis* etc., nitrogenase activity was found to be regulated by NH_4^+ (Paneque et al. 1987).

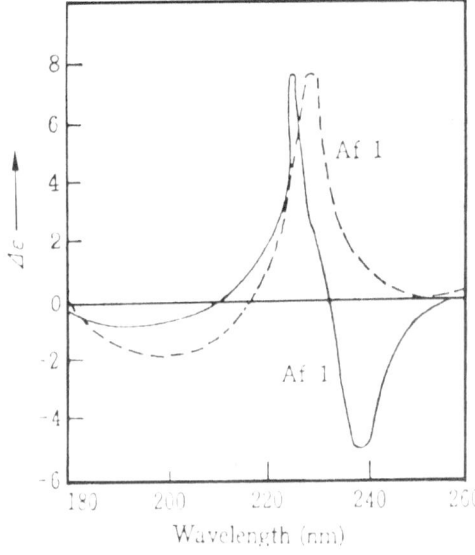

Fig. 2. UV-CD spectrum of MoFe protein from N_2-grown (Af1) and NH_4^+-grown (Af1) *Alcaligenes faecalis*

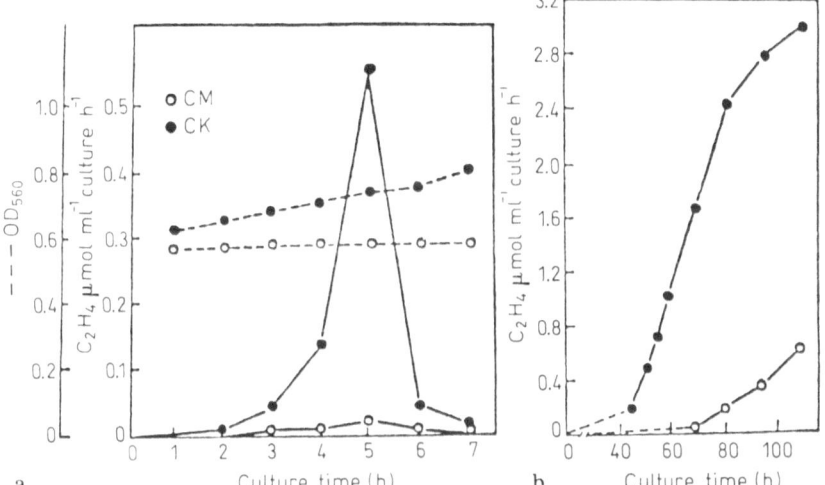

Fig. 3. The nitrogenase activity of *A. faecalis* A-15 in NH_4^+ culture after derepression. **a.** In liquid culture condition. **b.** In semi-solid culture condition. CK: without CM; CM: chloromycetin 100 µg/ml

Addition of ammonium ion to the culture of *A. faecalis* actively fixing dinitrogen caused an immediate inhibition of nitrogenase activity which was recovered after exhaustion of ammonium from the medium (You et al. 1983b). This inhibition seemed to involve the operation of ammonium assimilation system, because the process was prevented or reversed by (L-methionine-DL-sulfoximine) (MSX), an

Table 3. Effect of MSX on C_2H_2 reduction activity of *A. faecalis* A-15 inhibited by NH_4^+

Treatment	nm C_2H_4 formed/ml culture/h					
	10.0	11.5	13.0	15.0	17.0	19.0
3 ml culture + 0.135 mg $(NH_4)_2SO_4$ + 3.4 mg MSX	343.0	476.4	275.0	487.5	350.0	260.3
3 ml culture + 0.135 mg $(NH_4)_2SO_4$ + 6.8 mg MSX	505.3	570.5	300.0	663.3	386.5	310.0
3 ml culture + 0.135 mg $(NH_4)_2SO_4$ + 16.9 mg MSX	660.2	823.3	794.1	921.3	554.7	141.4
3 ml culture + 0.135 mg $(NH_4)_2SO_4$ (CK)	337.7	260.9	243.9	361.2	157.7	141.4
3 ml culture + 16.9 mg MSX (CK)	722.0	1000.0	930.6	1628.8	1247.8	534.4

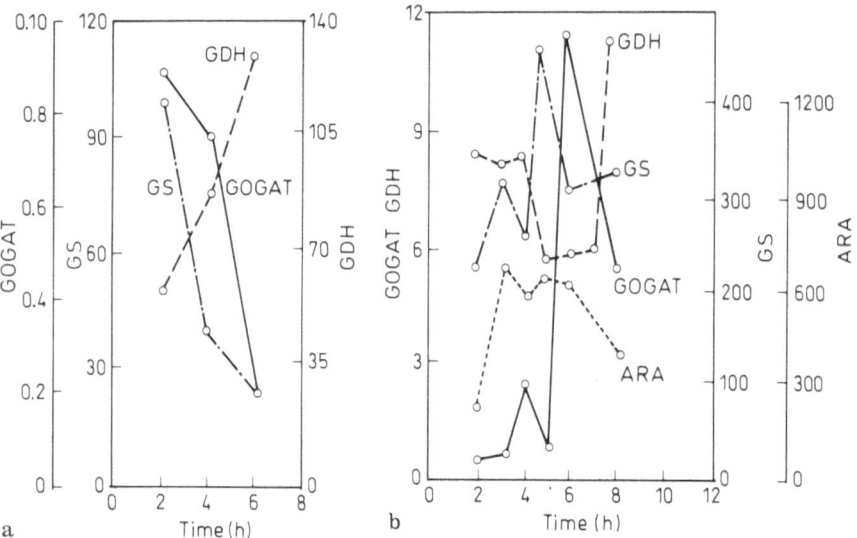

Fig. 4. Variation of GS, GOGAT and GDH activities in NH_4^+-grown (**a**), N_2-grown (**b**) cells f *A. faecalis* A-15. ARA: C_2H_2 µ mol $h^{-1}ml^{-1}$ culture; GS: Pi nmol min^{-1} mg^{-1} protein; GOGAT, GDH: NADPH µ mol min^{-1} mg^{-1} protein

inhibitor or glutamine synthetase (GS) (Table 3) (Rennie et al. 1982; You et al. 1983a).

The enzymes involved in the assimilation of NH_4^+ in bacteria were glutamate dehydrogenase (GHD), GS and glutamate synthase (GOGAT) (Postgate 1982). They were present in *A. brasilense* and *A. lipoferum* (Gauthier and Elmerich 1977), *A. faecalis* (Zhang et al. 1989) and most probably in all new rhizocoenotic diazotrophs. The GS and GOGAT activities in *A. faecalis* were higher in the cell grown in N_2 than in NH_4^+, while the GDH levels were higher in NH_4^+-grown cells (Fig. 4). The results indicate that assimilation of ammonium in nitrogen-fixing cells of *A. faecalis* proceeds via the GS-GOGAT pathway (You et al. 1988a; Zhang et al. 1989).

3.2.4 Nitrate and Nitrite Assimilation

Nitrate-dependent growth requires the presence of assimilatory nitrate and nitrite reductases (Brown et al. 1974). *A. faecalis* is capable of NO_3^--dependent growth under anaerobic condition (Lin and You 1987). However, *A. faecalis* predominantly utilizes NH_4^+ as the nitrogen source, when they grow aerobically in the culture medium containing both ammonium and nitrate. The presence of ammonium ions might delay the development of the nitrate uptake system until ammonium present was taken up by the cells, nitrate being then absolutely required for assimilatory nitrate uptake induction to take place (Paneque et al. 1987). Under anaerobic condition, nitrate serves as a terminal respiratory electron acceptor by *A. faecalis* (Lin and You 1987) and most strains of *A. brasilense* and *A. lipoferum* (Eskew et al. 1977; Scott et al. 1979). Nitrite inhibits the growth rate of *A. faecalis* even in its low concentrations (Lin and You 1987). Nitrite strongly inhibits nitrate assimilation, and the fact that both nitrate and nitrite suppress the biosynthesis of nitrogenase also has been reported (Paneque et al. 1987; Zimmer et al. 1987).

3.2.5 Denitrification

Under oxygen-limitation nitrate is converted to nitrite and/or gaseous NO and N_2 by *A. faecalis* (Lin and You 1987), *K. pneumoniae* and *E. aeroqenes* (Ji and Hollocher 1988). Denitrification by these diazotrophs may result in a loss of nitrate nitrogen normally available for plant growth. Thus, they may have an unfavorable effect in associating with rice when the nitrate content is high and the DOT is limited in paddy soils.

3.2.6 Nitrate-Dependent Nitrogenase Activity

A. faecalis and *E. cloacae* showed NO_3^--dependent nitrogenase activity under severe oxygen limitation (Lin and You 1987; et al. 1981), but nitrite inhibits this activity under the same conditions. Contradictory results were obtained by Nelson and Knowles (1978) and Neyra et al. (1977) and Scott et al. (1979) in azospirilla. It is doubtful whether this activity has any significance for diazotrophs growth and anaerobic nitrogen fixation, because the activity was much lower than that observed under aerobic and microaerobic conditions (Döbereiner and Pedrosa 1987).

3.3 Association with Rice Plants

Associations of azospirilla and other diazotrophs with cereals and forage grass have been described by many authors (Boddey and Döbereiner 1988; Dart 1986; Okon 1985), but the nature of these associations and the contribution of each partner have so far not been ascertained.

3.3.1 Chemotaxis of Bacteria Towards Rice Roots

Chemotactic reactions enable bacteria to carry out directed movement in a chemical gradient. In rhizosphere root effects are primarily due to the delivery of carbon substrates for microbial growth, and therefore a gradient effect should be expected that diminishes with the distance from the root. Also pH, CO_2 and nutrient gradient can be expected to occur under the influence of root system (Marschner 1985). Nutrient gradients provide advantageous conditions for the chemotaxis of bacteria (Pigram and Willians 1976). It has been found that they tend to attract *A. faecalis* towards the rice root. The capillary assay determining the mobility of bacteria shows chemotaxis of *A. faecalis* in response to root extract and exudate (Fig. 5). Only a few bacteria are chemotactic to some organic acids and could be used as a carbon source for their growth (Qui et al. 1984). After a short-term incubation with *A. faecalis*, the bacteria were absorbed and randomly accumulated on the root surface to form a layer ranging from 50–100 μm (Fig. 6). In a longer incubation, the bacteria colonized rice root epidermal surface with mucilaginous sheath (mucigel) (Fig. 7) (You et al. 1983b). However, when ammonium was added to the culture medium, the bacteria would be removed from the root surface, as in the case of *Azospirillum* (Yoshida and Yoneyama

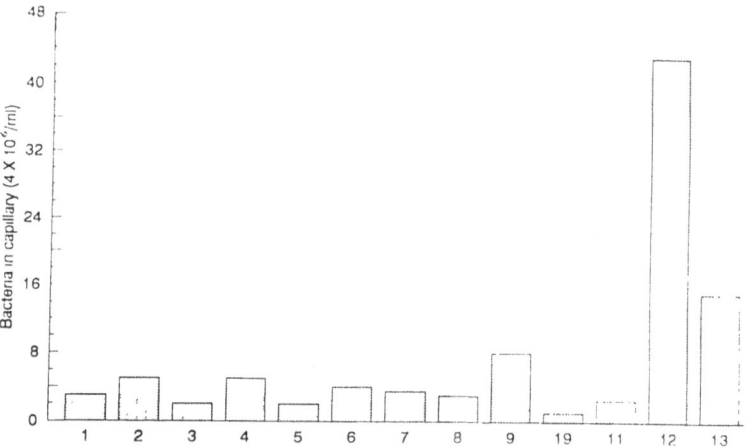

Fig. 5. Chemotaxis of *A. faecalis* A-15 to several materials 1. lactate 2. malate 3. succinate 4. oxalate 5. citrate 6. α-ketoglutarate 7. aconitate 8. fumarate 9. tartrate 10. benzoic acid 11. salicylic acid 12. extract of rice root 13. exudatant of rice root

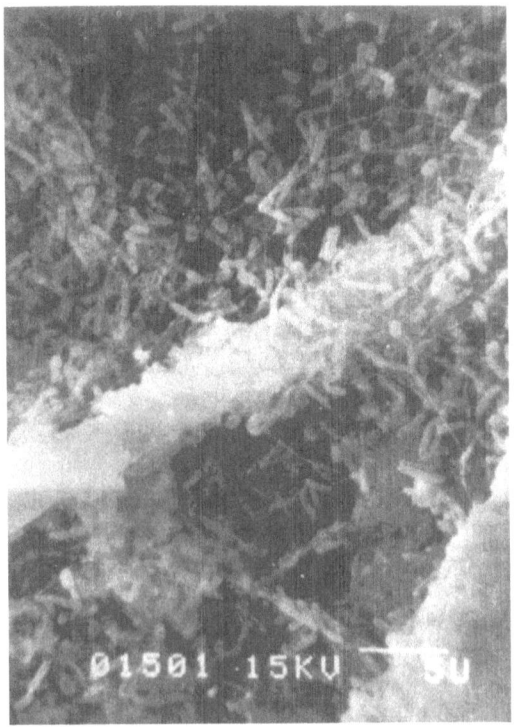

Fig. 6. SEM photograph of *A. faecalis* A-15 on the surface of rice root

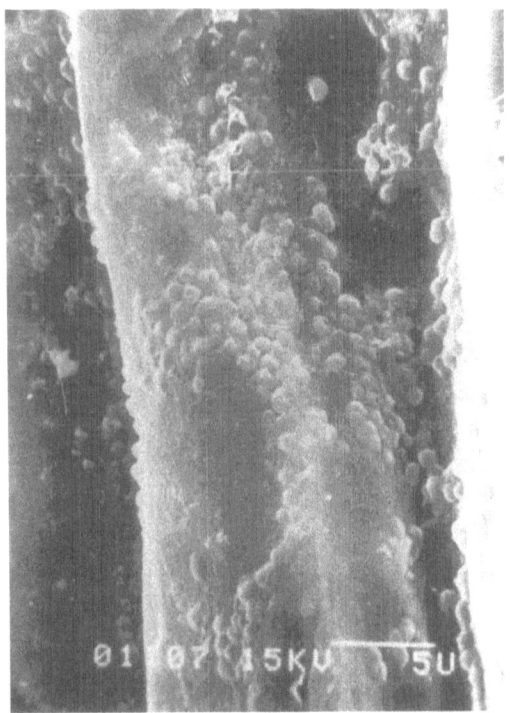

Fig. 7. SEM photograph of *A. faecalis* A-15 after 72 h incubation with rice root, the bacteria colonized root epidermal surface with mucigel

1981). When ammonium concentration reached up to 60 ppm in the medium, only a few bacteria were absorbed on the rice root surface (Zhang et al. 1984).

3.3.2 Interactions Between *A. faecalis* and Rice Plants

Using ^{15}N tracer assays, it was found that the associative N_2-fixing activity (ARA) of *A. faecalis* with rice was rather high. ARA of the associative system could reach 3140 nmol. C_2H_4/g dry root/h, and that *A. faecalis* excreted ammonium into the medium in the late log phase of its growth on N_2, and about $20 \sim 30\%$ of the total amount of nitrogen fixed by bacteria was excreted (You et al. 1983b). About 1/3 of nitrogen fixed by *A. faecalis* could translocate rapidly into the root and the leaves in 69 h (Table 1). *Enterobacter cloacae* produce lactate, a peptide-containing compound and gas mixture during its growth. The gas mixture consists of hydrogen, carbon dioxide, ammonia and other unknown gases (You et al. 1981). More recently, the exudates of rice root and *A. faecalis* have been estimated. About 9 kinds of amino acids and two kinds of phytohormones namely IAA (indoacetic acid) and GA, (gibberellin), and malic acid, but no carbohydrates were found in the pure culture of *A. faecalis* (Table 4). Rice root exudates consisted of 2.25 mg of organic acids, 1.27 mg carbohydrates and 2.61 mg of amino acids per gram fresh root. In the exudate the organic acids from high to low were in the order of citric acid, malic acid, succinic acid and lactic acid. Among the 15 kinds of amino acids the contents of the basic ones were the highest. The carbohydrates consisted of glucose, fructose and sucrose (Tables 5, 6). As to phytohormones only gibberellin-GA_3 was found in the noninoculated rice root exudates. The components and contents of rice root exudates were dependent on the variety of rice. Inoculation with *A. faecalis* stimulated the excretion of rice root and affected the components and contents of root exudates (Table 7). By using $^{14}CO_2$ tracer, 2.63% of photosynthate in noninoculated rice was excreted, while 3.52% was released from inoculated rice. Most of them were uptaken by *A.*

Table 4. Composition analysis of exudate from *A. faecalis*

Phytohormones (μg/ml)		Organic acids (μg/ml)		Amino acids (μg/ml)	
IAA	0.0097	Malic acid	2.152	Glx	0.80
				Ile	0.12
				Ala	0.98
GA_3	0.0315	Citric acid	0.000	Tyr	0.22
				Phe	0.17
				Leu	0.22
ABA	0.0000	Succinic acid	0.000	His	0.24
				Lys	0.86
				Arg	0.44

Table 5. Contents of organic acids and sugars in the culture media of rice non-inoculated and inoculated with *A. faecalis*

Rice Varieties		Organic acids (mg/g fresh root)				Carbon hydrates (mg/g fresh root)		
		Lactic acid	Succinic acid	Citric acid	Malic acid	Fructose	Glucose	Sucrose
Yuefu (*japonica*)	+	0.31	0.00	0.05	0.31	5.56	0.00	9.03
	−	0.00	0.72	0.67	0.71	5.85	0.00	9.14
Jingbai (*japonica*)	+	0.78	0.00	0.22	0.00	2.97	0.00	0.00
	−	0.46	1.19	1.83	1.00	3.39	1.01	9.83
Qinai (upland)	+	0.99	0.19	0.00	0.38	2.53	1.26	0.00
	−	0.53	0.00	0.64	0.44	8.13	1.01	1.01
Rongjing 1 (*indica*)	+	0.12	0.00	0.10	0.84	2.71	0.34	0.00
	−	0.28	0.00	0.38	0.17	9.26	0.00	2.14

+: inoculated with *A. faecalis*. −: non-inoculated

Table 6. Contents of amino acids (mg/g fresh root) in the culture media of rice varieties non-inoculated and inoculated with *A. faecalis*

Amino Acids	Rice Varieties							
	Yuefu		Jingbai		Qinai		Rongjing No. 1	
	+	−	+	−	+	−	+	−
Asx	0.032	0.067	0.051	0.092	0.016	0.073	0.000	0.057
Thr	0.013	0.062	0.065	0.043	0.113	0.052	0.000	0.000
Ser	0.013	0.052	0.015	0.043	0.073	0.077	0.000	0.000
Glx	0.026	0.084	0.014	0.068	0.100	0.105	0.000	0.042
Gly	0.004	0.052	0.015	0.081	0.050	0.056	0.000	0.076
Ala	0.008	0.045	0.015	0.080	0.043	0.063	0.000	0.066
Val	0.000	0.029	0.000	0.158	0.014	0.063	0.000	0.000
Met	0.000	0.000	0.000	0.054	0.000	0.000	0.000	0.000
ILe	0.085	0.070	0.120	0.057	0.109	0.094	0.136	0.243
Leu	0.139	0.063	0.207	0.174	0.161	0.203	0.127	0.201
Phe	0.147	0.703	0.181	0.147	0.167	0.167	0.142	0.228
Tyr	0.190	0.101	0.161	0.163	0.194	0.111	0.092	0.214
His	0.160	0.465	0.313	0.596	0.193	0.815	0.111	0.318
Lys	0.157	0.299	0.250	0.392	0.237	0.490	0.243	0.238
Arg	0.226	0.310	0.142	0.547	0.126	0.354	0.151	0.358

+: inoculated rice.
−: non-inoculated rice.

Table 7. Effect of *A. faecalis* on root exudation of rice seedlings

Rice types		Root exudation (dpm/g dry root)	
		24 hours	48 hours
Japonica rice	+	1.65×10^8	5.73×10^5
	−	3.58×10^6	2.42×10^5
Indica rice	+	1.66×10^6	5.77×10^5
	−	3.82×10^5	1.94×10^5
Upland rice	+	2.15×10^6	4.05×10^5
	−	5.96×10^5	2.96×10^5

Note: +: Inoculated   −: Non-inoculated

Table 8. Effect of rice root exudation on the nitrogen-fixing activity of *A. faecalis*

System of rice association N_2 fixers	Distribution of ^{14}C in plant		Root exudation (dpm/g dry root)	Nitrogenase activity (nmol C_2H_4/g dry root h)
	leaf (dpm/g dry leaf)	root (dpm/g dry leaf)		
I	6.30×10^5	1.29×10^4	1.95×10^5	1861.2
II	5.90×10^5	5.90×10^3	1.33×10^5	610.9
III	6.33×10^5	7.00×10^3	1.05×10^5	252.4

Note: I, II, III – The different rice plants of Jingbai variety.

faecalis in rhizosphere. The photosynthetic rate and excretory ability affected the nitrogenase activity in the rhizosphere. The higher the nitrogenase activity in the rhizosphere of the rice the greater the amount of root exudates was (Table 8) (Lin and You 1989).

3.3.3 Non-Nodular Endorhizospheric Nitrogen Fixation

A. faecalis not only accumulate on the rice root surface, but also could enter into the root itself. The bacteria were visible in the intercellular space of epidermal cells, cortical parenchyma and inside the root cells, when it was examined by microscopy after inoculation. A direct immunofluorescence assay also confirmed the presence of these bacteria inside the root cells, while FA-stained bacteria were not obeserved in the noninoculated control (Zhou and You 1988).

Additional evidence for bacteria endorhizocoenosis was obtained with ^{10}B α-tracking technique (You et al. 1985). The longitudinal sections (Fig. 8) of rice root inoculated with ^{10}B-labeled bacteria show that there were more tracks

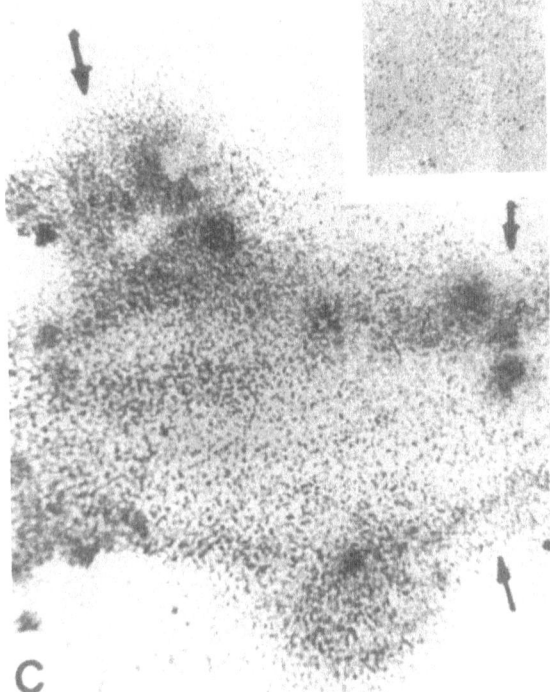

Fig. 8. ^{10}B α-Track of rice root inoculated with ^{10}B-labeled *A. faecalis* (longitudinal section). 150x. *INSERT*: Non-inoculated control, but the rice plants were grown in the normal culture medium containing H_3 $^{10}BO_3$

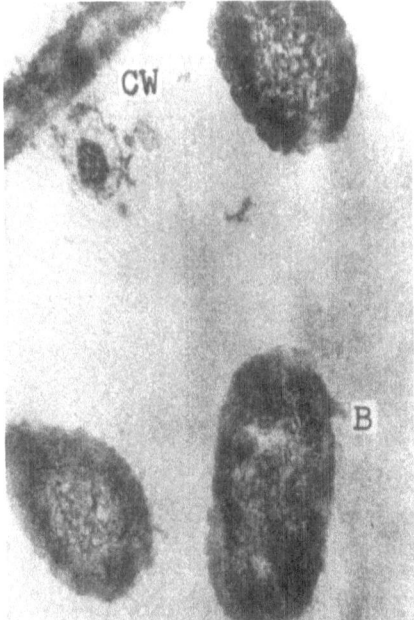

Fig. 9. Electron micrograph of ultrathin section of rice root inoculated with *A. faecalis* (40 000 X). CW, cell wall; B, bacterium

than in the control (Fig. 8 insert), and the tracks were concentrated on the surface of rice root. About 10% of the tracks was distributed inside the roots, and the density of tracks increased from inside the root to the outside.

An electronmicrograph of an ultrathin section of rice root inoculated with *A. faecalis* showed that some bacteria could be seen to have invaded the apparently intact root cells (Fig. 9).

Protoplasts isolated from rice root and callus which had been inoculated with *A. faecalis* were examined. The micrographs showed no bacteria on the protoplast surface (Fig. 10 A). However, many bacteria could be seen inside the protoplast broken by osmotic shock (Fig. 10 B), and in the ultrathin section of the protoplast (Fig. 10 C).

A. faecalis could live symbiotically with callus induced from rice root cell on N-free MS culture medium. This symbiont could fix N_2 (Table 9). *A. faecalis*

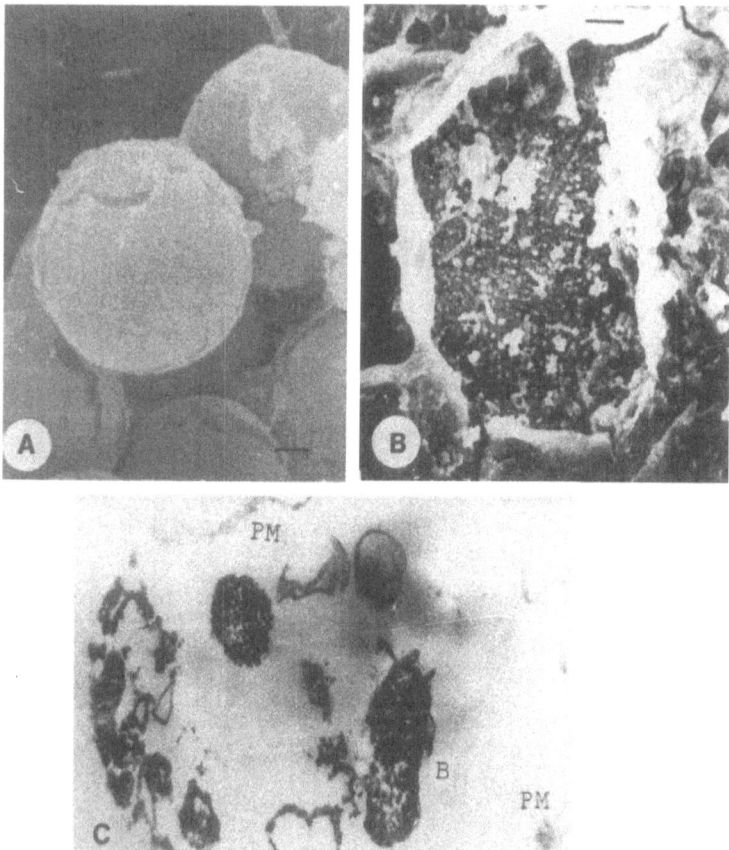

Fig. 10. Scanning and transmission electron micrographs of protoplasts prepared from rice root inoculated with *A. faecalis*. **A** Protoplast preparation (SEM; bar, 5 μm). No bacteria are visible on the surface. **B** Protoplasts broken by osmotic shock showing bacteria inside (SEM; bar 5 μm). **C** Ultrathin section of protoplasts from rice root callus (TEM: 16 000 x). PM, protoplast membrane; B, bacterium

Table 9. $^{15}N_2$ fixing rate by the symbiotic association between rice root callus and *A. faecalis*

Rice variety	Sample no.	Callus, g dry wt.	Total N mg	$^{15}N_2$ at. % excess	N fixed µg N/g dry wt. per day
		Non-inoculated calli			
Chunghua no. 1	1	0.098	3.12	0	0
	2	0.028	0.51	0	0
	3	0.039	0.62	0	0
Yufu	1	0.125	2.15	0	0
	2	0.045	1.13	0	0
	3	0.045	0.85	0	0
		Calli inoculated with *A. faecalis*			
Chunghua no. 1	1	0.045	1.12	2.09	18.94
	2	0.040	1.13	1.76	18.47
	3	0.044	1.32	1.79	19.83
Yufu	1	0.030	0.55	4.63	30.49
	2	0.052	1.17	—	—
	3	0.050	1.09	4.40	35.53

Note: The calli were grown in Erlenmeyer flasks on nitrogen-free MS medium. The cotton stopper were replaced by serum stoppers when the calli were grown for 24 h. Before this, calli were surface washed with water under strict aseptic conditions. The flasks were evacuated and filled with a gas mixture containing $^{15}N_2$ (abundance, approx. 99%)

could grow and multiply in the cell, as indicated by the fact that (i) bacteria were visible in all callus cells which were grown from a single cell inoculated with this bacterium, and (ii) *A. faecalis* could produce alkali during their growth and the pH value of the culture medium rose along with the biomass or OD_{560}, the presence of bromothymol blue in the medium could indicate increases in pH by changing the color from white to blue (You et al. 1983a).

As has been previously discovered (Qiu et al. 1981a), *A. faecalis* uses organic acids as sole carbon sources for their growth, but they utilize sugar poorly. *A. faecalis* cannot grow in the MS medium, because it contains only sucrose as the sole carbon source. On the other hand, the callus induced from rice root cell cannot grow in nitrogen-free MS medium. However, the bacteria-callus symbiotic association grew well in the MS medium, suggesting that the callus cells supported the bacteria with a carbon source for fixing nitrogen, and the bacteria provided the callus with fixed nitrogen for its growth. Hence, rice is thought to be in a very close association with *A. faecalis* (You and Zhou 1989).

4 Genetics

Genetic studies provide information to help understand how organisms perform their physiological and biochemical functions. *Klebsiella pneumoniae* is the best studied diazotroph in terms of *nif* genetics and is taken as the basic model for genetic studies on other diazotrophs. Most genetic studies on associative diazotrophs are on *A. brasilense* (Elmerich et al. 1988). Only a few results on the genetics of nitrogen fixation from rice root-associated diazotrophs, e.g. *E. cloacae*

(Zhu et al. 1986), *K. oxytoca* NG 13 (Uozumi et al. 1986), *A. faecalis* (Wang et al. 1989), have been reported.

4.1 Localization of *nif* Genes

By using a modified Casse method, the plasmids from several species of rice root-associated diazotrophs, including *A. faecalis* A15, *E. cloacae* strains E26 and EnSn, *K. planticola* DWUL2, *K. oxytoca* NG13 and *P. saccharaphila*, have been isolated and detected. Except *K. oxytoca* NG13, all bacteria described above were found to harbour plasmids. Among them *A. faecalis*, *E. cloacae* E26 and EnSn and *K. planticola* contained one or two large plasmids ranging from 30 to 200 MDa (Wang et al. 1989).

 E. cloacae nif fragments, covering the *nif*B-Y genes, and *R. leguminosarum nif* HD genes were used as probing for the presence of homologous genes in total DNA of *A. faecalis* A15, *E. cloacae* E26 and EnSn, *K. planticola* DWUL2, *K. oxytoca* NG13, and *P. saccharaphila*. ^{32}P-labeled plasmid pBG1 carrying the *nif*HD genes and plasmid pST1140 carrying *E. cloacae nif*B-Y genes, hybridized to the total DNA from bacteria described above (Wang et al. 1989). The size of *nif*-homologous *Eco*RI, *Hin*dIII and *Sal*I DNA fragments from various diazotrophs are shown in Table 10. The results implied that the *nif*HD and *nif* B-Y genes on the NH_4^+-grown cells of *A. faecalis* were comparable with those in the N_2-grown cells, despite the absence of nitrogenase activity.

4.2 Identification of *hup* Gene

The roles of uptake hydrogenase have been discussed in Sect. 3. Using plasmid pHVT116 containing *R. japonicum hup* gene as probe, homology between the

Table 10. Restriction enzyme analysis of *nif* genes of several associative nitrogen fixers by Southern hybridization

Enzyme	Strains	*nif*B–Y probe	*nif*DH probe
EcoR I	*A. faecalis* A15 *A. faecalis* A15 (NH_4^+-grown)	6.3*, 3.2	13.2
Hind III	*K. planticola* DWUL2	12.9, 5.9, 4.4	11.4, 5.8
Sal I	*A. faecalis* A15 *A. faecalis* A15 (NH_4^+-grown) *E. cloacae* EnSn *K. planticola* DWUL2 *P. saccharophila*	9.0, 3.0, 2.7 5.1, 2.9 9.0, 5.9, 4.3, 3.3	7.4, 5.6 4.5 11.0, 5.2, 3.9 2.8, 2.3

* fragments are indicated in kb

probe and restricted DNA of *A. faecalis* A15, *E. cloacae* EnSn, *K. planticola* DWUL2 and *P. saccharaphila* was found. *hup* gene of *A. faecalis* was located on chromosome. However, it was located on the large plasmid for *E. cloacae* EnSn. *nif* genes and *hup* gene are located on the same replicon (Yuan et al. 1990).

4.3 Construction of Genomic Library of *A. faecalis*

A genomic library of *A. faecalis* strain A15H1 which possesses rather high nitrogenase activity was constructed. The total DNA of *A. faecalis* A15H1 was partially digested with Sau3AI. Fragements in the range of 13–20 kb recovered from agarose gel were cloned in bacteriophage EMBL 4 vector. A total number of 1.2×10^6 recombinants was obtained. By using *nif*DHK and *nif*A genes of *K. pneumoniae* from plasmids pSA30 and pMC71A as probes, the clones containing consensus sequences were obtained (Fig. 11). A 3.5 kb fragment containing *nif*H gene was subcloned in plasmid pUC19, and a new plasmid pAFH1 was constructed. (Hai et al. 1990).

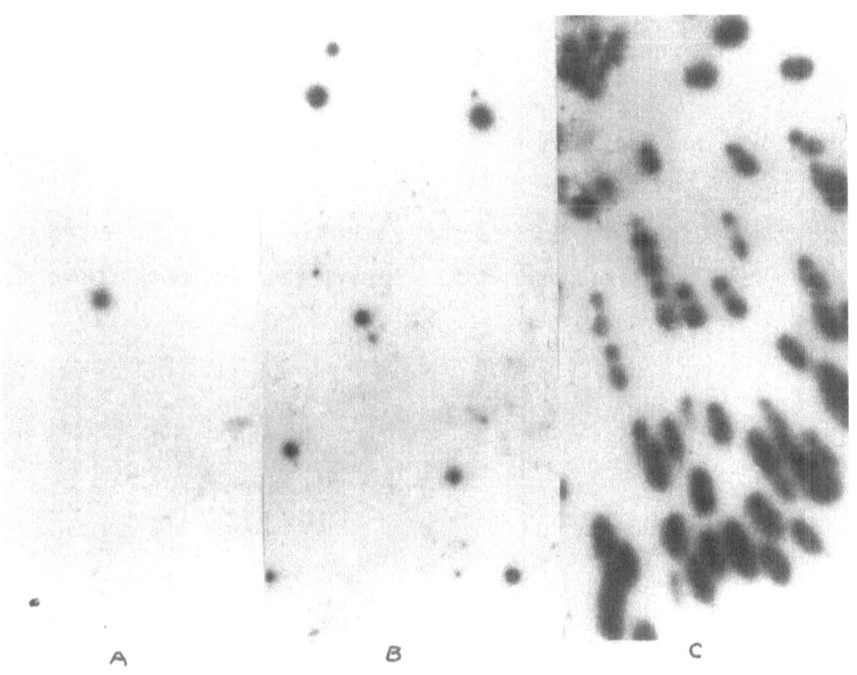

Fig. 11 A–C. Screening of *nif* HDK gene probe: ^{32}P-labelled *nif* HDK DNA of *K. pneumoniae*. **A.** The first round of screening. **B.** The second round of screening. **C.** The third round of screening

4.4 Transfer and Expression of Cloned *nif*A in *A. faecalis*

It is well known that ammonium effectively represses nitrogen-fixing activity (Postgate 1982). The *nif*A product, however, is an activator which is essential for the full expression of each *nif* promoter except that of its own operon (Dixon et al. 1980), and provides positive regulation of *nif* genes in the presence of ammonium and other fixed nitrogen (Ausubel 1984).

In order to get an ammonia-resistant strain of *A. faecalis*, tri-parental mating method was used to introduce pCK3 containing *nif*A fragement from *K. pneumoniae* and pCK5 containing *nif*A from *A. chroococcum* into *A. faecalis* A15H1. The resistance of kanamycin was the common genetic marker of pCK3 and pCK5, while the wild type of *A. faecalis* was sensitive to kanamycin. The results showed these aspects when pCK3 was used. They were quite similar in the case of *A. vinelandii* as a recipient. The *nif*A gene product of *K. pneumoniae* and *A. chroococcum* could activate the expression of *nif* genes in *A. faecalis*. The results also showed that the pCK3 and pCK5 plasmids both introduced nitrogenase activity in *A. faecalis* in the high levels of NH_4^+ (20 mM) (Cheng et al. 1990).

4.5 Transfer of *R. leguminosarum* Structural *nif* Genes into *A. vinelandii*

Recently, a heterogeneous *nif* gene recombination was tested to see whether homology existed and was sufficient to ensure a cross-over of nitrogenase structural genes from *Sym* plasmids of *R. leguminosarum* to genome of *A. vinelandii*. A transfer system was established and a Tn5 insertion mutant of *A. vinelandii* was obtained.

Plasmid pRK2509 carrying *R. leguminosarum nif*D::Tn5 was complemented *in trans* for conjugation by pRK2013 and transferred from *E. coli* to *A. vinelandii*. The transfer frequency was 7×10^{-6} per recipient. Transconjugants were detected by conjugation of pPH1J1, which was incompatible with pRK2509, into *A. vinelandii* strain containing the pRK2509 and simultaneous selection for kanamycin resistance (retention of Tn5) and gentamycin resistance (conferred by pPH1J1). After screening and subculturing the exconjugants a mutant strain, *A. vinelandii* strain W1, was obtained.

Site-directed Tn5 mutation applied to a *R. leguminosarum nif*D region was transferred into the wild-type *A. vinelandii* genome by homologous recombination (Fig. 12). Phenotype effects of Tn 5 mutation in the region of the structural *nif* genes were determined by a complementation of nitrogenase component and by immunological analysis of proteins of the mutant extract. The Tn5 mutant in *nif*D region was still able to accumulate a significant amount of active nitrogenase component II (Fe protein), but caused repression of *nif*DK products synthesis (Fig. 13) (Wang et al. 1988).

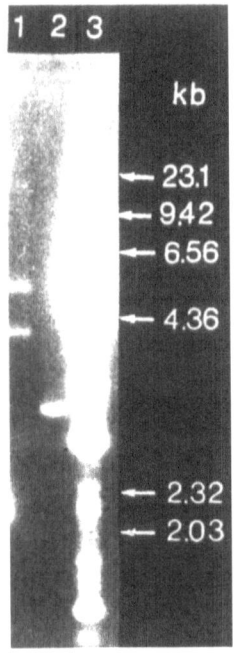

Lane 1: chromosomal DNA from
   A. vinelandii W1;

Lane 2: chromosomal DNA from
   A. vinelandii ATCC 478

Lane 3: plasmid pRK 2509.

On the right are shown *Hind*III
restriction fragments of phage λ
DNA with size in kb.

Fig. 12. Hybridization of ^{32}P-labelled *R. leguminosarum nif* D against chromosomal DNA from *A. vinelandii* digested by *Sal*I

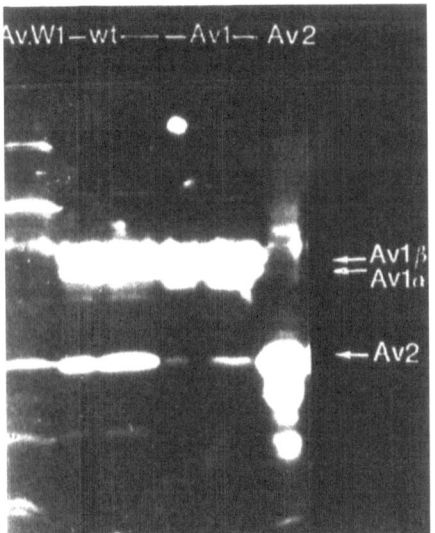

Fig. 13. Immunodetection of nitrogenase components I and II in *A. vinelandii* W1. Autoradiogram of a Western blot with cell-free extracts of mutant W1, wild type and pure nitrogenase components I and II. The position of Av1 α, Av1 β and Av2 bands are marked on the right

5 Agronomic Significance

Associations of diazotrophs with cereals can be exploited for the benefit of agriculture in many ways. The most immediate way seems to be the inoculation of field cereals with diazotrophs, but in many cases environmental factors and plant species affect the associations.

5.1 Rice Genotype Effects

As indicated above, through their exudates, rice varieties play an important role in the effectiveness of associative nitrogen fixation (Lin and You 1989). Five species of diazotrophs, namely *A. faecalis, K. planticola, K. oxytoca, E. cloacae* and *A. brasilense* have been selected as inoculants for eight species of wetland and upland rice subspecies, *indica, japonica* and hybrid rice. The results (Fig. 14) indicate that the different species of diazotrophs may have significant responses, e.g. the highest N_2-fixing activities were observed in the rice-*A. faecalis* associative system and the lowest ones in the rice-*A. brasilense* system. There was a significant variation in nitrogen-fixing activities among the other 3 rice-bacteria systems, suggesting that there might exist a close relationship between associative N_2-fixing bacteria and rice plants (Zhang et al. 1990).

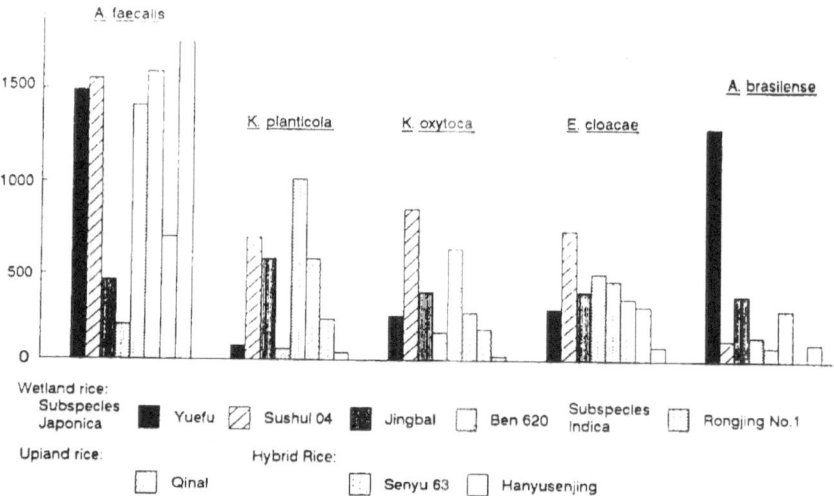

Fig. 14. Effect of rice genotypes on the supporting N_2 fixation reaction time: 72 h. ARA: C_2H_4 n mol./100 mg fresh root

Table 11. Effect of inocula on the yield and nitrogen content or rice (1988)

Inocula	Grain yield (kg/ha)	N content in rice (%)	Total N in rice (kg/ha)	N increase (%)
A. faecalis A15	9192.0	1.079	99.15	6.3
A. brasilense Sp245	8968.5	1.094	98.10	5.1
K. planticola DWUL2	8767.5	1.156	101.40	8.7
K. oxytoca NG13	8887.5	1.082	94.50	1.3
E. cloacae EnSn	8854.5	1.094	97.20	4.2
E. cloacae E26	8904.0	1.106	98.55	5.6
C K	8866.5	1.052	93.30	—

Table 12. Effect of inocula on the $^{15}N_2$-fixing activity in rice rhizosphere (1989)

	Inocula	Yield (kg/ha)	Total (%)	N atom % excess	Ndfa %	N fixed (kg/ha)	Yield of rice grain
	A. faecalis A15	15075.15	0.947	0.243	18.5	26.41	480.84
	E. cloacae E-26/73A	15362.55	0.966	0.257	13.8	20.47	492.50
Shoot part	E. cloacae E-26	14875.05	0.915	0.246	17.5	23.82	467.50
	K. oxytoca NG13/73A	14662.50	0.953	0.242	18.8	26.27	465.00
	K. oxytoca NG13	13937.53	0.855	0.261	12.4	14.78	464.17
	C K	13600.05	0.917	0.298	—	—	456.67
	A. faecalis A15	3101.55	0.743	0.291	5.83	13.43	
	E. cloacae E-26/73A	3364.20	0.823	0.301	2.58	7.14	
Root Part	E. cloacae E-26	3325.05	0.751	0.293	5.18	12.94	
	K. oxytoca NG13/73A	3350.10	0.721	0.303	1.94	4.68	
	K. oxytoca NG13	2575.05	0.753	0.327	7.12	13.81	
	C K	3400.05	0.669	0.309	—	—	

NG13/73A and E26/73A are transformants containing pCK3 plasmid.

5.2 Associated Rice Field Tests

Using ^{15}N dilution technique, dinitrogen-fixing rate of diazotrophs was estimated in field experiments during the rice growing seasons for 2 years. Rice seedlings were inoculated with diazotrophs: *A. faecalis*, *K. planticola*, *K. oxytoca* NG13, *E. cloacae* E26 and *K. oxytoca* NG13/73A and *E. cloacae* E26/73A. The latter two strains were transformants haboring plasmid containing *K. pneumoniae nif*A (Fuji et al. 1987). The results were presented in Tables 11 and 12. They indicate that the total nitrogen content in rice plants increased by up to 8%, total nitrogen fixed was about 20–30 kg/ha, and the grain yield was the same when nitrogen fertilizer was reduced by 1/3 (Zhang et al. 1990).

6 Prospects for the Future

Root-associated biological nitrogen fixation is one of the major source of nitrogen for wetland rice, estimated at 30 kg N/ha/crop, probably around 20%$_0$ of the total plant nitrogen (Boddey and Döbereiner 1984). However, the possible mechanisms of plant responses need to be further elucidated. The significant plant growth promotion by diazotrophs was caused probably not only by the nitrogen fixation (Van Verkum and Bohlool 1980), but also by mineralization of organic phosphorous compounds or solubilization of inorganic phosphorous compounds (Kloepper et al. 1989), or increased uptaking soil and fertilizer nitrogen (Kapulnik et al. 1985).

It has been confirmed that rice is still able to obtain significant N contributions from associative diazotrophs (App et al. 1980; Morris et al. 1985). However, the real benefit the rice could receive from the association will depend on: (1) the quantity of carbon substrate in the rhizosphere available to the diazotrophs; (2) the repression of ammonia and other fixed nitrogen on the nitrogenase activity of diazotrophs and (3) the availability of fixed nitrogen to rice.

It is apparent that research on associative nitrogen fixation should emphasize on two areas: the study of the mechanism of interaction between the rice plant and microorganism, and the study of the bacterial inocula which consist of naturally occurring bacteria and genetically engineered bacteria. Therefore, as a potential nitrogen source for rice cultivation, associative nitrogen fixation deserves serious consideration.

6 References

App AA, Watanabe I, Alexander M, Venture W, Daez C, Santigo T, De-Datta SK (1980) Non-symbiotic nitrogen fixation associated with the rice plant in flooded soil. Soil Sci 130: 283
Ausubel M (1984) Regulation of nitrogen fixation genes. Cell 37: 5

Boddey RM, Döbereiner J (1984) Nitrogen fixation associated with grasses and cereals. In Subba Rao NS (ed) Cur Developm Biol Nitrogen Fixation Oxford & IBH Publ, p 277

Boddey RM, Döbereiner J (1988) Nitrogen fixation associated with grasses and cereals: recent and perspectives for future research. Plant Soil 108: 53

Brown CM, McDonald-Brown DS, Meers JI (1974) Physiological aspects of microbial inorganic nitrogen metabolism. Adv Microbiol Physiol 11: 1

Burns RF, Hardy RWF (1975) Nitrogen Fixation in Bacteria and Higher Plants. Springer-Verlag, Berlin Heidelberg New York

Cheng Q, Hu CZ, Hai WL, You CB, Song W (1990) Transform action and expression of the cloned *nifA* gene of *Klebsiella pneumoniae* and *Azotobacter chroococcum* in *Alcaligenes faecalis*. In: A Treatise on Associative Nitrogen Fixation in Rice Rhizosphere (in press)

Dart PJ (1986) Nitrogen fixation associated with non-legumes in agriculture. Plant Soil 90: 303

Dawes EA, Senior PJ (1973) The role and regulation of energy reserve polymers in microorganisms. Adv Microbiol Physiol 10: 136

Dixon RA, Eady RR, Espin G, Hill S, Iaccarino M, Khan D, Merrick M (1980) Analysis of regulation of *Klebsiella pneumoniae* nitrogen fixation (nif) gene cluster with gene fusion. Nature (London) 276: 416

Döbereiner J, Day JM (1976) Associative symbioses in tropical grasses. In Newton WE, Nyman CJ (eds) Proc Ist Int Symp Nitrogen Fixation Washington State Univ Press, Pullman, p 518

Döbereiner J, Pedrosa FO (1987) Nitrogen-fixing bacteria in non-leguminous crop plants. Sci Tech Publ and Springer-Verlag, Berlin Heidelberg New York London Paris Tokyo

Eady RR (1981) Regulation of nitrogenase activity. In: Gibson AH, Newton WE (eds) Cur perspect nitrogen fixation (eds) Aust Acad Press, Canberra, p 172

Ela SW, Anderson A, Brill WJ (1982) Screening and selection of maize to enhance associative bacterial nitrogen fixation. Plant Physiol 70: 1564

Elmerich C, Galimandn M, Vielle C, Delorme F, De Zamaroczy M (1988) Nitrogen fixation genes of *Azospirillum*. In: Bothe H, de Bruijin FJ, Newton WE (eds) Nitrogen fixation: hundred years after Fischer, Stuttgart, New York, p 327

Eskew DL, Focht DD, Ting IP (1977) Nitrogen fixation, denitrification and pleomorphic growth in highly pigmented *Spirillum lipoferum*. Appl Envirn Microbiol 34: 582

Fuji T, Huang YD, Higashitani A, Nishimura Y, Iyama Y, Hirota Y, Yoneyama T, Dixon DA (1987) Effect of inoculation with *Klebsiella oxytoca* and *Enterobacter cloacae* nitrogen fixation by rice-bacteria associations. Plant Soil 103: 221

Gallon JR (1981) The oxygen sensitivity of nitrogenase: a problem for biochemists and micro-organisms. Trends Internatl Biochem Soc 6: 19

Gauthier D, Elmerich C (1977) Relationship between glutamine synthetase and nitrogenase in *Spirillum lipoferum*. FEMS Microbiol Lett 2: 101

Hai WL, Zheng HG, Wang B, You CB (1990) Construction of a genomic library of *Alcaligenes faecalis* and screening of positive clone containing *nif* genes. In A Treatise on Associative Nitrogen Fixation in Rice Rhizosphere (in press).

He FH, Zhang GQ, Kin SF, Jiang QG (1986) Isolation and identification of associative nitrogen-fixing bacteria in rhizosphere of rice, maize and sugarcane. Microbiol 13: 2

Huang SZ, Tang LF, Zhang WG, Liu CZ (1982) Observation of *Flavobacterium oryzae* sp nov M-sm-1612 in rice roots under electron mictoscope and the characteristics of nitrogen fixation associated with rice. Acta Microbiol Sin 27: 156

Ji XB, Hollocher TC (1988) Reduction of nitrate to nitric oxide by enteric bacteria. Biochem Biophys Res Comm 157: 106

Jia XM, Mo WY, Qian ZS (1989) Species and enumeration of nitrogen-fixing bacteria in rice root systems. Acta Agricul Univ Zhejiangensis 15: 57

Kapulnik Y, Feldman M, Okon Y, Hanis Y (1985) Contribution of nitrogen fixed by *Azospirillum* to the N nutrition of spring wheat in Israel. Biol Biochem 17: 509

Kloepper JW, Lifshitz R, Zablotowicz RM (1989) Free-living bacterial inocula for enhancing crop productivity. Trends Biotech 7: 39

Li X, Zhou FY, You CB (1989) H_2 uptake and carbon dioxide assimilation in *Alcaligenes faecalis*. Acta Phytophysiol Sin 15: 30

Lin M, You CB (1987) Denitrification and nitrogen fixation by *Alcaligenes faecalis*. Acta Agricul Nucl Sin 1: 3

Lin M, You CB (1989) Root exudates of rice (*Oryza sativa* L) and its interaction with *Alcaligenes faecalis*. Sci Agricul Sin 22(6): 6

Liu CZ, Tang LF, Huang SZ, Li JW, You CB (1980) Study on associative symbiotic nitrogen fixation of rice. Fujian Agricul Sci Tech (6): 1

Ludden PW, Roberts GP, Lowery RG, Fitzmaurice WP Saarill L, Lehman L, Lies D, Woehle D, Wirt H, Murrell SA, Pope MR, Kanemoto RH (1988) Regulation of nitrogenase activity by reversible ADP-ribosylation of dinitrogenase reductase. In: Bothe H, de Bruijin FJ, Newton WE (eds) Nitrogen fixation: Hundred years after. Fischer, Stuttgart, New York, p 157

Marschner M (1985) Nährstoffdynamik in der Rhizosphäre. Ber Deutch Bot Ges 98: 291

Morris DR, Zuberer DA, Weaver RW (1985) Nitrogen fixation by intact grass-soil cores using N-15 and acetylene reduction. Soil Biol Biochem 17: 89

Nelson LM, Knowles R (1978) Effect of oxygen and nitrate on nitrogen fixation and denitrification by *Azospirillum brasilense* grown in continuous culture. Can J Microbiol 24: 1395

Neyra CA, van Berkum P (1977) Nitrate reduction and nitrogenase in *Azospirillum lipoferum*. Can J Microbiol 23: 306

Okon Y (1985) *Azospirillum* as a potential inoculant for agriculture. Trends Biotechn 3: 223

Orme-Johnson WH (1985) Molecular basis of biological nitrogen fixation. Ann Rev Biophys Chem 4: 419

Paneque A, Cejudo FJ, Revilla E (1987) Nitrogen metabolism in heterotrophic bacteria. A comparative study of the short-term ammonium inhibition of dinitrogen fixation and nitrate assimilation in *Azotobacter chroococcum*. In: Ullrich WR, Apericio PJ, Syrett PJ, Castille F (eds) Inorg nitrogen metabolism. Springer-Verlag Berlin Heidelberg New York London Paris Tokyo, p 53

Partridge P, Yates MG (1982) Effect of chelating agents on hydrogenase in *Azotobacter chroococcum*.: Evidence that nickel is required for hydrogenase synthesis. Biochem J 204: 330

Pedrosa FO, Yates MG (1983) Effect of chelating agents and nickel ions hydrogenase activity in *Azospirillum brasilense*, *A. lipoferum* and *Derxia gummosa*. FEMS Microbiol Lett 17: 101

Pedrosa FO (1988) Physiology, biochemistry and genetics of *Azospirillum* and other root-associated nitrogen-fixing bacteria. CRC Critical Rev Plant Sci 6: 345

Postgate JR (1982) The fundamentals of nitrogen fixation. Cambridge Univ Press Cambridge London

Prigram NK, Williams FD (1976) Survival value of chemtaxis in mixed cultures. Can J Microbiol 22: 1771

Qian ZS, Mo WY, Chan SM, Jia XM (1981) Nitrogen-fixing activity in the rhizosphere of rice at the different growth stages. J Zhejiang Agricul Univ 7(2): 15

Qui YS, Zhou SP, Mo XZ, You CB, Wang DS (1980) Investigation of N_2-fixation bacteria isolated from rice rhizosphere. J Sci Monthly 25: 383

Qui YS, Zhou SP, Mo XZ, Wang DS, Hong JH (1981a) Study of nitrogen fixing bacteria associated with rice root. I Isolation and identification of organisms. Acta Microbiol Sin 21: 468

Qui YS, Zhou SP, Mo XZ, You CB, Wang DS (1981b) Investigation on nitrogen fixing bacteria in rice rhizosphere. In: Inst Soil Sci Acad Sin (ed) "Proc Sym Paddy Soil" Springer-Verlag Berlin, p 244

Qiu YS, Mo XZ, Zhang YL, Li X, You CB (1984) Some properties of the nitrogen-fixing associative symbiosis of *Alcaligenes faecalis* A-15 with rice plants. In: Veeger C, Newton WE (eds) Adv nitrogen fixation res PUDOC Wangeningen, p 64

Raju PN, Evans HJ, Seidler RJ (1972) An asymbiotic nitrogen-fixing bacterium from the root environment of corn. Proc Natl Acad Sci US 69: 3473

Rennie RJ, Freitas JR, Ruschel AP, Vose PB (1982) Isolation and identification of N_2 fixing bacteria associated with sugar cane (*Saccharum* sp.). Can J Microbiol 28: 462

scott DB, Scott CA, Döbereiner J (1979) Nitrogenase activity and nitrate respiration in *Azospirillum* spp., Arch Microbiol 121: 141

Song HY (1987) Nitrogen fixation of photosynthetic bacteria. In: You CB, Jiang YM, Song HY (eds) Biol nitrogen fixation Ch 13, Academic Press New York, p 246

Song W, You CD (1987) Isolation and purification and some characters of nitrogenase complex from *Azotobactre vinelandii*. Acta Phytophysiol Sin 13: 35

Song W, Zhang FR, You CB (1989) Synthesis and properties of MoFe protein of nitrogenase from N_2-grown and NH_4^+-grown *Alcaligenes faecalis*. Acta Phytophysiol Sin 15: 167

Tal S, Okon Y (1985) Production of the reverse material poly β-hydroxybutyrate and its function in *Azospirillum brasilense* Cd. Can J Microbiol 31: 608

Tibelius KH, Knowles E (1984) Hydrogenase activity in *Azospirillum brasilense* is inhibited by nitrite, nitric oxide, carbon monoxide and acetylene. J Bacteriol 160: 103

Uozumi T, Wang PL, Tonouchi N, Nam JH, Kim YM, Beppu T (1986) C loning and expression of the nifA of Klebsiella oxytoca in K. pneumoniae and Azospirillum lipoferum. Agricul Biol Chem 50: 1539

Van Berkum P, Bohlool BB (1980) Evalution of nitrogen bacteria in association with root of tropical grasses. Microbiol Rev 44: 491

Wang HX, You CB, Van den Bos RC (1988) Transfer of plasmid-borne Rhizobium leguminosarum nifD: Tn5 to Azotobacter vinelandii. Acta Agricul Nucl Sin 2: 47

Wang HX, Yuan HL, You CB (1989) Plasmid visualization and nif gene location in several nitrogen-fixing bacteria associated with rice plants. Acta Agricul Nucl Sin 3: 213

Wang ZF, Zeng KR, Yang YC (1987) Study on nitrogen fixing bacteria on rihizoplane of rice. Microbiol 14: 241

Watanabe I (1985) Nitrogen fixation associated with wetland rice. In "Nitrogen and Environment" Malik KA, Mujtaba Naqvi SH, Aleem MIH (eds) NIAB, p 185. Faisalabad Pakistan

Watanabe I (1986) Nitrogen fixation by non-legumes in tropical agriculture with special reference to wetland rice. Plant Soil 90: 343

Walterbury JB, Alloway CB, Turner KD (1983) A cellulolytic nitrogen-fixing bacterium cultured from the gland of Deshayes in shipworms (Bivalvia Teredinidae). Sci 221: 1401

Wu WL, Chen HQ (1986) A new variety of Derxia qummosa isolated from the rice seeds. J Fujian Agricul College 14: 323

Xu J, Wangn JW, Xin SY, Li JG (1987) Some properties of nitrogen fixation by Enterobacter sp 25 (E-25) and Klebsiella pneumoniae 12(K-12) isolated from rhizosphere of rice. Acta Agricul Bocerali-Sin 2: 27

Xu LS (1987) Hydrogenase and hydrogen metabolism in nitrogen fixing organisms. In: You CB, Jiang YM, Song HY (eds) Biol nitrogen fixation. Ch 8 Academic Press New York, p 129

Xu Q (1981) Cropping system in relation to fertility of paddy soil in China. In: Inst Soil Sci Acad Sin (ed) Proc Sym Paddy Soil Sin (ed) Springer-Verlag Berlin, p 220

Yoshida T, Yoneyama T (1981) Atmospheric N_2-fixation in flooded rice rhizospheres determined by ^{15}N isotope technique. In: Gibson AH, Newton WE (eds) Cur Persp Nitrogen Fixation. Aust Acad Press Canberra, p 496

Yoshida T, Rinaudo G (1982) Heterotrophic N_2 fixation in paddy soils. In "Microbiol Tropical Soils and Plant Productiv" Dommergues YR, Diem HG (eds) Nijhoff The Hague, p 75

You CB, Li JW, Song W, Li X (1979) Some properties of iron protein of nitrogenase from Azotobacter vinelandii (III). Acta Phytophysiol Sin 5: 215

You CB, Li JW, Song W, Zhang RJ, Zhou SP, YE SG (1981) Some physiological properties of nitrogen fixing bacteria Enterobacter cloacae. Acta Phytophysiol Sin 7: 43

You CB, Qiu YS (1982) Nitrogen fixation of Alcaligenes faecalis in association with rice seedlings. Sci Agricul Sin 15(6): 1

You CB, Li X, Wang YW, Zhu CZ, Hou JQ, Mo XZ, Lao JC, Qiu YS (1983a) Culture and physiological properties of nitrogen fixer Alcaligenes faecalis. Appl Atomic Energy in Agricul (4): 27

You CB, Li X, Wang YW, Qiu YS, Mo XZ, Zhang YL (1983b) Associative dinitrogen fixation of Alcaligenes faecalis with rice plants. Biol N_2 Fixation Newslett, 11: 92. Sydney Univ

You CB, Xiao JZ, Li X, Zhou FY, Wang YW (1984) Association of Alcaligenes faecalis with rice roots. Appl Atomic Energy in Agricul (1): 14

You CB, Xiao JZ, Zhou FY (1985) Determination of the distribution of nitrogen fixers by ^{10}B-track method. Plant Physiol Comm (1): 51

You CB (1987) Structure and function of nitrogenase. In: You CB, Jiang YM, Song HY (eds) Biol nitrogen fixation. Ch 2 Academic Press New York, p 20

You CB, Song W, Zhang DD, Zhou FY, Li X, Zhang FR (1988a) Biosynthesis of nitrogenase in NH_4^+-grown cells of Alcaligenes faecalis. In: Bothe H, de Bruijin FJ, Newton WE (eds) nitrogene fixation: Hundred years after. Fischer Stuttgart New York, p 141

You CB, Zhou FY, Zhang DD, Wang HX, Yuan HL (1988b) Association between Alcaliqenes faecalis and rice plant. ibid, p 802

You CB, Zhou FY (1989) Non-nodular endorhizosperic nitrogen fixation in wetland rice. Can J Microbiol 35: 403

You CB, Wang HX (1990) Rhizosphere nitrogen fixation in wetland rice. Acta Phytophysiol Sin 16: 209

Yuan HL, Wang HX, You CB (1990) Identification and analysis of hydrogen uptake (hub) genes of several associative nitrogen fixing bacteria with rice plant. Acta Agr Nucl Sin 4: 19

Zhang CS, Li JP, Ping SZ, Wang YD, Liu YZ, Wang HX, You CB (1990) Response of rice to inoculation with diazotrophic bacteria. In "A Treaties on Associative Nitrogen Fixation in Rice Rhizospher" (in press)

Zhang DD, Zhou FY, Li X, You CB (1989) Biosynthesis of nitrogenase in NH_4^+-grown cells of *Alcaligenes faecalis*. Acta Phytophysiol Sin 15: 35

Zhang YL, Mo XZ, Lao SH, Qiu YS, Li X, You CB (1984) Association of *Alcaligenes faecalis* A15 with rice root. Plant Physiol Comm (6): 32

Zhou FY, You CB (1988) Interaction between diazotrophic bacteria *Alcaligenes faecalis* and host plants rice. Sci Agricul Sin 21(4): 7

Zhou SP, Mo XZ, Ye SG, Cai XW, Qiu YS, Song W, You CB (1981) $^{15}N_2$ fixation of *Alcaligenes faecalis* and *Enterobacter cloacae*. Acta Agro Sin 7: 59

Zhu JB, Li ZG, Wang LW, Shen SS, Shen SC (1986) Temperature sensitivity of an *nif*A-like gene in *Enterobacter cloacae*. J Bacteriol 166: 357

Zimmer W, Roeben K, Dannerberg G, Bothe H (1987) The bacterial genus *Azospirillum* and its potential applications. In: Ullrich WR, Aparicio PJ, Syrett PJ, Castille F (eds) Inorg nitrogen metabolism. Springer-Verlag, Berlin Heidelberg New York London Paris Tokyo, p 177

CHAPTER 25

Symbiotic Nitrogen Fixation Resources: A Study on *Sinorhizobium fredii* and *Bradyrhizobium japonicum* and Their Applications

S.J. Wang, J.S. Lin, Z.W. Li, D.L. Xue, B. Qi,
G.M. Xu, and X.W. Zhang

The Nitrogen Fixation
and its Research in China
Editor: Guo-fan Hong
© Springer-Verlag Berlin Heidelberg 1992

1 Introduction

With the development of problems, such as that of food, population, energy resources and environmental pollution, scientists are now focusing on biological dinitrogen fixation in order to make a contribution to easing the situation. It is well known that nitrogen fertilizer from the chemical factories is approx. 45 million tons per year, that is about one third of that of the biological dinitrogen fixation on the earth each year. The amount of fixed dinitrogen through legume-*Rhizobium* symbiotic association is about 40 million tons a year. Therefore, the study of the resources of the legume-*Rhizobium* system will play an important role in agriculture, forestry and pasturage.

Chinese scientists have been paying special attention to soybean plants and their symbiotic partners not only because China is rich in the resource, but also the plants are rich in protein and lipids. It was Dr. Zhang Xianwu who first studied the soybean-*Rhizobium* association in China and published his paper "A Study of the Strain of *Bacillus radicicola* from the root nodules of Soybean" (Zhang 1937). In 1982 the isolation of fast-growing *Rhizobium* spp. from soybean nodules was reported for the first time. This ended the history that only slow-growing *Bradyrhizobium japonicum* strains can inoculate *Glycine max* plants. So far *Sinorhizobium fredii* strains have only been isolated in China.

2 Isolation, Identification, and Reservation of *Bradyrhizobium japonicum* and *Sinorhizobium fredii* Strains

More than twenty slow-growing *Bradyrhizobium japonicum* strains e.g. B15, B16, B17, B18, B19, B16-11c, and 345 fast-growing *Sinorhizobium fredii* strains e.g. QB11, QB13, QB46, QB112, QB113, QB1130, C3A, FR001, FR061, FR869, Sj881, Sj8813, Sj8855, Sj8866, Sj8911, Sj8912, Sj8918, SjSC2 have been isolated, identified and preserved in freeze-dried cultures (Wang Shujin 1985, 1989).

Fast-growing (FG) strains of *Rhizobium* are generally the microsymbionts of temperate legumes, whereas the slow-growing (SG) strains are in symbiosis with tropical legumes. Consequently, if a host is infected by the two types, FG strains may outcompete SG strains under certain environmental conditions such as at the lower temperature. Also, as acid producers the FG strains may be better adapted to growth in the alkaline soils and the greater tolerance of FG strains to NaCl is an advantage under conditions of salt stress. Our interest is to improve and construct better inoculars suitable to low temperature, boggy bottomland with alkaline/salt stress, because there are lands under such conditions in Northeast China.

Rhizobia were selected from Northeast China, where it has been thought is the origin of the species *G. max*. Wild species of the genus *Glycine* (as shown in Fig. 1) are ancestors of the cultivated soybean. The wild species had in recent

Fig. 1. The whole plant (*upper*) and root nodule (*lower*) of *Glycine soja* collected in Changbai Mountain, Northeast China

years been collected and cultured by the Tieling Institute of Agricultural Research. New varieties of wild soybean were reported, and the new Chinese name of "*G. gracilis* SKv. var. *nigra* SKv." was proposed and discussed (Fu 1986). We have recorded habitats and distributions of each taxon. So far 7286 strains of wild soybean have been found in China, of which 5853 strains were found in Northeast China. 450 strains of *Sinorhizobium fredii*, the symbiotic partners of the wild soybean, have been found, of which 331 strains were collected in Northeast China by us.

Symbiotic association between *G. soja* and *S. fredii* could be classified into five genotypes in accordance with the taxon of macrosymbiont: (1) *Glycine soja.*

(2) *Glycine soja f. lanceolata*, (3) *Glycine soja var. albiflora*, (4) *Glycine soja var. albiflora f. angustifolia*, (5) *Glycine gracilis* (Wang 1989). Although there are some similar characteristics among five genotypes, differences in serology, physiology and biochemistry are still obvious enough to distinguish one from another.

The soluble protein and lipase diagrams of the bacteria, with which the soybean roots were inoculated have been studied, together with the studies such as the utilization of carbon and nitrogen flagella, dyes reaction, tolerance to sodium chloride, etc. The similarities to the old strains were determined by the single-linkage method with computer. The results showed that the fast-growing soybean-*Rhizobium* group exhibits little similarity to other *Rhizobium* groups, including *Bradyrhizobium*. It was in agreement with Dr. Chen Wenxin's result who named this new genus *Sinorhizobium*.

3 Physiology and Biochemistry of *Sinorhizobium fredii*

A *Sinorhizobium fredii* strain QB1130 can inoculate effectively *Vigna unguiculata*, moan bean and American local prairie legumes (Davis, personal communication) besides *Glycine max* and *G. soja*. This is the first report that a *Sinorhizobium fredii* strain has a wide host range. As it has been known that slow-growing *Bradyrhizobium japonicum* has one flagellum only, whereas *S. fredii* has one to five flagella (Fig. 2). In addition to that the latter can utilize a wider range of carbon resource (Table 1) and tolerate much higher concentrations of NaCl (Table 2), whereas *B. japonicum* can not grow beyond 0.1 M NaCl (Wang 1987). No plasmids were detected in SG strains, while from 2 to 4 plasmids were found in *S. fredii* strains, e.g., QB1130, C3A and USDA191. Both QB1130 and C3A contained a large plasmid which migrated with the same mobility as the megaplasmid of *R. meliloti* strain JJ1. In addition, strain QB1130 contained three other plasmids, strain C3A contained two other plasmids. The plasmid in size of 295 kb has been reported in a number of *S. fredii* strains, including USDA191, and in each case it contained DNA sequences homologous to genes of *R. meliloti*. Southern blotting analysis demonstrated the hybridization of a *nif* probe with the 295 kb plasmid of QB1130, C3A and USDA191 (Fig. 3). In this regard, QB1130 resembles other *S. fredii* strains which have been previously described (Lin 1987).

The effectiveness of *S. fredii* strains on Peking demonstrated that these strains have potential agronomic importance. The ineffective nodulation of kent soybean with FG USDA205 is determined by a single dominant gene. Possibly, Peking could be used to introduce the character of compatibility with *S. fredii* into North American cultivars of soybean. Ertl and Fehr have shown that decreased yield of the progeny from a cross with *G. soja* can be overcome within three backcrosses with the North American parent.

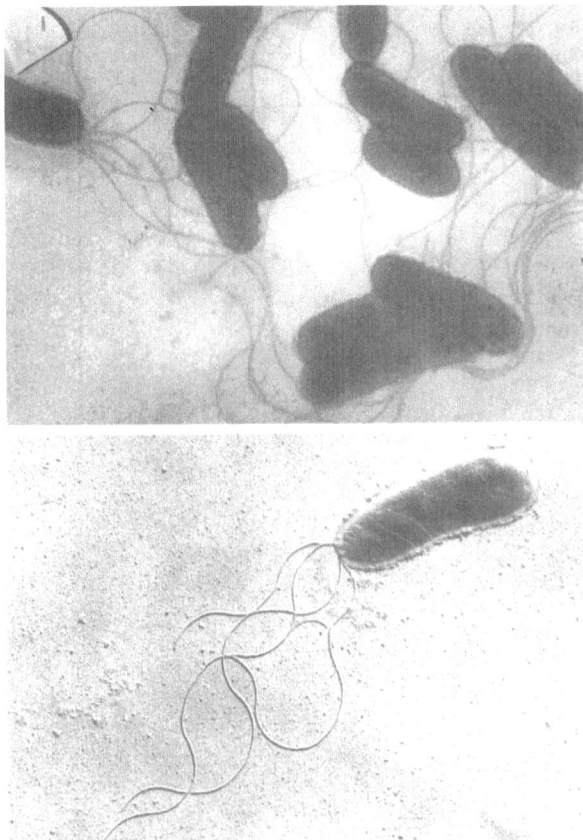

Fig. 2. The electron micrographs of *Sinorhizobium fredii*

We found that when the symbiosis was less effective, the nodule volume was getting smaller, less cells were infected per nodule, and less bacteroid density appeared per infected cell. Nitrogenase activity was determined by the bacteroid number. Therefore, the performance of the slow- and fast-growing rhizobia strains on the hosts are related to the ability of the bacteria to proliferate in the plant cells (Lin 1988, 1989).

Light and electron microscopy study showed that both *B. japonicum* and *S. fredii* strains can form excellent symbiosis with their host soybean cultivars, though FG rhizobia were found to predominate in some regions of China.

We have selected microsymbionts for the use in commercial soybean cultivars. They can form a good symbiotic association (Fig. 4), the micro- and macrosymbionts have the following characteristics (Xu 1989; Wang 1985, 1990):

1) Higher RuBPcase activity in the host plant leaves (6133 ^{14}C cpm vs. 2730 ^{14}C cpm in poor one.);

Table 1. Utilization of carbohydrates and organic acids by free-living SG (*Bradyrhizobium japonicum*) and FG (*Sinorhizobium fredii*) known to form effective nodules on *G. max* and *G. soja*

	S. fredii QB113-88	T428A	T487A	T431B	*B. japonicum* USDA110
Glycerol	+	+	+	+	+
D-Glucose	+	+	+	+	+
Galactose	+	+	+	+	+
D-Fructose	+	+	+	+	+
L-Arabinose	+	+	+	+	+
D-Mannitol	+	+	+	+	+
Gluconate	+	−	+	+	−
Na-Fumarate	−	−	+	+	+
Na-Malate	+	+	+	+	+
Na-Succinate	+	+	+	+	+
D-Mannose	+	+	+	+	−
Sucrose	+	+	+	+	−
Sorbitol	+	+	+	+	−
Maltose	+	+	+	+	−
Rhamnose	+	+	+	+	−
Trehalose	+	+	+	+	−
Inositol	+	+	+	+	−
Dulcitol	−	−	−	−	−

+: growth was supported
−: no growth

Table 2. Physiological and biochemical characteristics of *S. fredii* strains

Taxa of host plants	*G. soja*	*G. soja* f. *lanceolata*	*G. soja* var. *albiflora*	*G. soja* var. *albiflora* f. *angustifolia*
Strains	QB113-88	T428A	T487A	T431B
Suitable pH range	5.0–8.0	5.0–8.0	5.0–9.0	5.0–11.0
NaCl tolerance	0.3 M	0.3 M	0.5 M	0.6 M
BTB reaction	Produce acid	Produce acid	Produce acid slightly	Produce acid
BTB reaction under salt stress	Produce alkaline	Produce alkaline	Produce alkaline	Produce alkaline
Utilization of peptone broth	+	+	+	+
Litmus milk reaction	Form serum ring, acid	Mini, serum ring, alkaline	Mini serum ring, mini acid	Form serum ring Produce acid
Gelatin Hydrolysis	−	−	−	−
Nitrate reduction	+	+	+	+
H$_2$S reaction	−	−	−	−
Utilization of EtOH	+	+	+	+
Gram stain	−	−	−	−
3-Keto-lactose reaction	−	−	−	−

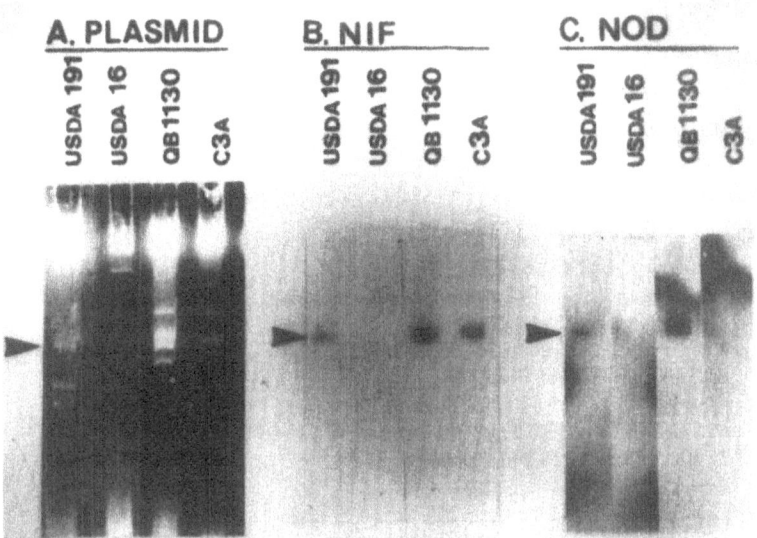

Fig. 3. Analysis of plasmid DNA from *S. fredii* strains USDA 191, QB1130 and C3A and *B. japonicum* strain USDA 16. **A** Agarose gel electrophoresis of plasmid DNA prepared by the method of Eckhardt (1978). Lanes: 1, USDA 191; 2, USDA 16; 3, QB1130 and C3A. The arrow indicates the position of the 295 kb plasmid. **B** Hybridization of the ^{32}P-labelled *K. pneumoniae nif* (HDK) probe against plasmid DNA. The same lane order as in *A*. The arrow indicates the position of the 295 kb plasmid. **C** Hybridization of the ^{32}P-labelled *R. meliloti* common nod probe against plasmid DNA. The lane order as in *A*. The arrow indicates the position of the 295 kb plasmid

2) Higher PEPcase activity in root nodules (6133 ^{14}C cpm vs. 586 ^{14}C cpm in poor one);
3) Higher photosynthesis activity in the leaves of host plant (14.9 mg dry weight $\cdot$ $dm^{-2} \cdot h^{-1}$ vs. 7.5 mg dry weight $\cdot dm^{-2} \cdot h^{-1}$);
4) Higher nitrogen-fixing activity (2079 nmol C_2H_4/gDW$\cdot$h$\cdot$vs. 575 nmol C_2H_4/gDW$\cdot$h in poor one);
5) Uptake hydrogenase positive in soybean nodules;
6) the recognition reaction between lectin of the soybean cultivar and the surface polysaccharide of microsymbiont take place much easier;
7) High content of uride in soybean plant xylem (3860–4260 µg/ml vs. 288 µg/ml in poor one);
8) More bacteroids within plant root nodules;
9) High seed yield (46.6 g/plant vs. 36.0 g/plant in poor one).

4 Recognition and Infection

Like *Bradyrhizobium japonicum*, *S. fredii* strains can infect roots of *Glycine soja* and *Glycine max*, and form nitrogen-fixing root nodules. The recognition and infection of soybean by *B. japonicum* has been studied by several groups, and

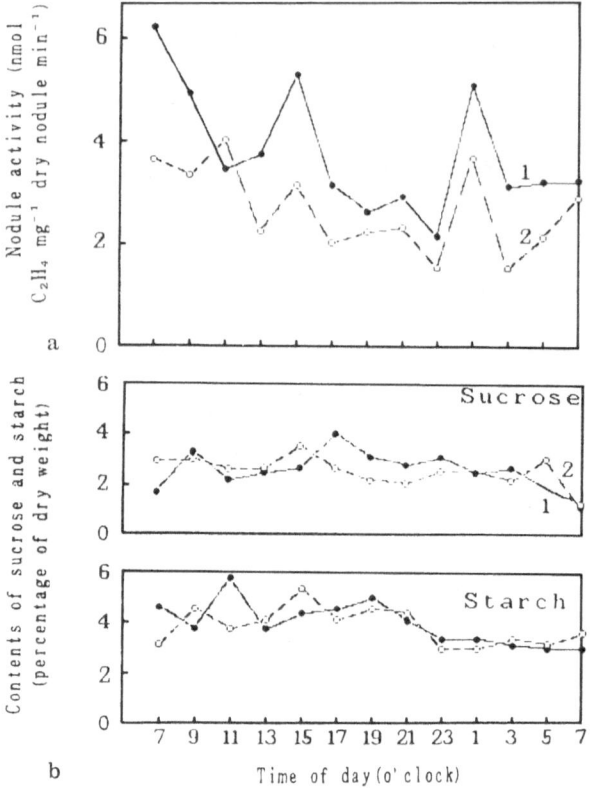

Fig. 4 a, b. Diurnal variations of the nodule activity (**a**) and the contents of sucrose and starch in the soybean root nodules (**b**) 1. Plants in the light 2. Plants in the shade

they found that the soybean lectin (SBL) and the bacterial extracellular poly-saccharides may play important roles in the same host, it does not necessarily mean that the molecular basis for recognition and infection is same. Mutation studies have revealed several specific properties observed in the recognition between *S. fredii* and soybean (Li 1989).

Twenty *S. fredii* strains from different sources were tested for their binding activity with FITC labeled SBL. All the strains were not able to bind SBL in their different culture stages, nor under the induction of soybean or wild soybean root exudates. Under the same conditions, however, all of the 10 strains of *Bradyrhizobium japonicum* tested expressed SBL binding activity in culture or under the induction of soybean or wild soybean root exudates (Fig. 5). Other rhizobia, such as *R. trifolii*, that do not nodulate *G. soja* and *G. max*, did not bind SBL in the same tests. *S. fredii* SC2 can bind to the root hairs of *G. soja*. Bindings occurred 12 h. after inoculation and were not inhibited by 60 mM of D-galactose. Exopolysaccharide (EPS)-deficient mutants of SC2 which reduced in EPS productivity or altered in EPS composition can still bind to and infect

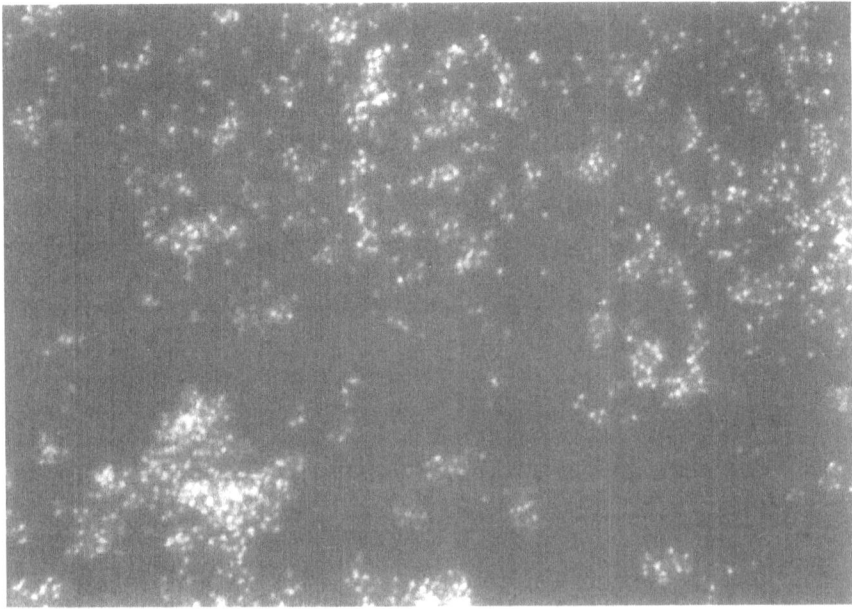

Fig. 5. The green fluorescence of FITC–SBL bound to *Bradyrhizobium japonicum* B16–11c (× 1000). FITC–SBL binding assay is as follows. *Rhizobium* and *Bradyrhizobium* strains were grown in the yeast extracrmanitol (YM) liquid medium. Bacterial cells at different culture stages were collected by centrifugation and then suspended in phosphate buffer (pH 7.2) FITC–SBL was added to a final concentration of 1.25 mg/ml, and the cell suspension was kept at room temperature for 30 min. The cells were washed in PBS and spread on slide. Dried cells were observed under Venox vertical UV-microscope. To check the influence of host plant on SBL binding, the bacterical cells were cultured in YM with or without soybean root exudate (SRE) or with or without wild soybean root exudate (WSRE), and then assayed for SBL binding as above

roots of *G. soja* (Fig. 6). In contrast, *B. japonicum* B16-11c was found to bind to the root hairs of *G. soja* in 1 hr. after inoculation, and the binding was inhibited by 60 mM of D-galactose. This indicates that the recognition mechanism of *S. fredii-G. soja* (*G. max*) is different with that of *B. japonicum-G. soja* (*G. max*).

S. fredii SC2 was mutagenized by transposon Tn5 and by nitrosoguanidine. In total, 29 Exo⁻ mutant strains deficient in EPS production were obtained. These Exo⁻ mutants showed 3 levels for their EPS productivity: 1) producing little EPS (about 1/10 of the parent strain, 5 strains); 2) producing micro-amount of EPS (about 1/30 of the parent strain, 6 strains); 3) EPSs are not detectable (18 strains). EPSs from some Exo⁻ mutant strains contain less uronic acid, no D-galactose, or more D-mannose, as compared with the EPS from the parent strain. This suggested a complicated pathway in the genetic control of EPS synthesis. All the Exo⁻ mutants, except Exo1, are capable of nodulating *G. soja* and *G. max* and fixing nitrogen at normal level with normal nodule structure. This implied that the recognition between *S. fredii* and *G. soja* may not require the production of EPS. The effect of EPS on nodulation may be associated

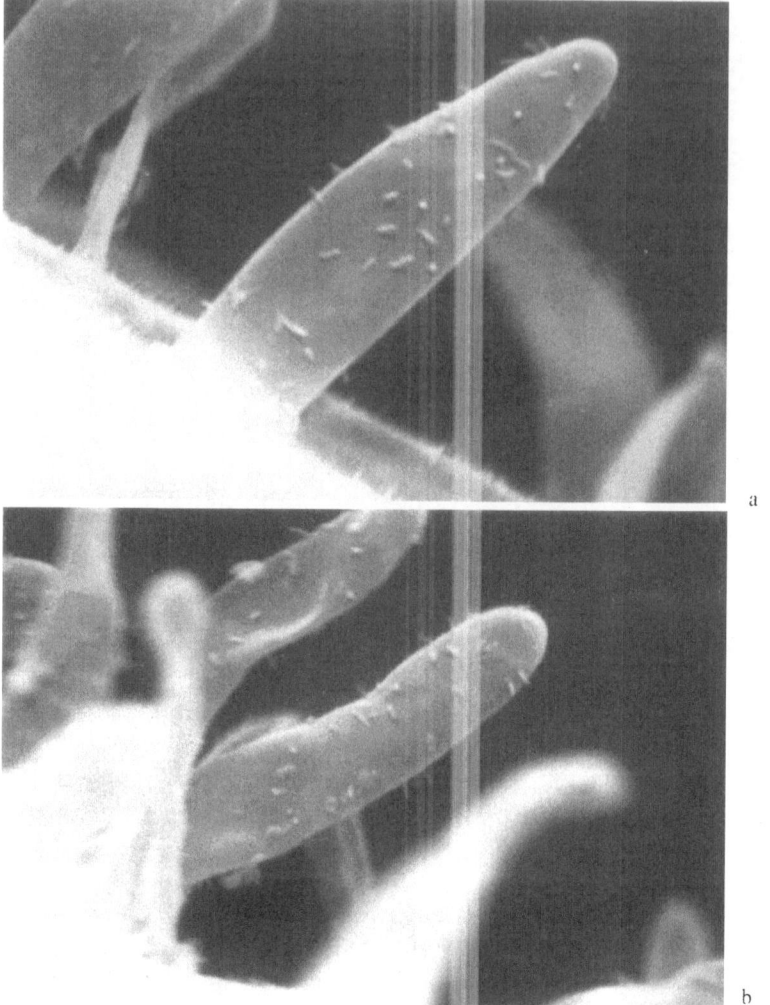

Fig. 6a, b. Binding of *Sinorhizobium fredii* to root hairs of *Glycine soja*. *G. soja* seedlings were cultured and inoculated with *S. fredii* as described by Stacey et al. (1980). Root fragments were dehydrated in a series of methyl cyanide concentrations up to 100%, and then further dried under low vacuum before coated with carbon and gold. Observation was made under Amray scanning electron microscope. *S. fredii* SC2 (**a**) and its EPS deficient mutant Exol (**b**) were found to be bound to root hairs of wild soybean in the end-on fasion

with host plant types. For instance, those plants forming apical nodules are very sensitive to the change in EPS. Those plants forming spherical nodules are relatively insensitive to EPS. As presented here, both soybean and wild soybean nodules were formed regularly when infected by the EPS deficient mutants. This does not exclude the possibility that in the latter case, special EPSs may be induced by plant factors and play important roles in the nodule development.

Exo1 nodulated *G. soja* at a reduced level (with nodulation efficiency about 10 percent of the parent strain). However, the nodules were fully developed and the nitrogen-fixing activity was expressed at the wild-type level. Exo1 could specifically bind to *G. soja* roots in the same way as the parent strain did (Fig. 6), but caused root hair curling at a significantly lower rate than the parent strain did. In culture, Exo1 produced no EPS, but made a considerable amount of CPS and LPS. The LPS of Exo1 contained less hexoses (39.9%) than that of the parent strain (79.2%). This may imply that *S. fredii* LPS may play a role in infection.

5 Application of *Sinorhizobium fredii* and *Bradyrhizobium japonicum* Strains in Agriculture

Zhang, Xu and Ding isolated some *Bradyrhizobium japonicum* strains in the 1950s, such as B15, B16, C18 etc. Those strains were widely used in Northeast China soybean fields before the 1960s, which increased the yield by 10% (Zhang Xianwu, 1957). Recently, Wang, Lin, and Xue improved some new *B. japonicum* and *S. fredii* strains, such as B16-11c, QB113 and QB1130, which have been used in nearly 30 thousand ha. soybean fields in the Sanjiang Plain, Helongjiang Province and increased the yield by 10–25 percent during 1983–1987. The reason for the increase of the soybean yield was the following:

1) Better symbiotic association between high-efficient photosynthesis soybean cultivars and nitrogen-fixing *Rhizobium* strains,
2) Ecological environment favorable for soybean growth.
3) Proper husbandry.

The soybean yield has been increased to 1950–2100 kg/ha from 982.5 kg/ha since the comprehensive reclamation of the boggy bottomland in Sanjiang Plain, Helongjiang Province during 1986–1989 (Wang 1990). The techniques of reclamation include: (1) irrigation and water conservancy engineering; (2) mechanical cultivation; (3) the introduction of new soybean breeds with high abilities of both photosynthesis and nitrogen fixation. The activities of nitrogen fixation of *Bradyrhizobium japonicum*, *Sinorhizobium fredii* and other N_2-fixers sharply increased after the reclamation. Deep ploughing with machines benefits the microbes as well as the root system. More than three *G. max* breeds with high photosynthesis rates were obtained and a strain of *S. fredii* Sj8855, which is tolerant to pesticide (MBC), with high N_2-fixation and suitable in boggy bottom 1 and have been developed (Table 3). We think that the above consideration may be applied in other boggy bottomlands to increase soybean yield.

Some *S. fredii* strains such as QB113, Sj8855 are of high efficiency in nitrogen fixation with good compatibility to Asian soybean cultivars and high tolerance under stress. They are now produced as commercial inoculants and widely used in Jilin, Helongjiang Province.

Table 3. Effect of high nitrogen fixation S. *fredii* strain Sj8855 inoculation and pesticide on growth, development and yield of soybean

| Treatment | Stalk height (cm) | Number of stalk nodes | Biomass (gDW/plant) | | | | Nodule number | 100 Seed weight (g) | Damping off of soybean | Yield (kg/ha) |
			Root	Stem	Leaf	Nodule				
Control	67.6	12.1	3.8	22.9	12.6	0.47	37.5	19.5	+ + +	2298.0
Pesticide (MBC) only	68.4	14.2	4.8	25.0	16.3	0.40	23.9	20.3	+	2390.0
Sj8855 inoculation only	72.6	17.8	7.2	40.6	25.0	0.65	40.3	20.1	+ +	2758.0
Sj8855 inoculation and MBC	74.2	14.7	5.4	34.5	18.8	0.48	43.3	20.8	+/−	2850.0

6 References

Fu P, and Chen Y (1986) Study on the classification of wild species of genus *Glycine* from Liaoning Province, Bull Bot Res 6(2): 117

Han J, Wang S, Lin J (1988) Functions of rhizosphere microorganism in agroecosystem and their effect on plant growth. Adv Ecol 5(2): 92

Li Z (1989) Studies on the interaction of *Rhizobium fredii* and *Glycine soja*, Ph.D dissertation, Inst of Appl Ecology. Academia Sinica

Lin J, Wang S, Qi B (1988) Present study on the communication between plant and microbe, Adv Ecol 5(4): 234

Lin J, Walsh KB, Johnson DA, Canvin DT, Wang S, Layzell DB (1987) Characterization of *R. fredii* QB1130, a strain effective on commercial soybean cultivars. Plant and soil 99: 441

Lin J, Walsh KB, Canvin DT, Layzell DB (1988) Structure and physiological bases for effectivity of soybean nodules formed by fast-growing and slow-growing bacteria. Can J Bot 66: 526

Stacey G, Paau AA, Brill WJ (1980) Host recognition in the *Rhizobium*-soybean symbiosis. Plant Physiol 66: 609

Wang S, He J (1956) Study on autumn soybean and peanut *Rhizobium*-soybean symbioses under different gaseous conditions. J Microbiol 1(3): 24

Wang S, Xue D, Qi B, Zhang X (1987) The ecological distribution of the fast-growing *R. japonicum* from different types of soil in Northeast China. J Microbiol 7(2): 18

Wang S et al. (1989) Ecological Distribution and identification of fast-growing R. fredii in different types of soil in Northest China. J Ecology 5: 51

Wang S et al (1989) Ecological Distribution and characteristics of *Glycine soja* plants and their microsymbiont, *Sinorhizobium fredii* in Northeast China area. J Microbiol 9(3): 35

Wang S, Lin J, Xue D, Xu G (1990) Effects of comprehensive reclamation of boggy bottomland on soil microbes and yield of soybean in Sanjiang Plain. J Appl Ecology 1(2): 120

Xu D, Shen Y, Wang S, Zhang X (1989) Studies on the relationship between photosynthesis and nitrogen fixation in the symbiotic system of soybean and nodule bacteria (*Rhizobium*). Acta Bot Sinica 31(2): 103

Zhang X (1937) Studies of the strains of *Bacillus radicicola* from the root nodules of soybean. Report of the Inst of Scientific Res Manchoukuo 1(5): 139

Zhang X, Xu G (1981) Study on *Rhizobium japonicum*. Bull Soil Microbiol 1: 25

Zhang X, Wang S et al. (1987) Study on the symbiotic association between *R. japonicum* mutant B16-11 and soybean plant mutant 18-79, 33-79, 55-79, 66-80, 77-80. KEXUE TONGBAO (A Montly J of Science) 6(6): 862

Zhang HY, Tang G, Xue D, Wang S, Zhang X (1985) Effect of the fast-growing *Rhizobium japonicum* QB113 symbiotic nitrogen fixation and yield of soybean. Soybean Science 4(4): 267

CHAPTER 26
Nitrogen-Fixing Blue–Green Algae

S.H. Li (S.H. Ley)

The Nitrogen Fixation
and its Research in China
Editor: Guo-fan Hong
© Springer-Verlag Berlin Heidelberg 1992

1 Introduction

Blue-green algae (Cyanobacteria) are a special group of prokaryotes. They have chlorophyll and phycobiliprotein and can fix carbon by oxygen-evolution photosynthesis like plants; and their genome contains nucleotide sequences which are comparable with that of 16S and 5S rRNA of eubacteria. Therefore, in tradition, they are called Cyanophyta (Myxophyta, Cyanophyceae), but in the recent decade, they are also called Cyanobacteria. (According to the "Manual of Systematic Bacteriology" Vol. 3 (1989), they can be called cyanobacteria or blue-green algae or Cyanophyceae). For convenience, I use BGA (blue-green algae) for these prokaryotes in this paper. BGA are widely distributed all over the earth. While fixing carbon from CO_2, certain BGA can fix dinitrogen from the atmosphere, and are called nitrogen-fixing BGA, including free living and symbiotic forms.

Because of their ability of photosynthesis and nitrogen fixation, BGA populate various environments from the Arctic to the Antarctic, and are very important in the nitrogen input in the global nitrogen cycle.

The water fern *Azolla* having the symbiont *Anabaena azollae*, a nitrogen-fixing BGA, has been used by Chinese farmers for centuries as a green manure to improve the nitrogen balance in rice paddy field and as green feeding for pigs.

Therefore, studies on the nitrogen-fixing BGA and their utilization are important in both basic and applied research (Carr et al. 1982; Fogg et al. 1973; Ley et al. 1959; Li et al. 1983; Roger et al. 1982; Venkstaramen 1972).

In this paper, we will describe part of our research carried out in the Department of Phycology, Institute of Hydrobiology, Chinese Academy of Sciences (called also Academia Sinica), Wuhan, 430072, China. I am very grateful to my colleagues Profs. QL Wang, HM Lin, MJ Yu, XK Zhang YQ Wang, YD Liu, ZR He, LF Dai, and LR Song for their useful discussions and the materials kindly offered.

2 Nitrogen-Fixing Blue-Green Algae

There are about 2000 taxa of BGA having been recorded in the world. Morphologically, they can be divided into three groups: single cell BGA, non-heterocystous filamentous BGA and heterocystous filamentous BGA. In general, the nitrogenase is located in the heterocysts, specially differentiated vagetative cells, and all the heterocystous forms are believed to be nitrogen-fixing. But many forms of non-heterocystous filamentous BGA and some of the unicellular forms are also showing the dinitrogen-fixing ability in microaerobic or/and anaerobic conditions. Totally, more than 150 species of 33 genera have been reported to be able to fix nitrogen (Li and Wang 1983).

In China, there are about one thousand taxa of BGA recorded. Among them many taxa are known as nitrogen-fixing forms. More than 20 species have been assayed for their nitrogen-fixing capacity.

In the Freshwater Algal Culture Collection of this Institute about one thousand strains are in collection, among them 232 strains are heterocystous, and some of them were assayed for their dinitrogen-fixing ability by the methods of either nitrogen analysis or acetylene reduction in culture.

According to the habitats, they are grouped into four categories: most of the strains were collected from the wetland or aquatic habitats, some from subarid or arid habitats, some from edaphic habitats and very few are symbionts. The ability of dinitrogen fixation is various with different strains and also with different environmental conditions. It is very interesting that the *Anabaena azollae*, isolated from *Azolla* and cultivated for a long time in the artificial nitrogen-free mineral medium, shows the highest acetylene reduction activity (Table 1). The wetland forms are moderate. The activity of the subarid or edaphic isolates are relatively low.

81 taxa of edaphic BGA are recorded, which were isolated from the soil samples from Anlu, Hubei, (Liu and Li 1989). Among them, 37 species are heterocystous, which accounted for 45.7% of the total taxa. 8 species (10 strains) grown in the soil were examined for their growth rate and nitrogen fixation in culture. They are *Anabaena oryzae*, *A. variabilis*, *Calothrix Braunii*, *Nostoc carneum*, *N. entophytum*, *N. piscinale*, *Scytonema schmidlei*, *Tolypothrix boutellei*. They could accumulate dry matter of 0.1–0.7 mg/ml, indicating 6.5–12.8 times increase in culture in 10 days. Growth coefficients (K) of *A. variablis*, *A. oryzae*, *N. entophytum*, *C. braunii* and *S. schmidlei* are 1.45–1.89, and the generation times are 16.2–12.3 hours. Their nitrogenase activities were shown in Table 2.

22 taxa of blue-greens are isolated from the subarid and arid soil samples from Qaidam Basin, Qinghai, and half of them (11 taxa) are hterocystous filamentous forms. Some have already been known as the nitrogen-fixing species (Hong et al., in press).

Table 1. Comparison of specific acetylene reduction in air of several blue-green algae

Species	Specific character	C_2H_2-reduction (nm mg^{-1} dry weight)
Anabaena azollae	symbiont of *Azolla*	219.1
Anabaena azotica	free-living, heterocystous filamentous	81.5
Anabaena cylindrica	free-living, heterocystous filamentous	68.5
Tolypothrix tnuis	free-living, heterocystous filamentous	69.2
Plectonema boryanum[a]	free-living, filamintous, non-heterocystous	0
Gloeocapsa LB 795	free-living, unicellular	48.9

[a] Fixing nitrogen under microaerobic conditions

Table 2. Comparison of nitrogenase activity of 10 strains isolated from rice-field in Anlu. 10 days after inoculation, culture medium HB 111[a]

Strains	Nitrogenase activity $(nM\ C_2H_4 \cdot mg^{-1}\ dry\ weight\ min^{-1})$
Anabaena oryzae	30.7
A. variabilis	44.9
Calothrix braunii	335.5
Nostoc carneum	117.5
N. entophytum str. 1	228.6
N. entophytum str. 2	90.5
N. entophytum str. 3	9.9
N. piscinale	13.4
Scytonema schmidlei	353.4
Tolypothrix bouteillei	200.8
Anabaena azotica	599.5

[a] Note: The formula of medium HB 111 see Li SH (S.H. Ley) 1981

3 Nitrogenase

Like the nitrogenase found in other nitrogen-fixing organisms, the nitrogenase of nitrogen-fixing BGA consists of two components, molybdenum–iron protein and iron protein, which are extremely oxygen-sensitive. Therefore the preparation of and reaction with the enzyme must be done in the anaerobic condition.

3.1 Isolation and Purification of Nitrogenase

We have established simplified and efficient procedures (Dai et al. 1985; Du et al. 1987; Lin et al. 1986) for isolation and purification of the two components of the nitrogenase of BGA.

(1) By using polyethyleneglycol instead of protamine and nucleases in the pretreatment, the process has been simplified.
(2) By using the preparative polyacrylamide gel electrophoresis instead of column chromatography, the process has been simplified and the purity of the component isolated was increased. The MoFe-protein and Fe-protein of BGA thus obtained have reached the electrophoretic purity (Fig. 1).

3.2 Some Properties of Nitrogenase of Blue–Green Algae

The basic properties of MoFe-protein and Fe-protein are as shown in Table 3.

They are quite similar with the nitrogenase components from other nitrogen-fixing organisms. However, the molecular weight of MoFe-protein is much

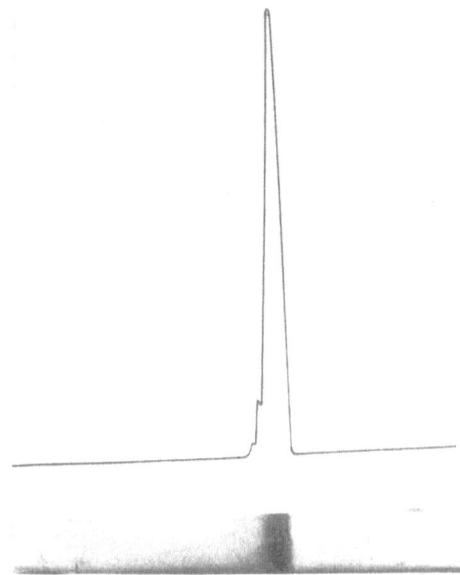

Fig. 1. Polyacrylamide gel electrophoretogram of Fe-protein of *Anabaena cylindrica* and its spectrum by microdensitometry

higher in our estimation. The nitrogenase from BGA is much more sensitive to temperature and oxygen. When the temperature raise to 55 °C, the nitrogenase of BGA will completely lose its activity within 3 min (with the exception of thermophilus BGA such as *Mastigocladus laminosum*). In the gas phase with 11% of oxygen, the nitrogenase loses its activity in 3 min (Du et al. 1987; Lin et al. 1986; LPIHB 1980).

The intensive inhibiting effect of the metal ions on the nitrogenase activity is increased by the order of $Na^+ > K^+ > Mg^{2+}$. In low concentration, $MgCl_2$

Table 3. Some properties of MoFe-protein and Fe-protein of nitrogenase from *Anabaena cylindrica*

Specific	MoFe-protein	Fe-protein
Subunit	4, homotype	2, homotype
Molecular weight (Daltons)	360,000	61,000
Metal content (g atom mol)		
Mo	1	—
Fe	18	3
Isoelectric point	5–5.5	
Asp + Glu (% of total aa)	19.8	22.3
Lys + Arg (% of total aa)	8.2	8.7
Km (vs. acetylene)	3.33×10^{-3}	
Optimal ratio for combination	1	2.5
Photospectrum absorption peak (nm)	258	280

may promote the nitrogenase activity. This may be due to its facilitating the binding of ATP with nitrogenase (Du et al. 1987; Lin et al. 1984; Lin et al. 1986). In higher concentration, Mg has conspicuous inhibiting effect. This is quite similar with Hallenback's result (Hallenback et al., 1979).

Ludden et al. (1979) found that Mn^{2+} and Mg^{2+} in high concentration had favorable effect on the nitrogenase activity of *Rhodospirillum rubrum*, and they concluded that there was an activating factor in the extract solution (Ludden et al. 1976). But we could not find such a factor in BGA.

3.3 Cross-Reaction of the Nitrogenase Components of *Anabaena cylindrica* and *Azobacter vinelandii*

In our experiments, we have demonstrated that the MoFe-protein from *A. cylindrica* recombined with Fe-protein of *A. vinelandii* shows a nitrogenase activity, and the Fe-protein from *A. cylindrica* recombined with MoFe protein from *A. vinelandii* also shows firstly a relatively strong nitrogenase activity. The specific activity of acetylene reduction and H_2-evolution of the latter cross recombination are 83.8% and 66.7% of those of the algal homologous components recombination, respectively. It was found in our kinetics study that the optimal molar ratio of Fe-protein to MoFe-protein for heterologous cross reaction is much higher (5:1) than that for the homologous complementary reaction, the time course for both reactions is similar (Fig. 2) (Du et al. 1987; He et al. 1985; Lin et al. 1986; LPIHB 1980a). Hallenbeck et al. (1979) considered that the Fe-protein of *A. cylindrica* is unable to recombine effectively with the MoFe-protein of *A. vinelandii*. However, our findings demonstrated that the Fe-protein of *A. cylindrica* can form an effective complex with the MoFe-protein of *A. vinelandii*.

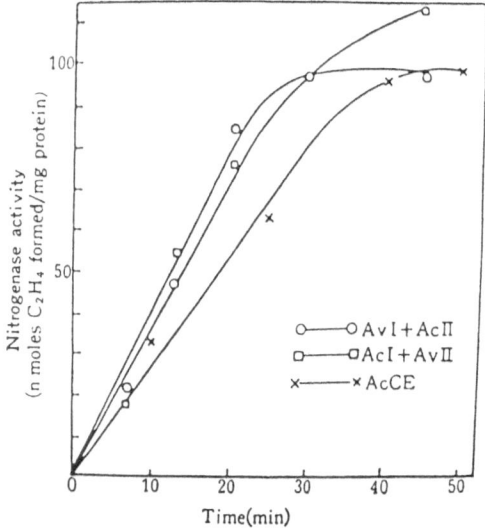

Fig. 2. Time course of the cross-reactivity of Fe-protein from *A. cylindrica* and MoFe-protein from *A. vinelandii*

It suggested that the components of nitrogenases from these two organisms are of high homology in their functions.

3.4 The Second Nitrogen-Fixing System of Blue–Green Algae

Since Bortels (1940) reported that the molybdenum is necessary for the growth of nitrogen-fixing BGA, 25 years after, Bulen and LeComte (1966) obtained for the first time MoFe-protein and Fe-protein from *Clostridium pasteurianum*, and 5 years later, Smith and Evens (1971) isolated also from *Anabaena cylindrica*. In 1947, Jensen and Spencer reported that *Clostridium butyricum* fixed N with molybdenum as well as vanadium. Becking (1962) claimed that in the strains of *Azobacter vinelandii* and *A. chroococcum*, vanadium can be substituted for molybdenum. From the *A. vinelandii* grown with V in place of Mo, V-containing nitrogenase was isolated (McKenna et al. 1970; Burns et al. 1971). In 1980, Bishop suggested that there is an alternative nitrogen-fixing system, and this idea was confirmed by Roson et al. of the Sussex group. We have recently found that there is a second nitrogen-fixing system (with VFe-protein and Fe-protein complex) in *Anabaena azollae* grown in autophototrophic conditions (Dai et al. 1990).

A. *azollae* were grown in the nitrogen-free media which either contain 0.5×10^{-6} M V_2O_5 or 1×10^{-6} M $Na_2Mo_2O_4$. The growth rates of both cultures are similar. Three days after, the specific acetylene reduction of V-grown algae is a little lower than that of the Mo-grown one. But the V-grown algae also produces methane, which is not produced by the Mo-grown culture. This is the key character of the V-nitrogenase. It suggests that there is an alternative nitrogen-fixing system in BGA. When the culture with V is put in the dark, no ethylene or methane are detected. It indicates that the production of ethylene and methane are light-dependent. When ethylene is used instead of acetylene in the reaction, no methane was produced. The mechanism where by methane is evolved from acetylene remains unknown.

It has been known that the H_2 evolution by reduction of H^+ in the N_2 or C_2H_2 reducing process is catalyzed by nitrogenase. The V-nitrogenase has greater capacity of H_2 evolution than nitrogenase prepared from the normal culture. It has also been proved that H_2-evolution is affected by C_2H_2. These results suggested that there may exist a second nitrogen-fixing system in the cells grown in V-containing culture.

4 Ecophysiology of Nitrogen-Fixing Blue–Green Algae

4.1 Carbon and Nitrogen Metabolism

In semi-steady-state continuous culture of *A. variabilis* ATCC 29413, it is found that owing to the CO_2-assimilation and nitrogen fixation, biomass increases

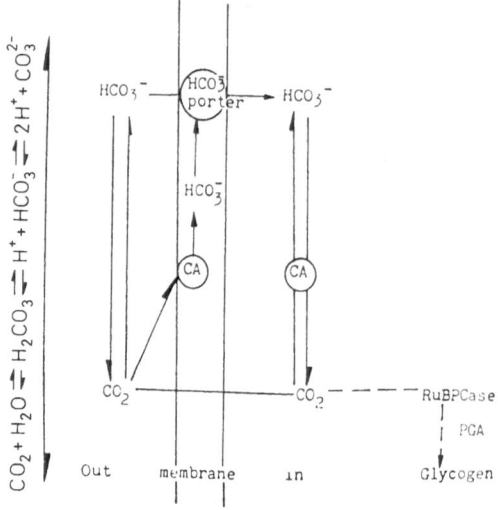

Fig. 3. A diagram for explaining the process of inorganic carbon absorption in cells of *Anabaena* spp., modified from the models proposed by Badger et al. (1985), Volokita et al. (1984), Abe et al. (1987)

during the light period, and owing to the continuous dilution and respiration, it decreases approximately at the same rate in the dark. The increase of the biomass in the light is mainly due to the synthesis and accumulation of glycogen. Glycogen accumulates up to 20% of the biomass at the end of the light period, then degrades down to 1–2% of dry weight in the dark. The synthesis rate of glycogen during light period is not uniform, being highest at the onset of illumination ($2.6 \, \mu g \cdot \mu g^{-1} \, chl \cdot h^{-1}$) (Liu 1988).

When CO_2 content in the gas phase increase to 1%, the biomass increases to 1.6–1.8 times. During the time of transition of CO_2 concentration from low to high level, the biomass increases sharply to $6.1 \, \mu g \cdot \mu g^{-1} \, chl \cdot h^{-1}$, and the glycogen is up to 56% of dry weight. Afterward, the glycogen content decreases gradually to a new semi-steady state. A possible model for the CO_2 and carbonate transport system of the cell is proposed (Fig. 3) (Liu 1988; Liu et al., in press).

Addition of ammonium to the culture stimulates the increase of biomass, but the accumulation of glycogen decreases. Supplying NH_4^+ and enriching the culture with CO_2 simultaneously lead to an enhancement of biomass but not for the accumulation of glycogen in cells. (Liu et al., in press).

Nitrogenase activity is low before the onset on light period, but increases rapidly in the following hours after the exposure to light. Light-induced nitrogenase activity reaches to its maximum in 4–6 h after light-on. Then drops due to the enzyme overturn. Nitrogenase synthesis in the heterocystous BGA *A. variabilis* only occurs in the early hours of light period.

Supplying ammonium and chloramphenicol to the culture at the onset of light, the nitrogen fixation is inhibited strongly. Throughout the diurnal cycle, nitrogenase activity in the dark is only one fifth in the light. No correlation is found of the dark activity with the glycogen content of the whole filaments. Glycogen derived from photosynthesis does not determine the nitrogenase activity in the dark.

The nitrogenase activity is also estimated from the increase of total nitrogen measured. The number of electrons necessary for ammonium formation indicates that no free hydrogen is produced. More than 90% of the nitrogen fixation occurs in the light period. The nitrogen assimilation of *A. variabilis* is essentially light-dependent (Liu 1988).

4.2 Regulation

4.2.1 Regulation by Ammonia

External ammonia rapidly and reversibly repress the nitrogenase activity of BGA. Although there is specific difference with various strains in ammonia repression, the difference is inconspicuous. 10 species of nitrogen-fixing BGA in our experiments were tested. No noticeable difference was found in their repression. Adding 1 mM NH_4Cl in the medium, all the strains lose 100% of their nitrogenase activity within 48 h. So it is very difficult to screen for the ammonia-resistant strains from the wild-types.

The mechanism of repression of ammonia on the nitrogenase activity has been studied with *Anabaena azotica*, *A. cylindrica* and *Plectonema* sp. (Dai et al. 1987).

(1) Relation of glutamine synthetase to the repression on ammonia on nitrogenase activity of BGA.

The addition of L-methionine-DL-sulphesimine (MSX) to the growing culture of *Anabaena azotica* results in releasing newly fixed NH_3 into the medium and inhibiting the glutamine synthetase (GS) activity, showing a negative correlation between GS activity and ammonia production.

Under N_2-fixing condition, the half-life time of the H_2ase activity is about 3.5 h, while that of GS is longer than 10 h. This suggests that the GS may not be a positive regulator of nitrogenase.

Glutamine (Gln) shows a high inhibition of the nitrogenase activity, but glutamine even in its high concentration does not inhibit GS activity. Stewart (1980) indicated that in BGA, GS could remarkably regulate directly or indirectly the N_2ase activity. In our experiments, the N_2ase is indirectly regulated by GS.

(2) The relation between the inhibition of N_2ase by ammonia and other physico-chemical factors.

The inhibition of N_2ase activity by ammonia and the elimination of this action with MSX are connected conspicuously with the algal culture. This is due to the amount of energy charge stored in the cells.

The effect of pH value on the repression of N_2ase activity by ammonia is also studied. Under alkaline condition, 3 h after adding 1 mM NH_4Cl to the culture, the N_2ase activity loses 87%, while in neutral condition, only 40% lost. It seems that in alkaline ammonia enter into cells more quickly than in neutral condition. Furthermore, adding $5\,\mu M$ MSX to the neutral culture medium causes complete loss of the GS activity within 30 min, but in high pH value, e.g. at pH 9, only 60% of the activity was lost.

(3) Effect of oxygen on the repression of N_2ase activity by NH_4^+.

In argon, nitrogen or helium + 0.05 atm O_2, the NH_4^+ dose not repress the N_2ase activity. The repression happens only in high pO_2 or in air. It seems that the repression of N_2ase activity by ammonia is O_2-dependent.

(4) Effect of amino acids pool on the N_2ase activity.

Of six amino acids tested, glutamic acid, aspartic acid, arginine and serine show little effect on the N_2ase activity, but the glutamine and asparagine shown remarkably repressive effect.

(5) The possible site where the N_2ase synthesis is inhibited by NH_4^+.

Chloromycetin and kanamycin were observed to reduce the N_2ase activity, but there is no such effect by actidione. The former two inhibitors may bind with 50s and 30s subunits of the prokaryote ribosome, to block the protein synthesis on 70s ribosome. It suggested that 70s ribosome may be the site where N_2ase synthesis was inhibited by NH_4^+.

The mechanism of the repression of the N_2ase activity by NH_4^+ is summarizing as follows: 1) The pathway of the NH_4^+ metabolism in *Richelia sinica* and *Anabaena azotica* is : Glutamine GS/GOGAT pathway. 2) The NH_4^+ does regulate not only the expression of the N_2-fixing genes of these two BGA (the synthesis of N_2ase), but also regulate the activity of the N_2ase synthesized. The regulation is of genetic and physiological levels. The possible site for the NH_4^+ regulated N_2ase synthesis should be on 70s ribosome. 3) The regulation of BGA N_2ase by NH_4^+ is indirect, and may be through the metabolites and other related factors.

4.2.2 Regulation by Oxygen

The effect of partial pressure of oxygen (pO_2) on the N_2ase activity of *Anabaena cylindrica* grown either in N_2-free or N_2-sufficient conditions respectively shows that the excessive oxygen in environment inhibits the N_2ase activity. In the strict anaerobic condition, the activity of N_2ase is limited. However, in the microaerobic condition (0.001–0.1 atm pO_2), the activity of N_2ase is promoted and can reach to a maximum. The N_2ase activity of cells grown in N_2-free condition is inhibited by 50% in 0.5 atm pO_2, but in the cells grown in N_2-sufficient media, the same degree of inhibition was observed by 0.35 atm pO_2. It shows that an increase in N_2ase activity and a decrease on oxygen sensitivity are found simultaneously in the cells of *A. cylindrica* grown in N_2-free condition (Heet et al. 1985). It suggests that the enhanced nitrogenase activity is possibly due to an increase in the respiration rate of the BGA.

4.3 Evolution of Hydrogen

The BGA are the organisms that evolove H_2, while fixing N_2. In the heterocystous BGA, the site of nitrogen fixation occurs mainly in heterocysts. When the nitrogenase catalyzes the reduction of N_2 to NH_3, the H^+ will also be reduced to H_2. There are two hydrogenase i.e. uptake hydrogenase and reversible hydrogenase.

The activity of H_2 evolution in light is possibly reaching to 10–15% of O_2-evolving activity in photosynthesis and corresponds to 50–80% of acetylene reduction activity.

The N_2-fixing and H_2-evolution of BGA are intensively light-dependent. Firstly, the light intensity for growth affects the proportion of heterocysts in heterocystous filamentous strains. The algae *Anabaena azotica* and *A. variabilis* grew under light of 1×10^3, 7×10^3 and 30×10^3 lux, the proportion of heterocysts in total cell countings is 1%, 5–7% and 8–11%, respectively. It means that the heterocyst number of the algae grown in the higher light intensity was about 10 times of that in the lower light intensity (Zhang et al. 1983a). In BGA, both nitrogen-fixing and H_2 evolution activities enhance with increase of light intensity. However, there is a light saturation point for nitrogen-fixing activity. With *Anabaena* spp.strains CA, the light saturation point closes to that of their photosynthesis, i.e. approximately 100 $\mu E.m^{-2}.s^{-1}$. But there is no such a light saturation point for H_2 evolution. H_2 production activity increases with light intensity up to 400 $\mu E.m^{-2}.s^{-1}$. (Zhang et al. 1983). The comparison studies on the phenotypic Hup^- mutant have shown that this is due to the activity of uptake hydrogenase. The uptake H_2ase activity is inversely related with light intensity for growth. In nitrogen-fixing BGA *Anabaena* spp. strains CA grown under 400 $\mu E.m^{-2}.s^{-1}$, the O_2-dependent H_2 uptake activity is close to zero (Zhang et al. 1986). Although the mechanism of this light regulation of H_2 uptake activity is still unknown, it has been clear that there is no remarkable relations with the O_2 content in the normal condition (unpublish data of Zhang).

It was found that the presence of Ni^{2+} and its concentration in the culture medium is the key factor regulating the net production of H_2 under the aerobic condition (Zhang et al. 1984). The uptake hydrogenase induction is strictly dependant on the Ni^{2+} concentration in the growth medium. When Ni^{2+} is as low as 10 nM, the activity of uptake H_2ase rises 8–10 times with the net H_2 production decreased about 80–90%. In the BGA grown in the medium containing Ni^{2+} (50 nM), two kinds of activity of uptake H_2ase were detected: the O_2-dependent H_2-uptake activity in dark is about 30% of the H_2 evolution activity: the light-dependent anaerobic H_2-uptake activity was about 50% H_2 evolution activity. The kinetic studies, using H_2 electrode to monitor the effects of Ni^{2+} and chloramphenicol on the H_2 uptake, show that Ni^{2+} may activate the H_2ase, and may also be required for its synthesis. The addition of EDTA to the growth media will markedly affect the net H_2 evolution. It is mainly due to the inhibition of the activity of uptake H_2ase by EDTA (Zhang et al. 1987).

4.4 Protection of Nitrogenase from Oxygen

Algal N_2ase is extremely sensitive to oxygen. The protection of nitrogenase from oxygen is very complicated in BGA. The protection against inactivation by oxygen involves the external air as well as the endogenous oxygen, evolved in the photosynthesis. Therefore, the ways for protection in blue–greens must be of multiple. In our laboratory, we have been tackling this problem in its four aspects.

4.4.1 Obstruction Protection

It is believed that nitrogenase is located in the heterocyst, which has a multilayer envelope and cell wall to obstruct the diffusion of oxygen into the cell. Due to its high respiratory rate, heterocysts have low oxygen tension.

Using ultraviolet radiation or NTG mutagenesis, a oxygen-sensitive mutant of heterocystous *Anabaena* 7120 was isolated. The mutant can not fix nitrogen in the air. Under microaerobic conditions, the mutant is capable of utilizing N_2 for growth with a low acetylene reduction activity. 1% O_2 (v/v) in the gas phase reduced the acetylene reduction activity by 15%, when the O_2 content raises to 5%, the activity decreases by 85%. The higher ability of the mutant to reduce 2,3,5-triphenyltetrazolium, the higher sensitivity to hydrophilous reagents such as SDS, triton-x-100 and the frailty of heterocysts in vitro demonstrates that the cell wall structure of the mutant is incomplete, therefore the permeability of oxygen into the cell wall increases. The defect in the cell wall structure of the mutant cell leads to the loss of its O_2-obstructing function. If the outer cell membrane of the wild strain is treated with EDTA and Tris to modify its structural completeness, its sensitivity to hydrophobic reagents increases. Alkaline phosphatase activity becomes higher, the permeability of the cells increases markably, and the nitrogen-fixing activity becomes lower in air. But under anaerobic conditions, the nitrogen-fixing activity was not affected. It shows that the structure of outer membrane is closely related to the protection of nitrogenase from oxygen (Wang et al. 1989).

4.4.2 The Oxygen-Scanvenging System in Cells

It was found that the addition of the subcellular components of sonicated cells of *Anabaena* or *Gloeocapsa* to suspension of the mutant cells leads to a 10-fold increase in nitrogenase activity, or more, in the microaerobic condition. These components could restore the acetylene-reducing activity at higher O_2 concentration. Similar protective effects have been observed in *Plectonema boryanum*, when the cell suspension is treated with subcellular components as above (Li (Ley) et al. 1981; Wang et al. 1981; Wang et al. 1982; Wang et al. 1983).

The subcellular components from sonicated cells are prepared by centrifugation at $15\,000 \times g$ and $80\,000 \times g$ in sequence, and then separated through the DEAE-cellulose column in 0.3 N NaCl, and the stable and dialysable active components were obtained. When the components were added to cell suspensions of *Anabaena* mutant or *Plectonema*, the oxygen was scavenged effectively, and the nitrogenase activity was expressed within 2–3 h under 0.8 atm O_2, while the control cells showed no activity even at as low oxygen as 0.25 atm O_2.

We consider that the active components may be the constructive components of the thylakoid membrane or granular inclusions of the cells, while are autoxidizable and involved in the photoreduction of oxygen. Thus the oxygen in the external environment could be scavenged rapidly, and the nitrogenase is protected from the inactivation.

It is likely that the oxygen scavenging systems are multiplex in BGA. The protective devices and their efficiency may be various in different groups of BGA, so that the nitrogenase activity will be expressed at different oxygen levels.

4.4.3 Enzymes Related to the Protection from Oxygen

We have compared the activities of superoxide dismutase, superhydroxidase and superoxidase of BGA which fix nitrogen in the air and under microaerobic conditions. The activity of superoxide dismutase is the highest in *Anabaena* 7120, and then *Gloeocapsa*, the lowest in *Plectonema* of about 35% to that of *Anabaena*. The superoxidase activity is also very low in *Plectonema*. But on the contrary, it has the highest activity of superhydroxidase. It seems that the superoxide dismutase may be more significant in protection of nitrogenase from oxygen in BGA

4.4.4 Relationship of Protection from Oxygen to Environmental Conditions

Nitrogen fixation of *Anabaena* has close relations to the light-dark alternation. The nitrogen fixation mainly takes place in the light period. The fixing activity decreases rapidly in the dark. But in the anaerobic condition, the decrease is rather slow. Comparison of the oxygen effects on the nitrogenase activity in the light and in the dark shows that the inactivation of nitrogenase in the dark is more rapid than that of in the light. The inactivation rate is in proportion with the oxygen content. This phenomenon does not result from the insufficient reductants and the energy supply. Because under the anaerobic conditions, no decrease in nitrogenase activity was found within 2.5 h in the dark, and 70% activity was retained even after 9 h in dark. It indicates that the nitrogenase is more easily and rapidly inactivated in the dark (Wang et al. 1987).

It is still unclear whether the oxygen is permeating into the heterocyst more easily or the decrease in oxygen scavenging ability of the cells happens in the dark. We have observed that the increase of the superoxide dismutase and superoxidase in *Anabaena* cells can be induced by the light intensity. These results show that the mechanism of protection of nitrogenase from oxygen is involving the process of light reaction. In the dark, the nitrogenase is inactivated and degraded by oxygen. Because of the synthesis of nitrogenase after the restore of light, the nitrogen fixation reappears. In this way the nitrogen fixation takes place.

The oxygen sensitivity of nitrogenase in *Anabaena* 7120 cells also relates to the temperature. In general, the sensitivity of nitrogenase to oxygen is increased in 38–40 °C. It is connected with the optimal growth temperature. The thermophilous strains of BGA shows the highest activity under such high temperature (Wang et al. 1983).

4.5 Differentiation of Akinete

Akinete differentiation in *A. azollae* begins at the late exponential phase of growth in batch culture. The process of akinete differentiation is heavily suppressed in the culture depleted of Fe or Ca ions, while phosphorus deficiency induces the formation of akinete. The induction by phosphorus deficiency can be enhanced by the increase of nitrate concentration. The akinete formation in the culture increases up to 41.3%, 42.2% and 68.8% in the presence of 1 mM 5-methyl-DL-tryptophan, 10 mM L-arginine and 10 mM DL-phenylanine, respectively.

Light limitation is probably one of the most important external factors triggering the initiation of akinete formation in *A. azollae*. The light and organic carbon source regulate the akinete differentiation. Under certain conditions, organic sources influence the process of differentiation of akinete in the same way as light does. It is suggested that the failure of akinete formation in some cells in the culture condition is caused by the lack of cell reserve accumulation.

The DNA content in akinete is three times more than that of vegetative cell (in akinetes of *A. sphaerica* 16 times more). The activities of glucose-6-phosphate-dehydrogenase, hexokinase and α-amylase declined sharply during the course of differentiation. The activity of catalasse changed gradually during the process.

We propose that the akinete differentiation in *A. azollae* is mainly controlled by the available energy in the cells (Song 1988).

5 Mutagenesis and Transformation: Genetic Approach

5.1 Mutagenesis

By using Ultraviolet (UV) or nitrosoguanidine (NTG) as mutagen, we have obtained various types of mutants, such as oxygen-sensitive heterocystous mutants of *Anabaena cylindrica*, nitrate reductase mutants, short filamentous heterocyst-minus mutant of *A. cylindrica*, tryptophan overproducing mutant of *A. spaerica*, etc. With regard to the efficiency of mutagenesis, NTG seems better than UV. As a mutagen, for example, after treatment with 100 µg ml^{-1} NTG

25% of *A. sphaerica* 1017 cells survive, the frequency of mutation for 6FT-resistance tryptophan overproducing strains is 4×10^{-6}, while no mutants were found in the UV treated culture. However, mutants induced by NTG were more likely subject to the spontaneous back mutation than that by UV. The frequency of spontaneous reverse mutation of the NTG induced oxygen-sensitive mutant *Anabaena*-71 is 1.9×10^{-6}, while the frequency for the UV-induced *Anabaena*-1 of the same phenotype is less than 10^{-8} (Li et al. 1987; Wang et al. 1981; Wang et al. 1987).

5.2 Transformation

The DNA isolated from the heterocystous *Anabaena* 7120 cells was used as a donor. Oxygen-sensitive nitrogen-fixing mutant *Anabaena*-1 was used as recipient. Of a large number of tests, only two of them showed positive results. The transformant are capable of growing in a medium without the combined N source in aerobic condition. Their specific activity for acetylene reduction is similar to that of the wildtype. The results showed that the oxygen-scavenging system was recovered by transformation of the oxygen-sensitive mutant *Anabaena*-1. The transformation frequency was about $10^{-6} - 10^{-5}$. But the transformation condition is not capable of replication (He et al. 1984).

Transformed by plasmid PACYC, *Anabaena*-1 strain was obtained, indicating that the chloramphenicol-resistant gene was expressed in the recipient, however, it is difficult to transform the chloramphenicol-resistant gene into the *Anabaena* widtype. It seems to be related with the structure of the cell wall. The transformation condition is also difficult for replication. It suggests that the transformation of *Anabaena* is more complicated than that of unicellular BGA (Wang ct al. 1988).

It was found that about half of the BGA strains isolated contained plasmids. We have also obtained sphaeroplast from *Anabaena cylindrica*, and found that the sphaeroplasts form regenerated algae colonies after 9 days. The cells regenerated may develop up to 20 cells filament (Kong 1987).

6 Physiological Ecology

Temperature, light intensity, pH, mineral nutrients, microguantity-elements, organic matters, pesticides, and population interactions were studied in the connection with the growth and nitrogen fixation of the nitrogen-fixing BGA, particularly of *Anabaena azotica* 686, both in the laboratory and field conditions.

Although the BGA can survive extremely severe environment, they proliferate mostly in the subtropics and tropics. The optimum temperature is $28-38\,^{\circ}\text{C}$ for the growth and nitrogen fixation of *A. azotica* 686. The algae grow slowly below 20 °C, however, if the temperature is higher than 45 °C, they usually become bleached and die out. In summer, the water temperature in the paddy field may reach as high as 40–45 °C at noon, the diurnal fluctuations of temperature allow the algae adapt themselves to grow in this habitat (Cui et al. 1983; Ley et al. 1959; Ley et al. 1963; Li (Ley) et al. 1982; Li (Ley) et al. 1983; Shen et al. 1990; Wang, MSS).

The light is a very important factor for growth and nitrogen fixation in BGA. Below the saturated light intensity, there is a close correlation between the light intensity and growth rate of *A. azotica* 686. The saturated light intensity of *A. azotica* is 30 000 lux. Up to 50 000 lux, the growth rate and nitrogen-fixing ability decrease but slightly. In summer, the light intensity may reach 100 000 lux with high temperature in the open field, and the algae may be scorched. The surface of the algal mat floating on the water becomes dry, whitish and crustlike. It suggests that part of the algae died. Therefore, algae growing in the summer bright day should be shaded to reduce the dosage of irradiation (Cui et al. 1983; Ley et al. 1963; Li (Ley) 1981; Li (Ley) 1984a; Li (Ley) et al. 1983). The effect on the colour change and pigment content is not so conspicuous as in *Tolypothrix tenuis*.

The BGA have been adapted to grow in neutral or slightly alkaline habitats. They can grow in media from pH 6 to 10. In the red soil area of South China, where the water and soil in paddy fields are more-or-less acid, the conditions are unfavorable to BGA. However, Chinese farmers are accustomed to using farm manures and lime to improve the soil, bringing the water in the paddy to around pH 7. If calcium-magnesium phosphate fertilizers and grass ashes are applied to the paddy during the inoculation of BGA, the algae grow well (Ley et al. 1963; Li (Ley) 1981a; Li (Ley) et al. 1982; Li (Ley) et al. 1983; Liu et al. 1989).

The range of phosphorus for the best growth of nitrogen-fixing BGA in the culture medium is 5–10 ppm. The higher nitrogenase activity and nitrogen content of the cells were observed when they grew at 10 ppm P. Phosphorus concentrations in the culture medium below or beyond 10 ppm would result in a rise in the value of the cellular C/N ratio. The external phosphorus concentration favours the formation of phosphorus compounds in the cells. But the cellular alkaline phosphatase activity would be sacrificed when the external phosphate or total cellular phosphorus level raises (Li (Ley) 1981a; Li (Ley) et al. 1983; Wang, MSS; Wang et al. 1963; Zhang 1985).

It was known that ammonium represses the nitrogenase activity. In the presence of ammonium, the nitrogen fixation of *A. azotica* 686 is inhibited completely, and the exogenous combined nitrogen would be the only source for the algal growth. Because of the strong absorption of ammonium by soil, the water contains little of it even in 1–3 days after application of the ammonium sulfate and ammonium bicarbonate as the base fertilizer in the paddy field. However, it does not affect the growth of BGA (Shen et al. 1983).

The metals and microquantity-elements required for growth and nitrogen fixation were shown in Table 4. It must be pointed out that the concentrations of each of the ions optimal for the algae are dependant upon the compounds

Table 4. Requirements of phosphorus, metals and boron by *Anabaena azotica* 686

Elements	Optimal conc. $(mg^{-1} l^{-1})$	Comments
P	10–15	—
Fe	1–1.5	Required for biosynthesis of iron sulfur protein, Fd, N_2ase
Na	5–10	—
K	10–15	—
Ca	20–30	More Ca^{2+} appeared to be required for growth on N
Mg	10–15	Constituent of chlorophyll
Cu	0.01–0.03	—
Mn	0.01–0.1	Required for oxygen production, Hill reaction
Mo	0.5	Constituent of nitrogenase
Co	$0.4–1.0 \mu g^{-1} l^{-1}$	—
B	0.01–0.1	—

they contain. For example, the optimal concentration of iron ions for *A. azotica* 686 is 1 ppm with ferrous sulfate, but 1.5 ppm with ferric chloride and 2 ppm with ferric citrate. And the organic metals seem to be more effective than their inorganic form (Li (Ley) et al. 1983).

A. azollae can grow heterotrophically in the dark. Fructose, glucose and sucrose supported the heterotrophic growth of *A. azollae*. Cultures adapted to phototrophic growth grow better in the dark when $NaNO_3$ is used as the nitrogen source. The concentration of chlorophyll *a* of the culture growth in the dark for 6 months decreases to 1/3–1/4 of that in the light. When *A. azollae* grow at 5500 lux, exogenous fructose and glucose still stimulate growth and increase nitrogen fixation (Jin 1884).

The population interaction between two species, *Tolypothrix tenuis* and *Anabaena variablilis* 1058 are studied in the axenic culture. The allelopathic action of the specific population is taking place through the extracellular products. In our experiments, three types of actions were found. (1) The lethal agents of these two species are different and they affect each other. (2) Inhibition of growth by each other is observed in one community. (3) The extracellular products of *A. variabilis* 1058 promote the growth of *T. tenuis*. Direct competitive interaction occurs between populations. This competition is controlled by the extracellular products which may regulate the structural composition and succession of the community (Liu et al. 1991).

7 Mass Culture and Utilization

The potential significance in agricultural applications of BGA is obvious. But how to obtain the huge amount of biomass is important for their exploitations. We have been, therefore, developing techniques for mass cultivation as well as

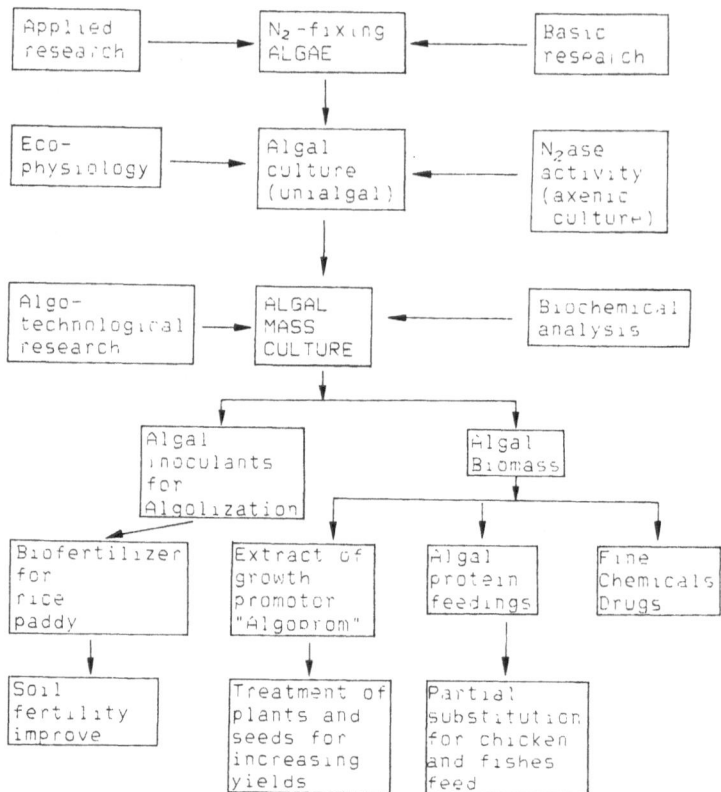

Fig. 4. A schematic representation of the general research programmes on nitrogen-fixing blue-green algae carried out in the Department of Phycology, Institute of Hydrobiology, Chinese Academy of Sciences, China

utilization since 1958. The whole program of our research work is shown schematically in Fig. 4.

7.1 Mass Production of Biomass

From several hundred samples collected from Hubei, Hunan and Guangdong provinces, we have selected the strains with high growth rate, high nitrogen-fixing ability and good adaptability to the environment by screening the unialgal culture in nitrogen-free medium followed by assaying the nitrogen-fixing capacity in the axenic culture. *Anabaena azotica* 686, *A. oscillarioides, A. sphaerica* 1017, *A. variabilis* 1058, *A. oryzae* 13, *Anabaena* sp. 1042, *Anabaena* sp. 1105, *Aulosira fertillissima, Tolypothrix tenuis* and *Richelia sinica* were further examined for potential inoculant strains. (Li (Ley) 1981a; Li (Ley) 1981b; Shen et al. 1990; Wang, MSS)

On the basis of the culture behaviours both in the laboratory and in the plastic sheet covered nursery bed in the open-air, we selected eight strains, and the mixture of them was used as an inoculant, "Mixed Algal Inoculant" (MAI), and MAI was cultivated in large scale. The MAI in mass culture adapted to the changing wheather. The different strains have their different physical requirements. Some strains in the MAI could grow in a broad range of weather conditions.

We have developed methods for mass cultivation and production of the MAI as the biofertilizer used in the late rice paddy. They are (Li (LEY) 1981a; Wang, MSS; Liu et al. 1990):

(1) Cultivation in a plastic sheet (PVC)-covered nusery bed in the field. The bed was inoculated with 50–100g fresh MAI per m². In general it could develop to 1500 g in 5 days, and may reach 2000g in three days at an optimal condition.

(2) Cultivation during the rest period between two crops. About 380–600 kg fresh MAI per ha or 22–37 kg dry MAI powder per ha have been used as inoculant usually. The BGA may grow up to 7500 kg per ha in 3–5 days, although up to 12 750 kg per ha has been recorded.

(3) Cultivation under PVC covered canopy. In recent years, we have established a microalgae culture and production base in Huangmei county (30 °N, 115 °E), Hubei province. There are 5170 m² shallow ponds under 50 canopies. The MAI are cultivated in those shallow ponds with 150–250 g fresh inoculant per m². It can be harvested in a day or two, with 100 g per m² algae left as inoculant for continuous growth. In this semi-continuous culture, 7–11 g dry weight $m^{-2}d^{-1}$ of biomass is harvested in average in warmer seasons. The maximal production is $22 g m^{-2} d^{-1}$. The cost per kilogram of dry algal biomass produced in this way is around 0.75–1.00 U.S. dollars.

7.2 Uses as Biofertilizer

Rice is the major crop in China and covers about 22.5 million ha. Most of the late rice paddy (second crop) is nitrogen dificient. The utilization of nitrogen-fixing BGA as biofertilizer for the late rice paddy has great significance in increasing crop yield. Long term plot tests and field trials have been done to evaluate possibility of using BAG as biofertilizer for the late rice paddy (Ley et al. 1959; Li (Ley) 1981; Li (Ley) 1981a; Li (Ley) et al. 1983; Tseng et al. 1959). The rice grew better, with more tillers following inoculation, and the grain output increased. The inoculation of fresh MAI (380–600 kg ha⁻¹) or dry MAI (22–37 kg ha⁻¹) was carried out three to six days after rice seedlings transplant. The BGA would grow up to 7500–12750 kg ha⁻¹ (fresh weight) and the algal mat covers fully the water surface in 3–5 days, we call it "Algalization" . And then the BGA mat will be put down into the soil by intertillage 10–14 days after the transplantation of rice seedlings.

The rates of nitrogen fixation are determined by using the acetylene reduction technique *in situ*. In general, the level of nitrogenase activity is 1.3–3.8 μmoles $C_2H_4g^{-1}$ dry alga min^{-1}, although up to 5.8 μmoles have been recorded. A study of the diurnal variations in activity shows that the peak activity appears at about 09:00 o'clock, and then decreases gradually during the rest of the day, and is undetectable from evening through dawn. Based on the growth rate and nitrogenase activity, we estimated that the fixed nitrogen during the late rice crop was 24.5–29 kg N ha^{-1}, the efficiency of which is comparable to that when ammonium sulfate was applied at the density of 72–144 kg ha^{-1}.

The total trial areas of the late rice paddy scattered within Hubei are 173 ha in 1976, 4333 ha in 1977, 20,000 ha in 1978, and 23,000 ha in 1979. On the average, there are about 15% increase in yield over that of the uninoculated fields.

7.3 Uses as Animal Feed

The blue–greens are rich in protein, essential amino acids, vitamins and other required nutrients for animal growth (LPIHB 1984).

The *Anabaena* mixtures cultivated in the Algal Culture Base have been used as mixed feeds for poultry and fishes. We have tried to feed ten thousands of chickens with *Anabaena* mixture powder as a substitute for part of their feed (about 67%–100% of fish neal in mixed feed). The chickens fed in this way grow as good as the control with nicer fresh colour, and the eggs of reddish yolk.

The *Anabaena* mixture powder is also used as part of the mixed feeds for carp and grass carp. The results showed that the *Anabaena* powder promoted the growth of carp and grass carp.

7.4 Growth-Promoting Extracts Used as Growth Factor for Crop Plants

In 1964, we found that the rice seedlings in the nursery bed inoculated with *Anabaena* grew much better than those without inoculating. Since then, a series of experiments done in the pot and paddy fields showed that there seemed to be growth factors in the *Anabaena* cells (LPIHB, MSS; Li (Ley) 1988). In the 1970s, an extract of algal mixture was used to soak the seeds and sprinkle the young plants. The seeds sprouted much better, and the growth rate of crop plants was also improved, compared with that of the control.

After the separation with various methods, we have obtained a heat-stable, chromatographic pure substance, with a light absorption peak at 210 nm. We call it "Algoprom".

In 1980s, in collaboration with the Heilongjian Institute of Agricultural Moderization, Chinese Academy of Sciences, we applied "Algoprom" to the seed of rice, spring wheat and maize with success. The increase of crop yield was by about 7–10%.

7.5 H₂ Evolution—New Source of Renewable Energy

Since the late 1970s, in an effort to search for renewable energy sources, we began to work on the H_2 evolution of blue–greens. But the results obtained were often contradictory. We think that besides the difference in strains used, this discrepancy was caused by the growth factors which we have mentioned above. For example, most cultures were used without EDTA, and Ni^{2+} contamination was likely which would make 10 times variations in H_2 evolution. So we have obtained two phenotypic Hup mutants N9A and 18A from the fast-growing wild-type *Anabaena* spp. CA. with high N_2ase activity. These two mutants are quite similar in physiological characters such as growth rate, chlorophyll content, photosynthetic O_2 evolution and even in C_2H_2-reduction. They evolve continuously H_2 at the same rate of acetylene reduction in the normal aerobic BGA growth condition, and no hydrogen uptake activity can be detected. It demonstrated that there may be some mechanism by which the N_2ase activity is protected, and sufficient energy is made available for N-fixation. (H_2ase) seems not to be involved in the nitrogen fixation of blue–greens.

The H_2-evolution is, therefore, a potential energy source (Zhang et al. 1986).

8 References

Becking JH (1961) Studies on nitrogen-fixing bacteria of the genus *Beijerinkia*. Plant and soil 14: 49

Bishop P, Jarleuski DML, Hetherington DR (1980) Evidence for an alternative nitrogen fixation system in *Azotobacter vinelandii*. Proc Natl Acad Sci USA 77: 7342

Bortels H (1940) Uber die Bedeutung des Mglybands fur stick-stoffbindende Nostocacean Arch Mikrobiol 11: 155

Bulen WA, LeComte JR (1966) The nitrogenase system from *Azotobacter*: two enzyme requirements for N_2 reduction, ATP dependent H_2 evolution and ATP hydrolysis. Proc Natl Acad Sci USA 56: 979

Burns RC, Fuchsman WH, Hardy RWF (1971) Nitrogenase from vanadium-grown *Azotobacter*: isolation, characteristics, and mechanistic implications. Biochem Biology Res Comm 42: 353

Carr NG, Whitton BA (1982) The biology of cyanobacteria. Univ of California Press, Berkeley, Los Angeles, p 688

Cui X, Wang Q, Li S (SH Ley) (1983) Diurnal variation of nitrogen-fixing activity of blue–green algae in rice field. Wuhan Bot Res 1: 269

Dai L, He H, Lin H (1985) An anaerobic preparative electrophoretic method for purification of nitrogenase component II (Fe-protein) from *Anabaena cylindrica*. Plant Phys Comm 1: 50

Dai Lingfen, He Hui, Lin Huimin (1987) The role of glutamine synthetase in regulating nitrogenase activity in *Anabaena Azotica*. Acta Hydrobiol Sinica 11: 344

Dai Lingfen, He Hui, Lin Huimin (1990) A second nitrogen fixation system in cyanobacterium *Anabaena azollae*. Acta Hydrobiol Sinica 14: 183

Du Daixian, Lin Huimin, He Zhenrong, Dai Lingfen, Xin Wusheng, Li Shanghao (SH Ley) (1987) Purification and some properties of Fe-protein of nitrogenase from *Anabaena cylindrica*. Oceanologia et Limnologia Sinica 18: 358

Fogg GE, Stewart WDP, Fay P, Walsby AE (1973) The blue–green algae. Academic Press, London, New York, p 459

Hallenbeck PC, Kostel P, Benemann R (1979) Purification and properties of nitrogenase from the cyanobacterium, *Anabaena cylindrica*. Eur J Biochem 98: 275

He Jiawan, Wang Yeqin, Li Shanghao (SH Ley) (1984) Heterocystous Anabaena 7120 DNA in transformation. Acta Genetica Sinica 11: 100

He Zhenrong, Lin Huimin, Du Daixian, Dai Lingfen, Xin Wusen, Li Shanghao (SH Ley) (1985) Cross-reactivity of Fe-protein of Anabaena cylindrica and MoFe-protein of Azotobacter vinelandii. Acta Hydrobiol Sinica 9: 241

He Zhenrong, Du Daixian, Lin Huimin, Dai Lingfen, Xin Wusen (1985) The respiration protecting nitrogenase against oxygen in Anabaena cylindrica. Acta Hydrobiol Sinica 9: 324

Hong Ying, Li Yaoying, Li Shanghao (SH Ley) Preliminary study on the blue-green algae community of aird soil in Qaidam Basin. Acta Bot Sinica (submitted)

Jin Chuanyin (1984) Physiology of heterotrophic growth if blue-green algae Anabaena azollae. Acta Hydrobiol Sinica 8: 443

Kong Fangxiang (1987) Isolation, cultivation and regeneration of sphaeroplasts of blue-green algae. Master thesis, Institute of Hydrobiology, Academia Sinica

Ley SH, Yeh TC, Liu FJ, Wang LM, Tsui SK (1959) The nitrogen fixation of some blue-green algae from Chinese ricefield. Acta Hydrobiol Sinica 429

Ley SH, Yeh TC, Liu FJ, Tsui SK (1959a) The effects of nitrogen-fixing blue-green algae on the yields of rice plant. Acta Hydrobiol Sinica 440

Ley SH, Wang QL, Tseng CM (1963) Studies on the growth and reproduction of nitrogen-fixing blue-green algae I Effects of temperature and pH on the growth and nitrogen fixation of Anabaena azotica. Comp Abst 30th Ann Symp Bot Soc China, p 25

Ley SH, Yeh TC, Liu FJ, Wang LM (1963a) Studies on the growth and reproduction of nitrogen-fixing blue-green algae. III Effects of microelements on the growth and nitrogen fixation of Anabaena azotica. Comp Abst 30th Ann Symp Bot Soc China, p 26

Li Qinsheng, Li Shanghao (SH Ley) (1981) Bacteria that lyse nitrogen-fixing blue-green algae. Acta Hydrobiol Sinica 7: 377

Li Qinsheng, Li Shanghao (SH Ley) (1984) Ecological factors restricting blue-green algae population. Acta Ecol Sinica 4: 310

Li Qinsheng, Wang Yeqin, Li Shanghao (SH Ley) (1985) Plasmids of Flexibacter chinenses, a bacterium lysing Calothrix. Symposium of Genetics in Hubeoi (Abstracts)

Li Shanghao (SH Ley) (1981) Dinitrogen-fixing algae and their role in crop yield of rice. In: Emejuaiwe SO et al. (eds) Global impacts of applied microbiology. Academic Press, New York, p 287

Li Shanghao (SH Ley) (1981a) Studies on the nitrogen-fixing blue-green algae as fertilizer in the late rice crop. Acta Hydrobiol Sinica 7: 417

Li Shanghao (SH Ley) (1981b) Role of nitrogen-fixing blue-green algae in rice cultivation in China. In: Gibson AH, Newton WE (eds) Current perspectives in nitrogen fixation. Australian Acad Sci, Canberra, p 500

Li Shanghao (SH Ley), He Tianfu, Cui Xiquan (1982) Physiological ecology of Nostoc sphaerioides. 1st Int Phycol Congress, Abstracts

Li Shanghao (SH Ley) (1988) Cultivation and application of microalgae in the People's Republic of China. In: Stadler et al. (eds) Algal biotechnology, Elsevier London, pp 41

Li Shanghao (SH Ley), Wang Qianlin (1983) Nitrogen fixing blue-green algae, a source of biofertilizer. In: Tseng CK (ed) Proc of the Joint China-US Phycology Symposium. Science Press, Beijing, p 479

Li Shanghao (SH Ley), Wang Yeqin, He Jiawan (1981) Agents protecting nitrogenase from oxygen in blue-green algae. In: Current prespectives in nitrogen fixation. Australian Acad Sci Canberra, p 451

Li Yansen, Wang Yeqin, Li Shanghao (SH Ley) (1987) Selection and analysis of tryptophan regulatory mutants of Anabaena sphaerica 1017. Research on Agriculture Modernization 11: 47

Lin Huimin, He Zhenrong, Du Daixian, Xin Wusen, Li Shanghao (SH Ley) (1984) Studies on nitrogenase of blue-green algae. In: Veeger C, Newton WE (eds) Advances in nitrogen fixation research. Nijhoff/Junk

Lin Huimin, He Zhenrong, Du Daixian, Dai Lingfen, Xin Wusen, Li Shanghao (SH Ley) (1986) Studies on nitrogenase Mo-Fe protein of blue-green algae. Acta Hydrobiol Sinica 10: 16

Liu Mei, Liu Qifan, Zhang Xiankong, Li Shanghao (SH Ley) (1985) Studies on chlorophyll-protein complexes isolated from N_2-fixing blue-green algae Anabaena sp 7120 and their spectrum character. Acta Hydrobiol Sinica 9: 1

Liu Shimei, Li Shanghao (SH Ley) (1991) Interaction between two populations of the blue-green algae. Acta Bot Sinica 23: 110

Liu Yongding (1988) Metabolism of carbon and nitrogen in nitrogen-fixing cyanobacteria. Inst of Hydrobiology, Acad Sinica, p 138. Dissertation (PhD)

Liu Yongding, Ernst A, Böger P (to be submitted) The influence of CO_2 limitation on glycogen accumulation in *Anabaena variabilis* (cyanobacterium) and its relationship with CO_2 uptake

Liu Yongding, Li Shanghao (SH Ley) (1989) Species composition and veritcal distribution of blue–green algae in rice field soils, Hubei, China. Nova Hedwigia 48: 55

Liu Yongding, Wang Qianlin, Shen Yinwu, Jin Chuanyin, Lu Jinshu, Zhu Jiamin, Li Shanghao (SH Ley) (1990) Ecophysiological problems and the responsive techniques in the mass cultivation of *Anabaena* spp. (blue–green algae) in large-scale. Third Natl Conf Chinese Society of Phycology (Abstracts)

Liu Yu (1987) Comparative study on the growth promoting substance activity from various blue–greens. Inst Hydrobiol, Acad Sinica, p 67. Thesis (Master degree)

LPIHB (Laboratory of Phycology, Institute of Hydrobiology, Academia Sinica) (1980) The nitrogenase activity and certain properties in several N_2-fixing blue–green algae. Acta Hydrobiol Sinica 7: 57

LPIHB (1980a) Complementary functioning of nitrogenase of *Anabaena azotica* and *Azotobacter vinelandii*. Acta Hydrobiol Sinica 7: 61

LPIHB (1984) Utilization of nitrogen-fixing blue–green algae as feed for chickens. Report, Institute of Hydrobiology, Academia Sinica, 1984

LPIHB, Growth-promoting substance from nitrogen-fixing blue–green algae. MSS

Ludden PW, Burris RH (1976) Activiting factor for the iron–protein of nitrogenase from *Rhodospirillum rubrum*. Science 194(4263): 424

Mckenna CE, Benemann JR, Traylor TC (1971) Vanadium-containing nitrogenase preparation: implications for the role of molybdenum in nitrogen fixation. Biochem Biophys Res Comm 41: 1501

Roger PA, Watanabe I (1982) Research on algae, blue–green algae, and phototrophic nitrogen fixation at IRRI (1963–81), summarization, problems and prospects. IRRI Research Paper Series Philippines 78, p 21

Shen Yinwu, Wang Qianlin, Cui Xiqun, Li Shanghao (SH Ley) (1983) Effects of NH_4-nitrogen on nitrogen-fixing activity by blue–green algae. Wuhan Bot Research 1: 275

Shen Yinwu, Wang Qianlin, Li Shanghao (SH Ley) (1990) Study on some characteristics of *Richelia sinica*. Ocenologia et Limnologia Sinica 21: 4

Smith RV, Evans MCW (1971) Nitrogenase activity in cell free extracts of the alga, *Anabaena cylindrica*. J Bact 105: 913

Song Lirong (1988) Studies on akinete differentiation in *Anabaena azollae*. Inst of Hydrobiol, Acad Sinica, p 131. Dissertation (PhD)

Stewart WDP (1980) Some aspect of structure and function in N-fixing cyanobacteria. Ann Rev Microbiol 34: 397

Tseng CM, Yeh TC, Ley SH, Liu FJ, Wang LM (1959) A method for conservation of the nitrogen-fixing blue–green algae for inoculation. Acta Hydrobiol Sinica 452

Venkatarmen GS (1972) Algal biofertilizer and rice cultivation. Today and Tomorrow, New Dehli

Wang LM, Liu FJ, Yeh TC, Tsui SK, Ley SH, Tseng CM, Yu CL, Shih YC (1959) Effect of some fungicides and insecticides on the growth of nitrogen-fixing blue–green algae. Acta Hydrobiol Sinica 456

Wang Qianlin, Mass culture of nitrogen-fixing blue–green algae. MSS

Wang Qinlin, Liu QF, Peng YF, Shi YZ, Ley SH (1963) Studies on the growth and reproduction of nitrogen-fixing blue–green algae. II. Effect of phosphorus on the growth and nitrogen fixation of *Anabaena azotica*. Comp Abst 30th Ann Symp Bot Soc China, p 26

Wang Yeqin, Feng Bo, Li Shanghao (SH Ley) (1989) Relation of integrity of outer membrane with its protection from oxygen in *Anabaena* cell. Acta Botanica Sinica 31: 265

Wang Yeqin, He Jiawan, Dai Lingfen, Li Shanghao (SH Ley) (1981) Oxygen sensitive nitrogen-fixing mutants of *Anabaena*. Acta Botanica Sinica 23: 288

Wang Yeqin, He Jiawan, Li Shanghao (SH Ley) (1982) The oxygen-scavenging system protecting nitrogenase from oxygen in cell in blue–green algae. Acta Botanica Sinica 24: 231

Wang Yeqin, He Jiawan, Li Shanghao (SH Ley) (1983) Relation of temperature and oxygen sensitivity of nitrogen fixation in *Anabaena*. Nitrogen Fixation Communication 4: 56

Wang Yeqin, He Jiawan, Zhang Deping, Li Shanghao (SH Ley) (1987) Relationship between oxygen and nitrogen fixation of *Anabaena* 7120 in light, darkness and at high temperature (38–40 °C). Acta Hydrobiol Sinica 11: 247

Wang Yeqin, Li Shanghao (SH Ley) (1983a) Nitrate reduction dependent on pH and its regulation on nitrogen fixation in *Anabaena*. Nitrogen Fixation Communication 4: 57

Wang Yeqin, Xu Xudong (1988) DNA transformation and foreign gene introduction in sensitive mutant of *Anabaena*. Symp Biotechnology (Abstracts) 18

Wang Yeqin, Yang Lin, Feng Bo (1987a) The regulation of heterocyst differentiation and changes of cell proteins in *Anabaena* 7120. Acta Hydrobiol Sinica 11: 41

Wealand JL et al. (1989) Changes in gene expression during nitrogen stravation in *Anabaena variabilis* ATCC 29413. J Bacteriol 171(3): 1284

Yang Lin, Wang Yeqin, Feng Bo (1987) Proteases and some of their characteristics in nitrogen fixation *Anabaena* 7120. Acta Hydrobiol Sinica 11: 203

Yang Shao (1989) Isolation and comparison of the cytoplasmic, thylakoid membrane and cell walls of the filamentous cyanobacterium *Anabaena* 7120 and its oxygen sensitive mutant. Inst of Hysrobiol Acad Sinica, Thesis (MSc).

Zhang Xiankong, Ben Haskell J, Tabita RR, van Baalen C (1983) Aerobic hydrogen production by the heterocystous cyanobacteria *Anabaena* spp. strains CA and IF. J Bact 156: 1118

Zhang Xiankong, Liu Qifang, Liu Mei (1987) Simultaneous photoharvesting of hydrogen and biomass from heterocystous cyanobacteria. In: Int Phycotalk Symposium Abstracts. Benares, India, p 39

Zhang Xiankong, Liu Qifang, Liu Mei, Wang Houle, Ley Shanghao (1983a) Effect of light intensity on pigmentation, photosythetic O_2-evolution and N_2-fixation in blue–green algae. Proc 1st Chinese Phycological Symposium, p 29

Zhang Xiakong, Tabita FR, van Baalen C (1984) Nickel control of hydrogen production and uptake in *Anabaena* spp. strains CA and IF. J G Microbiol 130: 1815

Zhang Xiankong, Tabita FR, van Baalen C (1986) The phenotypic Hup⁻ mutants of the nitrogen fixing blue–green alga *Anabaena* spp. strains CA. Acta Hydrobiol Sinica 10: 217

Zhang Xiaomin (1985) Phosphorus nutrition of *Anabaena variabilis*. Inst Hydrobiol Acad Sinica, Thesis (MSc)

CHAPTER 27

Nitrogen Fixation of *Azolla* and its Utilization in Agriculture in China

Chung-Chu Liu and Wei-Wen Zheng

The Nitrogen Fixation
and its Research in China
Editor Guo-fan Hong
© Springer-Verlag Berlin Heidelberg 1992

1 Introduction

The tiny aquatic fern, *Azolla* is an eukaryotic fern plant associated with a prokaryotic blue–green alga. In 1785, Lamarck named *Azolla* as a genus under *Saliviniaceae*. Wettstein separated *Azolla* from *Saliviniaceae* and set up a new family, *Azollaceae*. In 1873, a German, Strasburger found a kind of blue–green alga which belonged to *Nostocaceae*, and was located in the *Azolla* leaf cavities and named it *Anabaena azollae*. For this reason, the *Azolla* plant has been called *Azolla-Anabaena azollae* association in the scientific literature (Fig. 1).

China is thought to be the earliest country to grow and utilize *Azolla* in the World. According to legend, the farmers in Wengzhou area, Zhejiang Province, used *Azolla* as paddy green manure 700 years ago (Liu 1979a). Up to now, it is still difficult to examine the exact time when Chinese people started to grow and utilize *Azolla*. The earliest written record so far found is a report of "Heap *Azolla* in Paddy Field" which was published in the "Agricultural Journal" by Hong Bing Wen in Rui-an County, Zhenjiang Province in Guangxu the 24th year in the Qing Dynasty (Liu et al. 1986b). It can be seen from this newsprint that farmers in the Wengzhou area at that time made an attentive observation on *Azolla* multiplying speed and its effect on manuring paddy fields. They accumulated abundant experience on the collection of *Azolla* germplam, over-wintering, preservation and multiplication of *Azolla* inoculum. In addition to

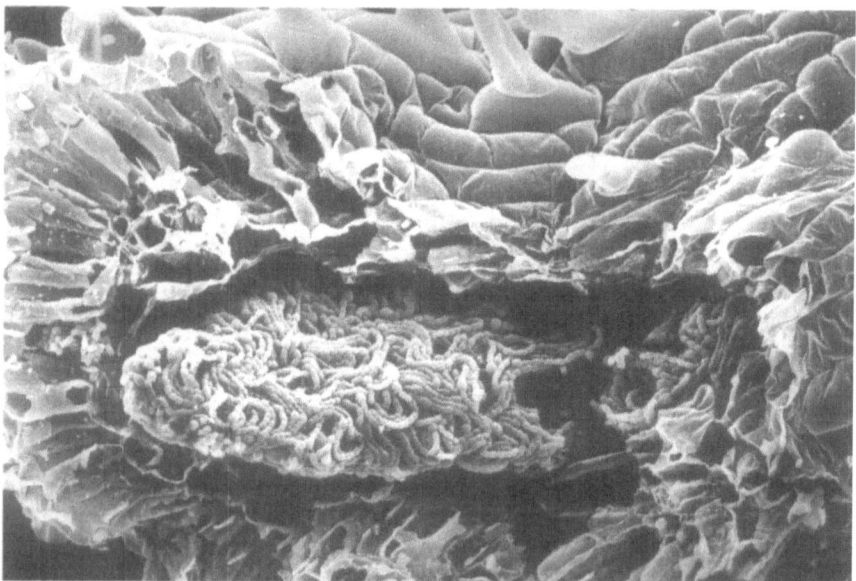

Fig. 1. Scanning electron micrograph of an *Azolla* leaf cavity, showing *Anabaena* filaments occupying the whole cavity. × 700

Zhejiang Province, in this country also there is a long history to grow and use *Azolla* in the agricultural production from Fujian, Hunan, Jiangshu, Guangxi Provinces. However, scientific research on *Azolla* and nation-wide popularization of *Azolla*-growing techniques have developed only within the past 40 years. Since the founding of the People's Republic of China, the concerned departments of the government have paid more attention to *Azolla* and the scientific research personnel in various provinces have made a significant contribution to *Azolla* cultivation and utilization in China. For example, the Fujian Academy of Agricultural Sciences started research on nitrogen fixation of *Azolla* and its utilization from the early period of the 1960s. In the late 1970s and early 1980s, C_2H_2 reduction and isotope *in situ* determining techniques were introduced. As the result, the determination of nitrogen fixation of *Azolla* tends to be more and more accurate. Up to now, the Fujian Academy of Agricultural Sciences has received several rewards in research achievements on *Azolla* from the State Ministry of Agriculture and Fujian Province. In 1984, the Chinese Ministry of Agriculture, Husbandy and Fishery and the Fujian Provincial Government jointly established the National *Azolla* Research Centre in Fuzhou City; henceforth, China has a special organization on *Azolla* research.

2 The Nitrogen-Fixing Activity of *Azolla* and its Effecting Factors

For the last several decades, researchers in different countries have determined and analysed the N-fixing capacity of *Azolla*. The Kejeldahl method was adopted in the early period. After entering the 1970s, the C_2H_2 reduction method was used to estimate the nitrogen amount fixed by *Azolla*. However, the above-mentioned results are mostly estimated under the special conditions in which *Azolla* continuously grows on the open water surface in a suitable environment. These conditions differ from the real ecological condition in which *Azolla* grows; furthermore, the results obtained under these conditions are quite different. In 1978, based on the fixed nitrogen amount measured by C_2H_2 reduction and N content in *Azolla* determined by the Kejeldahl method, combined with the real biomass of *Azolla* produced under field growth conditions, we calculated the annually fixed nitrogen amount of *Azolla* (Table 1).

Table 1. The annual fixed nitrogen amount of *Azolla* (Kg N/ha)

Azolla species by C_2H_2 reduction	Annual fixed nitrogen amount measured by C_2H_2 reduction	Annual fixed nitrogen amount determined by the Kejeldahl method
A. pinnata	160–188	435–496
A. filiculoides	249–292	604–689

It can be seen from Table 1 that the amount of fixed nitrogen as measured by the Kejeldahl method is obviously higher than that by the C_2H_2 reduction method. The C_2H_2 reduction method is a simple, rapid method, however, it is only a qualitative measure, because *Azolla* fixes nitrogen under field conditions which is very different from that in C_2H_2 reduction, and the indirect estimation of the C_2H_2 reduction method needs to be calibrated and converted to the real nitrogen fixed. Thus, a more direct and reliable isotope dilution method was adopted. The principle of this method is that *Azolla* plants and reference plants (*Anabaena*-free *Azolla* or *Lemna*) were grown simultaneously in the ^{15}N-containing solution, their individual N content and ^{15}N abundance were determined at the end of the experiment, and then the formula of the isotope dilution method was used to calculate the nitrogen fixed by *Azolla*.

Based on our observations, the amount of fixed nitrogen by *Azolla* in the C_2H_2 reduction method is lower than that measured by ^{15}N dilution (Table 2). Although, the results which have been obtained under different ecosystems are not the same, the conclusion that the annual amount of nitrogen fixed by *Azolla* ranges between 160–250 kg N/ha is now accepted by most investigators.

The N-fixing activity is affected by factors such as illumination, temperature, humidity, combined N concentration in the medium and the plant quality of *Azolla* itself.

In 1979, a comprehensive study on N-fixing activity of different *Azolla* species was conducted under various light intensity and temperature conditions. The results showed that the highest nitrogenase (N_2ase) activity appeared at 1000 lux, 25 °C, which decreased at 35 °C, then at 15 °C, N_2ase activity ceased. This fact illustrated that, under suitable light intensity, 25 °C is the appropriate temperature for nitrogen fixation of *Azolla* (Liu et al. 1979b).

The combined N concentration in medium has a significant effect on the nitrogen fixation of *Azolla*. Based on our investigation, the high concentration of combined N obviously depresses the N-fixing activity of *Azolla*. Among nitrogenous compounds, NH_2-N strongly harms the activity, NH_4-N has less effect, while NO_3-N does least harm. The action of different forms of nitrogen varies with its concentration. 7.5–15 ppm KNO_3 has a certain stimulation on *Azolla* growth and its N-fixing activity, the *Azolla* fronds still have some N-fixing activity when the concentration increased to 1500 ppm, but *Azolla* fronds died

Table 2. *Azolla* fixed nitrogen as measured by different methods (mg N/pot. day) (Liu et al. 1984)

Azolla species	C_2H_2 reduction	^{15}N dilution
A. pinnata var. *putian*	0.685 ± 0.160	1.299 ± 0.204
A. pinnata var. *indica*	0.488 ± 0.078	0.800 ± 0.190
A. microphylla	0.573 ± 0.99	0.468 ± 0.111

at 4500 ppm in large amount after 10 days of cultivation. High concentrations of ammonium sulfate are very harmful. N-fixing activity decrease sharply at 1000 ppm, the optimum concentration for *Azolla* growth and nitrogen fixation is at 10 ppm. The order of harmfulness is therefore from KNO_3 through NH_4NO_3 to $(NH_4)_2SO_4$ (You et al. 1981).

It is clear from the above results that the low concentration of combined nitrogen has a positive reaction on nitrogen fixation of *Azolla*. In other words, *Azolla*, under this condition can not only accept the combined nitrogen fixed by N-fixing *Anabaena azollae*, but also absorb some amount of nitrogen from medium via its root system to meet the requirement of its growth and development. Because the combined nitrogen exists more-or-less in the natural ecosystem, and nitrogen fixation consumes a certain amount of energy, it is probably an economical way for *Azolla* to absorb some nitrogen as a complement, while utilizing the fixed-N provided by its symbionts. This fact also elucidated that there is a subtle relation between fern and algae in controlling nitrogen metabolism.

It is more interesting that *Azolla* may excrete a part of the combined nitrogen via its root system while fixing nitrogen. In 1980, we used the N-tracing method to investigate the ^{15}N-excreting process of *A. filiculides* and *A. pinnata*. The result showed that, on the 1st, 2nd and 3rd day of the experiment, *Azolla* excretes 550, 930 and 4330 mg NH_3, respectively, while fixing ^{15}N. Furthermore, this N-excreting process is uninterrupted, and the excrement is supposed to be the metabolic product. This experiment also demonstrated that the excreted-N may be well absorbed by rice plants, which should be of certain significance in increasing rice production (Liu et al. 1980).

3 Nitrogen-Fixing *Anabaena azollae*: Classification and Identification

The actual N-fixer in *Azolla* is *Anabaena azollae* which are located inside the *Azolla* leaf cavity. *Anabaena azollae* is a filament which is linked by many cells and called algal trichome. This filament consists of two kinds of cells, one is small in size (2–5 μm long and 2–3 μm wide) and called vegetative cell, the other is large (5–8 μm long and 4–5 μm wide) and called heterocyst (Fig. 2). It is now proved that heterocysts contain N_2ase, they are the differentiated cells specially for nitrogen fixation. This morphological character of *Anabaena azollae* is very similar to that of free-living *Anabaena cylindrica* which belong to the *Cyanophyta*, *Cyanophyceae*, *Nostocales*. In many investigations, *Anabaena cylindrica* is always compared or placed on a par to *Anabaena azollae*.

However, for a long period, it has been argued whether the *Anabaena azolla* from different *Azolla* species belong to the same species. Their similarities and differences have also been explored. Many foreign investigators considered that the *Anabaena azollae* from different *Azolla* are the same although they are different in cell size, pigment distribution, etc.

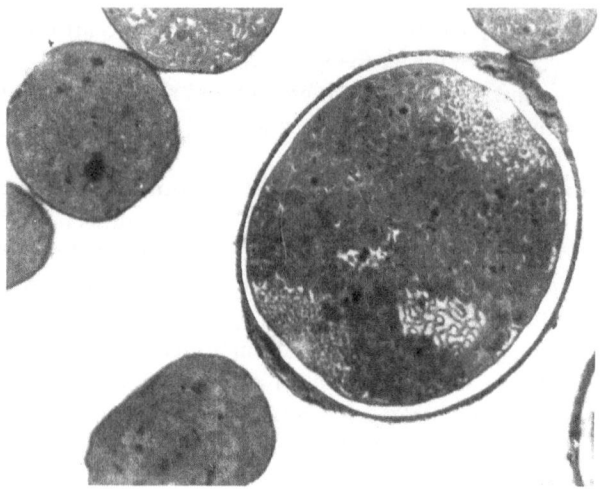

Fig. 2. Transmission electron micrograph of *Anabaena azollae* in *Azolla* leaf cavity. The big cell is a heterocyst, and the small ones are vegetative cells. × 10 000

Since 1986, the monoclonal antibody technique has been used to test the surface antigen of *Anabaena azollae*. The *Anabaena azollae* from the existing two subgenera and 7 *Azolla* species, i.e. *A. filiculoides*, *A. caroliniana*, *A. microphylla*, *A. mexicana*, *A. rubra*, *A. pinnata* and *A. nilotica* were isolated and purified. Then the purified *Anabaena azollae* were injected separately into BALB/C mice, the spleen cells from immunized mice were fused with SP_2/O myeloma cells of the same line mouse. 40 cell lines of hybridoma were successfully established. These hybridoma cell lines may excrete the antibodies which specially react with different symbiotic *Anabaena azollae* (Liu et al. 1986).

The next experiment used these 40 monoclonal antibodies (McAb) to test the symbionts from various *Azolla* species (or strains), *Tolypothrix tenus* and *Anabaena azotica* by indirect fluorescent antibody assay (IFA). The result showed that these 40 McAbs belong to 8 types of antibody (Table 3). Then the representative antibodies of 8 types were used for systematic classification and identification of *Anabaena azollae*.

It can be seen from Table 3 that, type I McAb reacts to all symbiotic *Anabaena azollae* from different *Azolla* species (or strains); type II McAb reacts to the symbionts from 7 *Azolla* species except *A. pinnata*; type III McAb only recognizes the symbionts from *Euazolla*; type IV McAb only recognize the symbionts from *Rhizosperma*; types V, VI McAb only react with the symbionts from one *Azolla* species in *Rhizosperma* respectively; type VII McAb reacts with the symbionts from 3 species *Azolla* in *Euazolla*; type VIII McAb recognizes the symbionts from both *Euazolla* and *Rhizosperma* subgenera. All 8 types of McAbs cannot react with *Tolypothrix tenus* and two cultures of free-living *Anabaena azotica*.

Table 3. The reaction character of 40 McAbs to *Anabaena azollae*, *Tolypothrix tenus* (Liu et al. 1987)

Type	Number of hybridoma cells	Source of *Anabaena azollae*								*Tolypo-thrix tenus*	*Anabaena azotica*	
		Ap	Ai	An	Ac	Ar	Af	Ami	Ame		I	II
I	17	+	+	+	+	+	+	+	+	−	−	−
II	4	−	+	+	+	+	+	+	+	−	−	−
III	12	−	−	−	+	+	+	+	+	−	−	−
IV	1	+	+	+	−	−	−	−	−	−	−	−
V	3		−	+	−	−	−	−	−	−	−	−
VI	1	+	+	−	−	−	−	−	−	−	−	−
VII	1	−	−	−	+	+	+	−	−	−	−	−
VIII	1	−	−	+	+	+	−	−	+	−	−	−

Note: Ai: *A. imbricata*, Ap: *A. pinnata*, An: *A. nilotica*, Ac: *A. caroliniana*, Ar: *A. rubra*, Af: *A. filiculoides*, Ami: *A. microphylla*, Ame: *A. mexicana*

A further experiment revealed that FA reactions of the 8 types of McAbs to the symbionts from different *Azolla* species during spring, summer, autumn and winter seasons are consistent. This means the basis for classifying and identifiying *Anabaena azollae* by using cell surface antigens is reliable.

Then, these 8 McAbs were used for FA reaction to symbionts from 195 *Azolla* strains. The results showed an obvious regularity. According to this basis, the symbionts from different *Azolla* species have been divided into 8 antigenic groups. The symbionts from *A. pinnata*, *A. imbricata* (a variety of *A. pinnata*), *A. nilotica* and *A. filiculoides* belong to its independent antigenic group respectively, this is in agreement with the *Azolla* classification. The differences are that the different species of *Azolla* in the classification, such as *A. rubra* and *A. caroliniana* were found to be of the same antigenic group of symbionts; the various *Azolla* strains in the same species, such as the *Azolla* strains from *A. mexicana*, whose original growing area is South and North America, get their symbionts from different antigenic groups. Because the present classification of *Azolla* strains is confusing, our result possesses a consultable value in *Azolla* classification.

To sum up, our investigations showed that the symbionts from various *Azolla* strains are different, this difference seems more significant between *Euazolla* and *Rhizosperma*. (Liu et al. 1989a).

4 The Acquirement of *Anabaena*-free *Azolla* and Reconstitution of *Anabaena*-free *Azolla* with *Anabaena azollae*

It is a long-cherished wish of many investigators to reconstitute *Anabaena*-free *Azolla* and *Anabaena azollae* or exchange the algae from one *Azolla* species to another by using an artificial method.

In order to conduct fern-algae recombinations, the first important thing is to acquire *Anabaena*-free *Azolla* and isolated *Anabaena azollae*. The *Anabaena*-free *Azolla* were obtained via stemtip culture method (Bai 1979). In this experiment, a new way to obtain *Anabaena*-free *Azolla* was achieved, i.e. by growing the *Azolla* megasporocarps without their indusia. The principle is that the *Anabaena azollae* symbionts enter into the sporocarps when *Azolla* plants produce megasporocarps. Finally, the algae transform to akinets and are located in the top area of the megasporocarp indusium (Peters 1983). For this reason, it is possible to remove the *Anabaena azollae* in the indusium by cutting it off from the megasporocarps. In practice, the apical membrane on top of the megasporocarp should be removed together with the indusium by a tiny scalpel under a stereoscope to avoid some akinets which remain on the membrane, then the decapitated megasporocarps are incubated under regular conditions, the resulting seedlings will become *Anabaena*-free *Azolla*. C_2H_2 reduction tests, whole plant-transparentizing and pressing method, continuous sectioning observation, electromicroscopic observation, and N-free solution cultivation methods all proved that the *Azolla* fronds obtained by decapitated megasporocarps are *Anabaena*-free *Azolla* indeed (Zheng 1987).

The source of *Anabaena azolla* used in recombination of *Azolla Anabaena* are:

1. the *Anabaena* (akinets) in megasporocarp indusia;
2. the isolated *Anabaena* strains obtained by Bai Ke zhi's (1979) or Newton's (1979) methods, which grow under 500 lux, 25–28 C for 1 2 months.

The method for reconstitution (i.e. inoculating *Anabaena azollae*) is the following:

1. to place the indusia from other *Azolla* species onto the decapitated megasporocarps (i.e. the apical membranes and indusia of the megasporocarps were removed);
2. to inoculate the cultural *Anabaena* cells by capillary to the decapitated megasporocarps. The inoculated megasporocarps are cultured in IRRI culture solution containing 50 ppm NH_4NO_3. The seedlings which possess 7 8 true leaves were transferred to N-free IRRI solution for further cultivation (Liu et al. 1988).

One of the main methods to identify if the reconstitution is successful or not is to use the monoclonal antibody technique, i.e. to confirm if the symbionts in recombinated *Azolla* possess the same antigenic reactivity with the inoculated *Anabaena azollae*.

The scanning electromicroscope has been used to reveal the reconstitution process of external *Anabaena* with new host *Azolla* plant; the monoclonal technique also proved that the symbionts in reconstituted *Azolla* are the original inocula of external *Anabaena azollae*. These results demonstrated that the symbionts from different *Azolla* species can be successfully exchanged or reconstituted by using the above-mentioned method (Liu et al. 1988).

The success of fern-algae recombination provides a new means to investigate *Azolla-Anabaena azollae* symbiosis, also providing a powerful tool to further expound theoretically the symbiosis mechanism, especially the establishment of the symbiosis process. It also opens a new way to breed *Azolla* strains. It is possible to breed new excellent *Azolla* species by selecting optimum fern-algae recombinations and exchanging techniques after understanding the characteristics of various *Azolla* species and their symbionts (Liu et al. 1988).

5 Utilization of *Azolla* Nitrogen Fixation in Agriculture in China

Essentially, the utilization of *Azolla* in agricultural production is to convert the cheap biological nitrogen which is acquired from nitrogen fixation of *Azolla* into edible food protein via various agronomic practices. In production, *Azolla* biomass is usually incorporated into soil as green manure or serves as fodder to increase livestock, poultry and fish products. In the last several decades, in Fujian Province and other provinces in China the main ways to improve the N-fixing capacity of *Azolla*, to prolong its N-fixing duration, and to increase its nitrogen transforming rate has been via *Azolla* breeding and selection, the improvement of *Azolla* culturing techniques and reformation of cropping systems.

5.1 Selection of Multi-Resistant Species and *Azolla* Breeding by Hybridization

In South China, *A. imbricata*, which serves as the main traditional strain of *Azolla* in China, cannot normally grow after April owing to the "three harmfulness", i.e. pests and disease, algae, and high temperature. The N-fixing potential of *Azolla* cannot be given full play, the growth and utilization of *Azolla* almost stops after June every year.

In order to improve this situation, in addition to a series of management practices in the field, attention was paid to collecting and selecting *Azolla* germplasm. In the last ten years or so, the *Azolla* collection in the National Azolla Research Center has covered 502 strains which come from more than 30 countries in 5 continents in the world including China. This center has bulit up the largest *Azolla* germplasm collection in the world.

Based on this collection, a more high pest-resistant, high temperature-resistant, shade-tolerant and more adaptive *A. caroliniana* has been selected after several year's laboratory and field comparisons. Since 1981, the multiple resistance of *A. caroliniana* has been proved by field cultivation in Fujian, Hunan, Zhejiang and Henan Provinces. For example, *A. caroliniana* was mix-grown with *A. filicuoides* in early rice fields in Fuqing County, Fujian Province.

After the beginning of May, *A. filiculoides* in paddy fields gradually died. However, a certain amount of *A. caroliniana* (745 kg/mu) still remain in paddy fields even after harvesting in July. After incorporating *Azolla* in late rice, a part of *Azolla* floats on the water surface and naturally grows in the paddy field. In addition to the 890 kg of *Azolla* which are naturally incorporated in one Mu paddy soil during the growth period of late rice, the residual *Azolla* biomass still has 10 200 kg/ha after the harvesting of late rice in November (Liu 1987).

The utilization of *A. caroliniana* increases rice grain yield by about 10% (Wei et al. 1982).

Since 1986, *A. microphylla* and *A. filicuoides* have been used as parent plants for sexual hybridization of *Azolla*. 4 excellent hybrid *Azolla* strains have been obtained (Wei et al. 1986). The significant differences between hybrid *Azolla* and their parent plants have been shown by zymographic analysis, morphological character identification and physiological–biochemical determination. The hybrid *Azolla* also showed superiority in its overall biomass, quality and tolerance over the present popularized *Azolla* species. These 4 hybrid *Azolla* are now becoming popular in Liaoning, Jilin and Helongjiang Provinces in North China.

In recent years, tissue culture, gene transformation and radiation techniques have also been used to improve *Azolla* species so that they possess high N-fixing capacity and tolerance.

5.2 The Reform of Cultivation Techniques

In the traditional *Azolla*-growing technique in paddy fields (i.e. dual growth of *Azolla* under rice canopy), the space in paddy fields and solar energy cannot be fully utilized owining to several limitations. *Azolla* in early rice fields produces only 3.5–5.25 kg per Mu (pure N basis). It has been known that the main limiting factors in *Azolla*-growing are the limited *Azolla*-growing space and insufficient solar energy. In order to achieve the goal to obtain both high grain yield and *Azolla* biomass, the traditional spacing method for rice seedlings needs to be reformed, and the practices adopted in growing *Azolla* and rice should also be reformed. Since 1976, the *Azolla*-growing technique called "double wide-narrow-rows spacing method" has been successfully applied (Liu 1979).

In this method two lines of rice seedlings are closely planted, leaving one *Azolla*-growing channel. The row distance between close lines is 13–14 cm, the plant distance for conventional rice varieties is 7–8 cm, for hybrid rice 8–10 cm; the width of the *Azolla*-growing channel for conventional rice is 35–50 cm, for hybrid rice 46–60 cm (Liu 1986b).

When using this new culturing technique, the rice plant structure in the paddy field was altered, and light, heat, and moisture conditions were improved. According to investigations in Hunan and Shangxi Provinces, the light intensity, light transmissivity and solar radiant heat in the "double wide-narrow-rows spacing" field were 30% increased over those in the conventional spacing field, whereas the relative humidity decreased by about 7%. The paddy ecosystem is favorable to the growth of the rice plant, thus the effective panicle rate and grain number increased by about 10%. On the other hand, it is also favorable to the growth and multiplication of *Azolla*, and the *Azolla*-growing period is extended by about 15–20 days in comparison with conventional paddy fields, the fresh biomass of *Azolla* in one Mu area reached 2000–3000 kg, the fixed nitrogen is 7–10 kg (on the basis of pure N). According to the data obtained in Hunan Province, this method also can effectively increase soil fertility, the dual growth of *Azolla* in early rice fields may leave plant residues in the field, the combined-N in the residual *Azolla* continues to promote the growth of rice seedlings in the late rice season, the increased rate reaching 735 kg/ha.

Another important reform in *Azolla* cultivation is to change water cultivation into moist soil cultivation. *Azolla* is an aquatic plant: it was generally considered that *Azolla* can only grow in the habitat with a covering water layer. We found ten years ago that some *Azolla* species such as *A. filiculoides.*, *A. caroliniana* possess moisture-growing character, i.e. it may normally grow and multiply on a moist soil surface where no water layer is available.

The investigations showed that the moist soil cultivation of *Azolla* can not only save water but also depress the harmfulness from pests and algae which use the water layer as their main habitat, while *Azolla* fronds grow vigorously. As a new technique, the moist soil cultivation of *Azolla* still needs to be further investigated and improved. It can be predicated that, following the breakthrough of moist soil cultivation, it will be applied in North China where there is less rainfall, but where a covenient irrigation system is available. In this way, *Azolla* can capture more solar energy and provide a source of cheap biological nitrogen (Liu 1987; Liu et al. 1986).

5.3 The Rice-*Azolla*-Fish System

Because the symbionts of *Azolla* provide sufficient combined nitrogen to meet the requirement of its growth and multiplication, the N-content in *Azolla* is higher than other green manure plants. Also, its crude protein content generally reaches 25%, sometimes up to 30%, it is therefore an excellent green feed. In South China, the history to use *Azolla* as fodder for pig, duck, chicken and fish has lasted for several centuries. Starting from the 1980s, following the popularization and utilization has been continuously increasing.

Since the reform and the opening door policy, following the development of rural commodity economy, the traditional manner of utilizing *Azolla* has faced a severe challenge. In order to further promote the economical benefit of *Azolla* cultivation and utilization, we comprehensively combined conventional "dual growth of *Azolla* in paddy fields" and "pisciculture in paddy fields" practices into a "Rice-*Azolla*-Fish system". In recent years, this scientific plot has been put into practice, the large-area field trials in Fujian, Zhejian, Hunan, Shichuang and Liaoning Provinces demonstrated that the "Rice-*Azolla*-Fish system" offers significant economical, soil and ecosystematic benefits and has been accepted by farmers. According to the incomplete statistics made in Fujian Province in 1989, the "Rice-*Azolla*-Fish system" was applied on 16 600 ha, rice grain yield increased 5500 tons and fish production increased about 7500 tons.

In the "Rice-*Azolla*-Fish system", *Azolla* as a fodder is taken up by fish, the fish dungs increase the organic matter in paddy soil and thus promote the multiplication of *Azolla*. On the other hand, the mutual promotion and restraint among the life in the "Rice-*Azolla*-Fish system" significantly decrease the harmfulness of *Azolla* pests. The biomass product in this system is generally higher than that of the conventional rice field. The results achieved from 1984–1986 in the Jianning demonstration area of the Fujian Academy of Agricultural Sciences showed that the annual *Azolla* biomass in the "Rice-*Azolla*-Fish System" in the double rice cropping area ranges 60 000–75 000 kg/ha. The data obtained by the Hunan Academy of Agricultural Sciences during 1984–1985 showed that the average *Azolla* biomass in this system in semilate rice field ranges 48 750–86 250 kg/ha with the maximum being 184 125 kg/ha. The result obtained at the Soil and Fertilizer Institute of Shichuang Academy of Agricultural Sciences illustrated that *Azolla* biomass in this system reached over 75 000 kg/ha, among which 1500 kg serves as fodder, 2500 kg as feed for pigs and the remaining 1000 kg serves as green manure in the paddy field (Ye et al. 1984; Liu 1987).

An experiment conducted by ^{15}N-labelled *Azolla* showed that in the "Rice-*Azolla*-Fish system" the rate of utilizing nitrogen in *Azolla* fronds reached as high as 67.76%; in the "Rice-Azolla System", when N-*Azolla* was used as basal manure, the rate was only 46.06%; when used as topdressing manure, 51.6% was obtained (Table 4).

Another ^{15}N-labelled trail showed that, after taking up [15]N-*Azolla*, the ^{15}N in *Tilapia nilotica* changed from 0.366% to 0.431%, the N-content in fish body is 2.41%. In other words, in the Rice-*Azolla*-Fish system, nitrogen fixed by *Azolla* was partially absorbed by fish and then transformed into fish protein, another part was utilized by rice plants. For this reason, the total N-utilizing rate is higher than in the Rice-*Azolla* system which contains no fish.

Table 4. The ^{15}N-utilizing rate of *Azolla* in various utilizing manners (Weng et al. 1987)

Treatment	^{15}N-utilizing rate (%)		
	By fish	By rice	Total
Rice-*Azolla*-Fish system	38.24	29.52	67.76
Rice-Azolla, [15]N-*Azolla* as basal manure		46.06	46.06
Rice-Azolla, [15]N-*Azolla* as topdressing manure		51.06	51.06

In recent years, the "Rice-*Azolla*-Fish system" has been developed into a new cropping system. The computer was used to establish a mathematical model with high output and low consumption. The management and manipulation in this system tend to be more standardized and scientific.

In 1989, the rice grain yield reached 11 250 kg/ha, fish production reached about 3750 kg/ha in the demonstrative station of the FAAS in Lianbang Fuzhou. The profit increased 3-fold compared with the conventional cropping system.

6 References

Bai K, Yu SL, Cheng WL, Yang SY, Cai C (1979) Isolation and pure culture of algae-free *Azolla* and *Anabaena*-azollae. Kexue Tong Bai (Chinese) 24(14): 664

Lin Cang, Liu Chung-Chu, Zheng De-Ying, Tang Long-Fei, Watanabe I (1988) The recombination between *Anabaena*-free *Azolla* and *Anabaena azollae*. Scientia Sinica (B) 7: 700

Liu Chung Chu (1979a) Use of *Azolla* in rice production in China. In: Nitrogen and Rice. Edited by IRRI. Philippines, p 376

Liu Chung Chu (1987) Reevaluation of Azolla Utilization in Agricultural production. In: IRRI (ed) *Azolla* utilization. Philippines, p 67

Liu Chung Chu (1984) Prospects of the utilization of *Azolla* in rice fields. Soil and Fertilizer (Chinese) 6: 16

Liu Chung Chu, Cheng Bing Huang, Wei Wen-Xiong, You Chung-Biao, Li Jia-Wei, Song Wei (1980) Preliminary exploration on the process of nitrogen excretion by *Azolla*, Scientia Agricult Sinica 4: 39

Liu Chung Chu, Cheng Youquan, Tong Long-Fei, Zheng Qi, Shong Tei-Ying, Chen Mingang, Li Yi-Ying, Lin Tian-Long (1989a) Studies on preparation monoclonal antibody against *Anabaena azollae* on its utilization. Scientia Sinica (B) 1: 44

Liu Chung Chu, Cheng Youquan, Tong Long-Fei, Zheng Qi, Shong Tie-Ying, Chen Mingan, Li Yi-Ying, Lin Tian-Long (1986) Studies on monoclonal antibody against *Anabaena azollae*. J Fujian Acad Agr Sci 1(2): 1

Liu Chung Chu, Lin Cang, Ren Zhujian, Cheng Qia-ken (1979a) Preliminary study on some physiological aspects of *Azolla*. Scientia Agriculture Sinica 2: 63

Liu Chung Chu, Lin Cang, Ren Zhu-jian (1979b) Preliminary study on some physiological aspects of *Azolla*. Scientia Agriculture Sinica 2: 63

Liu Chung Chu, Zheng Wei-Wen (1989b) *Azolla* in China. Agricultural Publ House. Beijing

Moore AW (1969) *Azolla*: biology and agronomic significance. Botanical Review 35: 17

Newton JW, Herman AI (1979) Isolation of cyanobacteria from the aquatic fern, *Azolla*. Arch Microbiol 120: 161

Peters GA, Calvert HE (1983) The *Azolla–Anabaena azollae* symbiosis. In: Goff EJ (ed) Alagal symbiosis. Cambridge University Press, New York, p 109

Wei Wen-Xiong, Chen Fuong-Ye, Lu Pei-Ji, Zheng Wei-Wen, Jin Guiying (1982) Studies on resistances of *A. caroliniana*. Construction of Soil

Wei Wen-Xiong, Jin Guiying, Zhang Ning (1986) Primary report of *Azolla* hybridization study. J Fujian Acad Agr Sci 1(1): 73

Weng Boqi, Cheng Bing-Huang, Tang Jianyang, Liu Chung Chu (1987) Study on utilization and effect of *Azolla* nitrogen with ^{15}N isotope labelled method in the Rice-*Azolla*-Fish cropping system. J Fujian Acad Agr Sci 2(1): 16

Ye Guo-Tiang, Chen Zhengnan, Jin Guiying, Jia Feng, Tang Long-Fei (1984) Study on Rice-*Azolla*-Fish System. Agricultural Sciences and Technology, Fujian (4): 24

You Chong-Biao, Liu Jing Wei, Song Wei, Liu Chung Chu, Wei Wen-Xiong, Che Bing-Huang (1981) Influence of nitrogen on the physiological properties of *Azolla*. 1. The effect of nitrate salt. Acta Physiolog Sinica (2): 97

Zheng Deying, Lin Cang, Tang Long-Fei, Liu Chung Chu (1987) The recombination of *Anabaena*-free *Azolla* with *Anabaena azollae*. 1. The acquirement and identification of *Anabaena*-free *Azolla*. J Fujian Acad Agriculture Sciences (2): 42

CHAPTER 28
Studies on Resource of Actinorhizal Trees and Their *Frankia* Endophytes

G.X. Chen, J.B. Huang, S.M. Shen, Z.Y. Zhao,
J.D. Jiang, H.F. Yang, H.C. Liu, and A.D. Yu

The Nitrogen Fixation
and its Research in China
Editor Guo-fan Hong
ⓒ Springer-Verlag Berlin Heidelberg 1992

1 Introduction

The actinorhizal tree-*Frankia* association is one of the important symbiotic nitrogen-fixing systems. The *Frankia* endophytes are of strong ability to nodulate various host plants of economic importance for forestry, and fix nitrogen. Actinorhizal trees can be used as a source of timber, fuelwood and fiber and increase the productivity of natural forests and plantations.

China lacks forest resources with per capita forest area and per capita forest volume being well below the world's average levels. At present forestry is still a weak link in her national economy. As a developing country, China can not afford the intensive use of fertilizers in forests. The use of nitrogen-fixing actinorhizal plants in managed forests and plantations may be an efficient way to increase timber, fiber and biomass production without the use of nitrogen fertilizer.

The research on nitrogen fixation by actinorhizal trees has been carried out at the Institute of Applied Ecology, Academia Sinica since 1978 and financially supported by the State Science and Technology Commission and Academia Sinica since 1983. More than 20 scientific researchers in the areas of forestry, forest ecology, plant physiology, microbiology and biochemistry have been involved in the research. In this paper, some of the results from studies on actinorhizal trees resources and their *Frankia* endophytes are briefly presented.

2 A Survey of Actinorhizal Trees Resources in China

A survey of actinorhizal tree resources in different parts of China has been carried out since 1978. Fourty-six species belonging to six genera are recorded so far (Shen et al. 1981; Huang et al. 1984; Zhao et al. 1990). Among them, 20 species (or varieties) are new records of actinorhizal trees. The species of actinorhizal trees in China are listed in Table 1.

The survey shows that China is rich in the resource with about 20% of the present tally of actinorhizal plants in more than 200 species. The nodules of all species identified were capable of reducing acetylene. Selected plant species were grown in combined nitrogen-free medium, and the plants were nodulated. Increased amounts of nitrogen were observed. The result is presented in Table 2, demonstrating that noculated plants accumulate a considerable amount of nitrogen from the atmosphere (Huang et al., unpublished data). In addition, the nodule micro-structure of the species were examined under both optical and electron microscopes, and the vesicles, which have been believed to be responsible for the nitrogen fixation, were observed in all these species.

The annual amount of nitrogen fixed under field conditions was measured by the acetylene reduction method. The amount fixed on *Hippophae rhamnoides* (2 years old), *Alnus sibirica* (4 years) and *A. sibirica* (5 years) was 23, 55 and 86 kg

Table 1. List of actinorhizal trees in China (*new record)

Elaeagnus:

E. angustifolia L., *E. umbellata* Thunb., **E. bockii* Diels, *E. commutata, E. conferta* Roxb., **E. crispa* Thunb., *E. glabra* Thunb., **E. gonyanthes* Benth., **E. henryi* Warb., *E. macrophylla* Thunb, **E. mollis* Diels, **E. moorcroftii* Wall., **E. multiflora* Thunb. var. *ovata,* **E. oldhami* Maxim., **E. oxycarpa* Schlecht., *E. pungens* Thunb., **E. stellipila* Rehd., **E. umbellata* var. pavifolia Servet., **E. virides* var. *delavayi* Lec. *E. conferta* Roxb., *E. griffithii* Serv.

Hippophae:

H. rhamnoides L., **H. rhamnoides* var. *porcera*

Alnus:

**A. cremastogyne* Burk., **A. ferdinandi* Makio, *A. formosana, A. glutinosa, A. hirsuta, A. incana, A. japonica* Sieb. et Zucc., *A. japonica* var. *microphylla,* **A. mandshurica* (Call.) Hand.-Mzt., *A. nepalensis* D. Don., *A. rugosa* var. *americana,* **A. sibirica* Fish., *A. tinctoria* Sarg., *A. traboculosa* Hand.-Mzt.

Myrica:

**M. sculenta* Buch-Ham., **M. nana* Sieb. et Zucc., *M. rubra* Sieb. et Zucc

Coriaria:

C. nepalensis Wall., **C. sinica* Maxim., **C. terminalis* Hemsl.

Casuarina:

C. equisetifolia L., *C. glauca, C. temhissima.*

---

per hectar, respectively. Poplar trees (*Populus pseudo-simonii*) grew much better in association with *Hippophae rhamnoides* than in pure stands. They increased in height and width by 58–169% and 106–328%, respectively (Huang et al. 1984).

The seasonal nodulation variations of the 3 selected tree species, *Alnus cremastogyme* Burk, *Elaeagnus oxycarpa* Schlecht and *Hippophae rhamnoides*, were monitored. The result (Fig. 1) (Jiang et al. 1982) showed that nodulation occurs in May, June, July and August with a peak in June. Nodulation almost does not occur after August.

The experiments on the inoculated *Hippophae rhamnoides* seedlings were carried out in the afforestation nursery during 1987–1988. The results obtained indicated that the nitrogenase activity of nodules and the biomass production of the seedlings were increased by 80% and 34%, respectively (Chen et al. 1990).

3 Isolation and Reinoculation of *Frankia*

Since Callaham et al. succeeded for the first time to isolate a *Frankia* strain from *Comptonia peregrina*, more and more actinorhizal endophytes have been isolated and classified in the genus *Frankia*. In 1980, we began to isolate *Frankia*

Table 2. The increased amount of nitrogen accumulated by nodulated plants cultured in water-washed sand*·**

Species	Type	Period of growth, days	Dry weight, g Whole plant	Nodules	Total N content of plant, mg
Alnus ferdinandi	Nod		1.33	0.029	15.06
Makio	Non-nod	103	0.37	–	3.58
Alnus cremastogyne	Nod		9.02	0.0268	140.25
Burk	Non-nod	103	1.34	—	10.87
Alnus sibirica	Nod		4.97	0.124	67.61
Fish	Non-nod	100	0.62	—	5.84
Hippophae rhamnoides	Nod		2.12	0.063	38.85
var. *porcera*	Non-nod	100	0.96	—	16.81
Elaeagnus oxycarpa	Nod		2.15	0.062	38.94
Schlecht	Non-nod	98	0.78	—	10.35
Elaeagnus moorcroftii	Nod		1.73	0.031	32.36
Wall	Non-nod	98	0.86	–	12.56
Elaeagnus mollis	Nod		1.06	0.024	17.66
Diels	Non-nod	98	0.14	–	1.01
Elaeagnus crispa	Nod		9.07	0.318	179.37
Thunb.	Non-nod	97	0.28	—	4.51
Coriaria sinica	Nod		16.91	0.123	143.00
Maxim	Non-nod	86	6.76	—	57.61

* Seeds were sown in April, and the seedlings were transplanted to sand which was thoroughly washed with water. The plants were inoculated by the aqueous suspensions of crushed nodules collected from the corresponding plant species, or by the aqueous suspensions of the habitat soil; the plants were watered with N-free Sideris-Young nutrient solution and harvested at the time as indicated.
** All the data shown in the table are means of 12 plant samples.

endophytes from nodules of actinorhizal trees, and in 1981 we obtained a strain Acc13 from nodules of *Alnus cremastogyne* Burk. Since then, *Frankia* spp. from 34 actinorhizal tree species were isolated and were cultured successfully (Table 3) (Jiang et al. 1984; Yang et al. 1984; Jiang et al. 1985; Jiang et al. 1987; Zhao et al. 1990). We have used different methods and to our knowledge, the dilution and direct isolation methods were simpler, thus more suitable.

The isolated strains share morphological characteristics. Hyphae are branching and septate and are 0.6–2.0 µm in diameter. They are Gram-positive.

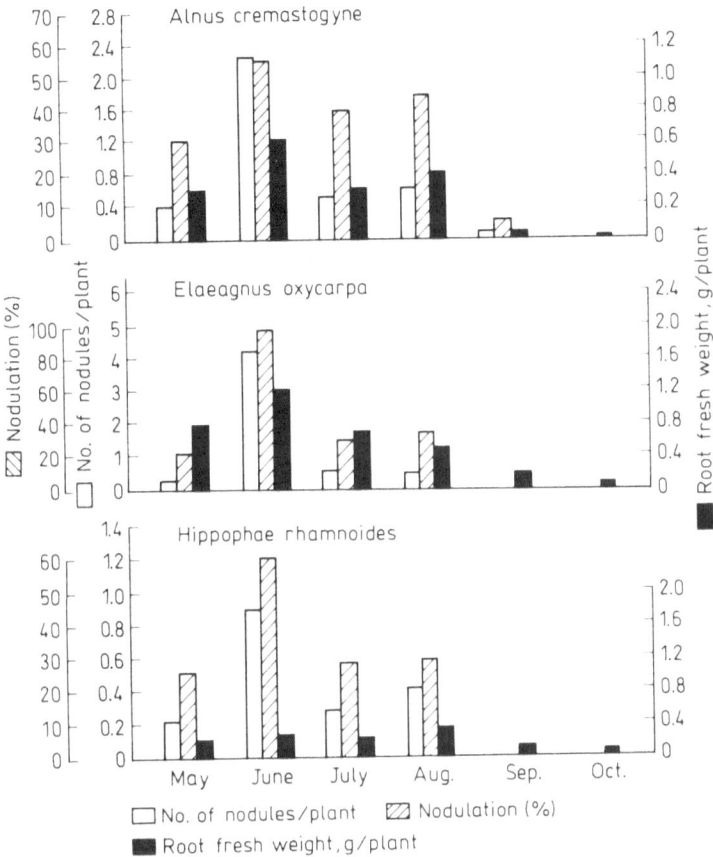

Fig. 1. Seasonal variations in nodulation

$$*\text{Nodulation } (\%) = \frac{\text{Nodulated plants}}{\text{Observed plants (10)}} \times 100\%$$

Sporangia are formed by means of intercalary or terminal filament swelling, irregular in shape with 3–35 μm in diameter. Mature sporangia contain spores, which are non-motile and irregular in shape and are 1–2 μm in diameter. In defined media, vesicles are formed. They are either spherical or spheroid at 2.5–4 μm in diameter and linked with filaments through a vesicle-stalk (Plate).

The isolated strains have been proved to be able to renodulate the host plants (Table 4).

Table 3. Isolated *Frankia* strains (*, new records)

Host species	Strains
*Alnus cremastogyne Burk	*Frankia* sp. Acc 13
	Frankia sp. Acc 8207
	Frankia sp. Acc 83201
A. hirsuta Turcz	*Frankia* sp. Ahc 8201
A. glutinosa	*Frankia* sp. Agc 8204
	Frankia sp. Agc 8404
A. japonica Sieb	*Frankia* sp. ajc 8206
A. glutinosa	*Frankia* sp. Acc-Ag 8209
A. nepalensis	*Frankia* sp. Anc 8301
A. tinctoria Sarg	*Frankia* sp. Atc 8405
	Frankia sp. Atc 205
A. kaloanel var. *Sibirio*	*Frankia* sp. Akc 8402
A. incana	*Frankia* sp. Aic 8403
A. japonica var. *microphylla*	*Frankia* sp. Ajc 904
*A. mandshurica Hand-Mazz.	*Frankia* sp. Amc 718
*Elaeagnus Oxycarpa Schlecht	*Frankia* sp. Eoc 85
	Frankia sp. Eoc 811
*E. multiflora Thunb var. *ovata*	*Frankia* sp. Emoc 1211
*E. gonyanthes Benth.	*Frankia* sp. Egc 107
E. angustifolia L.	*Frankia* sp. Eac 321
	Frankia sp. Eac 530
*E. stellipila Rehd.	*Frankia* sp. Esc 326
*E. crispa thunb.	*Frankia* sp. Ecc 327
*E. moorcroftii wall	*Frankia* sp. Emc 627
	Frankia sp. Emc 61
E. umbellata Thunb	*Frankia* sp. Euc 529
*E. bockii Diels	*Frankia* sp. Ebc 612
*E. mollis Diels	*Frankia* sp. Emc 7
	Frankia sp. Emc 626
*E. virides var. *delavayi* Lec,	*Frankia* sp. Evc 921
E. umbellata Thunb	*Frankia* sp. Euc 64
E. glabra Thunb	*Frankia* sp. Egc 8401
E. conferta Roxb	*Frankia* sp. 8815, 8912
E. griffithii Serv	*Frankia* sp. 8822
Hippophae rhamnoides L.	*Frankia* sp. Hrc 97
	Frankia sp. Hrc 922
	Frankia sp. Hrc 1213
*H. rhamnoides var. *porcera*	*Frankia* sp. Hrc 2
Casuarina equistifolia L.	*Frankia* sp. Cec 14
	Frankia sp. Cec 215
M. rubra Sieb. et Zucc.	*Frankia* sp. Mrc 8302
	Frankia sp. Mrc 221
	Frankia sp. Mrc 204

4 Cultivation of *Frankia* Endophytes

With the successful isolation of a number of *Frankia* endophytes, it was possible to study the comparative physiology and biochemistry of the bacteria both in their symbiosis with plants and in the free-living state.

Plate

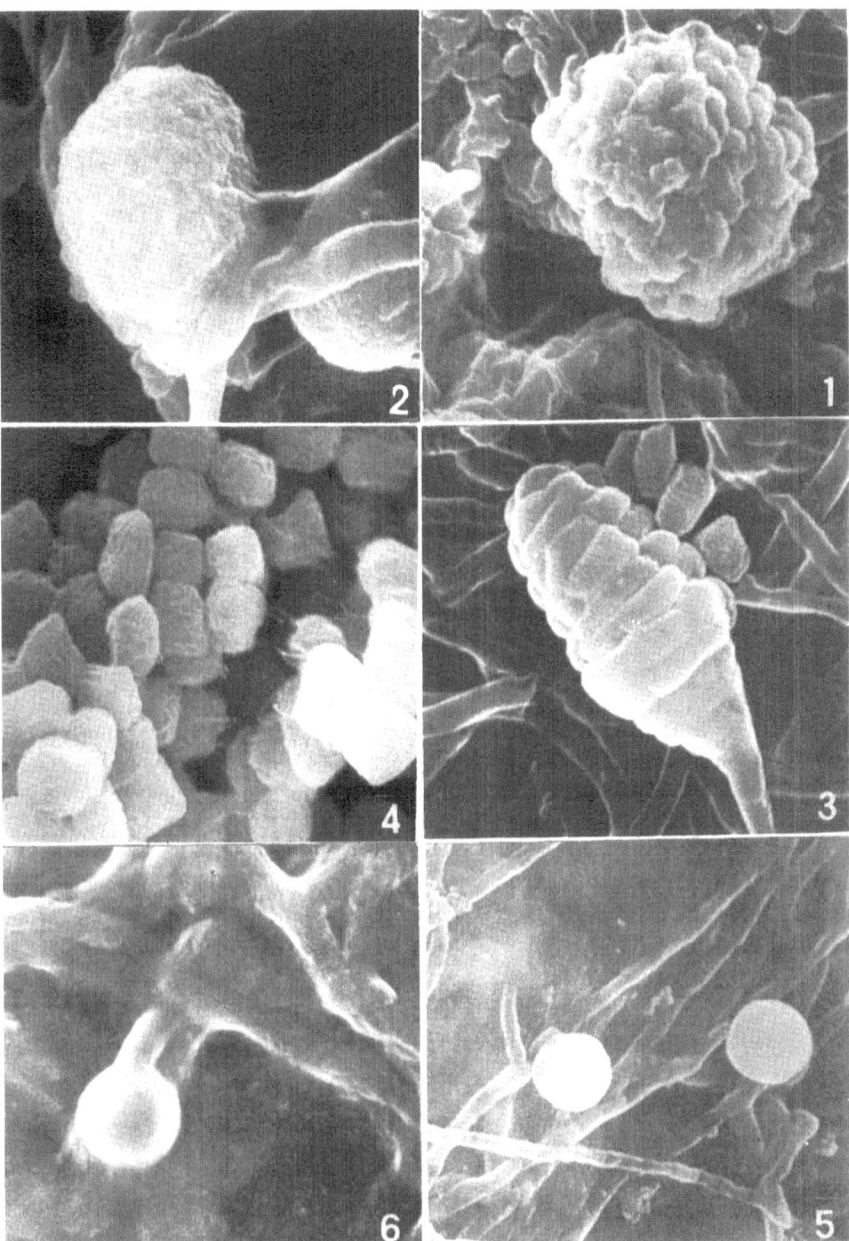

Scanning electron micrograph of Frankia strains
1–3 Sporangia: 1 Ahc8201(× 6,000), 2 Ajc8206(× 20,000), 3 Acc8207(× 9,000), 4 Spores
Ahc8201(× 20,000). 5 Vesicle Ahc8201(× 6,000), 6 Vesicle Egc107(× 12,000)

Table 4. Nodulation and growth of reinfected hosts

Host plants	Growth media	Strains	Type	No. of nodules per plant	Growth Height (cm)	Weight (g)ʳ	Acetylene reduction µ mol C_2H_4/h.g. fr
Alnus			Ino.			2.28	15.7
	Water	Acc13					
Cremastogyne			Non-ino.			0.85	0
Elaeagnus		Eoc85	Ino.	4–5	45	1.75	6.05
	Pearlite						
oxycarpa		Eoc811	Ino.	2–3	40	1.45	2.68
			Non-ino.	0	28	1.0	0
Elaeagnus			Ino.	6–7	45	7.5	7.15
	Pearlite	EmcR12					
multiflora							
var. *ovata*			Non-ino.	0	34	6.1	0
E. gonyanthes			Ino.	3–8		4.5	3.50
	Water	Egc107					
			Non-ino.	0		2.5	0
Hippophae			Ino.	9–11	53	5.2	0.6
	Water	Hrc97					
rhamnoides			Non-ino.	0	28	2.7	0

Considering the results on cross-inoculation and the potential utilization in the forest, we have isolated from *Alnus cremastogyne* and *Hippophae rhamnoides* nodules two strains, Acc13 and Hrc97, and examined in detail their growth behavior under different growth conditions, including different growth temperature, medium, pH and nutritional sources. The results (Chen et al. 1986) indicated that the strains can utilize a range of C- and N-sources for growth, higher protein yield produced with Tween-80, propionate and malic acid as C-sources and with casamino acid, ammonium chloride, yeast extract and aspartic acid as N-sources (Tables 5 and 6).

Table 5. Growth yields of *Frankia* Acc13 and Hrc97 on various carbon sources

Carbon sources (g/l)	µg protein/7 ml Acc13	Hrc97
Tween-80 (2)	433.8	318.7
Propionate (0.2) + Succinate (0.2)	334.1	504.9
Propionate (0.5)		357.1
Glucose (10)	169.9	413.0
Fructose (10)	118.0	491.8
Malic acid (0.5)	347.8	364.6
Succinate (0.5)	147.5	193.1
Acetate (0.5)	175.5	231.5
Tween-80 (1)	335.0	288.1
Tween-80 (1) + Propionate (0.2)	537.7	706.6

Table 6. Growth yields of *Frankia* strains Acc13 and Hrc97 on various nitrogen sources

Nitrogen sources (g/l)	µg protein/7 ml	
	Acc13	Hrc97
Casamino acids (5)	476.8	373.2
NH₄Cl (0.5)	345.4	381.4
Yeast extract (0.5)	251.2	412.4
Urea (0.5)	283.7	300.7
KNO₃ (0.5)	276.8	367.8
Casamino acids (5) + Yeast extract (0.5)	514.6	
NH₄Cl (0.5) + Yeast extract (0.5)	349.2	406.9
Glutamic acid (0.5)	359.2	152.7
Aspartic acid (0.5)	309.9	447.1

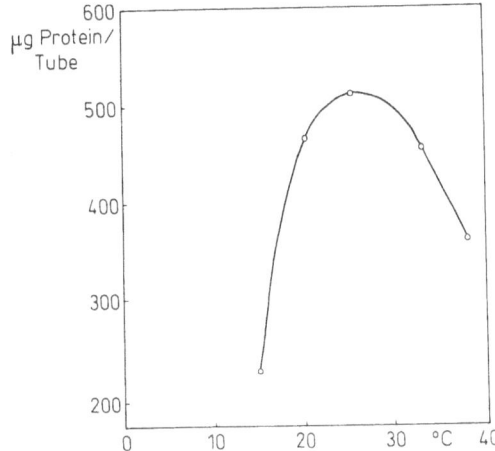

Fig. 2. Effect of temperature on growth yields of *Frankia* strain Hrc 97

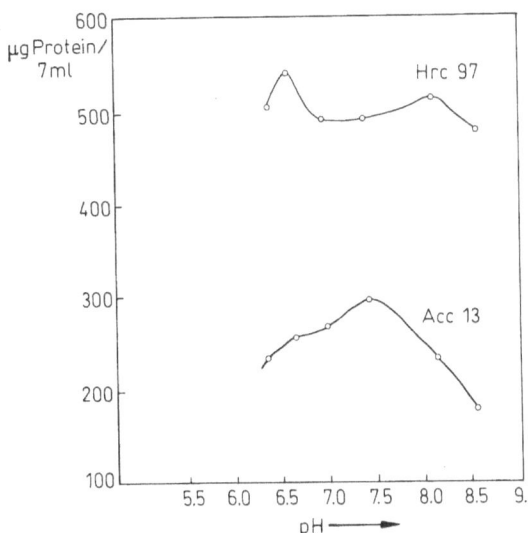

Fig. 3. pH effect on growth yields of *Frankia* strains

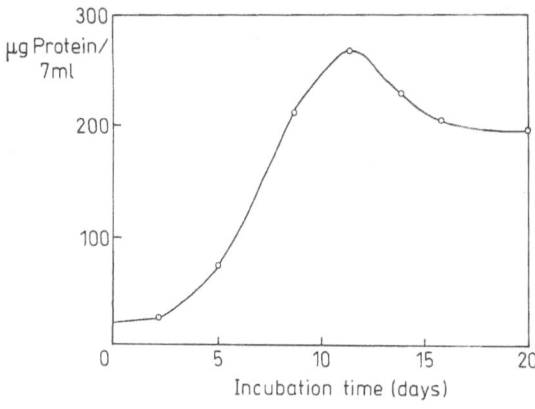

Fig. 4. Growth kinetics of Acc13 (in 10-l fermenter)

The optimum growth temperature for Acc13 strain was between 28–33 °C, while for the Hrc 97 strain it was 24–30 °C (Fig. 2) (Chen et al. 1986). The range of favorable pH for Hrc97 growth was between 6.5–8.0, and for strain Acc13 it was close to 7.0–7.5 (Fig. 3) (Chen et al. 1986). It was found that a cultivation in 10-fermenters had a doubling time of 50 h (Fig. 4) (Chen et al. unpublished data).

5 Host Specificity and Physiology of *Frankia* Endophytes

The bacterial genus *Frankia* is easily recognized both in vivo and in vitro. But as yet, there are not sufficient biochemical, morphological, or anatomical criteria for establishing species.

In order to learn more about the infectiveness and effectiveness of the actinomycete *Frankia* and provide basic information for taxonomic classification

Table 7. Host species tested

Host species	Plant material
Alnus cremastogyne Burk.	The same year seedlings
A. glutinosa (L.) Gaerth	The same year seedlings
A. hirsuta Turcz	The same year seedlings
A. japonica Sieb. et Zucc.	The same year seedlings
Elaeagnus gonyanthes Benth.	Cuttings
E. multiflora var. *ovata*	Cuttings
E. oxycarpa Schlecht	Seedlings and cuttings
E. umbellata var. *pavifolia*	Cuttings
E. viridis var. *delavayi*	Seedlings
Hippophae rhamnoides	The same year seedlings
M. rubra Sieb. et Zucc.	Seedlings
Casuarina equisetifolia	Seedlings

Table 8. *Frankia* strains used in cross-inoculation

Frankia isolate	Original host	Media
Acc13	*Alnus cremastogyne*	QMOD
Acc8207	*A. cremastogyne*	modified QMOD (Tween 80 as carbon source)
Agc8204	*A. glutinosa*	modified QMOD (Tween 80 as carbon source)
Ahc8201	*A. hirsuta*	modified QMOD (Tween 80 as carbon source)
Ajc8206	*A. japonica*	modified QMOD (Tween 80 as carbon source)
Egc107	*Elaeagnus gonyanthes*	modified QMOD (Tween 80 as carbon source)
Emoc1211	*E. multifora* var. *ovata*	modified QMOD (Tween 80 as carbon source)
Eoc85	*E. oxycarpa*	modified QMOD (Tween 80 as carbon source)
Hrc97	*Hippophae rhamnoides*	modified QMOD (Tween 80 as carbon source)
Cec14	*Casuarina equisetifolia*	modified QMOD (Tween 80 as carbon source)
Mrc8302	*Myrica rubra*	modified QMOD (Tween 80 as carbon source)

of *Frankia* strains at the species level and for selecting good strains for future use in forests, we carried out a series of cross-inoculation experiments with 10 pure, cultured strains and 12 host species (Huang et al. 1984). 10 strains were isolated from the root nodules on actinorhizal trees belonging to 10 species of 5 genera in 4 families; and the host species tested belong to 5 genera in 4 families (Tables 7 and 8).

The results (Table 9) indicate that *Frankia* strains from the nodules of *Alnus* did infect and nodulate *A. cremastogyne, A. japonica, A. hirsuta, A. glutinosa*

Table 9. Nodulation of host plant

Strain nodulated Host plants (%)	Acc 13	Acc 8207	Ahc 8201	Ajc 8206	Agc 8204	Emc 1211	Egc 107	Eoc 85	Hrc 97	Mrc 8302	Cec 14
Alnus cremastogyne	85	100	100	75	100	0	0	0	0		
A. japonica		100	100	50							
A. hirsuta Turcz			100								
A. glutinosa (L.) Gaerth		100									
Elaeagnus oxycarpa	0	0	0	0		100	89	100	75		
E. viridis var. *delavayi* Lec	0					14	5	0		0	
E. multiflora var. *ovata*			0	0		50	50	100	100	0	
E. umbelata var *pavifolia*				0					60		
E. gonyanthes		0		0			90	70	30	0	0
Hippophae rhamnoides L.									100		
Myrica rubra Sieb. et Zucc.	0	56	0	0						80	0
Casuarina equisetifolia L.	0	0	0							0	80

Table 10. Characteristics of *Frankia* sp.

Group	Sub-group	Strain	Original host	Sugar alcohol utilization	Pyruvate utilization	Diffusible pigment	pH
1	A	Ahc8201	*A. hirsuta* Turcz	–	+	–	5.5–7.5
		Agc8204	*A. glutinosa* (L.) Gaerth	–	+	–	5.5–7.5
		Agc8206	*A. japonica* Sieb. et Zucc.	–	+	–	5.5–7.5
		Acc8207	*A. cremastogyme* Burk	–	+	–	5.5–7.5
		AccAg8209	*A. glutinosa* (L.) Gaerth	–	+	–	5.5–7.5
		Acc13	*A. cremastogyne* Burk	–	+	–	5.5–7.5
	B	Mrc8302	*Myrica rubra*	+(few), poor growth	+	–	5.5–7.5
2		Ce414	*Casuarina equsielifolia* L.	+(few), poor growth	+	–	5.5–6.5
3	C	Emcl211	*E. multiflora*	+	–	a,b,c	5.5–8.5
		Egc107	*E. gonyanthes*	+	–		5.5–8.5
	D	Hrc97	*H. rhamnoides*	+–	+–(or –)	b,c	5.5–8.5
		Eoc85	*E. oxycarpa*	+–(or –)	+–(or –)		5.5–8.5

[a] light pink. [b] light brown. [c] brown

and *M. rubra*. However, no nodules were observed on the roots of *E. oxycarpa*, *E. viridis var. delavayi*, *E. multiflora var. ovata*, *E. umbelata var. pavifolia*, *E. gonyanthes* and *Casuarina* when they were inoculated with Acc13, Acc8207, Ahc8201 and Ajc8206. The pure, cultured strains isolated from *Elaeagnus* can infect and nodulate the host species of the same genus and family except *E. viridis var. delavayi*, which can only be poorly nodulated by a few strains from *Elaeagnus*. The strain Hrc97 from *Hippophae rhamnoides* did nodulate *Elaeagnus*, but not *E. viridis* var. *delavayi*. All of the strains isolated from *Elaeagnaceae* did not nodulate *Alnus* and *Casuarina*. Strain Cec14 isolated from *Casuarina equisetifolia* was able to infect and nodulate its own host plants. Based on these results, we suggested that *Frankia* endophytes can be divided into the following three host infectivity groups: *Alnus + Miryca* group, *Elaeagnaceae* group and *Casuarina* group. Similar results (Zhao et al. 1988) were obtained from physiological and biochemical studies (Table 10).

6 Induction of Nitrogenase in *Frankia*

Different C-source, N-source, way of incubation (shaking or stationary) and O_2-concentration in the gas phase were tested for establishing the optimum conditions for induction of nitrogenase in the in-vitro cultures of *Frankia*. The results obtained show that nitrogenase activities (acetylene reduction) occur only in the presence of O_2, optimal pO_2 at 15–20%, indicating nitrogenase in *Frankia* strains is more resistent to O_2 than that in *Rhizobium*. The fixed N-compounds (e.g. ammonium chloride) drastically repressed nitrogenase synthesis. The way of incubation (shaking or standing) also strongly affected the acetylene reduction activity. Little activity could be detected when the cells grew by standing at the optimal pO_2. We tested the specific activities of strains which ranged between 200–400 nmol C_2H_4 h·mg protein under the favorable conditions (Chen et al. 1984).

7 Nitrogen Fixation by Cell-free Extract from *Frankia* Pure Cultures

After establishing the conditions for *Frankia* growth and for the in vitro induction of nitrogenase on a large scale, we developed a technique for preparing cell-free extracts which have nitrogenase activity from *Frankia* pure cultures. The technique can be summarized as follows (Chen et al. 1987).

1. Preparation procedure

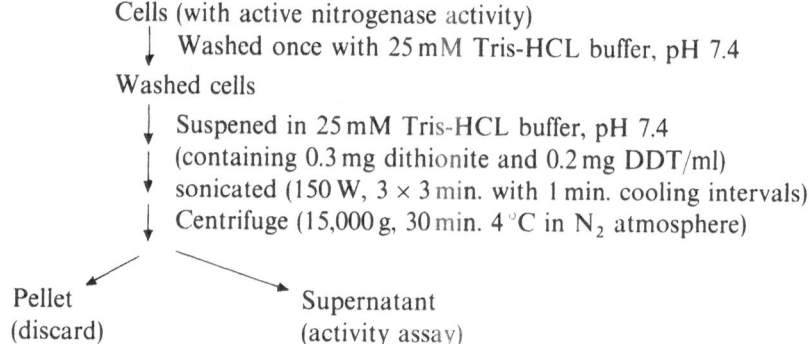

Cells (with active nitrogenase activity)
↓ Washed once with 25 mM Tris-HCL buffer, pH 7.4
Washed cells
↓ Suspened in 25 mM Tris-HCL buffer, pH 7.4
| (containing 0.3 mg dithionite and 0.2 mg DDT/ml)
↓ sonicated (150 W, 3 × 3 min. with 1 min. cooling intervals)
↓ Centrifuge (15,000 g, 30 min. 4 °C in N_2 atmosphere)

Pellet   Supernatant
(discard)   (activity assay)

2. 2 ml of activity assay contains
 8 µmol ATP
 15 µmol $MgCl_2$
 51 µmol creatine phosphate
 1.5 mg creatine kinase
 36 µmol sodium dithionite
 32 µmol Tris-HCL buffer (pH 7.4)
 Cell-free extract

The cell-free extract prepared by this procedure gave a specific activity of 22 nM C_2H_4/h·mg protein.

8 Uptake Hydrogenase in *Frankia* Endophytes

In all N_2-fixing organisms studied so far, while it reduces N_2, nitrogenase produces molecular hydrogen (H_2) simultaneously. Many N_2-fixing organisms, including the microsymbiont *Frankia* in actinorhizal plants, recycle much of this H_2 by uptake hydrogenase. In order to choose appropriate combinations between host plants and endobiont for the increased N_2-fixing efficiency, we monitored the uptake hydrogenase activities of thirteen *Frankia* strains isolated from *Alnus*, *Elaeagnus*, *Hippophae* and *Myrica*. Strain Acc-Ag8207 was found to have higher uptake hydrogenase activity. The expression of uptake hydrogenase was studied in relation to the concentrations of carbon and nitrogen sources, and the partial pressure of H_2, N_2, O_2 and CO_2 in the gas phase. The results (Yu et al. 1990) showed that the partial pressure of H_2 is the most effective factor. The concentrations of carbon and nitrogen sources also have considerable effects. The presence of H_2 is necessary for the expression of the gene for uptake hydrogenase. High oxygen concentration repressed hydrogenase gene expression. The uptake hydrogenase gene was highly expressed in *Frankia* strain Acc-Ag8209 in the medium which contains 0.15 mmol sodium propionate.

Studies on Resource of Actinorhizal Trees   553

9 References

Chen Guanxiong, Yu Adong, Wu Jie (1984) Study on nitrogenase activity in free-living *Frankia*. J Microbiol 4(3): 6

Chen Guanxiong, Yu Adong, Wu Jie (1986) Studies on conditions for growing *Frankia* endophytes. J Microbiol 6(1): 37

Chen Guanxiong, Yu Adong, Wu Jie, Wang Biao (1986) A new medium suitable for *Frankia* growth. J Microbiol 6(4): 55

Chen Guanxiong, Yu Adong, Wu Jie (1987) Nitrogen fixation by cell-free extract from *Frankia* pure cultures. KEXUE TONGBAO 32(9): 646

Chen Guanxiong, Yu Adong, Wu Jie, Shang Shuhui (1990) Effect of inoculation with *Frankia* on nitrogen fixation and biomass production of seedlings (in preparation)

Huang Jiabin, Shen Shanmin, Liu Huichang, Jiang Jiande, Yang Huifan, Zhao Zhenying, Yang Bifan (1984) An investigation of the resource of non-legumnious nodulating trees in China. In: Veeger C, Newton WE (eds.) Advances in nitrogen fixation research. Martinus Nijhoff, Amsterdam, 373

Huang Jiabin, Zhao Zhenying, Chen Guanxiong, Liu Huichang (1985) Host range of *Frankia* endophytes. Plant and Soil. 87: 61

Jiang Jiande, Zhang Yuhua (1982) Nodulation seasenal variation of some actinorhizal plants. Scientia Silvae Sinica 18(4): 425

Jiang Jiande, Yang Huifan, Huang Jiabin, Zhu Baoqin, Liu Huichang, Zhang Yuhua, Yang Bifang (1984) Isolation and cultivation of *Frankia* strain Acc13 from root nodules of *Alnus cremastogyne*. Acta Microbiol Sinica 24(1): 37

Jiang Jiande, Zhu Baoqin (1985) Isolation, cultivation and infectivity of actinorhizal endophyte of *Casuarina equisetifolia*. KEXUE TONGBAO 30(6): 833

Jiang Jiande, Zhao Zhenying, Zhu Baoqin (1987) Morphological observation and culture of *Myrica rubra* isolate. J North-East Forestry University 15(2): 114

Shen Shanmin, Huang Jiabin, Liu Huichang, Yang Huifan, Jiande, Yang Bifan (1981) A preliminary report on 8 new species and varieties of nodulated non-leguminous trees exhibiting nitrogen-fixation activity. Bull Inst Forestry and Pedology, Acad Sinica 4: 1

Yang Huifan, Jiang Jiande, Huang Jiabin, Zhu Baoqin, Zhang Yuhua, Liu Huichang, Yang Bifang, Zhao Zhenying (1984) Isolation and reinoculation of endophytes of *Hippophae rhamnoides* and *Elaeagnus*. Acta Microbiol Sinica 24(4): 315

Zhao Zhenying, Huang Jiabin, Liu Huichang, Zhu Baoqin, Yang Huifan (1988) Comparative studies on *Frankia* strains from different genera and families of actinorhizal plants. Acta Microbiol Sinica 28(3): 279

Zhao Zhenying, Xu Qingde, Ge Yinghua. A survey of resource of actinorhizal trees in Xishuangbanna (unpublished work)

Yu Adong, Chen Guanxiong, Liu Zhe, Li Hongbin, Gao Yan (1990) Study on uptake hydrogenase activity in *Frankia* strains. J Microbiol (in press)

CHAPTER 29
Studies on the Symbiotic Nitrogen-Fixing Actinomycete *Frankia* in China

J. Ding, Z.Z. Zhang, F.Y. Su, H.J. Sun, Y.L. Huang,
Q.D. Xu, W.H. Li, Z.W. Li, Y.H. Cui, H. Zou, Y. Wu,
D.H. Zhang, and Y.D. Li

The Nitrogen Fixation
and its Research in China
Editor: Guo-fan Hong
© Springer-Verlag Berlin Heidelberg 1992

1 Introduction

Nitrogen fixation by symbiotic and asymbiotic microorganisms has been studied for a number of years at the Laboratory of Soil Microbiology of the Institute of Applied Ecology. In 1978, Callaham and his colleagues for the first time isolated *Frankia* endophytes. Since, we have begun our research on the actinomycete *Frankia*. In this paper we will describe the advances made in this laboratory, and present a general picture of the field in this country with special emphasis on the following points:

(1) Actinorhizal plant resources in China.
(2) Improvement of isolation techniques of *Frankia* strains.
(3) Biological studies on *Frankia*.
(4) Achievement in utilization of actinorhizal plant resources in forest planting in China.
(5) Future directions.

2 Actinorhizal Plant Resources in China

In China, actinorhizal plants are very abundant, widespread and of various species. In 1978, the Institute of Applied Ecology made a survey of nitrogen-fixing resources among nonleguminous woody plants in China. According to the report of Huang Jiabin et al. (1984), about 44 species in 6 genera among 5 families of nonlegume woody plants have been reported to have actinomycete-induced nodules capable of nitrogen fixation. Among them about 20 are new species which enriched our collection of actinorhizal plant resources in the world (Huang Jiabin et al. 1984). Up to the present 4 species in *Casuarina* are added (Qin Min 1988).

Our microbiological research focuses on *Alnus*, *Hippophae*, *Casuarina* and *Myrica*, because of their ecological characteristics, and their economic importance. *Alnus* is widespread in China from Northeast to Southwest, capable of nodulation on humid soil, and used as a component of mixed forest and as fertilizer for soil improvement due to its active nitrogen fixation. *Hippophae*, widely distributed in semiarid regions, is drought resistant, and often used for rehabilitation of depauparate soils, as a component of mixed forest and for provision of fuel. *Casuarina* is used for plantation of protection forest in Southern China due to its quick growing under poor conditions. *Myrica* is an acidophil, especially widespread in yellow soil and is used for fruit production.

The ability of actinorhizal plants to grow vigorously on nitrogen-deficient soil and their value as soil-improvers depend on their capacity for symbiotic nitrogen fixation. Therefore, it is very important to isolate active endophytes from them.

3 Improvement of the Technique for Isolation of *Frankia* Strains

In 1978, Callaham isolated for the first time *Frankia* endophyte from *Comptonia penegraina* (Callaham et al. 1978). Since then there has been an increased interest in the group of actinorhizal plants. Baker et al. (1979) have successfully applied the sucrose layered-gradient method to isolate *Frankia* endophytes from actinorhizal plants. Lalonde et al. (1981) have established an isolation technique by pre-treatment of nodules with osmium tetroxide.

Based on a comparison of advantages and disadvantages of the above methods, we developed a simple isolation technique and successfully isolated a number of *Frankia* endophytes from Alnus, *Elaeagnus, Hippophae, Casuarina* and *Myrica*.

This method has many advantages, without any complex enzyme treatment process and without the harmful effects of OsO_4 to people. It can be used as a routine technique for isolating endophytes from various actinorhizal plants without high speed centrifugation. *Frankia* strains isolated by this method are summarized in Table 1.

At present this method is widely used in many laboratories due to its simplicity, reliability and reproducibility.

This method includes the following main steps:

Fresh nodules are collected from host plants in the field or in greenhouse pot-culture, and nodules are sterilized for 10 min in 0.1% $HgCl_2$ solution. Each nodule lobe is aseptically cut into small pieces to expose endophyte cells. Each piece is aseptically transferred to a test tube with Bap medium. Tubes are incubated at 28 °C until outgrowths from the nodule lobe piece are visible.

4 Biological Studies on *Frankia*

Morphological, physiological, ecological studies on *Frankia* were systematically carried out, and many interesting results have been obtained.

Table 1. *Frankia* strains isolated at the Institute of Applied Ecology

Strains	Host plant	Isolation method	Ref.
Acc13	*Alnus crematogyne*	Sucrose centrifugation	Jiang 1984
At4	*Alnus tinctoria*	Nodule pieces method	Li, Ding 1985
Emovrl1	*Elaeagnus multiflora* var. *ovata*	Nodule suspension method	Yang 1984
Cc01	*Casuarina Cunninghamiana*	Nodule pieces	Li, Ding 1986
Hr16	*Hippophae rhamnoides*	Nodule pieces method	Wu, Ding 1987
Mn12	*Myrica nana*	Nodule pieces method	Ding 1987

Table 2. Morphological characteristics of *Frankia* strains from different host plants (μm)

Strains	Host plant	Hyphae	Sporangia	Spores	Vesicles	Vesiclephores
At4	*Alnus tinctoria*	0.6–1.2	40–60	1.6–2.0	1.7–3.0	0.9–2.2
As2–8	*Alnus sibirica*	0.6–1.2	40–60	1.6–2.0	1.7–3.0	0.9–2.2
Ah1	*Alnus hirsuta*	0.6–1.0	35–60	1.4–1.7	2.2–3.9	1.3–2.0
HFCp11	*Comptonia peregrina*	0.6–1.1	30–45	1.7–2.4	2.0–3.8	1.5–4.0
Mn12	*Myrica nana*	0.4–0.7	15–35	1.3–3.0	3.7–4.0	3.0–4.4
Mg⁺	*Myrica gale*	0.6–1.0	15–30	1.4–1.7	1.7–3.5	0.9–2.2
MP13	*Myrica pensylvanica*	0.6–1.0	15–30	1.4–1.7	2.2–3.9	1.3–4.0
Cc01	*Casuarina cunninghamiana*	0.6–1.3	35–60	1.7–3.0	2.2–4.1	2.8–5.0
Ce37	*Casuarina equisetifolia*	0.6–1.3	35–60	1.7–3.0	2.2–4.1	2.8–5.0
ALLcl1	*Allocasuarina lehmanniana*	0.6–1.3	35–60	1.7–3.0	2.2–4.1	2.8–5.0
Hr16	*Hippophae rhamnoides*	0.4–0.9	15–30	1.4–2.0	1.9–2.4	1.0–1.9
Hr30	*Hippophae rhamnoides*	0.4–0.9	15–30	1.4–2.0	1.9–2.4	1.0–1.9
Em12	*Elaeagnus multiflor*	0.5–1.2	40–60	1.5–2.5	2.1–3.7	1.3–3.0

4.1 Morphology

All strains isolated from nodules of *Alnus, Elaeagnus, Hippophae, Casuarina* and *Myrica* have similar morphological characteristics. The actinomycete grows as a filamentous, septate and branched mycelial mat. In nitrogen-free Bap medium abundant vesicles are formed. Vesicles were of spherical form or pyriform, formed terminally on short parent hyphae branching from hyphae strands. In old culture, terminal or intercalary sporangia are formed. It is necessary to emphasize that endophytes isolated from *Casuarina* in old culture frequently appear as reproductive torulose hyphae (RTH), resulting from the enlargement and multiple segmentation of vegetative hyphae into torulose chains of sporelike cells. RTH differ from sporangia in morphology. RTH are not observed in strains isolated from *Alnus, Hippophae* or *Myrica*. Morphological studies also show that *Frankia* strains from different hosts have their own morphological characteristics in width of hyphae, loose or close arrangement of mycelium, size of sporangium, size and form of vesicle, length of vesiclephore. A comparison of morphological characteristics of *Frankia* strains from different host plants is given in Table 2 (Huang Yali et al. 1989).

4.2 Physiology

C-sources and N-sources were sysmatically studied. About 23 C-sources were examined. Simple organic acids were well used by *Frankia* strains. Mono-, di-, polysaccharides were not efficiently used. According to their C-source utiliza-

Table 3. Nitrogenase activity of *Frankia* strains from different host plants

Strains	Host plant	Nitrogenase activity nM C_2H_2/mg. protein, min.
Cc13	*Casuarina cunninghamiana*	4.76
Cc4	*Casuarina cunninghamiana*	1.62
Cc01	*Casuarina cunninghamiana*	2.00
Ce37	*Casuarina equisetifolia*	4.67
AllcI1	*Allocasuarina lehmanniana*	1.00
At4	*Alnus tinctoria*	10.16
At97	*Alnus tinctoria*	0.7
As28	*Alnus sibirica*	3.14
Ah1	*Alnus hirsuta*	5.22
CpI1	*Comptonia peregrina*	2.4
Mn35	*Myrica nana*	0.89
Hr16	*Hippophae rhamnoides*	2.44

tion spectrum, some C-sources could be used as a diagnostic characteristics for taxonomy. Acetic and propionic acid could be used by *Alnus, Myrica Frankia* strains. Pyruvic acid could be well used by *Hippophae Frankia* strains. All these three acids could be well used by *Casuarina Frankia* strains (Cui et al. 1988).

Nitrogenase activity of free-living *Frankia* strains were determined by gas chromatography in nitrogen-free medium. Results showed that nitrogenase activity varies with different host plants and under different culture conditions. The representative results from 8 species in 6 genera host plants are listed in Table 3 (Ding et al. 1986).

4.3 Cell Chemical Composition

The components of DAP and sugars in 41 *Frankia* strains isolated from 7 genera of host plants were analysed. The results showed that all 41 *Frankia* strains contain meso-DAP and an unknown hexose. In contrast, the hexose is absent in all other actinomycetes tested. This unknown hexose was found to be 2-*O*-methyl-D-mannose. The meso-DAP and 2-*O*-methyl-D-mannose may be used as diagnostic characteristics for *Frankia*. All 41 *Frankia* strains contain glucose, galactose, mannose, xylose, ribose, rhamnose and arabinose, etc. The sugar content in strains from different hosts is quite different: *Hippophae* contains more rhamnose than *Alnus, Comptonia* and *Myrica* do; *Allocasuarina cunninghamiana* contains arabinose and *Ceanothus americanus* contains fucose (Su 1987).

Amino acids in *Frankia* cells were determined by automatic amino acid analyzer. Results are listed in Table 4.

Results showed that glutamine or glutamic acid content in *Frankia* cells is very high, implying a high percentage of the protein in Frankia cells. The difference in amino acid composition in strains from different hosts are striking.

Table 4. Content of amino acids in cells of *Frankia* strains

Strains Amino acids	Avs13	Mn65	Hr30	Cc01
	content (μmg/mg protein)			
Asparagine, Aspartic acid or both	4.37	—	1.52	—
Threonine	5.67	8.36	2.95	29.46
Serine	16.24	—	—	—
Glutamine, Glutamic acid or both	78.17	619.50	197.27	299.78
Glycine	13.76	2.25	1.36	—
Alanine	15.91	21.28	5.68	19.57
Cysteine	5.02	6.40	7.42	—
Valine	6.32	6.92	4.92	—
Methionine	—	4.15	—	N.D.
Isoleucine	—	4.21	1.52	N.D.
Leucine	—	2.13	1.74	N.D.
Tyrosine	—	1.90	—	3.15
Phenylalanine	2.54	3.23	3.33	6.03
Lysine	3.84	4.38	3.41	8.53
NH_4^+	14.09	10.33	5.04	40.92
Arginine	—	7.84	—	7.66
Total amino acids	151.84	693.56	234.83	374.18
Gln, Glu or both content%	51.48	89.32	84.01	80.12

N.D. not determined

In *Frankia* strain AvsI there was no methionine, leucine, isoleucine, tyrosine, and arginine, in Frankia sp Hr30, serine, methionine, tyrosine, arginine were absent. In *Frankia* Mn65, asparagin, aspartic acid, serine were absent, in *Frankia* Sp. Cc01, asparagine, aspartic acid, serine, glycine, cysteine, valine were absent (Zou Hua et al.).

4.4 Host Range of Pure-Culture *Frankia* Strains and Their Serological Relationships

In order to understand their infectivity and host range of pure culture *Frankia* strains, cross-inoculation experiments with 11 representative *Frankia* strains and 17 host species were carried out successfully. Results show that *Frankia* strains have their host specificity groups: *Alnus* group, *Elaeagnus* group and *Casuarina* group (Huang Jianbin et al. 1985).

(1) *Alnus* group: strains which nodulate *Alnus*, *Myrica*.
(2) *Elaeagnus* group: strains which nodulate *Elaeagnus*, *Hippophae*.
(3) *Casuarina* group: strains which nodulate *Casuarina*.

These results are in agreement with those obtained in the serological study. The immunological study of 30 *Frankia* strains isolated from diverse host plants

Table 5. *Frankia* strains used in serological studies

Strains	Original host	Source
Ce4	*Casuarina cunninghamiana*	Guangdong, China
Ce24	*Casuarina equisetifolia*	Guangdong, China
Ce25	*Casuarina equisetifolia*	Guangdong, China
Ce28	*Casuarina equisetifolia*	Guangdong, China
Ce37	*Casuarina equisetifolia*	Guangdong, China
Cel3	*Casuarina equisetifolia*	USA
CeR43	*Casuarina equisetifolia*	USA
AllcI1	*Allocasuarina cunninghamiana*	USA
As28	*Alnus sibirica*	Changbai mountain, China
As23	*Alnus sibirica*	Changbai mountain, China
At4	*Alnus tinctoria*	Liaoning, China
At97	*Alnus tinctoria*	Liaoning, China
AgI	*Alnus glutinosa*	USA
ArI3	*Alnus rugosa*	USA
AvsI3	*Alnus virdis spp, sinuata*	USA
Hr16	*Hippophae rhamnoides*	Liaoning, China
Hr18	*Hippophae rhamnoides*	Liaoning, China
Hr24	*Hippophae rhamnoides*	Liaoning, China
Hr30	*Hippophae rhamnoides*	Liaoning, China
Hr32	*Hippophae rhamnoides*	Liaoning, China
Hr37	*Hippophae rhamnoides*	Liaoning, China
Mn7	*Myrica nana*	Yunnan, China
Mn8	*Myrica nana*	Yunnan, China
Mn12	*Myrica nana*	Yunnan, China
Mn13	*Myrica nana*	Yunnan, China
Mn35	*Myrica nana*	Yunnan, China
Mn65	*Myrica nana*	Yunnan, China
Mn89	*Myrica nana*	Yunnan, China
MpI	*Myrica pensylvanica*	USA
CpI1	*Comptonia penegrina*	USA

growing in different geographic area was undertaken to establish their taxonomic relationship. *Frankia* strains used in this test are listed in Table 5.

The SPA co-agglutination method was applied. Based on the SPA co-agglutination test, these *Frankia* strains can be divided into three major serogroups, which are listed in Table 6. The first serogroup includes strains isolated from *Alnus*, *Comptonia* and *Myrica* host plants, the second includes strains isolated from *Hippophae* host plant, and the third was isolated from *Casuarina*, *Allocasuarina* host plants (Sun 1987). The SPA co-agglutination is highly specific, sensitive and time-saving, therefore, it is a very convenient method for identification and group-division of *Frankia* strains.

In order to identify and characterize the numerous *Frankia* strains isolated, the method of electrophoretic whole-cell protein patterns was applied. 18 *Frankia* strains from different host plants, representing 8 genera, 16 species were used. These *Frankia* strains were determined by one-dimensional sodium sulfate-polyarcylamide gel electrophoresis.

Table 6. Serotypes of 30 *Frankia* strains from 5 genera of host plants

Serogroup	Strains	CpI1 Antiserum labelled SPA	At4 Antiserum labelled SPA	Mn35 Antiserum labelled SPA	Hr21 Antiserum labelled SPA	Ce37 Antiserum labelled SPA
I	As23	+ +	+ +	+ +	−	+
	As28	+ +	+ +	+ +	−	+
	At4	+ +	+ +	+ +	−	+
	At97	+ +	+ +	+ +	−	+
	Ag1	+ +	+ +	+ +	−	+
	ArI3	+ +	+ +	+ +	−	+
	ArsI3	+ +	+ +	+ +	−	+
	CpI1	+ +	+ +	+ +	−	+
	Mn7	+ +	+ +	+ +	−	+
	Mn8	+ +	+ +	+ +	−	+
	Mn12	+ +	+ +	+ +	−	+
	Mn13	+ +	+ +	+ +	−·	+
	Mn35	+ +	+ +	+ +	−	+
	Mn65	+ +	+ +	+ +	−	+
	Mn89	+ +	+ +	+ +	−	+
	MpI	+ +	+ +	+ +	−	+
II	HrI6	−	−	−	+ +	+
	HrI8	−	−	−	+ +	+
	Hr21	−	−	−	+ +	+
	Hr30	−	−	−	+ +	+
	Hr32	−	−	−	+ +	+
	Hr37	−	−	−	+ +	+
III	Cc4	+	+	+	+	+ +
	Ce24	+	+	+	+	+ +
	Ce25	+	+	+	+	+ +
	Ce28	+	+	+	+	+ +
	Ce37	+	+	+	+	+ +
	CeI3	+	+	+	+	+ +
	CcR43	+	+	+	+	+ +
	AllcI1	+	+	+	+	+ +

+ + strong coagglutination
+   trace coagglutination
−   no coagglutination

The results showed that polypeptide patterns vary greatly with different *Frankia* strains. According to the cluster analysis results, 18 *Frankia* strains are divided into three gel electrophoresis groups. The gel group A contains strains isolated from *Alnus*, *Myrica* and *Comptonia*. The gel group E contains strains isolated from *Elaeagnus* and *Hippophae*. The gel group C contains strains isolated from *Casuarina*, *Allocasuarina* and *Ceanothus*. Three gel electrophoresis groups were observed which matched with *Alnus*, *Elaeagnus* and *Casuarina* host-specificity group (Zhang Daohai et al., in press).

Hybridization between *Bam*HI-digested *Frankia* total DNAs and *nif* HDK probes from *Klebsiella pneumoniae* (pSA30) showed that the patterns of *nif*-hybridizing fragments were different among different compatible groups. At4 produced 3 kb, 1.9 kb and 0.8 kb fragments. Hr-16 produced 2.1 kb and 1.7 kb and Ccol produced a 5.5 kb fragment (Cui 1990).

Based on the results of cross-inoculation, serological tests, chemical taxonomy, and molecular hybridization, the *Frankia* strains we have tested may be classified into three groups.

4.5 Symbiotic Physiology and Environmental Conditions

Physiological studies showed that different lux illumination has a great influence on infection, nodulation and nitrogen fixation of the symbiotic system. Various host-symbionts demonstrate different sensitivity to lux illumination. High illumination intensity gives a good effect on nodulation of *Casuarina*, while *Alnus* is not very sensitive to the intensity of illumination. *Alnus* can be infected and nodulated under high (10 000 lux) and low (4 000 lux) illumination with wide adaptability to illumination intensity. The original habitat characteristics of *Alnus* in the mixed forest may be reflected in its wide requirement of illumination (Ding et al. 1987). A field survey also showed that acetylene reduction activity of nodules of the alder tree under full lighting conditions (21 000 lux) is more active (6.25 µM C_2H_4/g.h. fresh nodule) than nodules of alder under tree-crown conditions (8 000 lux, 1.12 µM C_2H_4/g.h. fresh nodule) (Yang et al. 1984).

Temperature tests showed that *Alnus* nodulation occurred most efficiently in the temperature ranging from 25–28 °C. Temperatures above 30 °C reduced the nodulation rate and the amount of nodules. Good nodulation occurred in the temperature range of 32–35 °C in *Casuarina*. The difference of the nodulating temperature under laboratory conditions among *Casuarina* and *Alnus* may reflect their requirements for the natural ecological environment (Ding et al. 1987).

Field monitoring results also showed that acetylene reduction activity of alder nodules at a soil temperature of 20 °C was two times higher than at a soil temperature of 8 °C (1.2 µM C_2H_4/g.h. fresh nodule) (Yang Sihe et al. 1984).

A series of experiments with *Frankia* sp.Hr32 on *Hippophae rhamnoides* were carried out in the laboratory. Seedlings of *Hippophae rhamnoides* were inoculated with *Frankia* sp.Hr32 and was grown in water culture. The effects of environmental factors such as salinity, pH value, combined-N and phosphorus supply on the infection, nodulation and plant growth were studied. Nodulation began at 0.2% NaCl level and high proportions of nodulation occurred at 0.1% NaCl level. Weak salt resistance of *Frankia* Hr32 led to non-nodulation at high salinity. The seeding of *Hippophae rhamnoides* attained the best nodulation at pH 7.0–8.0. No nodule was formed at pH 9.0 and 4.2. When the solution contained 10 ppm P_1 or more, 90% or higher nodulation was found, but low phosphorus supply limited the nodulation of the seedling. Sufficient phosphorus supply was necessary for plant growth and nodulation of *Hippophae rhamnoides*, also for nodule development and growth. 6 ppm or less combined-N supply improved the nodulation. No nodule was formed at 16 ppm combined-nitrogen. The nodulated seedlings of *Hippophae rhamnoides* had two times more drying weight than the control. Nodulation also delayed the fall of the leaves of *Hippophae rhamnoides* (Li et al. 1989).

5 Utilization of Actinorhizal Plant Resources in Forest Planting in China

In China the forest cover has been very low. Actinorhizal plants were successfully applied in forest plantings, which stimulated the forest development. Great socioeconomic benefits were derived from large-scale plantations of actinorhizal plants. In Southern China over 1 000 000 ha of *Casuarina* plantations have been established in about 3000 km of shelterbelt plantings along the coastal dunes fronting the South China Sea (Turnbull 1981). The great coastal windbreak plantation of *Casuarina* has stopped coastal and desert sand dune encroachment and increased crop yield. This achievement in construction of protection forests in prized as a Green Great Wall in Southern China. In Northern China sea-buckthorn resources are very abundant, widespread, occupying 940 000 hectares. In Northern China large-scale plantation of *Hippophae* is carried out due to their role in sand stabilization, soil conservation, rehabitation of depauperate soil, the provision of fuel, the construction of mixed forest and high economic benefit from comprehensive utilization of seabuckthorn. Therefore, seabuck-thorn is the gem of arid areas and is prized as a way from depauperation to wealth in poor regions (Liu 1986).

6 Future Directions

The success in isolating actinomycete Frankia has laid a foundation for its possible application. It is clear that, in the near future, substantial effort showed be focused on the following two aspects:

6.1 Selection and Construction of Fast-Growing Nitrogen-Fixing Actinomycetes

In recent years advances have been made both in afforestation and in the exploitation of seabuckthorn resources in China. Aerial seeding afforestation of seabuckthorn has succeeded. Therefore, selection and construction of fast-growing nitrogen fixing actinomycetes will not only promote *Frankia* research, but also be of great use in afforestation.

6.2 Trisymbiosis Research

Nitrogen-fixing plants have a relatively high demand for phosphorus. Ecological research showed that many actinorhizal plants are often infected with endo-mycorrhizas or ectomycorrhizas. Cooperation between mycorrhizas and nodulation can greatly affect plant growth, particularly on poor soils. Strengthening research in three-membered symbiosis will promote wide application of actinorhizal plants, especially on phosphorus-deficient soils. It will stimulate research in symbiotic physiology and below-ground ecology.

Acknowlegements. We would like to thank Professor John G. Torrey and Dr. Dwight D. Baker for their kindly providing American *Frankia* strains. This work was supported by the Natural Science Foundation of China.

7 References

Baker D, Torrey JG, Kidel GH (1979) Isolation by sucrose-density fractionation and cultivation in vitro of actinomycete from nitrogen-fixation root nodules. Nature 281: 76

Callaham D, Del Tredici P, Torrey JG (1978) Isolation and cultivation in vitro of the actinomycetes causing root nodulation in *Comptonia.* Science 199: 899

Cui Yuhai, Ding Jian (1988) Study of the C-source-utilizing spectra of six typical *Frankia* strains isolated from different hosts. J Microbiol 8(3): 30

Cui Yuhai (1990) Molecular cloning of symbiotic genes in *Frankia* PhD thesis Institute of Applied Ecology, Academia Sinica, Shenyang, China

Ding Jian, Zhang Zhongze, Li Zhongwei, Su Fengyan, Sun Huijun, Huang Yali, Wu Yang, Cui Yuhai, Xu Qingde, Li Weiguang (1986) Determination of nitrogenase activity of free-living *Frankia* strains by gas chromatography. J Microbiol 6(2): 33

Ding Jian, Zhang Zhongze, Sun Huijun, Su Fengyan, Huang Yali, Xu Qingde, Li Zhongwei, Wu Yang (1987) Comparative studies on biological characteristics of *Frankia* strains from *Alnus, Casuarina, Hippophae* and *Myrica.* J Microbiol 7(2): 16

Huang Jiabin, Shen Shanmin, Liu Huichang, Jiang Jiande, Yang Huifang, Zhao Zhenuing, Yang Binfang (1984) An investigation of the resource of nonleguminous noddulating trees in China. In: Veeger C, Newton WE (eds) Advances in nitrogen fixation research Academic Press, New York

Huang Jiabin, Zhao Zhenying, Chen Guanxiong, Liu Huichang (1985) Host range of *Frankia* endophytes. Plant and Soil 87: 61

Huang Yali, Ding Jian, Zhang Zhongze, Su Fengyan, Xu Qingde, Li Weiguang (1989) Studies on Morphological characteristics of Frankia strains from different host plants. J Microbiol 9(3): 29

Jiang Jiande, Yang Huifan, Huang Jiabin, Zhu Baoqin, Liu Huichang, Zhong Yuhua, Yang Bifang (1984) Isolation and cultivation of *Frankia* strains Ace13 from root nodules of *Alnus* cremastogyne. Acta Microbiol Sinica 24(1): 37

Lalonde M, Calvert HE, Pine S (1981) Isolation and use of *Frankia* strains in actinorhizae formation. In: Gibson AH, Newton WE (eds) Current perspectives in nitrogen fixation. p 296 Australian Acad Sci, Canbea

Li Zhongwei, Ding Jian (1985) Isolation of *Frankia* Sp.At4 from root nodules of *Alnus tinctoria,* cultivation in vitro and investigation on its infectivity. J Microbiol 5(3): 17

Li Zhongwei, Ding Jian (1986) Isolation of *Frankia* Sp.FSCc01 from root nodules of *Casuarina cunninghamiana* and its cultivation in vitro and nodulation conditions. Acta Microbiol Sinica 26(4): 295

Liu Kun (1986) Summary speech at National Meeting of Experience Exchange of Seabuckthorn Exploitation and Utilization (5–10 Sept.'86) Taiyuan, Shanxi in Seabuckthorn Abstracts, China Forestry Publishing House, p 125

Li Yuandan, Ding Jian (1989) Nitrogen-fixing symbiosis with *Frankia-Hippophae rhamnoides* I. Effects of the environmental factors on nodulation and growth of *Hippophae rhamnoides*. J Microbiol 9(3): 22

Qin Min (1988) Studies on *Frankia* spp. isolated from *Casurinaceae*. Master thesis Agricultural College of Guangxi

Su Fengyan, Ding Jian, Sun Huijun, Huang Yali, Xu Qingde, Zhang Zhongze, Li Zhongwei, Wu Yang, Li Weiguang, Wan Limin (1987) Analysis of chemical composition of 41 strains *Frankia* from different host plants. J Microbiol 7(2): 57

Sun Huijun, Ding Jian, Wu Yang, Xu Qingde, Huang Yali, Su Fengyan, Zhang Zhongze, Li Weiguang (1987) The serogroups of *Frankia* J Microbiol 7(2): 51

Turnbull JW (1981) The use of *Casuarina equisetifolia* for protection of forests in China. In: *Casuarina* ecology management and utilization. Proc Int Workshop, Canberra, Australia 17-21 August 1981 p 55

Wu Yang, Ding Jian (1987) Studies on biological charcteristics of *Frankia* sp. from *Hippophae rhamnoides*. Acta Microbiol Sinica 27(3): 227

Yang Huifan, Jiang Jiande, Huang Jiabin, Zhu Baogin, Zhang Yuhua, Liu Huichang, Yang Bifang, Zhao Zhenying (1984) Isolation and reinoculation of endophytes of *Hippophae rhamnoides* and *Elaeagnus*. Acta Microbiol Sinica 24(4): 315

Yang Sihe, Huang Jiabin, Lin Jihui, Liu Huichang (1984) Nitrogen fixation by nodules of alder and ecological condition in natural forest of Northeastern China. J Ecology 1: 10

Zhang Daohai, Ding Jian, Su Fengyan (in press) Grouping of *Frankia* strains by using sodium dodecyl sulfate-polyacrylamide gel electrophresis. J Microbiol

Zhang Zhongze, John G Torrey (1985) Studies of an effective strain of *Frankia* from *Allocasuarina lehmanniana* of the *Casuarinaceae*. Plant and Soil 87: 1

Zou Hua, Ding Jian (in press) Comparison of amino acids in the cell of *Frankia* strains from different host plants. J Microbiol

CHAPTER 30
Symbiotic Nitrogen Fixation in Tropical and Subtropical Leguminous Trees

W.N. Huang, K.Q. Cai, L.X. Cai, Y.D. Wu,
Z.W. Xu, G. Lan, and X.L. Zou

The Nitrogen Fixation
and its Research in China
Editor Guo-fan Hong
© Springer-Verlag Berlin Heidelberg 1992

1 Introduction

The utilization and development of nitrogen-fixing trees as resources has played an important role in establishing forests in denuded land, in producing more wood in industrial plantations, in conserving soil and water, and in improving ecosystems. A survey on the nodulation status of some tropical and subtropical woody leguminous species in Fujian Province and an investigation of the ecophysiology of symbiotic nitrogen fixation in several leguminous trees have been carried out in recent years.

2 A Survey on the Nodulation and Nitrogen Fixation of some Tropical and Subtropical Woody Leguminous Species of Fujian

Some 270 species of 46 genera of woody leguminous plants are distributed in China, some 90 species of 33 genera of leguminous trees are found in Fujian (FSTC 1987). A survey was carried out on the nodulation status of 29 species of Papilionoideae, 22 species of Mimosoideae and 16 species of Caesalpinioideae in Fujian (Huang et al. 1983). Among these species, 29 of the former, 21 of the following and 2 of the latter were nodulated. These species are listed in Table 1.

Table 1. The nodulation and nitrogen fixation of some woody leguminous species in Fujian

Genus	Species	Nodulation	C_2H_2-reducing activity (μmol C_2H_4/g FW/h)
Mimosoideae			
Acacia	auriculaeformis	+	3.015–11.791
	confusa	+	2.803–24.091
	cunninghamii	+	4.983–11.088
	farnesiana	+	6.493–8.170
	glauca	+	9.820–39.180
	mangium	+	9.449
	mearnsii	+	4.574–5.755
	nolosericea	+	2.275–3.609
Adenanthera	pavonina	–	
Albizia	chinensis	+	13.617
	falcataria	+	16.660–21.930
	julibrissin	+	2.094
	kalkora	+	6.016
	lebbeck	+	29.930–58.130
	odoratissima	+	6.207–19.180
	procera	+	5.375–27.848
	sp.	+	8.440
Enterolobium	contortisiliquum	+	7.760

Table 1. (*Continued*)

Genus	Species	Nodulation	C_2H_2-reducing activity (μmol C_2H_4/g FW/h)
Leucaena	leucocephala	+	10.980–24.990
	leucocephala cv. salvador	+	7.564–29.870
Mimosa	pionea	+	3.930
	pudica	+	3.810–8.535
Papilionoideae			
Amorpha	fruticosa	+	3.409–25.520
Cajanus	cajan	+	22.510
	cajan (L.) Millsp		
	f. flavus	+	10.290
Dalbergia	balansae	+	10.290
	hupeana	+	16.380–38.110
	obtusifolia	+	10.094
	odorifera	+	9.746
	sisso	+	35.630
	sp.	+	59.530–89.300
			35.632
Desmodium	gyroides	+	1.744
	heterocarpom	+	5.797
Erythrina	arborescens	+	22.239
	corallodendron	+	33.187
Indigofera	tinctoria	+	1.926–8.654
	suffruticosa	+	12.357
Lespedera	bicolor	+	0.874–2.007
	cuneata	+	38.032–49.356
	formosa	+	9.321
Moghania	macrophylla	+	2.354–4.228
Ormosia	henryi	+	4.172
	pinnata	+	3.082
Pterocarpus	indicus	+	17.500–22.600
Robinia	pseudoacacia	+	28.115
Sesbania	cannabina	+	6.856–7.395
	rostrata	+	10.905–18.785
Tephrosia	candida	+	8.900
	purpurea	+	3.099–6.978
Uraria	crinita	+	25.789
Wisteria	sinensis	+	0.196
Caesalpinioideae			
Caesalpinia	pulcherrima	–	
Cassia	laevigata	+	3.038
	mimosoides	+	2.319
	nodosa	–	
	occidentalis	–	
	siamea	–	
	surattensis	–	
	tora	–	
Delonix	regia	–	
Bauhinia	purpurea	–	
	variegata	–	
	faberi	–	
Cercis	chinensis	–	
Gleditsia	sinensis	–	
Gymnocladus	chinensis	–	
Peltophorum	pterocarpum	–	

3 Ecophysiology of Symbiotic Nitrogen Fixation in Leguminous Trees

3.1 Effect of Environmental Factors on Nodulation and Nitrogen Fixation of Leguminous Trees

Any factors in the environment which affect the growth of the host plants are likely to affect the development and function of the root nodules.

3.1.1 Seasonal Variations in Nodulation and Nitrogen Fixation of Woody Legumes

In the period 1982–1984, observations were made on the seasonal variations in nodulation and nitrogen fixation of *Acacia glauca* and *Leucaena leucocephala* grown in the Xiamen area (Huang et al. 1988, 1989a). Generally, more nodules formed and nitrogen-fixing activity was higher in summer and autumn than in

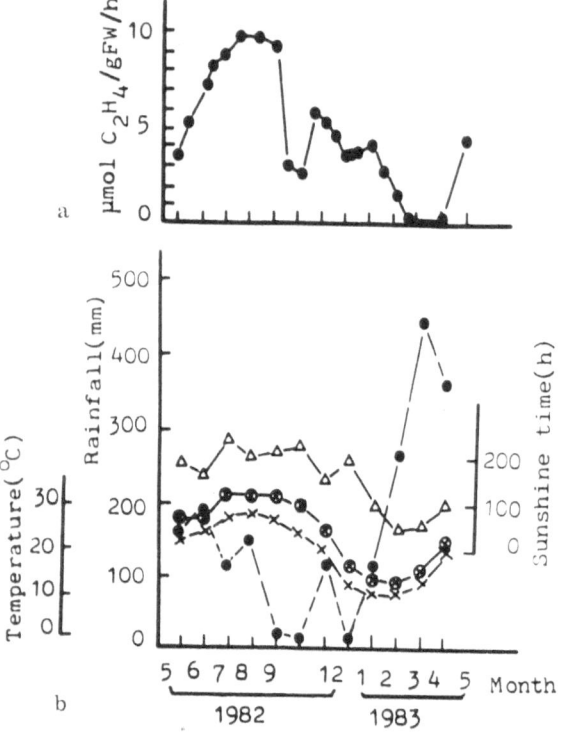

Fig. 1a. Seasonal variation in C_2H_2-reducing activity of root nodules on *A. glauca*; **b** variation of climatic factors for study period. $\times$——$\times$ air temperature, $\otimes$——$\otimes$ soil temperature at the depth of 10 cm, ·——· montly rainfall and $\triangle$——$\triangle$ montly sunshine time

the early spring and winter. The C_2H_2-reducing activity of nodules in *A. glauca* rose to a maximum in July and high rates of C_2H_2-reducing activity were maintained in August–September, and declined to a minimum during 10 to 20 October (Fig. 1). This might be a response of the desiccation of leguminous plants and their nodules at a time when both rainfall and soil moisture were relatively low. The second peak in activity of nodules occurred in November when new nodules formed. The activity of nodules declined after November, possibly due to the drop of both air and soil temperature. From 10 February to 27 April nodules were found to decay because of continual rain. The seasonal variations in nitrogen-fixing activity of nodules of *L. leucocephala* and the climatic variations in 1984 are shown in Fig. 2.

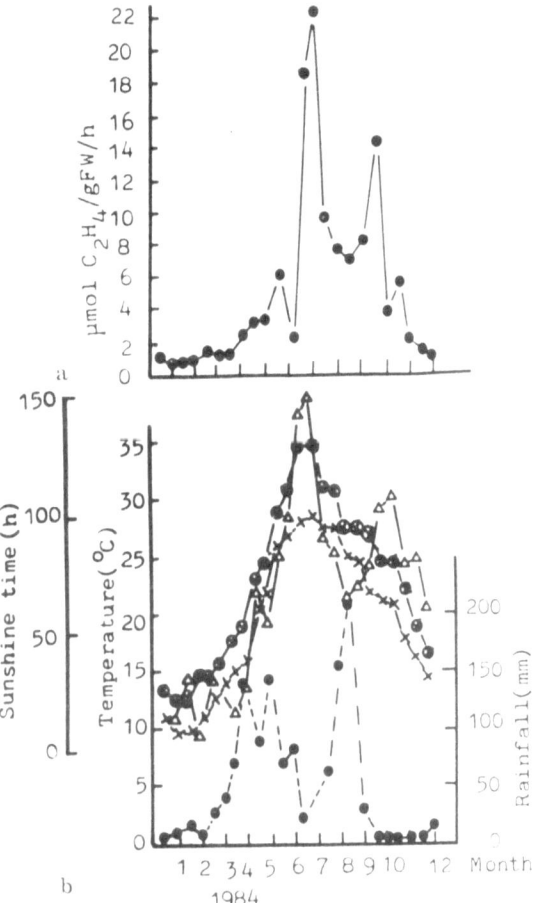

Fig. 2a. Seasonal variation in C_2H_2-reducing activity of root nodules on *L. leucocephala*; **b** variation of climatic factors for study period. ×——× air temperature, ⊗——⊗ soil temperature at the depth of 10 cm, ·——· montly rainfall and △——△ montly sunshine time

3.1.2 Effect on C_2H_2-Reducing Activity of Nodules on Woody Legumes

3.1.2.1 Temperature

The variation in air and soil temperatures was continually observed for 5 days before analysis of samples. Table 2 shows that temperature may affect C_2H_2-reducing activity in nodules of *A. glauca* grown in the field. Higher C_2H_2-reducing activity in nodules exhibited at air temperatures of 25–30 °C and at soil surface temperatures of 30–36 °C. The activity declined markedly below 20 °C.

Potted plants of *L. leucocephala, A. glauca, Albizia falcataria* and *Sesbania cannabina* were kept in the climatic chamber with different temperatures and

Table 2. Effect of temperature on C_2H_2-reducing activity in nodules of *A. glauca* grown in the field

Date determined	Air temp. (°C)	Soil temp. (°C)	C_2H_2-reducing activity (μmol C_2H_4/g FW/h)
27 July 1982	28.8	36.1	9.203
26 Aug. 1982	25.8	35.7	8.918
30 Nov. 1982	19.7	21.1	3.936
9 Feb. 1983	10.5	11.8	0.944

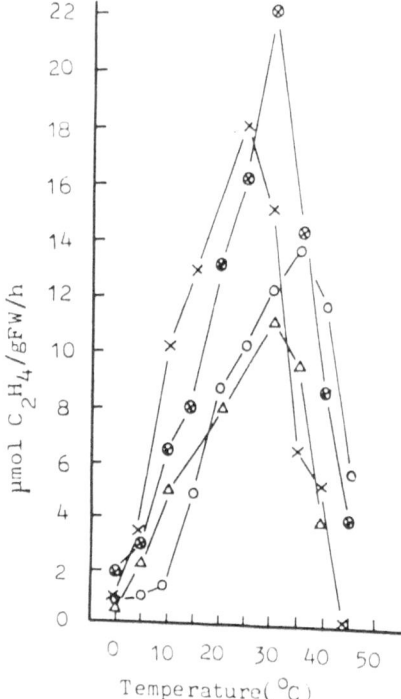

Fig. 3. Effect of temperature on C_2H_2-reducing activity in nodules of woody legumes. ×——× *L. leucocephala*, ○----○ *A. glauca*, ⊗—— ⊗ *S. cannabina* and △——△ *A. falcataria*

then exposed for 3 h and 24 h, respectively. Figure 3 shows that the response of each species of the trees to temperature was different (Huang et al. 1987, 1988, 1989a). The C_2H_2-reducing activity in nodules of *L. leucocephala* rose to a maximum at 25 °C, and rapidly reduced below 10 °C or over 30 °C, then almost reached zero at 0 °C or 45 °C. Maximum C_2H_2-reducing activity in nodules of *A. glauca* was at 35 °C, and that of both *A. falcataria* and *S. cannabina* at 30 °C. With the prolongation of unfavourable temperature, C_2H_2-reducing activity decreased (Fig. 4). When potted seedlings of *A. falcataria* were kept in the climatic chamber with lower temperature (0–5 °C) for 24 h, almost no

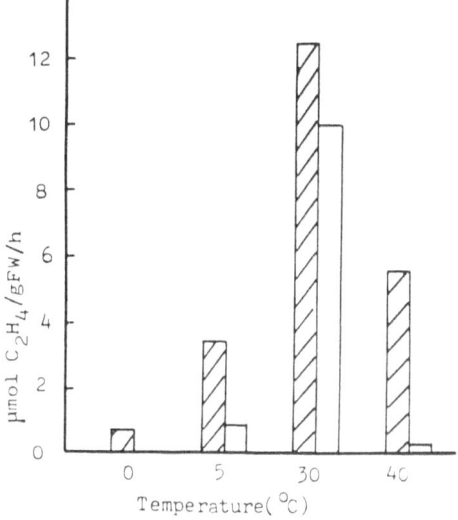

Fig. 4. Effect of low extended unfavourable temperature on C_2H_2-reducing activity in nodules of *L. leucocephala*. ▨3h and ▭24h

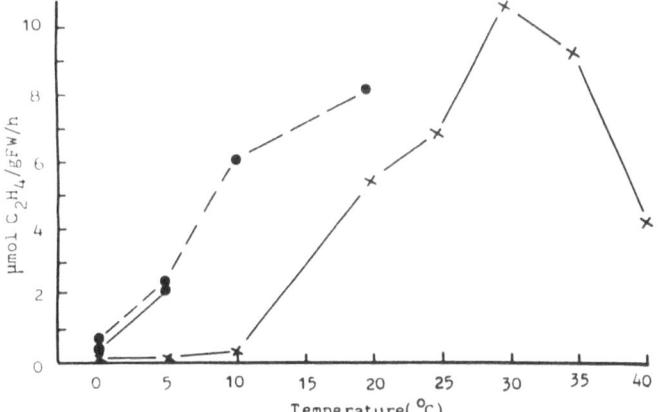

Fig. 5. Effect of low temperature on C_2H_2-reducing activity in nodules of *A. falcataria*. ·——· 3h, ×——× 24h and ·---· tested plants transplants to outdoor at 25 °C for 24 h

C_2H_2-reducing activity was detected. When potted plants were transplanted outdoors at 25 °C for the whole day, C_2H_2-reducing activity recovered to some extent as shown in Fig. 5.

3.1.2.2 Rainfall and Soil Moisture

Too little or too much water may limit or inhibit C_2H_2-reducing activity in nodules. Rainfall was continually observed for 15 days before analysis of nodular samples of *A. glauca* (Huang et al. 1989a). Table 3 shows that minimum C_2H_2-reducing activity was observed due to a period of drought in the first 12 days of October.

The results of soil moisture analysis during a period of drought indicated that soil moisture was only 3.5%–3.55% above 10 cm depth of the soil and 4.4%–4.47% between 10–30 cm depth. Under optimum conditions nodules of *A. glauca* contained 80.35% water, but at times of drought nodules had only 59% water and some of these nodules were wilted. At this time C_2H_2-reducing activity in nodules rapidly declined. Table 4 and Fig. 6 show that the effect of soil moisture on nitrogen fixing of nodules in *L. leucocephala* was marked.

Table 3. Effect of rainfall on C_2H_2-reducing activity in nodules of *A. glauca*

Date determined	Air temp. (°C)	Soil temp. (°C)	Rainfall (mm)	C_2H_2-reducing activity (μmol C_2H_4/g FW/h)
13 Sept. 1982	24.3	28.0	14.0	8.800
27 Sept. 1982	24.5	28.7	2.3	2.700
12 Oct. 1982	23.6	27.8	0	2.300

Table 4. Effect of rainfall on C_2H_2-reducing activity in nodules of *L. leucocephala*

Date determined	Air temp. (°C)	Soil temp. (°C)	Rainfall (mm)	C_2H_2-reducing activity (μmol C_2H_4/g FW/h)
24 June 1983	27.4	30.3	97.9	9.250
13 July 1983	28.5	35.3	5.0	1.100
12 Sept. 1983	27.2	29.2	57.6	5.140
26 Sept. 1983	27.4	33.1	6.5	0.400

3.1.2.3 Combined Nitrogen

The formation of nodules in seedlings of *L. leucocephala* was inhibited strongly by combined nitrogen (Cai, Huang 1987). Seedlings of *L. leucocephala* were grown in pots containing a nutrient solution with combined nitrogen at various concentrations (0, 26, 52, 78 ppm). The rate of nodulation in *L. leucocephala* seedlings reached 100% on the 7th day after inoculation in nitrogen-free nutrient

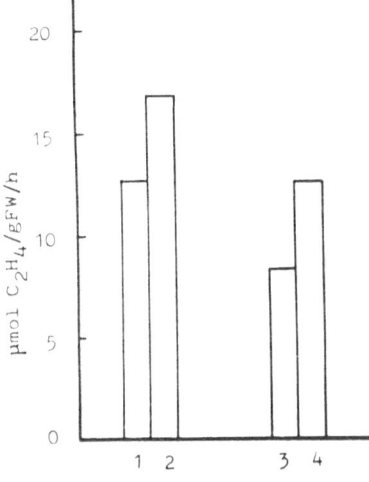

Fig. 6. Effect of soil moisture on C_2H_2-reducing activity in nodules of *L. leucocephala*; soil moisture content: 1, 10%; 2, 14%; 3, 9.8%; 4. 12.5%

solution, but it decreased with the increase of nitrogen concentration in the nutrient solution. On the 45th day after inoculation, the rate of nodulation was only 18.2% at 78ppm N, and the number of nodules and fresh weight per plant also decreased markedly (Table 5).

The C_2H_2-reducing activity in nodules was also inhibited by combined nitrogen. Table 6 shows that the higher the level of supplied nitrogen becomes, the lower the C_2H_2-reducing activity in nodules. However, all seedlings of *L. leucocephala* grown in nutrient solution with combined nitrogen had more dry

Table 5. Effect of combined nitrogen on nodulation of *L. leucocephala*

Treatment	No. of tested plants	No. of nodulated plants	Rate of nodulation (%)	Average No. of nodules (grain/plant)	Average fresh wt of nodules (mg/plant)
−N	45	45	100	51.5	148.4
26 ppm N	44	39	88.6	2.2	8.0
52 ppm N	44	28	63.6	1.2	4.5
78 ppm N	44	8	18.2	0.3	1.3

Table 6. Effect of combined nitrogen on C_2H_2-reducing activity in nodules of *L. leucocephala*

Treatment	C_2H_2-reducing activity (μmol C_2H_4/g FW/h)
−N	7.564
26 ppm N	6.072
52 ppm N	5.384
78 ppm N	4.683

weight than those grown in nitrogen-free nutrient solution, especially lower concentration of combined nitrogen markedly stimulated the growth of *L. leucocephala* seedlings. Our experimental results suggested that although seedlings of *L. leucocephala* might nodulate and fix nitrogen in poor soil, early supplementation with a little combined nitrogen was beneficial to the growth of seedlings.

3.1.2.4 Soil pH

The effect of pH in the nitrogen-free culture fluid on nodulation of *L. leucocephala* seedlings was investigated (Cai, Huang 1987; Cai et al. 1985; Huang et al. 1989b). Table 7 and Fig. 7 show that in culture fluid with a pH between 5.5 and 8.5, the rate of nodulation was 100% on the 7th day after inoculation, but the number of nodules per plant at pH 7.0 was the highest, and the number of nodules per plant at pH 8.5 was more than that at pH 5.5 on the 27th day after inoculation. Although *L. leucocephala* seedlings might nodulate in light acid soil, it was fitted to neutral and light alkaline soil. Formation of nodules was strongly inhibited in culture fluid with a pH between 3.6 and 4.5. At pH 3.6,

Table 7. Effect of pH in culture fluid on number of nodules in *L. leucocephala* (average number of nodules: grain/plant)

Days after inoculation	pH					
	3.6	4.0	4.5	5.5	7.0	8.5
7	0	0.14	2.57	11.13	12.13	8.50
17	0	1.43	8.43	16.63	19.90	16.00
27	0	2.14	10.00	17.25	23.63	18.88
37	0.29	3.43	10.29	17.25	25.88	19.13
47	0.29	3.71	11.43	18.63	31.50	21.63

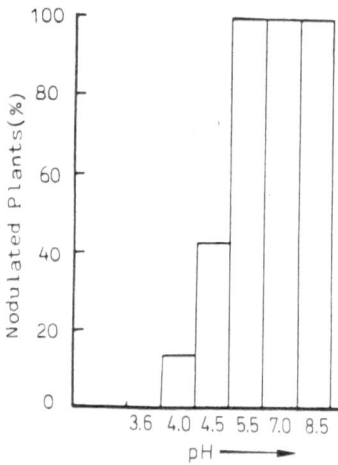

Fig. 7. Effect of pH in culture fluid on rate of nodulation in *L. leucocephala* (7 days after inoculation)

Table 8. Effect of different soil types on growth and nitrogen fixation of *L. leucocephala*

Location	Soil types	pH	Ages of trees	Height (M)	DBH (cm)	C_2H_2-reducing activity (μmol C_2H_4/g FW/h)
Xikou Nanan County	Riverside sandy loam	6.8	2	4.8	4.5	12.084
Tiaohu, Huian County	Red loam	4.6	2	1.7	1.5	2.828
Chihu, Huian County	Sandy loam	5.4	2	2.6	3.0	6.870
Heshan, Xiamen city	Red podzolic laterite	5.6	2	4.3	4.0	9.250

Table 9. Effect of different soil types on growth and nitrogen fixation of *A. auriculaeformis*

Location	Soil types	pH	Ages of trees	Height (M)	DBH[a] (cm)	C_2H_2-reducing activity (μmol C_2H_4/g FW/h)
Pinghe	Red loam	4.7	4	9.0	8.1	7.550
Zhaoan	Red loam	5.6	4	/	/	8.383
Xiamen	Red podzolic laterite	5.4	4	8.9	7.8	4.808

[a] Diameter breast high

no nodules were detected in seddlings of *L. leucocephala* on the 7th day after inoculation. On the 47th day after inoculation, the rate of nodulation only attained 28.6%, and the average number of nodules per plant was only up to 0.29 grains. At this time, the growth of the root system decreased and lateral roots were thin and short. In culture fluid at pH 4.0, the rate of nodulation was 14.3% on the 7th day after inoculation. The number of nodules per plant was less and 53% nodules formed on roots grown on the outside of the culture fluid. Although the rate of nodulation attained 100% on the 14th day after inoculation in the culture fluid at pH 4.5, the number of nodules per plant was only 42% of that grown in the culture fluid at pH 7.0. We found that *L. leucocephala* grew rapidly in soil with a pH between 6.0 and 7.5, but it did not grow well in acid soil at lower than pH 4.7 in the south of Fujian (Table 8).

The effect of soil pH on growth and nitrogen fixation of *A. auriculaeformis* was investigated. Table 9 shows that in the acid red soil with a pH between 4.7 and 5.6, *A. auriculaeformis* not only grew rapidly, but also the C_2H_2-reducing activity in nodules maintained at the increased level. Therefore, *A. auriculaeformis* is expected to perform much better on habitats of acid soil.

3.1.2.5 Light Intensity

The effect of light intensity on nodulation and nitrogen fixation was remarkable. When seedlings of three species *L. leucocephala*, *A. glauca* and *Dalbergia*

Table 10. Effect of light instensity on C_2H_2-reducing activity in nodules of woody legumes (μmol C_2H_4/g FW/h)

Plants	Natural light (75 000 Lux)		Dark within 24 h
	Cut stems	Whole plant	(80 Lux)
L. leucocephala	1.766	11.220	2.046
A. glauca	1.091	5.625	1.906
D. balansea	1.229	8.246	1.996

balansae) were transferred to the dark within 24 h, all C_2H_2-reducing activity in nodules declined rapidly. The C_2H_2-reducing activity in nodules of the three species in the dark represnted 1/5, 1/3 and 1/4, respectively, of that of the controls, which grew in natural light (Table 10). When the stems were cut, the C_2H_2-reducing activity in nodules of the three species represented only 1/6, 1/5 and 1/7, respectively, of that of the intact plants, which grew in natural light within 24 h.

3.1.2.6 Diurnal Variation

The pattern of the rate of photosynthesis of L. leucocephala and A. glauca varied greatly from sunrise to sunset during a fine day. Figure 8 shows that the low rate of photosynthesis at sunrise soon rose to a maximum at a mid-day and then declined at sunset. Two peaks appeared in diurnal variations in C_2H_2-reducing activity of nodules (Figs. 9–11). The first peak of C_2H_2-reducing activity in two species (Fig. 9 for L. leucocephala and Fig. 10 for A. glauca) appeared between 9 p.m. and 11 p.m. and the second peak occurred between

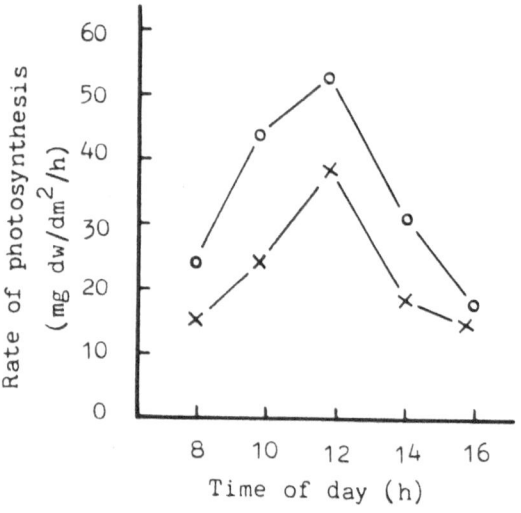

Fig. 8. Diurnal variations in rate of photosynthesis of A. glauca and L. leucocephala.
×———× A. glauca and ○----○ L. leucocephala

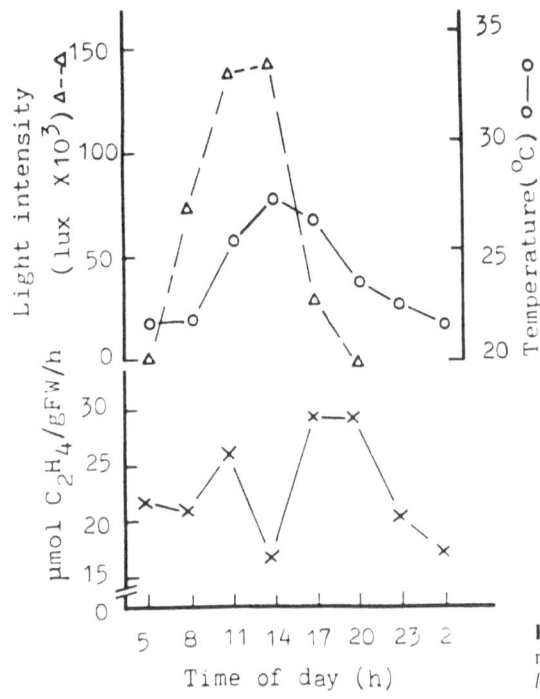

Fig. 9. Diurnal variations in C_2H_2-reducing activity in nodules of *L. leucocephala*

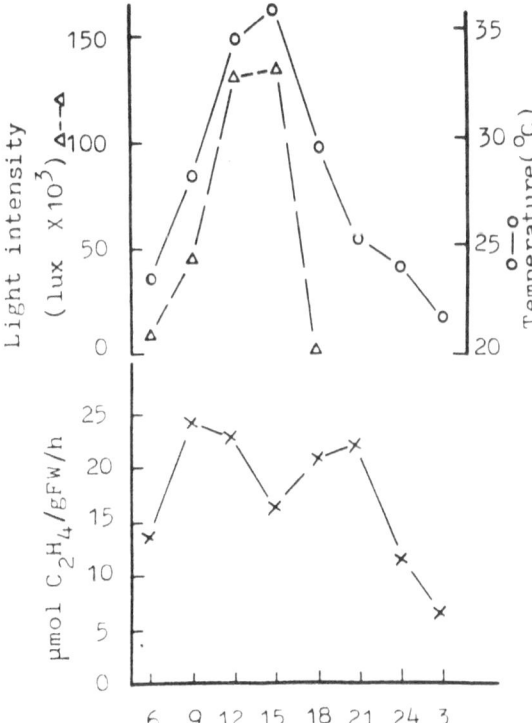

Fig. 10. Diurnal variations in C_2H_2-reducing activity in nodules of *A. glauca*

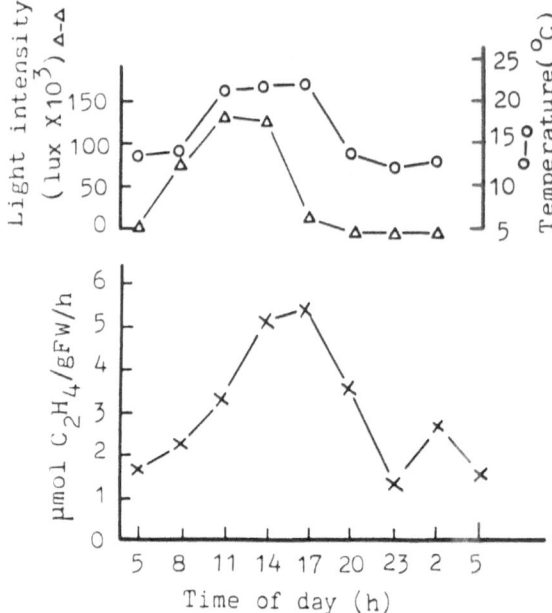

Fig. 11. Diurnal variations in C_2H_2-reducing activity in nodules of *A. falcataria*

Table 11. Effect of isolation time on C_2H_2-reducing activity of nodules on some woody legumens ($\mu mol\, C_2H_4/g\, FW/h$)

Plants	Time isolated (hours)			
	1	3	5	7
A. glauca	10.839	5.197	4.104	3.112
A. falcataria	16.296	12.428	/	/

17 a.m. and 21 a.m. The first peak of C_2H_2-reducing activity in *A. falcataria* (Fig. 11) occurred in the late afternoon (5 a.m.) and the second peak at 2 o'clock the next morning. Table 11 shows that C_2H_2-reducing activity of isolated nodules was related to the length of time during which they were isolated from the intact plants. The longer the time of isolation, the lower the C_2H_2-reducing activity of nodules. The results suggests that photosynthesis supply is the principle factor limiting nitrogen fixation.

3.2 H₂-Uptake and Nitrogen Fixation of Nodules on Woody Legumes

The ability to recycle H_2 evolved by nitrogenase is thought to be of importance in increasing the efficiency of N_2 fixation and to be a factor in increasing plant yield in symbiotic system (Abbrech et al. 1979; Evans et al. 1985; Wang, Song

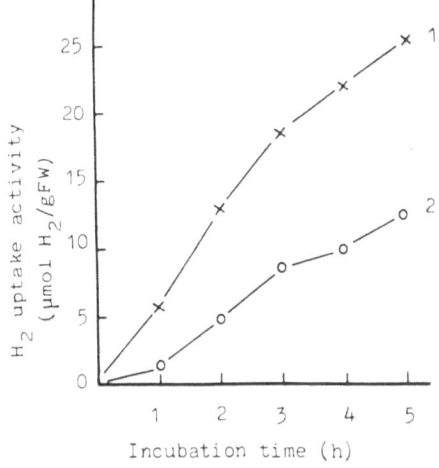

Fig. 12. H$_2$ uptake activity of nodules on *L. leucocephala* (*1*) and *A. auriculaeformis* (*2*) at different incubation times

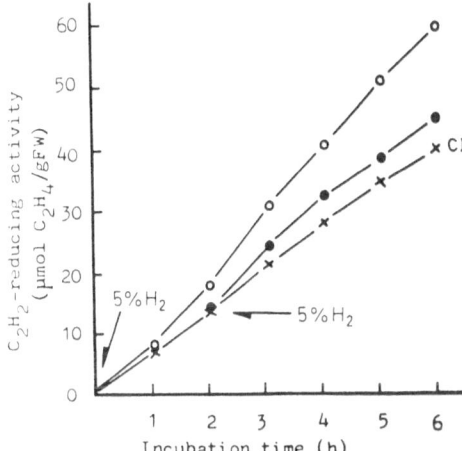

Fig. 13. Effect of additive H$_2$ on C$_2$H$_2$-reducing activity of nodules on *L. leucocephala*
○—○ 5% H$_2$ was added before reaction
·——· 5% H$_2$ was added after 2 h reaction
×——× No H$_2$ was added
At the time indicated by *arrows*, H$_2$ was added

Table 12. H$_2$ uptake activities and relative efficiencies of some woody leguminous nodules

Genus	Species	H$_2$ release (μmol H$_4$/g FW/h)	H$_2$ uptake (μmol H$_4$/g FW/h)	C$_2$H$_2$-reducing activity (μmol C$_2$H$_4$/g FW/h)	Relative efficiency
Acacia	*auriculaeformis*	0	2.07–3.33	4.80–8.30	1
	cunninghamii	0	1.30–2.15	2.58	1
	farnesiana	0	2.44–13.01	7.38	1
	glauca	0	8.72–10.46	11.13–16.49	1
Albizia	*falcataria*	0	1.15–1.83	4.03	1
	odoratissima	0	2.86–4.40	3.72	1
Enterolobium	*contortisiliqum*	0	1.14	1.78	1
Leucaena	*leucocephala*	0	7.31	8.21–10.25	1
Dalbergia	*hupeana*	0	3.70–5.53	5.41–14.60	1
Pterocarpus	*indicus*	0	1.10	1.57	1
Robinia	*pseudoacacia*	0	0.97	3.33	1

Table 13. Effect of H_2 on C_2H_2-reducing activity of nodules on *A. auriculaeformis* (μmol C_2H_4/g FW/h)

Time isolated (h)	1	2	3	5	7	9
CK[a]	1.800	1.826	1.889	1.818	1.809	1.702
+ 7.5% H_2[b]	2.549	2.481	2.479	2.383	2.332	2.224

[a] No H_2 was added
[b] 7.5% H_2 was added before reaction

1984). The H_2 uptake in nodules of some leguminous trees was investigated. Our results are listed in Table 12.

Figure 12 shows that expression of H_2 uptake activity was marked in nodules of *L. leucocephala* and *A. auriculaeformis*. Molecular hydrogen (H_2) marked a effect on the C_2H_2-reducing activity of nodules (Fig. 13 and Table 13) (Huang et al. 1990).

3.3 Relation Between the Development of Nodules and the Function of Nitrogen Fixation in Woody Legumes

The structure and function of nodules on *L. leucocephala* and *A. glauca* at different stages of development were investigated (Cai and Huang 1986; Huang et al. 1987; Wu et al. 1985a, 1985b, 1988). The life-span of root nodules in tropical leguminous trees might extend over winter. New nodules might occur on the top of old ones. The nodules changed their external shapes considerably during their growth, from single spheres or columns to branchy or multibranchy forms. A great difference was observed in external shapes of nodules which were taken at the same time (Fig. 14). However, all forms of nodules had a common

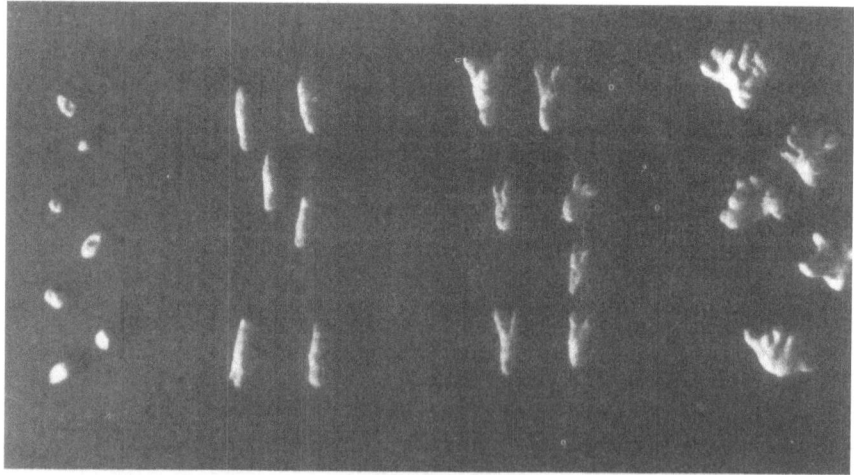

Fig. 14. Difference in external shapes of *L. leucocephala* nodules, which were taken at the same time

Table 14. Effect of different degrees of ripeness of nodules on C_2H_2-reducing activity woody legumens

Plants	Date determined	$\mu mol\, C_2H_4/g\, FW/h$		
		Juvenile	Mature	Senescent
L. leucocephala	6 Dec. 1983	8.069	12.207	1.580
A. falcataria	7 July 1986	10.190	16.558	7.696
A. auriculaeformis	12 Aug. 1989	7.681	11.791	3.123

Table 15. C_2H_2-reducing activity in different sections of mature nodules on *L. leucocephala*

Date determined	$\mu mol\, C_2H_4/g\, FW/h$		
	Top	Middle	Base
17 Aug. 1983	6.623	6.887	1.825
28 Aug. 1983	5.570	3.490	0.747
12 Sept. 1983	4.916	3.807	1.463

structure with bacteroid-containing tissues, vascular tissues, meristems and cortices.

The C_2H_2-reducing activity in nodules was closely related to the content of leghemoglobin. According to various shades of red in sections and the external shape, the nodules might be divided roughly into three physiological stages: juvenile, mature and senescent nodules. The highest C_2H_2-reducing activity, as determined by acetylene reduction assay, was observed in the mature nodules, the juvenile nodules ranked the second and the senescent ones followed as the last (Table 14). The mature nodules contained the greatest number of bacteroids (Wu et al. 1985a, b). The juvenile nodules had more bacteriods than the senescent nodules.

Different sections of the nodules have exhibited significant fluctuations in their C_2H_2-reducing activity. The C_2H_2-reducing activity in the top region of nodules was the highest, the middle of nodules ranked the second and the base followed as the last (Table 15).

4 Effect of Leguminous Trees on Soil Improvement

The red soil of hilly areas in southeast China, especially in Fujian Province, is very deficient in organic matter and mineral nutrients. Physically, it is compact and very sticky when wet, poor in aeration, slow in infiltration of water, and very acid in reaction.

Most of these areas have already been reforested, yet quite a sizeable area remains barren and needs to be properly exploited.

Table 16. Effect of leguminous trees on soil improvement

	Depth of soil layer (cm)	Soil aggregate %	Organic matter %	Total N %	Total P %	Total K %	Soil pH
Forestland	0–10	27.36	1.57	0.086	/	/	7.25
with 9-year-old	10–30	15.90	1.06	0.038	/	/	7.40
A. glauca							
Control (without	0–10	16.68	1.37	0.071	/	/	7.00
A. glauca)	10–30	14.60	0.71	0.029	/	/	7.10
Forestland with	0–30	/	4.389	0.123	0.031	0.403	5.70
6-year-old							
A. auriculaeformis							
Control (without	0–30	/	1.564	0.050	0.013	0.387	5.45
A. auriculaeformis)							
Forestland with	0–20	/	2.17	0.073	/	/	5.20
3-year-old	20–40	/	1.97	0.093	/	/	
L. leucocephala							
Control (without	0–20	/	1.04	0.051	/	/	5.60
L. leucocephala)	20–40	/	0.68	0.038	/	/	

The results of our investigation show that some leguminous trees are ideal as pioneer and fast-growing plants for afforestation, because of their special adaptation to barren land and their rapid growing ability with remarkably high nitrogen fixing activity of root nodules (Huang et al. 1987a, 1989a, 1989b; Huang, Huang 1989; Huang, Cai 1984).

The litter fall of 9-year-old *A. glauca*, 6-year-old *A. auriculaeformis* and 3-year-old *L. leucocephala* were investigated (Huang et al. 1989a; Huang, Huang 1989). The results show that total litter production per year of *A. glauca*, *A. auriculaeformis* and *L. leucocephala* was $600–800 \, g/M^2$, $610 \, g/M^2$ and $430–1200 \, g/M^2$, respectively. The nutrients of decomposable litter and the secretion of roots returned to soil, therefore, the fertility of soil was improved. The results from our analysis of the soil indicated that soil aggregate, organic matter and nutrients were all increased in the forest land (Table 16).

The role of nitrogen fixation of woody legumes in the field, forest and natural environment has drawn great attention in China. The results mentioned above regard our studies on the ecophysiology of symbiotic nitrogen fixation. Further research is under way.

5 References

Abbrecht SL et al. (1979) Hydrogenase in *Rhizobium japonicum* increases nitrogen fixation by nodulated soybeans. Science 203: 1255

Cai Keqiang, Huang Weinan (1986) Studies on the nitrogen fixation of root nodules of *Leucaena* seedlings. Plant Physiol Commun 3: 35

Cai Keqiang, Huang Weinan (1987) Effect of various nitrogen levels and H concentration on nodulation and nitrogen fixation of *L. leucocephala* seedlings. Subtropical Plant Res Commun 2: 16

Cai Longxiang et al. (1985) Isolation and reinoculation of Rhizobium of *L. leucocephala* and *A. glauca*. Subtropical Plant Res Commun 2: 4

Evans HJ et al. (1985) Biochemical characterization, evaluation and genetics of H_2 recycling in *Rhizobium*. In: Ludden PW, Burris JE (eds) Nitrogen fixation and CO metabolism. Elsevier, Amsterdam, New York, pp 3

Fujian Scientific and Technical Commission (1987) Flora Fujianica, Tomus 3. Science Press, Fujian

Huang Shengquan, Huang Weinan (1989) A preliminary study on introduction and cultivation of *A. auriculaeformis*. Subtropical Plant Res Commun 2: 21

Huang Weinan et al. (1983) Preliminary survey on the available symbiotic nitrogen fixing plants in Fujian Province. I. Distribution, growth habit, nodulation and nitrogen fixation and utilization of some leguminous trees. Subtropical Plant Res Commun 2: 15

Huang Weinan et al. (1987) Relation between the development of nodules in *L. leucocephala* and the nitrogenase activity. J Fujian Acad Agricult Sci 2(2): 48

Huang Weinan et al. (1987) Studies on nodulation and nitrogen fixation of *Albizia falcataria*. Subtropical Plant Res Commun 2: 5

Hung Weinan et al. (1988) Effect of climatic factor on nodulation and nitrogen fixation of *L. leucocephala*. Subtropical Plant Res Commun 2: 6

Huang Weinan et al. (1989) Growth and nitrogen fixation of *L. leucocephala* in different ecological environment and effect on soil improvement. Subtropical Plant Res Commun 1: 6

Huang Weinan et al. (1989) Nodulation and nitrogen fixation of *A. glauca* and effect on soil improvement. Subtropical Plant Res Commun 2: 1

Huang Weinan et al. (1990) Expression of H uptake and nitrogen fixation in nodules of *A. auriculaeformis*. Subtropical Plant Res Commun 1: 1

Huang Weinan, Cai Keqiang (1984) *Leucaena*-apromising fast-growing tree for agriculture, forestry and animal husbandry in South China. Subtropical Plant Res Commun 2: 43

Wang Hua, Song Hongyu (1984) AH-uptake hydrogenase in *Rhizobium astragali* in relaton to nitrogen fixation. Acta Phytophysiol Sinica 10(1): 63

Wu Yide, Huang Weinan et al. (1985) A study on the changing bacteroid perimembrane structures during the development of leguminous root nodules. J Fujian Agricult College 14(3): 185

Wu Yide, Huang Weinan et al. (1985) Studies on symbiotic nodules of tropical legume: 1. Morphogenesis and cellular structure observation. WuYi Science J 5: 307

Wu Yide, Huang Weinan et al. (1988) Ultrastructural study on the fusion and senescent disintegration of peribacteroid membrane in Acacia root nodules. Acta Agriculturae Nucleatae Sinica 2(3): 184

CHAPTER 31
Rhizobium Resources in the Arid Region of Xinjiang

G.L. Guan

The Nitrogen Fixation
and its Research in China
Editor: Guo-fan Hong
© Springer-Verlag Berlin Heidelberg 1992

1 Introduction

The Xinjiang region is situated at 36–48 °N and 71–96 °E in the middle of Eurasia and far from the ocean. Being surrounded by high mountains, its terrain is complicated. Rainfall is rare and evaporation is high causing the region to suffer from dryness. Annual and daily temperature fluctuations are drastic. This makes it a typical continental inland arid region. Affected by topographic features, many different ecologic conditions have been formed. Desertification and salinization of soils are widespread. However, in this great natural arid ecological system there are various kinds of nitrogen-fixing microorganisms, which have become fitted to the dry and saline conditions. Xinjiang is a vast land of about 1.65 million square kilometers, but the cultivated land only accounts for 2% of the whole area (Wen 1965). Therefore, the nitrogen-fixing microorganisms play a very important role in the nitrogen cycle in this ecosystem. *Rhizobium* research and its application in this region would help to improve the soil fertility, increase crop productivity and is also beneficial for energy-saving and environmental protection. According to Jordan and Allen (1974) so far, only about 0.3–0.4% of the *Rhizobium* species have been studied with respect to their symbiosis with plants. Most of the rhizobia have not been isolated and studied. To study *Rhizobium* resources in the Xinjiang arid area and identify new species or strains will certainly be useful for *Rhizobium* classification. Our research work in the period from 1982 to 1987 included testing the nitrogen-fixing ability of rhizobia in symbiosis, their resistance to unfavorable conditions and exploring their physiological and biochemical characteristics. This paper describes some of our research results.

2 Strains of Rhizobia in the Arid Region of Xinjiang

Through extensive investigations of leguminous plants, including those from natural grasslands, sandy desert, forest and saline-alkali soils, we have obtained 373 strains of rhizobia from 1208 nodule samples that belong to 109 species of 31 genera of leguminous plants (Table 1). Among them, 88 new *Rhizobium* strains from 39 species were isolated from the plants nodules (Table 2).

Taxonomic studies were carried out through nodulation tests. The species of leguminous plants were infected by homologous rhizobia, followed by cross-inoculation. The plants tested included, among others, *Medicago, Glycine max, Pisum sativum, Phaseolus vulgaris* and *Vicia sinesis.* Some rhizobia isolated from root nodules of *Astragalus* could not nodulate any one of the above 5 plants. But some rhizobia, such as those from *Glycine max, Vicia sinensis,* and *Phaseolus vulgaris* infected plants that were also infected by the rhizobia isolated

Table 1. *Rhizobium* strains in the Xinjiang arid region

Host plant (genera)	Number of species	Number of strains	*Rhizobium* strains (codes)
Alhagi	1	8	R001–R008
Ammopiptanthus	1	1	R009
amorpha	1	1	R010
Caragana	2	5	R011–R015
Cicer	1	4	R016–R019
Astragalus	34	65	R020–R084
Trigonella	5	30	R085–R114
Eremosparton	1	1	R123
Glycine	1	22	R124–R145
Hedysarum	5	14	R146–R159
Halimodendron	1	5	R160–R164
Glycyrrhiza	5	18	R165–R173 R366–R374
Lathyrus	4	17	R374–R187 R375–R377
Lotus	1	1	R188
Medicago	8	47	R189–R220 R345–R359
Melilotus	3	24	R221, R222 R225–R242 R360–R363
Onobrychis	2	13	R243–R254 E364
Orobus	1	5	R255–R259
Oxytropis	5	7	R260–R266
Pisum	1	2	R267, R268
Phaseolus	1	1	R295
Sophora	1	10	R296–R305
Thermopsis	1	1	R312
Trifolium	5	23	R313–R335
Crotalaria	1	5	R118–R122
Vicia	10	15	R270–R294
Sphaeropysa	1	9	R115–R117 R306–R311
Vigna	2	5	R336–R338 R340, R341
Turukhan	1	1	R339
Coronilla	1	1	R334
2Nd	2	2	R343, R344

from root nodules of *Oxytropis chrogossica* and *Swainsona salsula*. In view of the above facts, the laws described by Jordan (1984) and Yu et al. (1985) may not be appropriate for the classification of rhizobia isolated in this region. So the rhizobia isolated from root nodules of leguminous plants growing in this region have been temporarily named after their respective plants and been given code numbers.

Table 2. *Rhizobium* strains which were not reported in the literature*

Host plant	*Rhizobium* strains (codes)
Ammopiptanthus mongolicus	R009
Caragana intermedia	R011–R014
Astragalus oxyglottis	R020–R028
Astragalus schanginianus	R032–R034
Astragalus ammodytes	R035, R036
Astragalus grandiflorus	R043
Astragalus lasiophllus	R069
Astragalus pseudobrachytropis	R044–R049
Astragalus suidunesis	R052
Astragalus densiflorus	R058
Astragalus austrosibiricus	R059, R060
Astragalus amygdalinus	R061
Astragalus arpilobus	R062
Astragalus chorgossicus	R064
Astragalus kararensis	R050, R051
Astragalus penduliflorus	R066
Astragalus lehmannianus	R070
Astragalus hypogaeus	R071
Astragalus depauperatus	R074
Astragalus steinbergianus	R078
Astragalus follicularis	R080
Astragalus adsurgens	R081–R084
Astragalus stenoceras	R042
Cicer arietinum	R016–R019
Eremosparton songoricum	R123
Glycyrrhiza glabra	R170, R171
Hedysarum scoparium	R159
Medicago rivularis	R216–R218
Melilotus albus	R231–R238, R221
Oxytropis chorgossica	R260
Trigonella arcuata	R092–R103
Trigonella orthoceras	R089–R091
Trigonella tenuis	R106–R111
Vicia semenovii	R279–R281
Vicia sinkiangensis	R282–R286
Lathyrus quinquenerves	R185–R187
Oxytropis lappomica	R261, R262
Astragalus dshimensis	R065
Astragalus platyphyllus	R075

* In contrast to the report by ON Allen and WX Chen

3 Nitrogen Fixation in the Legume-*Rhizobium* Symbiosis

In suitable environments, rhizobia and their host plants can recognize each other and form nodules. On stimulation of leguminous root exudates, rhizobia propagate rapidly and after passing through the epidermis tissue form infectious threads. In the meantime they promote cell division of the root tissue, and

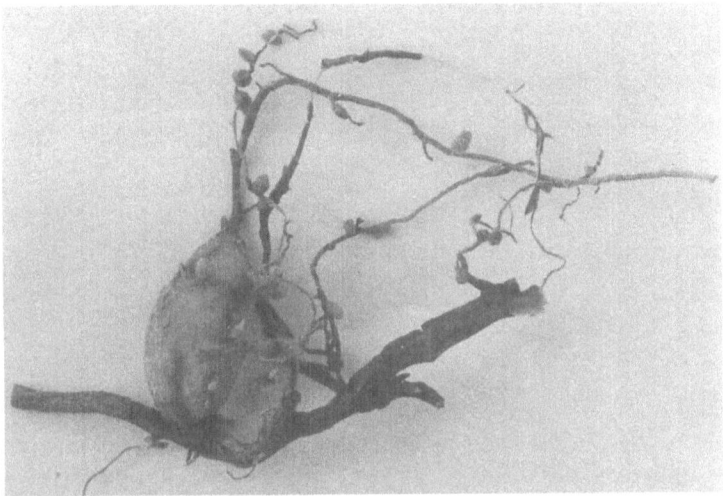

Fig. 1. Leguminous nodules of *Lathyrus tuberosus* (Xinjiang arid region)

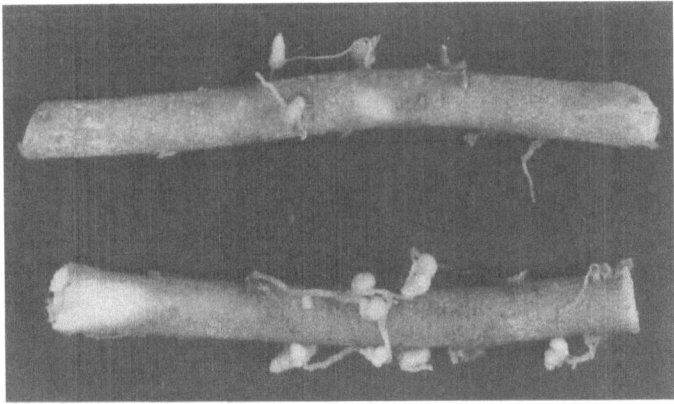

Fig. 2. Leguminous nodules of *Astragalus chorgossicus* (Xinjiang arid region)

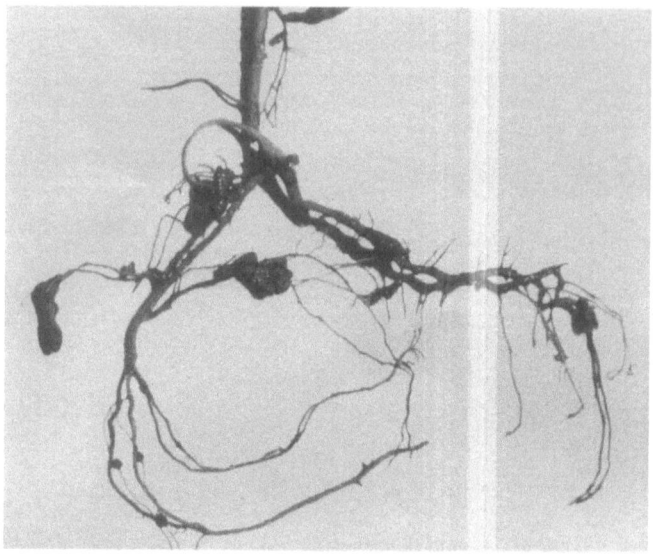

Fig. 3. Leguminous nodules of *Sophora alopecuroides* (Xinjiang arid region)

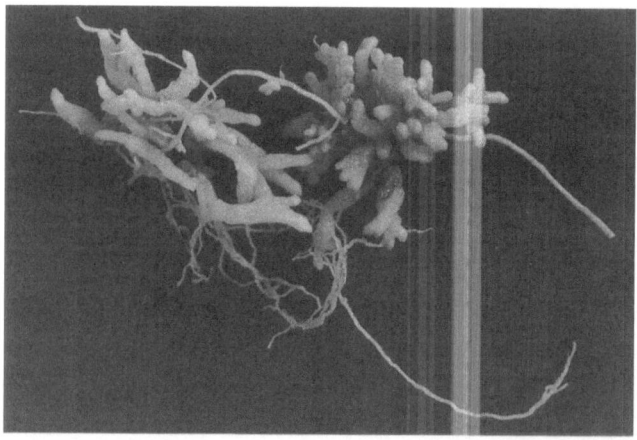

Fig. 4. Leguminous nodules of *Onobrychis viciifolia* (Xinjiang arid region)

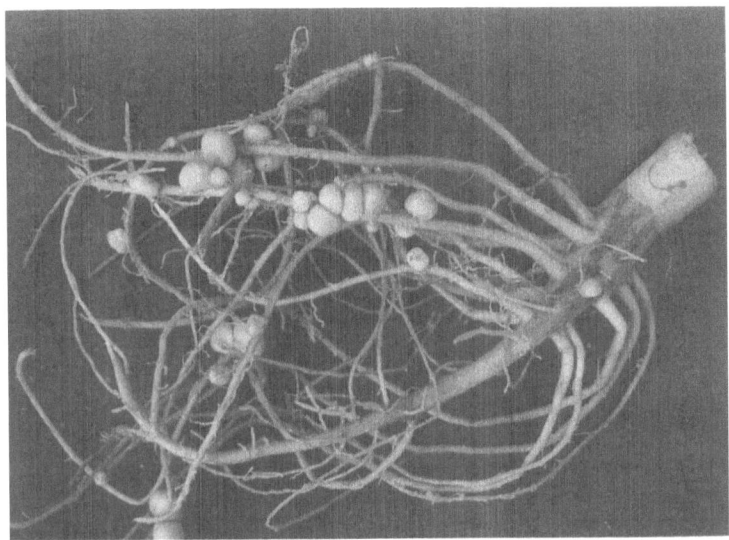

Fig. 5. Leguminous nodules of *Vigna* sp. (Xinjiang arid region)

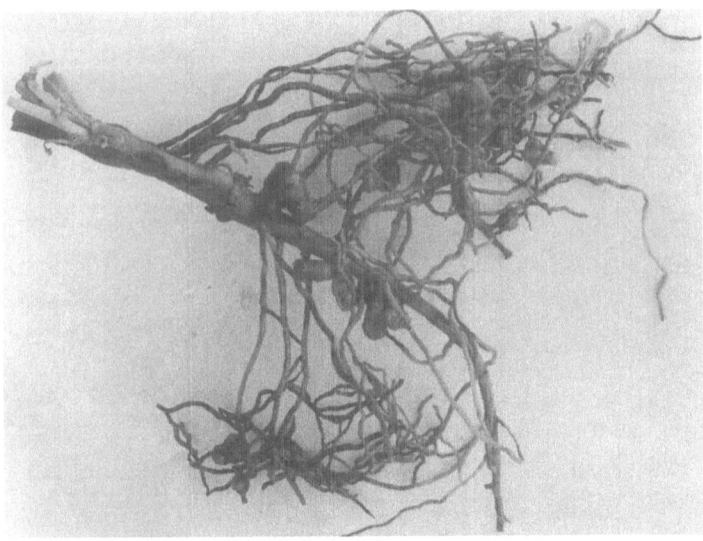

Fig. 6. Leguminous nodules of *Cicer arietinum* (Xinjiang arid region)

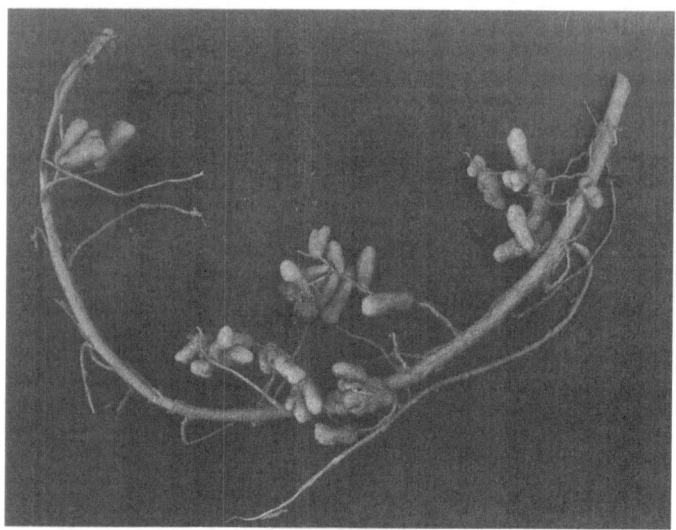

Fig. 7. Leguminous nodules of *Caragana intemedia* (Xinjiang arid region)

Fig. 8. Leguminous nodules of *Lathyrus quinuenervis* (Xinjiang arid region)

Fig. 9. Leguminous nodules of *Astragalus stenoceras* (perennial plant) (Xinjiang arid region)

Fig. 10. Leguminous nodules of *Astragalus ammodytes* (ephemeral plant) (Xinjiang arid region)

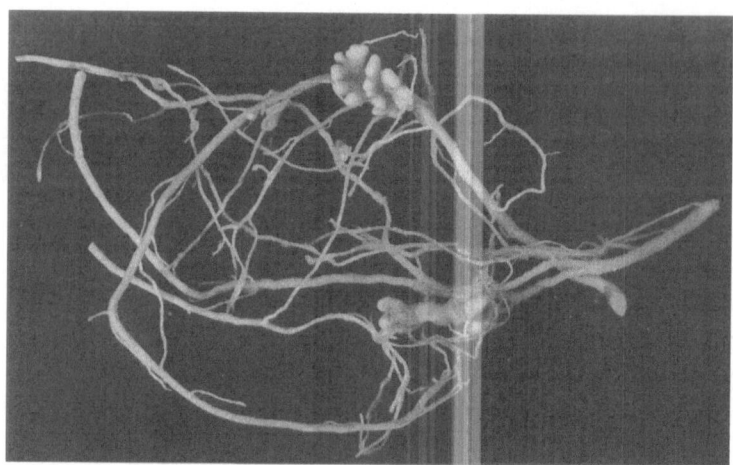

Fig. 11. Leguminous nodules of *Vicia villosa* (Xinjiang arid region)

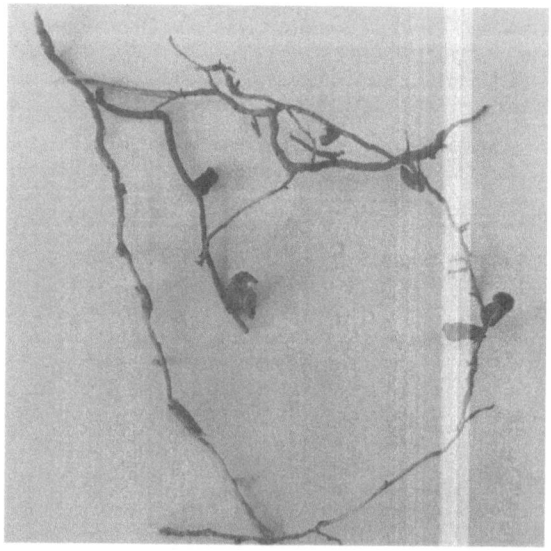

Fig. 12. Leguminous nodules of *Calophaca chinensis* (Xinjiang arid region)

nodules are then gradually formed. These leguminous root nodules vary in shape in the Xinjiang arid region (see Figs. 1–12). After entering the host cells, the rhizobia change into bacteroids, which carry out azofication. Infecting leguminous plants and forming nodules for nitrogen fixation is the remarkable characteristic of rhizobia which is different from other bacteria in the soil. In order to make a comprehensive investigation of the legume-*Rhizobium* symbiosis, more than 1,000 samples of root nodules were collected from over 100 species of legumes at different growing stages from budding to flowering; the nitrogen fixation activities of these nodule samples were detected with the acetylene reduction method as described by the Shanghai Institute of Plant Physiology. The results showed that 95% of the root nodules are effectively fixing nitrogen but the activities in the different species varies greatly (Table 3). *Astragalus* generally possessed high activities, at a maximum of 365.53 μmol $C_2H_4 g^{-1} h^{-1}$,

Table 3. Acetylene reduction activity, of nodules sampled in the Xinjiang arid region

Host plants	Sampling place and conditions	C_2H_4 reduction activity (μmol $C_2H_4 g^{-1} h^{-1}$)
Astragalus adsurgens	Mosuowan sands	63.23
Astragalus grandiflorus	Huocheng sand dune	19.12
Astragalus hypogaeus	Yili sand dune	16.78
Astragalus karkarensis	Tacheng grassland	365.53
Astragalus lasiphyllus	Urumqi desert hill	121.70
Astragalus lehmannianus	Urumqi desert hill	26.88
Astragalus oxyglottis	Fukang sand dune	29.98
Astragalus steinbergianus	Qitai sand dune	17.56
Astragalus stenoceras	Chabuchaer desert	119.81
Astragalus suiduensis	Huocheng sand dune	16.78
Alhagi sparsifolia	Turpan sands	2.48
Ammopiptanthus mongolicus	Turpan sands	12.63
Caragana intemedia	Turpan sands	113.40
Cicer arietium	Urumqi farmland	1.17
Caragana jubata	Turpan sands	10.64
Eremosparton songoricum	Qitai sand dune	10.31
Halimodendron halodendron	Mosuoiwan sands	10.25
Halimodendron mongolicum	Turpan sands	23.99
Hedysarum alpinum	Manasi forest	3.33
Glycine max	Shihezi farmland	8.75
Lathyrus humilis	Mosuowan sands	15.66
Onobrychis sp.	Qitai grassland	5.58
Pisum sativum	Misuowan farmland	0.10
Sophora alopecuroides	Shihezi farmland	0.06
Sphaerophysa salsula	Yili desert	28.27
Trifolium pratense	Taheng grassland	65.91
Trifolium repens	Taheng grassland	523.90
Trigonella concellata	Miquan desert hill	29.98
Trigonella foenum-graecum	Turpan farmland	85.98
Vicia sepium	Tacheng grassland	46.64
Vicia villosa	Urumqi farmland	0.90
Vicia sinensis	Turpan farmland	0.24
Medicago falcata	Manasi forest	83.64

42 times higher than for *Glycine max* nodules in the same region. The root nodules of other leguminous plants peculiar to arid regions such as *S. alopecuroides, H. mongolicum, C. intermedia* and an ephemeral plant *T. cancellata,* etc. also possessed higher nitrogen-fixing activities (more than $20\,\mu mol\ C_2H_4\,g^{-1}\,h^{-1}$). *Trifolium* plants, which are widely distributed in the world, are also found in Xinjiang. The root nodules of *T. repens* growing in the South Lake grassland were pink with high activities up to $523.90\,\mu mol\ C_2H_4\,g^{-1}\,h^{-1}$, 60 times more than that of *Glycine max.*

The factors affecting the nitrogen-fixing activity of nodules are very complex. They depend on genetic properties as well as environmental conditions. Energy, which is necessary for bacteroids in the root nodules to live and fix nitrogen, is supplied through the photosynthetic plant products. Therefore, the nitrogen fixation of nodules is regulated by the host plant. According to Bergenson (1982), near the phase of flowering, soybeans have fixed only 30–40% of their N, but when maturity is approached the value becomes over 80%. We have observed the nitrogen-fixing abilities during different developmental stages of one ephemeral plant, *T. jenuis,* growing in natural conditions and of three leguminous plants, *V. villosa, C. arietinum* and *M. sativa* which grow in artificially cultivated conditions. Root nodules were collected from the four plants and their nitrogen fixation activities were determined. The results are summarized in Table 4. It shows that the activities of nodules not only closely relate to the growth periods, but also to the plant habits.

The maximum activities appeared in the budding stage. The nitrogen fixation activities of *T. jenuis* nodules decreased with plant growth, the activity in the flowering period was only 30% as much as that in the budding stage. Perennial *M. sativa* nodules maintained relatively high activities in all the growth stages.

As described by Suttonetal (1981) (7), many features of bacteroid metabolism can be regarded as adaptations to the unusual environment in which they function. This environment is largely under the plant control, since all substances entering or leaving the bacteroid containing tissues. Bacteroids are characterized

Table 4. Relationship between acetylene reduction activity of root nodules and the growth periods of plants

Host plants	Living habits	C_2H_2 reduction activity ($\mu mol\ C_2H_4\,g^{-1}\,h^{-1}$)			
		in budding	in early flowering	in flowering	in podding
Trigonella fennis	Ephemeral plant	25.62	19.84	7.80	0.14
Vicia villosa	annual plant	19.14	14.34	8.10	0.27
Cicer arietinum	annual plant	3.04	3.00	10.98	0.08
Medicago sativa	perennial	16.62	16.32	14.46	13.32

Table 5. Effect of the detachment-time of root nodules from the host on acetylene reduction

Time (h)	Acetylene reduction activity (μmol $C_2H_4\,g^{-1}\,h^{-1}$)*		
	Trigonella foenum-graecum nodules	*Cicer arietinum* nodules	*Vigna* sp. nodules
1st	2.44	0.12	ND
2nd	2.50	2.82	1.26
3rd	7.56	1.14	1.56
4th	3.06	0.72	11.34
5th	1.14	0.60	30.54
6th	1.56	8.34	6.96
7th	1.50	0.78	8.76
8th	1.86	1.02	12.00

* Reaction temperature: 30 °C. ND: not determined

by high nitrogenase activity and are normally found in tissue with a high leg-hemoglobin content. Nitrogenase activity changes in the bacteroid. The nodule samples were collected during flowering period of *T. foenum-graecum*, *Vicia* sp. and *C. arietinum* and tested for the acetylene reduction activity. The results showed that the nitrogenase activities of the detachment nodules from different kinds of plants had their own peak value in different time (Table 5). The maximum nitrogenase activities of *T. foenum-graecum* nodules appeared in the third hour, but *C. arietinum* in the sixth, *Vicia* sp. in the fifth.

4 H$_2$-Evolution and -Uptake in the Legume-*Rhizobium* Symbiosis

It was reported by Evans et al. that cell-free nitrogenases from all known sources produce H_2 concomitant with N_2 reduction. This H_2 production during N_2 fixation usually accounts for between 25% and 35% of the electron flow through the nitrogenase reaction system. As described by Schubert et al. (1976) with most symbionts only 40–60% of the electron flow to nitrogenase system was transferred to nitrogen. The remainder was lost through hydrogen evolution. Therefore hydrogen evolution in the process of nitrogen fixation costs a lot of the energy. However, it was confirmed by many scientists that uptake hydrogenase systems were present in some legume nodules formed by some rhizobia. Evans (1980) listed the following advantages of the unidirectional hydrogenase system. 1) H_2 oxidation may utilize an excess of O_2 and thereby contribute toward the maintenance of O_2-labile nitrogenase; 2) H_2 removal may decrease H_2 accumulation to such a concentration that could be inhibitory to nitrogenase; and 3) evolution of H_2 from the nitrogenase system is ATP dependent and therefore represents energy wastage, but H_2 oxidation via the hydrogenase complex leads to ATP synthesis and therefore conserves some energy that otherwise would be lost. According to Maier et al. (1978), the genetic information for hydrogenase, like that of nitrogenase, appears to reside in the bacteria rather than in the legume. Eisbrenner and Evans (1983) found that most of the cowpea strains of *Rhizobium* examined are Hup$^+$, but not for strains of *R. meliloti* or *R. trifolii*. Uptake hydrogenase activities of the root nodules excised from 37 species that belong to 20 genera of legumes were detected in our laboratory by the method of Evans, only the nodules of *A. ammodytes* were unable to take up hydrogen, while the others had uptake hydrogenase activities: 3 species of the root nodules measured 1 μmol $H_2 g^{-1}h^{-1}$, 14 species had 1–5 μmol $H_2 g^{-1}h^{-1}$, 10 species had 5–10 μmol $H_2 g^{-1}h^{-1}$, 9 species had more than 10 μmol $H_2 g^{-1}h^{-1}$. The highest uptake hydrogenase activity was up to 68.65 μmol $H_2 g^{-1}h^{-1}$. The results show that most of the rhizobia including *R. meliloti* and *R. trifolii* isolated from the Xinjiang region generally possessed an uptake hydrogenase system. Nodules formed by these rhizobia emit little or no hydrogen into the air (Table 6). So they are more efficient. In an effort to evaluate the extent of energy loss as H_2 evolution from root nodules, Evans had utilized a relative efficiency estimate during nitrogen fixation. The relative efficiencies of nitrogen fixation

Table 6. Hydrogen-uptake activities and relative efficiency of nitrogen fixation of leguminous nodules

Host plant	H_2-uptake activity	H_2 evolution activity (μmol H.G.h.)	C_2H_2 reduction (μmol $C_2H_4 g^{-1}h^{-1}$)	Relative efficiency
Orobus luteus	7.19	0	1.61	1
Lathyrus humilis	6.28	0	29.28	1
Vicia cracca	15.65	0	3.24	1
Hadysarum austrosibiricum	6.35	0	1.11	1
Trifolium lupinaster	7.09	0	7.14	1
Thermopsis kuelunice	3.76	0	6.42	1
Halimodendron halodendron	3.89	0.04	2.82	0.92
Medicago falcata	3.12	0	3.18	1
Astragulus ammodytes	0	7.26	15.20	0.53
Melilotus officinalls	14.22	0	4.72	1
Glycine max	0.47	0.35	0.47	0.25
Crotalaria juncea	3.93	0	ND	ND
Trigonella arcuata	68.65	0	52.64	
Alhagi pseudalhagi	4.56	0	3.12	1
Oxytropis sp.	4.59	0	5.40	1
Glycyrrhiza glabra	0.70	0.07	ND	1
Sphaerophsa salsula	0.78	0	1.32	ND
Caragana intemedia	28.54	0	48.18	1
Onobrychis tanaitica	12.27	0	1.16	1
Astragalus atenoceras	7.48	0	10.98	1
				1

ND: not determined

Table 7. The relationship between hydrogen-uptake activities of leguminous nodules and growth periods of plants

Host plant	H_2-uptake activity (μmol $H_2 g^{-1} h^{-1}$)		
	in budding	in flowering	in podding
Trigonella arrcuata	68.65	21.26	1.88
Lathyrus quinquenervirus	9.33	0.18	0.04
Onobrychis sp.	1.33	0.54	0.09
Cicer arietium	2.16	0.76	0.09
Vicia sativa	4.31	1.61	0.22
Glycine max	4.73	0.23	0.03

Table 8. Hydrogen-uptake activities and acetylene reduction activities of *Trigonella arcuata** nodules in different growth periods

Growth periods	H_2-uptake activity	H_2-evolution activity	C_2H_2 reduction activity	Relative efficiency
	(μmol H_2/g.h.)		(μ mol $C_2H_4/g^{-1} h^{-1}$)	
In binding	24.55	0	52.64	1
In flowering	21.26	0	30.28	1
In podding	1.88	0	7.38	1

of the root nodules detached from 20 species of leguminous plants in the Xinjiang region were calculated by Evans formula (Table 6).

Maier has proved that the uptake hydrogenase activity in rhizobia may be expressed in the free-living condition, but poper medium and incubative conditions must be selected. The uptake hydrogenase expression of bacteriods in root nodules was affected by the nodules microenvironment and regulated by plant growth. Through measuring the uptake hydrogenase activity of the nodules from 6 species of legumes in different growth periods, the maximum activities were found in budding and the minimum in podding (Table 7). However, positive efficiency was found in different growth periods in the nodules from *T. arcuata* which only grows in the arid region (Table 8).

5 Rhizobia Resistance

Dryness, salinization, high attitude and the great and frequent changes in temperature are the striking characteristics of the ecological conditions in the Xinjiang arid region. The Turpan Basin is a typical arid region with high temperature. The air temperature is up to 45–50 °C during the summer, the soil

temperature (0–20 cm deep) in the daytime is 35–40 °C. There is 10 mm of precipitation. Many kinds of rhizobia found in this region can form symbioses with
legumes and fix nitrogen (Table 9), and some rhizobia from leguminous nodules
are able to resist low temperatures. The nodules of *A. pseudobrachytropis* and
T. arcuata were found at air temperatures of 10–17 °C and soil temperatures
of 6–12 °C (0–20 cm deep) in early spring near Urumqi desert hills; the acetylene
reduction activities were 13.20 µmol C_2H_4 $g^{-1}h^{-1}$ for *T. arcuata* and 20.50 µmol
C_2H_4 $g^{-1}h^{-1}$ for *A. pseudobrachytropis*.

Even though rainfall is rare and evaporation is severe in early summer and
the moisture percentage of soil (0–20 cm) decreased to 6.5–10.2%, some rhizobia
could still nodulate with *T. jenuis* and fix nitrogen with an activity of 18.00 µmol
C_2H_4 $g^{-1}h^{-1}$. According to Vincent (1977), the salts concentrations tolerable
to rhizobia are $CaCl_2$ 0.14 M KCl 0.32, NaCl 0.6 M, LiCl 0.14 M. The tolerance

Table 9. Acetylene reduction activities of leguminous nodules from Turpan Basin

Climatic* conditions	Sampling date	Host plant	C_2H_2 reduction activity (µmol C_2H_4/g.h.)
Annual	1984.6.05	*Glycyrrhiza glabra*	203
average	1984.6.05	*Caragana sp.*	0
temperature	1984.6.05	*Ammopiptanthus mongolius*	21
12–14 °C	1984.6.07	*Sphaerophsa salsula*	471
July: 30–34 °C	1982.7.19	*Astragalus adsurgens*	158
Jan: −16–10 °C	1983.6.16	*Vigna sinensis*	80
Annual	1983.6.16	*Crotalaria juncea*	177
precipitation	1983.6.16	*Sophora alopecuroides*	60
20–26 mm	1983.6.16	*Hedysarum scoparium*	0

* As described by ZW Wen (1965)

Table 10. Rhizobium growth on YMA medium with different concentrations of sodium chloride

Concentration of NaCl (mol/l)	*Rhizobium* strains (codes)	Percentage of examined strains
00.68	R152, R116, R262, R266, R115, R001, R327, R189 R146, R363, R023, R196, R022, R271, R276, R165 R267, R375, R003, R342, R164, R334, R092, R025 R339, R240, R009, R119, R171, R032, R278, R063 R305 These strains were isolated from 34 species of 19 genera of legumes.	47% of 87
0.86–1.03	R116, R115, R189, R146, R363, R023, R196, R022 R276, R335, R003, R342, R164, R334, R240, R009 R119, R171, R032, R298, R063, R305 These strains were isolated from 22 species of 14 genera of legumes.	29% of 87

Incubation temperature: 28 °C

Table 11. *Medicago rhizobia* growth on YMA medium with different concentrations of sodium sulphate

Concentration of Na$_2$SO$_4$(mol/l) strains (codes)	CK	0.46 0.91	0.64 1.27 1.46	1.64	1.32	2.00	2.18	2.36
R348	+	+		−	−	−	−	−
R351	+	+		−	−	−	−	−
R353	+	+		+	+	−	−	−
R352	+	+		+	+	+	+	+
R347	+	+		+	+	+	+	+
R355	+	+		+	+	+	+	+
R346	+	+		+	+	+	+	+
R345	+	+		+	+	+	+	+
R347	+	+		+	+	+	+	+
R349	+	+		+	+	+	+	+
R356	+	+		+	+	+	+	+

" + " shows growth; " − " shows no growth

Table 12. Growth of *Rhizobium* strains on YMA medium at different pH

pH of incubation	Growing strains (codes)	Percentage of examined strains
4	R270, R061, R031, R263, R266, R115, R286, R015 R312, R159, R244, R292, R189, R146, R243, R363 R023, R196, R277, R185, R022, R271, R165, R170 R257, R267, R011, R335, R342, R126, R334, R092 R025, R339, R099, R119, R171, R032, R190, R289 R063, R305 These strains were isolated from 42 species of 22 genera of legume.	67% of 68
10	R270, R152, R082, R116, R177, R061, R031, R291 R332, R266, R115, R286, R001, R123, R015, R159 R312, R327, R011, R188, R244, R292, R189, R146 R243, R296, R363, R023, R196, R160, R277, R185 R271, R276, R165, R176, R178, R257, R267, R340 R017, R003, R342, R126, R164, R334, R092, R025 R339, R119, R171, R032, R190, R289, R063, R305 These strains were isolated from 56 species of 25 genera of legume.	85% of 68
4-10	R334, R092, R025, R339, R119, R171, R032, R190 R290, R063, R305, R271, R165, R170, R257, R267 R017, R342, R126, R189, R146, R243, R363, R023 R196, R227, R185, R285, R015, R159, R312, R244 R292, R270, R061, R031, R266, R115 These strains were isolated from 38 species of 20 genera of legume.	60% of 68

Incubation temperature: 28 °C

to Na.K was 0.3 M for *R. trifolii*, 0.5–0.6 M for *R. meliloti*. Our experiments have shown 47% of rhizobia in the Xinjiang region of being able to grow on YMA medium with 0.68 M NaCl; 29% grow with 0.86–1.03 M NaCl (Table 10). 8 strains of 11 *R. meliloti* had even stronger tolerance and could grow in 2.36 M Na_2SO_4 (Table 11). The salt tolerance may be one of the causes why rhizobia can live in symbiosis in the Xinjiang soil under harsh conditions.

It was described by Jordan et al. (1984) that the optimum pH range for rhizobia, growth is 6–7, but some rhizobia can grow at pH 4.5 9.5, such as *R.*

Table 13. Growth of *R. meliloti* strains on YMA medium at different pH

Strains (codes)	pH 4	6	7	8	10	12	13
R345	−	+	+	+	+	+	−
R346	−	+	+	+	+	+	−
R347	−	+	+	+	+	+	−
R348	−	+	+	+	+	+	−
R349	−	+	+	+	+	+	−
R351	−	+	+	+	+	+	−
R378	−	+	+	+	+	+	−
R352	−	+	+	+	+	+	−
R353	−	+	+	+	+	+	−
R354	−	+	+	+	+	+	−
R355	−	+	+	+	+	+	−
356	−	+	+	+	+	+	−

Incubation temperature: 28 °C
"+" shows growth; "−" shows no growth.

Table 14. Growth of some *Rhizobium* strains at different temperatures

Incubation temperature	Growing strains (codes)	Percentage of examined strains
43 °C	R270, R080, R177, R061, R291, R332, R283 R001, R123, R015, R327, R011, R188, R244 R296, R363, R023, R312, R160, R022, R165 R241, R178, R243, R344, R092, R339, R091 R171, R118, R190. These strains were isolated from 31 species of 17 genera of legumes.	43% of 118
8 °C	R270, R152, R116, R177, R031, R291, R332 R263, R283, R266, R286, R159, R312, R189 R234, R023, R213, R160, R022, R271, R276 R165, R176, R178, R257, R267, R340, R017 R335, R033, R342, R126, R146, R025, R379 R099, R240, R009, R032, R298, R063, R247 R305, R118 These strains were isolated from 44 species of 22 genera of legumes	62% of 118

Rhizobium strains grow on YMA medium.

meliloti growing at pH 9.5. Our studies have exposed that rhizobia found in the Xinjiang region were able to grow in a wider pH range (Table 12). *R. meliloti* isolated in Xinjiang can grow even at pH 12 (Table 13).

According to Jordan, the optimum temperature range for rhizobia growth is 25–30 °C. A few of *R. meliloti* can grow at 42.5 °C, very few rhizobia could grow at 4 °C. The rhizobia in Xinjiang had a wider range of temperature for growth. 43% out of 118 examined strains were able to grow at 43 °C (Table 14). It is well known that microorganisms without spores generally can be killed at 60 °C in 2–3 min. According to the tests by Grahan et al. (1963), no strains of *Rhizobium* survived incubation at 60 °C for 5 min. However, 58% of the rhizobia isolated in the Xinjiang region could tolerate 60 °C for 10 min and after this treatment could still grow well at 28 °C (Table 15).

Table 15. Tolerance of *Rhizobium* strains to high temperature

Treatment by temperature	Survival strains	Percentage of examined strains
55 °C for 20 min	R001, R009, R011, R082, R061, R023, R123 R015, R118, R126, R165, R160, R159, R176 R189, R243, R231, R257, R291, R313, R298 R305 These strains were isolated from 22 species of 19 genera of legumes.	65% of 120
60 °C for 10 min	R061, R082, R011, R015, R001, R118, R123 R126, R165, R160, R159, R176, R189, R231 R234, R257, R291, R298, R305 These strains were isolated from 19 species of 18 genera of legumes.	58% of 120

Rhizobium: strains grow on YMA medium.

Table 16. *Rhizobium* strains with resistance to different concentrations of streptomycin

Concentration of streptomycin (r/ml)	Growing strains (codes)	Percentage of examined strains
500	R116, R177, R291, R263, R115, R188, R243 R296, R023, R022, R271, R165, R176, R178 R267, R340, R017, R003, R342, R025, R099 R190, R247 These strains were isolated from 17 genera of legumes.	41%
1000	R116, R291, R263, R115, R188, R243, R196 R023, R165, R176, R267, R017, R342, R339 These strains were isolated from 13 genera of legumes.	26%

Rhizobia strains grow on YMA medium.

Rhizobia are generally sensitive to a wide range of antibiotics. The germs of both Gram-positive and Gram-negative bacteria are inhibited by antibiotics, so are rhizobia. We found that some rhizobia isolated in this region had a strong resistance to streptomycin (Table 16).

Through several of our experiments above, we proved that rhizobia in the Xinjiang region are tolerant to such unfavorable factors as high temperature, salts, alkali, antibiotics, etc., implying that they have acquired these genetic properties under environmental pressure.

6 Physiological and Biochemical Properties of *Rhizobium* Bacteria

Jordan and Allen (1974) described the *Rhizobiaceae* family to include two genera, *Rhizobium* and *Agrobacterium*. They characteristically do not produce 3-ketolactose and do not utilize citrate. Our results show that random sampling

Table 17. Utilization of lactose and citrate by *Rhizobium* strains

Strains (codes)	Production of 3-keto-lactose	Utiliza-tion of citrate	Strains (codes)	Production of 3-keto-lactose	Utiliza-tion of citrate
R171	−	−	R276	−	−
R168	−	−	R180	−	+
R173	−	+	R316	−	−
R311	−	+	R326	−	+
R291	−	−	R324	−	−
R083	−	−	R331	−	+
R268	−	−	R146	−	+
R145	−	−	R259	−	+
R001	−	+	R188	−	+
R011	−	+	R342	−	+
R196	−	+	R270	−	+
R210	−	−	R038	−	+
R336	−	+	R023	−	+
R162	−	+	R059	−	−
R302	−	+	R061	−	+
R266	−	+	R050	−	+
R010	−	+	R043	−	+
R185	−	−	R058	−	+
R286	−	−	R058	−	+
R339	−	+	R096	−	−
R089	−	−	R252	−	+
R153	−	+	R119	−	−
R340	−	−	R153	−	+
R123	−	+	R312	−	+
R240	−	−	R231	−	−
R009	−	−	R159	−	+
R295	−	−	R175	−	+
R025	−	+	61A76*	−	−

* *R. japonicum* used as standard. "+" utilization; "−" no utilization.
We examined 105 strains which were isolated from 55 species of legumes.

Table 18. Reactions of Litmus–milk by *Rhizobium* strains

Strains (codes)	Colour	pH	Serum zone	Strains (codes)	Colour	pH	Serum zone
R168	red	5.0	+	R059	red	6.0	+
R173	blue	8.0	+	R118	red	5.0	+
R291	red	5.0	+	R115	blue	9.0	+
R001	red	6.0	+	R185	blue	8.8	+
R011	red	6.0	+	R276	red	5.0	+
R210	blue	8.5	+	R316	blue	8.5	+
R338	red	5.0	+	R326	red	6.0	+
R162	red	5.0	+	R324	blue	9.0	+
R266	red	5.0	+	R331	red	6.0	+
R010	blue	8.8	+	R188	red	6.0	+
R023	blue	9.0	+	R342	blue	8.0	+
R061	red	6.0	+	R270	red	6.0	+
R050	red	6.0	+	R123	red	6.0	+
R043	red	5.0	+	R175	blue	8.5	+
R053	red	6.0	+	R025	blue	9.0	+
R339	red	6.0	+	R096	red	6.0	+
R114	red	5.0	+	R252	red	5.5	+
61A76*	blue	9.0	+				

* *R. japonicum* used as standard. "+" shows serum zone formed.
We tested 52 strains which were isolated from 34 species of legumes.

of 105 strains of rhizobia from 55 species of leguminous plants were all uncapable of oxygenating lactose into 3-ketolactose, which was in line with the above characteristic. However, 34 of the above strains were able to utilize citrate, which was different from the second characteristic reaction described by Jordan (Table 17). Trinek reported that *Rhizobium leucaence* utilized citrate.

In carrying out a Litmus–milk test, the *Rhizobium* genus was divided into 2 groups (as described by Jordan and Allen, 1974) including 6 model species. Only *R. meliloti* could cause litmus–milk acid and formed a serum zone, others caused alkaline litmus–milk and formed a serum zone. In our tests, we continued observations for 45 days and found that 23 out of 34 species caused litmus–milk acid and formed a serum zone (Table 18).

B.T.B. tests were performed in our laboratory reveiled that 96% of 53 *Rhizobium* strains produced acid, which conformed to the litmus–milk reaction. 45 strains were tested for the utilization of 12 kinds of saccharides as carbon source. It was observed that monosaccharides, disaccharides and polysaccharides were all utilizable by these strains. Jordan and Allen (1974) reported that rhizobia could not liquefy or only slowly liquefy gelatin and could not utilize starch and casein. However, 76% of strains isolated in this region could liquefy gelatin, and among them, 33.6% could hydrolyze casein, and 35% could utilize starch (Tables 19, 20).

Based on the above studies, which we carried out using the method described by the Nanjing Institute of Soil Research, Chinese Academy of Sciences and also by Zhou, it was demonstrated that although the features of some rhizobia

Table 19. Hydrolyzation of casein and liquefaction of gelatin by *Rhizobium* strains

Strains (codes)	Hydroly-zation of casein	Liquefac-tion of gelatin	Strains (codes)	Hydroly-zation of casein	Liquefac-tion of gelatin
R171	−	−	R181	+ +	+ + +
R168	−	−	R318	+ + +	+ +
R173	−	+ +	R326	−	+ +
R311	−	−	R324	+	+ +
R291	−	+	R331	+ + +	+
R081	+ + +	+ + +	R146	+ + +	+ + +
R286	−	+	R259	+ +	+
R145	−	−	R188	−	+
R001	+ +	+ +	R342	+	+ +
R011	−	+	R270	+ + +	+ + +
R196	−	+	R038	−	+
R210	−	−	R023	+	+
R338	−	+	R059	−	+
R162	+ +	+ +	R061	+ +	+ + +
R302	−	+	R050	−	+
R266	+	+ +	R043	−	+
R010	−	+	R063	+ + +	+ +
R185	−	+ +	R058	−	−
R286	−	+	R339	+ +	+ + +
R276	−	+ +	R096	−	−
R114	−	−	R252	−	+ +
R254	+	+ +	R118	−	−
R340	−	+	R155	−	+
R123	+ + +	+ + +	R312	+ +	+
R240	−	−	R231	−	−
R009	−	−	R159	−	+
R295	−	−	R175	+ +	+ + +
R018	−	+	61A76*	−	−

* *R. japonicum* used as standard. " + " shows the ability to hydrolyze casein or liquefy gelatin, " + + " and " + + + " indicate the increasing degrees of the ability, we have tested 105 strains which were isolated from 56 species of legumes.

isolated in the Xinjiang region conformed to those common to rhizobia, some strains were not in line with the properties described in Bergey's Manual of Determinative Bacteriologgy (8th edn).

7 Prospects of *Rhizobium* Application in Xinjiang

We have selected strains with high activities of nitrogen fixation, great infectious ability and resistance to harsh conditions by means of tube culture and basin planting. Thye were then inoculated on leguminous plants in the field. The results showed that the grass crop of *Medicago*, *Glycyrrhiza glabra* root

Table 20. Utilization of starch by *Rhizobium* strains

Strains (codes)	Utilization of starch	Strains (codes)	Utilization of starch
R171	−	R266	−
R168	−	R010	−
R173	+ +	R185	−
R311	−	R286	+
R291	−	R181	+
R083	−	R316	−
R268	−	R326	+
R126	−	R332	+
R001	+	R324	+
R011	+	R146	−
R196	+	R259	−
R210	−	R188	+
R164	−	R342	−
R302	−	R270	+
R050	+	R038	−
R063	−	R025	−
R058	+	R023	−
R059	−	R061	+
R339	−	R096	−
R114	−	R259	+
R253	−	R119	−
R340	−	R115	−
R123	+	R312	−
R231	−	R159	−
R043	+	61A76*	−

* *R. japonicum* used as standard. "+" shows utilization of the starch; "−" shows no utilization.

and *Glycine max* seed crop increased by 30%, 40%, and 10%, respectively. In addition, we have noticed that some rhizobia in Xinjiang could produce pink and gold-yellow pigments, a large amount of mucopolysaccharides and proteinase. These products may be used in the fermentation industry. 373 strains of rhizobia isolated from 109 species of legumes were a representation of the whole *Rhizobium* population in this region. To our knowledge, some strains we described here were not previously reported in the literature. The discovery of new *Rhizobium* strains adds a taxonomic value. Our experiments have proved that some rhizobia isolated in Xinjiang possess distinct physiological, biochemical and ecological characteristics, and may serve as valuable research materials for molecular gene-tics of nitrogen fixation.

Acknowledgements. This research was supported by the National Science Foundation of China. Help was offered in this research by Professor H.Y. Song is sincerely appreciated. Li Zhong-yuan, Wang Wei-wei, Yang Yu-suo, Guo Pei-xin et al. participated in this research.

8 References

Allen ON, Allen EK (1981) The Leguminosae. A source book of characteristics, uses, and nodulation. Univ. of Wisconsin Press, p 707

Bergersen FJ (1982) Root nodules of legumes: structure and function. Wiley, New York, p 9

Chen WX, Wu BH (1987) Survey and taxonomy of leguminous root nodule bacteria of Xinjiang region. Chinese Agricult Sci 20(6): 22

Eisbrenner G, Evans HJ (1983) Aspects of hydrogen metabolism in nitrogen-fixing legumes and other plant-microbe associations. Ann Rev Plant Physiol 34: 105

Evans HJ, Emerich DW et al. (1980) Hydrogen metabolism in the legume-*Rhizobium* symbiosis. In: Newton WE, Orme–Johnson WH (eds) Nitrogen fixation. Vol 2. Symbiotic associations and cyanobacteria. p 69

Evans HJ, Emerich DW et al. (1980) The role of hydrogenase in nodule bacteroids and free-living rhizobia. In: Stewent WDP, Gllon, JR (eds) Nitrogen fixation. Academic Press, London, p 55

Grahan PH, Parker CA, et al. (1963) Spore formation and heat resistance in *Rhizobium*. J Appl Bacteriol 86: 1353

Jordan DC, Allen ON (1974) Family III *Rhizobiaceae* conn 1938. In: Bachanan RE, Gibbons NE et al. (eds) Bergey's Manual of Determinative Bacteriology 8th edn. Williams and Wilkins, Baltimore, p 261

Jordan DC (1984) Family III Rhizobiaceae conn 1938. In: Krieg NR, Holt JG (eds) Bergey's Manual of Systematic Bacteriology, 1st edn. William and Wilkins, Baltimore, p 235

Maier RJ, Campbell ER et al. (1978) Expression of hydrogenase activity in free-living *Rhizobium japonicum*. Proc Natl Acad Sci USA 75(7): 3258

Nanjing Institute of Soil Research, Chinese Academy of Science (1985) Research methods of soil microbiology. Science Press, Beijing, p 90

Sutton WD, Pankhurst CE, Craig AS (1981) The *Rhizobium* Bacteriod State. In: Giles KL, Atherly AG (eds) Biology of the *Rhizobiaceae*. Academic Press, p 149

Shanghai Institute of Plant Physiology. Research Academy of Science (1974) A sample and easy method of the acetylene reduction for nitrogen fixation. Botanica Sinica 16: 383

Schubert KR, Evans HJ (1976) Hydrogen evolution: A major factor affecting the efficiency of nitrogen fixation in nodulated symbionts. Proc Natl Acad Sci USA 73(4): 1207

Tinick MJ (1980) Relationships amongst the fast-growing *Rhizobia* of *Lablab purpureus*, *Leucaena*, *Leucocephala*, *Mimosa* spp., *Acacia farnesiana* and *Sesbania grandiflora* and their affinities with other rhizobial groups. J Appl Bacteriol 49: 39

Vincent JH (1970) A manual for the practical study of the root-nodule bacteria. Blackwell Scientific, Oxford, p 1

Vincent JM (1977) *Rhizobium* general microbiology. In: Hardy RWF, Silver WS (eds) A treatise on dinitrogen fixation, Sect. 3: Biology. Wiley, New York, p 277

Wen ZW (1965) The soil geography of Xinjiang. Science Press, Beijing, p 3, 190

Yang CY, Han YL (1985) Claves Plantarum Xinjiang gensium. Xinjiang People's Publ House, p 1

Yu DF, Li GA (1985) Microbiology. Science Press, Beijing, p 87

Zhang LY (1985) A preliminating study on the ephemerals in the Mosowan District, Xinjiang. Acta Phytoecol et Geobotanica Sinica 9(3): 213

Zhou DQ (1983) Research methods of microbiology. Science Technology Press, Shanghai, p 97